Multispectral and Hyperspectral Remote Sensing Data for Mineral Exploration and Environmental Monitoring of Mined Areas

Multispectral and Hyperspectral Remote Sensing Data for Mineral Exploration and Environmental Monitoring of Mined Areas

Editors

Amin Beiranvand Pour
Basem Zoheir
Biswajeet Pradhan
Mazlan Hashim

MDPI • Basel • Beijing • Wuhan • Barcelona • Belgrade • Manchester • Tokyo • Cluj • Tianjin

Editors
Amin Beiranvand Pour
Universiti Malaysia Terengganu (UMT)
Malaysia

Basem Zoheir
University of Kiel
Germany

Biswajeet Pradhan
University of Technology Sydney
Australia

Mazlan Hashim
Universiti Teknologi Malaysia
Malaysia

Editorial Office
MDPI
St. Alban-Anlage 66
4052 Basel, Switzerland

This is a reprint of articles from the Special Issue published online in the open access journal *Remote Sensing* (ISSN 2072-4292) (available at: https://www.mdpi.com/journal/remotesensing/special_issues/hyperspectral_mining).

For citation purposes, cite each article independently as indicated on the article page online and as indicated below:

LastName, A.A.; LastName, B.B.; LastName, C.C. Article Title. *Journal Name* **Year**, *Volume Number*, Page Range.

ISBN 978-3-0365-1264-8 (Hbk)
ISBN 978-3-0365-1265-5 (PDF)

© 2021 by the authors. Articles in this book are Open Access and distributed under the Creative Commons Attribution (CC BY) license, which allows users to download, copy and build upon published articles, as long as the author and publisher are properly credited, which ensures maximum dissemination and a wider impact of our publications.

The book as a whole is distributed by MDPI under the terms and conditions of the Creative Commons license CC BY-NC-ND.

Contents

About the Editors . . . . **vii**

Amin Beiranvand Pour, Basem Zoheir, Biswajeet Pradhan and Mazlan Hashim
Editorial for the Special Issue: Multispectral and Hyperspectral Remote Sensing Data for Mineral Exploration and Environmental Monitoring of Mined Areas
Reprinted from: *Remote Sens.* **2021**, *13*, 519, doi:10.3390/rs13030519 . . . . **1**

Lida Noori, Amin Beiranvand Pour, Ghasem Askari, Nader Taghipour, Biswajeet Pradhan, Chang-Wook Lee and Mehdi Honarmand
Comparison of Different Algorithms to Map Hydrothermal Alteration Zones Using ASTER Remote Sensing Data for Polymetallic Vein-Type Ore Exploration: Toroud–Chahshirin Magmatic Belt (TCMB), North Iran
Reprinted from: *Remote Sens.* **2019**, *11*, 495, doi:10.3390/rs11050495 . . . . **7**

Arindam Guha, Yasushi Yamaguchi, Snehamoy Chatterjee, Komal Rani and Kumranchat Vinod Kumar
Emittance Spectroscopy and Broadband Thermal Remote Sensing Applied to Phosphorite and Its Utility in Geoexploration: A Study in the Parts of Rajasthan, India
Reprinted from: *Remote Sens.* **2019**, *11*, 1003, doi:10.3390/rs11091003 . . . . **31**

Amin Beiranvand Pour, Yongcheol Park, Laura Crispini, Andreas Läufer, Jong Kuk Hong, Tae-Yoon S. Park, Basem Zoheir, Biswajeet Pradhan, Aidy M. Muslim, Mohammad Shawkat Hossain and Omeid Rahmani
Mapping Listvenite Occurrences in the Damage Zones of Northern Victoria Land, Antarctica Using ASTER Satellite Remote Sensing Data
Reprinted from: *Remote Sens.* **2019**, *11*, 1408, doi:10.3390/rs11121408 . . . . **53**

Basem Zoheir, Ashraf Emam, Mohamed Abdel-Wahed and Nehal Soliman
Multispectral and Radar Data for the Setting of Gold Mineralization in the South Eastern Desert, Egypt
Reprinted from: *Remote Sens.* **2019**, *11*, 1450, doi:10.3390/rs11121450 . . . . **95**

Lei Sun, Shuhab Khan and Peter Shabestari
Integrated Hyperspectral and Geochemical Study of Sediment-Hosted Disseminated Gold at the Goldstrike District, Utah
Reprinted from: *Remote Sens.* **2019**, *11*, 1987, doi:10.3390/rs11171987 . . . . **125**

Basem Zoheir, Mohamed Abd El-Wahed, Amin Beiranvand Pour and Amr Abdelnasser
Orogenic Gold in Transpression and Transtension Zones: Field and Remote Sensing Studies of the Barramiya–Mueilha Sector, Egypt
Reprinted from: *Remote Sens.* **2019**, *11*, 2122, doi:10.3390/rs11182122 . . . . **145**

Amin Beiranvand Pour, Tae-Yoon S. Park, Yongcheol Park, Jong Kuk Hong, Aidy M Muslim, Andreas Läufer, Laura Crispini, Biswajeet Pradhan, Basem Zoheir, Omeid Rahmani, Mazlan Hashim and Mohammad Shawkat Hossain
Landsat-8, Advanced Spaceborne Thermal Emission and Reflection Radiometer, and WorldView-3 Multispectral Satellite Imagery for Prospecting Copper-Gold Mineralization in the Northeastern Inglefield Mobile Belt (IMB), Northwest Greenland
Reprinted from: *Remote Sens.* **2019**, *11*, 2430, doi:10.3390/rs11202430 . . . . **179**

Seyed Mohammad Bolouki, Hamid Reza Ramazi, Abbas Maghsoudi, Amin Beiranvand Pour and Ghahraman Sohrabi
A Remote Sensing-Based Application of Bayesian Networks for Epithermal Gold Potential Mapping in Ahar-Arasbaran Area, NW Iran
Reprinted from: *Remote Sens.* **2020**, *12*, 105, doi:10.3390/rs12010105 219

Laura Tuşa, Mahdi Khodadadzadeh, Cecilia Contreras, Kasra Rafiezadeh Shahi, Margret Fuchs, Richard Gloaguen and Jens Gutzmer
Drill-Core Mineral Abundance Estimation Using Hyperspectral and High-Resolution Mineralogical Data
Reprinted from: *Remote Sens.* **2020**, *12*, 1218, doi:10.3390/rs12071218 253

Milad Sekandari, Iman Masoumi, Amin Beiranvand Pour, Aidy M Muslim, Omeid Rahmani, Mazlan Hashim, Basem Zoheir, Biswajeet Pradhan, Ankita Misra and Shahram M. Aminpour
Application of Landsat-8, Sentinel-2, ASTER and WorldView-3 Spectral Imagery for Exploration of Carbonate-Hosted Pb-Zn Deposits in the Central Iranian Terrane (CIT)
Reprinted from: *Remote Sens.* **2020**, *12*, 1239, doi:10.3390/rs12081239 277

Hodjat Shirmard, Ehsan Farahbakhsh, Amin Beiranvand Pour, Aidy M Muslim, R. Dietmar Müller and Rohitash Chandra
Integration of Selective Dimensionality Reduction Techniques for Mineral Exploration Using ASTER Satellite Data
Reprinted from: *Remote Sens.* **2020**, *12*, 1261, doi:10.3390/rs12081261 311

Tomás Martín-Crespo, David Gómez-Ortiz, Silvia Martín-Velázquez, Pedro Martínez-Pagán, Cristina de Ignacio-San José, Javier Lillo and Ángel Faz
Abandoned Mine Tailings Affecting Riverbed Sediments in the Cartagena–La Union District, Mediterranean Coastal Area (Spain)
Reprinted from: *Remote Sens.* **2020**, *12*, 2042, doi:10.3390/rs12122042 341

Robert Jackisch, Sandra Lorenz, Moritz Kirsch, Robert Zimmermann, Laura Tusa, Markku Pirttijärvi, Ari Saartenoja, Hernan Ugalde, Yuleika Madriz, Mikko Savolainen and Richard Gloaguen
Integrated Geological and Geophysical Mapping of a Carbonatite-Hosting Outcrop in Siilinjärvi, Finland, Using Unmanned Aerial Systems
Reprinted from: *Remote Sens.* **2020**, *12*, 2998, doi:10.3390/rs12182998 361

Baodong Ma, Xuexin Li, Ziwei Jiang, Ruiliang Pu, Aiman Liang and Defu Che
Dust Dispersion and Its Effect on Vegetation Spectra at Canopy and Pixel Scales in an Open-Pit Mining Area
Reprinted from: *Remote Sens.* **2020**, *12*, 3759, doi:10.3390/rs12223759 393

About the Editors

Amin Beiranvand Pour is currently an associate professor at University Malaysia Terengganu (UMT). He was a post-doctoral researcher and senior lecturer (Assistant Professor) in Geoscience and Digital Earth Centre (INSTeG) at University Technology Malaysia (UTM) from 2013 to 2016, and a researcher at the Korea Polar Research Institute (KOPRI) from 2017 to 2020. He earned his B.Sc. and M.Sc. in geology and economic geology from Shiraz University, Iran. He received his Ph.D. in remote sensing with specialty in mineral exploration from University Technology Malaysia (UTM) in 2012. He was project leader of several mineral exploration projects using satellite remote sensing technology in arid and semi-arid terrains, Antarctic, Arctic and tropical environments. Dr. Beiranvand Pour is one the most dynamic researchers in the field of mineral exploration using remote sensing data and has published 140 research papers since 2011 (total citation: 2324; H-index: 29; i10-index: 46). Presently, he is conducting several Special Issues in remote sensing, minerals and mining journals.

Basem Zoheir is a professor of Mineralogy and Economic Geology at Benha University (Egypt), and is now a researcher staff member and an AvH fellow at Kiel Univ (CAU-Germany). He obtained his Ph.D. from the University of Munich (Germany), and was a post-doctorate at the Universities of Tubingen, TU Clausthal, Geneva, Stockholm, Graz, TU Lulea. As a Fulbrighter, Basem Zoheir conducted research at the USGS DFC-Denver, and has published > 60 research articles in international ISI journals. Basem Zoheir serves now as an AE for Ore Geology Reviews and acted as an international referee/examiner for a number of research funding foundations/institutions. Besides his academic work, he has provided consultations to national and international mining companies. His expertise and research highlights exploration targeting of mineral deposits, with emphasis on temporal/spatial distribution of orogenic gold deposits in the Arabian-Nubian Shield.

Biswajeet Pradhan is an internationally established scientist in the field of Geospatial Information Systems (GIS), remote sensing and image processing, complex modeling/geo-computing, machine learning, and soft-computing applications, natural hazards and environmental modeling, and remote sensing of Earth observation. He is also a distinguished professor at the University of Technology, Sydney. He was listed as the World's Most Highly Cited Researcher by Clarivate Analytics Report for five consecutive years: 2006–2020 as one of the world's most influential minds. In 2018–2020, he was awarded as World Class Professor by the Ministry of Research, Technology and Higher Education, Indonesia. He is a recipient of the Alexander von Humboldt Research Fellowship from Germany. In 2011, he received his habilitation in "Remote Sensing" from Dresden University of Technology, Germany. Between February 2015 and January 2020, he served as "Ambassador Scientist" for Alexander Humboldt Foundation, Germany. Professor Pradhan has received 55 awards since 2006 in recognition of his excellence in teaching, service, and research. Out of his more than 550 articles (total citation: 32,383; H-index: 97; i10-index: 399), more than 500 have been published in science citation index (SCI/SCIE) technical journals. He has written eight books and thirteen book chapters.

Mazlan Hashim is the Director of Research Institute of Sustainable Environment, Universiti Teknologi Malaysia (UTM). He is a Fellow of the Academy Sciences Malaysia. Professor Hashim is also a Senior Fellow at the Geoscience & Digital Earth Centre (INSTeG), Faculty of Built Environment & Surveying, UTM. He has held several positions as Visiting Professor/Scientist at various international university. He is an expert in remote sensing applications with geomatics engineering background. He has published his research works in more than 120 high impact articles, and he has recently been listed among the World's Top 2% Scientists for 2019 citations in the Engineering field.

Editorial

Editorial for the Special Issue: Multispectral and Hyperspectral Remote Sensing Data for Mineral Exploration and Environmental Monitoring of Mined Areas

Amin Beiranvand Pour [1,*], Basem Zoheir [2,3], Biswajeet Pradhan [4,5,6,7] and Mazlan Hashim [8,9,10]

1. Institute of Oceanography and Environment (INOS), Universiti Malaysia Terengganu (UMT), Kuala Nerus 21030, Malaysia
2. Department of Geology, Faculty of Science, Benha University, Benha 13518, Egypt; basem.zoheir@ifg.uni-kiel.de
3. Institute of Geosciences, University of Kiel, Ludewig-Meyn Str. 10, 24118 Kiel, Germany
4. Centre for Advanced Modelling & Geospatial Information Systems (CAMGIS), Faculty of Engineering and Information Technology, University of Technology Sydney, Sydney, NSW 2007, Australia; Biswajeet.Pradhan@uts.edu.au
5. Department of Energy and Mineral Resources Engineering, Sejong University, Choongmu-gwan, 209 Neungdong-ro Gwangjin-gu, Seoul 05006, Korea
6. Center of Excellence for Climate Change Research, King Abdulaziz University, P.O. Box 80234, Jeddah 21589, Saudi Arabia
7. Earth Observation Center, Institute of Climate Change, Universiti Kebangsaan Malaysia, UKM Bangi 43600, Malaysia
8. Geoscience and Digital Earth Centre (INSTeG), Research Institute for Sustainable Environment, Universiti Teknologi Malaysia, Johor Bahru 81310, Malaysia; mazlanhashim@utm.my
9. Faculty of Built Environment & Surveying, Universiti Teknologi Malaysia, Johor Bahru 81310, Malaysia
10. Department of Tourism Science, Graduate School of Urban Environmental Sciences, Tokyo Metropolitan University, Minami-Osawa 1-1, Hachiouji, Tokyo 192-0397, Japan
* Correspondence: beiranvand.pour@umt.edu.my; Tel.: +60-9-6683824; Fax: +60-9-6692166

Keywords: mineral exploration; multispectral and hyperspectral data; mining; remote sensing

1. Introduction

In recent decades, multispectral and hyperspectral remote sensing data provide unprecedented opportunities for the initial stages of mineral exploration and environmental hazard monitoring. Increasing demands for minerals because of industrialization and exponential growth in population emphasize the necessity for replenishing exploited reserves by exploration of new potential zones of mineral deposits. Identification of host-rock lithologies, geologic structural features, and hydrothermal alteration mineral zones are the most conspicuous applications of multispectral and hyperspectral remote sensing satellite data for mineral exploration in the metallogenic provinces and frontier areas around the world [1–11]. Numerous ore deposits such as orogenic gold, porphyry copper, carbonatite, massive sulfide, epithermal gold, podiform chromite, uranium, magnetite, and iron oxide copper-gold (IOCG) deposits have been successfully prospected and discovered using multispectral and hyperspectral remote sensing satellite imagery [12–22].

The Advanced Spaceborne Thermal Emission and Reflection Radiometer (ASTER), Landsat data series, the Advanced Land Imager (ALI), Worldview-3, Hyperion, HyMap and the Airborne Visible/IR Image Spectrometer (AVIRIS) remote sensing data serve as low-cost tools for ore mineral exploration [3,7,11–13,20,23]. Additionally, Synthetic Aperture Radar (SAR) data contains a high potential for structural mapping and lineament extraction. The Phased Array type L-band Synthetic Aperture Radar (PALSAR) satellite remote sensing data are particularly used for mapping structurally controlled orogenic gold mineralization in the arid and tropical environments due to its penetration capability [7,18,24–29].

Several advanced image processing algorithms and machine learning techniques can be successfully used to extract essential information related to hydrothermal alteration minerals and lithological units at pixel and sub-pixel levels for indicating high potential zones of ore mineralizations. Different types of image processing algorithms have been used to extract spectral information from multispectral and hyperspectral remote sensing data for instance (i) band-ratio, indices, and logical operator based methods; (ii) principal components and transformation based methods—such as principal component analysis (PCA), independent component analysis (ICA), and minimum noise fraction (MNF); (iii) shape-fitting based algorithms—such as spectral angle mapper (SAM), matched-filtering (MF), and mixture-tuned matched-filtering (MTMF); and (iv) partial unmixing and target detection methods—such as linear spectral unmixing (LSU), constrained energy minimization (CEM), orthogonal subspace projection (OSP), and adaptive coherence estimator (ACE) [2,8]. Machine learning techniques are developing progressively crucial to unravel several image processing challenges in the coming future. Although the techniques are subject to scientific interest for the remote sensing mineral exploration community, but generic implementation is still in initial stages.

Furthermore, human-induced changes—in the form of mine excavation, open-pit mining, transportation, mine tailing, mineral processing in mining zones, mine waste, dust pollution, and acid runoff—necessitate a proper monitoring of mining areas by remote sensing observations. Environmental pollution mapping and monitoring of mined areas are the main challenges that need to be addressed for future sustainability and environmental management in metallogenic provinces and surrounding areas. Consequently, a special issue entitled "Multispectral and Hyperspectral Remote Sensing Data for Mineral Exploration and Environmental Monitoring of Mined Areas" is proposed, which is expected to particularly motivate researchers for presenting the latest achievements in the field of geological remote sensing for mineral exploration and environmental monitoring.

A total of 20 manuscripts have been submitted to this special issue, which were evaluated by professional guest editors and reviewers. Subsequently, 14 papers attained the level of quality and novelty anticipated by *Remote Sensing* and finally were revised, accepted, and published in the special issue. The achievements of articles presented in this special issue are summarized in the following section.

2. Summary of Papers Presented in This Special Issue

Noori et al. [3] compared different image processing algorithms for mapping hydrothermal alteration zones associated with polymetallic vein-type mineralization using ASTER data in the Toroud–Chahshirin Magmatic Belt (TCMB), North Iran. Selective principal component analysis (SPCA), band ratio matrix transformation (BRMT), spectral angle mapper (SAM), and mixture tuned matched filtering (MTMF) were implemented and compared to map hydrothermal alteration minerals at the pixel and sub-pixel levels. Subtle differences between altered and non-altered rocks and hydrothermal alteration mineral assemblages were detected and mapped in the study area. Results indicate several high potential zones of epithermal polymetallic vein-type mineralization in the northeastern and southwestern parts of the study area, which can be considered for future systematic exploration programs. Guha et al. [30] used emittance spectroscopy and ASTER broadband thermal remote sensing data to map phosphorite associated with carbonate-rich sediments of the Aravalli Super Group, Rajasthan, India. In this study, a relative band depth (RBD) image using selected emissivity bands of ASTER (bands 11, 12, and 13) was developed for mapping and delineating phosphorite from the dolomite or carbonate host-rock lithologies. Additionally, the RBD is capable to differentiate low-grade phosphorite exposures from high-grade phosphorite zones. The authors recommended that the RBD of broadband ASTER thermal infrared (TIR) bands can be used for targeting phosphorite occurring under similar geological systems around the world.

Pour et al. [10] mapped listvenite occurrences in the damage zones of northern Victoria Land, Antarctica using ASTER Data. Principal component analysis (PCA)/independent component analysis (ICA) fusion technique, linear spectral unmixing (LSU), and constrained energy minimization (CEM) algorithms were implemented to extract spectral information for detecting alteration mineral assemblages and listvenites. Mineralogical assemblages containing Fe^{2+}, Fe^{3+}, Fe-OH, Al-OH, Mg-OH, and CO3 spectral absorption features were detected by applying PCA/ICA fusion to visible and near infrared (VNIR) and shortwave infrared (SWIR) bands of ASTER. Silicate lithological groups were mapped and discriminated using PCA/ICA fusion to TIR bands of ASTER. Goethite, hematite, jarosite, biotite, kaolinite, muscovite, antigorite, serpentine, talc, actinolite, chlorite, epidote, calcite, and dolomite were detected using LSU and CEM algorithms. Several potential zones for listvenite occurrences were identified, typically in association with mafic metavolcanic rocks (Glasgow Volcanics) in the Bowers Mountains. Zoheir et al. [18] utilized Landsat 8-Operational Land Imager (OLI), ASTER, PALSAR and Sentinel-1 satellite data coupled with field and microscopic investigations to unravel the setting and controls of gold mineralization in the Wadi Beitan–Wadi Rahaba area in the South Eastern Desert of Egypt. Band ratios, RBD and mineralogical indices are used to extract the representative pixels form Landsat 8-OLI and ASTER bands. Lineaments were manually and automatically extracted from PALSAR and Sentinel-1 data. The data fusion approach was used and showed no particular spatial association between gold occurrences and certain lithological units but indicates a preferential distribution of gold–quartz veins in zones of chlorite–epidote alteration overlapping with high-density intersections of lineaments. A priority map with zones defined as high potential targets for undiscovered gold resources were generated for the Wadi Beitan–Wadi Rahaba area in this study.

Sun et al. [31] integrated ground-based hyperspectral imaging and geochemistry data for resource exploration and exploitation of sediment-hosted disseminated Gold at the Goldstrike District, UT, USA. Ground-based hyperspectral imaging was applied to study a core drilled in the Goldstrike district covering the basal Claron Formation and Callville Limestone. The integration of remote sensing and geochemistry data helped to identify an optimum stratigraphic combination of limestone above and siliciclastic rocks below in the basal Claron Formation, as well as decarbonatization, argillization, and pyrite oxidation in the Callville Limestone, that are related with gold mineralization. Zoheir et al. [17] used multi-sensor satellite imagery data, including Sentinel-1, PALSAR, ASTER, and Sentinel-2, for mapping the regional structural control of orogenic gold mineralization in the Barramiya–Mueilha sector. Feature-oriented principal component selection (FPCS) was applied to polarized backscatter ratio images of Sentinel-1 and PALSAR datasets for regional structural mapping and identification of potential dilation loci. The PCA and band ratioing techniques are applied to ASTER and Sentinel-2 datasets for lithological and hydrothermal alteration mapping. The radar and multispectral satellite data abetted a better understanding of the structural framework and unraveled settings of the scattered gold occurrences in the study area.

Pour et al. [11] utilized Landsat-8, ASTER and WorldView-3 multispectral remote sensing imagery for prospecting copper-gold mineralization in the Northeastern Inglefield Mobile Belt (IMB), Northwest Greenland at regional, local, and district scales. Hydrothermal alteration minerals such as iron oxide/hydroxide, Al/Fe-OH, Mg-Fe-OH minerals, silicification (Si-OH), and SiO_2 mineral groups were mapped using directed principal components analysis (DPCA) technique, Linear spectral unmixing (LSU) and adaptive coherence estimator (ACE) algorithms. Several high potential zones for Cu-Au prospecting were identified in the IMB, Northwest Greenland, including (i) the boundaries between the Etah metamorphic and meta-igneous complex rocks and sedimentary successions of the Franklinian Basin in the Central Terrane, (ii) orthogneiss in the northeastern part of the Cu-Au mineralization belt adjacent to Dallas Bugt, and (iii) the southern part of the Cu-Au mineralization belt nearby Marshall Bugt. Bolouki et al. [12] investigated a remote sensing-based application of Bayesian networks for epithermal gold potential mapping in Ahar-Arasbaran area, NW Iran. Landsat Enhanced Thematic Mapper+ (Landsat-7 ETM+), Landsat-8, and ASTER datasets were used to detect hydrothermal alteration zones associated with epithermal gold mineralization using band ratio, relative absorption band depth (RBD) and PCA techniques. The Bayesian network classifier was used to synthesize the thematic layers of hydrothermal alteration zones. Many new potential zones of epithermal gold mineralization were identified in the Ahar-Arasbaran region.

Tuşa et al. [32] estimated mineral abundance in drill-core samples collected from Bolcana porphyry copper-gold deposit by employing hyperspectral short-wave infrared (SWIR) data and scanning electron microscopy-based image analyses using a mineral liberation analyzer (SEM-MLA). Machine learning algorithms were executed to combine the two data types and upscale the quantitative SEM-MLA mineralogical data to drill-core scale. Quasi-quantitative maps over entire drill-core samples were acquired. Sekandari et al. [13] used Landsat-8, Sentinel-2, ASTER, and WorldView-3 spectral imagery for exploration of carbonate-hosted Pb-Zn deposits in the Central Iranian Terrane (CIT). Band ratios and PCA techniques were adopted and implemented to map alteration minerals and lithologies. Fuzzy logic modeling was applied to integrate the thematic layers produced by the image processing techniques for generating mineral prospectivity maps. The most favorable/prospective zones for hydrothermal ore mineralizations and carbonate-hosted Pb-Zn mineralization in the study region were particularly mapped and indicated at regional and district scales. Shirmard et al. [33] integrated selective dimensionality reduction techniques such as PCA, ICA, and minimum noise fraction (MNF) for mineral exploration using ASTER satellite data. The fuzzy logic model was used for integrating the most rational thematic layers derived from the techniques for mineral prospectivity mapping in the Toroud-Chahshirin range, Central Iran.

Martín-Crespo et al. [34] presented the results of the geo-environmental characterization of La Matildes riverbed, affected by mine tailings in the Cartagena–La Unión district, Murcia (southeast Spain) using geophysical and geochemical techniques. Two electrical resistivity imaging (ERI) profiles were carried out to obtain information about the thickness of the deposits and their internal structure. The geochemical composition of borehole samples from the riverbed materials shows significantly high contents of As, Cd, Cu, Fe, Pb, and Zn being released to the environment. Results demonstrated that surface extraction in three open-pit mines have changed the summits of Sierra de Cartagena–La Unión and rock and metallurgical wastes have altered the drainage pattern and buried the headwaters of ephemeral channels. Jackisch et al. [35] integrated drone-borne photography, multi- and hyperspectral imaging, and magnetics data for mapping a carbonatite-hosting outcrop in Siilinjärvi, Finland. Structural orientations and lithological units are deduced based on high-resolution, hyperspectral image-enhanced point clouds. Unmanned aerial system (UAS)-based magnetic data allow an insight into their subsurface geometry through modeling based on magnetic interpretation. A geologic map is resulted discriminating between the principal lithologic units and distinguishes ore-bearing from waste rocks. Ma et al. [36] investigated the dust dispersion characteristics in Kuancheng mining area, Hebei Province, North China using the American Meteorological Society (AMS) and the U.S. Environmental Protection Agency (EPA) regulatory model (AERMOD). The spectral characteristics of vegetation canopy under the dusty condition were simulated, and the influence of dustfall on vegetation canopy spectra was studied based on the three-dimensional discrete anisotropic radiative transfer (DART) model. The experimental results show that the dust pollution along a haul road was more severe and extensive than that in a stope. Taking dust dispersion along the road as an example, the variation of vegetation canopy spectra increased with the height of dust deposited on the vegetation canopy. The findings would be beneficial to decision-makers or researchers for the remote sensing application to mapping and assessing the dust effect in mining areas.

3. Concluding Remarks

The sympathetic and judicious comments delivered by the reviewers enhanced each of the papers published in this special issue, which came to fruition only because they were willing to volunteer their time and attention. We hope that the investigations published in this special issue will assist mineral exploration communities and mining companies about the application and integration of multispectral and hyperspectral remote sensing data for mineral exploration and environmental monitoring of mined areas.

Author Contributions: All authors have read and agreed to the published version of this manuscript.

Funding: This research received no external funding.

Acknowledgments: The guest editors would like to thank the authors who contributed to this special issue and the reviewers who helped to improve the quality of the special issue by providing constructive recommendations to the authors. We would like to express our appreciation to Quenby Qu (assistant editor), all authors and reviewers who contributed their time, research, and specialty for this special issue. We wish to extend our sincere gratitude to Quenby Qu (assistant editor) and MDPI editorial team for supporting the guest editors in efficiently processing each manuscript.

Conflicts of Interest: The authors declare no conflict of interest.

References

1. Mars, J.C.; Rowan, L.C. Spectral assessment of new ASTER SWIR surface reflectance data products for spectroscopic mapping of rocks and minerals. *Remote Sens Environ.* **2010**, *114*, 2011–2025. [CrossRef]
2. Pour, B.A.; Hashim, M. The application of ASTER remote sensing data to porphyry copper and epithermal gold deposits. *Ore Geol. Rev.* **2012**, *44*, 1–9. [CrossRef]
3. Noori, L.; Pour, B.A.; Askari, G.; Taghipour, N.; Pradhan, B.; Lee, C.-W.; Honarmand, M. Comparison of Different Algorithms to Map Hydrothermal Alteration Zones Using ASTER Remote Sensing Data for Polymetallic Vein-Type Ore Exploration: Toroud–Chahshirin Magmatic Belt (TCMB), North Iran. *Remote Sens.* **2019**, *11*, 495. [CrossRef]
4. Ninomiya, Y.; Fu, B. Thermal infrared multispectral remote sensing of lithology and mineralogy based on spectral properties of materials. *Ore Geol. Rev.* **2019**, *108*, 54–72. [CrossRef]
5. Pour, A.B.; Park, Y.; Park, T.S.; Hong, J.K.; Hashim, M.; Woo, J.; Ayoobi, I. Regional geology mapping using satellite-based remote sensing approach in Northern Victoria Land, Antarctica. *Polar Sci.* **2018**, *16*, 23–46. [CrossRef]
6. Pour, A.B.; Hashim, M.; Park, Y.; Hong, J.K. Mapping alteration mineral zones and lithological units in Antarctic regions using spectral bands of ASTER remote sensing data. *Geocarto Int.* **2018**, *33*, 1281–1306. [CrossRef]
7. Pour, A.B.; Park, T.S.; Park, Y.; Hong, J.K.; Zoheir, B.; Pradhan, B.; Ayoobi, I.; Hashim, M. Application of multi-sensor satellite data for exploration of Zn-Pb sulfide mineralization in the Franklinian Basin, North Greenland. *Remote Sens.* **2018**, *10*, 1186. [CrossRef]
8. Pour, A.B.; Hashim, M.; Hong, J.K.; Park, Y. Lithological and alteration mineral mapping in poorly exposed lithologies using Landsat-8 and ASTER satellite data: North-eastern Graham Land, Antarctic Peninsula. *Ore Geol. Rev.* **2019**, *108*, 112–133. [CrossRef]
9. Pour, A.B.; Park, Y.; Park, T.S.; Hong, J.K.; Hashim, M.; Woo, J.; Ayoobi, I. Evaluation of ICA and CEM algorithms with Landsat-8/ASTER data for geological mapping in inaccessible regions. *Geocarto Int.* **2019**, *34*, 785–816. [CrossRef]
10. Pour, A.B.; Park, Y.; Crispini, L.; Läufer, A.; Kuk Hong, J.; Park, T.-Y.S.; Zoheir, B.; Pradhan, B.; Muslim, A.M.; Hossain, M.S.; et al. Mapping Listvenite Occurrences in the Damage Zones of Northern Victoria Land, Antarctica Using ASTER Satellite Remote Sensing Data. *Remote Sens.* **2019**, *11*, 1408. [CrossRef]
11. Pour, A.B.; Park, T.-Y.; Park, Y.; Hong, J.K.; Muslim, A.M.; Läufer, A.; Crispini, L.; Pradhan, B.; Zoheir, B.; Rahmani, O.; et al. Landsat-8, Advanced Spaceborne Thermal Emission and Reflection Radiometer, and WorldView-3 Multispectral Satellite Imagery for Prospecting Copper-Gold Mineralization in the Northeastern Inglefield Mobile Belt (IMB), Northwest Greenland. *Remote Sens.* **2019**, *11*, 2430. [CrossRef]
12. Bolouki, S.M.; Ramazi, H.R.; Maghsoudi, A.; Beiranvand Pour, A.; Sohrabi, G. A Remote Sensing-Based Application of Bayesian Networks for Epithermal Gold Potential Mapping in Ahar-Arasbaran Area, NW Iran. *Remote Sens.* **2020**, *12*, 105. [CrossRef]
13. Sekandari, M.; Masoumi, I.; Beiranvand Pour, A.M.; Muslim, A.; Rahmani, O.; Hashim, M.; Zoheir, B.; Pradhan, B.; Misra, A.; Aminpour, S.M. Application of Landsat-8, Sentinel-2, ASTER and WorldView-3 Spectral Imagery for Exploration of Carbonate-Hosted Pb-Zn Deposits in the Central Iranian Terrane (CIT). *Remote Sens.* **2020**, *12*, 1239. [CrossRef]
14. Mars, J.C.; Rowan, L.C. Regional mapping of phyllic and argillic-altered rocks in the Zagros magmatic arc, Iran, using Advanced Spaceborne Thermal Emission and Reflection Radiometer (ASTER) data and logical operator algorithms. *Geosphere* **2006**, *2*, 161–186. [CrossRef]
15. Duuring, P.; Hagemann, S.G.; Novikova, Y.; Cudahy, T.; Laukamp, C. Targeting iron Ore in banded iron formations using ASTER data: Weld Range Greenstone Belt, Yilgarn Craton, Western Australia. *Econ. Geol.* **2012**, *107*, 585–597. [CrossRef]
16. Ducart, D.F.; Silva, A.M.; Toledo, C.L.B.; Assis, L.M. Mapping iron oxides with Landsat-8/OLI and EO-1/Hyperion imagery from the Serra Norte iron deposits in the Carajás Mineral Province, Brazil. *Braz. J. Geol.* **2016**, *46*, 331–349. [CrossRef]
17. Zoheir, B.; El-Wahed, M.A.; Pour, A.B.; Abdelnasser, A. Orogenic Gold in Transpression and Transtension Zones: Field and Remote Sensing Studies of the Barramiya–Mueilha Sector, Egypt. *Remote Sens.* **2019**, *11*, 2122. [CrossRef]

18. Zoheir, B.; Emam, A.; Abdel-Wahed, M.; Soliman, N. Multispectral and Radar Data for the Setting of Gold Mineralization in the South Eastern Desert, Egypt. *Remote Sens.* **2019**, *11*, 1450. [CrossRef]
19. Sekandaril, M.; Masoumi, I.; Beiranvand Pour, A.M.; Muslim, A.; Hossain, M.S.; Misra, A. ASTER and WorldView-3 satellite data for mapping lithology and alteration minerals associated with Pb-Zn mineralization. *Geocarto Int.* **2020**, in press. [CrossRef]
20. Moradpour, H.; Rostami Paydar, G.; Pour, A.B.; Kamran, K.V.; Feizizadeh, B.; Muslim, A.M.; Hossain, M.S. Landsat-7 and ASTER remote sensing satellite imagery for identification of iron skarn mineralization in metamorphic regions. *Geocarto Int.* **2020**, in press. [CrossRef]
21. Safari, M.; Maghsoudi, A.; Pour, A.B. Application of Landsat-8 and ASTER satellite remote sensing data for porphyry copper exploration: A case study from Shahr-e-Babak, Kerman, south of Iran. *Geocarto Int.* **2018**, *33*, 1186–1201. [CrossRef]
22. Beygi, S.; Talovina, I.V.; Tadayon, M.; Pour, A.B. Alteration and structural features mapping in Kacho-Mesqal zone, Central Iran using ASTER remote sensing data for porphyry copper exploration. *Int. J. Image Data Fusion* **2020**, in press. [CrossRef]
23. Rani, K.; Guha, A.; Kumar, K.V.; Bhattacharya, B.K.; Pradeep, B. Potential Use of Airborne Hyperspectral AVIRIS-NG Data for Mapping Proterozoic Metasediments in Banswara, India. *J. Geol. Soc. India* **2020**, *95*, 152–158. [CrossRef]
24. Pour, A.B.; Hashim, M. Structural geology mapping using PALSAR data in the Bau gold mining district, Sarawak, Malaysia. *Adv. Space Res.* **2014**, *54*, 644–654. [CrossRef]
25. Pour, A.B.; Hashim, M.; Marghany, M. Exploration of gold mineralization in a tropical region using Earth Observing-1 (EO1) and JERS-1 SAR data: A case study from Bau gold field, Sarawak, Malaysia. *Arab. J. Geosci.* **2014**, *7*, 2393–2406. [CrossRef]
26. Pour, A.B.; Hashim, M. Integrating PALSAR and ASTER data for mineral deposits exploration in tropical environments: A case study from Central Belt, Peninsular Malaysia. *Int. J. Image Data Fusion* **2015**, *6*, 170–188. [CrossRef]
27. Pour, A.B.; Hashim, M. Structural mapping using PALSAR data in the Central Gold Belt, Peninsular Malaysia. *Ore Geol. Rev.* **2015**, *64*, 13–22. [CrossRef]
28. Pour, A.B.; Hashim, M.; Makoundi, C.; Zaw, K. Structural Mapping of the Bentong-Raub Suture Zone Using PALSAR Remote Sensing Data, Peninsular Malaysia: Implications for Sediment-hosted/Orogenic Gold Mineral Systems Exploration. *Resour. Geol.* **2016**, *66*, 368–385. [CrossRef]
29. Pour, A.B.; Hashim, M.; Park, Y. Gondwana-Derived Terranes structural mapping using PALSAR remote sensing data. *J. Indian Soc. Remote Sens.* **2018**, *46*, 249–262. [CrossRef]
30. Guha, A.; Yamaguchi, Y.; Chatterjee, S.; Rani, K.; Vinod Kumar, K. Emittance Spectroscopy and Broadband Thermal Remote Sensing Applied to Phosphorite and Its Utility in Geoexploration: A Study in the Parts of Rajasthan, India. *Remote Sens.* **2019**, *11*, 1003. [CrossRef]
31. Sun, L.; Khan, S.; Shabestari, P. Integrated Hyperspectral and Geochemical Study of Sediment-Hosted Disseminated Gold at the Goldstrike District, Utah. *Remote Sens.* **2019**, *11*, 1987. [CrossRef]
32. Tuşa, L.; Khodadadzadeh, M.; Contreras, C.; Rafiezadeh Shahi, K.; Fuchs, M.; Gloaguen, R.; Gutzmer, J. Drill-Core Mineral Abundance Estimation Using Hyperspectral and High-Resolution Mineralogical Data. *Remote Sens.* **2020**, *12*, 1218. [CrossRef]
33. Shirmard, H.; Farahbakhsh, E.; Beiranvand Pour, A.; Muslim, A.M.; Müller, R.D.; Chandra, R. Integration of Selective Dimensionality Reduction Techniques for Mineral Exploration Using ASTER Satellite Data. *Remote Sens.* **2020**, *12*, 1261. [CrossRef]
34. Martín-Crespo, T.; Gómez-Ortiz, D.; Martín-Velázquez, S.; Martínez-Pagán, P.; de Ignacio-San José, C.; Lillo, J.; Faz, Á. Abandoned Mine Tailings Affecting Riverbed Sediments in the Cartagena–La Union District, Mediterranean Coastal Area (Spain). *Remote Sens.* **2020**, *12*, 2042. [CrossRef]
35. Jackisch, R.; Lorenz, S.; Kirsch, M.; Zimmermann, R.; Tusa, L.; Pirttijärvi, M.; Saartenoja, A.; Ugalde, H.; Madriz, Y.; Savolainen, M.; et al. Integrated Geological and Geophysical Mapping of a Carbonatite-Hosting Outcrop in Siilinjärvi, Finland, Using Unmanned Aerial Systems. *Remote Sens.* **2020**, *12*, 2998. [CrossRef]
36. Ma, B.; Li, X.; Jiang, Z.; Pu, R.; Liang, A.; Che, D. Dust Dispersion and Its Effect on Vegetation Spectra at Canopy and Pixel Scales in an Open-Pit Mining Area. *Remote Sens.* **2020**, *12*, 3759. [CrossRef]

Article

Comparison of Different Algorithms to Map Hydrothermal Alteration Zones Using ASTER Remote Sensing Data for Polymetallic Vein-Type Ore Exploration: Toroud–Chahshirin Magmatic Belt (TCMB), North Iran

Lida Noori [1], Amin Beiranvand Pour [2], Ghasem Askari [1], Nader Taghipour [1,*], Biswajeet Pradhan [3,4], Chang-Wook Lee [5,*] and Mehdi Honarmand [6]

[1] School of Earth Sciences, Damghan University, Damghan 3671641167, Iran; lida.noori7161@yahoo.com (L.N.); gh.askari@du.ac.ir (G.A.)
[2] Korea Polar Research Institute (KOPRI), Songdomirae-ro, Yeonsu-gu, Incheon 21990, Korea; beiranvand.amin80@gmail.com
[3] Center for Advanced Modelling and Geospatial Information Systems (CAMGIS), Faculty of Engineering and Information Technology, University of Technology Sydney, 2007 New South Wales, Australia; Biswajeet.Pradhan@uts.edu.au or Biswajeet24@gmail.com
[4] Department of Energy and Mineral Resources Engineering, Choongmu-gwan, Sejong University, 209 Neungdong-ro Gwangjin-gu, Seoul 05006, Korea
[5] Division of Science Education, Kangwon National University, 1 Kangwondaehak-gil, Chuncheon-si 24341, Korea
[6] Department of Ecology, Institute of Science and High Technology and Environmental Sciences Graduate University of Advanced Technology, Kerman 7631133131, Iran; mehonarmand167@gmail.com
* Correspondence: taghipour@du.ac.ir (N.T.); cwlee@kangwon.ac.kr (C.-W.L.); Tel.: +98-913340-9359 (N.T.); +61-29514-7937 (C.-W.L.)

Received: 31 January 2019; Accepted: 22 February 2019; Published: 1 March 2019

Abstract: Polymetallic vein-type ores are important sources of precious metal and a principal type of orebody for various base-metals. In this research, Advanced Spaceborne Thermal Emission and Reflection Radiometer (ASTER) remote sensing data were used for mapping hydrothermal alteration zones associated with epithermal polymetallic vein-type mineralization in the Toroud–Chahshirin Magmatic Belt (TCMB), North of Iran. The TCMB is the largest known goldfield and base metals province in the central-north of Iran. Propylitic, phyllic, argillic, and advanced argillic alteration and silicification zones are typically associated with Au-Cu, Ag, and/or Pb-Zn mineralization in the TCMB. Specialized image processing techniques, namely Selective Principal Component Analysis (SPCA), Band Ratio Matrix Transformation (BRMT), Spectral Angle Mapper (SAM) and Mixture Tuned Matched Filtering (MTMF) were implemented and compared to map hydrothermal alteration minerals at the pixel and sub-pixel levels. Subtle differences between altered and non-altered rocks and hydrothermal alteration mineral assemblages were detected and mapped in the study area. The SPCA and BRMT spectral transformation algorithms discriminated the propylitic, phyllic, argillic and advanced argillic alteration and silicification zones as well as lithological units. The SAM and MTMF spectral mapping algorithms detected spectrally dominated mineral groups such as muscovite/montmorillonite/illite, hematite/jarosite, and chlorite/epidote/calcite mineral assemblages, systematically. Comprehensive fieldwork and laboratory analysis, including X-ray diffraction (XRD), petrographic study, and spectroscopy were conducted in the study area for verifying the remote sensing outputs. Results indicate several high potential zones of epithermal polymetallic vein-type mineralization in the northeastern and southwestern parts of the study area, which can be considered for future systematic exploration programs. The approach used in this research has great implications for the exploration of epithermal polymetallic vein-type mineralization in other base metals provinces in Iran and semi-arid regions around the world.

Keywords: Toroud–Chahshirin Magmatic Belt (TCMB); remote sensing; ASTER; hydrothermally altered zones; polymetallic vein-type mineralization

1. Introduction

Since the Bronze Age, polymetallic vein-type ores have been important sources of precious metal and established a main type of deposit for various base-metals [1–5]. Polymetallic vein-type ore deposits precipitated in the geological structures such as faults, fractures, brecciated rocks, and porous layers, where the pressure, temperature, and several other chemical factors are suitable for the precipitation [1,6]. Moreover, during the ore mineral precipitation processes, hydrothermal fluids react with the mineral constituents of lithological units they are passing and produce hydrothermal alteration zones with distinctive mineral assemblages [7]. The presence of intrusive rocks and hydrothermal alteration zones associated with polymetallic vein-type deposits provide an important guide for exploring this type of ore mineralization especially by the application of advanced satellite remote sensing data [8,9].

The Advanced Spaceborne Thermal Emission and Reflection Radiometer (ASTER) contains appropriate spectral and spatial resolution to detect spectral absorption features of hydrothermal alteration minerals and lithological units [10–16]. ASTER datasets can be used to identify and remotely map hydrothermal alteration zones associated with polymetallic vein-type ore deposits in vegetated regions and well-exposed terrain, especially in a semi-arid environment [8,9,17]. ASTER measures reflected radiation in three bands in the 0.52- to 0.86 μm (the visible and near-infrared (VNIR)), six bands in the 1.6- to 2.43 μm (the shortwave infrared (SWIR)), and five bands of emitted radiation in the 8.125- to 11.65 μm (the thermal infrared (TIR)) with 15, 30, and 90 meter resolution, respectively [18,19]. Hydrothermal alteration zones associated with various ore deposits such as porphyry copper, orogenic gold, epithermal gold, massive sulfide, iron, and chromite deposits have been successfully detected and mapped using ASTER imagery in metallogenic provinces around the world [20–27]. Specifically, some studies used ASTER data for the exploration of polymetallic vein-type ore deposits. Mahanta and Maiti [8] used ASTER VNIR+SWIR spectral data for mapping alteration zones such as kaolinization, ferruginization, silicification, phosphatization, and sulphidation associated with the polymetallic vein-type mineralization in the South Purulia Shear Zone (SPSZ), East India. In a recent work, Ahmadirouhani et al. [9] identified and mapped hydrothermal alteration zones, including propylitic, phyllic, argillic, and gossan with Cu-Fe-Au vein-type mineralization using ASTER VNIR+SWIR spectral bands in the Bajestan region, northern sector of the Lut Block, East Iran.

The Toroud–Chahshirin Magmatic Belt (TCMB) is located in Semnan province, central-north Iran (Figure 1A) and contains numerous occurrences of epithermal polymetallic vein-type mineralization (Figure 1B). The TCMB is the largest known goldfield and base metals province in central-north Iran [28,29]. Hydrothermally altered zones are reportedly associated with polymetallic vein-type mineralization in this belt [29–33]. Propylitic, phyllic, argillic, and advanced argillic alteration and silicification occur generally with Au-Cu, Ag, and/or Pb-Zn mineralization. Therefore, mapping and identification of hydrothermal alteration mineral assemblages using ASTER satellite remote sensing data in the TCMB can be considered as a cost-effective and applicable tool for targeting and prospecting epithermal polymetallic vein-type mineralization. In this research, the Moaleman region of the TCMB was selected (Figure 1B). This region has a high potential for epithermal polymetallic vein-type mineralization, particularly anomal Cu-Au values associated with altered dacite and dacite-andesite and volcaniclastics rocks [30,34,35]. Since no report on a comprehensive remote sensing investigation is available for epithermal Cu-Au exploration in the Moaleman region of the TCMB, results of an ASTER remote sensing mapping are necessary for future systematic exploration projects. The main objectives of this study are (i) to detect hydrothermal alteration mineral zones and assemblages using VNIR and SWIR spectral bands of ASTER data by application of specialized image processing techniques, including Selective Principal Component Analysis (SPCA) [36,37], Band Ratio Matrix Transformation (BRMT) [16], Spectral Angle Mapper (SAM) [38], and Mixture Tuned Matched Filtering (MTMF) [39–41]; (ii) to compare the results derived from SPCA and BRMT transformation algorithms and SAM and MTMF spectral mapping algorithms to map alteration minerals at the pixel and sub-pixel levels and (iii) to prospect high potential zones of epithermal Cu-Au mineralization for future systematic exploration programs in the study area.

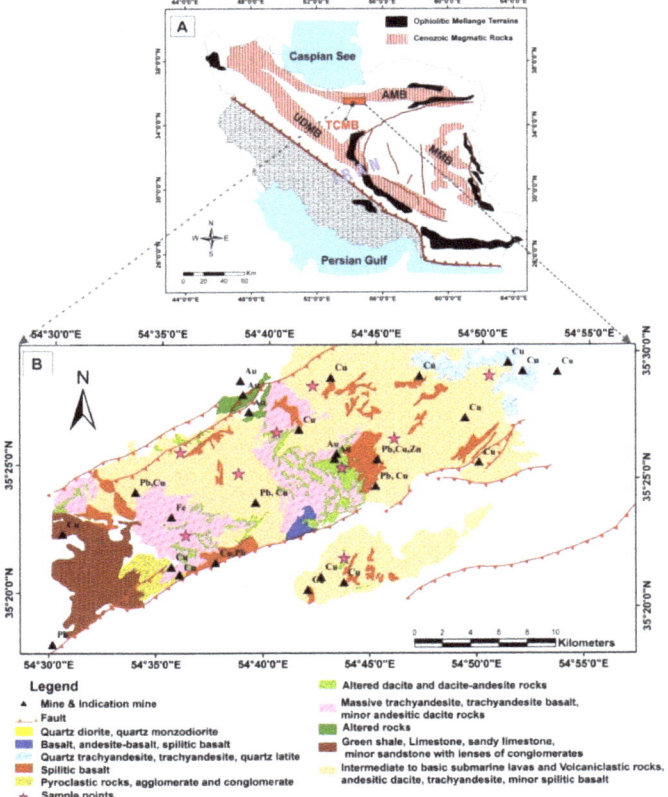

Figure 1. (**A**) The geographical location of the Toroud–Chahshirin Magmatic Belt (TCMB) in North of Iran (red rectangle). (**B**) Simplified geological map of the Moaleman region showing the distribution of polymetallic vein-type mineralization (modified from [42]). Abbreviations: UDMB: Urumieh–Dokhtar Magmatic Belt; MMB: Makran Magmatic Belt; AMB: Alborz Magmatic Belt.

2. Geology of the TCMB and Mineralization

The TCMB is situated in the central to eastern Alborz Magmatic Belt (AMB) and lies in the northern part of the Central Iran Structural Zone (CISZ) (Figure 1A). It is restricted between the E–W trending Toroud fault in the south and the E–W trending Anjilou fault in the north [29]. This magmatic arc has a complex tectonic, magmatic, and stratigraphy history [29,30]. Alavi et al. [43] and Alavi [44] and proposed that Torud–Chahshiran range and adjacent volcanic rocks are displaced to Eocene magmatism in the CISZ to the south. The Magmatic arc contains mainly of igneous rocks of Tertiary age, while there are also scattered outcrops of metamorphosed Paleozoic and Mesozoic rocks [45]. Most of the magmatic activities in the TCMB occurred in the middle to late Eocene and have been divided into three stages from oldest to youngest, including (i) explosive volcanic activity represented by rhyolite to rhyodacite tuffs and locally andesitic lava flows, with subordinate marls, tuffaceous marlstones, and sandstones; (ii) lava flows and pyroclastic rocks of andesite, trachyandesite, and basaltic andesite composition; and (ii) subordinate dacitic-rhyodacitic rocks and hypabyssal intrusive rocks [46].

In the magmatic belt, many of the known mined deposits (gold and base metals) are associated with hydrothermally altered zones [30,31]. These deposits mostly include epithermal veins such as Kuhe Zar (Au-Cu), Abolhassani (Pb-Zn-Ag-Au), Pousideh (Cu), Dian (Cu), Cheshmeh Hafez (Pb-Zn), Gandy (Au-Ag, Pb-Zn), Chahmussa (Cu), Darestan (Cu), and Robae (Fe-Cu) mineralization zones. Moreover, Baghu (placer gold, turquoise) and Challu (Cu), Khanjar (Pb-Ag-Zn) are other types of mineralization in the sedimentary rocks of the TCMB [30,47–50]. Hydrothermal alterations such as propylitic, phyllic, argillic and advanced argillic, and silicification are reported in this magmatic arc, generally where alkaline to sub-alkaline plutonic rocks such as andesite are intruded into hosted volcano pyroclastic rocks [29,47]. For instance, mineralization in the Gandy deposit occurs in quartz-sulfide veins and breccias and is associated with alteration halos of argillic and propylitic (approximately 4 km) as well as narrow supergene jarosite, kaolinite, and iron hydroxide [29,31]. The granodiorite rocks are intruded in pyroclasitic and andesitic lavas in the Kuh-Zar deposit and Au and Cu rich hydrothermal fluids affected the host rocks which were followed by propylitic, phyllic, argillic and silicification alterations [50]. In the Moaleman area, mineralization is controlled by major faults. Numerous ore mineralizations such as Cu, Au, Pb, Zn, and Fe were reported in this area (Figure 1B). Although these deposits have a small size, most of them contain valuable economic mineralization [35].

3. Materials and Methods

3.1. Pre-processing of the Remote Sensing Data

A cloud-free level 1B ASTER in hierarchical data format (HDF) that was acquired on 25 March 2003 was used for remote spectral analyses in this study. This image was pre-georeferenced to UTM zone 40 North projections with using the WGS-84 data. The SWIR bands re-sampled to the spatial resolution of VNIR so that all pixels of nine bands (VNIR+SWIR) with 15*15 m^2 pixel size were stacked. The level 1B data product measures radiance at the sensor without atmospheric corrections [18]. Therefore, atmospheric correction is necessary before image processing analysis for converting radiance-calibrated data to apparent reflectance. The Log-residual atmospheric correction technique was applied to the ASTER image in this study. The Log Residuals calibration is capable to remove atmospheric transmittance, topographic effects, solar irradiance, and albedo effects [51,52]. It produces a pseudo reflectance image, which is highly applicable for detecting absorption features related to alteration minerals. Additionally, Crosstalk correction was applied to the ASTER dataset [53]. We have performed this correction by Cross-Talk correction software that is available from www.gds.aster.ersdac.or.jp. The ENVI (Environment for Visualizing Images, http://www.exelisvis.com) version 5.2 and ArcGIS version 10.3 software (Esri, Redlands, CA, USA) packages were used to process the remote sensing datasets.

3.2. Image Processing Algorithms

The main target of the specialized image processing techniques adopted in this study was to apply image processing techniques that are capable of mapping hydrothermal alteration minerals at the pixel and sub-pixel levels using VNIR+SWIR spectral bands of ASTER data. Therefore, subtle differences between altered and non-altered rocks and hydrothermal mineral assemblages could be feasible by implementing specialized image processing techniques as follows.

3.2.1. Principal Component Analysis (PCA)

The PCA is a multivariate statistical technique that used to reduce the dimensionally of input data and reduces the additional frequency among the data, as a result, the possibility of useful data loss is minimized. This information in terms of quantity is a very small part of the overall information content available in original bands. Spectral bands are selected that contain absorption and reflection features of alteration minerals. In this way, a new image (PC) is generated on the axes with the new coordinate system [36,37]. The resulting component is more interpretable than are the original images. A PC image contains the unique contribution of eigenvector loadings (magnitude and sign) for absorption and reflection bands of alteration mineral or mineral group is able to enhance the mineral or mineral group. If the loading is positive in the reflective band of a mineral the image tone will be bright, and if it is negative, the image tone will be dark for the enhanced target mineral [37]. The PCA technique has been applied to multispectral remote sensing data such as ASTER for highlighting spectral responses related to specific hydrothermal alteration minerals associated with porphyry copper mineralization [54–59].

In this research, the Selective Principal Component Analysis (SPCA) [60], known also as Directed Principal Component Analysis (DPCA) [61], was applied on VNIR+SWIR bands for mapping the specific alteration zones associated with polymetallic vein-type mineralization in the study area. The basic difference between the PCA and SPCA is that in the SPCA only a subgroup of bands is selected depending on the aims that plan to be achieved. In this study, according to the known ASTER band indices for hydrothermal alteration mineral mapping [62–65], some subsystems (specific bands) were selected for SPCA analysis. Bands 1, 2, and 4 were selected for mapping iron oxides/hydroxide minerals (Table 1A). Bands 4, 5, and 6 were designated for argillic alteration mapping (Table 1B). Bands 5, 6, and 7 were nominated to specify phyllic zone (Table 1C) and bands 7, 8, and 9 were used to map propylitic alteration zones.

Table 1. Eigenvector loadings matrix calculated using Selective Principal Component Analysis (SPCA) for selected Advanced Spaceborne Thermal Emission and Reflection Radiometer (ASTER) bands. (**A**) Iron oxides/hydroxides minerals; (**B**) Argillic alteration; (**C**) Phyllic alteration; and (**D**) Propylitic alteration.

A	Band1	Band2	Band4	B	Band4	Band5	Band6
PC1	−0.57	−0.57	−0.58	PC1	0.58	−0.57	−0.57
PC2	−0.42	−0.39	0.81	PC2	−0.81	0.35	0.46
PC3	−0.7	−0.14	−0.14	PC3	−0.61	−0.73	0.67
C	Band5	Band6	Band7	D	Band7	Band8	Band9
PC1	−0.57	−0.57	−0.57	PC1	−0.58	−0.57	−0.57
PC2	−0.42	−0.39	0.81	PC2	−0.73	0.06	0.67
PC3	−0.69	0.71	0.02	PC3	0.35	−0.81	0.45

3.2.2. Band Ratio Matrix Transformation (BRMT)

The Band Ratio Matrix Transformation (BRMT) is a semiautomatic lithological-mineralogical mapping technique, which is proposed and established for sedimentary strata discrimination [16]. This analytical method extracts key spectral characteristics using VNIR and SWIR spectral bands of ASTER. Although BRMT methodology is proposed for sedimentary rocks discrimination, the effectiveness of this technique for mapping hydrothermal alteration zones in igneous rocks background is promising. The robustness of BRMT arises from a combination of statistical factors, including variance percent (V%), positive and negative correlation averages (±r), correlation averages (±rk), and contribution percent (C%). It maps key spectral characteristics using RGB color composites and rule classifier of the band ratio (BR) and band transform (BT) bands. In this study, the BRMT method was applied to VNIR+SWIR bands of ASTER data for detailed mapping of hydrothermal alteration zones in the study area. Table 2 shows positive and negative correlation and contribution percent of the band ratios n1–n36 used for BRMT transformation. Table 3 shows eigenvalues, variance percent (%V), positive and negative correlation averages (+rk > 0.1 and −rk < 0.1), and correlation percent of BT1 to BT10, which hold maximum spectral properties extracted from the image. The selected BT bands containing specific spectral properties (negative and positive contribution percent >3%) are listed in Table 4.

Table 2. Positive and negative correlation (±r) and contribution percent (C%) of the band ratios n1–n36 used for Band Ratio Matrix Transformation (BRMT).

Band Ratio	Negative		Positive		Band Ratio	Negative		Positive	
	$-r$	C%	$+r$	C%		$-r$	C%	$+r$	C%
n1	−0.34	1.86	0.40	4.81	n19	−0.58	3.13	0.12	1.49
n2	−0.86	4.65	0.36	4.38	n20	−0.60	3.24	0.13	1.61
n3	−0.54	2.91	0.28	3.42	n21	−0.37	2.01	0.12	1.40
n4	−0.98	5.32	0.12	1.39	n22	−0.82	4.44	0.22	2.66
n5	−0.56	3.06	0.12	1.50	n23	−0.36	1.96	0.31	3.71
n6	−0.60	3.26	0.12	1.45	n24	−0.36	1.98	0.35	4.23
n7	−0.63	3.42	0.13	1.57	n25	−0.37	2.03	0.33	4.02
n8	−0.45	2.46	-	-	n26	−0.51	2.75	0.52	6.22
n9	−0.32	1.75	0.32	3.87	n27	−0.44	2.37	0.21	2.56
n10	−0.55	3.01	0.15	1.79	n28	−0.41	2.25	0.31	3.68
n11	−0.55	2.96	0.11	1.36	n29	−0.43	2.33	0.34	4.07
n12	−0.41	2.23	-	-	n30	−0.36	1.94	0.56	6.74
n13	−0.98	5.32	0.13	1.62	n31	−0.32	1.75	0.33	4.04
n14	−0.40	2.18	0.11	1.30	n32	−0.37	2.02	0.32	3.85
n15	−0.48	2.62	-	-	n33	−0.38	2.06	0.39	4.69
n16	−0.54	2.94	0.18	2.21	n34	−0.23	1.27	0.40	4.83
n17	−0.95	5.19	0.26	3.17	n35	−0.52	2.84	0.18	2.18
n18	−0.39	2.11	0.16	1.98	n36	−0.44	2.38	0.18	2.22

Table 3. Eigenvalues, variance percent (%V) and positive and negative correlation averages ($+r_k > 0.1$ and $-r_k < 0.1$) of the forward BRMT for BT1–BT10.

BT Number	1	2	3	4	5	6	7	8	9	10
Eigenvalue	0.12300000006606	0.00928	0.00643	0.00335	0.00255	0.00125	0.00098	0.00083	0.00005	0.00002
V%	83.24	6.28	4.35	2.27	1.72	0.85	0.66	0.56	0.03	0.01
$+\bar{r}_k$	0.40	0.25	0.17	0.22	0.23	0.27	0.21	0.24	-	-
$+\bar{r}_k\%$	20.07	12.70	8.51	11.02	11.39	13.67	10.58	12.06	-	-
$-\bar{r}_k$	−0.92	−0.42	−0.38	−0.36	−0.27	−0.24	−0.27	−0.26	-	-
$-\bar{r}_k\%$	29.48	13.36	12.08	11.54	8.78	7.62	8.82	8.32	-	-

Table 4. The selected band transform bands (BTs) contain specific spectral properties (negative and positive contribution percent > 3%).

Contribution > 3%	BT													
Positive	n30	n26	n34	n1	n33	n2	n24	n29	n31	n25	n9	n32	n23	n28	n3	n17
Negative	n10	n5	n19	n20	n6	n7	n22	n2	n17	n13	n4	-	-	-	-	-

3.2.3. Mixture Tuned Matched Filtering (MTMF)

The MTMF is a partial unmixing, hybrid method based on the combination of well-known signal processing methodologies and liner mixture theory [39–41]. MTMF consist of two phases, an MF calculation for abundance estimation and a mixture tuning calculation for the identification and rejection of false positives [6–68]. In this study, the endmembers were extracted from the image using the Minimum Noise Fraction (MNF), Pixel Purity Index (PPI), and n-dimensional visualization techniques [69,70]. The MNF method was accomplished to separate noise from the data. The PPI detects the highest purity pixels in the image. The n-dimensional visualization was used to locate, identify, and cluster the purest pixels and the most extreme spectral responses (endmembers) in the VNIR+SWIR ASTER dataset. The threshold of 2.5 was applied to PPI. The output of PPI represents as bright pixels (more purity) and dark pixels (less spectral purity). After applying the n-dimensional visualization method, 10 n-D classes (endmembers) were extracted, which indicate distinctive absorption features related to alteration minerals. Figure 2 shows the extracted endmember spectra (n-D classes) for the study area. Comparison of the absorption characteristics of the extracted endmembers (n-D classes) with the USGS spectral library is considered for identification of alteration minerals.

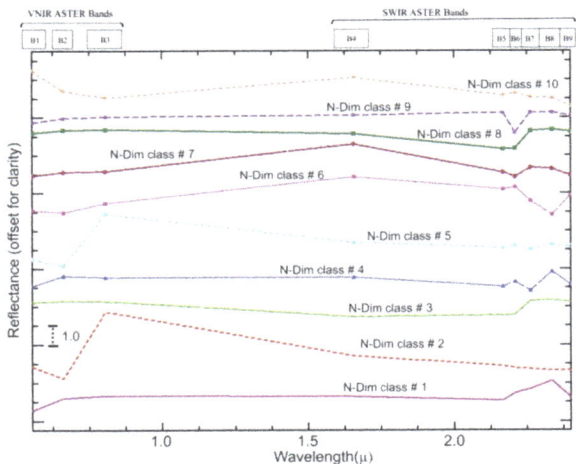

Figure 2. The endmember mineral extracted from the ASTER dataset using of n-Dimensional visualization method. The reference spectra were resampled to the response function of ASTER bands. The ASTER band center positions are also shown.

3.2.4. Spectral Angle Mapper (SAM)

The SAM is a classification method that classifies the pixels based on similar spectral properties with reference spectra [38]. It uses endmember spectra that can be extracted directly from the image or spectral library. This process of calculating the spectral angle between the reference spectra and image determines the similarity of these two groups. The outputs of the algorithm are angles between zero and one. The large angle shows less similarity and the smaller angle more similarity [71,72]. In this analysis, the SAM was applied to VNIR+SWIR ASTER bands with default value 0.1 (radians) as a threshold, and the spectra for running the algorithm were obtained from the USGS library as reference spectra [73]. Figure 3 shows end-member minerals extracted from the USGS library, including hematite, jarosite, montmorillonite, illite, muscovite, and chlorite.

3.3. Fieldwork Data, Laboratory Analysis and Verification

For collecting field data, 10 zones showing the high spatial distribution of hydrothermal alteration minerals in the Moaleman region were selected. Global positioning system (GPS) survey was acquired using a GPS (Garmin eTrix 10, Nanjing Sifang Mapping Equipment Ltd., Jiangsu, China) with an average accuracy of 5m to find the exact location of the selected zones in the ASTER scene. Field photos were taking from alteration zones. Seventy rock samples were collected from the alteration zones and lithological units for laboratory analysis, including thin section preparation, X-ray diffraction (XRD) analysis, and analytical spectral devices (ASD) spectroscopy. The XRD analysis was applied using Advance-D8 XRD Bruker model at Central Laboratory of Damghan University, Damghan, Iran. The exposure time of powder samples (1 g) was about one hour by a monochromatic ray in the wavelength of 5.4 Å and ranging angles between 5 to 70 degrees. The step of diffraction was set as 0.2 degrees to guarantee detection of clay minerals which are detectable in low angles. The copper anode is used to generate X-rays with a voltage of 25 kilovolts (kV). The spectra of the representative samples from altered zones were measured using a FieldSpec3®spectroradiometer operating in the 0.35-2.5 μm spectral range at the Department of Ecology, Institute of Science and High Technology and the Environmental Sciences Graduate University of Advanced Technology, Kerman, Iran. The fore-optics were at a small distance from the surface under observation. Spectralon of Labsphere which is made of polytetrafluoroethylene (PTFE) and sintered halon G-80 was used as a white reference panel.

About 10 measurements were performed per spot. Moreover, the Kappa coefficient was calculated using a Matlab code developed by Askari et al. [74] for SAM and MTMF results for accuracy assessment (Table 5A,B).

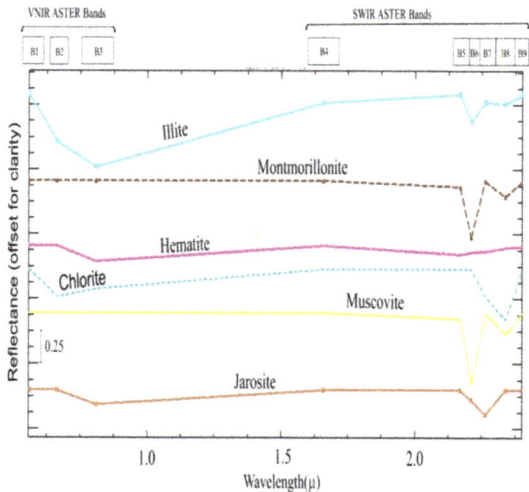

Figure 3. Reference endmember spectra of hematite, jarosite, montmorillonite, illite, muscovite, and chlorite obtained from the USGS spectral library for mapping alteration zones in the study area. The reference spectra were resampled to the response function of ASTER bands. The ASTER band center positions are also shown.

Table 5. (A) The accuracy assessment for the Mixture Tuned Matched Filtering (MTMF) method based on GPS survey collected during the field study. (B) The accuracy assessment for the Spectral Angle Mapper (SAM) method based on GPS survey collected during the field study.

	(A) Ground Truth Samples			
Class	Argillic	Phyllic	Propylitic	User Acc. (Percent)
Unclassified	1	15	1	—
Argillic	57	0	0	98.2
Phyllic	0	84	0	84.8
Propylitic	0	0	49	98
Prod. Acc.	100	100	100	
Over. acc. (Percent)			Kappa Coef. (Percent)	
95.7			0.93	
	(B) Ground Truth Samples			
Class	Argillic	Phyllic	Propylitic	User Acc. (Percent)
Unclassified	1	2	9	-
Argillic	53	25	6	62
Phyllic	4	72	0	92.3
Propyllitic	0	0	35	79.5
Prod. Acc.	92.9	74.2	85.3	-
Over. acc. (Percent)			Kappa Coef. (Percent)	
84.4			0.78	

4. Results and Discussion

4.1. Alteration Mapping Results Derived from ASTER Data

Considering statistical results derived from SPCA (Table 1A–D), it evident that hydrothermally altered rocks were mapped in the SPC images that contain a unique contribution of components (eigenvectors) related to spectral characteristics of the alteration minerals. Table 1A shows eigenvector values for mapping iron oxide/hydroxide minerals in the study area. Analyzing the eigenvector loadings shows that the SPC2 contains a strong to moderate contribution of band 1 (−0.42) and band 2 (−0.39) with negative signs and a strong contribution of band 4 (0.81) with a positive sign. Iron oxide minerals have absorption features in bands 1 and 2 and reflectance features in band 4 of ASTER, respectively [54,62,75]. Therefore, the SPC2 is able to enhance oxide/hydroxide minerals as bright pixels due to opposite signs of the eigenvector loadings in the absorption bands (negative signs in bands 1 and 2) and reflection band (positive sign in band 4) (Figure 4A).

Looking at the eigenvector loadings in Table 1B for mapping argillic alteration indicates that the SPC3 has a strong contribution of band 4 (−0.61) and band 5 (−0.73) with negative signs and a strong contribution of band 6 (0.67) with a positive sign, respectively. Kaolinite and alunite are main constituents of argillic alteration that normally exhibit Al-OH absorption features in bands 5 and 6 of ASTER [63]. Al-OH minerals show maximum reflectance features in band 4 of ASTER that covers the spectral region of 1.6 μm [64]. Thus, the argillic alteration zone appears as dark pixels in the SPC3 image because of a negative sign in band 4 (reflection band). Dark pixels in the SPC3 image were inverted to bright pixels by multiplication to −1 (Figure 4B).

Table 1C shows the eigenvector loadings for mapping phyllic alteration zone. The SPC3 shows strong eigenvector loadings for band 5 (−0.69) and band 6 (0.71) with opposite signs, while band 7 (0.02) has a very small contribution in the SPC3. The phyllic zone composed of illite/muscovite (sericite) produces an intense Al-OH absorption feature at band 6 of ASTER [52]. Phyllic alteration zone in the study area manifests in bright pixels in the SPC3 image (Figure 4C). Considering the eigenvector loadings in Table 1D for identification of propylitic alteration zone in the study area, the SPC3 contains strong loadings of band 8 (−0.81) and moderate contribution of bands 7 (0.35) and 9 (0.45) with opposite signs, respectively. The propylitic alteration zone consisting of epidote, chlorite, and calcite display strong absorption features in band 8 of ASTER [64]. For that reason, the SPC3 image depicts propylitic alteration zone as bright pixels in the study area (Figure 4D).

Propylitic, phyllic, argillic, and advanced argillic alteration zones were reported as dominated alteration zones with epithermal polymetallic vein-type mineralization in the study area [30]. The surface distribution pattern of iron oxide/hydroxide minerals, argillic alteration zone, phyllic zone, and propylitic zone is almost similar and mostly concentrated in the central and southwestern parts of the study area (see Figure 4A–D). However, argillic zone and iron oxide/hydroxide minerals show more similar spatial distribution and strong surface abundances compare to phyllic and propylitic zones. Accordingly, a Red-Green-Blue (RGB) color composite was assigned to the SPC3 of argillic alteration, SPC3 of phyllic alteration, and SPC3 of propylitic alteration images for providing a false color-based classification image of the detected pixels. Figure 5 shows the resultant image for the study area. Argillic alteration zone appears in red and yellow colors, it is evident that the red zone can be considered as the advanced argillic and yellow zone is a combination of Argillic and phyllic alteration zones (Figure 5). The phyllic alteration zone is represented in green and cyan colors. The mixture of phyllic and propylitic alteration zones depict as cyan color. Propylitic alteration zone manifests as blue color (Figure 5). Comparison with the geological map of the study area (see Figure 1B), most of the ore mineralizations are concentrated in argillic and advanced argillic alteration zones (see Figure 5), which are associated with dacite and dicite-andesite, spilitic basalt, trachyandesite basalt, and quartz trachyandesite lithological units. Several advanced argillic (red mixed with yellow pixels) zones are observable in the southwestern and northeastern parts of the study area, which could be considered as prospective zones.

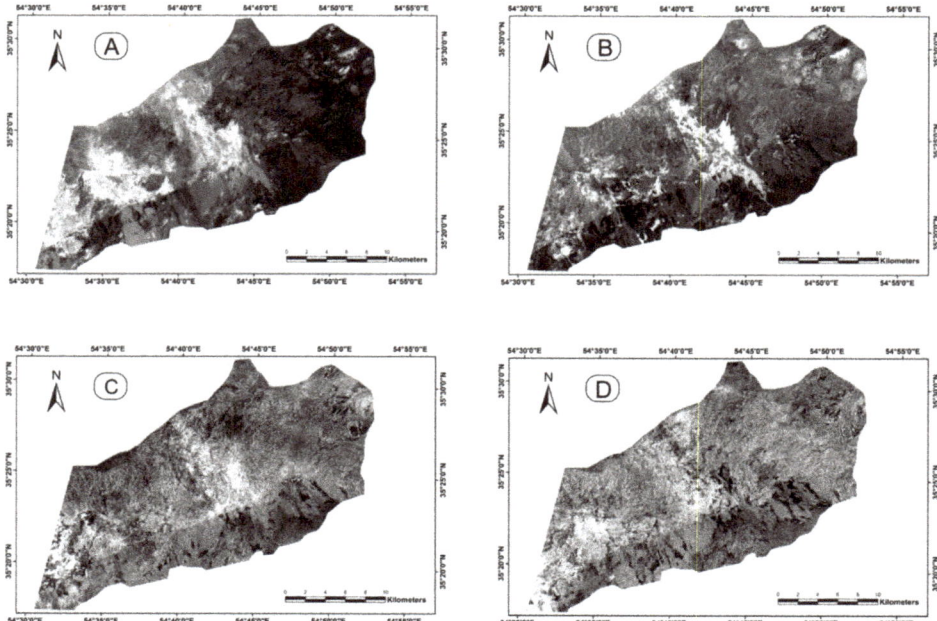

Figure 4. SPC images derived from SPCA analysis. (**A**) SPC2 image showing oxide/hydroxide minerals as bright pixels; (**B**) SPC3 image showing the argillic alteration zone as bright pixels; (**C**) SPC3 image showing the phyllic alteration zone as bright pixels; and (**D**) SPC3 image showing the propylitic alteration zone as bright pixels.

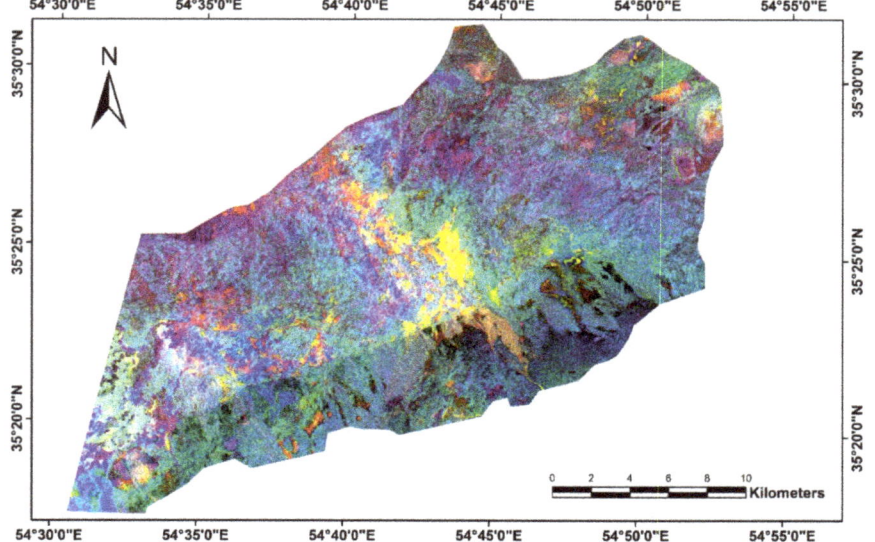

Figure 5. RGB false color composite of the SPC3 of argillic alteration (R), SPC3 of phyllic alteration (G), and SPC3 of propylitic alteration (B) images covering the study area.

Considering of statistical results calculated for the BRMT algorithm (Tables 2–4), the eigenvalues, Vi%, positive and negative correlation averages ($+\bar{r}_k > 0.1$ and $-\bar{r}_k < 0.1$) for the BT1, BT2, and BT3 are considerable (Table 3). High number of eigenvalues, Vi%, and $+\bar{r}_k\%$ and $-\bar{r}_k\%$ were estimated in the BT1, BT2, and BT3. The BT1 contain the highest eigenvalue (0.123) Vi% (83.24), $+\bar{r}_k\%$ (20.7), and $-\bar{r}_k\%$ (29.48). The BT2 shows eigenvalue of 0.00928, Vi% of 6.28, $+\bar{r}_k\%$ of 12.70, and $-\bar{r}_k\%$ of 13.36. The BT3 has an eigenvalue of 0.00643, Vi% of 4.35, $+\bar{r}_k\%$ of 8.51, and $-\bar{r}_k\%$ of 12.08 (Table 3). It shows that these BTs contain most of the spectral information that was extracted by the BRMT algorithm from the image. Therefore, the BTs were used for producing RGB false color composite to reveal the most spectrally dominated hydrothermal alteration zones and lithological units in the study area (Figure 6). With reference to SPCA results, it is discernable that argillic, phyllic, and advanced argillic alteration zones are most spectrally dominated alteration zones in the study area, which appear as magenta color in Figure 6. These alteration zones are typically concentrated in the central, southwestern, and northeastern parts of the study area, which contain mineralogically interesting zones for ore exploration.

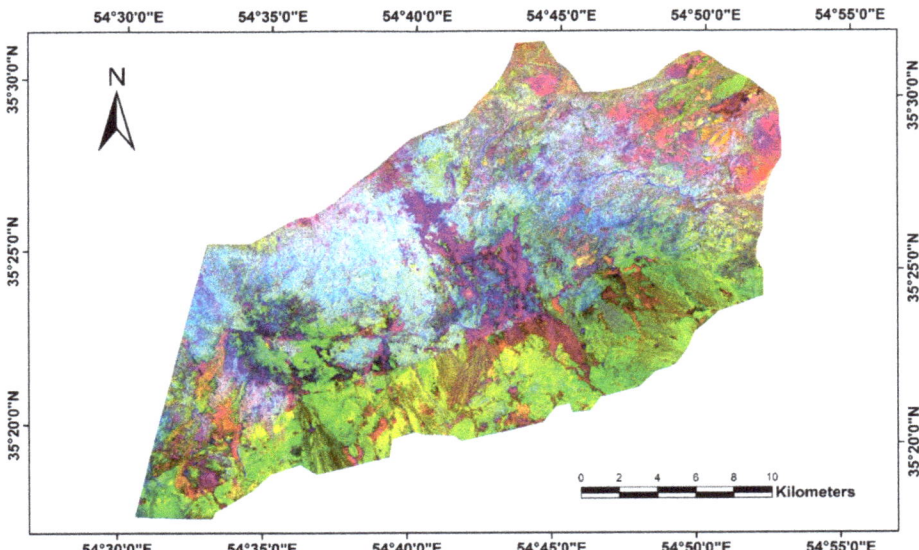

Figure 6. RGB false color composite of the BT1 (R), BT2 (G), and BT3 (B) derived from the BRMT algorithm for the study area.

The analysis of the negative and positive contribution percent > 3% for BTs (see Tables 2 and 4) indicates that the BTs contain a negative contribution holding more specific spectral properties attributed to lithological units and hydrothermal alteration zones. Therefore, the BTs containing a negative contribution > 3% (Table 4), such as n10, n5, n19, n20, n6, n7, n22, n2, n17, n13, and n4, were used for running the rule image classifier. Figure 7 shows the resultant BRMT classification map for the study area. Eight classes (C1 to C8) were identified. The class C1 (red pixels) can be considered a moderate propylitic alteration zone that is combined with unaltered volcaniclastic rocks. This class (C1) mainly covers the eastern and northeastern parts of the study area. The class C2 (blue pixels) includes the advanced propylitic alteration zone, which is generally concentrated in the central and southwestern part of the study area (Figure 7). The advanced argillic alteration zone is depicted in class 3 (green pixels), which covers typically central and western parts and many other small exposures in the whole of the study area. Class C4 (cyan pixels) is a combination of argillic, iron oxide, and propylitic mineral assemblages and sedimentary rocks that weathered and transferred to the alluvial fan. Class 5 (brown pixels) is an accumulation of iron oxide minerals that are within the highly altered parts of argillic and phyllic zones. Sandstone and alluvium can be considered in class 6 (light yellow pixels), while class 7 (magenta pixels) might be an admixture of some weathered rocks of classes 1, 2, and 3. Class 8 (mustard pixels) represents unknown units that may consist of some weathered and transferred sedimentary rocks. By using the geological map of the study area (see Figure 1) as a reference, it is obvious that the most of the reported ore mineralizations in the study area are located in the interior of class 3 (advanced argillic alteration zone) of the BRMT classification map (Figure 7). Thus, some perspective zones (green pixels) could be considered in the southwestern and northeastern sectors.

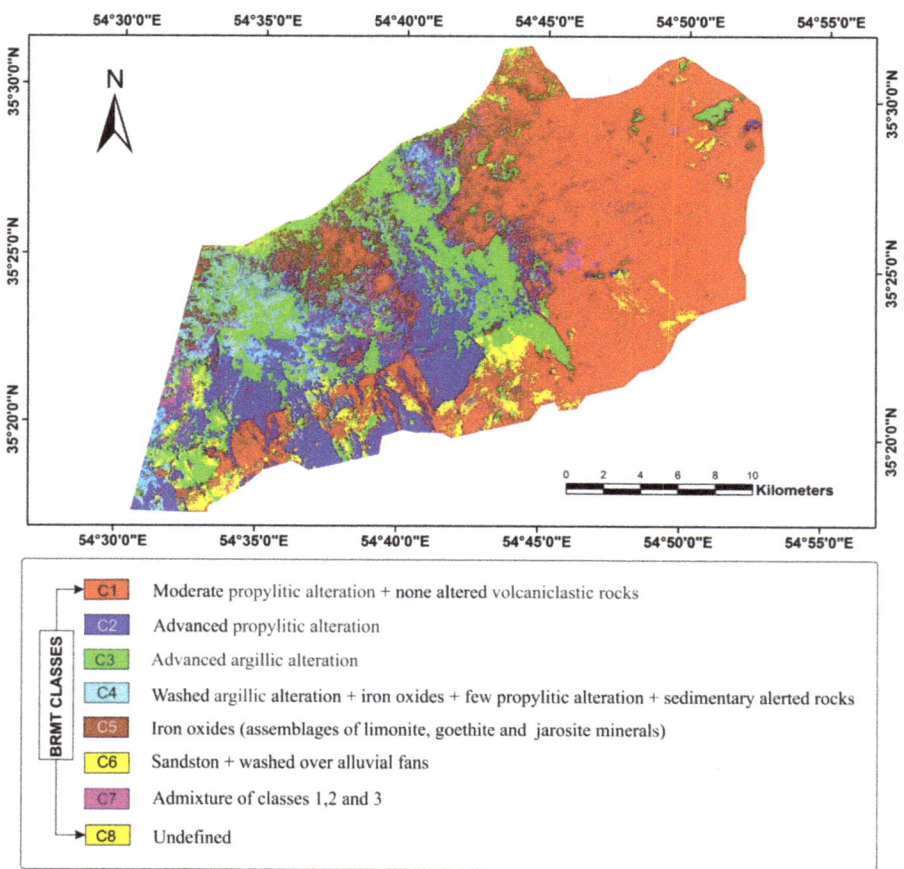

Figure 7. BRMT classification map for the study area.

Within the VNIR+SWIR interval, for identification of the n-D classes (#1 to #10) derived from the n-Dimensional visualization analysis technique (see Figure 3), diagnostic absorption features related to electronic process intensities and crystal field transitions due to Fe^{2+}, Fe^{3+}, and REE transitional metals in the VNIR [76,77] and vibrational overtones of the fundamental Al-OH/Fe-OH/Mg-OH and CO_3 in the SWIR [78,79] should be taken into consideration. Generally, dominant spectral patterns correspond to the most abundant minerals or spectrally dominant mineral groups. Subordinate spectral patterns produce spectral variability due to spectrally less active or less abundant phases in the rock [80]. Therefore, the wavelength shift of the distinctive absorption features directly or inversely depends on the abundance of spectrally active mineral groups within the rocks. Comparison of the absorption characteristics of the n-D classes extracted for the study area with the USGS spectral library indicate some spectrally distinctive mineralogical phases (see Figure 2). The n-D class #1 does not contain any distinctive absorption features related to alteration minerals and could be considered as an unaltered/unknown class. The n-D class #2 represents diagnostic absorption features related to Fe3+ transitional metals, coinciding with bands 2 and 3 of ASTER. Thus, it contains iron oxide/hydroxide absorption features, which are attributable to hematite. There are no recognizable absorption features in the n-D class #3, which can be classified as unaltered/unknown.

Analyzing the n-D class #4 and n-D class #5 indicates spectral signatures attributed to jarosite, which correspond with bands 2, 3, and 7 of ASTER (see Figure 2). In fact, band 7 of ASTER is able to detect Fe-OH absorption features caused by jarosite and/or Fe-muscovite [81]. So, the n-D class #4 might contain spectral signatures of other mineral groups with subordinate spectral patterns. The n-D class #6 represents chlorite due to major Mg, Fe-OH absorption properties in band 8 of ASTER (2.30–2.360 μm). A major Al-OH absorption feature positioned in band 6 of ASTER is obvious in the n-D class #7, which reflects the spectral signatures of illite (see Figures 2 and 3). The n-D class #8 might be considered as mixed spectral signatures of illite, muscovite, and montmorillonite. The n-D class #9 is characterized by a strong absorption feature centered at 2.20 μm (coinciding with band 6 of ASTER), which is attributable to muscovite/montmorillonite (see Figures 2 and 3). The n-D class #10 exhibits mixed spectral signatures of hematite and jarosite (see Figure 2).

Fraction images of end-members (the n-D classes #1 to #10) resulting from MTMF analysis appear as a series of greyscale rule images (one for each extracted end-member) for the study area. High digital Number (DN) values (bright pixels) in the rule image represents the subpixel abundance of the end-member mineral in each pixel and map its location. The pseudo-color ramp of greyscale rule images was generated to illustrate high fractional abundance (high DN value pixels) of end-members (the n-D classes) in the study area (Figure 8). It helps to distinguish the contrast between subpixel targets and surrounding areas. This contrast expresses the fractional abundance of the target mineral present in the rule image. It should be noted here that unaltered/unknown class (the n-D class #1 and n-D class #3) was omitted during the production of Figure 8.

Figure 8 shows the pseudo-color ramp of the n-D class fraction images derived from the MTMF algorithm for the study area. Considering the fractional abundance of detected endmember minerals, muscovite, montmorillonite, and illite spectrally governed the study area, while hematite, jarosite, and chlorite have less contribution in total mixed spectral characteristics. Spatial distribution of the minerals with similar spectral features such as muscovite and montmorillonite (absorption features near 2.20 μm) and hematite and jarosite (absorption features near 0.48 μm to 0.85 μm) is comparable. It derives from the fact that ASTER multispectral signatures contain some limitations for detecting subtle differences between analogous absorption characteristics especially when mixture occurs. However, the ASTER VNIR and SWIR bands are sufficiently positioned to detect spectral feature differences between important key minerals [82]. Referencing geological map of the study area (see Figure 1), muscovite/montmorillonite/illite mineral assemblages are typically concentrated in the central, southwestern, and northeastern parts associated with dacitic and andesitic units. Nevertheless, hematite/jarosite/chlorite mineral assemblages are present in low surface abundance (Figure 8). The high concentration of iron oxide/hydroxide minerals is noticeable in the western and southwestern part of the study area associated with trachyandesite basalt units. In this part, the fractional abundance of muscovite/montmorillonite/illite is not high (Figure 8). Accordingly, the southwestern and northeastern parts of the study area contain a number of mineralogically interesting zones for ore mineralizations and holding high potential zones for future systematic exploration program.

SAM classification technique was implemented for mapping the spatial distribution of prevalent minerals such as hematite, jarosite, montmorillonite, illite, muscovite, and chlorite in the alteration zones. Hematite contains absorption features in bands 1 and 3 of ASTER. Jarosite shows absorption characteristics in bands 1, 3, and 7 of ASTER. Montmorillonite displays weak absorption features in band 5 and strong absorption features in band 6 of ASTER. Illite exhibits distinctive absorption properties in bands 5 and 6 of ASTER. Muscovite has absorption features in bands 1, 2, and 6 of ASTER. Chlorite contains absorption characteristics in bands 1, 2, and 3 and diagnostic absorption features in band 8 of ASTER (see Figure 3). Therefore, these spectral absorption signatures of alteration minerals in the VNIR+SWIR bands of ASTER can be used for detecting subtle differences between alteration minerals by running the SAM algorithm. Figure 9 shows the SAM classification map for the study area. Detailed spatial distribution of the selected minerals was mapped within the alteration zones. Montmorillonite, muscovite, and illite are the most dominated minerals in the argillic, phyllic, and advanced argillic alteration zones. However, hematite and jarosite demonstrate moderate surface distribution and chlorite has very low abundance in the argillic, phyllic, and advanced argillic alteration zones. High concentration of chlorite is mapped only in the propylitic alteration zone that is associated with hematite and jarosite. Several concentrations of muscovite and illite are mapped in the southwestern and northeastern parts of the study area, which previously deliberated as high potential zones for ore mineralizations (Figure 9).

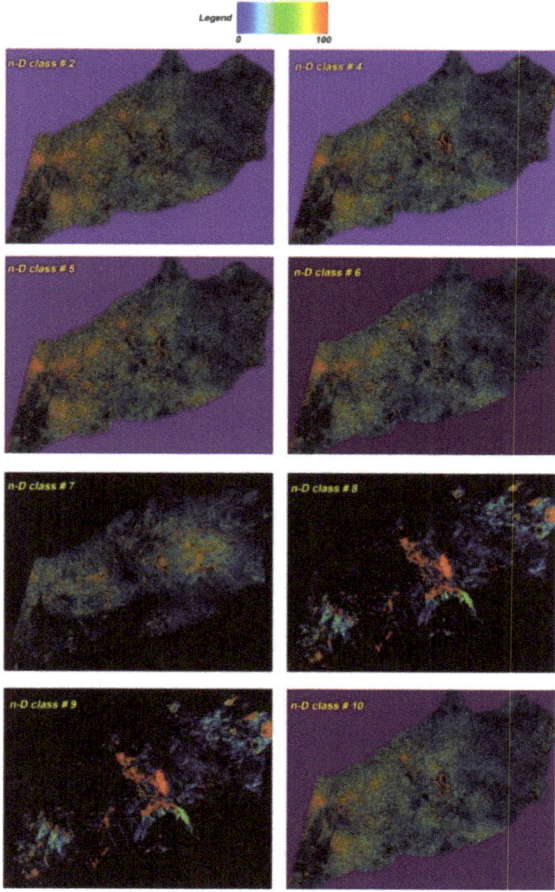

Figure 8. Fraction images of the selected n-D classes derived from MTMF algorithm for the study area. Pseudo-color ramp was applied to greyscale rule images. n-D class #2: hematite; n-D class #4 and n-D class #5: jarosite; n-D class #6: chlorite; n-D class #7: illite; n-D class #8: mixed spectral signatures of illite, muscovite, and montmorillonite; n-D class #9: muscovite/montmorillonite; n-D class #10: mixed spectral signatures of hematite and jarosite.

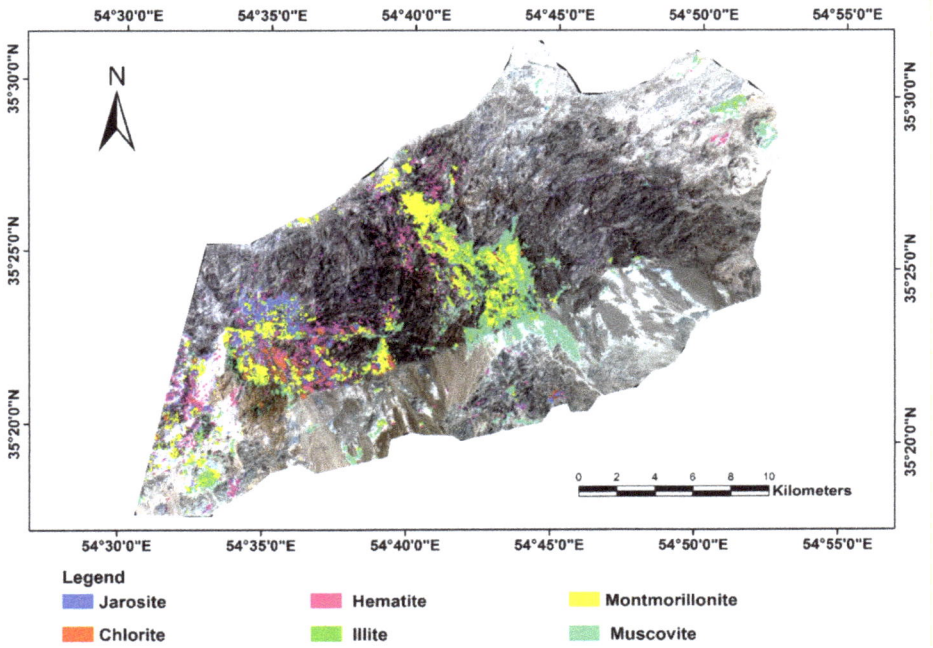

Figure 9. Spectral Angle Mapper (SAM) classification map for the study area.

4.2. Fieldwork, Laboratory Analysis and Verification Results

Comprehensive geological fieldworks were carried out in the study area especially in the detected hydrothermal alteration zones. A number of prospects and mineralogically interesting zones were visited. The precise location of the alteration zones was recorded using GPS survey. The lithological units and alteration zones were checked and samples were collected. Part of the hand specimen was split off for a thin section and the rest was crushed to a grain size of less than 2 mm for XRD analysis. In particular, the study area contains a significant concentration of advanced argillic, argillic, and phyllic alteration zones (Figure 10A–C). However, hematite-rich altered oxidized zones and propylitic alteration zones cover also large parts of the study area (Figure 10D–F). Typically, hematite-rich altered oxidized zones and phyllic alteration zones show the close spatial relationship in many parts of the study area (See Figure 10D).

Petrographic studies were carried out on thin sections of the collected rock samples. Thin section observations indicate the transformation of primary silicate minerals (feldspars) to secondary altered minerals (sericite, clay minerals, calcite and epidote) (Figure 11A–F). Plagioclase is typically replaced by sericite, clay minerals, calcite, epidote, and quartz in the most of alteration zones. Veins and subhedral grains of opaque minerals are more observable in the thin sections of advanced argillic, argillic, and phyllic alteration zones (see Figure 11A–D). In the propylitic zone, the original minerals are fully replcaed by secondary minerals (calcite and epidote) (see Figure 11E,F). Minerals identified in the collected rock samples from hydrothermal alteration zones using XRD analysis include montmorillonite, illite, goethite, hematite, muscovite, albite, orthoclase and quartz in advanced argillic and argillic zones (Figure 12A,B); muscovite, illite, hematite, magnetite, albite, epidote, calcite, montmorillonite, and quartz in the phyllic zone (Figure 12C,D); epidote, calcite, chlorite, albite, anorthite, and quartz in the propylitic zone (Figure 12E,F).

Figure 10. Field photographs of the hydrothermal alteration zones in the study area. (**A**) A panoramic view of argillic alteration zones; (**B**) a view of the advanced argillic alteration zone; (**C**) a regional view of the phyllic alteration zone; (**D**) a regional view of hematite-rich altered oxidized zones in association with phyllic alteration zones; (**E**) a view of the propylitic zone; (**F**) a close up of a specimen from the propylitic alteration zone.

The ASD spectroscopy is sensitive to detect the presence of alteration minerals with strong absorption features in the mineralogically interesting zones. Figure 13 shows the average reflectance spectra of phyllic, gossan (hematite-rich altered oxidized zone), argillic and propylitic rock samples. The reflectance spectra from phyllic samples show three prominent absorption features near 1.40 µm, 1.90 µm, and 2.20 µm, due to vibrational overtone and combination tones involving OH-stretching modes [83,84]. The absorption features near 1.40 µm and 1.90 µm in the phyllic samples can be attributed to OH stretches occurring at about 1.4 µm and the combination of the H-O-H bend with OH stretches near 1.90 µm [78]. The feature near 2.20 µm is due to a combination of the OH-stretching fundamental with Al-OH bending mode [84,85]. These spectral characteristics exhibit similarities to the spectra of muscovite (dominant absorption features located around 2.20 µm), which is a main alteration mineral in the phyllic zone.

Strong absorption features near 0.50 μm and 1.0 μm in the gossan samples are due to electronic transitions in iron ions (ferric and ferrous ions) [79,83]. Moreover, the feature near 2.27 μm is attributed to a combination of OH stretch and Fe-OH bend [78]. The two strong absorption features at about 0.50 μm and 1.0 μm are normally considered for hematite and goethite in the VNIR regions [86]. Jarosite has a diagnostic absorption feature at 2.27 μm [78]. The OH and H2O vibrational bands near 1.40 μm and 1.90 μm are also commonly seen in iron oxides/hydroxides spectra [86]. Thus, hematite, goethite, and jarosite are major alteration mineral constituents in the gossan zone.

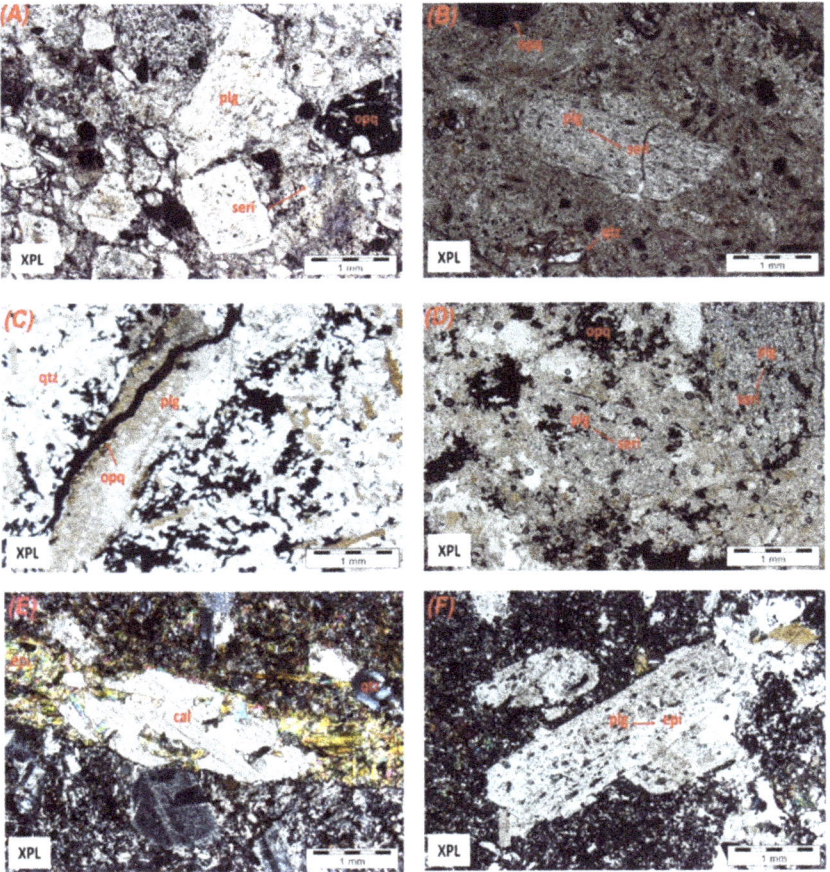

Figure 11. Different types of alteration mineralogy in the alteration zones. Microphotographs of (**A**) argillic zone: plagioclase has been replaced by sericite and clay mineral groups; (**B**) advanced argillic zones: plagioclase crystals topotactically replaced by sericite and clay mineral groups; (**C**) phyllic zone: vein of opaque minerals and relicts of plagioclase that replaced by clay mineral groups and quartz; (**D**) phyllic zone: relicts of plagioclase replaced by sericite; (**E**) propylitic zone: completely replaced original mineralogy by calcite, epidote, and quartz; (**F**) propylitic zone: variolitic to sub-ophytic texture of plagioclase phenocrysts replaced by epidote. Abbreviation: plg = plagioclase, seri = sericite, opq = opaque minerals, qtz = quartz, epi = epidote, cal = calcite.

The reflectance spectra of argillic samples display two overall absorption features at about 1.40 μm and 1.90 μm (due to the OH and H2O vibrational bands) and also consist of maximum absorption near 2.17 μm and 2.20 μm (Figure 13). Montmorillonite, kaolinite, illite, and alunite show spectral characteristics consisting of a major absorption feature at around 2.20 μm associated with a secondary feature between 2.16 and 2.18 μm, which are related to stretching vibration of the inner and outer hydroxyl groups (Al-OH bending mode) [85,87]. The spectra from Propylitic samples show also characteristics absorption features near 0.50 μm and 1.0 μm (due to ferric and ferrous iron ions) and the OH and H2O vibrational bands (about 1.40 μm and 1.90 μm) (Figure 13). The absorption feature near 2.35 μm in the samples is due to a combination of OH-stretching fundamental with the Mg-OH bending mode, and the feature near 2.50 μm can be attributed to combination and overtone bands of CO3 fundamentals, respectively [78,83]. Chlorite and epidote show distinctive absorption features at around 2.35 μm [85]. Carbonates (calcite, aragonite, and dolomite) have diagnostic absorption features between 2.30 and 2.50 μm [88].

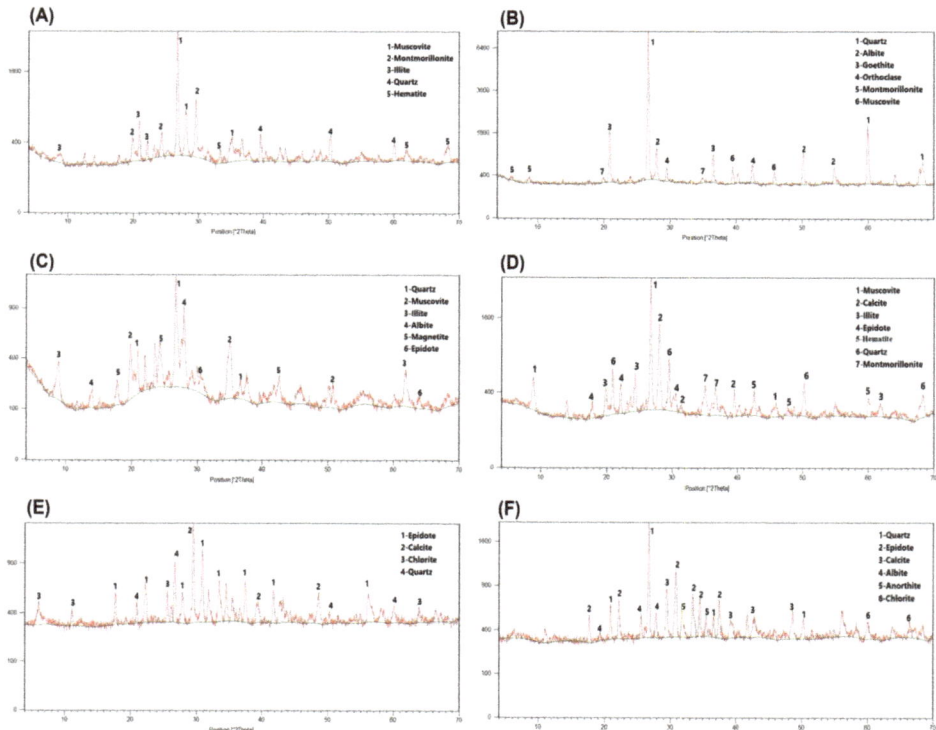

Figure 12. Results of XRD analysis shows minerals of representative samples collected from (**A**) and (**B**) advanced argillic and argillic zones; (**C,D**) phyllic zone; and (**E,F**) propylitic zone.

Comparing the ASD spectroscopy results with XRD analysis indicates most diagnostic spectral features in the phyllic zone derived from muscovite, in the argillic zone resulting from montmorillonite and illite, and in the propylitic zone associated with chlorite, epidote, and calcite. Considering the XRD and ASD analysis, iron oxide/hydroxide minerals are associated with the alteration mineral assemblages in advanced argillic and argillic, phyllic, and propylitic alteration zones. Table 5A,B shows the accuracy assessment results for the MTMF and SAM methods based on GPS survey collected during fieldwork. Analysis of the statistical factors indicate that the overall accuracy and Kappa Coefficient for MTMF and SAM are 95.7 and 0.93 (see Table 5A) and 84.4 and 0.78 (see Table 5B), respectively. The assessment emphasizes that both MTMF and SAM methods provide accurate mapping results in the study area. However, the MTMF method was capable of providing more accurate results for mapping the surface distribution of hydrothermal alteration minerals.

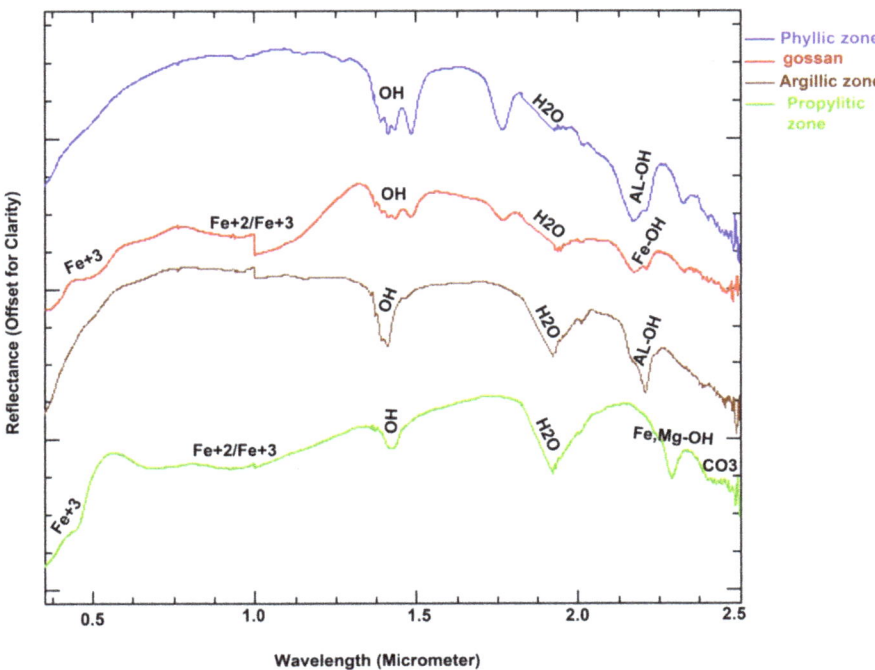

Figure 13. Laboratory reflectance spectra (average) of altered rock samples collected from phyllic, gossan, argillic, and propylitic zones. The main absorption feature spectra attributed to Fe^{+2}, Fe^{+3}, OH, H_2O, Fe-OH, Al-OH, Fe, Mg-OH, and CO_3 are annotated. A sensor-shift could be seen around 1.0 µm especially for gossan and argillic zone.

5. Conclusions

ASTER remote sensing data processing provides maps of surface alteration mineralogy for the Moaleman region of the TCMB, which illustrate several high potential zones of polymetallic vein-type mineralization. The image processing algorithms implemented in this analysis, including the SPCA, BRMT, SAM, and MTMF provided a consistent way to identify and map hydrothermal alteration zones in the study area. The SPCA, BRMT spectral transformation algorithms used in this study were capable of mapping the surface distribution of hydrothermally altered rocks and lithological units. The surface distribution pattern of iron oxide/hydroxide minerals, argillic alteration zone, phyllic zone, and propylitic zone was mapped in the study area using a unique contribution of eigenvector loading in the SPC3. Iron oxide/hydroxide minerals and propylitic zones showed similar spatial distribution, while strong analogous surface distribution patterns were more obvious for argillic, phyllic, and iron oxide/hydroxide minerals zones. The BT1, BT2, and BT3 hold most of the spectral information in the image, which was extracted by the BRMT algorithm. The most spectrally dominated hydrothermal alteration zones, including argillic, phyllic, and advanced argillic alteration zones were revealed. Additionally, the BTs contain a negative contribution > 3% providing comprehensive information as eight different lithological/alteration/weathered classes (C1 to C8) for the study area. Therefore, more alteration/lithological information can be obtained from the BRMT algorithm compare to the SPCA.

Ten endmembers (n-D classes) were extracted using n-dimensional visualization method, which indicates distinctive absorption features related to alteration minerals in the study area. Spectrally distinctive mineralogical phases were identified, such as hematite, jarosite, chlorite, illite, muscovite, and montmorillonite, which may have some spectral signatures of subordinate mineralogical phases. Fraction images of end-members derived from the MTMF algorithm showed the similar sub-pixel distribution for minerals contain analogous spectral features, which can be attributed to some limitations of ASTER data for detecting subtle differences between equivalent absorption characteristics especially when mixture occurs. Detailed spatial distribution of prevalent minerals including hematite, jarosite, montmorillonite, illite, muscovite, and chlorite in the alteration zones was mapped using the SAM algorithm. Results indicate that montmorillonite, muscovite, and illite are the most dominated minerals in the argillic, phyllic, and advanced argillic alteration zones. However, hematite and jarosite demonstrate moderate surface distribution and chlorite has very low abundance in these alteration zones. A high concentration of chlorite was mapped only in the propylitic alteration zone, which was also associated with hematite and jarosite. The MTMF results confirmed the presence of alteration minerals and their spatial distribution at the subpixel level in the study area, while SAM mapped subtle differences between alteration minerals in the alteration zones. The accuracy assessment results show the MTMF method is proficient to be responsible for more accurate outcomes for mapping the surface distribution of hydrothermal alteration minerals. Hydrothermal alteration zones and minerals that mapped using ASTER data processing demonstrate good correspondence with the results of field survey, petrography, XRD analysis, and spectral measurements acquired by ASD spectrometer. A number of high potential zones of epithermal polymetallic vein-type mineralization were identified in the study area, particularly in the northeastern and southwestern sectors, which can be considered for future systematic exploration programs. The approach developed in this study can be used for the exploration of epithermal polymetallic vein-type mineralization in other base metals provinces in semi-arid regions around the world.

Author Contributions: L.N. performed experiments and field data collection; A.B.P. wrote the manuscript and analyzed the data; G.A. supervised and contributed to writing and analysis; N.T. supervised; C.-W.L. and B.P. edited, restructured, and professionally optimized the manuscript including the funding acquisition; M.H. provided ASD data.

Funding: This research is supported by the UTS under grant numbers 323930, 321740.2232335 and 321740.2232357. In addition, this research was supported by a grant from the National Research Foundation of Korea, provided by the Korea government (No. 2017R1A2B4003258, 2018).

Acknowledgments: We acknowledge the School of Earth Sciences, Damghan University for providing all facilities during this research. We are also thankful to Korea Polar Research Institute (KOPRI), Faculty of Engineering and Information Technology, the University of Technology, Sydney, and Department of Ecology, University of Advanced Technology, Kerman.

Conflicts of Interest: The authors declare no conflict of interest.

References

1. Cox, D.P. Descriptive model of polymetallic veins. In *Mineral Deposit Models*; Cox, D.P., Singer, D.A., Eds.; U.S. Geological Survey Bulletin: Reston, VI, USA, 1986; Volume 1693, p. 125.
2. Haynes, S.J. Vein-type ore deposits: Introduction. *Ore Geol. Rev.* **1993**, *8*, 205–211. [CrossRef]
3. Ono, S.; Hirai, K.; Matsueda, H.; Kabashima, T. Polymetallic Mineralization at the Suttsu Vein-type Deposit, Southwestern Hokkaido, Japan. *Resour. Geol.* **2004**, *54*, 453–464. [CrossRef]
4. Gharesi, M.; Karimi, M. Vein type mineralization and related alterations of Narigun polymetalli deposit, East of Yazd, Central Iran. In *Proceedings of the 10th International Congress for Applied Mineralogy (ICAM)*; Springer: Berlin/Heidelberg, Germany, 2012; pp. 229–236.

5. Tu, W.; Du, Y.S.; Wang, G.W.; Lei, Y.P. Cordilleran vein type Pb-Zn-polymetallic deposits of the Xidamingshan district, Guangxi, SW China: Fluid inclusion and geochemical studies. *Geol. Ore Depos.* **2013**, *55*, 494–502. [CrossRef]
6. Box, S.E.; Bookstrom, A.A.; Zientek, M.L.; Derkey, P.D.; Ashley, R.P.; Elliot, J.E.; Peters, S.G. *Assessment of Undiscovered Mineral Resources in the Pacific Northwest: A Contribution to the Interior Columbia Basin Ecosystem Management Project*; USGS Open-File Report OF 95-682; USGS: Reston, VI, USA, 1996.
7. Lowell, J.D.; Guilbert, J.M. Lateral and vertical alteration-mineralization zoning in porphyry ore deposits. *Econ. Geol.* **1970**, *65*, 373–408. [CrossRef]
8. Mahanta, P.; Maiti, S. Regional scale demarcation of alteration zone using ASTER imageries in South Purulia Shear Zone, East India: Implication for mineral exploration in vegetated regions. *Ore Geol. Rev.* **2018**, *102*, 846–861. [CrossRef]
9. Ahmadirouhani, R.; Rahimi, B.; Karimpour, M.H.; Malekzadeh Shafaroudi, A.; Pour, A.B.; Pradhan, B. Integration of SPOT-5 and ASTER satellite data for structural tracing and hydrothermal alteration mineral mapping: Implications for Cu–Au prospecting. *Int. J. Image Data Fusion* **2018**, *9*, 237–262. [CrossRef]
10. Pour, A.B.; Hashim, M.; Hong, J.K.; Park, Y. Lithological and alteration mineral mapping in poorly exposed lithologies using Landsat-8 and ASTER satellite data: North-eastern Graham Land, Antarctic Peninsula. *Ore Geol. Rev.* **2017**. [CrossRef]
11. Pour, A.B.; Hashim, M.; Park, Y.; Hong, J.K. Mapping alteration mineral zones and lithological units in Antarctic regions using spectral bands of ASTER remote sensing data. *Geocarto Int.* **2018**, *33*, 1281–1306. [CrossRef]
12. Pour, A.B.; Park, T.S.; Park, Y.; Hong, J.K.; Zoheir, B.; Pradhan, B.; Ayoobi, I.; Hashim, M. Application of multi-sensor satellite data for exploration of Zn-Pb sulfide mineralization in the Franklinian Basin, North Greenland. *Remote Sens.* **2018**, *10*, 1186. [CrossRef]
13. Pour, A.B.; Park, Y.; Park, T.S.; Hong, J.K.; Hashim, M.; Woo, J.; Ayoobi, I. Regional geology mapping using satellite-based remote sensing approach in Northern Victoria Land, Antarctica. *Polar Sci.* 16, 23–46. [CrossRef]
14. Pour, A.B.; Park, Y.; Park, T.S.; Hong, J.K.; Hashim, M.; Woo, J.; Ayoobi, I. Evaluation of ICA and CEM algorithms with Landsat-8/ASTER data for geological mapping in inaccessible regions. *Geocarto Int.* **2018**. [CrossRef]
15. Testa, F.J.; Villanueva, C.; Cooke, D.R.; Zhang, L. Lithological and hydrothermal alteration mapping of epithermal, porphyry and tourmaline breccia districts in the Argentine Andes using ASTER imagery. *Remote Sens.* **2018**, *10*, 203. [CrossRef]
16. Askari, G.; Pour, A.B.; Pradhan, B.; Sarfi, M.; Nazemnejad, F. Band Ratios Matrix Transformation (BRMT): A Sedimentary Lithology Mapping Approach Using ASTER Satellite Sensor. *Sensors* **2018**, *18*, 3213. [CrossRef] [PubMed]
17. Mars, J.C. *Regional Mapping of Hydrothermally Altered Igneous Rocks along the Urumieh-Dokhtar, Chagai, and Alborz Belts of Western Asia Using Advanced Spaceborne Thermal Emission and Reflection Radiometer (ASTER) Data and Interactive Data Language (IDL) Logical Operators—A Tool for Porphyry Copper Exploration and Assessment*; U.S. Geological Survey Scientific Investigations Report 2010-5090-O; USGS: Reston, VI, USA, 2014; 36p.
18. Abrams, M. The Advanced Spaceborne Thermal Emission and Reflaction Radiometer (ASTER). Data Products for the high spatial resolution imager on NASA Terra Platform. *Int. J. Remote Sens.* **2000**, *21*, 847–859. [CrossRef]
19. Fujisada, H. Design and performance of ASTER instrument. In *Advanced and Next-Generation Satellites*; Breckinridge, J.B., Ed.; International Society for Optics and Photonics: Bellingham, WA, USA, 1995; Volume 2583, pp. 16–25.
20. Kurata, K.; Yamaguchi, Y. Integration and Visualization of Mineralogical and Topographical Information Derived from ASTER and DEM data. *Remote Sens.* **2019**, *11*, 162. [CrossRef]
21. Rowan, L.C.; Hook, S.J.; Abrams, M.J.; Mars, J.C. Mapping hydrothermally altered rocks at Cuprite, Nevada, using te Advanced Spaceborne Thermal Emission and Reflection Radiometer (ASTER), A new satellite-imaging system. *Econ. Geol.* **2003**, *98*, 1019–1027. [CrossRef]
22. Carrino, T.A.; Crósta, A.P.; Toledo, C.L.B.; Silva, A.M.; Silva, J.L. Geology and hydrothermal alteration of the Chapi Chiara prospect and nearby targets, Southern Peru, using ASTER data and reflectance spectroscopy. *Econ. Geol.* **2015**, *110*, 73–90. [CrossRef]

23. Cudahy, T. Mineral Mapping for Exploration: An Australian Journey of Evolving Spectral Sensing Technologies and Industry Collaboration. *Geosciences* **2016**, *6*, 52. [CrossRef]
24. Sheikhrahimi, A.; Pour, B.A.; Pradhan, B.; Zoheir, B. Mapping hydrothermal alteration zones and lineaments associated with orogenic gold mineralization using ASTER remote sensing data: A case study from the Sanandaj-Sirjan Zone, Iran. *Adv. Space Res.* **2019**, in press. [CrossRef]
25. Rajendran, S.; Nasir, S. Characterization of ASTER spectral bands for mapping of alteration zones of volcanogenic massive sulphide deposits. *Ore Geol. Rev.* **2017**, *88*, 317–335. [CrossRef]
26. Rajendran, S.; Thirunavukkarasu, A.; Balamurugan, G.; Shankar, K. Discrimination of iron ore deposits of granulite terrain of Southern Peninsular India using ASTER data. *J. Asian Earth Sci.* **2011**, *41*, 99–106. [CrossRef]
27. Rajendran, S.; Al-Khirbasha, S.; Pracejusa, B.; Nasira, S.; Al-Abria, A.H.; Kusky, T.M.; Ghulam, A. ASTER detection of chromite-bearing mineralized zones in Semail Ophiolite Massifs of the northern Oman Mountains: Exploration strategy. *Ore Geol. Rev.* **2012**, *44*, 121–135. [CrossRef]
28. Rastad, E.; Tajeddin, H.; Rashidnejad-Omran, N.; Babakhani, A. Genesis and gold (copper) potential in Darestan-Baghou mining area. *Iran. Geosci. J.* **2000**, *36*, 60–79. (In Persian)
29. Fard, M.; Rastad, E.; Ghaderi, M. Epithermal Gold and Base Metal mineralization at Gandy Deposite, North of central Iran and the Role of Rhyolitic Intrusions. *J. Sci. Islamic Repub. Iran* **2006**, *17*, 327–335.
30. Shamanian, G.H.; Hedenquist, J.W.; Hattori, K.H.; Hassanzadeh, J. The Gandy and Abolhassani epithermal deposits in the Alborz magmatic arc, Semnan Province, northern Iran. *Econ. Geol.* **2004**, *99*, 691–712. [CrossRef]
31. Ziaii, M.; Abedi, A.; Kamkar, A.; Zendahdel, A. GIS modelling for Au-Pb-Zn potential mapping in Torud-Chahshirin area Iran. *Int. J. Min. Environ. Issues* **2010**, *1*, 17–27.
32. Moghadam, H.S.; Li, X.H.; Stern, R.J.; Santos, J.F.; Ghorbani, G.; Pourmohsen, M. Age and nature of 560–520 Ma calc-alkaline granitoids of Biarjmand, northeast Iran: Insights into Cadomian arc magmatism in northern Gondwana. *Int. Geol. Rev.* **2016**, *58*, 1492–1509. [CrossRef]
33. Moghadam, H.S.; Li, X.H.; Santos, J.F.; Stern, R.J.; Griffin, W.L.; Ghorbani, G.; Sarebani, N. Neoproterozoic magmatic flare-up along the N. margin of Gondwana: The Taknar complex, NE Iran. *Earth Planet. Sci. Lett.* **2017**, *474*, 83–96. [CrossRef]
34. Mehrabi, B.; Siani, M.G. Mineralogy and economic geology of Cheshmeh Hafez polymetallic deposit, Semnan province, Iran. *J. Econ. Geol.* **2010**, *2*, 1–20, (In Persian with English Abstract).
35. Ghorbani, G. Chemical composition of minerals and genesis of mafic microgranular enclaves in intermediate-acidic plutonic rocks from Kuh-e-Zar area (south of Semnan). *Iran. J. Crystallogr. Mineral.* **2007**, *2*, 293–310.
36. Crosta, A.P.; Desouza Filho, C.; Azevedo, F.B. Targeting key alteration mineral in epithermal deposits in Patagonia, Argentin, using ASTER imagery and principal component Analysis. *Int. J. Remote Sens.* **2003**, *24*, 4233–4240. [CrossRef]
37. Loughlin, W.P. Principal Components Analysis for alteration mapping. *Photogramm. Eng. Remote Sens.* **1991**, *57*, 1163–1169.
38. Kruse, F.A.; Lefkoff, A.B.; Boardman, J.B.; Heidebrecht, K.B.; Shapiro, A.T.; Barloon, P.J.; Goetz, A.F.H. The Spectral Image Processing System (SIPS)—Interactive Visualization and Analysis of Imaging spectrometer Data. *Remote Sens. Environ.* **1993**, *44*, 145–163. [CrossRef]
39. Boardman, J.W. Leveraging the high dimensionality od AVIRIS data for improved sub-pixel target unmixing and rejection of false positives: Mixture Tuned Matched filtering. In Proceedings of the Summaries Workshop Seventh Annual JPL Airborne Geoscience Workshop, Pasadena, CA, USA, 12–16 January 1998; Green, R.O., Ed.; 1998; pp. 55–56.
40. Harsanyi, J.C.; Farrand, W.H.; Chang, C.I. Detection of Subpixel Signatures in hyperspectral image sequences. Proceeding of the 1994 ASPRS Annual Conference, Reno, NV, USA, 25–28 April 1994; pp. 236–247.
41. Boardman, J.W.; Green, R.O. Exploring the spectral variability of the Earth as Measured by AVIRIS in 1999. In *Proceedings of the Summaries of the Ninth Annual JPL Airborne Geosciences Workshop*; Jet Propulsion Special Publication: Pasadena, CA, USA, 2000; Volume 8, p. 10.
42. Eshraghi, S.A.; Jalali, A. *Geologycal Map of Moaleman, 1:100,000*; Geology Survey: Tehran, Iran, 2006.
43. Alavi, M. Tectonic of zagros orogenic belt of Iran; new data and interpretation. *Tectonophysics* **1994**, *229*, 211–238. [CrossRef]

44. Alavi, M.; Vaziri, H.; Seyed-Emami, K.; Lasemi, Y. The Triassic and associated rocks of the Nakhlak and Aghdarband areas in central and northeastern Iran as remnants of the southern Turanian active continental margin. *Geol. Soc. Am. Bull.* **1997**, *109*, 1563–1575. [CrossRef]
45. Alavi, M. Structures of the Zagros fold-thrust belt in Iran. *Am. J. Sci.* **2007**, *307*, 1064–1095. [CrossRef]
46. Hushmandzadeh, A.R.; Alavi, M.; Haghipour, A.A. *Evalution of Geological Phenomenon in Toroud Area (From Precambrian to Recent) of Iran*; Report No. H5; GSI.IR: Tehran, Iran, 1978; 136p.
47. Akhyani, M.; Kharqani, M.; Rahimi, M.; Sereshki, F. Alteration Zones Detection of Troud-Chah Shirin Volcanic—Plutonic Belt using different processing methods of ASTER images. *Eng. Environ. Geol.* **2015**, *24*, 107–116.
48. Liaghat, S.; Sheykhi, V.; Najjaran, M. Petrology, gheochemistry and genesis of Baghu turquoise, Damghan. *J. Sci.* **2008**, *34*, 133–142, (In Persian with English Abstract).
49. Nahidifar, E.; Fardoost, F.; Rezaii, M. Mineralogy of Dian Copper Deposit (South of Damghan). In Proceedings of the 17th Symposium of Geological Survey of Iran, Tehran, Iran, (In Persian)29–31 October 2013.
50. Roohbakhsh, P.; Karimpour, M.H.; Malekzadeh, A. Geology, mineralization, geochemistry and petrology of intrusions in the Kuh Zar Au-Cu deposit, Damghan. *J. Econ. Geol.* **2018**, *10*, 1–2.
51. Green, A.A.; Craig, M.D. Analysis of aircraft spectrometer data, with logarithmic residuals. In *Proceedings of the Airborne Imaging Spectrometer Data Analysis Workshop*; Vane, G., Goetz, A., Eds.; JPL: Pasadena, CA, USA, 1985; pp. 111–119.
52. Research Systems, Inc. *ENVI Tutorials*; Research Systems, Inc.: Boulder, CO, USA, 2008.
53. Iwasaki, A.; Tonooka, H. Validation of a crosstalk correction algorithm for ASTER/SWIR. *IEEE Trans. Geosci. Remote Sens.* **2005**, *43*, 2747–2751. [CrossRef]
54. Pour, B.A.; Hashim, M. Identification of hydrothermal alteration minerals for exploring of porphyry copper deposit using ASTER data, SE Iran. *J. Asian Earth Sci.* **2011**, *42*, 1309–1323. [CrossRef]
55. Pour, B.A.; Hashim, M. Spectral transformation of ASTER and the discrimination of hydrothermal alteration minerals in a semi-arid region, SE Iran. *Int. J. Phys. Sci.* **2011**, *6*, 2037–2059.
56. Pour, B.A.; Hashim, M. Identifying areas of high economic-potential copper mineralization using ASTER data in Orumieh–Dokhtar Volcanic Belt, Iran. *Adv. Spaceborn Res.* **2012**, *49*, 753–769. [CrossRef]
57. Pour, B.A.; Hashim, M. The application of ASTER remote sensing data to porphyry copper and epithermal gold deposits. *Ore Geol. Rev.* **2012**, *44*, 1–9. [CrossRef]
58. Honarmand, M.; Ranjbar, H.; Shahabpour, J. Application of Principal Component Analysis and Spectral Angle Mapping in the mapping of hysrothermal alteration in the Jebal–Barez area, southeastern Iran. *Resour. Geol.* **2012**, *62*, 119–139. [CrossRef]
59. Khaleghi, M.; Ranjbar, H.; Shahabpour, J.; Honarmand, M. Spectral angle mapping, Spectral information divergence and Principal component analysis of the ASTER SWIR data for exploration of porphyry copper mineralization in the Sarduiyeh area, Kerman province, Iran. *Appl. Geomat.* **2014**, *6*, 49–58. [CrossRef]
60. Siljestrom, P.; Moreno, A.; Vikgren, K.; Cáceres Puro, L. The application of selective principal components analysis (SPCA) to a Thematic Mapper (TM) image for the recognition of geomorphologic configuration. *Int. J. Remote Sens.* **1997**, *18*, 3843–3852. [CrossRef]
61. Fraser, S.J.; Green, A. A software defoliant for geological analysis of band ratios. *Int. J. Remote Sens.* **1987**, *8*, 525–532. [CrossRef]
62. Abdelsalam, M.; Stern, R. Mapping gossan in arid regions with landsat TM and SIR-C images, the Beddaho Alteration Zone in northern Eritrea. *J. Afr. Earth Sci.* **2000**, *30*, 903–916. [CrossRef]
63. Rowan, L.C.; Mars, J.C. Lithologic mapping in the Mountain Pass, California area using Advanced Spaceborne Thermal Emission and Reflection Radiometer (ASTER) data. *Remote Sens. Environ.* **2003**, *84*, 350–366. [CrossRef]
64. Mars, J.C.; Rowan, L.C. Regional mapping of phyllic- and argillic-altered rocks in the Zagros magmatic arc, Iran, using Advanced Spaceborne Thermal Emission and Reflection Radiometer (ASTER) data and logical operator algorithms. *Geosphere* **2006**, *2*, 161–186. [CrossRef]
65. Rowan, L.C.; Robert, G.S.; John, C. Distribution of hydrothermally altered rocks in the Peko Diq, Pakistan mineralized area based on spectralysis of ASTER data. *Remote Sens. Environ.* **2006**, *104*, 74–87. [CrossRef]
66. Yang, C.; Everitt, J.H.; Bradford, J.M. Yield estimation from hyperspectral imagery using spectral angle mapper (SAM). *Am. Soc. Agric. Biol. Eng.* **2008**, *51*, 729–737.

67. Hosseinjani, M.; Tangestani, M. Mapping alteration minerals using sub-pixel unmixing of ASTER data in the Sarduyeh area, SE Kerman, Iran. *Int. J. Digit. Earth* **2011**, *4*, 487–504. [CrossRef]
68. Dennison, P.E.; Roberts, D.A. Endmember selection for multiple endmember spectral mixture analysis using endmember average RMSE. *Remote Sens. Environ.* **2003**, *87*, 123–135. [CrossRef]
69. Boardman, J.W.; Kruse, F.A. *Automated Spectra Analysis: A Geologic Example Using AVIRIS Data, North Grapevine on Geologic Remote Sensing*; Environmental Research Institute of Michigan: Arbor, MI, USA, 1994; pp. 407–418.
70. Green, A.A.; Berman, M.; Switzer, P.; Craig, M.D. A transformation for ordering multispectral data in terms of image quality with implications for noise removal. *IEEE Trans. Geosci. Remote Sens.* **1988**, *26*, 65–74. [CrossRef]
71. Jensen, J.R. *Introductory Digital Image Processing*; Pearson Prentice Hall: Upper Saddle River, NJ, USA, 2005.
72. Sabins, F.F. Remote sensing for mineral exploration. *J. Geol. Rev.* **1999**, *14*, 157–183. [CrossRef]
73. Kokaly, R.F.; Clark, R.N.; Swayze, G.A.; Livo, K.E.; Hoefen, T.M.; Pearson, N.C.; Wise, R.A.; Benzel, W.M.; Lowers, H.A.; Driscoll, R.L.; et al. *USGS Spectral Library Version 7*; U.S. Geological Survey Data Series 1035; USGS: Reston, VI, USA, 2017; 61p.
74. Askari, G.; Li, Y.; Moezzi Nasab, R. An adaptive polygonal centroidal voronoi tessellation algorithm for segmentation of noisy SAR images. *Int. Arch. Photogramm. Remote Sens. Spat. Inf. Sci.* **2014**, *XL-2/W3*, 65–68. [CrossRef]
75. Velosky, J.C.; Stern, R.J.; Johnson, P.R. Geological control of massive sulfide mineralization in the Neoproterozoic Wadi Bidah shear zone, southwestern Saudi Arabia, inferences from orbital remote sensing and field studies. *Precambrian Res.* **2003**, *123*, 235–247. [CrossRef]
76. Hunt, G.R. Electromagnetic radiation: The communication link in remote sensing. In *Remote Sensing in Geology*; Siegal, B.S., Gillespie, A.R., Eds.; John Wiley & Sons: New York, NY, USA, 1980; pp. 5–45.
77. Burns, R.G. *Mineralogical Applications of Crystal Field Theory*; Cambridge University Press: Cambridge, UK, 1993; 551p.
78. Clark, R.N. Spectroscopy of rocks and minerals, and principles of spectroscopy. In *Manual of Remote Sensing*; Rencz, A., Ed.; Wiley and Sons Inc.: New York, NY, USA, 1999; Volume 3, pp. 3–58.
79. Hunt, G.R. Spectral signatures of particulate minerals in the visible and near-infrared. *Geophysics* **1977**, *42*, 501–513. [CrossRef]
80. Sgavetti, M.; Pomilio, L.; Meli, S. Reflectance spectroscopy (0.3–2.5 µm) at various scales for bulk-rock identification. *Geosphere* **2006**, *2*, 142–160. [CrossRef]
81. Di Tommaso, I.; Rubinstein, N. Hydrothermal alteration mapping using ASTER data in the infiernillo porphyry deposite, Argentina. *Ore Geol. Rev.* **2007**, *32*, 275–270. [CrossRef]
82. Kruse, F.A.; Perry, S.L. Mineral mapping using simulated Worldview-3 short-wave-infrared imagery. *Remote Sens.* **2013**, *5*, 2688–2703. [CrossRef]
83. Hunt, G.R.; Evarts, R.C. The use of near-infrared spectroscopy to determine the degree of serpentinization of ultramafic rocks. *Geophysics* **1980**, *46*, 316–321. [CrossRef]
84. Hunt, G.R. Near-infrared 1/3–2/4 (µm) spectra of alteration minerals potential for use in remote sensing. *J. Geophys.* **1979**, *44*, 1974–1986. [CrossRef]
85. Bishop, J.L.; Lane, M.D.; Dyar, M.D.; Brwon, A.J. Reflectance and emission spectroscopy study of four groups of phyllosilicates: Smectites, kaolinite-serpentines, chlorites and micas. *Clay Miner.* **2008**, *43*, 35–54. [CrossRef]
86. Cudahy, T.J.; Ramanaidou, E.R. Measurement of the hematite: Goethite ratio using field visible and near-infrared reflectance spectrometry in channel iron deposits, Western Australia. *Aust. J. Earth Sci.* **1997**, *44*, 411–420. [CrossRef]
87. Frost, R.L.; Johansson, U. Combination bands in the infrared spectroscopy of kaolins—A drift spectroscopic study. *Clays Clay Miner.* **1998**, *46*, 466–477. [CrossRef]
88. Gaffey, S.J. Spectral reflectance of carbonate minerals in the visible and near-infrared (0.35–2.55 microns): Calcite, aragonite, and dolomite. *Am. Mineral.* **1986**, *71*, 151–162.

© 2019 by the authors. Licensee MDPI, Basel, Switzerland. This article is an open access article distributed under the terms and conditions of the Creative Commons Attribution (CC BY) license (http://creativecommons.org/licenses/by/4.0/).

Article

Emittance Spectroscopy and Broadband Thermal Remote Sensing Applied to Phosphorite and Its Utility in Geoexploration: A Study in the Parts of Rajasthan, India

Arindam Guha [1,*], Yasushi Yamaguchi [2], Snehamoy Chatterjee [3], Komal Rani [1] and Kumranchat Vinod Kumar [1]

1. Geosciences Group, National Remote Sensing Centre, Indian Space Research Organization (ISRO) Balanagar, Hyderabad 500037, India; pasrichakomal@gmail.com (K.R.); vinodkumar_k@nrsc.gov.in (K.V.K.)
2. Graduate School of Environmental Studies, Nagoya University, Nagoya 464-8601, Japan; yasushi@nagoya-u.jp
3. Department of Geological and Mining Engineering and Sciences, Michigan Technological University, Houghton, MI 49931, USA; schatte1@mtu.edu
* Correspondence: arindamguha.1976@gmail.com

Received: 11 April 2019; Accepted: 24 April 2019; Published: 27 April 2019

Abstract: The contrast in the emissivity spectra of phosphorite and associated carbonate rock can be used as a guide to delineate phosphorite within dolomite. The thermal emissivity spectrum of phosphorite is characterized by a strong doublet emissivity feature with their absorption minima at 9 μm and 9.5 μm; whereas, host rock dolomite has relatively subdued emissivity minima at ~9 μm. Using the contrast in the emissivity spectra of phosphorite and dolomite, data obtained by the thermal bands of Advanced Spaceborne Thermal Emission and Reflection Radiometer (ASTER) sensor were processed to delineate phosphorite within dolomite. A decorrelation stretched ASTER radiance composite could not enhance phosphorite rich zones within the dolomite host rock. However, a decorrelation stretched image composite of selected emissivity bands derived using the emissivity normalization method was suitable to enhance large surface exposures of phosphorite. We have found that the depth of the emissivity minima of phosphorite gradually has increased from dolomite to high-grade phosphorite, while low-grade phosphate has an intermediate emissivity value and the emissivity feature can be studied using three thermal bands of ASTER. In this context, we also propose a relative band depth (RBD) image using selected emissivity bands (bands 11, 12, and 13) to delineate phosphorite from the host rock. We also propose that the RBD image can be used as a proxy to estimate the relative grades of phosphorites, provided the surface exposures of phosphorite are large enough to subdue the role of intrapixel spectral mixing, which can also influence the depth of the diagnostic feature along with the grade. We have validated the phosphorite pixels of the RBD image in the field by carrying out colorimetric analysis to confirm the presence of phosphorite. The result of the study indicates the utility of the proposed relative band depth image derived using ASTER TIR bands for delineating Proterozoic carbonate-hosted phosphorite.

Keywords: ASTER; emissivity; emissivity normalization method; dolomite; phosphorite; relative band depth (RBD)

1. Introduction

Phosphorites are known as the source rock of fertilizer, and its exploration is very important for agriculture-dependent countries like India that have huge populations. New methods are essential for exploring additional pockets of phosphorite as India imports 85% of the phosphorite used to make fertilizer [1]. In this study, we propose a simple, easily reproducible thermal remote sensing-based method for the delineation of phosphorite as the input for detailed exploration. Phosphate minerals are known to have diagnostic absorption features within the spectral domain of 8.3 to 11.25 micrometers [2]. However, there are no records available on the attempts made to analyze and utilize the emissivity spectra of phosphorite (constituted with phosphate-bearing minerals) for geological exploration. This study aims to delineate phosphorite within host rocks based on the processing of broadband thermal multispectral data using the emissivity contrast of phosphate and dolomite. Phosphorite or sedimentary phosphate deposit is a sedimentary rock which is constituted varieties of phosphate minerals such as apatite, fluroapatite, etc. [3]. In phosphorite, fluorine in phosphate minerals are often replaced by hydroxyl, chlorine irons [3]. On the other hand, host rock or associated rock of phosphorite is primarily dolomite. Dolomite is sedimentary rock primarily constituted with different carbonate minerals like dolomite (predominant), calcite, quartz, etc. Main constituent minerals of phosphorite (i.e., fluroapatite) and its host rock (predominantly dolomite) have their diagnostic emissivity features within the spectral domain of 8.3 to 11.25 micrometers [3,4]. The study is relevant for the exploration of Paleoproterozoic phosphorite. The phosphorites of Paleoproterozoic age are primarily hosted by dolomite, dolomitic limestone, and associated carbonate rocks in different parts of the world, for example, the Irece Basin (eastern-central Brazil), Simian series of rocks (central Gujhao, China), and Heerapur (Madhya Pradesh, India) [5,6]. These phosphorites occur as bands, patches of different size within the dolomite. These patches could be resolvable in the thermal images using the emissivity spectra of phosphate minerals. Emissivity spectra of the target are the result of atomic and molecular vibration. Vibrational spectroscopy in the thermal infrared (TIR) domain is sensitive to the molecular structure and chemical composition of minerals [7]. All rock-forming minerals display spectral signatures in their emissivity spectra due to the different vibrational modes resulting from the stretching and bending vibrations of bonds in their crystal lattices [7,8]. Most importantly, fundamental vibrational absorptions of geologic materials occur within the spectral range of 3 to 50μm [7,8]. However, the records of mapping geological material using thermal spectra were limited to quartz-rich igneous intrusive rocks, other silicate mineral-dominant rock types and metasedimentary rock units [9–14]. Records are also limited on the use of broadband thermal spectroscopy to study economically important rocks, like phosphorite.

At present, no spaceborne thermal sensor is operative which can collect emissivity spectra with a fine spectral resolution (for example, with a spectral resolution of 10 to 30 nm). However, the spectral domain of 8.0 to 11.5 μm is being used for space-based broadband thermal spectroscopy as it is within the atmospheric window and suitable for detecting various silicate minerals [9–12,14]. In the present context, five thermal spectral bands of the Advanced Spaceborne Thermal Emission and Reflection Radiometer (ASTER) sensor operative within the spectral domain of 8.125 to 11.625 μm [15,16] are suitable for carrying out the study of phosphorite mapping using thermal bands as different geological targets can be targeted using quantitative mineralogical parameter such as emissivity [17,18]. ASTER was launched with the Terra satellite in 1999 and gained popularity as spectral bands of ASTER are capable for detecting spectrally sensitive minerals like calcite, different clay minerals, mica, etc. [15,16]. This popularity is especially true for bands in the shortwave infrared (SWIR) spectral domain and TIR spectral domain [19–24]. ASTER TIR bands have also been widely used for delineating feldspar-rich intrusive, e.g., albite granitoids, alkali granite, and different granitoid systems, and also for delineating mafic igneous complexes from their silicic counterparts in geological mapping [10,11,25–29]. However, the spectral dimensionality of ASTER TIR bands is an issue as these bands are also known to display striping noise and a poor signal to noise ratio (SNR) [28]. Furthermore, the spatial resolutions of spaceborne thermal sensors are coarse (ASTER has 90-meter spatial resolution for its TIR bands). Therefore, the scene within the pixels of spaceborne thermal sensors is also heterogeneous. This heterogeneity hinders the detailed characterization of the target using its emissivity spectrum as the emissivity of a pixel (which is containing different target) is often different to that of the target unless the pure target occupies a considerable portion of the pixel.

There has been an attempt to delineate phosphorite from dolomite using the contrast of their reflectance spectra [30]. This spectral contrast in the visible near infrared (VNIR) and SWIR domain is due to the presence of a secondary vibrational feature (overtone and combination) of carbonate mineral in dolomite and absence of such feature in the phosphorite [31,32]. Therefore, the detection of phosphorite in the VNIR-SWIR domain is an indirect approach. On the other hand, the TIR domain is characterized by the doublet vibrational feature of phosphate-bearing mineral (i.e., fluroapatite) of phosphorite [2]. There are no records on the use of emissivity spectroscopy to delineate carbonate and phosphate and corresponding upscaling to the broadband emissivity of ASTER bands. Here, attempt has been made to derive an image enhanced product to delineate phosphorite and also use the same product as the proxy to find the relative grade variation in phosphorite under specific condition. Therefore, an approach is proposed for targeting Palaeoproterozoic phosphorite based on the spectral contrast of phosphorite and the host rock dolomite in the TIR bands of ASTER. The potential of broadband emissivity feature as a proxy to process the ASTER TIR bands to delineate low and high-grade phosphorite from the host dolomite rock has been analyzed in this study. In this regard, we studied the emissivity contrast of dolomite and phosphorite in their laboratory spectra and also in ASTER TIR sensor resampled counterparts. Further, we compared the image-based emissivity spectra of phosphorite pixels with their corresponding ASTER resampled laboratory spectra to ensure that the spectral features of phosphorite have been translated from ASTER resampled laboratory spectra to their image spectra. After confirming the translation, the relative spectral emissivity bands of ASTER (derived from ASTER level 1B data) were processed to derive an appropriate index image to delineate phosphorite. Potential of the relative emissivity extraction method in preserving the shape of the emissivity spectra of minerals and rocks, provides the scope of deriving a simple and reproducible method to delineate phosphorite using their broadband emissivity spectra [12,33–35].

2. Study Area and Geology

The study area is located 18 km southeast of Udaipur, one of the major towns of Rajasthan, the largest state of India (Figure 1).

Table 1. ASTER thermal infrared (TIR) data specifications (here visible–near-infrared (VNIR) bands are used for preparing the study area map while data processing is restricted to TIR bands). Band 6 to Band 9 are part of ASTER short wave infrared(SWIR) region. These SWIR spectral bands are not used in this study.

Sensor Type	Band Number	Spectral Width or Wavelength Range (μm)	Spatial Resolution (meter)	Radiometric Resolution (in bits)
Visible Infrared (VNIR)	1	0.52–0.60	15	8
	2	0.63–0.69	15	8
	3N	0.78–0.86	15	8
Thermal Infrared Sensor (TIR)	Band 10	8.125–8.475	90	12
	Band 11	8.8475–8.825		
	Band 12	8.925–9.275		
	Band 13	10.25–10.95		
	Band 14	10.95–11.65		

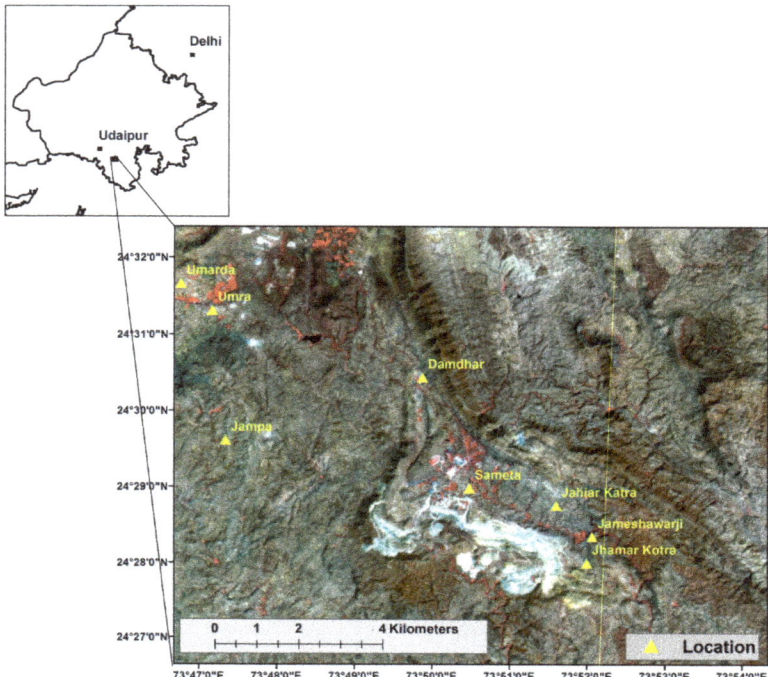

Figure 1. The extent of the study area is shown on the Advanced Spaceborne Thermal Emission and Reflection Radiometer (ASTER) false color composite image prepared using visible and near-infrared bands of ASTER. In this false color composite image, Red = Band 3 of ASTER visible near infrared (VNIR) sensor, Green =Band 2 of VNIR sensor, Blue =Band 1 of VNIR sensor of ASTER. For specification of ASTER VNIR bands, please refer Table 1.

In the study area, phosphorite is associated with carbonate-rich sediments of the Aravalli Super Group [36]. A geological map of the study area is presented in Figure 2. Although the grade of the regional metamorphism is low for these sediments (metamorphosed under green-schist facies), at places, higher grade metamorphism is also reported [37]. The distribution of different depositional facies, particularly dolomite with stromatolitic phosphorite, is controlled by paleo sea floor topography suggesting the presence of an epicontinental sea during the deposition of phosphorites [36]. It has also been suggested that the deposition of phosphorite and dolomite were triggered in the Paleoproterozoic period after the ephemeral relief in the platform part of the paleo sea was obliterated by intense weathering [37,38]. Dolomitic marble, dolomite, stromatolitic limestone, cherty quartzite phyllite, and quartzite are the major sedimentary rock units of the Aravalli Group while metavolcanics are intermittent volcanic units of a volcano-sedimentary sequence of the lower Proterozoic period in the study area. Granite gneiss, amphibolites, migmatites, etc. are part of the basement rock above which Aravalli sediments were deposited (Figure 2).

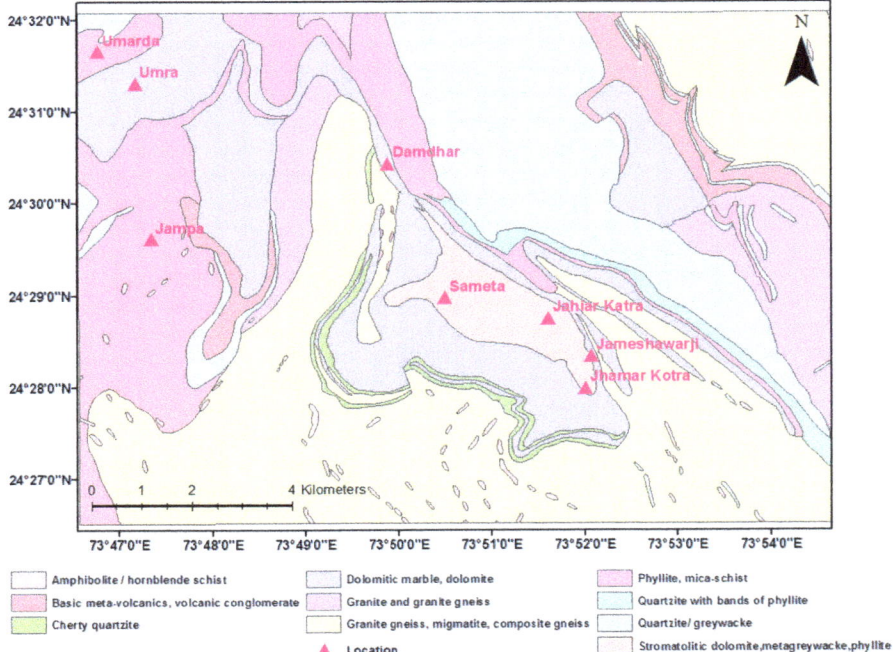

Figure 2. Lithological Map of the area (Source: Geological Survey of India (GSI); Unpublished). Triangles are important locations where phosphorites are exposed.

3. Materials and Methods

3.1. Materials

3.1.1. Rock Samples

In the study area, we have identified two major types of phosphorite: one type is low-grade phosphorite with P_2O_5 content 10–15% and another variant is high-grade Stromatolite bearing phosphorite with P_2O_5 content 28–30%. Low-grade phosphorite is massive and rich in dolomite mineral with fluroapatite. High-grade phosphorite is stromatolitic, easily weathered, and constituted with primarily fluroapatite. In these stromatolitic phosphorites, we either have both dolomite and fluroapatite as dominant minerals, or it is predominantly constituted with fluroapatite. Mode of occurrence of the samples in the field (dolomite, low-grade, and high-grade phosphorite) is shown in Figure 3. The dolomite and associated phosphorites are exposed on the denudational hills and pediment surface. Samples are collected from the surface exposures of these geomorphic units for the spectral analysis. Initially, large samples were broken from the surface exposures of rocks. Each representative sample was broken into few fragments. One sample fragment was used for spectral analysis, and other two fragments were used for X-Ray diffraction (XRD) and X-Ray fluorescence (XRF) analysis, respectively.

Figure 3. (**a**). Surface exposures of dolomite. (**b**) Surface exposures of low-grade rock phosphate (**c**). Surface exposures of high-grade phosphorite.

3.1.2. ASTER Data

ASTER has nine bands in the VNIR and SWIR spectral domain and five bands in the TIR domain [15,16,39,40] (Table 1). The TIR subsystem of the ASTER sensor operates within the spectral domain of 8.125 to 11.65 µm and is characterized by five spectral bands. These channels collect spectral data using a single telescope with a spatial resolution of 90 m. It has a "whiskbroom" scanning mirror [40]. Each band uses ten mercury–cadmium–telluride detectors that are cooled to 80 K using a mechanical split Sterling cycle [40]. We have used ASTER Level 1B data for the spatial mapping of phosphorite rich zones within the study area. The reason for selection of ASTER Level 1B data was guided by the previous study results obtained by different researchers using the same type of dataset (ASTER Level 1B) on the derivation of the mineralogically sensitive geological index [11,12,25]. Use of Level 1B data would also help in reducing the uncertainties resulted from the implementation of atmospheric correction algorithms on the radiance data [10,11]. Further, we assumed that the role of the atmosphere would be minimal in the recorded radiance of the thermal multispectral sensor, which has its spectral bands in atmospheric window of TIR domain. Further, records are available on the derivation of relative emissivity from Level 1B data of ASTER TIR bands [33,34] Emissivity normalization is one of such proven method which is known to be effective in deriving the shape of emissivity spectra [41,42].

3.1.3. Spectral Data

Emissivity spectra of representative samples of dolomite and major variants of phosphorite were collected in the laboratory. We have collected thermal emissivity spectra within the spectral domain of 8 to 12 µm using a portable Fourier-transformed (FT)-infrared (IR) spectrometer manufactured by D&P Instruments, United States of America [43]. The spectrometer has a functional spectral range of 2 to 16 micrometers. However, we have not processed and analyzed the emissivity spectra of the rock samples for the spectral domain of 2 to 8 µm as this domain is beyond the spectral range of ASTER TIR bands. The spectral resolution of the spectrometer is 4 cm^{-1} wave number within the spectral range of 8.125 to 11.67µm (i.e., the spectral range of ASTER). The spectrometer is composed of a nitrogen-cooled indium–gallium arsenide/mercury cadmium telluride detector [40] and worked on the principle of Fourier-transform [44]. Therefore, it can collect spectral data for very large wavelength domains.

3.1.4. Mineralogical and Chemical Data

XRD data of representative samples of phosphorites are used to estimate the minerals present in the samples to understand how mineralogy (i.e., dominant constituent minerals) contributes to shaping the Emissivity spectra of phosphorite samples. We used a specialized diffractometer system (6E-XRD 3003 TT automated system) to carry out diffraction studies of the powdered samples (200 mesh size) using the characteristic CuK(α) radiation (crystal monochromated).

A wavelength dispersive XRF instrument (MagiX Pro PW 244-PANalytical model) was used to estimate the major oxides with the primary emphasis on understanding the variation of P_2O_5 content in the phosphorite samples. Sample preparation method followed for collecting XRD and XRF data is similar to the method; which has been discussed in the literature [30].

3.2. Methods

3.2.1. Spectral Data Collection and Analysis

We have collected representative samples of dolomite and major variants of phosphorite samples from the study area. These samples were cut into rectangular blocks of 4-inch x 5-inch size to 5-inch x 6-inch size range. Samples were placed under the optics of the spectrometer and viewed with the optics to ensure that the emittance of the samples was collected from the sample surface only. Emissivity spectra of representative samples of phosphorite and host rock were derived from the collected emitted radiance of the sample and black body in the laboratory using an FTIR spectrometer (Figure 4). An emissivity spectrum is a plot of emissivity as a function of wavelength. Here, emissivity is derived by estimating the ratio of the emitted radiance of the sample to the emitted radiance of a blackbody at a specified temperature. We collected the emitted radiance of the sample after elevating its temperature using a heater to ensure that the emissivity spectra are collected with a high signal to noise ratio (SNR). We also maintained the isothermal condition during the measurement of emitted radiance by keeping the sample above a low conductive unit so that it would not conduct its accumulated heat fast. Before measuring the emitted radiance of the sample, we calibrated the instrument by measuring the emitted radiance of a blackbody at two different temperatures. One measurement was taken at 10 °C (lower than ambient temperature), and another was taken a few degrees centigrade higher than the temperature of the hot sample [44]. Measured emitted radiance of a black body at two different temperatures (one higher than the sample temperature and the other lower than the sample temperature) was helpful to estimate the emitted radiance of a blackbody at the same temperature with that of the sample [44]. Black body emitted radiance was used to normalize the emitted radiance of the sample to derive the emissivity of the sample temperature. The instrument performs satisfactorily within the temperature range of 5° to 40° centigrade [43].

Table 2. Results of XRD and X-ray fluorescence (XRF) analysis of representative samples of phosphorite and dolomite to identify the presence of the dominant mineral phases and their relative proportion along with the P_2O_5 Content [30].

Nature of Samples	Sample No. and Details	Major Dominant Minerals Identified Using XRD Data (Arranged as Per the Decreasing Order of Relative Abundance)	P_2O_5 Content in % (XRF)
Phosphorite and dolomite	P1(low-grade phosphorite)	Fluroapatite, dolomite	13.94
	P6 (Dolomite)	Dolomite, quartz	—
	P9 (Dolomite bearing high-grade phosphorite)	Fluroapatite, dolomite,	38.53
	P11 (Dolomite depleted high-grade phosphorite)	Fluroapatite, quartz	39.11

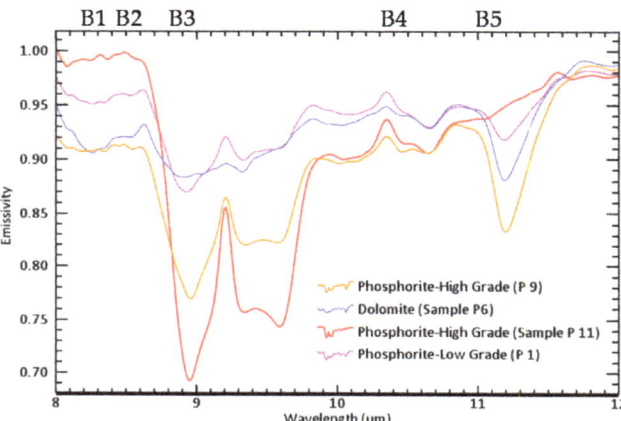

Figure 4. Continuum removed thermal emissivity spectra of representative samples of dolomite and phosphorites (P1 is low-grade; P11 and P9 are medium-to-high-grade phosphorite; P6 is dolomite). Broad mineralogy of the samples and P_2O_5 content of the samples are given in Table 2. Wavelength positions of five TIR bands of ASTER are also shown (B1 to B5).

While collecting emissivity spectra, blackbody temperature is regulated using electrostatic heating [44]. The facility to perform electrostatic heating is in-built in the spectrometer. We used a gold plate having a low emissivity (0.2) to measure the contribution of downwelling radiance [44]. Downwelling radiance is subtracted from the emitted radiance of the sample and blackbody before their respective emittance/emitted radiance values are used to derive emissivity. The conceptual framework and methodology to collect emissivity spectra have been discussed in the literature [44,45].

In this study, the emissivity spectra of different grades of phosphorite (broadly two grades: 10–15% and 38–40% P_2O_5 content) were compared with the emissivity spectrum of dolomite (Figure 4). Higher-grade phosphorites are represented by two specific types: one is dolomite-bearing phosphorite and the other is stromatolite without the presence of dolomite as a constituent mineral. We compared spectral contrast of major targets (different phosphorite and dolomite) and also analyzed the spectral contrast of different samples of each variant of phosphorite and dolomite to understand the inter-rock and intrarock variation of the emissivity spectra (Figures 4 and 5). Further, the emissivity spectra of specified grade phosphorite samples and host rock were compared with the emissivity spectra of respective dominant constituent minerals to understand how mineralogy influences the emittance spectra of rocks [45,46] (Figure 6). Laboratory emissivity spectra of rock samples were also resampled to the ASTER bandwidth to understand how spectral contrast of dolomite and host rock was translated from sample to ASTER bandwidth (Figure 7). An attempt was also made to understand how spectral contrast between dolomite and phosphorite would be gradually reduced if the spectra of both the end members are linearly mixed within a pixel of ASTER TIR sensor, which is of 90-m size (Figure 8). In order to derive spectra of pixels mixed with dolomite and phosphorite in different proportion, we added their respective pure spectra with different weights. Weights are assigned based on the assumption that spatial extent (in terms of fraction) of the pixel is occupied by only these two targets (i.e., dolomite and phosphorite) in different fractions. This was required to understand the role of other factors such as intrapixel mixing in reducing the spectral contrast of dolomite and phosphorite except for the grade (i.e., P_2O_5) of phosphorite. Detailed results related to the analysis of spectral data are discussed in the results section.

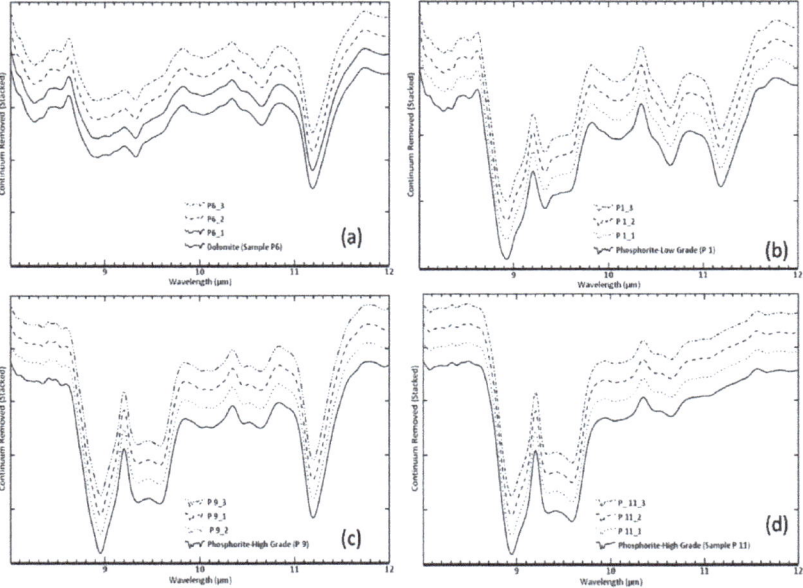

Figure 5. (**a**) Emissivity spectra of different samples of host rock dolomite(different samples of same rock type are denoted P6_1, P6_2, and so on). (**b**) Emissivity spectra of different samples of low-grade phosphorite (different samples of same rock type are denoted P1_1, P1_2, and so on). (**c**) Emissivity spectra of medium to high-grade phosphorite (different samples of same rock type are denoted P9_1, P9_2, and so on). (**d**) Emissivity spectra of high-grade phosphorite (different samples of same rock type are denoted P11_1, P11_2, and so on).

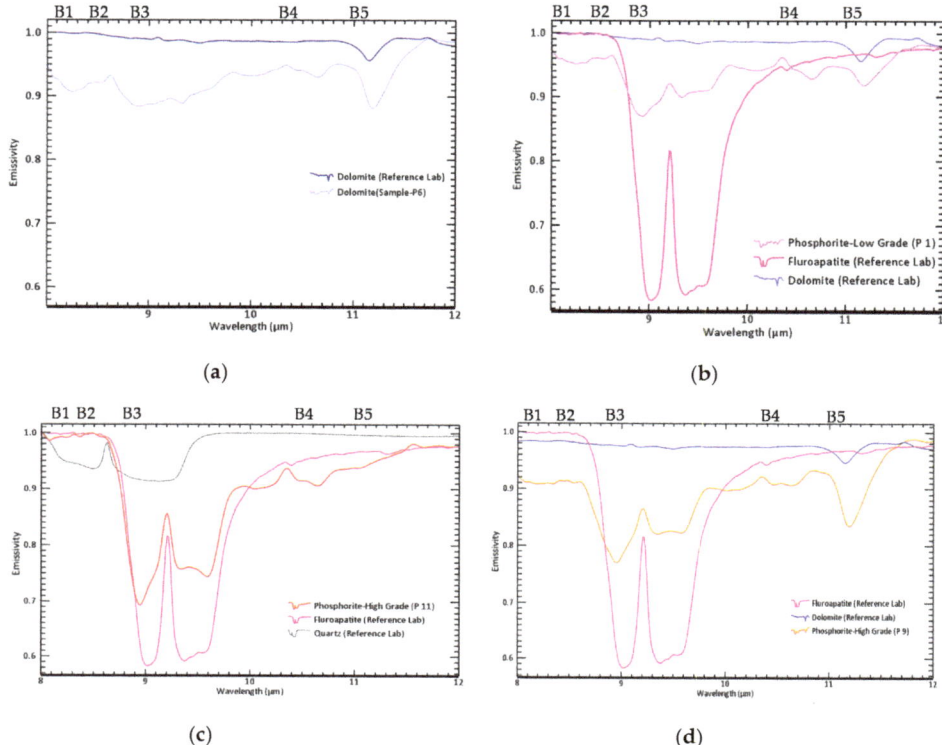

Figure 6. (**a**). Continuum removed thermal spectrum of dolomite (P6) and the spectra of the dominant constituent mineral are plotted. (**b**) Continuum removed thermal spectrum of low-grade phosphorite (P1), and the spectra of dominant constituent minerals are plotted. (**c**) Continuum removed thermal spectrum of medium to high-grade phosphorite (P11) and the spectra of dominant constituent minerals are plotted (**d**). Continuum removed thermal spectrum of medium to high-grade phosphorite (P9) and the spectra of dominant constituent minerals are plotted. Mineral spectra is collected from United States Geological Survey (USGS) spectral library. The spectral data base is available with ENVI 5.1 software package. Wavelength positions of five TIR bands of ASTER are also shown (B1 to B5).

We also compared the shape of the broadband emissivity spectrum of phosphorite of ASTER pixel with the ASTER resampled counterpart (Figure 9). This was required to confirm the fact that image-based emissivity spectra of phosphorite could preserve the broadband emissivity feature of the phosphorite that has been observed in the ASTER resampled laboratory spectra. Image-based emissivity spectra were collected from a few regions of interests (ROI) distributed above the known phosphorite exposures around Jhamar\Kotra and Sameta areas. This has made the basis for ASTER TIR image processing for delineating phosphorite within its host rock.

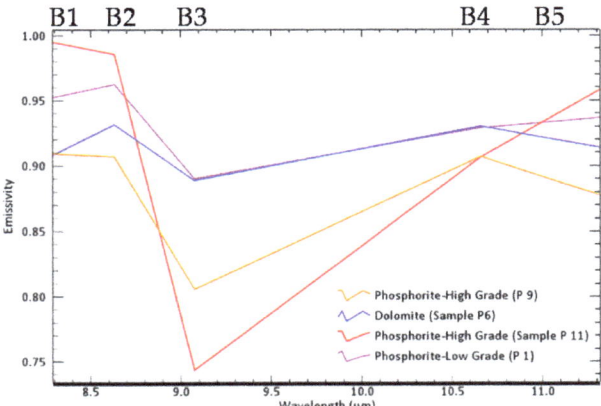

Figure 7. ASTER resampled spectra of representative samples of phosphorite and dolomite. Wavelength positions of five TIR bands of ASTER are also shown (B1 to B5).

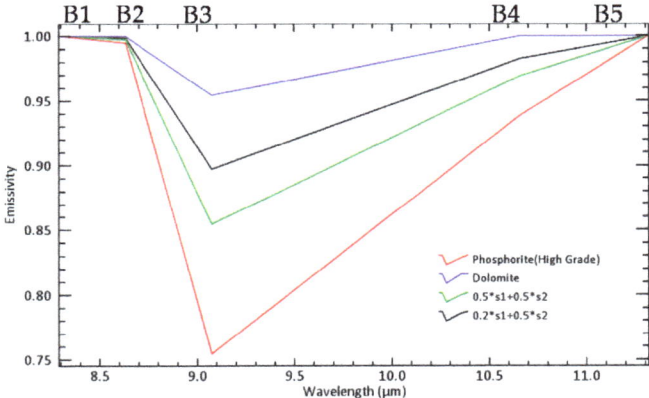

Figure 8. Comparison between phosphorite, dolomite and their intermediate mixed variants. Intermediate mixed variants are derived by linear mixing Emissivity spectra of dolomite (S1) and phosphorite (S2) to understand the role of intrapixel mixing in the detection of phosphorite. Wavelength positions of five TIR bands of ASTER are also shown (B1 to B5).

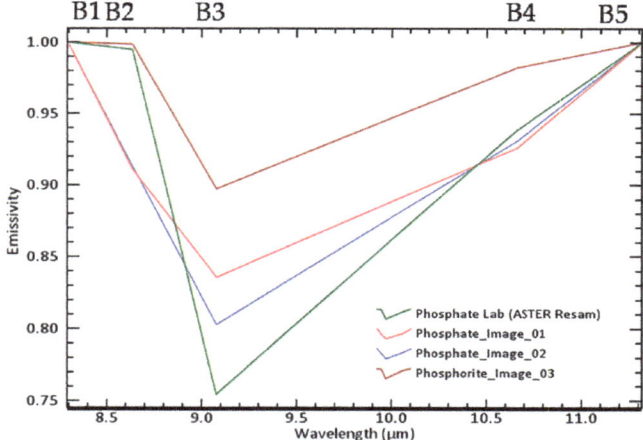

Figure 9. Comparison between the ASTER TIR band resampled mean high-grade phosphorite spectrum and with the few image spectra of phosphorite pixels. These pixels are the known exposures of phosphorite around the Jhamar Kotra mining area. Wavelength position of ASTER five TIR bands are also shown (B1 to B5).

3.2.2. Mineralogical and Chemical Analysis

Original representative samples of phosphorite and dolomite were broken into three pieces to carry out mineralogical and elemental analysis along with the analysis of emissivity spectra. One piece of the sample of each type of rock was used for the collection and analysis of emissivity spectra. The other two pieces of the same sample were used for XRF and XRD analysis respectively. A wavelength dispersive XRF instrument was used to estimate the dominant oxides. Representative samples of phosphorite and dolomite were pulverized and consequently sieved using 200-mesh sizes for XRF samples. The results of XRF analysis are given in Table 2. In this study, P_2O_5 content is only used to relate this chemical data with the depth of the broadband diagnostic emissivity feature of different phosphorite samples.

We used the diffractometer system (6E-XRD 3003 TT automated system) to utilize the CuK(α) radiation (crystal monochromated) for X-ray diffraction studies of the powdered samples (200 mesh size). Before performing XRD measurements, phosphorite rock samples were pulverized and consequently sieved using 80-mesh sieves. After obtaining the desired grain size, we separated the lighter mineral phases (e.g., phosphate and dolomite) from the heavier mineral fractions using methylene iodide solution. We have separately analyzed powdered fractions of dolomite and apatite minerals using the diffractometer to avoid the overlapping of peaks of different mineral phases in the diffractograms of the samples. Predominant minerals (i.e., fluorapatite, dolomite, and quartz) were only identified using XRD analysis based on their peaks at a specified incident angle (Figure 4). The mineralogy of samples with their relative abundance derived from XRD study was used as a reference to analyze the emissivity spectra of rocks.

3.2.3. ASTER Data Analysis and Field Validation

We analyzed ASTER Level 1B "at sensor radiance" data with reference to the geological map of the study area. Paleoproterozoic phosphorites are primarily associated with dolomite or other carbonate rocks. Therefore, we processed the ASTER data for a portion of the study area; which was covering the spatial extent of the host rock—dolomite. In this regard, we used the lithological boundary of dolomite as delineated in the geological map of the Geological Survey of India to spatially subset the ASTER TIR image. The spatial subsetting of the ASTER image is useful in reducing the number of targets to be delineated in ASTER thermal bands. This may enhance the mapping accuracy (as number of unknown to be detected will be few) as ASTER TIR bands are known for poor intrinsic dimensionality due to the presence of striping noise and the poor signal received at band 14 of the sensor [28,29].

Based on the broadband emissivity features of phosphorite, we prepared image composite using three radiance bands after attempting decorrelation stretching (Figure 10). We derived the radiance image composite after extracting the desired portion from the ASTER scene using the lithological outline of dolomite as "region of interest" (Figure 11). Further, we enhanced the separability of surface exposures of phosphorite rich zones from host rock by deriving the emissivity image composite. In this regard, we derived emissivity using the "emissivity normalization" method. The image-based emissivity product derived using the above method is known for preserving the overall shape of emissivity spectra and the wavelength of diagnostic emissivity features of the target [33,34,42]. In this method, temperature values of the pixels were derived for each spectral band using a fixed emissivity value (0.96) [42]. Consequently, we derived spectral emissivity for each pixel using the highest temperature value of each pixel (i.e., from the set of different temperature values in different bands) derived using fixed emissivity in the first instance. Before deriving relative emissivity from the ASTER thermal radiance bands, we implemented "in-scene atmospheric correction" on ASTER radiance bands based on the assumption that atmosphere remained uniform over the scene and a black body is present in the scene [47]. In this method, we also assumed that the atmosphere is single layered and downwelling component of atmosphere is absent. Atmosphere calibration is made based on deriving gain and offset by regressing measured and theoretical radiance of blackbody at specified wavelength. In the scene, pixels which record highest temperature are assumed as the mathematical approximation of black body [47]. After calibrating the scene using thermal atmospheric correction, emissivity normalization method was implemented. For delineating phosphorite, we also used three emittance bands to derive decorrelation stretched emissivity composite image to delineate phosphorite within dolomite (Figure 12).

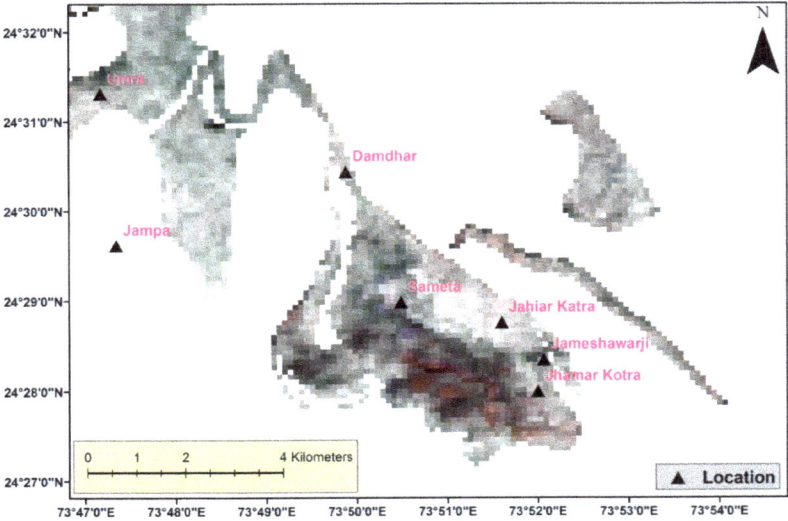

Figure 10. Decorrelation stretched false color composite prepared using spectral bands of ASTER level 1B radiance data. In this image composite, Red = Band 13, Green =band 12, and Blue = band 11 of ASTER thermal infrared sensor. Please refer to Table 1 for ASTER TIR band nuber detail.

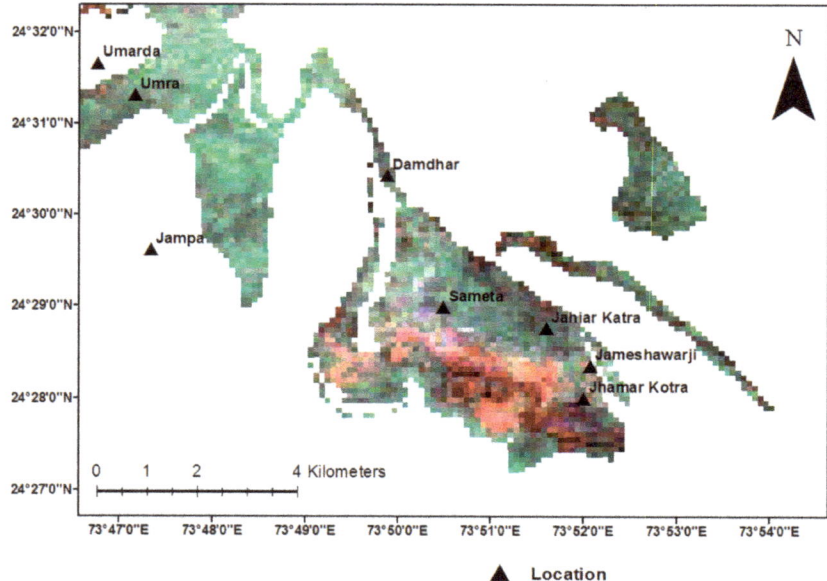

Figure 11. Decorrelation stretched emissivity normalization composite derived using red = emissivity of band 13, green = emissivity of band 12, and blue = emissivity of band 11. In this figure, Phosphorite pixels are enhanced with a red and pinkish-magenta color. Please refer to Table 1 for ASTER TIR band nuber detail.

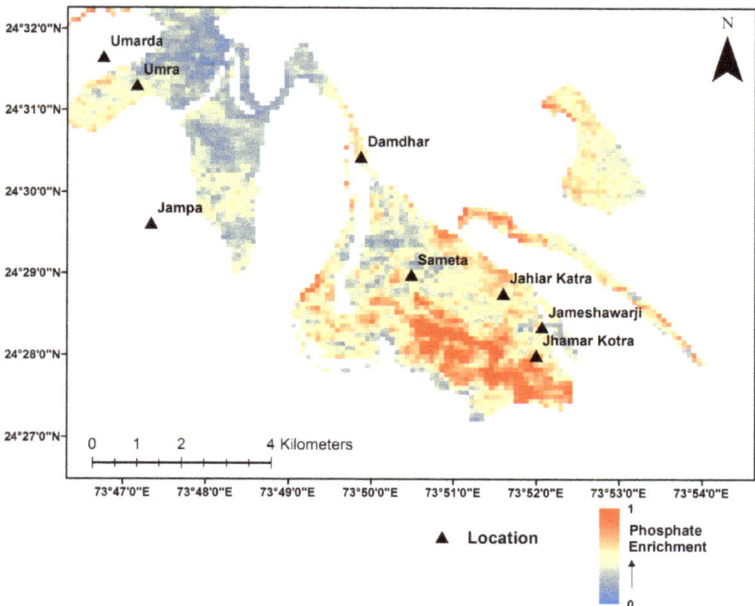

Figure 12. Density sliced relative band depth image derived using three emissivity bands of ASTER (Band 13+Band 11)/Band 12. Relative subpixel abundance of phosphate is increasing with yellow to red color, while blue zones are dolomite.

Further, a relative band depth (RBD) image was derived using emissivities of bands 11, 12, and 13 (derived using the emissivity normalization method) to enhance the strong emissivity minima on emissivity spectra of phosphorite (Figure 12). This RBD image was derived as per the contrast observed in the ASTER resampled laboratory emissivity spectra of dolomite and phosphorite and their mixed variants (Figures 8 and 9). We derived a relative band depth (RBD) image to delineate rock phosphate using a single band product. Finally, we validated the TIR image enhanced products (RBD image and emissivity image composite) by visiting the field locations of phosphorite pixels and confirming the presence of phosphorite based on rapid colorimetric analysis of samples collected from the rock exposures (small rectangles) (Figure 13). In this regard, the pulverized rock samples (rock samples were broken and manually pulverized) were mixed with an acidic ammonium molybdate solution to rapidly identify the presence of phosphate in the sample.

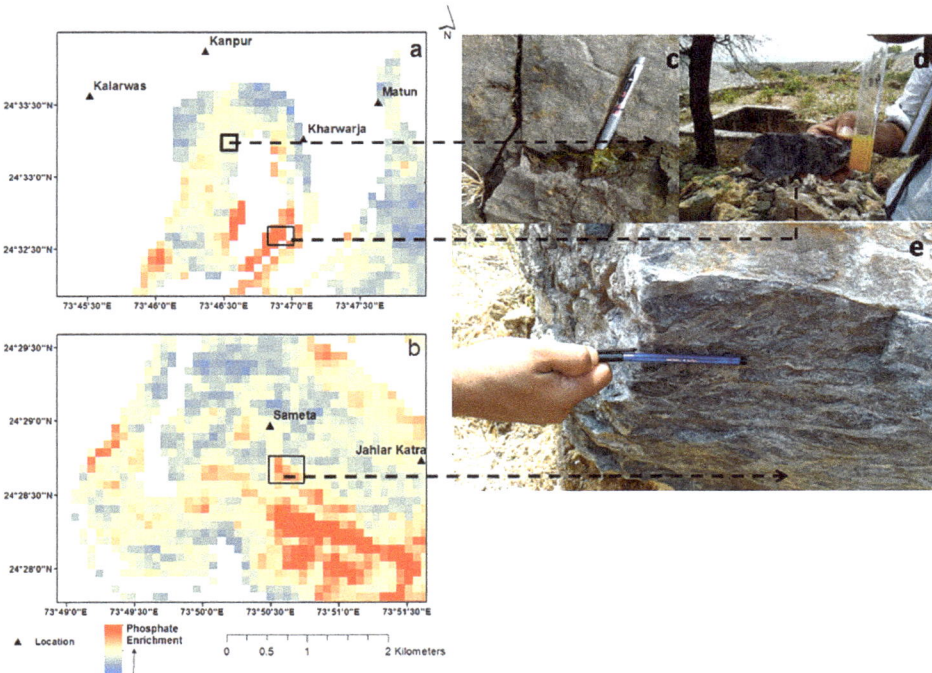

Figure 13. (**a**). Density sliced relative band depth image for Kharawarja (Matun). (**b**). Sameta and the Jhmar Kotra mine. High phosphorite rich zones have been demarcated with red color. Field check areas beyond the present mine are shown with a rectangle. Field validation in the east of the Kharwarja area (**c**) and south of the Kharwarja area (**d**) and Sameta area (**e**).

4. Results and Discussion

We analyzed the emissivity spectra of representative samples of different rocks to understand inter-rock emissivity variation (Figure 4) of phosphorite and dolomite. We also analyzed the contrast in the emissivity spectra of different samples of same rock type (Figure 5). We found emissivity spectra of different samples of the same rock were similar to each other and overlapping one over the other. Therefore, emissivity spectra of different samples of same rock were stacked and plotted (Figure 5a–c). It is quite evident that the emissivity contrast of different samples of same rock would be indistinguishable when we would analyze their ASTER resampled counterpart or their image spectra as the broad emissivity spectra would be further more generalized. Therefore, we concentrated our study to identify the inter-sample emissivity contrast in this study and used the inter-rock emissivity variation as a reference for processing ASTER TIR bands. In the study area, we found one low-grade and two high-grade variants of phosphorite. Details of these samples were discussed in the Section 3.1. We further analyzed the mineralogical significance of emissivity spectra (and related contrast) of dolomite and phosphorites using the XRD data of these samples as the reference (Table 2). Analysis of Emissivity spectra is the basis for processing of the ASTER data. Therefore, we analyzed the emissivity spectra of low-grade and high-grade phosphorite and host rock with reference to the emissivity spectra of their constituent minerals (Figure 6). While analyzing the emissivity spectra of these samples, it was observed that phosphorite had a strong doublet with minimal at 9 μm and 9.5 μm, while emissivity spectrum of dolomite sample was characterized with emissivity minima at 11.2 μm [2] (Figures 4 and 6a). Similar to dolomite, a diagnostic emissivity minima (i.e., at 11.2 μm) was also observed in the low-grade phosphorite (i.e., P1), but the same feature is obscured in the emissivity spectra of dolomite depleted, high to moderate grade phosphorite samples (P11) (Figures 4 and 6a–c). We, however, could identify the same feature in the emissivity spectra of dolomite-bearing high-grade phosphorite (Figure 6d). The fundamental vibration of carbonate bonds governs the emissivity minima at 11.2 μm [48]. In contrast, the emissivity doublet of phosphorite is absent in the emissivity spectra of dolomite samples (sample P6) (Figure 4), and subdued emissivity minima are identified in the dolomite emissivity spectrum (Figure 6a). The above details on the spectral contrast of dolomite and phosphorite are subdued in the ASTER resampled emissivity spectra of these samples (Figure 8).

However, we observed that the ASTER resampled emissivity spectra of phosphorite samples had the stronger depth of emissivity minima with respect to the dolomite emissivity feature at 9.2 μm (Figure 7). Depth of the emissivity feature of high-grade variant was more than the low-grade variant of phosphorite.

It is known that the emissivity minima of emissivity spectra of rocks and constituent minerals can be studied to estimate relative grade or compositional variation based on the assumption that the grain size has a negligible influence on the quantitative parameters (i.e., depth and width) of emissivity features [48,49]. In the study, this assumption may be true as dolomite, and different phosphorites had similar grain size. This provided us a scope to relate emissivity minima with the grade or P_2O_5 content. We found that the depth of broadband emissivity minima of phosphorite correlates broadly with two phosphorite grade or variants present in the study area (Figure 8 and Table 2). This assumption is made based on the observation that high-grade phosphorite (high P_2O_5 content) has a larger depth of its diagnostic emissivity feature with respect to lower grade phosphorite. However, in addition to grade, the spectral purity of the pixels of the ASTER image may also influences the depth of emissivity minima. The depth of the broad emissivity feature gradually reduces (i.e., the depth will be reduced) if the dolomite is linearly mixed with different proportions of phosphorite within the spatial extent of the large pixels of the ASTER bands. The role of intrapixel mixing in reducing the depth of emissivity feature has been analyzed based on deriving mixed pixel spectra by linearly combining ASTER resampled emissivity spectra of dolomite and phosphorite (high-grade variants) with different weights (proportional to their respective spatial abundance in the pixel) (Figure 8). We found that the depth of the emissivity feature of phosphorite would be gradually reduced with the decrease in the relative spatial abundance of phosphorite within the pixel. Therefore, delineation of phosphorite within the dolomite would be difficult if the patchy or very small phosphorite exposures are mixed with the exposure of dolomite occupying the major portion of the pixel. However, broadband emissivity minima of phosphorite can be used to derive ASTER based image products to delineate large exposures of phosphorite and also can be used to relatively delineate the grade of phosphorite exposures from the low-grade variants provided the exposures are of larger size.

Further, we also confirmed that the shape of the image spectra of phosphorite pixels was comparable with the ASTER resampled laboratory counterpart of emissivity spectrum of phosphorite. This was ascertained by cross-comparing the spectra of phosphate mines at selected places with the ASTER resampled emissivity spectrum of high-grade phosphorite (Figure 9). The comparative assessment of image spectra of phosphorite with respect to ASTER resampled laboratory counterpart is essential to ascertain the spectral consistency of target from ground to sensor.

Consequently, we processed the ASTER TIR bands based on the analysis of the emittance spectra of dolomite and the two major variants of phosphorite, as discussed in the previous section. ASTER data were further processed for the portions covering the spatial extent of host rock dolomite as the main or large exposures of phosphorite occur within dolomite/dolomitic marble. We prepared the ASTER radiance image composite using selected radiance bands after attempting the decorrelation stretching. In the radiance-composite image of ASTER TIR bands (prepared using bands 11–13) we could not delineate phosphorite exposures effectively (except some patches on Jhamarkotra mine (Figure 10). Consequently, we prepared an emissivity composite to delineate large surface exposures of phosphorite using the emissivities of bands 11–13. In this image product, phosphorite exposures were enhanced with pinkish-red color (Figure 11). We also derived a relative band depth (RBD) image (Figure 12) using the emissivity patterns defining the shoulders (bands 11 and 13) and absorption minima (band 12) of the broadband emissivity feature of phosphorite (Figure 7). A higher value of this RBD image was indicative of progressive phosphorite enrichment within the dolomite as the depth of the emissivity feature was broadly related to the phosphate grade (i.e., higher grade phosphorite had a larger depth than the lower grade variant), provided the phosphorite exposures were large (Figures 7, 8 and 12). In this colour density sliced RBD image, "red colored" pixels were indicative of high-grade phosphorite while yellow colored pixels were indicative of low-grade phosphorite. Blue pixels indicated the presence of dolomite.

We carried out field validation to clarify the results of RBD images at the selected sites. We collected rock samples from the exposures identified with red or yellow pixels in the RBD image. Most of the areas are phosphate bearing. Phosphorite-bearing rock powders were found to have changed color from colorless to yellow as ammonium molybdate reacts with phosphate to precipitate "yellow colored" ammonium phosphate [50].

In this regard, we identified phosphorite within dolomite by injecting acidic ammonium molybdate solution in the pulverized sample of rock collected from the rock exposure at the west of Kharwarja. The RBD image shows an intermediate value with yellow color at this site (shown with rectangle). The phosphorite is associated with the dolomite, and the reported grade was low [37] (Figure 13a,c). We could also identify the presence of stromatolitic phosphorite at a site occurring at the south of Kharwarja, The site had relative high value in RBD image as it was above the red pixels of RBD image (Figure 13a,d). Stromatolitic phosphorites are high-grade phosphorite in the study area [37]. Similarly, stromatolitic phosphorite was also identified in a site situated to the north of Sameta (Figure 13b,e). We showed the field location of this site with a small rectangle on the RBD image and site was above the high-value pixels (i.e., it is red colored) of RBD image.

The proposed method of phosphorite mapping using TIR bands would be applicable for any Proterozoic dolomite hosted phosphorite deposit in any part of the world. The proposed RBD image can also be used to identify the relative grade variations in the phosphorite exposures provided the exposures are large enough to make band depth values of RBD image invariant to the modifications; which could be due to intrapixel mixing and the mapping was attempted within the spatial extent of host rock of phosphorite. The proposed method is rapid and can be used to identify areas with high phosphorite content for the detailed exploration of rock-phosphate. Index-based delineation of phosphorite using ASTER TIR bands is guided by characteristic absorption feature of phosphorite. Therefore, index images derived from ASTER relative emissivity bands would not only help in the delineation of phosphorite, but also would help in relatively estimating grade of phosphorite with large and homogeneous surface exposures. Similar grade estimation may not be possible in VNIR-SWIR spectral domain as grade sensitive spectral feature is absent in the VNIR-SWIR domain. However, small patchy exposures of phosphorite which can be detected using spectral contrast of dolomite and phosphorite in ASTER SWIR band product based on the implementation of subpixel mapping approach can be subdued in the TIR band based product proposed for phosphorite [30]. This is due to intratarget spectral mixing of dolomite and small patchy exposure of phosphorite, which would hinder the detection of small exposures in broadband, coarse resolution thermal bands of ASTER. However, the proposed approach of TIR band based mapping of phosphorite can be supplemented with a geophysical survey like caliper logging or gammy ray logging (phosphate is often associated with radioactive minerals) for the detailed exploration of identified anomaly [51].

5. Conclusions

Based on the methodology adopted and the results obtained, the following conclusions are derived from the present study.

a. Emissivity spectra of dolomite and phosphorite are distinct from each other. A strongly emissive doublet characterizes the emissivity spectra of phosphorite samples while dolomite is devoid of such emissivity minima.
b. The spectral contrast of dolomite and phosphorite has been further generalized in the ASTER image spectra, and ASTER resampled laboratory spectra (Figure 6). The contrast in the emissivity is limited to the depth variation of the emissivity feature at 9.1 μm for dolomite and phosphorite. Phosphorite emissivity spectra have a larger depth with respect to the emissivity spectrum of dolomite having negligible depth at the wavelength.
c. We proposed an RBD image-based on the emissivity contrast of dolomite and phosphate. The proposed RBD image of ASTER TIR bands can be used to delineate phosphorite provided the spatial mapping using the RBD image is restricted within the spatial extent of the host rock, i.e., dolomite or carbonate rocks. In this study, low-grade phosphorite exposures have intermediate value (yellow color), while high-grade phosphorite have high value (identified with yellow pixels)
d. The proposed approach of broadband TIR band based phosphorite mapping is simple, reproducible and can be used for targeting phosphorite occurring under similar geological setups. Many important carbonate phosphorite deposits in the world have a similar geological setup.

Author Contributions: The details of the contribution made by all the authors in different segment of work are as follows: (a) Conceptualisation: A.G. and K.V.K.; (b) Methodology: A.G., K.R.; Validation, A.G. and K.R.; Formal Analysis with suggestions to improve part of the analysis of the final result: A.G., Y.Y., S.C.; Writing-Original Draft Preparation: A.G., Y.Y., S.C.; Writing-Review & Editing: A.G., S.C., Y.Y.; Supervision: K.V.K.; Funding Acquisition: K.V.K., A.G. (for internal fund of Indian Space Research Organisation, Department of Space, Govt. of India) and Yashushi Yamaguchi(Publication Charge).

Funding: This research received no external funding.

Acknowledgments: The authors are thankful to the Director of the National Remote sensing Centre for his overall guidance. The authors are thankful to the authorities of the Atomic Mineral Directorate for exploration and research, Hyderabad, India (AMD) for providing analytical support. The authors are also thankful to the Officers of the AMD and Geological Survey of India for their guidance during validation of result.

Conflicts of Interest: The authors declare no conflict of interest.

References

1. Choudhuri, R. *Two Decades of Phosphorite Investigations in India*; Geological Society: London, UK, 1990; Volume 52, pp. 305–311.
2. Lane, M.D.; Dyar, M.D.; Bishop, J.L. Spectra of phosphate minerals as obtained by visible-near infrared reflectance, thermal infrared emission, and Mössbauer laboratory analyses. In Proceedings of the Lunar and Planetary Science Conference, League City, TX, USA, 12–16 March 2007; Volume 38, p. 2210.
3. Tucker, M.E. *Sedimentary Petrology: An Introduction to the Origin of Sedimentary Rocks*, 3rd ed.; John Wiley & Sons: Hoboken, NJ, USA, 1 April 2009.
4. Buettner, K.J.; Kern, C.D. The determination of infrared emissivities of terrestrial surfaces. *J. Geophys. Res.* **1965**, *70*, 1329–1337. [CrossRef]

5. Misi, A.; Kyle, J.R. Upper Proterozoic carbonate stratigraphy, diagenesis, and stromatolitic phosphorite formation, Irecê Basin, Bahia, Brazil. *J. Sediment. Res.* **1994**, *64*, 299–310.
6. Cook, P.T.; Shergold, J.H. *Phosphate Deposits of the World, Proterozoic and Cambrian Phosphorites*; Cambridge University Press: Cambridge, UK, 1986; Volume 1, p. 386.
7. Salisbury, J.W.; Walter, L.S. Thermal infrared (2.5–13.5 µm) spectroscopic remote sensing of igneous rock types on particulate planetary surfaces. *J. Geophys. Res.* **1989**, *94*, 9192–9202. [CrossRef]
8. Salisbury, J.W.S.; D'Aria, D.M. Emissivity of terrestrial materials in the 8–14 µm atmospheric windows. *Remote Sens. Environ.* **1992**, *42*, 83–106. [CrossRef]
9. Ninomiya, Y.; Fu, B.; Cudahy, T.J. Detecting lithology with advanced spaceborne thermal emission and reflection radiometer (ASTER) multispectral thermal infrared "radiance-at-sensor" data. *Remote Sens. Environ.* **2005**, *99*, 127–139. [CrossRef]
10. Ding, C.; Liu, X.; Liu, W.; Liu, M.; Li, Y. Mafic and ultramafic and quartz-rich rock indices deduced from ASTER thermal infrared data using a linear approximation to the planck function. *Ore Geol. Rev.* **2014**, *60*, 161–173. [CrossRef]
11. Ding, C.; Li, X.; Liu, X.; Zhao, L. Quartzose–mafic spectral feature space model: A methodology for extracting felsic rocks with ASTER thermal infrared radiance data. *Ore Geol. Rev.* **2015**, *66*, 283–292. [CrossRef]
12. Guha, A.; Kumar, V. New ASTER derived thermal indices to delineate mineralogy of different granitoids of Archaean Craton and analysis of their potentials with reference to Ninomiya's indices for delineating quartz and mafic minerals of granitoids-an analysis in Dharwar Craton, India. *Ore Geol. Rev.* **2016**, *74*, 76–87.
13. Rani, K.; Guha, A.; Pal, S.K.; Vinod Kumar, K. Comparative analysis of potentials of ASTER thermal infrared band derived emissivity composite, radiance composite and emissivity-temperature composite in geological mapping of Proterozoic rocks in parts Banswara, Rajasthan. *J. Indian Soc. Remote Sens.* **2019**. [CrossRef]
14. Van der Meer, F.D.; Van der Werff, H.M.A.; Van Ruitenbeek, F.J.A.; Hecker, C.A.; Bakker, W.H.; Noomen, M.F.; Van der Meijde, M.; Carranza, E.J.M.; De Smeth, J.B.; Woldai, T. Multi-and hyperspectral geologic remote sensing: A review. *Int. Appl. Earth Observ. Geoinf.* **2012**, *14*, 112–128. [CrossRef]
15. Yamaguchi, Y.; Kahle, A.B.; Tsu, H.; Kawakami, T.; Pniel, M. Overview of advanced spaceborne thermal emission and reflection radiometer (ASTER). *IEEE Trans. Geosci. Remote Sens.* **1998**, *36*, 1062–1071. [CrossRef]
16. Abrams, M. The advanced spaceborne thermal emission and reflection radiometer (ASTER): Data products for the high spatial resolution imager on NASA's Terra platform. *Int. J. Remote Sens.* **2000**, *21*, 847–859. [CrossRef]
17. Becker, F.; Li, Z.L. Surface temperature and emissivity at various scales: Definition, measurement and related problems. *Remote Sens. Rev.* **1995**, *12*, 225–253. [CrossRef]
18. Tang, H.; Li, Z.L. *Quantitative Remote Sensing in Thermal Infrared: Theory and Applications*; Springer: Heidelberg, Germany, 2014.
19. Hubbard, B.E.; Crowley, J.K. Mineral mapping on the Chilean–Bolivian Altiplano using co-orbital ALI, ASTER and Hyperion imagery: Data dimensionality issues and solutions. *Remote Sens. Environ.* **2005**, *99*, 173–186. [CrossRef]
20. Hewson, R.D.; Cudahy, T.J.; Mizuhiko, S.; Ueda, K.; Mauger, A.J. Seamless geological map generation using ASTER in the Broken Hill-Curnamona province of Australia. *Remote Sens. Environ.* **2005**, *99*, 159–172. [CrossRef]
21. Chen, X.; Warner, T.A.; Campagna, D.J. Integrating visible, near-infrared and short-wave infrared hyperspectral and multispectral thermal imagery for geological mapping at Cuprite, Nevada. *Remote Sens. Environ.* **2007**, *110*, 344–356. [CrossRef]
22. Bell, J.H.; Bowen, B.B.; Martini, B.A. Imaging spectroscopy of jarosite cement in the Jurassic Navajo Sandstone. *Remote Sens. Environ.* **2010**, *114*, 2259–2270. [CrossRef]
23. Brandmeier, M. Remote sensing of Carhuarazo volcanic complex using ASTER imagery in Southern Peru to detect alteration zones and volcanic structures–a combined approach of image processing in ENVI and ArcGIS/ArcScene. *Geocarto Int.* **2010**, *25*, 629–648. [CrossRef]
24. Bedini, E. Mineral mapping in the Kap Simpson complex, central East Greenland, using HyMap and ASTER remote sensing data. *Adv. Space Res.* **2011**, *47*, 60–73. [CrossRef]

25. Ninomiya, Y.; Matsunaga, T.; Yamaguchi, Y.; Ogawa, K.; Rokugawa, S.; Uchida, K.; Muraoka, H.; Kaku, M. A comparison of thermal infrared emissivity spectra measured in situ, in the laboratory, and derived from thermal infrared multispectral scanner (TIMS) data in Cuprite, Nevada, USA. *Int. J. Remote Sens.* **1997**, *18*, 1571–1581. [CrossRef]
26. Aboelkhair, H.; Ninomiya, Y.; Watanabe, Y.; Sato, I. Processing and interpretation of ASTER TIR data for mapping of rare-metal-enriched albite granitoids in the Central Eastern Desert of Egypt. *J. Afr. Earth Sci.* **2010**, *58*, 141–151. [CrossRef]
27. Matar, S.S.; Bamousa, A.O. Integration of the ASTER thermal infra-red bands imageries with geological map of Jabal Al Hasir area, AsirTerrane, the Arabian Shield. *J. Taibah Univ. Sci.* **2013**, *7*, 1–7. [CrossRef]
28. Yajima, T.; Yamaguchi, Y. Geological mapping of the Francistown area in north-eastern Botswana by surface temperature and spectral emissivity information derived from advanced spaceborne thermal emission and reflection radiometer (ASTER) thermal infrared data. *Ore Geol. Rev.* **2013**, *53*, 134–144. [CrossRef]
29. Son, Y.S.; Kang, M.K.; Yoon, W.J. Lithological and mineralogical survey of the Oyu Tolgoi region, Southeastern Gobi, Mongolia using ASTER reflectance and emissivity data. *Int. J. Appl. Earth Observ. Geoinf.* **2014**, *26*, 205–216. [CrossRef]
30. Guha, A.; Vinod Kumar, K.; Porwal, A.; Rani, K.; Singaraju, V.; Singh, R.P.; Khandelwal, M.K.; Raju, P.V.; Diwakar, P.G. Reflectance spectroscopy and ASTER based mapping of rock-phosphate in parts of Paleoproterozoic sequences of Aravalli Group of rocks, Rajasthan, India. *Ore Geol. Rev.* **2018**. [CrossRef]
31. Gaffey, S.J. Spectral reflectance of carbonate minerals in visible and near infrared: Anhydrous carbonate minerals. *J. Geophys. Res.* **1987**, *92*, 1429–1440. [CrossRef]
32. Gaffey, S.J. Spectral reflectance of-carbonate minerals in the visible and near infrared (0.35–2.55 microns): Calcite, aragonite, and dolomite. *Am. Mineral.* **1986**, *71*, 151–162.
33. Li, Z.L.; Becker, F.; Stoll, M.P.; Wan, Z. Evaluation of six methods for extracting relative emissivity spectra from thermal infrared images. *Remote Sens. Environ.* **1999**, *69*, 197–214. [CrossRef]
34. Li, Z.L.; Tang, B.H.; Wu, H.; Ren, H.; Yan, G.; Wan, Z.; Trigo, I.F.; Sobrino, J.A. Satellite derived land surface temperature: Current status and perspectives. *Remote Sens. Environ.* **2013**, *131*, 14–37. [CrossRef]
35. Guha, A.; Vinod Kumar, K. Integrated approach of using aster derived emissivity and radiant temperature for delineating different granitoids—a case study in parts of Dharwar Craton, India. *Geocarto Int.* **2015**, *31*, 860–869. [CrossRef]
36. Roy, A.B.; Paliwal, B.S.; Shekhawat, S.S.; Nagori, D.K.; Golani, P.R.; Bejarniya, B.R. Stratigraphy of the Aravalli Supergroup in the type area. *Geol. Soc. India Mem.* **1988**, *7*, 121–138.
37. Banerjee, D.M.; Schidlowski, M.; Arneth, J.D. Genesis of upper proterozoic Cambrian phosphorite deposits of India: Isotopic inferences from carbonate fluroapatite, carbonate and organic carbon. *Precambrian Res.* **1986**, *33*, 239–253. [CrossRef]
38. Roy, A.B.; Paliwal, B.S. Evolution of lower Proterozoic epicontinental deposits: Stromatolite-bearing Aravalli rocks of Udaipur, Rajasthan, India. *Precambrian Res.* **1981**, *14*, 49–74. [CrossRef]
39. Abrams, M.; Tsu, H.; Hulley, G.; Iwao, K.; Pieri, D.; Cudahy, T.; Kargel, J. The advanced spaceborne thermal emission and reflection radiometer (ASTER) after fifteen years: Review of global products. *Int. J. Appl. Earth Observ. Geoinf.* **2015**, *38*, 292–301. [CrossRef]
40. NASA. ASTER. Available online: https://asterweb.jpl.nasa.gov/ (accessed on 1 January 2016).
41. Hook, S.J.; Gabell, A.R.; Green, A.A.; Kealy, P.S. A comparison of techniques for extracting emissivity information fromthermal infrared data for geologic studies. *Remote Sens. Environ.* **1992**, *42*, 123–135. [CrossRef]
42. Kealy, P.S.; Hook, S.J. Separating temperature and emissivity in thermal infrared multispectral scanner data: Implications for recovering land surface temperatures. *IEEE Trans. Geosci. Remote Sens.* **1993**, *31*, 1155–1164. [CrossRef]
43. D&P Instruments. Available online: http://www.dpinstruments.com/ (accessed on 15 June 2018).
44. Ruff, S.W.; Christensen, P.R.; Barbera, P.W.; Anderson, D.L. Quantitative thermal emission spectroscopy of minerals: A laboratory technique for measurement and calibration. *J. Geophys. Res.* **1997**, *102*, 14899–14913. [CrossRef]
45. Christensen, P.R.; Bandfield, J.L.; Hamilton, V.E.; Howard, D.A.; Lane, M.D.; Piatek, J.L.; Ruff, S.W.; Stefanov, W.L. A thermal emission spectral library of rock-forming minerals. *J. Geophys. Res. Planets* **2000**, *105*, 9735–9739. [CrossRef]

46. Kokaly, R.F.; Clark, R.N.; Swayze, G.A.; Livo, K.E.; Hoefen, T.M.; Pearson, N.C.; Wise, R.A.; Benzel, W.M.; Lowers, H.A.; Driscoll, R.L.; et al. *USGS Spectral Library Version 7*; No. 1035; US Geological Survey: Reston, VA, USA, 2017.
47. Johnson, B.R.; Young, S.J. *In-Scene Atmospheric Compensation: Application to SEBASS Data Collected at the ARM Site, Technical Report, Space and Environment Technology Center*; The Aerospace Corporation: El Segundo, CA, USA, 1998.
48. Hamilton, V.E.; Christensen, P.R.; McSween, H.Y., Jr. Determination of Martian meteorite lithologies and mineralogies using vibrational spectroscopy. *J. Geophys. Res. Planets* **1997**, *102*, 25593–25603. [CrossRef]
49. Hamilton, V.E. Thermal infrared emission spectroscopy of the pyroxene mineral series. *J. Geophys. Res. Planets* **2000**, *105*, 9701–9716. [CrossRef]
50. Mission, G. Colorimetric estimation of phosphorus in steels. *Chemiker Zeitung* **1908**, *32*, 633.
51. Wynn, J.C.; Bazzari, M.; Bawajeeh, A.; Tarabulsi, Y.; Showail, A.; Hajnoor, M.O.; Techico, L.; Wynn, J.P. *Phosphate Content Derived from Well Logging, Al Jalamid Phosphate Deposit, Northern Saudi Arabia*; U.S. Geological Survey Mission Data File Report IR-869; U.S. Geological Survey: Reston, VA, USA, 1994; 9p.

© 2019 by the authors. Licensee MDPI, Basel, Switzerland. This article is an open access article distributed under the terms and conditions of the Creative Commons Attribution (CC BY) license (http://creativecommons.org/licenses/by/4.0/).

Article

Mapping Listvenite Occurrences in the Damage Zones of Northern Victoria Land, Antarctica Using ASTER Satellite Remote Sensing Data

Amin Beiranvand Pour [1,2,*], Yongcheol Park [1], Laura Crispini [3], Andreas Läufer [4], Jong Kuk Hong [1], Tae-Yoon S. Park [1], Basem Zoheir [5,6], Biswajeet Pradhan [7,8], Aidy M. Muslim [2], Mohammad Shawkat Hossain [2] and Omeid Rahmani [9]

1. Korea Polar Research Institute (KOPRI), Songdomirae-ro, Yeonsu-gu, Incheon 21990, Korea; ypark@kopri.re.kr (Y.P.); jkhong@kopri.re.kr (J.K.H.); typark@kopri.re.kr (T.-Y.S.P.)
2. Institute of Oceanography and Environment (INOS), University Malaysia Terengganu (UMT), 21030 Kuala Nerus, Terengganu, Malaysia; shawkat@umt.edu.my (M.S.H.); aidy@umt.edu.my (A.M.M.)
3. DISTAV—University of Genova—Corso Europa 26, 16132 Genova, Italy; laura.crispini@unige.it
4. Federal Institute for Geosciences and Natural Resources (BGR), Stilleweg 2, 30655 Hannover, Germany; andreas.laeufer@bgr.de
5. Department of Geology, Faculty of Science, Benha University, Benha 13518, Egypt; basem.zoheir@ifg.uni-kiel.de
6. Institute of Geosciences, University of Kiel, Ludewig-Meyn Str. 10, 24118 Kiel, Germany
7. Centre for Advanced Modelling and Geospatial Information Systems (CAMGIS), Faculty of Engineering and Information Technology, University of Technology Sydney, New South Wales 2007, Australia; Biswajeet.Pradhan@uts.edu.au
8. Department of Energy and Mineral Resources Engineering, Choongmu-gwan, Sejong University, 209 Neungdong-ro Gwangjin-gu, Seoul 05006, Korea
9. Department of Natural Resources Engineering and Management, School of Science and Engineering, University of Kurdistan Hewlêr (UKH), Erbil, Kurdistan Region 44001, Iraq; omeid.rahmani@ukh.edu.krd
* Correspondence: Amin.Beiranvand@kopri.re.kr; Tel.: +82-3-27605472

Received: 20 May 2019; Accepted: 10 June 2019; Published: 13 June 2019

Abstract: Listvenites normally form during hydrothermal/metasomatic alteration of mafic and ultramafic rocks and represent a key indicator for the occurrence of ore mineralizations in orogenic systems. Hydrothermal/metasomatic alteration mineral assemblages are one of the significant indicators for ore mineralizations in the damage zones of major tectonic boundaries, which can be detected using multispectral satellite remote sensing data. In this research, Advanced Spaceborne Thermal Emission and Reflection Radiometer (ASTER) multispectral remote sensing data were used to detect listvenite occurrences and alteration mineral assemblages in the poorly exposed damage zones of the boundaries between the Wilson, Bowers and Robertson Bay terranes in Northern Victoria Land (NVL), Antarctica. Spectral information for detecting alteration mineral assemblages and listvenites were extracted at pixel and sub-pixel levels using the Principal Component Analysis (PCA)/Independent Component Analysis (ICA) fusion technique, Linear Spectral Unmixing (LSU) and Constrained Energy Minimization (CEM) algorithms. Mineralogical assemblages containing Fe^{2+}, Fe^{3+}, Fe-OH, Al-OH, Mg-OH and CO_3 spectral absorption features were detected in the damage zones of the study area by implementing PCA/ICA fusion to visible and near infrared (VNIR) and shortwave infrared (SWIR) bands of ASTER. Silicate lithological groups were mapped and discriminated using PCA/ICA fusion to thermal infrared (TIR) bands of ASTER. Fraction images of prospective alteration minerals, including goethite, hematite, jarosite, biotite, kaolinite, muscovite, antigorite, serpentine, talc, actinolite, chlorite, epidote, calcite, dolomite and siderite and possible zones encompassing listvenite occurrences were produced using LSU and CEM algorithms to ASTER VNIR+SWIR spectral bands. Several potential zones for listvenite occurrences were identified, typically in association with mafic metavolcanic rocks (Glasgow Volcanics) in the Bowers Mountains. Comparison of the remote sensing results with geological investigations in the study area demonstrate invaluable implications of the remote sensing approach for mapping poorly exposed lithological units, detecting possible zones of listvenite occurrences and discriminating subpixel abundance of alteration mineral assemblages in the damage zones of the Wilson-Bowers and Bowers-Robertson Bay terrane boundaries and in intra-Bowers and Wilson terranes fault zones with high fluid flow. The satellite remote sensing approach developed in this research is explicitly pertinent to detecting key alteration mineral indicators for prospecting hydrothermal/metasomatic ore minerals in remote and inaccessible zones situated in other orogenic systems around the world.

Keywords: Bowers Terrane; listvenite; hydrothermal/metasomatic alteration minerals; damage zones; ASTER; Northern Victoria Land; Antarctica

1. Introduction

Hydrothermal/metasomatic alteration mineral assemblages are one of the significant indicators for ore mineralizations in the damage zones of lithotectonic units in orogenic systems [1–3]. They can be detected and mapped by the application of multispectral satellite remote sensing data [4–12]. Listvenite is a metasomatic rock composed of variable amounts of quartz, magnesite, ankerite, dolomite, sericite, calcite, talc and sulfide minerals. It is formed by interaction of mafic and ultramafic rocks with low to intermediate temperature CO_2- and S-rich fluids, and is commonly found along the major fault and shear zones at terrane boundaries or major tectonic units in orogenic systems [13–16]. As such, listvenite is spatially associated with ophiolites, greenstone belts and suture zones in orogenic belts [17–24]. Listvenite occurrences are considered to represent key indicators for certain mineral associations connected with ore mineralizations such as gold and other hydrothermal deposits like Ag, Hg, Sb, As, Cu, Ni, Co, as well as magnesite and talc [14,18,20,25,26].

Advanced Spaceborne Thermal Emission and Reflection Radiometer (ASTER) multispectral remote sensing satellite data provide appropriate spatial, spectral and radiometric resolutions suitable for mapping hydrothermal/metasomatic alteration mineral assemblages [6–12,27–31]. Iron oxide/hydroxide, hydroxyl-bearing and carbonate mineral groups present diagnostic spectral absorption features due to electronic processes of transition elements (Fe^{2+}, Fe^{3+} and REE) and vibrational processes of fundamental absorptions of Al-OH, Mg-OH, Fe-OH, Si-OH, CO_3, NH_4 and SO_4 groups in the visible and near infrared (VNIR) and shortwave infrared (SWIR) regions [32–35]. These mineral groups can be detected using three VNIR (from 0.52 to 0.86 μm; 15-m spatial resolution) and six SWIR (from 1.6 to 2.43 μm; 30-m spatial resolution) spectral bands of ASTER [36,37]. Additionally, thermal infrared bands (TIR; 8.0–14.0 μm; 90-m spatial resolution) of ASTER are capable of discriminating silicate lithological groups due to different characteristics of the emissivity spectra derived from Si–O–Si stretching vibrations in the TIR region [36,38–42]. Accordingly, ASTER remote sensing satellite datasets are particularly useful for the detection of listvenites and alteration mineral assemblages occurring in damage zones of terrane and major tectonic boundaries in orogens around the world. However, only a few studies exist, which used ASTER remote sensing data for the regional mapping of listvenite occurrences, such as Rajendran et al. [43], who used ASTER VNIR+SWIR spectral bands for the detection of listvenites along the serpentinite–amphibolite interface of the Semail Ophiolite in the Sultanate of Oman.

In Northern Victoria Land (NVL) of Antarctica, the widespread occurrence of listvenites was documented as one of the main types of hydrothermal/metasomatic fault-related rocks in the damage zones between the Wilson Terrane (WT) and the Bowers Terrane (BT) (Figures 1 and 2) [44]. Thisboundary coincides with the Lanterman suture zone, where mafic and ultramafic rocks with indications of UHP (ultrahigh pressure) metamorphism implies ancient subduction processes at the palaeo-Pacific active continental margin of East Gondwana during the Late-Ediacaran to Early Paleozoic Ross Orogeny [45–48]. Structural investigations in the poorly exposed damage zones of the terrane boundaries reveal that the dominant features are (i) steeply dipping reverse and strike-slip faults; (ii) diffuse veining; and (iii) hydrothermal/metasomatic alteration mineral assemblages and listvenites (Figure 2) [47,49–51].

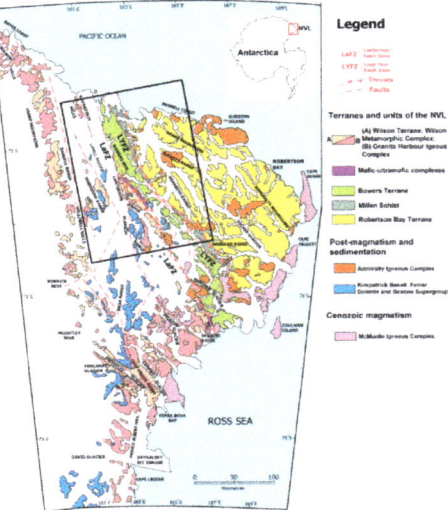

Figure 1. Geological map of Northern Victoria Land (NVL). Modified from [7,48] based on the GIGAMAP series [52]. The black rectangle shows the coverage of ASTER images used in this study.

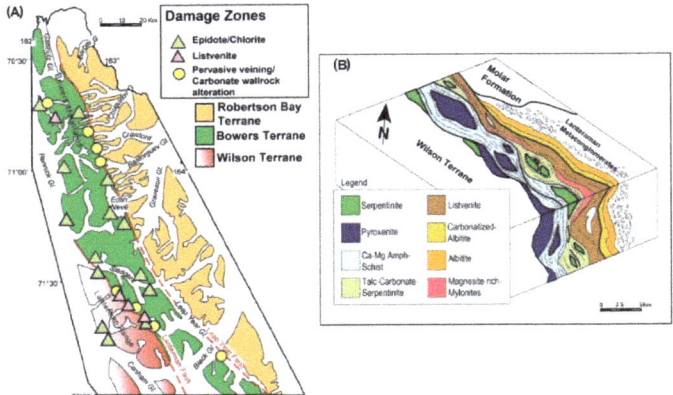

Figure 2. (**A**) Sketch map with the location of the fault (damage) zones where the hydrothermal alteration and veining is more intense (modified from [51]). (**B**) Sketch map of lithological-mineralogical sequences in the damage (fault) zone at the boundary of WT and BT (modified from [44]).

Another occurrence of listvenites was reported from the upper Dorn Glacier at the eastern side of the northern Bowers Mountains only a few km west of the BT-RBT boundary represented by the Millen Schist Belt [1,50]. In the aforementioned paper, the authors reported gold mineralization in the listvenite bodies and the surrounding hydrothermal alteration mineral zones. The gold-bearing quartz veins (Dorn lodes) are hosted by hydrothermally altered rocks (approximately up to 300 m wide), which are mainly characterized by variable amounts of ankerite, muscovite, chlorite, pyrite and arsenopyrite. The mineralization is located in a brittle-ductile reverse high-strain shear zone in lower greenschist facies metavolcanic and metasedimentary rocks [1].

Recently, remote sensing studies have been conducted by Pour et al. [6,7] for regional-scale lithological mapping in NVL and local-scale alteration mineral mapping in the Morozumi Range and Helliwell Hills areas of the WT. However, no comprehensive remote sensing study is yet available for the boundary region between the WT and the BT further to the east, where listvenite bodies, in particular, occur in the damage zones. Consequently, the main objectives of this study are: (i) to extract spectral information directly from ASTER image spectra at pixel and sub-pixel levels for detecting alteration mineral assemblages and listvenites in the damage (fault) zones particularly of the WT-BT boundary using a series of specialized/standardized image processing algorithms; (ii) to prospect listvenite occurrences in the poorly exposed damage zones of the WT-BT boundary; (iii) to verify and compare ASTER image processing results with available field and laboratory data collected from the fault zones of the WT-BT boundary; and (iv) to test the results of the listvenites along the WT-BT terrane on particular listvenite occurrences in intra-BT fault zones in the northern Bowers Mountains and thus its general potential in detecting similar synorogenic mineral alteration zones in NVL and other orogens around the world.

2. Geological Setting of NVL and the Bowers Terrane

Northern Victoria Land (NVL) is composed of three NW-trending litho-tectonic units or terranes of late Proterozoic–Ordovician age, which are from west to east (i) the Wilson Terrane (WT), (ii) the Bowers Terrane (BT), and (iii) the Robertson Bay Terrane (RBT) (Figures 1 and 2) [53–55]. They are generally interpreted to have formed during west-directed subduction processes at the palaeo-Pacific active continental margin of East Gondwana during the Latest Ediacaran to Early Palaeozoic Ross Orogeny [45]. The WT encompasses polydeformed low- to high-grade (up to granulite facies) metasedimentary sequences intruded by the Granite Harbour Igneous Complex (calc-alkaline plutons with magmatic arc affinity of the latest Ediacaran to Cambro-Ordovician age) [45,56–58]. The BT comprises very low-grade to low-grade (prehnite-pumpellyite to lower-greenschist facies) metavolcanic and metasedimentary rocks, which are considered to be an intra-oceanic arc complex [59,60] or a fore-arc volcanic complex at the Ross-orogenic active continental margin [54]. The RBT is a very low- to locally in its western part low-grade (zeolite to prehnite-pumpellyite facies) turbidite sequence, which is interpreted as a synorogenic sedimentary pile in an accretionary environment [61–63]. The tectonic boundary between the WT and the BT is generally referred to as the Lanterman Fault Zone (LaFZ), whereas the Leap Year Fault Zone (LYFZ) separates the BT from the RBT by the strongly sheared Millen Schist Belt (Figures 1 and 2) [47,64,65]. These fault zones represent long-lived structures in the structural edifice of NVL, which were repeatedly reactivated [66].

After the Ross orogeny, the three terranes were intruded by Devonian/Carboniferous calcalkaline intrusions (Admiralty Intrusives) and associated felsic volcanics (Gallipoli Volcanics) [67]. A subsequent erosion/exhumation phase produced a regional peneplain surface, on which the Late Carboniferous-Early Jurassic terrestrial sedimentary sequence of the Beacon Supergroup was deposited [68,69]. This was followed by tholeiitic magmatism of late Early Jurassic age (Ferrar Dolerite and Kirkpatrick Basalt) [70]. The Cenozoic tectonics in NVL is predominantly linked to the development of the West Antarctic Rift System (WARS), which involved repeated reactivation of the former Paleozoic discontinuities of the NVL crust [66,71,72]. The Cenozoic transtensional structures influenced the emplacement of alkaline magmatic rocks of the McMurdo Igneous Complex (see Figure 1) [60].

The BT consists of three lithostratigraphic sequences of likely Cambrian to Ordovician in age, including from bottom to top the Sledgers, the Mariner, and the Leap Year groups, respectively [73]. The Sledgers Group includes: (i) the Glasgow Volcanics with basalts, spilites, volcanic breccia and tuffs and (ii) the Molar Formation with sandstone, conglomerate, mudstone and subordinate limestone. The Mariner Group comprises a regressive sequence of shallow marine sandstone, subordinate conglomerate, mudstone and limestone. The Leap Year Group consists of quartz-rich sandstone and conglomerate [74–76]. Figure 2B shows a schematic section of the WT-BT boundary, the strip close to the contact with the metaconglomerates on the BT side is characterized by xenolithic blocks of eclogite, amphibolite, serpentinite and listvenite as the main lithologies [77]. The metamorphic grade in the WT ranges from greenschist to amphibolite and up to the granulite facies, which is notably higher and more diverse than the entire BT. In the BT, higher metamorphic grades, including HP metamorphism, are confined only to the proximity of the Lanterman suture zone at the tectonic boundary between the WT and the BT [45,46,78].

3. Materials and Methods

3.1. ASTER Data Characteristics and Pre-processing

Twelve ASTER level 1T (Precision Terrain Corrected Registered At-Sensor Radiance) scenes obtained from U.S. Geological EROS (http://glovis.usgs.gov/) were used for mapping the poorly exposed damage zones of the WT-BT boundary. Table 1 shows the dataset characteristics of the ASTER images used in this research. Atmospheric correction was applied to the ASTER data using Fast Line-of-sight Atmospheric Analysis of Spectral Hypercube (FLAASH) algorithm [79,80], which is available in the ENVI (Environment for Visualizing Images, http://www.exelisvis.com) version 5.2 software package. The Sub-Arctic Summer (SAS) atmospheric and the Maritime aerosol models were used for running the FLAASH algorithm [81]. Crosstalk correction is required before processing of ASTER data to remove the influences of energy overflow from band 4 into bands 5 and 9 [82]. Hence, it was implemented with respect to the ASTER SWIR bands used in this research. The ASTER images were pre-georeferenced to the UTM zone 58 South projection using the WGS-84 datum and rotated to the north up to the UTM projection. Furthermore, the 30-m-resolution SWIR bands were re-sampled to have spatial dimensions of 15 m (corresponding to the VNIR 15-m resolution) using the nearest neighbour re-sampling technique for producing a stacked layer of VNIR+SWIR bands. A masking procedure was used to remove the snow/ice, cloud and shadow by applying the Normalised Difference Snow Index (NDSI) [7,30,83,84]. Radiance at the sensor TIR data without atmospheric corrections was used in this analysis for retaining the original radiance signature. The digital number (DN) value of each pixel in level 1T data at band i (i = 10–14) is linearly converted to radiance registered at the sensor (L_{sen}) (Wm^{-2}·Sr^{-1}·μm^{-1}) [41,42], by the application of Equation (1):

$$L_{sen}^i = cof^i \times (DN^i - 1) \qquad (1)$$

where, $cof^{10} = 0.006882$, $cof^{11} = 0.006780$, $cof^{12} = 0.006590$, $cof^{13} = 0.005693$, and $cof^{14} = 0.005225$.

Table 1. The dataset characteristics of ASTER images used in this research.

Granule ID	Date and Time of Acquisition	Path/Row	Cloud Coverage	Sun Azimuth	Sun Elevation
AST_L1T_00312282003214501	2003/01/01, 21:45:01	66/110	1%	55.417	34.625
AST_L1T_00301012003215056	2003/01/01, 21:51:05	66/110	2%	56.520	34.555
AST_L1T_00301012003215105	2003/01/01, 21:51:05	66/111	1%	56.972	34.077
AST_L1T_00301012003215114	2003/01/01, 21:51:14	66/111	1%	57.568	33.573
AST_L1T_00301192003213854	2003/01/19, 21:38:54	66/111	3%	61.896	29.830
AST_L1T_00301192003213845	2003/01/19, 21:38:45	66/111	1%	61.357	30.402
AST_L1T_00312022006215521	2006/12/02, 21:55:21	67/110	1%	52.247	34.707
AST_L1T_00312022006215530	2006/12/02, 21:55:30	67/110	3%	52.693	34.114
AST_L1T_00312022006215539	2006/12/02, 21:55:39	67/111	5%	53.417	33.845
AST_L1T_00301022005221335	2005/01/02, 22:13:35	68/110	1%	52.821	35.641
AST_L1T_00301022005221344	2005/01/02, 22:13:44	68/110	2%	53.395	35.254
AST_L1T_00301022005221353	2005/01/02, 22:13:53	68/111	4%	53.756	35.665

3.2. Image Processing Algorithms

Spectral information for detecting alteration mineral assemblages and listvenites in the damage zones of the WT-BT boundary were extracted at pixel and sub-pixel levels using specialized/standardized image processing algorithms. For regional-scale mapping of the WT-BT boundary, pixel-based algorithms, including Principal Component Analysis (PCA) and Independent Component Analysis (ICA) were used. Sub-pixel-based algorithms, namely Linear Spectral Unmixing (LSU) and Constrained Energy Minimization (CEM), were applied for detailed mapping in some selected subsets of the damage zones at local scale. Figure 3 shows an overview of the BT and surrounding area (the coverage of ASTER images used in this study), along with six selected subsets of the fault zones (zones 1–6) for detailed mineral mapping. A flowchart of the methodology used in this research is shown in Figure 4. For processing the datasets, the ENVI (Environment for Visualizing Images, http://www.exelisvis.com) version 5.2 and ArcGIS version 10.3 (Esri, Redlands, CA, USA) software packages were used.

Figure 3. ASTER mosaic of the Bowers Terrane (BT) and surrounding area (the coverage of ASTER images used in this study). Magenta rectangles indicate selected subsets of fault zones (zones 1–6) for detailed mineral mapping.

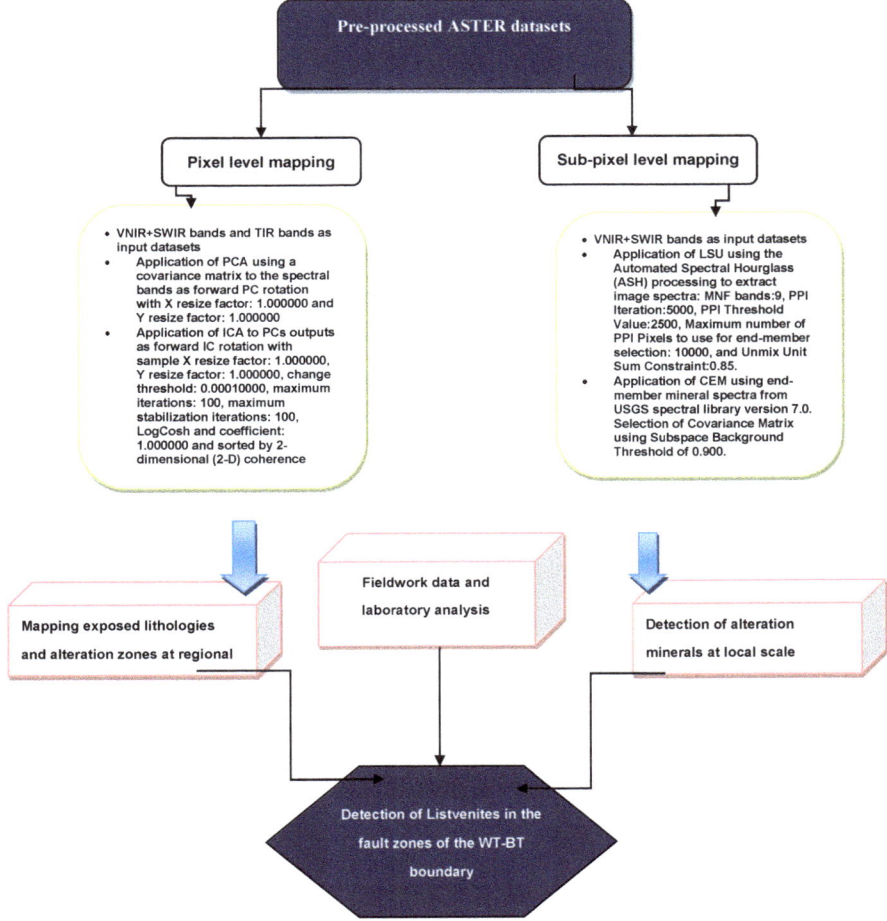

Figure 4. An overview of the methodological flowchart used in this study. Characterizations of the techniques can be found in ENVI Tutorials. Research Systems, Inc., Boulder, CO [81].

3.2.1. Spectral Information Extraction at the Pixel Level

A fusion of PCA and ICA algorithms was implemented to the ASTER VNIR+SWIR and TIR datasets for extracting image spectra at the pixel level to map poorly exposed lithologies and alteration mineral assemblages. The PCA technique selects uncorrelated linear combinations (Eigenvector loadings) of variables in such a way that each component successively extracted linear combination and contains a smaller variance [85–88]. The Eigenvector loadings contain important information for identifying hydrothermal/metasomatic alteration mineral assemblages. By computing the correlation of each band k with each component p, it is possible to determine how each band "loads" or are associated with each principal component (Equation (A1), see Appendix A) [86].

A PC image with moderate to high eigenvector loading for the indicative bands (reflection and/or absorption bands) of the mineral with opposite signs enhance that mineral [89]. If the loading is positive in the reflection band of the mineral, the enhanced pixels related to the mineral will manifest as bright pixels. On the contrary, if the loading is negative in the reflection band of the mineral, the enhanced pixels related to the mineral will manifest as dark pixels [90]. Consequently, eigenvector loadings in each PCA will identify the PC image in which the spectral information of the specific alteration mineral is loaded as bright or dark pixels [91]. The ICA is a statistical and computational technique for array processing and data analysis, aiming at recovering unobserved signals or 'sources' from observed mixtures, exploiting only the assumption of mutual independence between the signals [92,93]. It is a method for separating the combinations with the most non-Gaussian possible probability density functions from the more Gaussian signal mixtures [94,95]. These are identified as the "independent components" (ICs) of the observations [96].

For pixel-based image spectra extraction, PCA analysis can remove correlations, but it is not capable of omitting higher-order dependence. However, the ICA removes both correlations and higher-order dependence. Accordingly, a fusion of PCA and ICA has great capability to identify pixels related to poorly exposed lithologies and alteration minerals in the background of extensive snow/ice cover (Antarctic environments). The PCA can be used to give weight to the components and remove the correlation before applying ICA for revealing hidden factors. As a result, this fusion has a great performance to identify the pixels containing the spectral signature of the alteration minerals or mineral groups that are maximally independent of each other. In this study, four spatial subset scenes (include six selected damage zones; see Figure 3) covering exposed lithologies in the WT-BT boundary and intra-BT fault zones in the northern Bowers Mountains were selected for implementing the PCA/ICA fusion technique. The performance characteristics of the technique are summarized in Figure 4. The image eigenvectors were obtained for the PCA analysis using a covariance matrix of VNIR+SWIR and TIR bands (Table A1; Table A2, see Appendix A). Subsequently, a forward ICA rotation was applied to the PCs images (see Figure 4). The IC images were statistically examined for each of the selected spatial subset scenes (Tables A1 and A2, see Appendix A). The ICs contain maximally independent pixels of alteration minerals or mineral groups were selected to produce Red-Green-Blue (RGB) color composite image maps.

3.2.2. Spectral Information Extraction at the Sub-pixel Level

The LSU and CEM algorithms were applied to VNIR+SWIR bands for detailed mapping of alteration minerals and listvenites at the sub-pixel level in the six selected subsets of fault zones (see Figure 3). The LSU is a sub-pixel sampling algorithm [97–99], the reflectance at each pixel of the image is assumed to be a linear combination of the reflectance of each material (or end-member) present within the pixel. The LSU is used to determine the relative abundance of end-members within a pixel based on the end-members' spectral characteristics. In this algorithm, it is assumed that the observed pixel reflectance can be modeled as a linear mixture of individual component reflectance multiplied by their relative proportions. Mathematically, the LSU can be represented as Equation (A2) [100] (see Appendix A).

In this study, the Automated Spectral Hourglass (ASH) technique [81,101,102] was employed to extract reference spectra directly from the ASTER image for producing fraction images of end-members using the LSU. The ASH technique uses the spectrally over-determined data for finding the most spectrally pure pixels (end-members) to map their locations and estimates their sub-pixel abundances [81]. It includes several steps, namely: (i) the Minimum Noise Fraction (MNF) [103,104]; (ii) the Pixel Purity Index (PPI) [105]; and (iii) automatic end-member prediction from the n-Dimensional Visualizer [104,106]. The performance characteristics of MNF, PPI and the n-D Visualizer used in this study are shown in Figure 4. The continuum-removal process was applied to the extracted end-members for isolating their spectral features and putting them on a level playing field so they may be intercompared. Continuum-removal and feature comparison is the key to successful spectral identification [34,107,108].

The CEM is a target detection algorithm [109–111] that implements a partial unmixing of spectra to estimate the abundance of user-defined end-member materials from a set of reference spectra (either image or laboratory spectra) [112]. It specifically constrains the desired target spectra using a Finite Impulse Response (FIR) filter [113,114], while minimizing effects caused by unknown background signatures [115,116]. Mathematical details of the CEM performance can be found in Chang et al. [112] and Manolakis et al. [111]. For running the CEM, only prior knowledge of desired target spectra (end-member materials) is needed. In this study, the reference spectra of some hydrothermal alteration minerals (typically associated with listvenites) were selected from the USGS spectral library (version 7.0; [117]) for executing the CEM. The performance characteristics of the CEM are summarized in Figure 4. End-member spectra of goethite, hematite, jarosite, biotite, kaolinite, muscovite, antigorite, serpentine, talc, actinolite, chlorite, epidote, calcite, dolomite, siderite and chalcedony were selected and convolved to response functions of ASTER VNIR+SWIR bands (Figure 5). The end-member spectra of the target minerals were used to generate fraction images of prospective alteration minerals associated with listvenites in the poorly exposed fault zones along the WT-BT boundary and intra-BT fault zones in the northern Bowers Mountains.

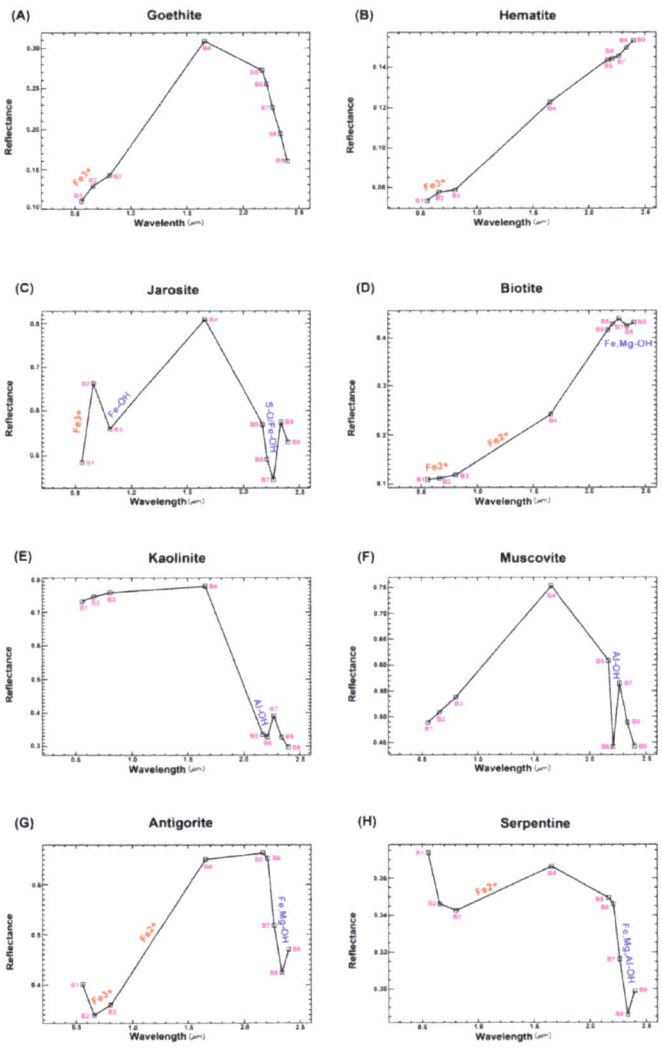

Figure 5. *Cont.*

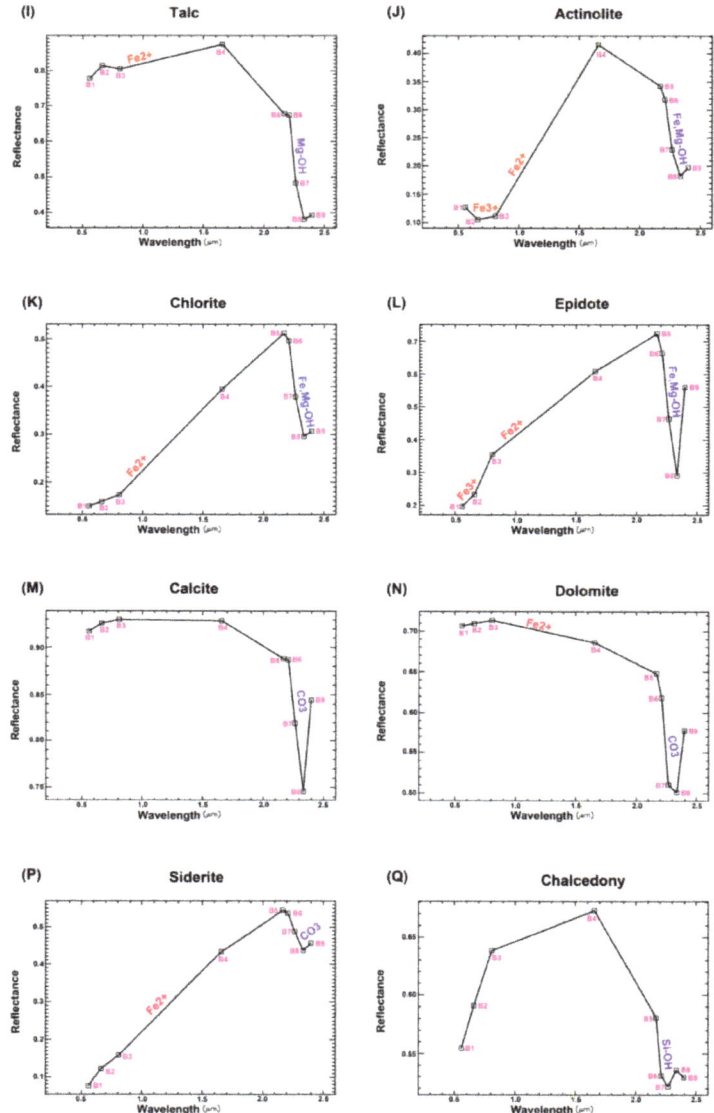

Figure 5. Laboratory reflectance spectra of the selected alteration minerals resampled to response functions of ASTER VINR+SWIR bands [117]. Cubes show the location of the ASTER VINR+SWIR bands (B1 to B9) from 0.50 μm to 2.5 μm. (**A**) goethite; (**B**) hematite; (**C**) jarosite; (**D**) biotite; (**E**) kaolinite; (**F**) muscovite; (**G**) antigorite; (**H**) serpentine; (**I**) talc; (**J**) actinolite; (**K**) chlorite; (**L**) epidote; (**M**) calcite; (**N**) dolomite; (**P**) siderite; and (**Q**) chalcedony.

3.3. Fieldwork Data and Laboratory Analysis

Different locations of exhumed fault systems that are systematically associated with hydrothermal alterations in the poorly exposed damage zones along the WT-BT and BT-RBT boundaries, as well as along major faults and shear zones within the BT and in the easternmost WT along the Rennick Glacier, were investigated. Geological field work was performed during various Italian (PNRA) and

German Scientific Expeditions (GANOVEX) in Northern Victoria Land (NVL) and was particularly conducted at several major outcrop sites in the Lanterman Range area, the Molar Massif, and the Bowers Mountains, the Explorers Range on the Rennick Glacier side to the western part facing the Lillie Glacier. Locations of alteration zones and listvenites were recorded using a Garmin Montana 608t handheld GPS with an average accuracy of 5 m and an iPhone SE using GPS plus GLONASS (Table A3, see Appendix A). Field photos were taken of exposed lithologies, hydrothermally fault-bounded altered rocks and listvenites occurrences during most recent expeditions (2015–2016 and 2016–2017 summer season) and several rock samples were also collected for laboratory analysis. The rock samples were examined by optical microscopy of thin sections and investigated by X-ray diffraction (XRD) analysis for determining their mineralogical composition. Mineral phases were investigated by a Philips PW3710 X-Ray diffractometer (current: 20 mA, voltage: 40 kV, range 2θ: 5–80°, step size: 0.02° 2θ, time per step: 2 s) at DISTAV (University of Genova, Italy), which mounted a Co-anode, as in [49]. Acquisition and processing of the XRD data were carried out using the Philips High Score software package. Additionally, confusion matrix (error matrix) and Kappa Coefficient were calculated for LSU classification mineral maps versus field data (Table A4, see Appendix A).

4. Results

4.1. Regional Overview of the BT and Surrounding Areas

A regional view of the poorly exposed lithological units was generated for the BT and surrounding areas using a mosaic of ASTER images (Figure 6). ASTER Fe-MI = (band 4/band 3) × (band 2/band 1), Al-OH-MI= (band 5 × band 7)/(band 6 × band 6) and Fe, Mg-OH-MI= (band 7 × band 9)/(band 8 × band 8) spectral-band ratio indices [7] were assigned to RGB color composite, respectively. Spatial distribution of iron oxide/hydroxide minerals, Al-OH minerals and Fe,Mg-O-H minerals is manifested by a variety of false color composite in the exposures (Figure 6). Exposed lithologies with a high content of iron oxide/hydroxide minerals are represented as red, magenta and orange colors. The Bowers Mountains and many other parts of the BT and neighbouring areas such as Morozumi Range, Helliwell Hills and Lanterman Range contain a high surface abundance of iron oxide/hydroxide minerals (red, magenta and orange pixels) (Figure 6). Regarding the geological maps of the region, the lithological units in these zones mostly consist of the Wilson Terrane metamorphic rocks, Granite Harbour Igneous Complex, metavolcanic rocks (Glasgow Volcanics), Admiralty Intrusives, Ferrar Dolerite and Kirkpatrick Basalt.

Most of the sedimentary rock units in the study area comprise the Beacon Supergroup, Robertson Bay Group, Molar Formation, Mariner Group and Leap Year Group, which appear in green and blue colors (Figure 6). Generally, sedimentary rocks contain large amounts of Al-OH and Fe,Mg-O-H mineral assemblages (detrital clay minerals). Very poorly exposed outcrops adjacent to the Morozumi Range, Helliwell Hills and ANARE Mountains are represented as cyan color (Figure 6). The exposures in the central part of the Mirabito Range, the southern part of the Alamein Range and the central-northern part of the Everett Range are depicted as yellow color (Figure 6). This indicates that these exposures mostly contain iron oxide/hydroxide minerals in association with Al-OH minerals. Some of the exposed rocks in the southern and western parts of the Mirabito Range and the northern part of the ANARE Mountains are manifested in purple color (Figure 6) due to the admixture of iron oxides/hydroxides with Fe,Mg-O-H mineral groups.

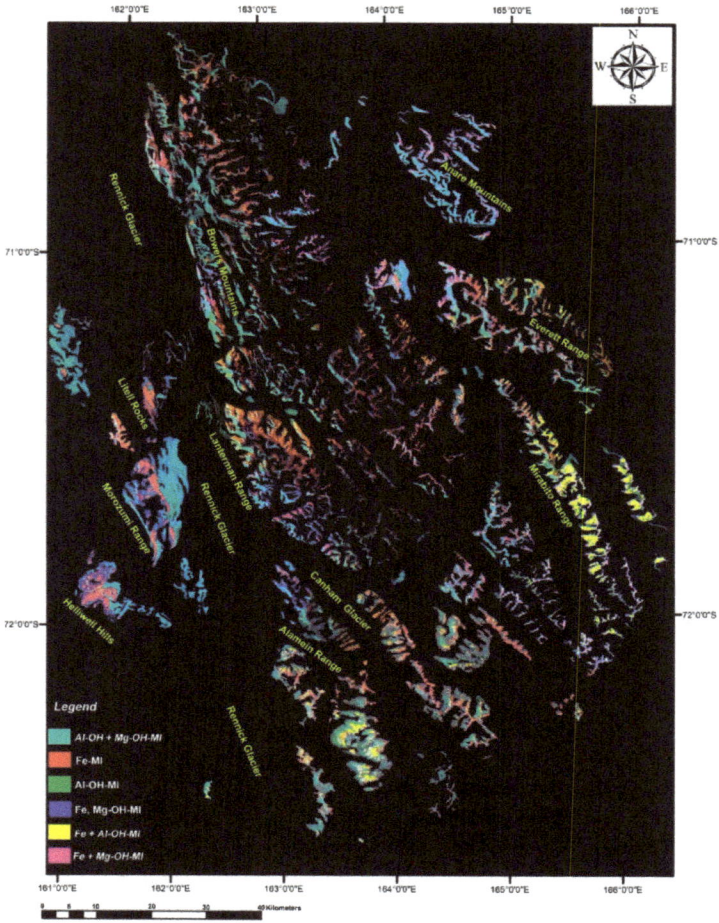

Figure 6. ASTER mosaic image of Fe-MI, Al-OH-MI and Fe,Mg-OH-MI as RGB color composite. It shows exposed lithologies in the Bowers Terrane (BT) and surrounding areas. The locations of some large mountain ranges are shown.

4.2. Alteration and Lithological Mapping in the Fault Zones at Regional Scale

The analysis of the eigenvector matrix for VNIR+SWIR bands (Table A1) and the visual examination of the output IC images indicate the existence of maximally independent pixels related to alteration minerals or mineral groups in some PCAs that contain unique contributions (magnitude and sign of eigenvector loadings) of the minerals. The eigenvector loadings calculated for the spatial subset covering zone 1 and the surrounding areas (Table A1A) reveal that PCA5 has a strong contribution of band 1 (0.571038) and band 3 (−0.731120) with opposite signs. PCA6 shows moderate loadings of band 5 (−0.281197), band 6 (0.308307) and band 7 (0.264687) and strong loadings of band 8 (−0.553969) and band 9 (0.588366), each with opposite signs. PCA7 contains strong weightings of band 5 (0.497821), band 6 (−0.618179) and band 7 (0.507327) with opposite signs, and moderate to low weightings of band 8 (−0.246778) and band 9 (0.048065).

Iron oxide/hydroxide minerals (hematite, goethite, jarosite and limonite) exhibit diagnostic absorption characteristics in band 1 (0.52–0.60 μm), band 2 (0.63–0.69 μm) and band 3 (0.78–0.86 μm) of ASTER [31,118]. The Al-OH minerals (kaolinite, alunite and muscovite) contain absorption features in band 5 (2.145–2.185 µm), band 6 (2.185–2.225 µm) and band 7 (2.235–2.285 µm) of ASTER [119]. On the other hand, the Fe,Mg-O-H and CO_3 minerals (chlorite, epidote and calcite) have distinctive absorption features in band 8 (2.295–2.365 µm) and band 9 (2.360–2.430 µm) of ASTER [120,121]. Accordingly, PCA5, PCA6 and PCA7 have great potential to hold maximally independent pixels related to iron oxide/hydroxide minerals, Al-OH minerals and Fe,Mg-O-H and CO_3 minerals, which can be specifically revealed using ICA analysis. Examination of the Z-Profiles (interactively plot the spectrum for the pixel under the cursor [81]) of the output IC images indicated that the identified pixels are independently and spectrally related to the indicated minerals.

PCA5, PCA7 and PCA6 were assigned to the RGB color composite for mapping iron oxide/hydroxide minerals, Al-OH minerals and Fe,Mg-O-H and CO_3 minerals, respectively. Figure 7A shows the resultant image map for the spatial subset covering zone 1 and the surrounding areas. Magenta, red, yellow and light yellow pixels predominate in the exposed zones, and green and blue pixels are less in abundance (Figure 7A). Thus, iron oxide/hydroxide minerals have high surface abundance in the exposed lithologies in zone 1. However, Al-OH minerals and Fe,Mg-O-H and CO_3 mineral assemblages have low surface abundance and are generally associated with iron mineral groups as magenta, yellow and light yellow pixels. With reference to the geological map of zone 1 and the surrounding areas, the exposed lithological units mostly consist of Wilson Terrane metamorphic rocks, Granite Harbour Igneous Complex, Beacon Supergroup and Ferrar Dolerite. Surface distribution of iron oxide/hydroxide minerals (red and magenta pixels) is typically associated with the Ferrar Dolerite and Granite Harbour Igneous Complex (Figure 7A), for instance, the exposures of Ferrar Dolerite in the northern part of the Alamein Range and the exposed zones of the Granite Harbour Igneous Complex along the Hunter Glacier in the southern part of the Lanterman Range. The Al-OH and Fe,Mg-O-H and CO_3 minerals (green, blue, yellow and light yellow pixels) are mainly concentrated in exposures of the Beacon Supergroup and Wilson Terrane metamorphic rocks (particularly amphibolite-facies metasedimentary rocks).

Table A1B shows the eigenvector loadings for the spatial subset covering zones 2 and 3. PCA3 shows strong contributions of band 1 (−0.470308) and band 3 (0.711152) with opposite signs. PCA6 contains strong loadings of band 8 (0.440341) and band 9 (−0.347880) with opposite signs. PCA7 has strong to moderate contributions of band 5 (0.612809), band 6 (−0.398723) and band 7 (0.186662) with opposite signs, and strong loadings of band 8 (0.403665) and band 9 (−0.344018). Considering the eigenvector loadings in PCA7, this PCA contains the contribution of both Al-OH and Fe,Mg-O-H and CO_3 mineral groups. ICA rotation is able to separate maximally independent pixels related to two different mineral groups. Therefore, PCA3, PCA7 and PCA6 were selected for ICA rotation and subsequent generation of an RGB color composite to detect iron oxide/hydroxide, Al-OH and Fe,Mg-O-H and CO_3 minerals, respectively. Figure 7B displays the resultant image map for the spatial subset covering zones 2 and 3. Several types of mineral assemblages are detected. Prevalent distribution of iron oxide/hydroxide minerals (red and magenta pixels) is associated with most of the exposures, while Al-OH minerals (green pixels) and Fe,Mg-O-H and CO_3 minerals (blue pixels) are specifically predominant in some exposed zones. The admixture of the mineral groups (yellow and cyan pixels) is also observable in some small exposures (Figure 7B). Comparison with the geological map of the study zones indicates that iron oxide/hydroxide minerals are typically associated with exposures of the Granite Harbour Igneous Complex of the WT, Glasgow metavolcanic rocks of the BT, Robertson Bay Group of the RBT and Admiralty Intrusives. The Al-OH minerals and Fe,Mg-O-H and CO_3 minerals characterize the exposed zones of metasedimentary rocks (the Molar Formation, Mariner Group and Leap Year Group) of the BT, Wilson Terrane metamorphic rocks and Ferrar Dolerite.

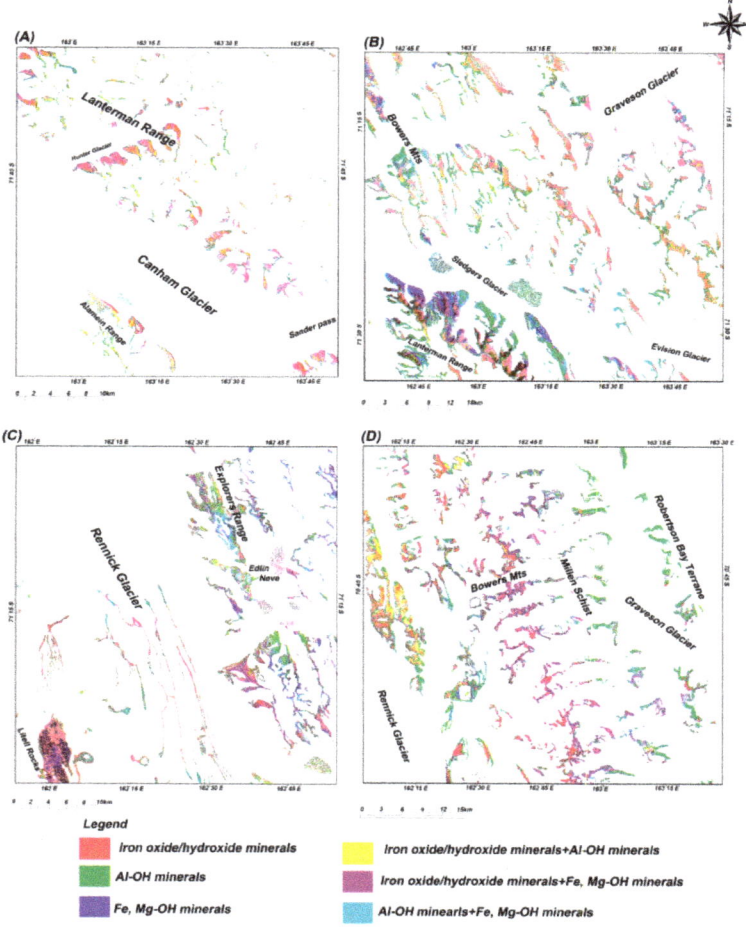

Figure 7. ASTER image maps derived from the ICA rotation and subsequent RGB color composite to VNIR+SWIR bands. (**A**) Spatial subset covering zone 1 and surrounding areas; (**B**) Spatial subset covering zones 2 and 3; (**C**) Spatial subset covering zone 4 and surrounding areas; and (**D**) Spatial subset covering zones 5 and 6.

Considering eigenvector loadings for the spatial subset covering zone 4 and surrounding areas (Table A1C), it is evident that PCA4, PCA5 and PCA6 contain spectral information to identify iron oxide/hydroxide, Al-OH and Fe,Mg-O-H and CO_3 mineral groups at the pixel level. PCA4 shows strong weightings of band 1 (0.646205) and band 3 (−0.741104) with opposite signs. PCA5 has strong to moderate contributions of band 5 (0.617458), band 6 (0.257579) and band 7 (−0.353971) with opposite signs. PCA6 contains strong loadings of band 8 (0.722429) and band 9 (−0.484494) with opposite signs (Table A1C). Therefore, the ICA rotation and subsequent RGB color composite were applied to PCA4, PCA5 and PCA6 for mapping the iron oxide/hydroxide, Al-OH and Fe,Mg-O-H and CO_3 mineral groups, respectively. Figure 7C displays the resultant image map for the spatial subset covering zone 4 and the surrounding areas. The iron oxide/hydroxide mineral group appears as red and magenta color pixels, which mainly predominate in exposed lithologies associated with the Litell Rocks region (consists of Kirkpatrick Basalts/Ferrar Dolerite) in the south-western part of the image. The Al-OH mineral group is represented as green and cyan color pixels, which are associated with exposed rocks in the Bowers Mountains (Figure 7C). Conceivably, they are exposures of the Molar Formation and the Mariner Group that contain a high content of clay minerals. The Fe,Mg-O-H and CO_3 minerals manifest in blue color pixels, which have low surface abundance in the study zone. However, they are generally associated with iron oxide/hydroxide minerals in several exposed zones (Figure 7C). The exposures could possibly contain a high content of Fe,Mg-O-H and iron oxide/hydroxide minerals in the Bowers Mountains are Glasgow metavolcanic rocks of the BT. The Autobahn Moraine is a prominent and linear large moraine stretching over several 10s of kilometres along the Rennick Glacier [122]. The iron oxide/hydroxide and Al-OH mineral groups are mapped in the Autobahn Moraine (the central part of Figure 7C).

Analysing the eigenvector loadings for the spatial subset covering zones 5 and 6 (Table A1D) indicate that PCA4 has strong contribution of band 1 (−0.576526) and band 3 (0.760114), PCA6 contains strong loadings of band 8 (−0.676196) and band 9 (0.342499) and PCA7 shows strong weightings of band 5 (0.454540), band 6 (−0.555810) and band 7 (0.600908). Thus, PCA4, PCA7 and PCA6 were selected for the ICA rotation and were subsequently assigned to an RGB color composite for identifying iron oxide/hydroxide, Al-OH and Fe,Mg-O-H and CO_3 mineral groups. Figure 7D shows the resultant image map for the spatial subset covering zones 5 and 6, which covers the Bowers Mountains, Millen Schist and Robertson Bay Terrane (RBT) from west to east. Magenta, red, yellow and green pixels govern most of the exposures, while blue pixels are much less abundant. Hence, iron oxide/hydroxide and Al-OH mineral groups and their admixture are dominant mineral assemblages in the study zone. The Glasgow Volcanics, the Molar Formation, Millen Schist, Robertson Bay Group and Admiralty Intrusives are exposed in zones 5 and 6 with reference to the geological map. It is discernible that the exposures of Glasgow Volcanics and Admiralty Intrusives contain a high surface distribution of iron oxide/hydroxide with some admixture of Al-OH minerals, which appear as magenta, red and yellow pixels in Figure 7D. The Molar Formation, Millen Schist and Robertson Bay Group appear in green color pixels due to the high content of Al-OH mineral assemblages (Figure 7D).

TIR radiation has been stated to be a function of temperature and emissivity [38,123]. In the ASTER TIR dataset, bands 10, 11 and 12 comprise spectral emissivity and temperature information, while in bands 13 and 14, surface temperature dominates the spectral emissivity information [7,123]. Additionally, ultramafic/mafic rocks (gabbro, dolerite and dunite; 40–45% SiO_2) show high spectral emissivity in bands 10 to 12 (8–9 µm) and low emissivity in bands 13 and 14 (10–12 µm), as well as high surface temperature due to low albedos and low thermal inertia [41,42,123–125]. Felsic rocks (granite and granitoid; 60–80% SiO_2) show low spectral emissivity in bands 10 to 12 and high emissivity in bands 13 and 14, as well as low surface temperatures attributable to high albedos [41,42,123,124]. In view of this, ASTER TIR bands incorporate maximally independent pixels related to ultramafic-to-mafic, intermediate and felsic lithological units, which can be specifically detected using PCA/ICA analysis.

The analysis of the eigenvector matrix for TIR bands in Table A2 shows unique contributions of eigenvector loadings (magnitude and sign) in some specific PCAs. The Z-Profiles of detected pixels indicate distinctive spectral emissivity features related to different lithological units (ultramafic, mafic, intermediate and felsic rocks) in the output IC images. Table A2A shows eigenvector loadings for the spatial subset of zone 1 and the surrounding areas. PCA2 shows strong contributions of band 12 (−0.793719) and band 14 (0.566669) with opposite signs. PCA3 contains strong loadings of band 10 (−0.639764) and band 11 (−0.511314) with negative signs and band 14 (0.432023) with a positive sign. PCA4 has strong weightings of band 10 (0.630731) and band 11 (−0.744700) with opposite signs. Considering the characteristics appearing in the eigenvector loadings (Table A2A), PCA2 contains maximally independent pixels related to ultramafic-to-mafic units, while PCA3 and PCA4 include the independent pixels of intermediate and felsic units.

The RGB color composite was generated for the zone 1 and the surrounding areas after ICA rotation of PCA2, PCA4 and PCA3 for mapping maximally independent pixels of ultramafic/mafic, felsic and intermediate lithological units, respectively (Figure 8A). The image map contains red, magenta, yellow, light yellow and green pixels. Comparison with the geological map indicates that red and magenta pixels generally match the Wilson Terrane metamorphic rocks and Ferrar Dolerite, which are composed of ultramafic-to-mafic lithological units. The yellow and light yellow pixels match well with exposures of the Granite Harbour Igneous Complex (Figure 8A). The green pixels match the Beacon Supergroup (arkosic quartz sandstone). The intermediate trend can be expected with the quartz-rich/feldspar-rich rocks (granite and granitoid), while the felsic trend is correlated with quartz-rich rocks (quartzose sedimentary rocks) due to the high content of SiO_2 and spectral property of quartz [42,124].

Based on the eigenvector loadings for the spatial subset of zones 2 and 3 (Table A2B), PCA2 shows strong positive loading of band 12 (0.733503) and strong negative loading of band 14 (−0.567605); PCA3 contains strong negative contribution of band 10 (−0.739002) and strong positive contribution of band 12 (0.467340); and PCA4 has strong negative weighting of band 10 (−0.528835) and strong positive weighting of band 11 (0.806255). Hence, PCA2, PCA3 and PCA4 images comprise maximally independent pixels related to ultramafic-to-mafic, intermediate and felsic lithological units, respectively. These PCA images were selected for ICA rotation and a false-color composite image map was generated by allocating red color to ultramafic-to-mafic rocks (the ICA2 image), green color to felsic rocks (the ICA4 image) and blue color to intermediate rocks (the ICA3 image) (Figure 8B). Green, cyan, yellow and red color pixels are depicted in the image map. Yellow pixels refer to combined mafic and felsic trends, while cyan pixels contain felsic in the intermediate signature. With reference to the geological map of the study area, red pixels are mainly associated with exposures of the Glasgow metavolcanic rocks of the BT and Ferrar Dolerite; yellow pixels are mostly considered with exposed zones of the Granite Harbour Igneous Complex and Wilson Terrane metamorphic rocks; green and cyan pixels are generally represented in the exposures of metasedimentary rocks (the Molar Formation, Mariner Group and Leap Year Group) of the BT, Robertson Bay Group and Admiralty Intrusives.

Table A2C shows the eigenvector matrix for the spatial subset covering zone 4 and the surrounding areas. PCA2 has high eigenvector loadings in band 12 (0.734577) and band 14 (−0.594568) with opposite signs. PCA3 shows a strong negative contribution of band 10 (−0.749770) and strong positive contribution of band 12 (0.439774). PCA4 contain strong positive loading of band 10 (0.532694) and high negative loading of band 11 (−0.814304). After ICA rotation, red, green and blue colors were used to produce a false-color composite image map of the ICA2 (ultramafic-to-mafic rocks), ICA4 (felsic rocks) and ICA3 (intermediate rocks). Figure 8C shows the resultant image map for zone 4 and the surrounding areas. Green, cyan, blue and red color pixels are observable. The concentration of red pixels mostly corresponds with the Litell Rocks region (Ferrar Dolerite) and Glasgow metavolcanic rocks, while green and cyan pixels are likely associated with metasedimentary rocks (Molar Formation, Mariner Group and Leap Year Group) of the BT. Blue pixels are probably the admixture of the metavolcanic and metasedimentary rocks in the Bowers Mountains, which show an intermediate trend in the image map (Figure 8C). The green, red and magenta pixels that appear as long linear/curve patterns in the central part of the image (corresponding with the Rennick Glacier) seem to be the Autobahn Moraine [122].

Considering the eigenvector loadings for the spatial subset covering zones 5 and 6 (Table A2D), PCA2 shows high weighting with a negative sign in band 12 (−0.765517) and great contribution with a positive sign in band 14 (0.585714). PCA3 has strong loading of band 10 (0.667689) and band 12 (−0.419996) with opposite signs. PCA4 contains a strong contribution of band 10 (0.605752) with a positive sign and high loading of band 11 (−0.766295) with a negative sign. The ICA rotation was applied to PCA2, PCA3 and PCA4. Accordingly, the ICA2, ICA4 and ICA3 were assigned to RGB false-color composite for mapping ultramafic-to-mafic rocks, felsic rocks and intermediate rocks, respectively. Figure 8D shows the resulting image map of the selected zone. Red and magenta pixels are mainly associated with the exposure of the Glasgow Volcanics, which are mainly mafic in composition. Yellow pixels with an intermediate composition likely correspond to the exposed zones of Admiralty Intrusives and the admixture of the metasedimentary and metavolcanic rocks of the Bowers Mountains.

Green and cyan pixels reflect felsic lithologies that typically characterize the Molar Formation, Millen Schist and Robertson Bay Group exposures (Figure 8D).

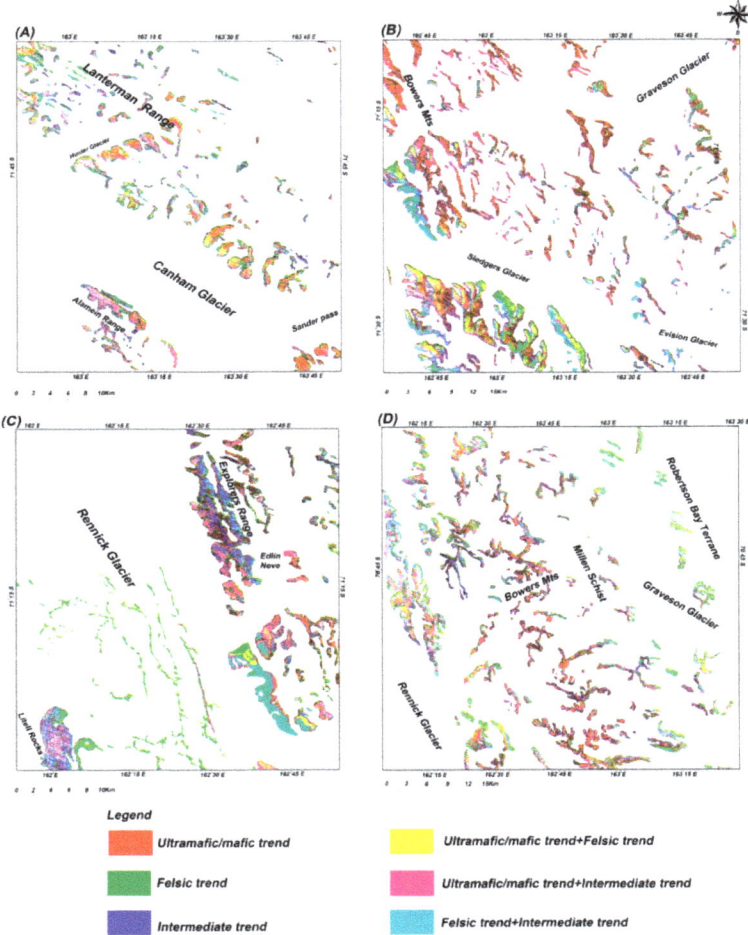

Figure 8. ASTER image maps derived from the ICA rotation and subsequent RGB color composite to TIR bands. (**A**) Spatial subset covering zone 1 and surrounding areas; (**B**) Spatial subset covering zones 2 and 3; (**C**) Spatial subset covering zone 4 and surrounding areas; and (**D**) Spatial subset covering zones 5 and 6.

4.3. Detection of Hydrothermal Alteration Minerals and Prospecting Listvenites in the Selected Subset of Damage Zones

The LSU and CEM algorithms were used for sub-pixel level mapping of alteration mineral assemblages and prospecting listvenites in the six selected subsets of damage (fault) zones (zones 1–6) (see Figure 3). A set of unique pixels (corresponding to a pure end-member) was defined using the n-Dimensional analysis technique for each selected subset. Figure 9 shows end-member (mean) spectra extracted for each selected subset of the six study zones. Subsequently, end-member spectra extracted for each selected subset from the apparent reflectance data were used to act as end-members for LSU spectral mapping. Comparison with selected end-member reflectance spectra of minerals from the

USGS spectral library that resampled to response functions of ASTER VINR+SWIR bands (see Figure 5) indicates that the extracted end-members or a subset of the extracted end-members can be considered for LSU classification.

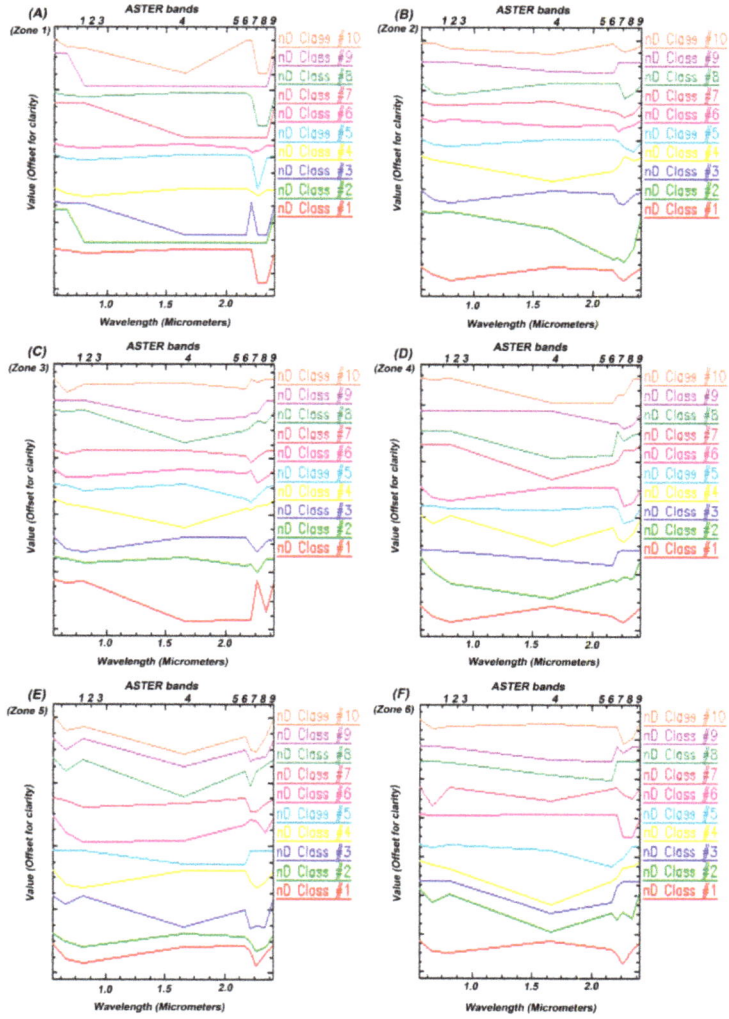

Figure 9. End-member (mean) spectra extracted from ASTER VINR+SWIR bands using the n-Dimensional analysis technique for the six selected subsets of the damage (fault) zones. (**A**) Zone 1; (**B**) Zone 2; (**C**) Zone 3; (**D**) Zone 4; (**E**) Zone 5; and (**F**) Zone 6. ASTER band center positions are shown for the selected zones.

The spectral patterns are classified based on the relative absorption intensities. The most dominant spectral patterns often characterize the most abundant minerals (spectrally dominant species) in the rock, which could also be influenced by subordinate phases (spectrally less active or less abundant groups). Generally, in a mineral mixture or association, the identification of different mineralogical phases is complicated when more than one spectrally active mineral group is present [126]. As a result, the extracted end-member spectra for the study zones could be classified to seven spectrally active groups based on Fe^{2+}, Fe^{3+}, Fe-OH, Al-OH, Mg-OH and CO_3 spectral absorption features, namely (i) goethite/hematite/jarosite group; (ii) kaolinite group; (iii) biotite/muscovite group; (iv) chlorite/epidote/actinolite group; (v) serpentine/antigorite/talc group; (vi) calcite/dolomite/siderite group; and (vii) unaltered/unknown mineral group.

Figure 9A shows extracted end-member spectra for the subset zone (1). Some obvious distinctions between spectral signatures for a variety of minerals are recognizable especially for iron oxide/hydroxide, kaolinite and biotite/muscovite groups. n-D class #1, n-D class #2, n-D class #8, n-D class #9 and n-D class #10 typically represent iron oxide/hydroxide absorption characteristics. Goethite, hematite and jarosite show strong Fe^{3+} and Fe-OH absorption features at 0.48 μm, 0.83–0.97 μm and 2.27 μm, coinciding with bands 1, 3 and 7 of ASTER, respectively [6,33]. Thus, n-D class #1, n-D class #8 and n-D class #10 can be attributed to jarosite (strong 2.27 μm; Fe-OH absorption) and n-D class #2 and n-D class #9 could be considered for goethite and hematite (0.48 μm and 0.83–0.97 μm; Fe^{3+} absorption) (Figure 9A). n-D class #3 and n-D class #7 can be grouped as unaltered/unknown minerals because these classes do not show any distinctive absorption features related to hydrothermal alteration minerals. n-D class #4 represents biotite due to major Mg,Fe-OH absorption near 2.30 μm and minor Fe^{3+} absorption features at 0.85 μm [108,120], which correspond with bands 3 and 8 of ASTER (see Figure 5D). n-D class #5 exhibits a distinct Al-OH absorption feature at 2.2 μm attributable to muscovite, coinciding with band 6 of ASTER [32,119,121]. n-D class #6 contains Al-OH absorption features of kaolinite at 2.17 μm and 2.2 μm that correspond with bands 5 and 6 of ASTER [33,108,120].

The fraction images of end-members resulting from LSU analysis for zone (1) appear as a series of greyscale rule images, one for each extracted end-member. High digital Number (DN) values (bright pixels) in the rule image represent the subpixel abundance of the target mineral in each pixel and map its location. Considering the resultant fraction images and extracted end-member spectra for zone (1), it is evident that goethite, hematite, jarosite, biotite, muscovite and kaolinite are the dominant minerals. For post classification of the fraction images, the rule image classifier tool was applied using a maximum value option. It should be noted that in spectral mixture analysis, a material with a spectral signature similar, but not identical, to that of an end-member can be modeled along with that end-member and be mapped in that end-member's fraction image [127]. For that reason, the rule image classifier tool is not capable of assigning all end-member minerals into their different classes. However, spectral signatures different from the background and other minerals can certainly be discriminated and classified. Thus, the red color class was designated for the goethite/hematite/jarosite group, the yellow color class was assigned to the biotite/muscovite group and the green color class was selected for the kaolinite group. The unaltered/unknown mineral group was not considered for post classification in the present study.

Figure 10A shows the LSU classification mineral map for zone (1). The results indicate that zone (1) is spectrally governed by the goethite/hematite/jarosite group, while the biotite/muscovite and kaolinite groups have a smaller contribution to the total mixed spectral characteristics. In zone (1), the Wilson Terrane metamorphic rocks, Granite Harbour Igneous Complex, Beacon Supergroup and Ferrar Dolerite are exposed. Therefore, a high surface abundance of iron oxide/hydroxide minerals is related to the crystal-field transitions of iron ions (Fe^{+2} and Fe^{+3}) in the primary mafic minerals (olivine, pyroxenes and plagioclase) and/or the alteration of primary mafic minerals within mafic rock units such as the Wilson Terrane metamorphic rocks and Ferrar Dolerite. Kaolinite high abundance zones are mostly associated with the detrital clay minerals of the Beacon Supergroup. The biotite/muscovite group seems to be a phyllic alteration zone associated with the Granite Harbour Igneous Complex. Accordingly, the presence of listvenite bodies in zone (1) is slightly feasible.

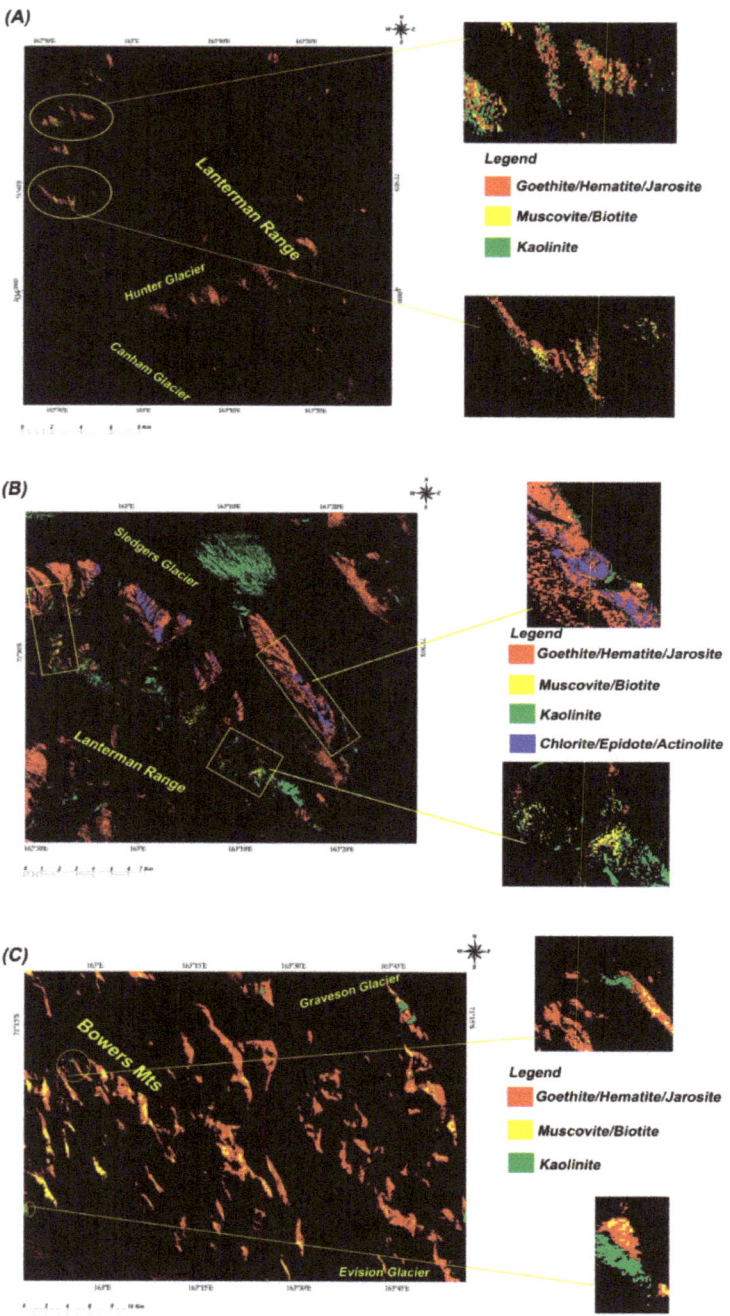

Figure 10. *Cont.*

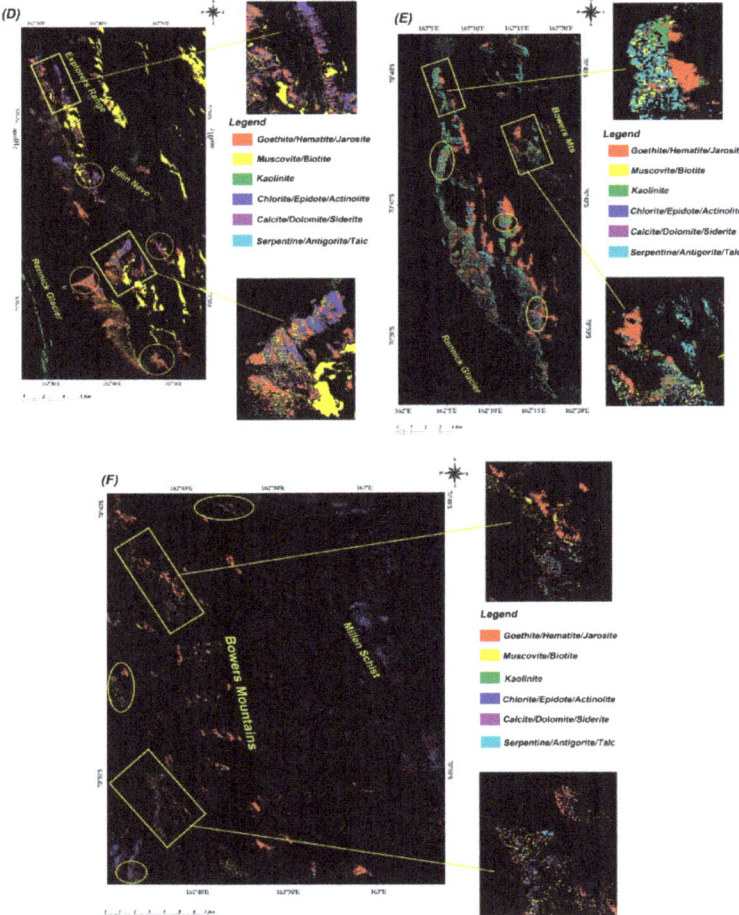

Figure 10. LSU classification mineral maps derived from fraction images of the extracted end-members. (**A**) Zone 1; (**B**) Zone 2; (**C**) Zone 3; (**D**) Zone 4; (**E**) Zone 5; and (**F**) Zone 6. Spectrally dominant mineral groups (concentration more than 10%) are depicted as colored pixels.

Analysis of the extracted end-member spectra for the subset zone (2) (Figure 9B) indicates several distinguishable spectral signatures for the alteration minerals. Hematite, jarosite, kaolinite, biotite, muscovite, actinolite, epidote and chlorite are recognizable (Figure 9B). n-D class #1 shows the spectral characteristics of jarosite and n-D class #2 shows the distinctive absorption features of kaolinite. n-D class #3 represents actinolite due to the iron absorption features in bands 2 and 3, and Mg, Fe-OH absorption in bands 7 and 8 of ASTER (see Figure 5J). n-D class #4 and n-D class #9 lack any distinct spectral features related to the alteration minerals, which can be considered an unaltered/unknown mineral group in zone (2). n-D class #5 and n-D class #8 represent epidote and chlorite because they display a slight iron absorption at 0.85–0.97 µm and a major Mg, Fe-OH absorption at 2.30–2.35 µm [33,108,128], corresponding to bands 2, 3 and 8 of ASTER data, respectively [121]. n-D class #6 and n-D class #7 characterize muscovite and biotite, respectively. n-D class #10 can be considered as mixed spectral features of goethite, hematite and jarosite.

Fraction images of end-members for zone (2) were classified using the rule image classifier tool. The blue color class was assigned for chlorite/epidote/actinolite group and added to the previous classes for producing the LSU classification mineral map. Figure 10B shows the resultant classification map for the subset of zone (2). Goethite/hematite/jarosite and chlorite/epidote/actinolite groups show high surface abundance, whereas kaolinite and biotite/muscovite groups exhibit moderate to less spectral contribution and spatial distribution in zone (2). The association of goethite/hematite/jarosite and chlorite/epidote/actinolite groups is observable with exposures of the Glasgow metavolcanic rocks and metasedimentary rocks (Molar Formation, Mariner Group and Leap Year Group). Kaolinite, biotite/muscovite and iron oxide/hydroxide assemblages are associated with the Granite Harbour Igneous Complex. The Wilson Terrane metamorphic rocks and Ferrar Dolerite exposures are generally dominated by goethite/hematite/jarosite group, while the Beacon Supergroup is governed by kaolinite group. For that reason, listvenite bodies could be located in a zone that spatially contains all of the four alteration mineral assemblages. Please note that the large greenish-colored spot in the area of the Sledgers Glacier is a dense crevasse field, which may contain contamination of clay mineral groups (Figure 10B). Several prospect zones for listvenite bodies are identifiable in zone (2); some of them are delimited by yellow rectangles in Figure 10B.

In zone (3), some spectral signatures related to alteration minerals such as goethite, hematite, jarosite, kaolinite, biotite and muscovite are discernible (Figure 9C). n-D class #1, n-D class #4, n-D class #8 and n-D class #9 can be considered to be an unaltered/unknown mineral group. They do not illustrate any significant absorption features related to the target alteration minerals. n-D class #2 and n-D class #3 represent jarosite. Spectral signatures related to kaolinite, biotite and muscovite are apparent in n-D class #5, n-D class #6 and n-D class #7, respectively. n-D class #10 can be characterized as having mixed spectral features of hematite and goethite. Figure 10C displays the LSU classification mineral map for zone (3). The goethite/hematite/jarosite and biotite/muscovite groups are spectrally predominant. The kaolinite group shows very low surface abundance. Iron oxide/hydroxide and biotite/muscovite mineral groups are concentrated in the exposures of the Glasgow metavolcanic rocks and metasedimentary rocks in the Bowers Mountains. Small exposures of the Granite Harbour Igneous Complex and Admiralty Intrusives contain kaolinite group minerals associated with iron oxide/hydroxide and biotite/muscovite mineral groups, which can be attributed to the alteration products of argillic and phyllic alteration zones. Hence, zone (3) has very low potential for containing listvenites.

Typical spectral signatures for a variety of mineral assemblages are decipherable for zone (4) (Figure 9D). n-D class #1 represents antigorite, which contains Fe^{3+} absorption features and Mg,Fe-OH absorption near 2.30 µm [128,129], coinciding with bands 2, 3 and 8 of ASTER (see Figure 5G). n-D class #2 shows 0.48 µm and 0.83–0.97 µm Fe^{3+} absorption and seems to be a combined spectral signature of goethite and hematite. The unaltered/unknown mineral group can be assigned to n-D class #3. Serpentine shows spectral characteristics related to crystal-field transitions in the Fe^{2+} near 0.4 and 0.5 µm and Fe^{3+} near 0.65 µm and a combination of OH-stretching fundamental with the Al-OH-bending and Fe,Mg-OH-bending modes near 2.20 and 2.30 µm [128,130,131]. Therefore, it seems that n-D class #4 has a serpentine spectral signature, which displays distinctive absorption features in bands 2, 3, 7 and 8 of ASTER (see Figure 5H). n-D class #5 appears to be a combined spectral signature of muscovite and biotite. n-D class #6 is characterized by chlorite and epidote spectral features. n-D class #7 can be considered as a mixed spectral signature for calcite and dolomite. Broad Fe^{2+} absorption features occur in calcite and dolomite spectra near 0.9–1.2 µm and vibrational processes of CO_3 radical cause absorption properties near 2.30–2.35 µm [132,133], which are equivalent to bands 4, 7 and 8 of ASTER. n-D class #8 exhibits siderite due to absorption characteristics related to Fe^{2+}, Fe^{3+} and CO_3 in bands 2, 3, 7 and 8 of ASTER (see Figure 5P). n-D class #9 exhibits the kaolinite spectral signature, and n-D class #10 can be considered to be a mixed spectral signature of hematite and jarosite.

Figure 10D shows the LSU classification mineral map for zone (4). Biotite/muscovite, goethite/hematite/jarosite and chlorite/epidote/actinolite mineral groups are spectrally significant and have high surface abundance. However, kaolinite, serpentine/antigorite/talc (cyan color class) and calcite/dolomite/siderite (purple color class) mineral assemblages are weakly distributed in the exposed zones. Since zone (4) covers the Bowers Mountains, biotite/muscovite mineral assemblages (yellow pixels) are likely associated with the Molar Formation and the Mariner Group metasedimentary sequences that contain high contents of detrital clay minerals. Glasgow Metavolcanic rocks contain goethite/hematite/jarosite and chlorite/epidote/actinolite mineral groups because of the alteration of primary mafic mineral within basalts, spilites, volcanic breccia and tuffs. The Autobahn Moraine in the Rennick Glacier [122] is mapped as green pixels in the southwestern part of zone (4) (Figure 4D). Erratic metasedimentary or sedimentary rocks and detritus delivered to this moraine from the Bowers Mountains and/or the Lanterman Range could possibly contain large amounts of kaolinite mineral groups. Zone (4) is a highly likely prospective area for listvenites, where the most of the alteration mineral groups are specifically concentrated and are associated particularly with the Glasgow Volcanics. Some of the prospects in zone (4) are shown by yellow rectangles and circles in Figure 10D.

Deciphering the extracted end-member spectra for zone (5) reveals several distinct spectral signatures for the alteration minerals (Figure 9E). n-D class #1 and n-D class #2 obviously represent chlorite and epidote, respectively. n-D class #3 seems to be the combination of antigorite and talc spectral signatures. Mixed spectra of biotite and muscovite can be seen in n-D class #4. Diagnostic absorption features related to alteration minerals could not be realized in n-D class #5 (unaltered/unknown mineral group). n-D class #6 characterizes actinolite spectral properties. n-D class #7 has the combined spectra of hematite and jarosite. It seems that the spectral signatures of kaolinite are exhibited in n-D class #8. Serpentine, antigorite and talc might display mixed spectral properties in n-D class #9. n-D class #10 represents serpentine. Figure 10E shows the LSU classification mineral map for zone (5). The goethite/hematite/jarosite, kaolinite and serpentine/antigorite/talc mineral groups are considerably predominant. Conversely, biotite/muscovite and chlorite/epidote/actinolite mineral assemblages show low spatial distributions and abundances in this zone. The association of iron oxide/hydroxide, serpentine/antigorite/talc, chlorite/epidote/actinolite mineral assemblages are mostly concentrated in the exposures of Glasgow Volcanics due to the alteration of mafic minerals within meta-basalts, spilites, volcanic breccia and tuffs. Kaolinite is mostly associated with the Molar Formation because of the high content of conglomerate and mudstone. Zone (5) may comprises several listvenite occurrences especially associated with Glasgow Volcanics, where a high surface distribution of serpentine/antigorite/talc group is detected with biotite/muscovite and iron oxide/hydroxide mineral groups. Yellow rectangles and circles show some of the prospective zones (Figure 10E).

Analyzing the spectral signatures related to alteration minerals for zone (6) indicates the presence of goethite, hematite, jarosite, biotite, muscovite, chlorite, actinolite, serpentine and talc (Figure 9F). n-D class #1 shows chlorite. n-D class #2 and n-D class #4 may have mixed spectral signatures related to goethite, hematite and jarosite. n-D class #3 and n-D class #8 do not have any indicative absorption features related to alteration minerals and can be considered an unaltered/unknown mineral group. Biotite and muscovite spectral characteristics can be found in n-D class #5 and n-D class #6, respectively. n-D class #7 represents actinolite. n-D class #9 and n-D class #10 characterize serpentine and talc. The LSU classification mineral map for zone (6) is shown in Figure 10F. Iron oxide/hydroxides (goethite, hematite and jarosite) and chlorite/epidote/actinolite mineral assemblages are spectrally major components, while biotite/muscovite and serpentine/antigorite/talc mineral groups are minor components in this zone. The Glasgow Volcanics show high content of iron oxide/hydroxides minerals, which are locally associated with chlorite/epidote/actinolite, biotite/muscovite and serpentine/antigorite/talc mineral groups. Listvenite bodies could be found in the local association of these mineral assemblages within Glasgow Volcanics exposures in the Bowers Mountains. Several locations could be taken into consideration for listvenite occurrences, some of which are demarcated by yellow rectangles and circles in Figure 10F.

The CEM algorithm was implemented to produce fraction images of selected end-member spectra from the USGS spectral library, including goethite, hematite, jarosite, biotite, kaolinite, muscovite, antigorite, serpentine, talc, actinolite, chlorite, epidote, calcite, dolomite, siderite and chalcedony (see Figure 5). The subset of the zones (2), (4) and (5) was selected for running the CEM because they exhibit a variation of mineral assemblages and several highly likely prospective zones for listvenite occurrences in the LSU classification mineral maps (see Figure 10B,D,E). The spatial subset of zone (2) covering the Lanterman Range and surrounding areas was selected to present in this paper. Fraction images of sixteen selected end-member minerals were produced using the CEM algorithm. Fractional abundance of target end-member minerals appears as a series of greyscale rule images, one for each selected mineral. Pseudo-color ramp of greyscale rule images was generated to illustrate high fractional abundance (high DN value pixels) of the target minerals in zone (2) (Figure 11). This comfortably distinguishes the contrast between subpixel targets and surrounding areas. This contrast expresses the fractional abundance of the target mineral present in the rule image.

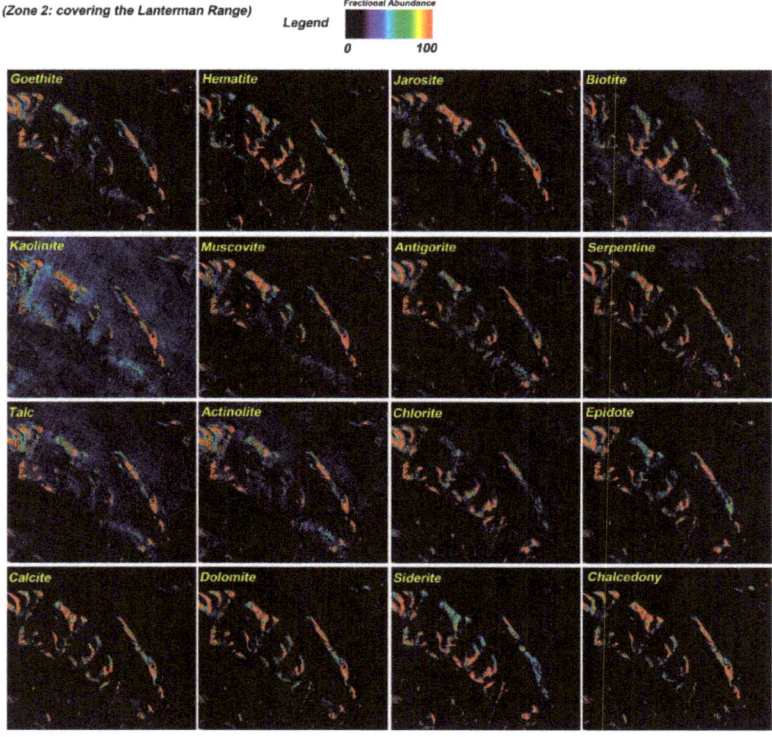

Figure 11. Fraction images of the selected end-member minerals derived from CEM algorithm for zone (2) covering the Lanterman Range and surrounding areas. Pseudo-color ramp was applied to greyscale rule images.

Considering Fe^{2+}, Fe^{3+}, Fe-OH, Al-OH, Mg-OH, Si-OH and CO_3 spectral absorption characteristics of target minerals and the limitations of VNIR+SWIR ASTER spectral bands, seven groups of mineral assemblages could also be discernible here. There might be some confusion for distinctive separation of the absorption features (electronic transitions and molecular vibrational overtones) present in alteration minerals using VNIR+SWIR spectral bands of ASTER, particularly when mixtures occur [134,135]. Therefore, some of the target alteration minerals may have a similar manifestation of the fractional abundances in the CEM rule images. For iron oxide/hydroxide mineral group, fractional abundances of goethite and hematite show almost similar appearance and high mixtures, while jarosite shows low mixtures with them and different spatial fractional abundance in some parts of zone (2) (Figure 11). Examination of the end-member spectra extracted from ASTER VINR+SWIR bands using n-Dimensional analysis for zone (2) indicates the presence of jarosite spectral characteristics as n-D class #1 and mixed spectral features of goethite, hematite and jarosite in n-D class #10 (see Figure 9B). Comparison of the fraction images of biotite and muscovite shows different spatial abundances and low combinations (Figure 11). Looking at the n-Dimensional analysis results for zone (2) shows the identification of spectral signatures for muscovite and biotite in n-D class #6 and n-D class #7, respectively (see Figure 9B).

Fractional abundance of kaolinite exhibits comparable manifestation with the spatial abundance of muscovite in many parts (Figure 11). n-D class #2 derived from n-Dimensional analysis for zone (2) is considered to be kaolinite (see Figure 9B); however, the CEM results indicate that this n-D class can be a mixture of kaolinite and muscovite. Antigorite, serpentine and talc display fractional abundance analogous to actinolite (Figure 11). n-D class #3 extracted from n-Dimensional analysis for zone (2) represents actinolite due to some distinctive spectral signatures (see Figure 9B), which are very similar to spectral features of serpentine/antigorite/talc group (see Figure 5G–J). Hence, there may be some confusion between these minerals when using VINR+SWIR bands of ASTER, and their spectral signatures could appear in only one n-D class (n-D class #3) (see Figure 9B). Fractional abundances of chlorite and epidote show some small spatial differences and high combination. Considering n-D classes for zone (2) (see Figure 9B), n-D class #5 and n-D class #8 are determined to be epidote and chlorite, respectively. Fraction images of calcite, dolomite, siderite and chalcedony exhibit similar spatial abundance and high mixtures with each other, although siderite has a lower distribution in some places (Figure 11). With respect to n-D classes for zone (2) (see Figure 9B), there is a high probability that n-D class #4 can be considered to be the mixed spectral signatures of the calcite/dolomite/siderite group, which was previously considered an unaltered/unknown mineral group in zone (2). Please note that a high degree of spectral contrast is required to distinguish the end-member mineral from the background materials [127]. n-D class #9 (see Figure 9B) might have mixed spectral signatures of chalcedony with other minerals, which was formerly assigned to the unaltered/unknown mineral group in zone (2). This is due to the fact that all mixed-pixel spectra always lie on the line that connects the component spectra [111].

4.4. Petrography and Mineralogy of Hydrothermal Alteration Minerals and Listvenites

Petrographic studies were carried out on samples of hydrothermally altered rocks from the major damage zones of the terrane boundaries and along intra-terrane faults or shear zones. The country rocks comprise low-grade metamorphic basalts and volcaniclastic rocks, low- high-grade metamorphic ultramafic and mafic igneous rocks and granitoids. The observed types of alteration can be divided into the following main groups: (i) Mg-Ca-Fe carbonation and/or silicification of metavolcanic rocks associated with syntectonic carbonate coatings on fault planes, hydraulic brecciation, and quartz-carbonate veining occurring along intra-BT faults and shear zones in the Bowers Mountains and along the BT-RBT boundary damage zone; (ii) epidote-bearing slickensides, epidote veining and epidotization occurring in low-grade metabasalts and in amphibolites in the Lanterman Range and the Bowers Mountains; (iii) epidote-chlorite-prehnite-bearing cataclastites, ultracataclastites, and indurated gouges in fault cores, as well as saussuritization of K-feldspar in the wallrock of granitoid rocks in the western Lanterman Range [136]; (iv) foliated listvenites with Mg-Ca carbonates, quartz, Cr-phengite, Cr-chlorite, magnesite, and talc derived from carbonation of mafic and ultramafic rocks along brittle-ductile shear zones within the WT-BT boundary damage zone and within the BT [1,137] (see Table A3). Figure 12A–H shows some exposures of typical altered rocks and listvenites.

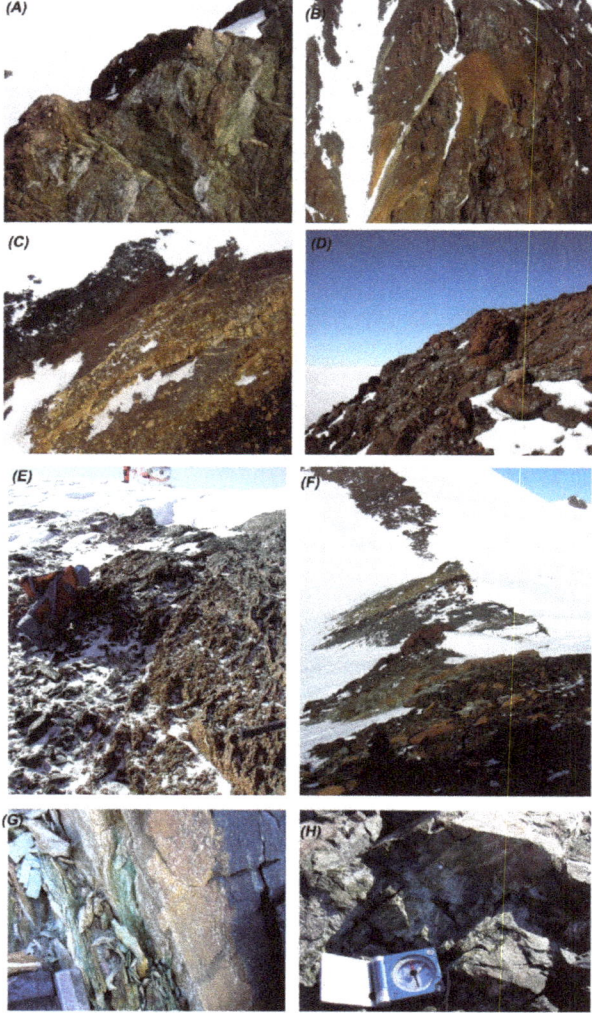

Figure 12. Field photographs of the typical altered rocks and listvenites. (**A**) Helicopter view of the fault zone characterized by epidote–prehnite–quartz coatings, and epidotization in Glasgow Volcanics country rocks, Mt Gow, Bowers Mountains; (**B**) Helicopter view of the reddish (iron oxide/hydroxide minerals) to greenish (chlorite-epidote minerals) alterations of Glasgow Volcanics along fault zones, NE slopes of Mt Gow, Bowers Mountains; (**C**) Helicopter view of the metasomatic alteration (carbonate dominated) of Glasgow Volcanics around a quartz–carbonate fault vein system (Dorn Glacier, for details see [1]; (**D**) Close view of reddish alteration (iron oxide/hydroxide minerals dominated) along fault zones in Glasgow Volcanics, NE slopes of Mt Gow, Bowers Mountains; (**E**) Close view of shear zone with magnesite-talc-quartz mylonite derived from mafic and ultramafic rocks at the WT-BT boundary, Lanterman Range; (**F**) View of layers of alternating foliated ultramafite and amphibolite within mylonitic shear zone characterised by listvenite and magnesite-talc-quartz rich mylonite, WT-BT boundary, Lanterman Range; (**G**) Close view of listvenites from a shear zone at the WT-BT boundary, Lanterman Range. The foliated green rock is rich in Cr-muscovite and chlorite, the light brown part is rich in Fe-Mg carbonates; (**H**) Close view of a damage zone with intense epidotization (greenish to pinkish color) and silicification of the host Glasgow Volcanics Mt Gow, Bowers Mountains.

Petrographic studies of the upper Dorn Glacier at the eastern side of the northern Bowers Mountains indicated the transformation of primary mafic minerals such as pyroxene (augite), olivine, plagioclase, ilmenite and amphibole to secondary altered minerals such as muscovite, Mg,Fe-chlorite, titanite, calcite, epidote-prehnite, hematite/limonite and siderite/ankerite in the hydrothermal alteration zones [1]. Figure 13A–D shows thin sections of epidote+prehnite+chlorite alteration on a fault in Glasgow Volcanics (A), carbonated basalt (B) and foliated listvenites (C and D). Primary mafic minerals in fine-grained green metabasalt (Glasgow Volcanics) replaced by epidote-chlorite-prehnite-bearing cataclastites in the fault zone. The phenocrysts of epidote and prehnite display a cataclastic fabric, while chlorite forms thin veinlets in the backgmass (Figure 13A). Mg-Ca-Fe carbonation has completely replaced the primary mafic minerals of the metabasalt. Some of the carbonates are phenocrystalline and anhedral (Figure 13B). Muscovite, Mg-Ca carbonates, quartz in foliated listvenites are characterized by very fine-grained brittle-ductile deformation along microfractures, and also disseminated in the background (Figure 13C–D). The mineral contents of some selected samples were also examined using XRD analysis (Figure 14A–E). The XRD analysis demonstrated that the predominant minerals in the altered rocks and listvenites were ankerite, muscovite, magnesite, kaolinite, hematite, dolomite, epidote, spinel, talc, clinochlore and quartz.

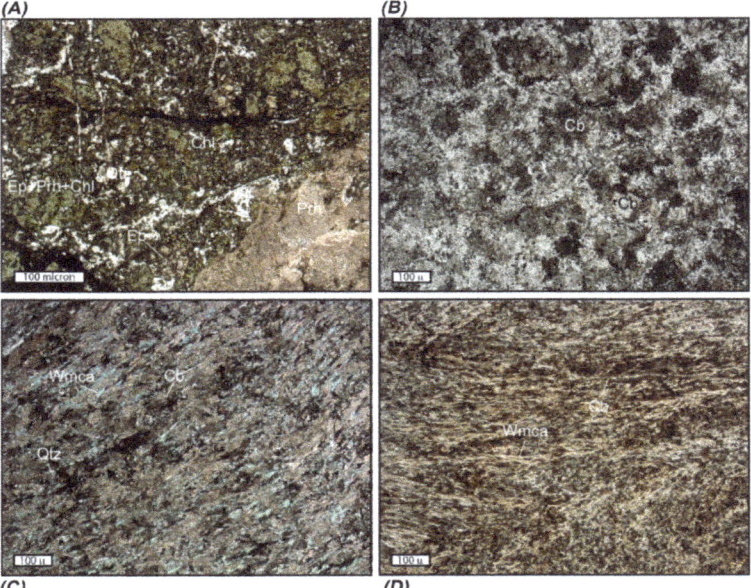

Figure 13. Different types of alteration mineralogy in hydrothermally altered rocks. Microphotographs of (**A**) epidote+prehnite+chlorite-dominated alteration in the damage zone of a fault in Glasgow Volcanics (plane-polarized light; #09.12.03GL 9); (**B**) carbonated basalt in the damage zone of a shear zone (Dorn Glacier—plane-polarized light; #08.12.05GL6); and (**C,D**) foliated listvenites from two brittle-ductile shear zones at the WT-BT boundary (**C**: crossed-polarized light, #23.12.96 GL 6; **D**: plane-polarized light, #19.12.96 GL 6). Abbreviation: Ep = epidote, Prh = prehnite, Chl = chlorite, Cb = carbonate, Wmica = white mica, Qtz = quartz.

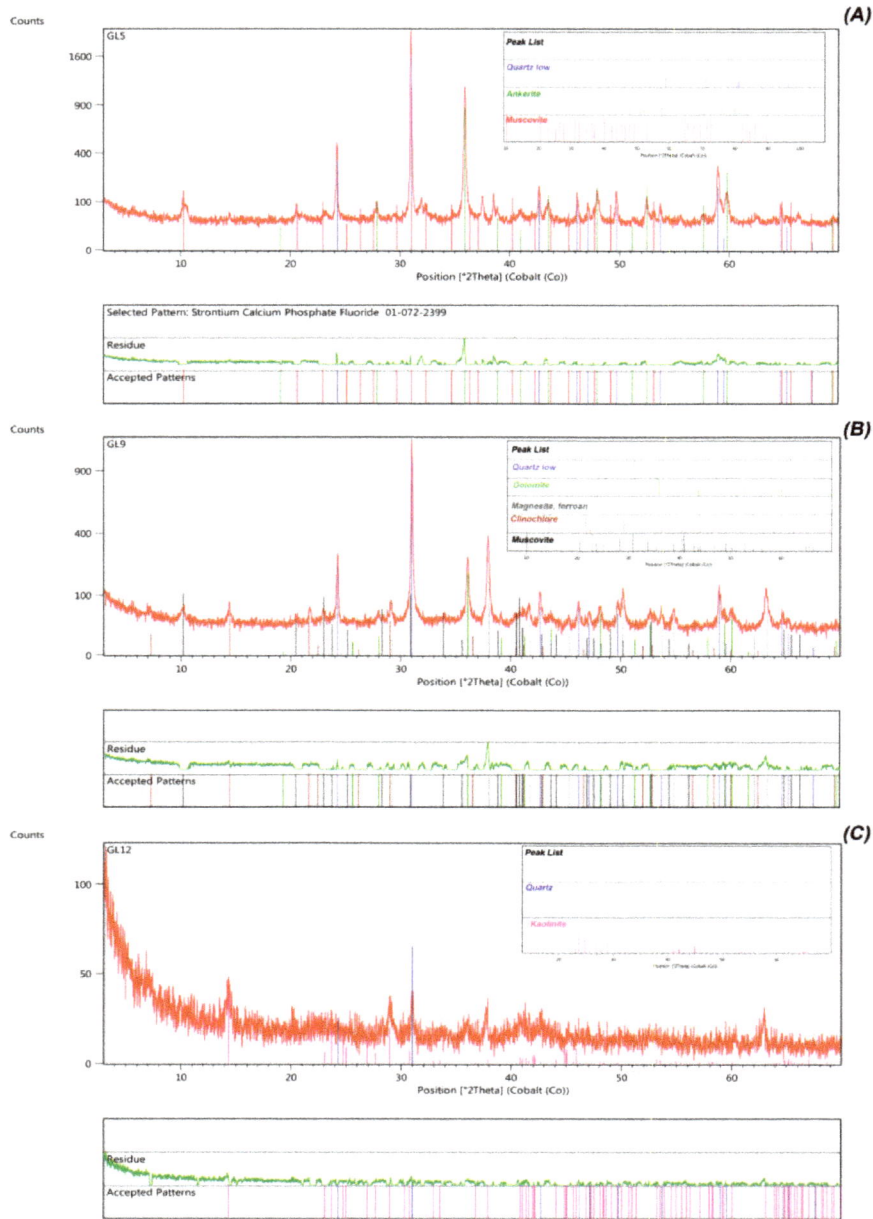

Figure 14. Cont.

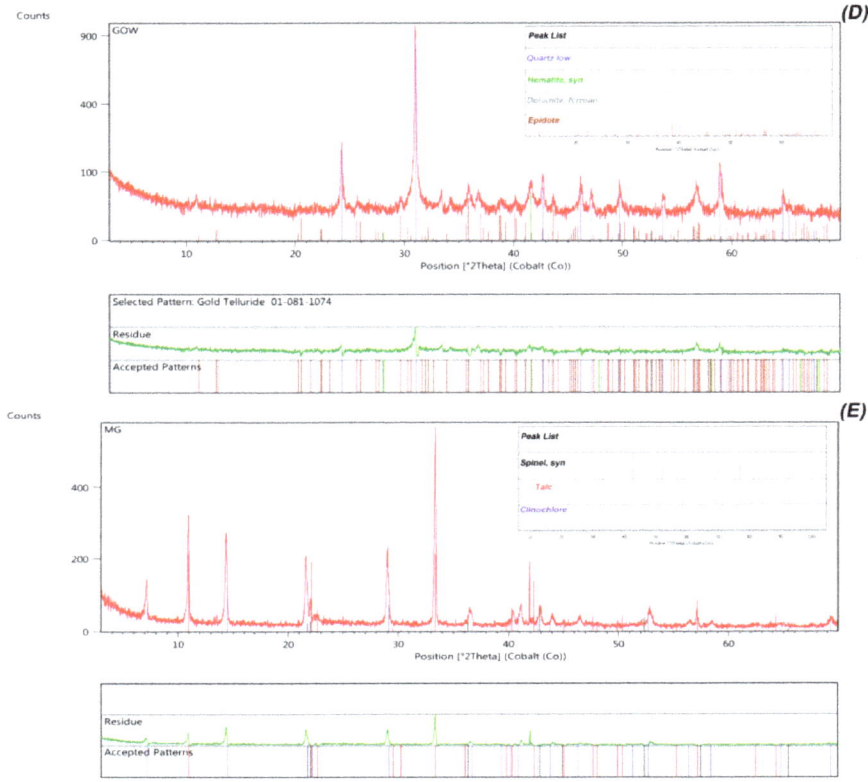

Figure 14. Representative XRD analysis of samples collected from listvenites (**A–C**) and altered Glasgow volcanics (reddish to greenish alterations) at Mt Gow, Bowers Mountains (**D–E**).

5. Discussion

Terrane boundaries are major tectonic discontinuities that often display a complex and polyphase structural evolution, and a variety of ore mineralizations were reported in these boundaries [2,3]. Listvenite, a carbonate-silica-dominated rock that forms as a result of CO_2- and K-metasomatism of ultramafic/mafic rocks, represents a key indicator for the occurrence of ore mineralizations along shear zones or major tectonic boundaries [20–23,138–140]. The ability to map hydrothermal alteration mineral and listvenites using spectral bands of ASTER satellite remote sensing data has been successfully established in many parts of the world [6–12,27–29,31,43]. Features in the spectra of lithological units are dependent upon their constituent minerals and intensities. The intensities are a function of the concentration, absorption coefficient and accessibility to the recording radiation of each constituent mineral [128]. Thus, the extraction of mineralogical information of interest from the remote sensing imagery required standardized/specialized image processing techniques to detect spectrally active minerals and revealing hidden key end-member spectra.

In this study, implementation of robust and standardized/specialized image processing techniques consisting of PCA/ICA fusion, LSU and CEM analysis to ASTER VNIR+SWIR and TIR bands provides a consistent way to detect spectrally active alteration mineral assemblages and lithological units at both pixel and sub-pixel levels in the poorly exposed damage zones of the WT-BT and BT-RBT boundaries and along intra-terrane faults and shear zones in NVL of Antarctica. Fusion technique of PCA and ICA was used to extract the image spectra at the pixel level for mapping alteration mineral assemblages and lithological units. The resultant image maps derived from PCA/ICA fusion identified the pixels containing the spectral signature of the alteration minerals or mineral groups that are maximally independent of each other using VNIR+SWIR and TIR bands of ASTER. PCA images contain high eigenvector loadings and opposite signs in the indicative bands (in the VNIR and SWIR regions) for oxide/hydroxide minerals, Al-OH minerals and Fe,Mg-O-H and CO_3 minerals were identified and used for ICA rotation to extract maximally independent pixels of the minerals. Maximally independent pixels related to ultramafic-to-mafic, intermediate and felsic lithological units (in the TIR region) particularly detected using PCA/ICA analysis. Sub-pixel abundances of spectrally pure pixels (end-members) related to alteration minerals were mapped using ASH technique and LSU classification, which directly derived from ASTER image reference spectra (in the VNIR+SWIR bands). Fraction images of end-members (extracted from the USGS library) obtained by CEM analysis were similar to LSU classification mineral maps, though some spectral mixing and confusion between minerals contain absorption features with subtle differences are undoubted.

Considering image maps of alteration minerals (Figure 7A–D), iron oxide/hydroxides minerals are spectrally dominated in the study zones. The exposures of Granite Harbour Igneous Complex, Glasgow metavolcanic rocks, Admiralty Intrusives and Ferrar Dolerite contain a high surface abundance of iron oxide/hydroxides minerals. Iron oxide spectral signatures are produced due to crystal-field transitions of iron ions (Fe^{+2} and Fe^{+3}) in the primary mafic minerals (olivine, pyroxenes and plagioclase) or caused by the alteration of the primary minerals to iron oxide/hydroxide mineral (hematite, goethite, jarosite and limonite) [126,128]. Therefore, the iron spectral signature can be mapped in both altered and unaltered exposures. The (meta-)sedimentary sequences of the Wilson Terrane metamorphic rocks, Molar Formation, Mariner Group, Leap Year Group, Robertson Bay Group and Beacon Supergroup predominantly show vibrational processes of fundamental absorptions of Al-OH, Fe,Mg-O-H and CO_3 modes. These groups of spectral features appear to be quite diagnostic of detrital clay minerals in the (meta-)sedimentary units. Furthermore, some of the alteration minerals are the products of regional metamorphic processes in the WT, BT and RBT. However, these mineral assemblages could also be weathering products of rocks taking into account the paleo-latitude of the rocks at the time of deposition, e.g., in the case of the redbeds of the Leap Year Group. Also, other metasedimentary rocks deposited in the accretionary environment (like the Robertson Bay Group) would have received detritus rich in Fe- or Al-rich oxides/hydroxides in a particular weathering environment as for these rocks in the early Ordovician (deep-reaching, intensive weathering in equatorial latitudes). Accordingly, only the exposures contain a strong combination of absorption features and intensities for iron oxide/hydroxide, Al-OH and Fe,Mg-O-H and CO_3 modes can be considered as hydrothermal/metasomatic alteration mineral zones, specifically in the Bowers Mountains and Lanterman Range (Figure 7A–D).

The stretching vibration of Si–O–Si tetrahedra in TIR regions was determined to discriminate silicate rocks by applying PCA/ICA fusion (Figure 8A–D). An ultramafic/mafic trend was mapped mostly in the exposures of the Glasgow Volcanics and Ferrar Supergroup (Ferrar Dolerite/Kirkpatrick Basalts). The Wilson Terrane metamorphic rocks, the Granite Harbour Igneous Complex and the Admiralty Intrusives show an intermediate trend. The (meta-)sedimentary rocks of the Molar Formation, Mariner Group, Leap Year Group, Robertson Bay Group and Beacon Supergroup were classified in the felsic trend (Figure 8A–D). Spatial distributions of spectrally active minerals containing Fe^{2+}, Fe^{3+}, Fe-OH, Al-OH, Mg-OH and CO_3 spectral absorption features, including goethite/hematite/jarosite, kaolinite, biotite/muscovite, chlorite/epidote/actinolite, serpentine/antigorite/talc and calcite/dolomite/siderite, were detected using n-Dimensional analysis technique and LSU classification in the six selected subsets of damage zones (Figure 10A–F). Goethite, hematite, jarosite, biotite, muscovite, kaolinite, actinolite, epidote and chlorite were the predominating minerals, while serpentine, antigorite, talc, calcite, dolomite and siderite were in a minority in the selected subsets. The results derived from LSU (Figure 10A–F) show good agreement with the PCA/ICA mapping results (Figure 7A–D). However, alteration minerals were detected systematically in the fraction images of end-members for the selected subsets. Several prospective zones for listvenite occurrences were identified in zones (2), (4), (5) and (6)

(Figure 10B,D–F), which were mostly concentrated and associated with the Glasgow Volcanics in the Bowers Mountains.

The fraction images derived from the CEM algorithm for detecting alteration minerals using the end-member spectra from the USGS spectral library confirmed the capability of ASTER VNIR+SWIR spectral bands to identify some important alteration minerals for prospecting listvenite occurrences. However, some confusion between the alteration minerals may occur due to spectral and spatial limitations of the ASTER multispectral bands, especially when mineral spectral signatures are mixed and/or contain subtle differences. Confusion between minerals with similar absorption bands near 0.9 µm, 2.2 µm and 2.30 µm, such as goethite, hematite, kaolinite, muscovite, chalcedony, serpentine, chlorite, epidote and calcite, is high. As a consequence, CEM fraction images of the alteration minerals show comparable surface abundances in zone (2) (Figure 11). Comprehensive fieldwork data, microscopy-based petrographic studies and XRD analysis in the major fault zones of the terrane boundaries and along the intra-terrane fault or shear zones verified the occurrence of Mg-Ca-Fe carbonation, epidotization and listvenites, especially in Glasgow Volcanics in the Bowers Mountains and Lanterman Range. Comparison of the LSU classification mineral maps with field data using a confusion matrix approach and Kappa Coefficient shows a very good match, which indicates the overall accuracy of 71.42% and the Kappa Coefficient of 0.57, respectively (Table A4, see Appendix A). Consequently, the remote sensing approach developed in this study is explicitly applicable for mapping alteration minerals and lithological units associated with ore mineralizations in remote and inaccessible areas such the Antarctic and Arctic regions, where access to field data is challenging.

6. Conclusions

Application of ASTER multispectral remote sensing data for detecting hydrothermal alteration mineral assemblages, and particularly listvenites, in the poorly exposed damage zones of the WT-BT boundary, the BT-RBT boundary and within the BT and in the eastern WT (NVL, Antarctica) confirmed alteration mineral patterns at several locations, including the predominance of iron oxide/hydroxide, biotite/muscovite, chlorite/epidote/actinolite, Mg-Ca-Fe carbonates and listvenites. The PCA/ICA fusion successfully helped to map and discriminated alteration minerals containing Fe^{2+}, Fe^{3+}, Fe-OH, Al-OH, Mg-OH and CO_3 spectral absorption features in the VNIR+SWIR bands, and silicate lithological units contained different characteristics of emissivity spectra derived from stretching vibration of Si–O–Si tetrahedra in TIR bands. n-Dimensional analysis technique confirmed the presence of several end-member alteration minerals such as goethite/hematite/jarosite, kaolinite, biotite/muscovite, chlorite/epidote/actinolite, serpentine/antigorite/talc, calcite/dolomite/siderite in the selected damage zones, the fraction abundances and spatial distributions of which were subsequently mapped by the LSU classification. The results showed that listvenite occurrences are mostly associated with metavolcanic rocks of Glasgow Volcanics in zones (2), (4), (5) and (6) in the Bowers Mountains, while the association of alteration minerals is generally associated with other sedimentary and igneous lithological units in the study zones. The CEM results verified the identification of some important alteration minerals for detecting listvenite occurrences, although some confusion between the alteration minerals may present using ASTER VNIR+SWIR spectral bands. Furthermore, geological fieldwork and laboratory investigations proved essential implications of the remote sensing data analysis for detecting alteration minerals in remote and inaccessible (Antarctic) environments. The mapping results indicate that ASTER data processing using the pixel/sub-pixel algorithms can provide an efficient approach to map hydrothermal alteration minerals, lithological units (ultramafic, mafic, intermediate and felsic) and listvenite occurrences in inaccessible parts of the NVL and can be broadly applicable in other tectonic boundaries or major tectonic fault zones around the world.

Author Contributions: A.B.P. wrote the manuscript and analyzed the data; Y.P. supervision and funding acquisition; L.C. and A.L. writing, editing, performed experiments and field data collection; J.K.H. and T.-Y.S.P supervision; B.Z. and BP writing and editing; A.M.M and M.S.H. image processing; O.R. editing and analysis.

Funding: This study was conducted as a part of KOPRI research grant PE19050.

Acknowledgments: We are thankful to the Korea Polar Research Institute (KOPRI) for providing all the facilities for this investigation. We acknowledge the logistic support of the Italian National Antarctic Research Programme (PNRA), the German GANOVEX (German Antarctic North Victoria Land Expedition) programme of the Federal Institute for Geosciences and Natural Resources (BGR) and the excellent helicopter support of Helicopters NZ during several expeditions that provided ground-truth data for the satellite-based remote sensing data. Special thanks to the colleagues of the Antarctic campaigns who shared with us the data from field investigations. L.C. thanks PNRA_2013/AZ2.02 and PNRA16_00040_REGGAE projects for financial support. A.L. acknowledges the financial support of the German Research Foundation (DFG) for some parts of this study (grant Kl429/18-1-3).

Conflicts of Interest: The authors declare no conflicts of interest.

Appendix A

PCA computing equation (Equation (A1)):

$$R_{KP} = \frac{a_{KP} \times \sqrt{\lambda_P}}{\sqrt{Var_K}} \tag{A1}$$

where, a_{kp} = eigenvector for band k and component p; $\lambda_P = P^{th}$ eigenvalue; and Var_K = variance of Kth band in the covariance matrix [86]. This computation results in a new n × n matrix filled with factor loadings. This data is typically represented in quantitative terms, which is a very small fraction of the total information content of the original bands. It is expected that the loaded information is indicated in the spectral signature of the desired mineral or mineral group [91].

Table A1. Eigenvector matrix for VNIR+SWIR bands of ASTER. (**A**): Spatial subset covering zone 1 and surrounding areas; (**B**): Spatial subset covering zones 2 and 3; (**C**): Spatial subset covering zone 4 and surrounding areas; and (**D**): Spatial subset covering zones 5 and 6.

(A)									
Eigenvector	Band 1	Band 2	Band 3	Band 4	Band 5	Band 6	Band 7	Band 8	Band 9
PCA 1	0.467805	0.509694	0.512981	0.217468	0.176799	0.246614	0.248711	0.197151	0.134638
PCA 2	−0.300431	−0.288970	−0.263316	0.338230	0.386615	0.342808	0.283244	0.363021	0.404354
PCA 3	0.157609	0.161323	−0.103135	0.006767	0.226125	−0.358278	−0.615519	0.073713	0.612094
PCA 4	0.044681	0.011848	−0.069971	0.090291	−0.633560	−0.438259	0.344247	0.463711	0.240442
PCA 5	0.571038	0.143893	−0.731120	−0.148991	0.161062	0.025766	0.172102	0.073443	−0.186704
PCA 6	0.085901	−0.004662	−0.067071	−0.301431	−0.281197	0.308307	0.264687	−0.553969	0.588366
PCA 7	−0.076892	−0.036997	0.134514	−0.155125	0.497821	−0.618179	0.507327	−0.246778	0.048065
PCA 8	0.068572	0.059992	−0.169358	0.829896	−0.117622	−0.150630	0.013344	−0.485437	−0.044016
PCA 9	0.565673	−0.777720	0.257238	0.065265	0.008201	−0.026038	−0.057528	0.007880	0.024873
(B)									
Eigenvector	Band 1	Band 2	Band 3	Band 4	Band 5	Band 6	Band 7	Band 8	Band 9
PCA 1	−0.479262	−0.517939	−0.509952	−0.225048	−0.160278	−0.238705	−0.238099	−0.191397	−0.123923
PCA 2	0.288066	0.267422	0.266615	−0.388911	−0.371102	−0.336818	−0.273033	−0.365435	−0.404865
PCA 3	−0.470308	−0.335548	0.711152	−0.063002	−0.061806	0.155815	0.278307	−0.089651	−0.207174
PCA 4	0.089149	0.081563	−0.378102	0.066965	−0.262842	0.327757	0.541766	0.014714	−0.606397
PCA 5	0.005198	−0.011714	0.013230	−0.289525	−0.571447	−0.297938	0.372870	0.457705	0.389860
PCA 6	−0.143194	0.033396	0.100700	0.637725	−0.213061	−0.346181	−0.285287	0.440341	−0.347880
PCA 7	0.048756	−0.037063	−0.000810	−0.381573	0.612809	−0.398723	0.186662	0.403665	−0.344018
PCA 8	0.019200	−0.050602	0.037382	−0.368910	−0.118495	0.574367	−0.492425	0.505200	−0.133907
PCA 9	−0.659478	0.731979	−0.091948	−0.132052	0.046537	0.019598	−0.007121	−0.002060	0.028260
(C)									
Eigenvector	Band 1	Band 2	Band 3	Band 4	Band 5	Band 6	Band 7	Band 8	Band 9
PCA 1	0.436596	0.474941	0.478323	0.259162	0.202467	0.288422	0.293737	0.233995	0.150469
PCA 2	−0.361523	−0.339702	−0.304584	0.341346	0.339529	0.323638	0.270272	0.338853	0.369764
PCA 3	−0.113406	−0.131491	0.013472	−0.038953	−0.199768	0.301020	0.615036	−0.041133	−0.676513
PCA 4	0.646205	0.069871	−0.741104	0.063857	0.083737	0.043079	0.029404	0.011287	−0.119854
PCA 5	−0.081856	0.003684	0.073346	0.258757	0.617458	0.257579	−0.353971	−0.466145	−0.361612
PCA 6	−0.081361	0.054497	0.026637	0.134626	0.188319	−0.340347	−0.252269	0.722429	−0.484494
PCA 7	−0.038421	0.062948	0.062948	0.128723	−0.501280	0.651852	−0.509411	0.194283	−0.050948
PCA 8	0.065332	−0.071443	0.035478	−0.833862	0.356670	0.335182	−0.066825	0.222638	0.021186
PCA 9	−0.478726	0.790450	−0.345997	−0.117250	0.039925	0.012703	0.092941	−0.045796	0.007618

Table A1. Cont.

				(D)					
Eigenvector	Band 1	Band 2	Band 3	Band 4	Band 5	Band 6	Band 7	Band 8	Band 9
PCA 1	0.457469	0.496121	0.484720	0.241961	0.190767	0.264428	0.267634	0.222557	0.153665
PCA 2	−0.318221	−0.311611	−0.309208	0.348166	0.377922	0.352360	0.263388	0.321127	0.381217
PCA 3	0.050573	0.100867	0.065376	−0.074127	0.348897	−0.190996	−0.603883	−0.184339	0.648564
PCA 4	−0.576526	−0.199530	0.760114	−0.121149	−0.074678	−0.014245	0.117300	−0.037045	0.120188
PCA 5	0.062048	0.065969	−0.120679	−0.198503	−0.582283	−0.293206	0.201001	0.452767	0.517116
PCA 6	0.076383	0.014735	−0.110901	−0.152342	−0.282329	0.477741	0.275588	−0.676196	0.342499
PCA 7	0.082735	−0.021608	−0.059222	−0.230230	0.454540	−0.555810	0.600908	−0.236367	0.060691
PCA 8	0.018744	0.033687	−0.043334	−0.827212	0.260085	0.381570	−0.039539	0.303178	−0.075067
PCA 9	−0.581066	0.774987	−0.234825	0.012884	0.021499	−0.022616	0.039495	−0.044352	−0.044133

Table A2. Eigenvector matrix for TIR bands of ASTER. (**A**): Spatial subset covering zone 1 and surrounding areas; (**B**): Spatial subset covering zones 2 and 3; (**C**): Spatial subset covering zone 4 and surrounding areas; and (**D**): Spatial subset covering zones 5 and 6.

			(A)		
Eigenvector	Band 10	Band 11	Band 12	Band 13	Band 14
PCA 1	−0.387065	−0.400716	−0.418903	−0.501105	−0.512857
PCA 2	0.124757	−0.148619	−0.793719	0.106037	0.566669
PCA 3	−0.639764	−0.511314	0.328506	0.186276	0.432023
PCA 4	0.630731	−0.744700	0.169973	0.079875	−0.111041
PCA 5	0.165821	0.036321	0.240257	−0.834608	0.465714

			(B)		
Eigenvector	Band 10	Band 11	Band 12	Band 13	Band 14
PCA 1	−0.393525	−0.404088	−0.412766	−0.496019	−0.515209
PCA 2	0.068174	0.244757	0.733503	−0.274307	−0.567605
PCA 3	−0.739002	−0.349952	0.467340	0.209083	0.263226
PCA 4	−0.528835	0.806255	−0.213160	−0.138353	0.075548
PCA 5	0.121244	−0.065522	0.166595	−0.784770	0.580852

			(C)		
Eigenvector	Band 10	Band 11	Band 12	Band 13	Band 14
PCA 1	−0.387916	−0.409696	−0.430823	−0.491519	−0.504450
PCA 2	0.059723	0.202526	0.734577	−0.249602	−0.594568
PCA 3	−0.749770	−0.351003	0.439774	0.248107	0.244302
PCA 4	0.532694	−0.814304	0.224886	0.010234	0.049680
PCA 5	−0.006039	0.069557	0.175536	−0.796521	0.574340

			(D)		
Eigenvector	Band 10	Band 11	Band 12	Band 13	Band 14
PCA 1	−0.386572	−0.407245	−0.432094	−0.491312	−0.506577
PCA 2	0.007595	−0.170349	−0.765517	0.204563	0.585714
PCA 3	0.667689	0.466491	−0.419996	−0.168193	−0.363168
PCA 4	0.605752	−0.766295	0.140070	0.132069	−0.093781
PCA 5	0.194321	-0.017612	0.176809	−0.819167	0.509541

LSU mathematical equation (Equation (A2)):

$$R_i = \sum_{i=1}^{N} F_e R_e + E_i \quad (A2)$$

where, R_i = Surface reflectance in band i of the sensor; F_e = Fraction of end-member e; R_e = Reflectance of end-member e in the sensor wave band i; N = Number of spectral end-members; and E_i = Error in the sensor band i for the fit of N end-members [100].

Table A3. Locations of alteration zones and listvenites recorded by Global Positioning System (GPS) readings.

Altered Rock Types	Coordinates
Listvenites	71°36.026′S–163°16.302′E
Listvenites	71°36.011′S–163°16.364′E
Carbonitization	70°47.459′S–162°39.438′E
Carbonitization	70°47.349′S–162°38.688′E
Listvenites + sulfides+ albitites	71°33.638′S–163°11.716′E
Epidotization + fault zone	71°23.702′S–162°47.968′E
Reddish to greenish alteration zone	71°19.056′S–162°35.167′E
Epidotization + chlorite in granitoids	71°31.135′S–162°59.086′E
Epidotization + prehnite in granite	71°49.569′S–161°18.145′E
Epidotization + prehnite in granite	71°29.123′S–162°38.567′E
Epidotization in granite	71°44.753′S–162°59.358′E
Epidote + serpentine + talc in high grade mafic-ultramafic rocks	71°27.635′S–162°53.098′E
Listvenite + serpentine in ultramafic rocks	71°34.546′S–163°11.185′E
Listvenite + carbonates in volcanoclastic rocks	71°32.738′S–163°31.602′E
Listvenite + carbonates + talc	71°36.768′S–163°16.407′E
Carbonitization + silica in volcanoclastic rocks	71°33.075′S–163° 31.949′E
Epidotization in Glasgow Volcanics	71°11.159′S–163°00.378′E
Epidote + prehnite + quartz in volcanoclastic rocks	71°23.048′S–162°48.796′E
Listvenite + quartz + hydraulic breccia in Glasgow Volcanics	71°27.078′S–163°26.689′E
Epidote coating fault in Glasgow Volcanics	71°11.292′S–162°35.369′E
Listvenites in ultramafic rocks	71°38.525′S–162°24.458′E

Table A4. Confusion matrix for LSU classification mineral maps versus field data.

Class	Listvenites	Carbonitization	Epidotization	Totals (Field data)	User's Accuracy
Listvenites	5	0	1	6	83.33%
Carbonitization	1	5	1	7	71.42%
Epidotization	1	2	5	8	62.50%
Totals (LSU maps)	7	7	7	21	
Producer's Accuracy	71.42%	71.42%	71.42%		

Overall accuracy = 71.42%, Kappa Coefficient = 0.57

References

1. Crispini, L.; Federico, L.; Giovanni, C.; Talarico, F. The Dorn gold deposit in northern Victoria Land, Antarctica: Structure, hydrothermal alteration, and implications for the Gondwana Pacific margin. *Gondwana Res.* **2011**, *19*, 128–140. [CrossRef]
2. Goldfarb, R.J.; Taylor, R.D.; Collins, G.S.D.; Goryachev, N.A.; Orlandini, O.F. Phanerozoic continental growth and gold metallogeny of Asia. *Gondwana Res.* **2014**, *25*, 48–102. [CrossRef]
3. Zaw, K.; Meffre, S.; Lai, C.K.; Burrett, C.; Santosh, M.; Graham, I.; Manaka, T.; Salam, A.; Kamvong, T.; Cromie, P. Tectonics and metallogeny of mainland Southeast Asia-A reviewand contribution. *Gondwana Res.* **2014**, *26*, 5–30.
4. Zoheir, B.; Emam, A.; El-Amawy, M.; Abu-Alam, T. Auriferous shear zones in the central Allaqi-Heiani belt: Orogenic gold in post-accretionary structures, SE Egypt. *J. Afr. Earth Sci.* **2017**, *146*, 118–131. [CrossRef]
5. Hewson, R.D.; Robson, D.; Carlton, A.; Gilmore, P. Geological application of ASTER remote sensing within sparsely outcropping terrain, Central New South Wales, Australia. *Cogent Geosci.* **2017**, *3*, 1319259. [CrossRef]
6. Pour, A.B.; Park, Y.; Park, T.S.; Hong, J.K.; Hashim, M.; Woo, J.; Ayoobi, I. Evaluation of ICA and CEM algorithms with Landsat-8/ASTER data for geological mapping in inaccessible regions. *Geocarto Int.* **2019**, *34*, 785–816. [CrossRef]

7. Pour, A.B.; Park, Y.; Park, T.S.; Hong, J.K.; Hashim, M.; Woo, J.; Ayoobi, I. Regional geology mapping using satellite-based remote sensing approach in Northern Victoria Land, Antarctica. *Polar Sci.* **2018**, *16*, 23–46. [CrossRef]
8. Pour, A.B.; Park, T.S.; Park, Y.; Hong, J.K.; Zoheir, B.; Pradhan, B.; Ayoobi, I.; Hashim, M. Application of multi-sensor satellite data for exploration of Zn-Pb sulfide mineralization in the Franklinian Basin, North Greenland. *Remote Sens.* **2018**, *10*, 1186. [CrossRef]
9. Pour, A.B.; Hashim, M.; Park, Y.; Hong, J.K. Mapping alteration mineral zones and lithological units in Antarctic regions using spectral bands of ASTER remote sensing data. *Geocarto Int.* **2018**, *33*, 1281–1306. [CrossRef]
10. Testa, F.J.; Villanueva, C.; Cooke, D.R.; Zhang, L. Lithological and hydrothermal alteration mapping of epithermal, porphyry and tourmaline breccia districts in the Argentine Andes using ASTER imagery. *Remote Sens.* **2018**, *10*, 203. [CrossRef]
11. Sheikhrahimi, A.; Pour, B.A.; Pradhan, B.; Zoheir, B. Mapping hydrothermal alteration zones and lineaments associated with orogenic gold mineralization using ASTER remote sensing data: A case study from the Sanandaj-Sirjan Zone, Iran. *Adv. Space Res.* **2019**, *63*, 3315–3332. [CrossRef]
12. Noori, L.; Pour, B.A.; Askari, G.; Taghipour, N.; Pradhan, B.; Lee, C.-W.; Honarmand, M. Comparison of Different Algorithms to Map Hydrothermal Alteration Zones Using ASTER Remote Sensing Data for Polymetallic Vein-Type Ore Exploration: Toroud–Chahshirin Magmatic Belt (TCMB), North Iran. *Remote Sens.* **2019**, *11*, 495. [CrossRef]
13. Spiridonov, E.M. Listvenites and zodites. *Int. Geol. Rev.* **1991**, *33*, 397–407. [CrossRef]
14. Halls, C.; Zhao, R. Listvenite and related rocks: Perspectives on terminology and mineralogy with reference to an occurrence at Cregganbaun, Co. Mayo, Republic of Ireland. *Mineral. Deposita* **1995**, *30*, 303–313. [CrossRef]
15. Uçurum, A. Listwaenites in Turkey: Perspectives on formation and precious metal concentration with reference to occurrences in East-Central Anatolia. *Ofioliti* **2000**, *25*, 15–29.
16. Akbulut, M.; Piskin, O.; Karayigit, A.I. The genesis of the carbonatized and silicified ultramafics known as listvenites: A case study from the Mihaliccik region (Eskisehir) NW Turkey. *Geol. J.* **2006**, *41*, 557–580. [CrossRef]
17. Buisson, G.; Leblanc, M. Gold bearing listwaenites (carbonatized ultramafic rocks) inophiolite complexes. In *Metallogeny of Basic and Ultrabasic Rocks*; Gallagher, M.J., Ixer, R.A., Neary, C.R., Prichard, H.M., Eds.; The Institution of Mining and Metallurgy: London, UK, 1986; pp. 121–132.
18. Buisson, G.; Leblanc, M. Gold in mantle peridotites from Upper Proterozoic ophiolites in Arabia, Mali, and Morocco. *Econ. Geol.* **1987**, *82*, 2091–2097. [CrossRef]
19. Hansen, L.D.; Dipple, G.M.; Gordon, T.M.; Kellett, D.A. Carbonated serpentinite (listwanite) at Atlin, British Columbia: A geological analogue to carbon dioxide sequestration. *Can. Mineral.* **2005**, *43*, 225–239. [CrossRef]
20. Zoheir, B.; Lehmann, B. Listvenite-lode association at the Barramiya gold mine, Eastern Desert, Egypt. *Ore Geol. Rev.* **2011**, *39*, 101–115. [CrossRef]
21. Azer, M.K. Evolution and economic significance of listwaenites associated with Neoproterozoic ophiolites in south Eastern Desert, Egypt. *Geol. Acta* **2013**, *11*, 113–128.
22. Kuzhuget, R.V.; Zaikov, V.V.; Lebedev, V.I.; Mongush, A.A. Gold mineralization of the Khaak-Sair gold-quartz ore occurrence in listwanites (western Tuva). *Russ. Geol. Geophys.* **2015**, *56*, 1332–1348. [CrossRef]
23. Belogub, E.V.; Melekestseva, I.Y.; Novoselov, K.A.; Zabotina, M.V.; Tret'yakov, G.A.; Zaykov, V.V.; Yuminov, A.M. Listvenite-related gold deposits of the South Urals (Russia): A review. *Ore Geol. Rev.* **2017**, *85*, 247–270. [CrossRef]
24. Abdel-Karim, A.A.M.; El-Shafei, S.A. Mineralogy and chemical aspects of some ophiolitic metaultramafics, central Eastern Desert, Egypt: Evidences from chromites, sulphides and gangues. *Geol. J.* **2017**, *53*, 580–599. [CrossRef]
25. Falk, E.S.; Kelemen, P.B. Geochemistry and petrology of listvenite in the Samail ophiolite, Sultanate of Oman: Complete carbonation of peridotite during ophiolite emplacement. *Geochim. et Cosmochim. Acta* **2015**, *160*, 70–90. [CrossRef]

26. Ferenc, S.; Uher, P.; Spišiak, J.; Šimonová, V. Chromium- and nickel-rich micas and associated minerals in listvenite from the Muránska Zdychava, Slovakia: Products of hydrothermal metasomatic transformation of ultrabasic rock. *J. Geosci.* **2016**, *61*, 239–254. [CrossRef]
27. Pour, B.A.; Hashim, M. The application of ASTER remote sensing data to porphyry copper and epithermal gold deposits. *Ore Geol. Rev.* **2012**, *44*, 1–9. [CrossRef]
28. Zoheir, B.; Emam, A. Field and ASTER imagery data for the setting of gold mineralization in Western Allaqi-Heiani belt, Egypt: A case study from the Haimur. *J. Afr. Earth Sci.* **2014**, *66*, 22–34. [CrossRef]
29. Gabr, S.S.; Hassan, S.M.; Sadek, M.F. Prospecting for new gold-bearing alteration zones at El-Hoteib area, South Eastern Desert, Egypt, using remote sensing data analysis. *Ore Geol. Rev.* **2015**, *71*, 1–13. [CrossRef]
30. Pour, A.B.; Hashim, M.; Hong, J.K.; Park, Y. Lithological and alteration mineral mapping in poorly exposed lithologies using Landsat-8 and ASTER satellite data: North-eastern Graham Land, Antarctic Peninsula. *Ore Geol. Rev.* **2019**, *108*, 112–133. [CrossRef]
31. Noda, S.; Yamaguchi, Y. Estimation of surface iron oxide abundance with suppression of grain size and topography effects. *Ore Geol. Rev.* **2017**, *83*, 312–320. [CrossRef]
32. Hunt, G.R. Spectral signatures of particulate minerals in the visible and near-infrared. *Geophysics* **1977**, *42*, 501–513. [CrossRef]
33. Hunt, G.R.; Ashley, R.P. Spectra of altered rocks in the visible and near-infrared. *Econ. Geol.* **1979**, *74*, 1613–1629. [CrossRef]
34. Clark, R.N. Spectroscopy of rocks and minerals, and principles of spectroscopy. In *Manual of Remote Sensing*; Rencz, A., Ed.; Wiley and Sons Inc.: New York, NY, USA, 1999; Volume 3, pp. 3–58.
35. Cloutis, E.A.; Hawthorne, F.C.; Mertzman, S.A.; Krenn, K.; Craig, M.A.; Marcino, D.; Methot, M.; Strong, J.; Mustard, J.F.; Blaney, D.L.; et al. Detection and discrimination of sulfate minerals using reflectance spectroscopy. *Icarus* **2006**, *184*, 121–157. [CrossRef]
36. Abrams, M.; Hook, S.J. Simulated ASTER data for geologic studies. *IEEE Trans. Geosci. Remote Sens.* **1995**, *33*, 692–699. [CrossRef]
37. Abrams, M.; Hook, S.; Ramachandran, B. ASTER User Handbook, Version 2. Jet Propulsion Laboratory, California Institute of Technology, 2004. Available online: http://asterweb.jpl.nasa.gov/content/03_data/04_Documents/aster_guide_v2.pdf (accessed on 21 September 2015).
38. Salisbury, J.W.; D'Aria, D.M. Emissivity of terrestrial material in the 8–14 μm atmospheric window. *Remote Sens. Environ.* **1992**, *42*, 83–106. [CrossRef]
39. Salisbury, J.W.; Walter, L.S. Thermal infrared (2.5–13.5 μm) spectroscopic remote sensing of igneous rock types on particulate planetary surfaces. *J. Geophys. Res.* **1989**, *94*, 9192–9202. [CrossRef]
40. Ninomiya, Y. Quantitative estimation of SiO_2 content in igneous rocks using thermal infrared spectra with a neural network approach. *IEEE Trans. Geosci.Remote Sens.* **1995**, *33*, 684–691. [CrossRef]
41. Ninomiya, Y.; Fu, B. Thermal infrared multispectral remote sensing of lithology and mineralogy based on spectral properties of materials. *Ore Geol. Rev.* **2019**, *108*, 54–72. [CrossRef]
42. Ninomiya, Y.; Fu, B. Regional lithological mapping using ASTER-TIR data: Case study for the Tibetan Plateau and the surrounding area. *Geosciences* **2016**, *6*, 39. [CrossRef]
43. Rajendran, S.; Nasir, S.; Kusky, T.M.; Ghulam, A.; Gabr, S.; El-Ghali, M.A.K. Detection of hydrothermal mineralized zones associated with listwaenites in Central Oman using ASTER data. *Ore Geol. Rev.* **2013**, *53*, 470–488. [CrossRef]
44. Crispini, L.; Capponi, G. Albitite and listvenite in the Lanterman Fault Zone (northern Victoria Land, Antarctica). In *Antarctica at the Close of a Millennium*; Gamble, J., Skinner, D.N.B., Henrys, S., Eds.; Ministry of Education: Wellington, New Zealand, 2002; Volume 35, pp. 113–119.
45. Goodge, J.W.; Fanning, C.M.; Norman, M.D.; Bennet, V. Temporal, Isotopic and Spatial Relations of Early Paleozoic Gondwana-Margin Arc Magmatism, Central Transantarctic Mountains, Antarctica. *J. Petrol.* **2012**, *53*, 2027–2065. [CrossRef]
46. Godard, G.; Palmeri, R. High-pressure metamorphism in Antarctica from the Proterozoic to the Cenozoic: A review and geodynamic implications. *Gondwana Res.* **2013**, *23*, 844–864. [CrossRef]
47. Crispini, L.; Federico, L.; Capponi, G. Structure of the Millen Schist Belt (Antarctica): Clues for the tectonics of northern Victoria Land along the paleo-Pacific margin of Gondwana. *Tectonics* **2014**, *33*, 420–440. [CrossRef]
48. Estrada, S.; Läufer, A.; Eckelmann, K.; Hofmann, M.; Gärtner, A.; Linnemann, U. Continuous Neoproterozoic to Ordovician sedimentation at the East Gondwana margin - Implications from detrital zircons of the Ross Orogen in northern Victoria Land, Antarctica. *Gondwana Res.* **2016**, *37*, 426–448. [CrossRef]

49. Capponi, G.; Crispini, L.; Meccheri, M. The metaconglomerates of the eastern Lanterman Range (northern Victoria Land, Antarctica): New constraints for their interpretation. *Antarct. Sci.* **1999**, *11*, 215–225. [CrossRef]
50. Crispini, L.; Vincenzo, G.D.; Palmeri, R. Petrology and 40Ar–39Ar dating of shear zones in the Lanterman Range (northern Victoria Land, Antarctica): Implications for metamorphic and temporal evolution at terrane boundaries. *Mineral. Petrol.* **2007**, *89*, 217–249. [CrossRef]
51. Federico, L.; Crispini, L.; Capponi, G. Fault-slip analysis and transpressional tectonics: A study of Paleozoic structures in northern Victoria Land, Antarctica. *J. Struct. Geol.* **2010**, *32*, 667–684. [CrossRef]
52. Pertusati, P.C.; Ricci, C.A.; Tessensohn, F. German-Italian Geological Antarctic Map Programme – the Italian Contribution. Introductory Notes to the Map Case. *Terra Antart. Rep.* **2016**, *15*, 1–15.
53. Kleinschmidt, G.; Tessensohn, F. Early Paleozoic westward directed subduction at the Pacific continental margin of Antarctica. In *Gondwana Six: Structure, Tectonics, and Geophysics*; McKenzie, G., Ed.; AGU Geophysical Monograph Series; American Geophysical Union: Washington, DC, USA, 1987; Volume 40, pp. 89–105.
54. Federico, L.; Capponi, G.; Crispini, L. The Ross orogeny of the transantarctic mountains: A northern Victoria Land perspective. *Int. J. Earth Sci.* **2006**, *95*, 759–770. [CrossRef]
55. Federico, L.; Crispini, L.; Capponi, G.; Bradshaw, J.D. The Cambrian Ross Orogeny in northern Victoria Land (Antarctica) and New Zealand: A synthesis. *Gondwana Res.* **2009**, *15*, 188–196. [CrossRef]
56. Stump, E. *The Ross Orogen of the Transantarctic Mountains*; Cambridge University Press: New York, NY, USA, 1995.
57. Paulsen, T.; Deering, C.; Sliwinski, J.; Bachmann, O.; Guillong, M. A continental arc tempo discovered in the Pacific-Gondwana margin mudpile? *Geology* **2016**, *44*, 915–918. [CrossRef]
58. Menneken, M.; John, T.; Läufer, A.; Giese, J. Zircons from the Granite Harbour Intrusives, northern Victora Land, Antarctica. In Proceedings of the POLAR 2018, Open Science Conference, Davos, Switzerland, 19–23 June 2018.
59. Jordan, H.; Findlay, R.; Mortimer, G.; Schmidt-Thome, M.; Crawford, A.; Muller, P. Geology of the northern Bowers Mountains, North Victoria Land, Antarctica. *Geol. Jahrb.* **1984**, *60*, 57–81.
60. Rocchi, S.; Capponi, G.; Crispini, L.; Di Vincenzo, G.; Ghezzo, C.; Meccheri, M.; Palmeri, R. Mafic rocks at the WilsoneBowers terrane boundary and within the Bowers Terrane: Clues to the Ross geodynamics in northern Victoria Land, Antarctica. In Proceedings of the 9th International Symposium on Antarctic Earth Sciences, Potsdam, Germany, 8–12 September 2003.
61. Wright, T.O.; Ross, R.J., Jr.; Repetski, J.E. Newly discovered youngest Cambrian or oldest Ordovician fossils from the Robertson Bay terrane (formerly Precambrian), northern Victoria Land, Antarctica. *Geology* **1984**, *12*, 301–305. [CrossRef]
62. Roland, N.W.; Läufer, A.L.; Rossetti, F. Revision of the Terrane Model of Northern Victoria Land (Antarctica). *Terra Antart.* **2004**, *11*, 55–65.
63. Goodge, J.W. Metamorphism in the Ross orogen and its bearing on Gondwana margin tectonics. In *Convergent Margin Terranes and Associated Regions: A Tribute to W.G.*; Cloos, M., Carlson, W.D., Gilbert, M.C., Liou, J.G., Sorensen, S.S., Eds.; Geological Society of America Special Paper: Boulder, CO, USA, 2007; Volume 419, pp. 185–203.
64. Ricci, C.A.; Tessensohn, F. The Lanterman-Mariner suture: Antarctic evidence for active margin tectonics in Paleozoic Gondwana. In *Geologisches Jahrbuch*; Tessensohn, F., Ricci, C.A., Eds.; Schweizerbart Science Publishers: Stuttgart, Germany, 2003; pp. 303–332.
65. Phillips, G.; Läufer, A.; Piepjohn, K. Geology of the Millen Thrust System, northern Victoria Land, Antarctica. *Polarforschung* **2014**, *84*, 39–47.
66. Rossetti, F.; Storti, F.; Läufer, A.L. Brittle architecture of the Lanterman Fault and its impact on the final terrane amalgamation in north Victoria Land, Antarctica. *J. Geol. Soc.* **2002**, *159*, 159–173. [CrossRef]
67. Borg, S.G.; Stump, E. Paleozoic magmatism and associated tectonic problems of Northern Victoria Land, Antarctica. In *Gondwana Six: Structure, Tectonics and Geophysics*; McKenzie, G., Ed.; Geophysical Monograph Series; American Geophysical Union: Washington, DC, USA, 1987; Volume 40, pp. 67–76.
68. Collinson, J.W. The palaeo-Pacific margin as seen from East Antarctica. In *Geological Evolution of Antarctica*; Thomson, M.R.A., Crame, J.A., Thomson, J.W., Eds.; Cambridge University Press: New York, NY, USA, 1991; pp. 199–204.
69. Schöner, R.; Bomfleur, B.; Schneider, J.; Viereck-Götte, L. A Systematic Description of the Triassic to Lower Jurassic Section Peak Formation in North Victoria Land (Antarctica). *Polarforschung* **2011**, *80*, 71–87.

70. Grindley, G.W. The geology of the Queen Alexandra Range, Beardmore Glacier, Ross Dependency, Antarctica; with notes on the correlation of Gondwana sequences. *N. Z. J. Geol. Geophys.* **1963**, *6*, 307–347. [CrossRef]
71. Rossetti, F.; Lisker, F.; Storti, F.; Läufer, A. Tectonic and denudational history of the Rennick Graben (North Victoria Land): Implications for the evolution of rifting between East and West Antarctica. *Tectonics* **2003**, *22*, 1016. [CrossRef]
72. Kleinschmidt, G.; Läufer, A.L. The Matusevich Fracture Zone in Oates Land, East Antarctica. In *Antarctica: Contributions to Global Earth Sciences*; Fütterer, D.K., Damaske, D., Kleinschmidt, G., Miller, H., Tessensohn, F., Eds.; Springer: Berlin/Heidelberg, Germany; New York, NY, USA, 2006; pp. 175–180.
73. Laird, M.G. Evolution of the Cambrian-Early Ordovician Bowers Basin, North Victoria Land, and its relationships with the adjacent wilson and Robertson Bay Terrane. *Mem. Della Soc. Geol. Ital.* **1987**, *33*, 25–34.
74. Laird, M.G.; Bradshaw, J.D. Uppermost Proterozoic and lower Paleozoic geology of the Transantarctic Mountains. In *Antarctic Geosciences*; Craddock, C., Ed.; University of Wisconsin Press: Madison, WI, USA, 1982; pp. 525–533.
75. Weaver, S.D.; Bradshaw, J.D.; Laird, M.G. Geochemistry of Cambrian volcanics of the Bowers Supergroup and implications for the Early Paleozoic tectonic evolution of northern Victoria Land. *Antarct. Earth Planet. Sci. Lett.* **1984**, *68*, 128–140. [CrossRef]
76. Wodzicki, A.; Ray, J.R.R. Geology of the Bowers Supergroup, central Bowers Mountains, northern Victoria Land. In *Geological Investigation in Northern Victoria Land*; Stump, E., Ed.; Antarctic Research Series; AGU: Washington, DC, USA, 1986; Volume 46, pp. 39–68.
77. Capponi, G.; Crispini, L.; Meccheri, M. Structural history and tectonic evolution of the boundary between the Wilson and Bowers terranes, Lanterman Range, northern Victoria Land. *Antarct. Tectonophys.* **1999**, *312*, 249–266. [CrossRef]
78. Ricci, C.A.; Talarico, F.; Palmeri, R.; Di Vincenzo, G.; Pertusati, P.C. Eclogite at the Antarctic palaeo-Pacific active margin of Gondwana (Lanterman Range, northern Victoria Land, Antarctica). *Antarct. Sci.* **1996**, *8*, 277–280. [CrossRef]
79. Cooley, T.; Anderson, G.P.; Felde, G.W.; Hoke, M.L.; Ratkowski, A.J.; Chetwynd, J.H.; Gardner, J.A.; Adler-Golden, S.M.; Matthew, M.W.; Berk, A.; et al. FLAASH, a MODTRAN4-based atmospheric correction algorithm, its application and validation. *IEEE Int. Geosci. Remote Sens. Symp.* **2002**, *3*, 1414–1418.
80. Kruse, F.A. Comparison of ATREM, ACORN, and FLAASH Atmospheric Corrections using Low-Altitude AVIRIS Data of Boulder, Colorado. In Proceedings of the 13th JPL Airborne Geoscience Workshop, JPL Publication 05-3, Jet Propulsion Laboratory, Pasadena, CA, USA, 31 March–2 April 2004.
81. Research Systems, Inc. *ENVI Tutorials*; Research Systems, Inc.: Boulder, CO, USA, 2008.
82. Iwasaki, A.; Tonooka, H. Validation of a crosstalk correction algorithm for ASTER/SWIR. *IEEE Trans. Geosci. Remote Sens.* **2005**, *43*, 2747–2751. [CrossRef]
83. Gupta, R.P.; Haritashya, U.K.; Singh, P. Mapping dry/wet snow cover in the Indian Himalayas using IRS multispectral imagery. *Remote Sens. Environ.* **2005**, *97*, 458–469. [CrossRef]
84. Hall, D.K.; Riggs, G.A.; Salomonson, V.V.; DiGirolamo, N.E.; Bayr, K.J. MODIS snow-cover products. *Remote Sens. Environ.* **2002**, *83*, 181–1194. [CrossRef]
85. Singh, A.; Harrison, A. Standardized principal components. *Int. J. Remote Sens.* **1985**, *6*, 883–896. [CrossRef]
86. Jensen, J.R. *Introductory Digital Image Processing*; Pearson Prentice Hall: Upper Saddle River, NJ, USA, 2005.
87. Chang, Q.; Jing, L.; Panahi, A. Principal component analysis with optimum order sample correlation coefficient for image enhancement. *Int. J. Remote Sens.* **2006**, *27*, 3387–3401.
88. Gupta, R.P. *Remote Sensing Geology*, 3rd ed.; Springer: Berlin, Germany, 2017; pp. 180–190.
89. Loughlin, W.P. Principal components analysis for alteration mapping. *Photogramm. Eng. Remote Sens.* **1991**, *57*, 1163–1169.
90. Crosta, A.; Moore, J. Enhancement of Landsat Thematic Mapper imagery for residual soil mapping in SW Minais Gerais State, Brazil: A prospecting case history in Greenstone belt terrain. In Proceedings of the 7th ERIM Thematic Conference: Remote Sensing for Exploration Geology, Calgary, AB, Canada, 2–6 October 1989; pp. 1173–1187.
91. Gupta, R.P.; Tiwari, R.K.; Saini, V.; Srivastava, N. A simplified approach for interpreting principal component images. *Adv. Remote Sens.* **2013**, *2*, 111–119. [CrossRef]
92. Hyvarinen, A.; Oja, E. Independent component analysis: Algorithms and applications. *Neural Netw.* **2000**, *13*, 411–430. [CrossRef]

93. Hyvärinen, A. Independent component analysis: Recent advances. *Philos. Trans. A Math. Phys. Eng. Sci.* **2013**, *371*, 20110534. [CrossRef] [PubMed]
94. Hyvärinen, A.; Zhang, K.; Shimizu, S.; Hoyer, P.O. Estimation of a structural vector autoregression model using non-Gaussianity. *J. Mach. Learn. Res.* **2010**, *11*, 1709–1731.
95. Shimizu, S. Joint estimation of linear non-Gaussian acyclic models. *Neurocomputing* **2012**, *81*, 104–107. [CrossRef]
96. Hyvärinen, A.; Karhunen, J.; Oja, E. *Independent Component Analysis*; A Wiley-Interscience Publication; John Wiley & Sons, Inc.: New York, NY, USA, 2001; pp. 1–12.
97. Boardman, J.W. Inversion of imaging spectrometry data using singular value decomposition. In Proceedings of the IGARSS'89 12th Canadian Symposium on Remote Sensing, Vancouver, BC, Canada, 10–14 July 1989; Volume 4, pp. 2069–2072.
98. Boardman, J.W. Sedimentary Facies Analysis Using Imaging Spectrometry: A Geophysical Inverse Problem. Ph.D. Thesis, University of Colorado, Boulder, CO, USA, 1992; p. 212.
99. Adams, J.B.; Smith, M.O.; Gillespie, A.R. Imaging spectroscopy: Interpretation based on spectral mixture analysis. In *Remote Geochemical Analysis: Elemental and Mineralogical Composition*; Pieters, C.M., Englert, P.A.J., Eds.; Cambridge University Press: New York, NY, USA, 1993; pp. 145–166.
100. Adams, J.B.; Sabol, D.E.; Kapos, V.; Filho, R.A.; Roberts, D.A.; Smith, M.O.; Gillespie, A.R. Classification of multispectral images based on fractions of endmembers: Application to land-cover change in the Brazilian Amazon. *Remote Sens. Environ.* **1995**, *52*, 137–154. [CrossRef]
101. Kruse, F.A.; Bordman, J.W.; Huntington, J.F. Comparison of airborne hyperspectral data and EO-1 Hyperion for mineral mapping. *IEEE Trans. Geosci. Remote Sens.* **2003**, *41*, 1388–1400. [CrossRef]
102. Kruse, F.A.; Perry, S.L. Regional mineral mapping by extending hyperspectral signatures using multispectral data. *IEEE Trans. Geosci. Remote Sens.* **2007**, *4*, 1–14.
103. Green, A.A.; Berman, M.; Switzer, P.; Craig, M.D. A transformation for ordering multispectral data in terms of image quality with implications for noise removal. *IEEE Trans. Geosci. Remote Sens.* **1988**, *26*, 65–74. [CrossRef]
104. Boardman, J.W.; Kruse, F.A. Automated spectral analysis: A geologic example using AVIRIS data, north Grapevine Mountains, Nevada. In Proceedings of the Tenth Thematic Conference on Geologic Remote Sensing, Environmental Research Institute of Michigan, Ann Arbor, MI, USA, 9–12 May 1994; pp. I-407–I-418.
105. Boardman, J.W.; Kruse, F.A.; Green, R.O. Mapping target signatures via partial unmixing of AVIRIS data. In *Summaries, Fifth JPL Airborne Earth Science Workshop*; JPL Publication: Pasadena, CA, USA, 1995; Volume 1, pp. 23–26.
106. Boardman, J.W. Automated spectral unmixing of AVIRIS data using convex geometry concepts. In *Summaries, Fourth JPL Airborne Geoscience Workshop*; JPL Publication: Pasadena, CA, USA, 1993; Volume 1, pp. 11–14.
107. Milliken, R.E.; Mustard, J.F. Estimating the water content of hydrated minerals using reflectance spectroscopy I. Effects of darkening agents and low-albedo materials. *Icarus* **2007**, *189*, 550–573.
108. Bishop, J.L.; Lane, M.D.; Dyar, M.D.; Brwon, A.J. Reflectance and emission spectroscopy study of four groups of phyllosilicates: Smectites, kaolinite-serpentines, chlorites and micas. *Clay Miner.* **2008**, *43*, 35–54. [CrossRef]
109. Farrand, W.H.; Harsanyi, J.C. Mapping the distribution of mine tailings in the Coeur d'Alene River Valley, Idaho, through the use of a constrained energy minimization technique. *Remote Sens. Environ.* **1997**, *59*, 64–76. [CrossRef]
110. Chang, C.I.; Heinz, D.C. Constrained subpixel target detection for remotely sensed imagery. *IEEE Trans. Geosci. Remote Sens.* **2000**, *38*, 1144–1159. [CrossRef]
111. Manolakis, D.; Marden, D.; Shaw, G.A. Hyperspectral Image Processing for Automatic Target Detection Applications. *Linc. Lab. J.* **2003**, *14*, 79–116.
112. Chang, C.I.; Liu, J.M.; Chieu, B.C.; Ren, H.; Wang, C.M.; Lo, C.S.; Chung, P.C.; Yang, C.W.; Ma, D.J. Generalized constrained energy minimization approach to subpixel target detection for multispectral imagery. *Opt. Eng.* **2000**, *39*, 1275–1281.
113. Oppenheim, A.V.; Willsky, A.S.; Young, I.T. *Signals and Systems*; Prentice-Hall, Inc.: Englewood Cliffs, NJ, USA, 1983; p. 256, ISBN 0-13-809731-3.
114. Johnson, S. Constrained energy minimization and the target-constrained interference-minimized filter. *Opt. Eng.* **2003**, *42*, 1850–1854. [CrossRef]

115. Harsanyi, J.C. Detection and Classification of Subpixel Spectral Signatures in Hyperspectral Image Sequences. Ph.D. Thesis, Department of Electrical Engineering, University of Maryland, Baltimore County, Baltimore, MD, USA, 1993.
116. Harsanyi, J.C.; Farrand, W.H.; Chang, C.I. Detection of subpixel signatures in hyperspectral image sequences. In Proceedings of the American Society of Photogrammetry & Remote Sensing, Reno, NV, USA, 25–28 April 1994; Lyon, J., Ed.; pp. 236–247.
117. Kokaly, R.F.; Clark, R.N.; Swayze, G.A.; Livo, K.E.; Hoefen, T.M.; Pearson, N.C.; Wise, R.A.; Benzel, W.M.; Lowers, H.A.; Driscoll, R.L.; et al. *USGS Spectral Library Version 7*; U.S. Geological Survey Data Series; United States Geological Survey: Reston, VA, USA, 2017; Volume 61, p. 1035. [CrossRef]
118. Cudahy, T. *Satellite ASTER Geoscience Product Notes for Australia*; CSIRO: Collingwood, Australia, 2012; Volume 1, ISBN EP-30-07-12-44.
119. Mars, J.C.; Rowan, L.C. ASTER spectral analysis and lithologic mapping of the Khanneshin carbonate volcano, Afghanistan. *Geosphere* **2011**, *7*, 276–289. [CrossRef]
120. Mars, J.C.; Rowan, L.C. Spectral assessment of new ASTER SWIR surface reflectance data products for spectroscopic mapping of rocks and minerals. *Remote Sens. Environ.* **2010**, *114*, 2011–2025. [CrossRef]
121. Mars, J.C.; Rowan, L.C. Regional mapping of phyllic- and argillic-altered rocks in the Zagros magmatic arc, Iran, using Advanced Spaceborne Thermal Emission and Reflection Radiometer (ASTER) data and logical operator algorithms. *Geosphere* **2006**, *2*, 161–186. [CrossRef]
122. GANOVEX Team. Geological map of North Victoria Land, Antarctica, 1:500,000—Explanatory notes. *Geol. Jahrb. B* **1987**, *66*, 7–79.
123. Yajima, T.; Yamaguchi, Y. Geological mapping of the Francistown area in northeastern Botswana by surface temperature and spectral emissivity information derived from Advanced Spaceborne Thermal Emission and Reflection Radiometer (ASTER) thermal infrared data. *Ore Geol. Rev.* **2013**, *134*, 134–144. [CrossRef]
124. Ninomiya, Y.; Fu, B.; Cudahy, T.J. Detecting lithology with Advanced Spaceborne Thermal Emission and Reflection Radiometer (ASTER) multispectral thermal infrared radiance-at-sensor data. *Remote Sens. Environ.* **2005**, *99*, 127–139. [CrossRef]
125. Ramakrishnan, D.; Bharti, R.; Singh, K.D.; Nithya, M. Thermal inertia mapping and its application in mineral exploration: Results from Mamandur polymetal prospect, India. *Geophys. J. Int.* **2013**, *195*, 357–368. [CrossRef]
126. Sgavetti, M.; Pomilio, L.; Meli, S. Reflectance spectroscopy (0.3–2.5 µm) at various scales for bulk-rock identification. *Geosphere* **2006**, *2*, 142–160. [CrossRef]
127. Farrand, W.H.; Harsanyi, J.C. Discrimination of poorly exposed lithologies in imaging spectrometer data. *J. Geophys. Res.* **1995**, *100*, 1565–1575. [CrossRef]
128. Hunt, G.R.; Evarts, R.C. The use of near-infrared spectroscopy to determine the degree of serpentinization of ultramafic rocks. *Geophysics* **1980**, *46*, 316–321. [CrossRef]
129. Van der Meer, F. Estimating and simulating the degree of serpentinization of peridotites using hyperspectral remotely sensed imagery. *Nonrenew. Res.* **1995**, *4*, 84–98. [CrossRef]
130. King, T.V.V.; Clark, R.N. Spectral characteristics of chlorites and Mg-serpentines using high-resolution reflectance spectroscopy. *J. Geophys. Res.* **1989**, *94*, 13997–14008. [CrossRef]
131. Evans, B.W. Control of the products of serpentinization by the $Fe^{2+}Mg_{-1}$exchange potential of olivine and orthopyroxene. *J. Petrol.* **2008**, *49*, 1873–1887. [CrossRef]
132. Crowley, J.K. Visible and near-infrared (0.4–2.5 µm) reflectance spectra of playa evaporate minerals. *J. Geophys. Res.* **1991**, *96*, 16231–16240. [CrossRef]
133. Gaffey, S.J. Spectral reflectance of carbonate minerals in the visible and near-infrared (0.35–2.55 microns): Calcite, aragonite, and dolomite. *Am. Mineral.* **1986**, *71*, 151–162.
134. Kruse, F.A.; Perry, S.L. Mineral mapping using simulated Worldview-3 short-wave-infrared imagery. *Remote Sens.* **2013**, *5*, 2688–2703. [CrossRef]
135. Rowan, L.C.; Mars, J.C. Lithologic mapping in the Mountain Pass, California area using Advanced Spaceborne Thermal Emission and Reflection Radiometer (ASTER) data. *Remote Sens. Environ.* **2003**, *84*, 350–366. [CrossRef]
136. Malatesta, C.; Crispini, L.; Laufer, A.; Lisker, F.; Federico, L. Effects of hydrothermal alteration during cycles of deformation along fault zones in granitoids (northern Victoria Land, Antarctica). *AGU Fall Meet. Abstr.* **2018**.

137. Crispini, L.; Capponi, G.; Laufer, A.; Lisker, F. Fault-controlled ancient hydrothermal systems in North Victoria Land, Antarctica. In Proceedings of the POLAR 2018–Where the Poles Come Together, Open Science Conference, Davos, Switzerland, 19–23 June 2018.
138. Kelemen, P.B.; Matter, J.; Streit, E.E.; Rudge, J.F.; Curry, W.B.; Bluztajn, J. Rates and mechanisms of mineral carbonation in peridotite: Natural processes and recipes for enhanced, in situ CO_2 capture and storage. *Ann. Rev. Earth Planet. Sci.* **2011**, *39*, 545–576. [CrossRef]
139. Klein, C.; Hurlbut, C.S.J. *Manual of Mineralogy*; Dana, J.D., Ed.; John Wiley and Sons: New York, NY, USA, 1985; p. 596.
140. Likhoidov, G.G.; Plyusnina, L.P.; Shcheka, Z.A. The behavior of gold during listvenitization: Experimental and theoretical simulation. *Dokl. Earth Sci.* **2007**, *415*, 723–726. [CrossRef]

© 2019 by the authors. Licensee MDPI, Basel, Switzerland. This article is an open access article distributed under the terms and conditions of the Creative Commons Attribution (CC BY) license (http://creativecommons.org/licenses/by/4.0/).

Article

Multispectral and Radar Data for the Setting of Gold Mineralization in the South Eastern Desert, Egypt

Basem Zoheir [1,2,*], Ashraf Emam [3], Mohamed Abdel-Wahed [4] and Nehal Soliman [5]

1. Department of Geology, Faculty of Science, Benha University, Benha 13518, Egypt
2. Institute of Geosciences, University of Kiel, Ludewig-Meyn Str. 10, 24118 Kiel, Germany
3. Geology Department, Faculty of Science, Aswan University, Aswan 81528, Egypt; emam99@aswu.edu.eg
4. Geology Department, Faculty of Science, Cairo University, Giza 12613, Egypt; mawahed@sci.cu.edu.eg
5. National Authority for Remote Sensing & Space Sciences (NARSS), Cairo 1564, Egypt; nehal.abdelrahman@narss.sci.eg
* Correspondence: basem.zoheir@fsc.bu.edu.eg or basem.zoheir@ifg.uni-kiel.de; Tel.: +2-106-279-2092

Received: 14 May 2019; Accepted: 16 June 2019; Published: 18 June 2019

Abstract: Satellite-based multi-sensor data coupled with field and microscopic investigations are used to unravel the setting and controls of gold mineralization in the Wadi Beitan–Wadi Rahaba area in the South Eastern Desert of Egypt. The satellite-based multispectral and Synthetic Aperture Radar (SAR) data promoted a vibrant litho-tectonic understanding and abetted in assessing the regional structural control of the scattered gold occurrences in the study area. The herein detailed approach includes band rationing, principal component and independent component analyses, directional filtering, and automated and semi-automated lineament extraction techniques to Landsat 8- Operational Land Imager (OLI), Advanced Spaceborne Thermal Emission and Reflection Radiometer (ASTER), Phased Array L-band Synthetic Aperture Radar (PALSAR), and Sentinel-1B data. Results of optical and SAR data processed as grayscale raster images of band ratios, Relative Absorption Band Depth (RBD), and (mafic–carbonate–hydrous) mineralogical indices are used to extract the representative pixels (regions of interest). The extracted pixels are then converted to vector shape files and are finally imported into the ArcMap environment. Similarly, manually and automatically extracted lineaments are merged with the band ratios and mineralogical indices vector layers. The data fusion approach used herein reveals no particular spatial association between gold occurrences and certain lithological units, but shows a preferential distribution of gold–quartz veins in zones of chlorite–epidote alteration overlapping with high-density intersections of lineaments. Structural features including en-echelon arrays of quartz veins and intense recrystallization and sub-grain development textures are consistent with vein formation and gold deposition syn-kinematic with the host shear zones. The mineralized, central-shear quartz veins, and the associated strong stretching lineation affirm vein formation amid stress build-up and stress relaxation of an enduring oblique convergence (assigned as Najd-related sinistral transpression; ~640–610 Ma). As the main outcome of this research, we present a priority map with zones defined as high potential targets for undiscovered gold resources.

Keywords: multispectral and radar data; data fusion; gold mineralization; Wadi Beitan–Wadi Rahaba; structural control; Najd Fault System; South Eastern Desert; Egypt

1. Introduction

Remote-sensing applications and the recently made free-of-charge or low-cost satellite data boosted scientific research and helped the industry to understand features controlling mineral resources. The ideal cases are areas uncovered by vegetation and characterized by arid or semi-arid atmospheric conditions. The Nubian Shield is a typical example of well-exposed crystalline basement rocks with historically known resources, i.e., gold and copper. The South Eastern Desert (SED) terrane, a part of the

Nubian Shield, is underlain by Neoproterozoic crystalline rock belts of mainly dismembered ophiolites, island arc metavolcanic/metasedimentary rocks, and less abundant poorly dated gneissic and schistose metasedimentary rocks. The tectonic build-up of the SED terrane was mainly shaped during the final assembly of East- and West-Gondwanas [1–3]. Accretion-related structures document two principal episodes of shortening at ca. 715–700 Ma and 685–665 Ma [4]. Deformation was initiated by northeast (NE)–southwest (SW) pure shear and progressed to simple shearing along the east-northeast (ENE)–west-southwest (WSW) direction. The first event is expressed by SW-verging intrafolial tight and overturned folds and thrusts, while the second event is documented by NW–NNW-trending large-scale open folds, piggyback thrusts strongly segmented by N–S and NW–SE to WNW–ESE sinistral strike–slip faults [5]. The SED terrane encompasses three major structural systems, namely the NW–WNW-trending Allaqi-Heiani suture, N–S Hamisana zone, and NW–SE Wadi Hodein–Wadi Kharit shear corridor (Figure 1) [1,2,6,7]. Deformation along the Wadi Hodein–Wadi Kharit shear corridor includes sinistral shearing and northwest-directed thrusting, extension, and tectonic escape, collectively resulting in a dominant northwesterly structural trend. The latter is considered the western extension of the Najd fault system in the Arabian Shield into the Nubian Shield [8].

Several gold occurrences are located in the SED terrane, mostly along second- or third-order shear zones assigned to as post-accretionary structures (Figure 1) [9,10]. The gold–quartz veins and related hydrothermal alteration halos in the SED occur in distinctive geologic/structural settings [10], namely (i) steeply or moderately dipping silicified, carbonated brittle–ductile shear zones between allochthonous listvenized ophiolitic blocks and island arc-metavolcanic and metasedimentary rocks, i.e., where acid and intermediate dykes and flat extensional granophyre dykes are abundant (e.g., Hutit, El-Beida, and El-Anbat deposits), (ii) steeply dipping anastomosing ductile shear zones wrapped around or cutting syn- or late-orogenic granitoid intrusions (e.g., Korbiai, Madari, Romite, Egat), and (iii) in ductile shear zones in highly deformed ophiolitic or island arc terranes cutting locally carbonaceous pelitic or volcanogenic metasedimentary rocks (e.g., Um El-Tuyor, Betam, Seiga, Shashoba, Um Garayat, and Haimur).

In the Wadi Rahaba–Wadi Beitan area, auriferous quartz veins are controlled mainly by NNW–SSE shear zones, but gold occurrences along the ENE–WSW or NE–SW fault/fractures zone are also less commonly observed. Generally, gold occurrences in the Wadi Rahaba–Wadi Beitan area are small-scale, with a few mineralized quartz veins that are generally < 100 m long. However, shafts, dumps, buildings and leaching basins left behind by ancient miners refer to extensive mining activities.

Band combination and false color composite (FCC) images are used in geological applications based on known optical and Synthetic Aperture Radar (SAR) characteristics in specific wavelength regions [11–21]. The satellite imagery spectral data used for mineral exploration span the visible to infrared regions. As hydrothermal alteration is a typical associate with hydrothermal mineral deposits, mapping or detecting the hydrothermal alteration zones is a prime focus of mineral exploration programs using the remote-sensing data. Most of the hydrothermal alteration mineral species have distinctive features in the shortwave infrared (SWIR) region, making the Advanced Spaceborne Thermal Emission and Reflection Radiometer (ASTER) and Landsat-8 Operational Land Imager (OLI) sensors important free-of-charge data sources for mineral mapping, particularly in arid geographical regions. Several techniques developed for the analysis of the SWIR data include band-rationing (BR), principal component analysis (PCA), the relative absorption band depth (RBD), and the spectral mineralogical indices that are proven effective in lithological and hydrothermal alteration mapping if integrated with field data [22–31].

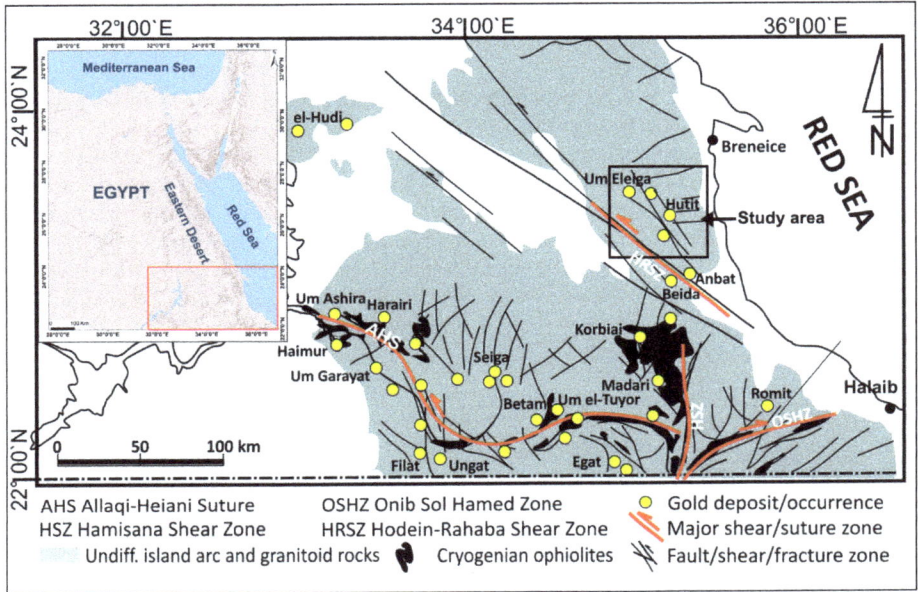

Figure 1. Simplified geological map of the South Eastern Desert (SED) of Egypt with the major deformation systems and gold occurrences shown.

The Phased Array Type L-Band Synthetic Aperture Radar (PALSAR) acquires data in the quad mode, i.e., horizontal-horizontal (HH), horizontal-vertical (HV), vertical-horizontal (VH), and Vertical-Vertical (VV). Also, a European radar imaging satellite Sentinel-1 is equipped with a synthetic aperture radar (SAR-C). The band math operator and directional filtering are used to generate backscatter images from SAR data, where the resultant images serve as a suitable data source for manual and automated techniques [32,33]. Extraction of structural lineaments from satellite-based imagery data is accomplished using edge enhancement, directional filtering, and manual digitizing techniques [34,35]. The automatic lineament extraction (ALE) technique is done using software algorithms, i.e., Canny Edge Detection [36]. However, the reference of the trustworthiness of the ALE with this algorithm is the manual extraction of lineaments that are verified by fieldwork.

The present work examines the spatial relationship between gold mineralization and structural elements, providing a meaningful hypothesis in relating gold metallogeny to the structures and evolution of the SED. Lithological and structural mapping is based on integrated field, microscopic, and multi-sensor imagery data (Landsat-8 OLI, ASTER, PALSAR, and Sentinel-1). Interpretations of the outcrop-scale and microscopic-scale structural elements are detailed herein to provide valuable information about the known occurrences for the sake of promoting the determination of new exploration targets along the regional structures.

2. Geologic Setting

The Wadi Beitan–Wadi Rahaba area comprises two major lithologic units: (1) autochthonous gneisses and migmatites in the west, and (2) allochthonous ophiolitic mélange and island arc-metavolcanic assemblage rocks in the east (Figure 2). The gneissic rocks are mainly hornblende–biotite granodiorite gneisses. They are light gray, coarse-grained, with feldspar and quartz as the main constituents in addition to less common hornblende, biotite, and garnet. Regionally, these gneisses are affected by extensive stromatic migmatization and show a well-developed banding. The ophiolitic mélange rocks (admixed serpentinite, metabasalt, chert, and carbonaceous metasedimentary rocks) occupy the central part of the study area as two elongated NNW-trending thrust sheets separated by a belt of island arc-metavolcanic/metavolcaniclastic rocks. The regional foliation generally strikes NW–SE and dips to the NE or to SW at moderate angles.

The ophiolitic mélange and arc assemblages are intruded by gabbro–diorite, granodiorite, and younger gabbro/granite, and also dissected by later dykes, quartz veins, and plugs. The gabbro–diorite intrusions are heterogeneous in composition, commonly of diorite, foliated with minor folds at their peripheries. The mylonitized syn-tectonic granodiorite is confined to the gneisses/ophiolitic mélange contact in the western part of the study area. They exhibit a well-developed mylonitic foliation, and are composed essentially of stretched orthoclase and quartz porphyroclasts, set in a fine- to medium-grained sheared matrix of orthoclase and quartz with a lower percentage of plagioclase and mafic minerals. The post-tectonic granites are common in the eastern part of the study area, but also exposed as dispersed blocks in the southwestern part (Figure 2). These rocks vary from buff and white coarse-grained granite to pink, fine-grained aplitic leucogranite.

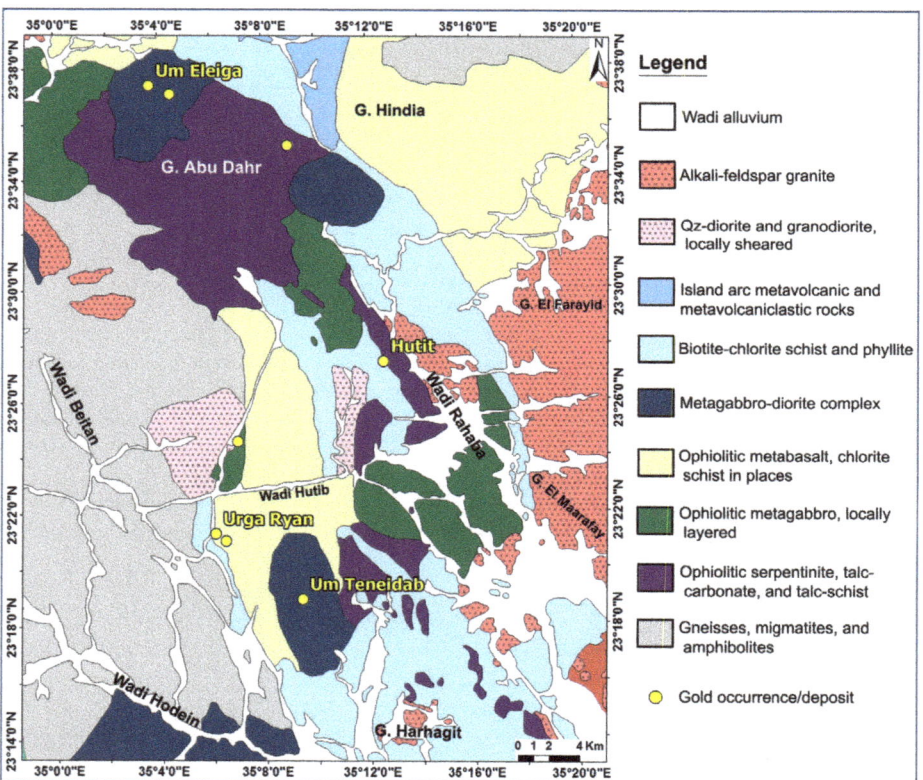

Figure 2. Geological map of Wadi Beitan–Wadi Rahaba area in South Eastern Desert (SED) of Egypt. Modified from Conoco Coral [37]. The unlabeled yellow filled circles are either insignificant or yet not studied occurrences.

The Wadi Hodein area is 20 km west of the Red Sea coast in the SED and is occupied by a Neoproterozoic greenstone belt of variably deformed ophiolite and island arc-volcanic/plutonic rocks. These metamorphic rocks are intruded by syn-tectonic granites and late- to post-tectonic gabbros and granites. Moreover, the Neoproterozoic rocks are unconformably overlain by Cretaceous sandstones and intruded by Tertiary basalt [38]. The ophiolite rocks (serpentinite, metagabbro, and less common pillow metabasalt) form discontinuous NW–SE-trending sheets (10–20 km in length) overthrusted on the island arc-metavolcanic/volcaniclastic rocks along steeply east-dipping thrust/strike–slip fault structures. Serpentinite and talc–carbonate rocks form large NW–SE-trending elongated masses. Moreover, small elongated slices (<100 m) of highly sheared serpentinite and talc–carbonate schist are tectonically incorporated within the island arc-metavolcanic/volcaniclastic terrains [1,39,40]. The island arc-metavolcanic/volcaniclastic assemblages form NNW–SSE- and NW–SE-trending belts of highly foliated and sheared basic–intermediate to acidic metavolcanic/volcaniclastic rocks [39–42].

Deformation history of the Wadi Hodein wrench includes multi-deformational stages D1–D4 [38,40,42–44]. The early NNE–SSW crustal shortening (D1) led to the creation of WNW–ESE tight intrafolial or overturned folds (F_1) and axial planar foliations (S_1) parallel to the primary (lamination) bedding (S_o) of the island arc-metavolcanic/volcaniclastic rocks. The second event (D2) involved subsequent sinistral transpression along thrust segments at the base of the ophiolitic blocks and formed steeply dipping, left-lateral NNW–SSE ductile shear zones and major faults that deformed most of the exposed lithologies. Macroscopic and mesoscopic NNW–SSE folds (F_2), verging commonly to the WSW, are associated with pervasive axial planar crenulation cleavage (S_2) in the metavolcanic/volcaniclastic rocks.

Gold-bearing quartz lodes are commonly observed where S_2 is superimposed on S_1. Finally, during the youngest deformational stage, a weak brittle deformation led to ENE–WSW dextral strike–slip faults cutting across the syn-orogenic granodiorite and pre-existing rocks. The brittle deformation event might be attributed to the Red Sea rift [42].

3. Remote-Sensing Data Processing

3.1. Data Characteristics and Methods

Fusion of satellite optical and microwave data is applied herein for comprehensive lithologic mapping and geological structures to inspect the regional and mine-scale setting of gold mineralization in the study area. The ASTER data (Table 1) cover a wide spectral range of 14 spectral bands, measuring reflected radiation in three bands between 0.52 and 0.86 µm (visible-near infrared, VNIR) with 15-m resolution, and six bands from 1.6 to 2.43 µm (shortwave Infrared, SWIR) with 30-m resolution. The emitted radiation is measured at 90-m resolution in five bands through the 8.125–11.65-µm wavelength region (thermal infrared; TIR). The ASTER standard data also include topographic information digital elevation models (DEM) [28]. The Landsat-8 OLI data comprise nine spectral bands from which seven bands measure the reflected VNIR and SWIR radiation with 30-m spatial resolution for bands 1–7 and 9, while the panchromatic band 8 has 15-m resolution. The ultra-blue band 1 is operative in coastal and aerosol targets, whereas band 9 is valued for cloud detection. The TIR bands collects two thermal bands (10 and 11) that measure the emitted radiation through the 10.6–12.5 µm wavelength region (TIR) with 100-m spatial resolution [29]. The applied projection method for the ASTER and Landsat-8 OLI/TIRS data is UTM (Universal Transverse Mercator), Zone N36 according to the WGS-84 datum.

In this study, subsets of cloud-free level 1B ASTER (AST_L1B_ 00312252006082422, acquired on 25 December 2006) and cloud-free level 1T OLI (Terrain-corrected, LC81730442019067LGN00, Path 173/Row 44, acquired on 8 March 2019) VNIR–SWIR data were processed using the ENVI software, version 5.1, provided by ITT Visual Information Solutions (now Exelis Visual Information Solutions). The spectral response curves of some selective minerals, acquired from U.S. Geological Survey spectral library [45], were employed to evaluate the diagnostic spectral features of rock forming minerals and assess lithological mapping and delineating the major mylonitic zones.

The active synthetic aperture radar (SAR) data were integrated with ASTER and ASTER–GDEM (global digital elevation model) data for structural enhancement, lineament extraction, and mapping of major fault/shear zones controlling the distribution of gold occurrences in the study area. The microwave SAR data are an important source of professional data used for mapping of geological structures [46].

The PALSAR sensor is an L-band SAR, with fully polarized (HH, HV, VH, and VV) and multi observation modes (fine, polarimetric, and ScanSar) with 10-, 30-, and 100-m spatial resolution, respectively [47,48]. The C-band SAR sensors of Sentinel-1 satellites have a dual-polarization (co-polarized VV or HH, and cross-polarized VH or HV), interferometric wide-swath (IW) mode and a spatial resolution of 5 × 20 m [49]. Subsets of Fine Beam Dual (FBD) HH + HV polarization, level 1.5 ALOS_PALSAR scenes (ALPSRP075090450 and ALPSRP077570445, acquired on 22 June and 9 July 2007, respectively) and Sentinel-1B data (S1B_IW_GRDH_1SDV_20180712t033727-20180712t033752, acquired on 12 July 2018) were processed for lineament extraction and structural mapping in this study.

Table 1. Summarized characteristics of the ASTER and Landsat-8 OLI/TIRS data.

Aster				Landsat-8 OLI/TIRS			
Bands	Spectral Region	Wavelength (μm)	Resolution (m)	Bands	Spectral Region	Wavelength (μm)	Resolution (m)
Band 1	VNIR	0.52–0.60	15	Band 1	Coastal	0.433–0.453	30
Band 2		0.63–0.69		Band 2	Blue	0.450–0.515	
Band 3		0.78–0.86		Band 3	Green	0.525–0.600	
Band 4	SWIR	1.60–1.70	30	Band 4	Red	0.630–0.680	
Band 5		2.145–2.185		Band 5	NIR	0.845–0.885	
Band 6		2.185–2.225		Band 6	SWIR	1.560–1.660	
Band 7		2.235–2.285		Band 7		2.100–2.300	
Band 8		2.295–2.365		Band 8	Panchromatic	0.500–0.680	15
Band 9		2.360–2.430		Band 9	Cirrus	1.360–1.390	30
Band 10	TIR	8.125–8.475	90	Band 10	TIR	10.60–11.19	100
Band 11		8.475–8.825		Band 11		11.50–12.51	
Band 12		8.925–9.275					
Band 13		10.25–10.95					
Band 14		10.95–11.65					

Methods applied for rigorous analysis of the SWIR data include band-rationing (BR), principal component analysis (PCA), independent component analysis (ICA), and the relative absorption band depth (RBD). Ninomiya's mineralogical indices were also used to effectively help in lithological and hydrothermal alteration mapping, integrated with fieldwork data (Figure 3) [22–31]. Adaptive filtering, band math, PCA, and directional filtering were applied to PALSAR and Sentinel-1 data for lineament extraction by manual and automated means. The latter was achieved though edge enhancement of the microwave data using the LINE extraction algorithm tool in the PCI-Geomatica software package (version 2017). The LINE algorithm includes edge detection, thresholding, and curve extraction [36,50,51]. The algorithm merges pairs of facing and close parallel polylines controlled by defining specific parameters. The final polylines could be exported as a vector layer and imported into the ArcGIS environment. Figure 3 illustrates the flowchart and techniques applied for the different data types in this study.

3.2. Pre-Processing of Satellite Data

Removing the atmospheric effects is an important pre-processing step needed for qualitative and quantitative analysis of surface reflectance data [52,53]. The atmospheric correction involves re-scaling and conversion of the radiance data collected by optical sensors to surface reflectance data; therefore, the obtained reflectance spectra can be calibrated directly with the standard reflectance spectra collected in the laboratory and field. In the present study, the FLAASH (fast line-of-sight atmospheric analysis of spectral hypercube) algorithm was employed for the atmospheric correction and reflectance conversion of Landsat-8 OLI and ASTER VNIR–SWIR data [54]. The VNIR and SWIR bands of ASTER and OLI sensors were subjected to radiometric calibration, layer-stacked, and re-sampled to 30-m spatial resolution. The fast line-of-sight atmospheric analysis of spectral hypercubes (FLAASH) algorithm [54] was then applied to the radiometrically calibrated radiance data with the band interleaved by line (BIL) format. Thermal atmospheric correction and emissivity normalization parameters were applied to the TIR bands.

The side-looking architecture of the microwave SAR sensors resulted in foreshortening, layover, and shadowing as the main distortions in the SAR images [55,56]. Foreshortening is predominant in the elevated mountainous geographic terrains, especially if the side-looking architecture is steep and the mountains slant toward the sensor. Layover occurs in the terrains of sufficiently steep slopes, where the layover zones appear with bright tone. Shadows in the SAR images occur as dark regions tarnished by thermal noise. In this study, such distortions were minimized by applying ortho-rectification of SAR subsets using the 12.5-m digital elevation model (DEM) of ALOS–PALSAR data. The SAR images are commonly corrupted by speckles that diminish the differentiation ability between objects and reduce the accuracy of feature extraction. In this study, we applied Lee adaptive filtering to the PALSAR and Sentinel-1B data to remove the radar speckles while preserving edges, lineaments, and structural features. The Lee filter enables smoothing of the speckled data based on statistics computed in filter windows, where the calculated pixel values replace the original pixel values using the surrounding pixels [57,58].

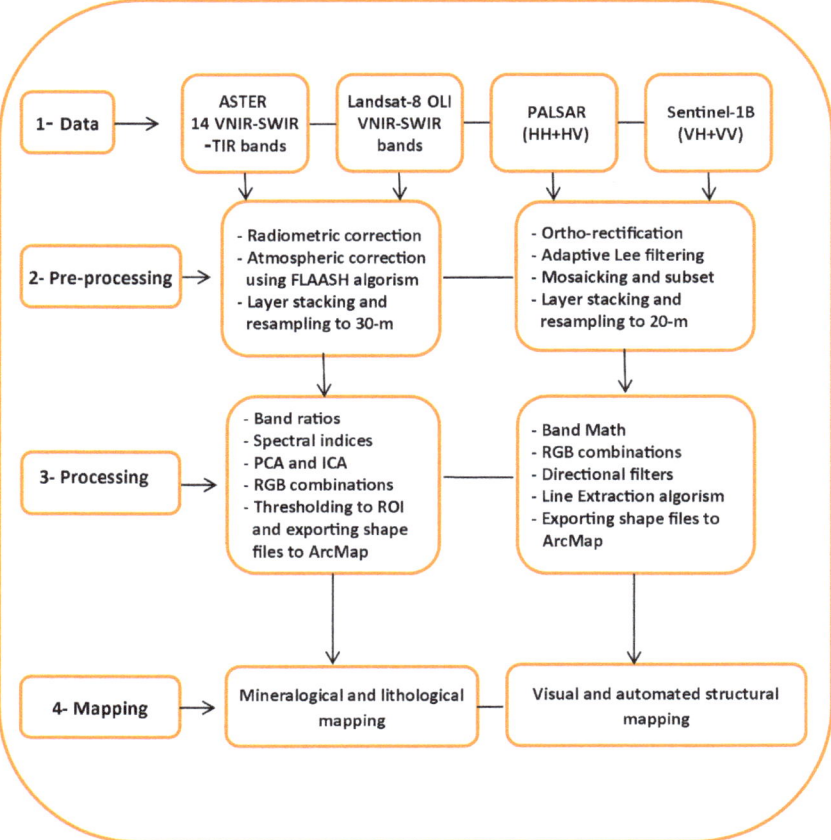

Figure 3. Flowchart of methods and data types and the succession of processing steps applied for the present study.

3.3. Optical Data-Based Lithological Mapping

3.3.1. Band Ratios and Relative Absorption Band Depth

Band-rationing, a powerful image processing technique, enhances the spectral variations between surface materials and highlights target anomalies, while suppressing some information such as the difference in albedo and topographic slope effects [59]. Band ratios are mathematical transformations in which the digital numbers (DN) of one band are divided by those of another band. They signify differences between DN values of diagnostic reflectance and adsorption features in the spectral bands. The relative absorption band depth (RBD) is a useful three-band math transformation in which the sum of the two shoulders having reflectance maxima is divided by the third band including the absorption feature minima [31]. Accordingly, this technique typifies the relative absorption depth of absorption peaks, and is utilized commonly for identifying rock-forming minerals that have diagnostic Fe, Mg-OH, and CO_3 absorption features. The OLI-band ratios (6/7, 6/2, 5/6, 4/2, 6/5 × 4/5) and the ASTER–RBD (6 + 9/8, 5 + 7/6, 6 + 9/8^2, 5 + 7/6^2) values were processed to improve the identification of the Fe, Mg-OH, Al-OH-bearing, and carbonate rocks in the study area.

The OLI-band ratio 6/7 (1.61/2.20 µm) is equivalent to the ASTER-band ratio 4/6 (1.656/2.209 µm) and the Enhanced Thematic Mapper (ETM)-band ratio 5/7 (1.65/2.22 µm) analyzed for CO_3 and OH-bearing minerals [28,60–62]. The grayscale image of this ratio brings out clay minerals, serpentine, and many alteration zones with a bright image signature (Figure 4A). The OLI-band ratio 6/5 (1.610/0.865) coincides with the ASTER-band ratio 4/3 (1.656/0.807 µm) and the ETM-band ratio 5/4 (1.650/0.825 µm). This ratio converts the ferrous iron (Fe^{+2})–silicate-bearing minerals (i.e., olivine and pyroxene) into bright pixels and clearly distinguishes the mafic from non-mafic rocks [62,63]. The OLI-band ratio 4/2 (0.6550/0.4825) approaches much of the ASTER-band ratio 2/1 (0.661/0.556 µm) and the ETM-ratio 3/1 (0.6600-0.4825 µm) that are used to highlight rocks rich in hematite [61]. The FCC image of three OLI-band ratios (6/7 in red (R), 6/5 in green (G), 4/2 in blue (B)) was used to characterize the mixed serpentinite, quartz–carbonate (listvenite), and talc–carbonate schist in yellow and reddish pixels (Figure 4B). The highly tectonized metavolcanic rocks and carbonaceous metasedimentary rocks appear in purple and lemon colors, respectively. The island arc-metavolcanic and metavolcaniclastic rocks appear as dark-green pixels, while granitic rocks exhibit a bluish-green spectral signature. The FFC OLI-band ratio image (R: 6/7, G: 6/2, B: 6/5 × 4/5) highlights the talc–carbonate schist, serpentinite, and graphite-bearing metasedimentary rocks as rose, red, and magenta pixels, respectively (Figure 5A). This image enables the distinction between mafic rocks as blue pixels and felsic rocks as green ones. The ASTER-based RBD (6 + 9/8^2) image (Figure 5B) and (7 + 9/8) image (not shown) intensify the spectral signature of amphibole, chlorite, and other Fe–Mg-OH-bearing minerals and show them as bright pixels, whereas the RBD (5 + 7/6) discriminates the Al-OH-bearing silicate minerals, i.e., muscovite, sericite, and kaolinite, commonly described for altered felsic rocks.

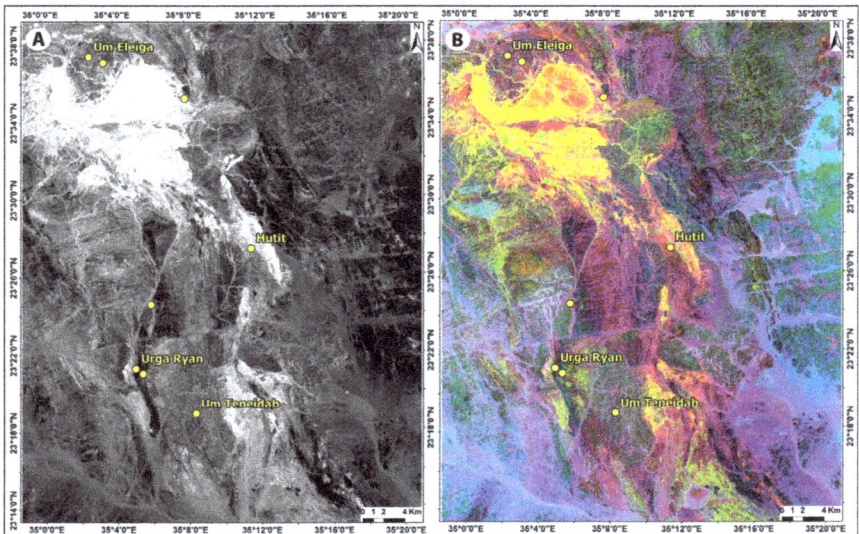

Figure 4. Spectral discrimination of ophiolitic rocks by a processed Landsat-8 OLI scene of the study area. (**A**) Grayscale band ratio (6/7) image; (**B**) false color composite (FCC) band ratio image (red (R): 6/7, green (G): 6/5, blue (B): 4/2) of the study area. The ophiolitic rocks are highlighted by yellow or lemon-yellow pixels and clearly display signs of shearing and dislocation. Note that the ophiolitic rocks are rather abundant along the eastern shear zone which accommodates the Hutit mine. The yellow-filled circles are gold occurrences and/or locations of old gold mines as in the geological map (Figure 2).

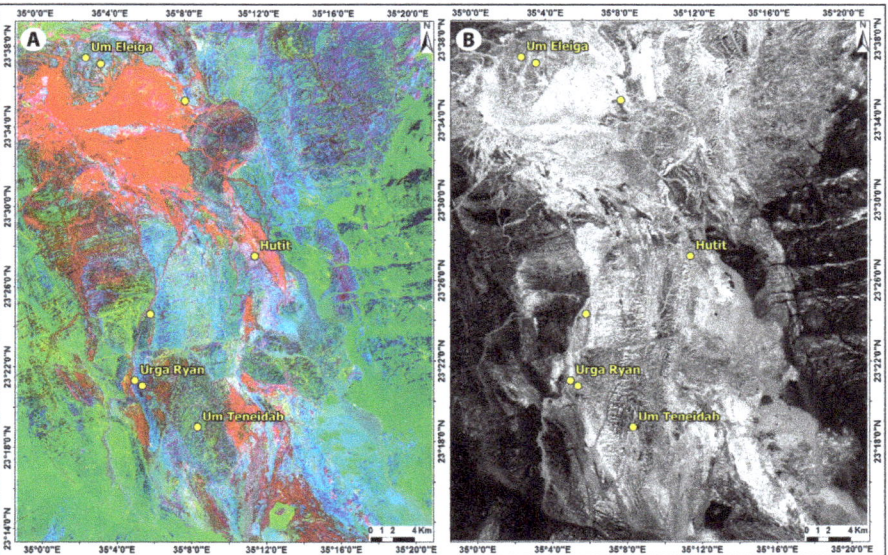

Figure 5. Ophiolitic mélange mapping promoted by processing OLI and ASTER data. (**A**) The FCC OLI-band ratio (R: 6/7, G: 6/2, B: 6/5 × 4/5) image discriminates terranes of the mafic rocks in the study area; (**B**) the grayscale ASTER RBD $(6 + 9/8^2)$ image clearly highlight the ophiolitic mélange terrane in the central part of the study area and is surrounded by felsic rocks (granites and granitic gneisses) in the eastern and western parts of the image.

3.3.2. Mineralogical Indices

The spectral mineralogical indices are reflectance combinations of two or more spectral bands signifying the relative abundance of target objects. Accordingly, the spectral indices of rock-forming minerals are mathematical expressions and band ratios used to facilitate the mapping of lithology and hydrothermally mineral alteration zones. Using the six ASTER–SWIR bands, four spectral mineralogical indices (OH-bearing mineral index, OHI; kaolinite index, KLI; alunite index, ALI; calcite index, CLI) were developed by Ninomiya [64] for mapping the hydrothermally altered zones and mineral anomalies. The OHI is calculated as (band 7/band 6) × (band 4/band 6), the KLI is developed by way of (band 4/band 5) × (band 8/band 6), and the ALI is formulated as (band 7/band 5) × (band 7/band 8), while the CLI is calculated as (band 6/band 8) × (band 9/band 8). Three spectral indices using ASTER TIR bands (quartz index, QI; carbonate index, CI; mafic index, MI) were advocated to facilitate the mapping of quartz, carbonate, and mafic–ultramafic rocks [65]. New spectral indices (Fe-mineral index, Fe-MI; Al-OH-bearing mineral index, Al-OH-MI; Fe–Mg-OH-bearing mineral index, Fe-Mg-OH-MI) were developed to evaluate the abundance of iron oxide/hydroxide minerals, Al-OH, CO_3, and Fe–Mg-OH-bearing alteration minerals [21]. For the Landsat-8 VNIR–SWIR data, the Fe-MI is calculated as (band 6/band 5) × (band 4/band 3), and the Al-OH-MI as (band 6/band 7) − (band 4). For ASTER VNIR–SWIR data, the Fe-MI is calculated through (band 4/band 3) × (band 2/band 1), the Al-OH-MI by means of (band 5) × (band 7/band 6), and the Fe-Mg-OH-MI as a result of (band 7) × (band 9/band 8). In this study, the QI, CI, and MI of Ninomiya et al. [65] and the Fe-Mg-OH-MI of Pour et al. [21] were processed using the ASTER SWIR–TIR bands. A better differentiation between the mafic and non-mafic rocks was attained by the MI and QI grayscale images, while the ultramafic rocks and associating carbonate and Fe–Mg-OH-bearing alteration zones were clearly differentiated in the grayscale images of CI and Fe-Mg-OH-MI (Figure 6A,B).

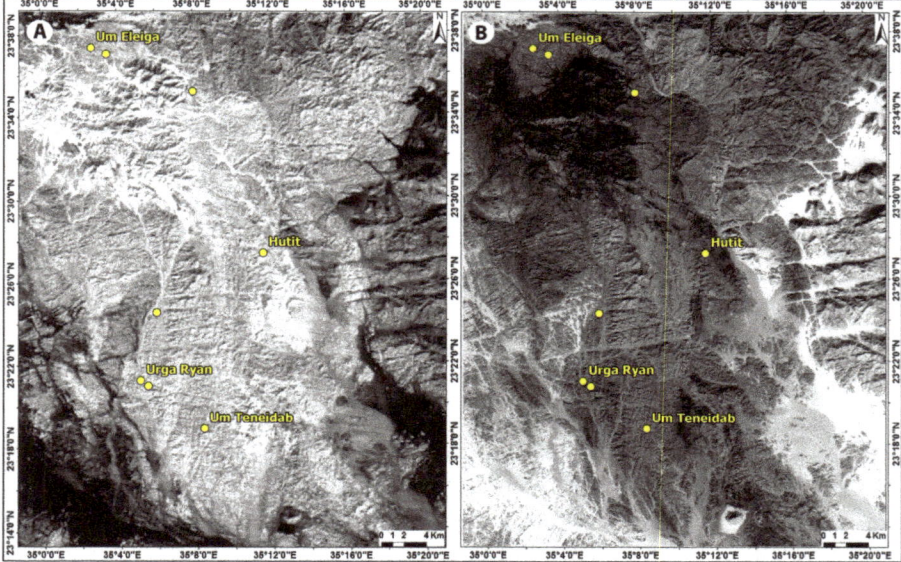

Figure 6. Mineralogical indices images based on the ASTER–SWIR data of the study area. (**A**) Grayscale image of the carbonate index (CI) of Ninomiya [65]; (**B**) grayscale image of the Fe-Mg-OH-MI after Pour et al. [21].

3.3.3. Principal and Independent Component Analyses (PCA and ICA)

The loading of Operational Land Imager (OLI) and ASTER multispectral data is attained to reduce dimensionality of correlated data and suppress the redundant information [66]. The resultant principal components are subjected to a nonlinear band generation method of the ICA transformation for finer classification and to remove correlation and detect the self-determining components. The eigenvector loadings of principal component 1 (PC1) show high contributions (0.480 and −0.490) for the ASTER relative band depth $(6 + 9)/8^2$ and Fe-Mg-OH MI, respectively (Table 2). The PC2 loadings indicate strongly negative contributions (−0.618 and −0.746) for the QI and relative band depth $(5 + 7)/6$, respectively, while PC3 shows highly positive loading (0.691) for the OLI-band ratio (5/6). The eigenvector loadings of PC5 show strongly positive (0.631) and negative (−0.596) contributions for the OLI-band ratios (5/6) and (6/7), respectively. The RGB composite combination of PC5, PC2, and PC3 distinguishes the undifferentiated mafic metavolcanic rocks of Wadi Rahaba from the felsic/mafic island arc metavolcanic rocks of G. El-Urga.

Table 2. Eigenvector loadings of principal component analysis (PCA) for OLI-band ratios and ASTER–RBD mineralogical indices. PC—principal component, QI—quartz index, Var.%—variation in percentage.

Eigenvector	Fe–Mg-OH	QI	(5 + 7)/6	$(6 + 9)/8^2$	(6/7)	(4/5)	(4/2)	(5/6)	Var.%
PC1	−0.492	−0.276	−0.013	0.480	0.313	0.375	−0.419	0.193	44.4
PC2	0.011	−0.618	−0.746	−0.173	−0.001	−0.084	0.081	−0.135	19.8
PC3	0.135	0.027	−0.078	−0.068	0.537	−0.017	0.452	0.691	16.3
PC4	−0.240	−0.101	0.176	0.135	0.398	−0.822	−0.041	−0.218	7.8
PC5	−0.202	−0.158	0.046	−0.052	−0.596	−0.367	−0.202	0.631	5.7
PC6	−0.041	0.651	−0.626	0.372	−0.048	−0.172	−0.101	0.048	2.9
PC7	0.016	−0.209	0.100	0.677	−0.304	−0.001	0.623	−0.088	1.9
PC8	−0.800	0.198	−0.046	−0.343	−0.074	0.113	0.416	−0.103	1.2

The PC5, PC2, and PC3 image highlights the post-tectonic granites of Wadi Rahaba, and the granodiorite gneisses and migmatites of Wadi Beitan (Figure 7A). The ophiolitic serpentinite, talc–chlorite schist, and talc–carbonate altered zones appear as indigo blue and pink pixels, whereas chlorite–tremolite–actinolite schists and Fe-OH-bearing alteration zones appear as green pixels. The RGB combination of independent components IC3, IC1, and IC2 efficiently discriminates the gabbroic rocks and mafic metavolcanic rocks with light-blue and cyan image signatures, while the island arc-metavolcanic rocks and chlorite–amphibole-bearing schists appear as pink pixels (Figure 7B). The ophiolitic serpentinite, talc–chlorite schist, and talc–carbonate alteration zones have yellow signatures, and the stream sediments of Wadi Na'am and Wadi Beitan appear with a lemon image signature.

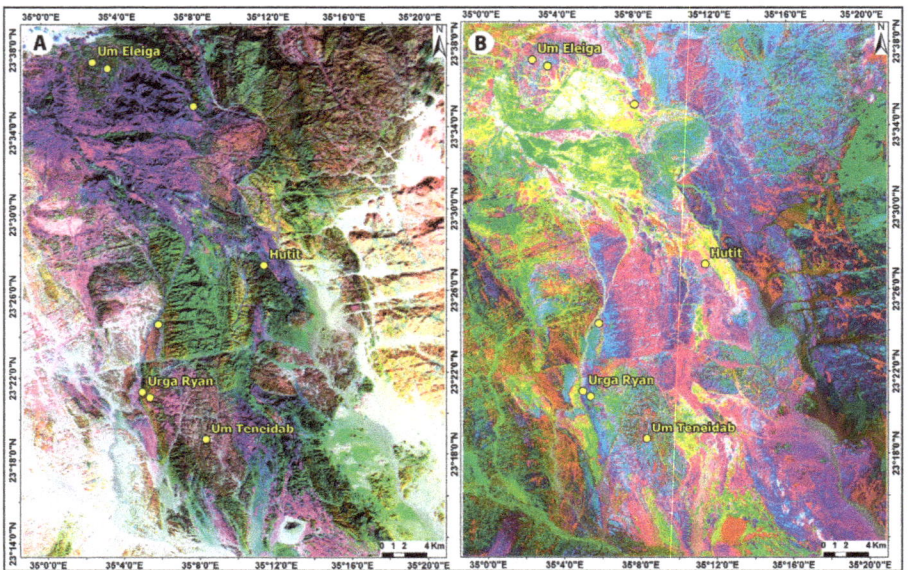

Figure 7. Principal and independent component analysis (PCA and ICA) images of OLI-band ratios, ASTER–RBD, and mineralogical indices. (**A**) FCC principal component (PC) image (R: PC5, G: PC2, B: PC3); (**B**) FCC independent component (IC) image (R: IC3, G: IC1, B: IC2).

3.4. SAR Data-Based Lineament Extraction

The band math operator was used to generate backscatter images from the Sentinel-1B (VV + VH, VV − VH, VV/VH, VV + VH/VH, VV − VH/VH) and PALSAR (HH + HV, HH − HV, HH/HV, HH + HV/HV, HH − HV/HV) bands. The resultant images were statistically analyzed using the PCA transformation (Tables 3 and 4). The PALSAR PC2 and Sentinel PC3 show the highest positive eigenvector loadings (0.894 and 0.810) for the subtraction and addition images (HH − HV and VV + VH, respectively). Meanwhile, the PALSAR PC3 has a strongly positive loading (0.874) for the addition image (HH + HV) and moderately negative loading (−0.448) for the subtraction image (HH − HV). In contrast, the Sentinel PC2 exhibits moderately positive loading (0.463) for the addition image (VV + VH) and highly negative loading (−0.653) for the ratio image (VV − VH/VH). The PALSAR PC4 and PC5 have highly negative eigenvector values (−0.761 and −0.805) for the ratio images (HH/HV and HH + HV/HV, respectively), while PC4 shows a strongly positive value (0.637) for the ratio image (HH − HV/HV). The Sentinel PC4 and PC5 have highly negative (−0.664) and positive (0.709) loadings for the ratio images (VV − VH/VH and VV/VH, respectively).

Because of its ability to outline structural lineaments, the backscatter PALSAR PC2 and Sentinel PC3 images were subjected to four directional filters (0°, 45°, 90°, and 135°) to enhance the linear features and realize the major structural trends (N–S, NE–SW, E–W, and NW–SE, respectively). Also, the false color composite images of PALSAR PC (PC2, PC5, and PC4), Sentinel PC (PC2, PC4, and PC3), and directional filters (90°, 135°, and 45°) were employed in RGB mode to improve the visualization of the extensive lineaments (Figure 8A,B). Using Coral Draw and ArcMap software, the structural lineaments were manually traced to produce a vector lineament layer. In the PCI Geomatica, the user-defined parameters (Table 5) of the ALE algorithm LINE were applied to the directional filters of PALSAR PC2 and Sentinel PC3 (Figure 9A,B). Hence, afterward, manual editing was carried out to improve the accuracy of lineament auto-detection by adding new or relocating existing segments.

Table 3. Eigenvector loadings of PCA for Sentinel-1B band math images.

Eigenvector	VV + VH	VV − VH	VV/VH	VV + VH/VH	VV − VH/VH	Var.%
PC1	−0.030	0.567	−0.568	−0.480	0.353	47.29
PC2	0.463	0.353	−0.356	0.329	−0.653	30.28
PC3	0.810	−0.222	0.209	−0.492	0.093	20.05
PC4	−0.358	−0.083	0.071	−0.648	−0.664	2.34
PC5	0.005	0.706	0.709	−0.010	−0.006	0.05

Table 4. Eigenvector loadings of PCA for PALSAR band math images.

Eigenvector	HH + HV	HH − HV	HH/HV	HH + HV/HV	HH − HV/HV	Var.%
PC1	0.205	−0.013	0.562	0.569	0.564	60.19
PC2	0.441	0.894	−0.054	−0.045	−0.040	20.59
PC3	0.874	−0.448	−0.109	−0.106	−0.113	17.64
PC4	−0.003	−0.010	−0.761	0.121	0.637	1.20
PC5	0.005	−0.002	0.300	−0.805	0.512	0.39

Table 5. Parameters used for the LINE extraction algorism applied to the SAR data.

Parameter	Value (pixel)
Edge filter radius	5
Edge gradient threshold	20
Curve length threshold	30
Line fitting error threshold	2
Angular difference threshold	15
Linking distance threshold	40

The basic statistical parameters of extracted lineaments are shown in Table 6. The manually extracted lineaments have smaller count (691) and greater maximum length (11.96 km) compared to those of automated lineaments extracted from PALSAR (count = 6324, max length = 3.78 km) and Sentinel-1 (count = 7062, max length = 3.47 km). The manually and automatically extracted lineaments were imported as data layers to the ArcMap environment, and the lineament density maps were, thus, generated using the line density module in the spatial analyst toolbox (Figure 10A,B). The manually extracted lineaments are overlain on the directionally filtered FCC images (R: 90°, G: 135°, B: 45°) of the Sentinel-1B PC3 (Figure 11A) and the automatically extracted lineaments on the directionally filtered PALSAR PC2 FCC (R: 90°, G: 135°, B: 45°) image (Figure 11B). The comparison weighs the manually extracted and directionally filtered Sentinel-1B results.

Table 6. Basic statistics of extracted lineaments from SAR data.

Method	Visual Extraction	Automated Extraction	
Used Data	RGB PC Images and Filters	Filtered PALSAR PC2	Filtered Sentinel PC3
Count	691	6324	7062
Minimum	0.49	0.38	0.34
Maximum	11.96	3.78	3.47
Sum	1949.07	4307.17	4448.77
Mean	2.82	0.68	0.63
Standard Deviation	1.74	0.33	0.30

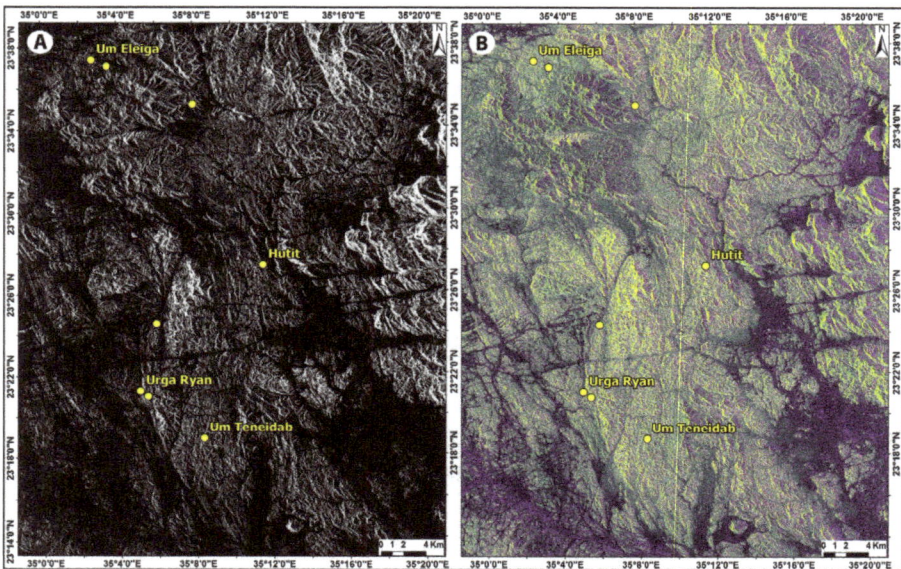

Figure 8. Structural mapping promoted using PALSAR and Sentinel-1B PCA data. (**A**) Grayscale image of PALSAR PC2; (**B**) FCC image of Sentinel-1 PCs (R: PC2, G: PC4, B: PC3).

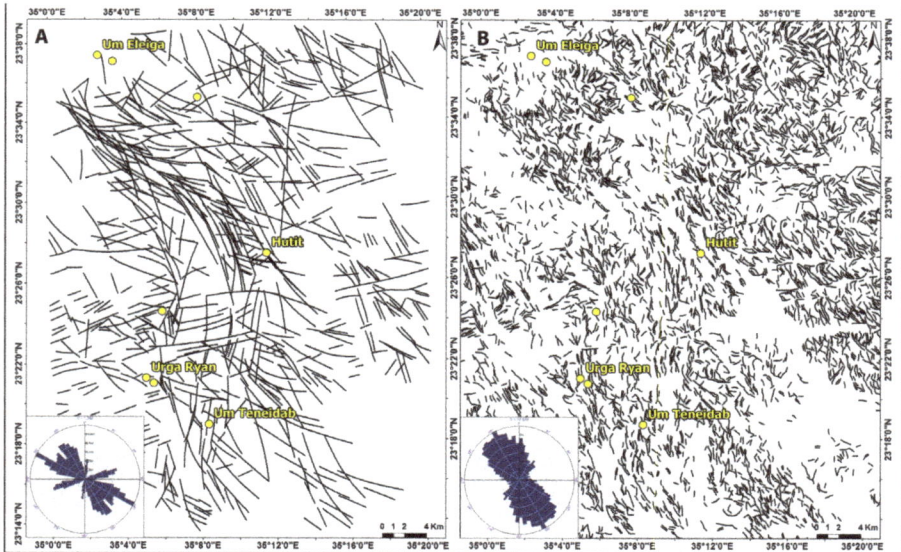

Figure 9. Geological lineament extracted from the SAR data of the study area. (**A**) Manually extracted lineaments aided by edge enhancement techniques of PALSAR and Sentinel-1 PCA images verified in the fieldwork; (**B**) results of the automated lineament extraction technique (ALE) using the same images as in Figure 8 after applying the directional filters. Insets are rose diagrams showing the main structural trends in the study area. The manually extracted lineament results obviously show the two main northwest (NW)–southeast (SE) and north-northwest (NNW)–south-southeast (SSE) structural trends, whereas the automated extraction shows only the prevailing NNW–SSE trend.

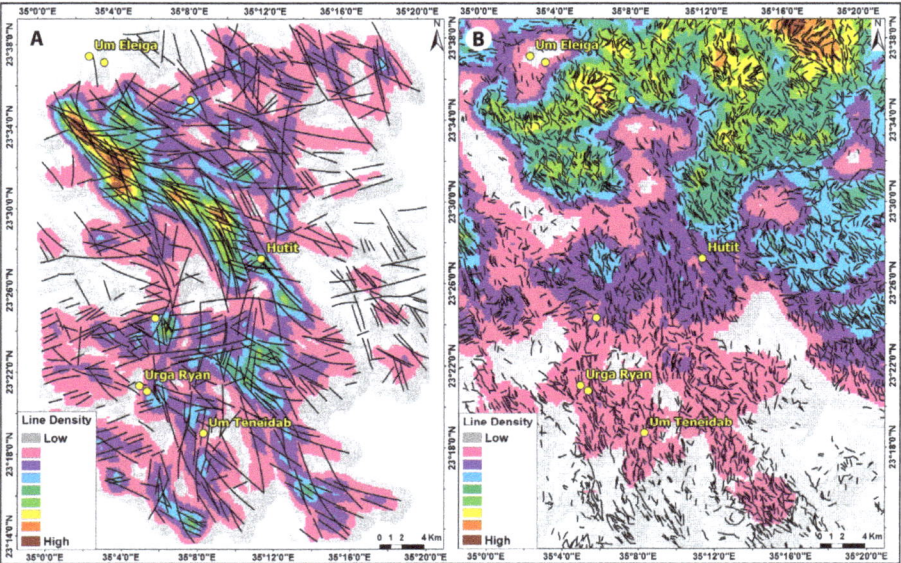

Figure 10. Density maps of the extracted structural lineaments in the study area. (**A**) Line density map of manually extracted lineaments aided by edge enhancement and directional filtering techniques; (**B**) line density map of lineaments extracted automatically by the LINE algorithm (see the text for more details).

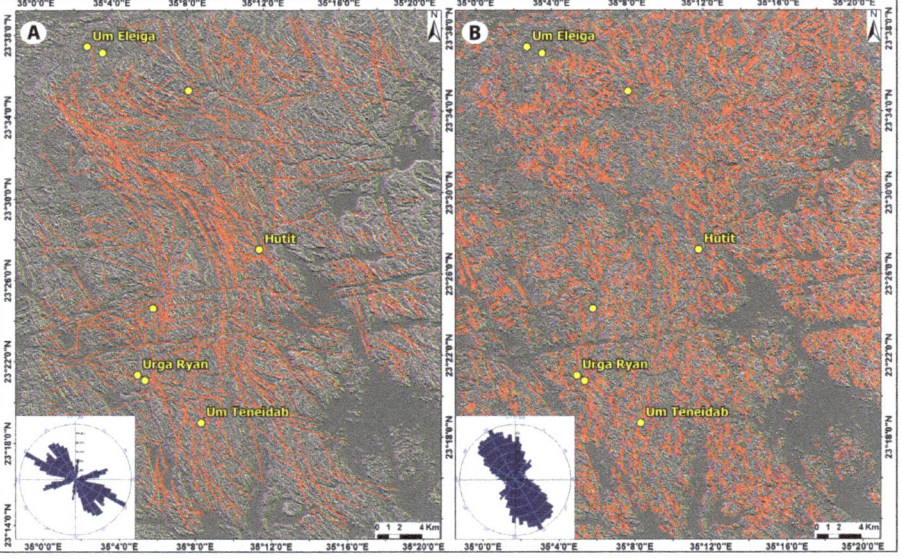

Figure 11. (**A**) Manually extracted lineaments overlaying a FCC image of directional filters (R: 90°, G: 135°, B: 45°) on the Sentinel-1B PC3; (**B**) automatically extracted lineaments superimposed on an FCC (R: 90°, G: 135°, B: 45°) image directionally filtered with PALSAR PC2.

4. Gold Occurrences

In this section, details of the significant gold occurrences in the study area are given. Aside from the comprehensive description of the geological setting of each of the gold occurrences, a pivotal emphasize is given to the deformation fabrics exhibited by the mineralized quartz veins and the structural elements observed in the mine areas and at the regional map scale.

4.1. Um Eleiga Gold Deposit

The Um Eleiga area, ~45 km west of the Red Sea coast, hosts a historic gold deposit in the SED. The gold mine lies along the NNW–SSE-trending Wadi Um Eleiga (Figure 12A), where traces of placer mining and shallow pits, dumps, ancient mining camps, stone anvils, hammers, and grinding date back to the Roman–Byzantine and early Islamic times [67–69]. Gold contents in quartz dumps collected from shallow pits grade as high as 28 g/t [69], whereas the fine-grained alluvium at the base of the Wadi deposits/terraces yield anomalous concentrations of up to 36 g/t Au [40,70].

The Um Eleiga mine area is hosted by an elliptical, zoned intrusive complex (ca. 32 km^2) encompassing quartz–gabbro, diorite, tonalite, and granodiorite (Figure 12A). The intrusive complex cuts through allochthonous ophiolitic blocks of serpentinite–chromitite (Gebel Abu Dahr) embedded in a highly tectonized matrix of pelitic and carbonaceous metasedimentary and metavolcanic rocks (see Figures 5A and 7A). The different rock varieties in the complex are separated from each other by gradational contacts. WNW-, NNW-, and N-trending fault/fracture sets densely dissect the complex. The central part of the complex features olivine-, pyroxene-, and/or hornblende-rich gabbros, whereas diorite surrounds the gabbroic core and locally shows a distinct porphyritic texture. Tonalite and granodiorite form the outer parts of the complex. A small body of albitized microdiorite and sets of lamprophyre and andesite dykes cut the western parts of the complex. In the eastern part, the highly tectonized gabbro and diorite contain intensely kaolinitized and oxidized zones in which massive and disseminated goethite, malachite, and azurite are abundant.

The mineralized quartz veins trend mainly NE–SW or ENE–WSW and cut the gabbroic rocks in the central part of the complex and extend beyond the gabbro–diorite boundary (Figure 12B). A later barren generation of N-, NW-, and E-trending quartz veins are restricted to fault intersections and tension gashes in the highly deformed gabbroic rocks. Zoheir et al. [71] suggested that fault/joint intersections are the main structural control of intensely hydrothermal alteration zones and high gold contents in the central part of the Um Eleiga complex. The E-trending fractures cutting the gabbroic rocks are locally associated with chlorite–calcite and chalcedonic quartz alteration assemblage. Sulfide-bearing quartz veins (5–40 cm thick) cutting the gabbro–diorite complex are scarce, and cannot be compared with the extensive old workings. It is herein considered that the old miners worked quartz veins, scattered quartz blocks, and the friable wadi alluvium underneath the consolidated terraces. These milky quartz veins have brecciated borders in which quartz fragments are cemented by chalcedonic quartz, calcite, chlorite, and sulfides (Figure 12C).

Gold-bearing quartz dumps enclose chlorite–sericite–calcite selvages and pyrite–malachite–limonite gossans (Figure 12D,E). The main ore minerals are pyrite, chalcopyrite, sphalerite, pyrrhotite, and gold. Pyrrhotite and sulfarsenide form scarce primary inclusions, whereas pyrite, chalcopyrite, sphalerite, and free gold are late in the paragenetic sequence. Pyrite forms disseminated euhedral to subhedral grains with pyrrhotite inclusions. Chalcopyrite and sphalerite are intergrown with subhedral pyrite grains and also occur as inclusions in pyrite grains. Free-milling gold blebs and specks occur along fine ribbons and selvages of wallrock enclosed in the quartz veins. Hydrothermal alteration phases associated with gold–sulfide quartz veins include fine-grained quartz, chlorite, calcite, sericite, rutile, and sulfides. These hydrothermal minerals are clearly late relative to the igneous paragenesis of the host zoned intrusion. Pyrite, chalcopyrite, and sphalerite are disseminated in domains of pervasive quartz–sericite–chlorite alteration.

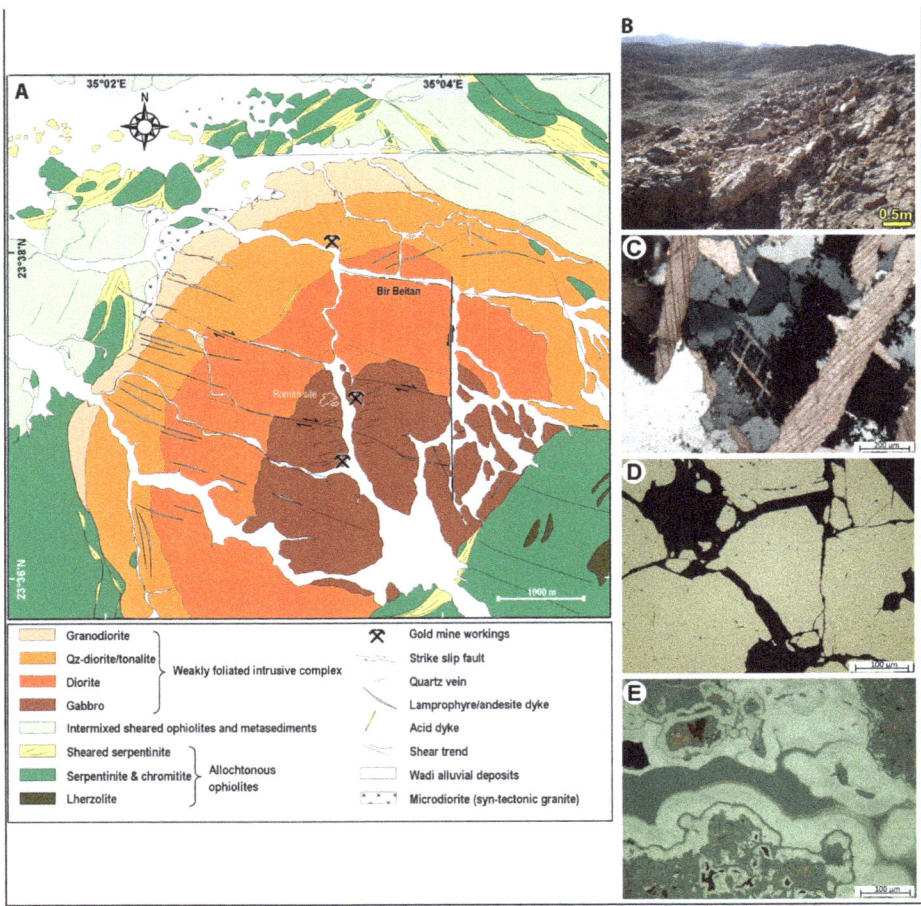

Figure 12. (**A**) Geological map of the Um Eleiga gold mine, modified from Zoheir et al. [71]; (**B**) milky quartz vein cuts across the gabbroic central part of the Um Eleiga intrusive complex; (**C**) transmitted, crossed polar light photomicrograph of the mineralized quartz veins showing dominant calcite, chlorite, and sulfide minerals; (**D**,**E**) Reflected light photomicrographs of the mineralized quartz veins showing disseminated euhedral to subhedral pyrite, and botryoidal malachite replacing chalcopyrite or other Cu-sulfides.

4.2. Hutit Gold Occurrence (Also Known as Huzama or Rahaba Mine)

The Hutit gold occurrence lies between the head of Wadi Huzama and Wadi Hutit, a small tributary of Wadi Rahaba (Figure 13A). The mine area is occupied by conspicuous, high to moderately elevated hills of serpentinite, mafic metavolcanics, and pelitic/psammopelitic metasedimentary rocks. Old mining in the area dates back to the early 20th century, but millstones from diorite and gabbro point to old workings, possibly from the Islamic times (seventh to eighth century). Recently, exploration mapping, structural survey, rock chip, and trench sampling, complemented by a ~30,000-m diamond core drilling program completed by Thani Ashanti (now Thani Stratex Resources Ltd.) between 2009 and 2013, indicate an in-house, non-Joint Ore Reserves Committee Code (non-JORC) resource estimate of ~0.5 Moz gold (http://thanistratex.com/projects/projects-overview/).

The old miners extracted the ore bodies from two main (northern and southern) mines. The mine area was mapped at the 1:1000 scale and the mine area was subdivided into northern and southern [69]. Abundant remains of grinding and separation plants are observed in the northern mine (Figure 13B). Nevertheless, the preserved crusher stages, leaching basins, and loading station in the southern mine reflect significant mining activities in the past. In both mines, a main entrance through a horizontal ~E–W adit leads to the veins at a distance of 20 or 35 m.

The gold-bearing quartz veins occur along the contact between elongate allochthonous serpentinite masses and successions of intercalated metavolcanic and metasedimentary rocks. Field criteria indicate that these rocks are tectonically intermixed and intercalated with graphite-bearing schists forming a distinct mélange unit, in which serpentinite blocks are embedded. Contacts between the serpentinite blocks and the underlying rocks are zones of intensive shearing, grain size reduction and abundant talc-, quartz-, and carbonate- rich rocks. Serpentinite is composed essentially of antigorite, relict olivine, subordinate talc, calcite, tremolite, and minor chromite and magnetite. In the sheared horizons, no relics of olivine or pyroxene are found, where the rocks are composed mainly of antigorite and talc. Blocks of ophiolitic metabasalt and metagabbro are embedded in the sheared matrix and stretched parallel to the NW–SE foliation (Figure 13B).

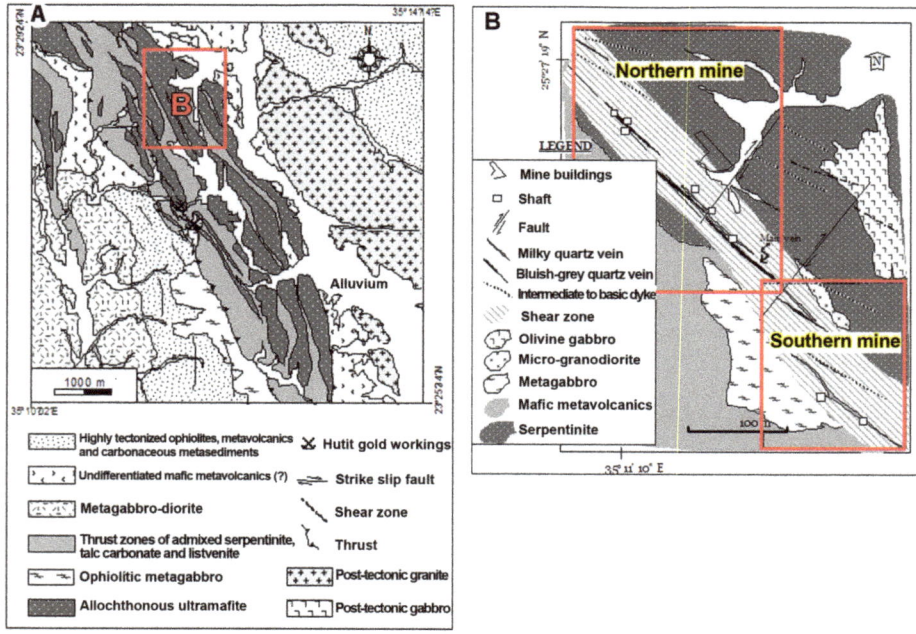

Figure 13. (**A**) Geological map of the Hutit gold mine and surroundings, compiled from field work and remote-sensing data interpretations and adapted from Hassan and El-Manakhly [69]; (**B**) detailed geological map of the main lode in the northern mine and southern mine areas.

Alternating mafic metavolcanic and metasedimentary rocks prevail in the northern mine. The mélange rocks are characterized by moderate to high deformation, especially in proximity to the large faults. The elongation of clasts within the mélange matrix locally defines a moderately north-northwest plunging lineation. The metavolcanic rocks are mainly dark-colored, foliated, and slightly or intensively contorted. These rocks occupy the western part of the mine area and form an NW–SE-trending belt. The intact massive blocks of these rocks assume basaltic and ultramafic protoliths (Figure 13C,D). The schistose varieties are mainly tremolite–actinolite and chlorite schists. They are intercalated locally with bands of metasedimentary rocks (i.e., metasiltstone and metamudstone), composed essentially of chlorite ± biotite, epidote, and quartz. These schistose rocks are graphite-bearing, especially in the northern part of the mine area.

An NNW–SSE elongate intrusion of gabbro cuts into the tectonized serpentinite in the northern mine. Small masses of micro-granodiorite dyke-like bodies (NW–SE), and dykes with different compositions cut the country serpentinite, metavolcanic, and gabbro rocks in the northern mine. Most of the dykes strike NW–SE, but a small number of the mafic dykes are NE-trending. Rhyodacite, dacite, and andesite dykes are generally porphyritic with tabular plagioclase and rhombic hornblende, embedded in a fine- to very fine-grained groundmass. The mafic dykes, mainly basalt to basaltic andesite, are notably abundant in the northern mine.

NW–SE thrust segments bound the ophiolitic serpentinite masses and dip moderately or steeply to the NE. Emplacement of the serpentinite slices is interpreted as being from east to west, constrained from moderately to steeply east-northeast dipping shear planes and consistently NW-trending stretching lineation, generally consistent with the W-directed tectonic transport of ophiolitic rocks in Wadi Ghadir area, north of the present study area [7]. A kilometer-scale shear zone striking in an NW-SE direction is superimposed on the thrust zone and related fabrics. Analyzing the shear planes, asymmetrical fabrics, and slickensides indicates that this shear zone is a reverse fault zone (Figure 14A), which accommodates a left-lateral displacement. Although nearly parallel to the thrust segments, this shear zone dips steeply to the west (Figure 14B). Conjugate joints and faults are common in the northern mine. Fractures in the country rocks follow two main trends, N 40° E and N 50–60° W. No direct cross-cutting relationship was observed between the quartz veins and dykes.

The mineralized quartz veins occur along a 150-m-wide shear zone, where quartz veins have anastomosing and undulating morphologies, both down-dip and along the strike. The shear zone, quartz veins, and associated hydrothermal alteration overprint the metamorphic mineral assemblage and fabrics in the host metavolcanic and serpentinite rocks. Two types of gold-bearing quartz veins are reported in the mine area, including bluish-gray and milky quartz veins. In the northern mine, a 180-m-long bluish-gray quartz vein varies in thickness from less than 30 up to 150 cm [69]. It strikes parallel to the shear zone (NW–SE) and dips 80° SW in the northern mine. This vein is made up mainly of gray quartz, carbonate, and subordinate colorless quartz and rare sulfides. The main entrance, along an adit from east to west, was used to work out this vein. A milky quartz vein of 20–50 cm thickness and 90 m length occurs in the vicinity of the main quartz vein. The bluish quartz veins are common in the northern mine, whereas milky quartz veins are rather dominant in the southern mine. The banded appearance of the bluish quartz veins (Figure 14C) and their association with intensively altered host rocks, with abundant signs of strain and the absence of gashes and tensional gaps, suggest that these veins were formed under a compressional stress regime through formation of the shear zone. Asymmetric bent quartz lenses (Figure 14E) provide signs of left-lateral shearing, but sub-vertical slickensides along the vein walls also corroborate the reverse nature of the shear zone. Field observations indicate that the milky quartz veins are younger than the bluish-gray quartz veins, on the basis of cross-cutting relationships (Figure 14F).

The bluish-gray quartz veins are surrounded by carbonated, ferruginated, and less commonly kaolinitized wallrocks. The milky quartz veins in the southern mine are commonly surrounded by sericite–chlorite and less commonly epidote where they cut through metavolcanic rocks. In both types of quartz veins, signs of plastic and brittle deformation are abundant. Sub-grain development is the most characteristic feature of zones where the quartz veins are narrow and branchiate. Ribbon-shaped grain formation and less commonly mortar texture are also observed in the quartz veins. All interstitial quartz grains show undulatory extinction, deformation bands, and minor development of tiny equidimensional recrystallized grains around grain margins.

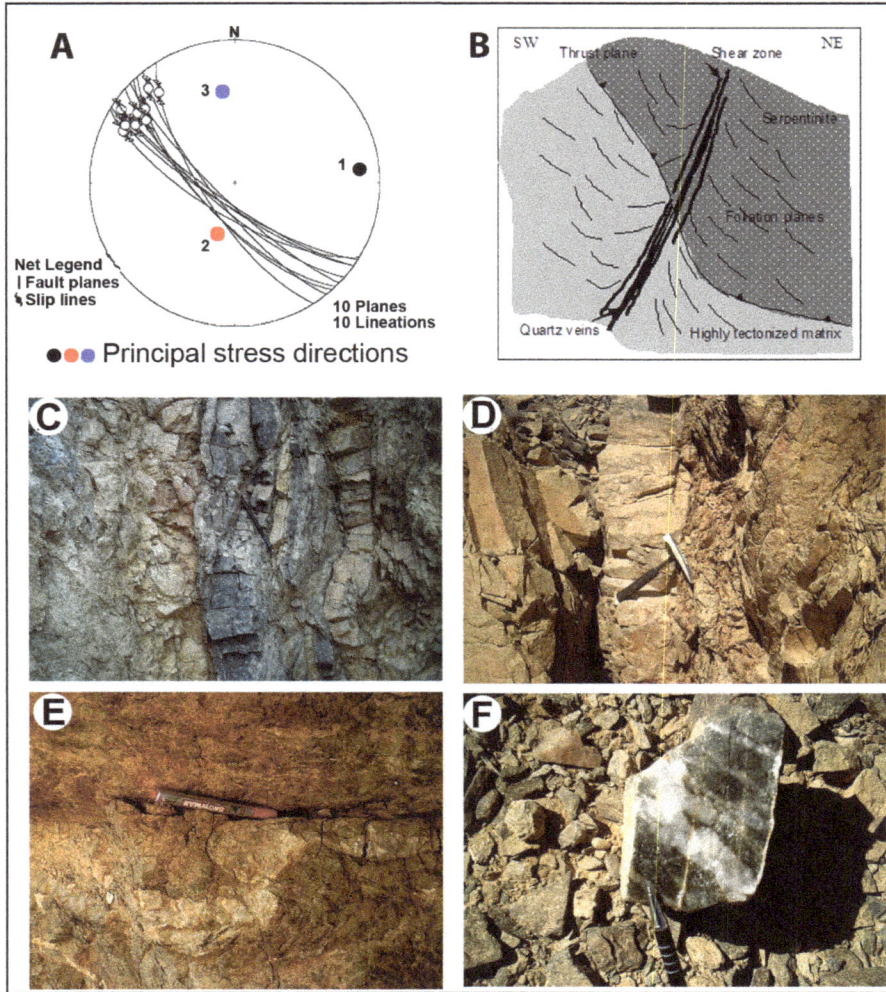

Figure 14. (**A**) Equal area stereographic projection (lower hemisphere) of planes of the shear zone and poles to the associated slickensides; (**B**) sketch drawing explaining the geometrical relationship between the quartz veined shear zone in the Hutit mine and the thrust structures. Note the deflection of the foliation about the shear zone (based on field observations); (**C**) nearly vertical, bluish-gray quartz veins associated with highly sheared serpentinite in the northern mine (looking NW); (**D**) milky quartz vein cutting metavolcanic rocks in the southern mine (looking N); (**E**) S-shaped quartz vein indicating a sinistral sense of shearing (looking W); (**F**) grayish quartz vein traversed by milky veinlets. The sample is from the northern mine lode.

The mineralogy of the quartz veins also includes arsenopyrite, pyrite, and less commonly gold. Both arsenopyrite and pyrite are usually altered into goethite. In this case, appearance of tiny gold, streaky or wire-like particles along the rhythmic zones of goethite is common. This indicates that oxidation led to remobilization of structure-bound gold from pyrite and arsenopyrite to form native gold in secondary sites. Data concerning the ore grade include fire assay concentrations of some samples from the grayish and milky quartz veins. Gabra [70] reported 1–40 g/t in quartz veins from the northern mine and 1–36 g/t in samples from quartz veins intercalated with sheared rocks in the southern mine. Takla et al. [72] analyzed samples from the two different types of quartz veins and reported an average of 20 g/t. They also investigated the hydrothermal alteration zone for its gold content and indicated that the adjacent altered wallrocks contain 8 g/t Au on average [72].

4.3. Um Teneidab Gold Mine (Also Known as Um Kalieb or Um Kalieba Mine)

The Um Teneidab mine is situated 3 km west of Gebel Um Teneidab and 13 km southwest of the Hutit mine. The Um Teneidab mine area is underlain by gabbroic rocks that are cut by abundant offshoots of fine-grained granite (Figure 15A). The contact between granite and the gabbroic host rocks is irregular and sharp (Figure 15B). The area was mapped at the 1:1000 scale and these rocks are assigned as metagabbro–diorite and late-orogenic granite [69]. Takla et al. [72] discussed the features in detail, implying that these rocks belong to the younger gabbroic rocks of the Egyptian basement complex. In the present work, we agree with Hassan and El-Manakhly [69], and classify the gabbroic rocks in the mine area as an island arc-metagabbro–diorite complex. This interpretation is based on some local foliated textures, a corona texture with brown hornblende bounding hypersthene crystals, and presence of more differentiated bosses with diorite composition.

Chlorite is common as an alteration mineral after the ferromagnesian mineral constituents of the host gabbros. The granitic rocks are composed of andesine, orthoclase, quartz, and intensively chloritized biotite. Approaching the quartz veins, plagioclase is more or less completely replaced by sericite and kaolinite. Pyrite is most common as alteration mineral disseminated in the hydrothermally altered granite and gabbroic rocks. Alteration is pervasive where the tectonized gabbro is densely seamed with granitic offshoots.

Structurally, the Um Teneidab mine area is traversed by conjugate NW–SE and NE–SW fault sets. Faults with no obvious lateral displacement dissect the granite body and offshoots in the western part of the mine area. Stretching lineation and slickensides along the quartz vein walls suggest that the shear zone experienced also little ductile deformation (Figure 15C). Formation of this shear zone is attributed to the competence heterogeneity between coarse-grained gabbro and fine-grained granite. Granularity gives additional cohesion contrast that might proceed to a discontinuity zone or plane between these two different lithologies. Abundant quartz veins and felsic dykes are controlled by NW–SE shear/fault sets, but show no direct cross-cutting relations. Deformation is intense in zones where the granite offshoots traverse the gabbroic rocks.

Gold in the Um Teneidab mine area is related to a system of 10–40-cm-thick milky quartz veins extending for more than 200 m along a wrenched shear zone (Figure 15A). These veins are NW- or NNW-trending and are commonly sub-vertical. The main lode is a zone of stockwork of veinlets (70 cm wide) bounded by hydrothermally altered wallrocks forming together a ~2-m-wide mineralization zone. The granite is notably sericitized and silicified close to the quartz veins (Figure 15D). The latter are massive, composed of coarse-grained quartz crystals locally fractured and filled with newly formed quartz, characteristically colorless (less than 3-cm-wide veinlets).

Most quartz veins in the mine area, particularly the thin ones, are completely recrystallized. Porphyroclasts embedded in less recrystallized, mosaic-like, strain-free quartz are observed along the flanks of quartz veins. The quartz porphyroclasts are lensoid and show strong undulose extinction, deformation lamellae, and sub-grain development; they contain numerous fluid inclusions of various generations. The boundaries of the shears are sharp, and the sulfides are clearly confined to zones of shearing and alteration.

Gold is disseminated as flakes in altered pyrite, associated with galena or as fillings in the microfractures of the quartz veins. Gold is also present in the hydrothermally altered wallrocks, i.e., the altered granite (quartz–sericite rocks). Less commonly, relics of pyrite are seen disseminated in the quartz veins, whereas gold wires along rhythmic zones in goethite are accidently seen. Gold occurs as native globules disseminated in the quartz veins, mostly along the grain boundaries. The quartz veins contain from <1 up to 30 g/t gold, with an average of 8 g/t in the altered wallrocks [72].

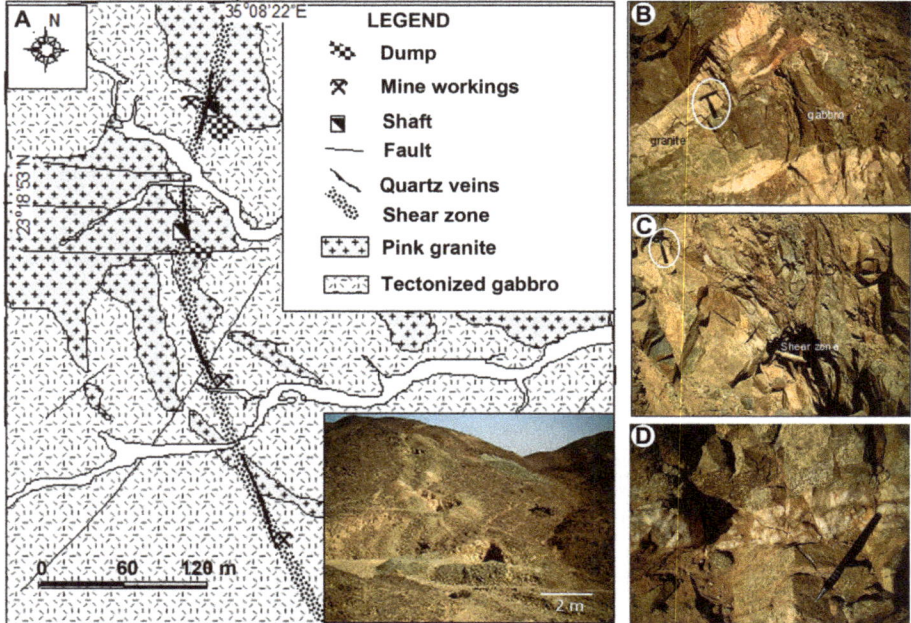

Figure 15. (**A**) Geologic map of the Um Teneidab gold mine, modified from Hassan and El-Manakhly [69]. The inset photo shows the mine shafts along the ~N–S line as in the map (photo looking to N); (**B**) granite offshoots cut the host gabbroic rocks and are associated with zones of discoloration in the mine area (photo looking to W); (**C**) shear zone with quartz veins between tectonized gabbro and fine-grained granite (photo looking to N); (**D**) milky quartz vein cutting across granite at the Um Teneidab gold mine. Notice the color bleaching of the wallrocks.

4.4. Urga Ryan Gold Occurrence

The Urga Ryan gold occurrence is located 17 km SW of the Hutit mine, ~10 km west of Gebel Um Teneidab. The location is ca. 2.5 km south of the intersection of the E–W Wadi Hutib and the nearly N–S Wadi Urga Ryan along the main wadi. The area surrounding the Urga Ryan occurrence is underlain by island arc-metavolcanic rocks, dominated by metaandesite and epidote–chlorite schist (Figure 16A). The metavolcanic sequence is locally affected by a several kilometer-scale shear system that led to intense shearing in an NNW–SSE direction overprinting the WNW–ESE schistosity of the metavolcanic rocks. In the eastern part of the mine area, a large granitoid intrusion of granodiorite or quartz diorite composition cuts the metavolcanic rocks. The granitoid rocks are slightly foliated and tapered along the foliation in the metavolcanic rocks and enclose elongated enclaves parallel to the metavolcanic rock schistosity. The old mine workings are situated in a low hill terrane that is underlain by sheared metavolcanic rocks (Figure 16A) along the main Wadi Urga Ryan, whereas mine houses spread over many tributaries around the area, likely reflecting considerable mine activities. The mineralization is, however, limited to small locations, particularly where the shearing is intense and quartz lenses are abundant.

The mineralization is confined to a local NNW–SSE shear zone, which dips steeply to westward, cutting across the sheared metavolcanic rocks (Figure 16B). The main lode is composed of boudinaged quartz veins and lenses, ranging in thickness from less than 5 cm to 30 cm and extending along the strike for more than 40 m. The host shear zone exhibits features of brittle and ductile regimes manifested by mylonitization, asymmetric boudinaged quartz lenses, and partial recrystallization

(Figure 16C). The sense of shear along this shear zone is derived from the lensoidal quartz pockets that point to a left-lateral movement concurrent with vein emplacement (Figure 16D). This observation suggests a spatial and temporal relationship between the shear zone and gold-bearing quartz veins.

The local dynamic recrystallization of the host metavolcanic rocks is assumed to have been strongly catalyzed by fluid flow through dilatant zones and promoted ductility-enhancing mineral reactions. These high-fluid-pressure features likely develop in rocks buried at great depths, indicating mesothermal conditions typical of orogenic gold deposits. In the mine area, hydrothermal alteration is confined to narrow zones of sheared wallrocks bounding the quartz veins and veinlets. The quartz veins gave a gold content ranging from 1–7 g/t [70].

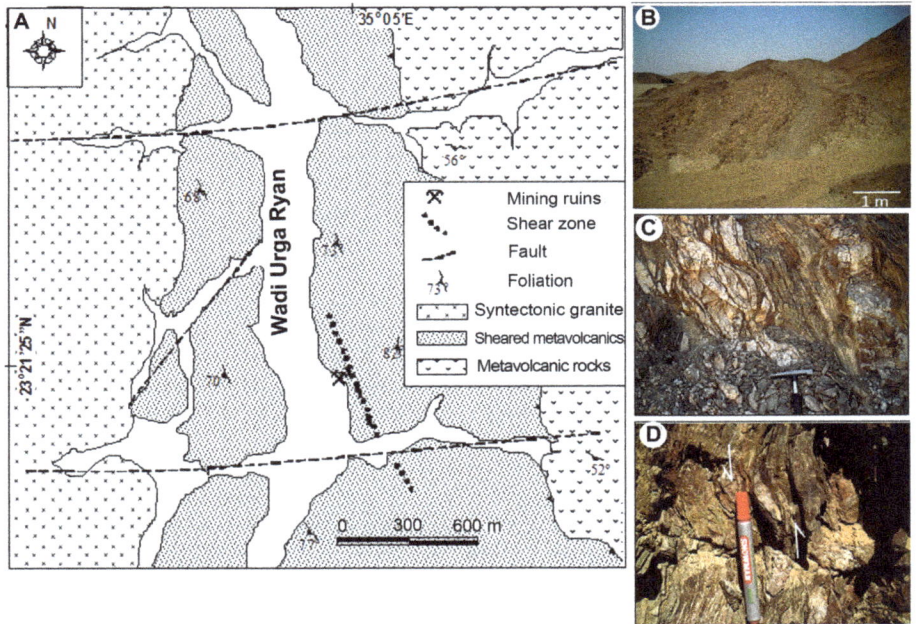

Figure 16. (**A**) Simplified geological map of the Urga Ryan gold occurrence (compiled from satellite-imagery data coupled with our fieldwork), (**B**) moderately dipping, strongly foliated, and sheared metavolcanic rocks at the Urga Ryan gold mine with zones of discoloration due to variable hydrothermal alteration distal and proximal to the gold-bearing veins; (**C**) boudinaged quartz lenses along the shear zone surrounded by ferruginated, sericitized wallrocks; (**D**) lensoid (sigmoidal) quartz lenses indicating sinistral sense of shear along the host shear zone. Inclined slickensides on the vein wall surfaces suggest that movement was oblique with lateral and vertical components.

5. Discussion

The ASTER and Landsat-8 OLI band ratio images showed the effective absorption features of the mafic rock-forming minerals and their metasomatic products. Carbonate minerals had diagnostic absorption features in the wavelength region 2.1–2.5 µm corresponding to the OLI–SWIR band 7 and ASTER–SWIR bands 6, 7, and 8. On the contrary, these minerals showed high-reflectance features in the wavelengths of OLI–SWIR band 6 and ASTER–SWIR band 4. Talc, antigorite, and other Mg-OH minerals showed diagnostic absorption responses at the wavelengths (1.39 and 2.3 µm) matching with the OLI–SWIR bands 6 and 7 and ASTER–SWIR bands 4 and 8. Such minerals exhibited strong reflectance through the wavelengths of OLI band 4 and ASTER bands 2 and 6. Kaolinite, muscovite, and other Al-OH-bearing minerals exhibited diagnostic absorption features within the wavelength range 2.1–2.2 µm equivalent to the OLI–SWIR band 7 and ASTER–SWIR bands 5 and 6. The diagnostic absorption features of Fe–Mg-OH-bearing minerals including amphiboles, chlorite, and epidote appeared through the wavelengths of ASTER–SWIR band 8, while their high-reflectance features were recorded within the ASTER–SWIR band 6. The OLI-band ratio (6/7), CI and ASTER–RBD (6 + 9/8^2 and 5 + 7/6^2) images were effective in discrimination of carbonate and clay-rich zones, talcous serpentinite, and highly tectonized Mg-OH-bearing rocks. Alternatively, the OLI-band ratio (6/5) and the mafic index (MI) were used to separate the mafic–ultramafic rocks from felsic rocks, while the RBD (6 + 9/8 and 7 + 9/8) and Fe-Mg-OH-MI were utilized to amplify the spectral signature of amphibole-, chlorite-, and epidote-bearing zones (Figures 4–7).

The automated extraction technique produces short dense lineaments that are difficult to relate to tectonically significant structures. Dray valleys (wadis) along the weakness zones correspond to the major faults in the automated lineament extraction results, such as Wadi Hutib, W Urga Ryan, and Urga Atshan. These results are not necessarily accurate. The manual extraction of lineaments results in long lineaments corresponding to the major structures deforming the lithological units in the study area (Figure 9A). The line density map represents the number of lineaments per square unit area. The line density maps show a high concentration of lineaments in the central and northern parts, where the NW–SE and NNW–SSE structural trends dominate (Figure 10A).

Thresholding the resultant grayscale images from band ratios, RBD, and mineralogical indices allowed the extraction of representative pixels (regions of interest), which were converted into vector shape files and then added to the ArcMap environment. Finally, the extracted lineaments were laid over the vector data layers of ratios and mineralogical indices to engender a final semi-automated image on which the litho-structural relationships and the spatial distribution of gold occurrences and lineament intersections were best presented (Figure 17). The high deformation and fracturing zones bounding the sheared ophiolitic belt contain most of the shear-associated gold occurrences (see Figure 4A,B). Such zones are propitious features, indicating high probability of zone potential for undiscovered gold resources.

The Wadi Beitan shear zone (Figure 17) discriminates the gneissic granite and gabbro–diorite complex from strongly foliated metavolcanic and metavolcaniclastic rocks in the western part of the map area. Shearing along this zone is oblique with a high-angle dip component along moderately to steeply eastward planes and a left-lateral strike (lateral) component. Widespread shearing bands, pervasive sericite, and carbonate alteration are observed along this zone. Together with the Wadi Khashab shear zone further south, the Wadi Beitan shear zone is part of an extensive (~100-km-long) shear corridor accommodating several, although generally small-scale, gold occurrences. Quartz veins showing signs of ductile shearing are locally sulfide-bearing and are associated with green malachite alteration zones. The Wadi Rahaba shear zone is a steeply east-dipping left-lateral brittle–ductile shear zone separating tectonically mixed ophiolitic rocks from schistose island arc rocks. The ophiolitic and metavolcanic rocks on both sides of the shear zone show S-shaped fault–drag folds, consistent with the sinistral movement. The Hutit gold mine occurs along the Wadi Rahaba shear zone. The Bir Beitan fault is a steeply dipping, ESE-trending strike–slip fault, with local silicification and hydrothermal breccia zones. Hydrothermal brecciation and silica alteration are locally coinciding zones with disseminated sulfide and local malachite staining.

The NNW–SSE shear zones seem to have rather high potential for hydrothermal alteration and quartz veining where intersected by the extensive WNW–ESE shear zones (Figure 17). These shear zones accommodate a prominent sinistral displacement, measured 5 km farther south [42]. A good example is the location of the Urga Ryan occurrence. Younger ~E–W or ENE–WSW faults cut the regional structural trend and dislocate the lithologic boundaries with apparent dextral displacements of 1–2 km. The mixed kinematic shear sense indicators (sigmoidal lenses and stretching lineation) along the WNW–ESE faults may imply that early left-lateral shearing along these faults was obliterated later by a right-lateral displacement, or as a result of rejuvenation of shears antithetic to the master NNW–SSE sinistral shear zones.

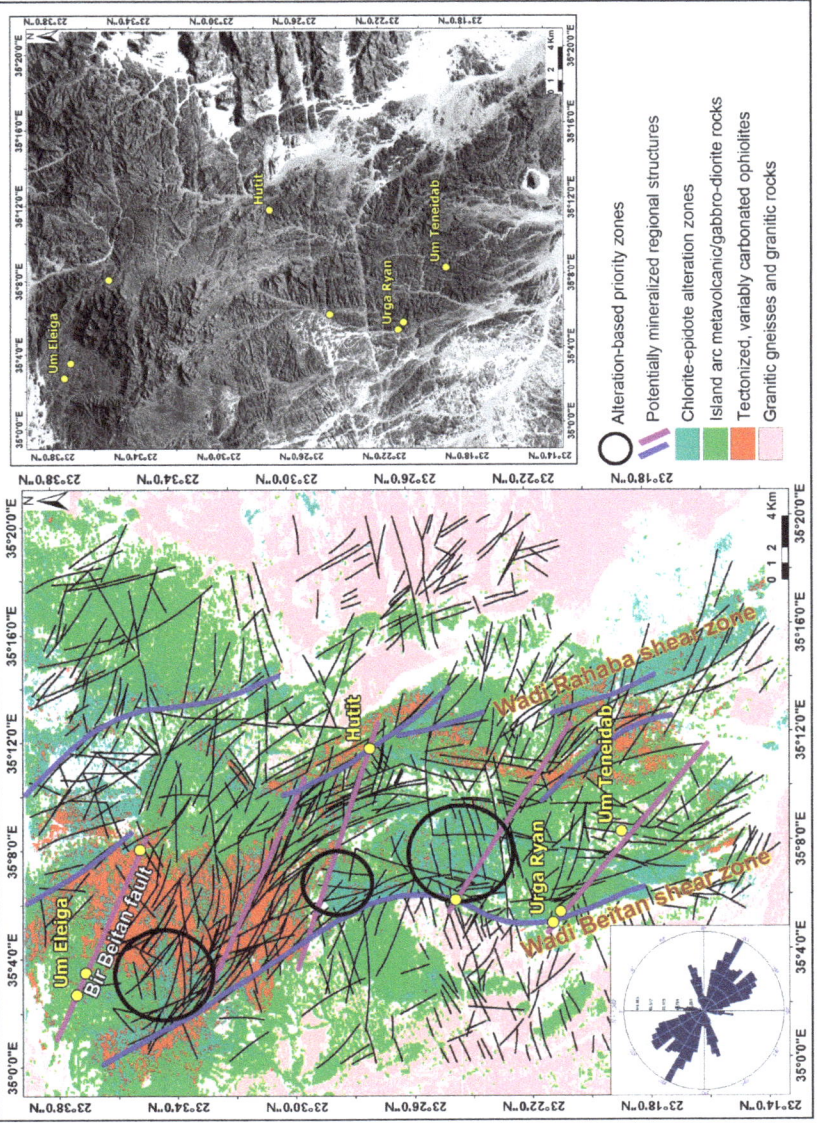

Figure 17. A fused lithological–structural map generated by merging the extracted structural lineaments and mineralogical index vector images of the study area. The inset is a rose diagram of the main structural lineaments. To the right is a grayscale ASTER-band 1 image of the study area. Zones with potential for hidden resources, considered as priority exploration targets, are highlighted.

Fusion of the field remote-sensing results indicates a prominent difference in the structural setting of gold occurrences. Occurrences confined to the main shear zones in the central parts of the shear corridor (i.e., Hutit and Urga Ryan occurrences) show abundant signs of plastic deformation and tinge with the main NNW–SSE shear trend. Gold occurrences in gabbro–diorite complexes (Um Eleiga and Um Teneidab) show a weaker association with shear zones, but occupy instead zones where shear splays fringe apart. The Um Eleiga and Um Teneidab occurrences are apparently controlled by the older WNW–ESE or NW–SE shear/fault zones that were later overprinted by the more prominent NNW–SSE shear zones. This could bear a temporal relationship suggesting that these two occurrences and other occurrences in identical settings predate the main shearing and the related gold occurrences. In other words, the spatial association revealed from the remote-sensing data and the structural observations during fieldwork may support the idea of gold introduction to the study area at different time episodes. The extracted mineral maps interpreted in terms of hydrothermal alteration and lithological controls on the gold mineralization emphasize the role played by the mylonitic zones (zones of finer grain size and carbonate and Fe-OH-mineral species) in controlling the distribution of the scattered gold occurrences in the study area. In accordance with this observation, we see that domains where the NW–SE and NNW–SSE shear/fault zones intersect are very likely zones of fluid focusing through the time of the shear system (Najd-related) development. Verification made through the mineralogical indices showed overlapping carbonate and hydrous minerals in such lineament intersection zones.

6. Conclusions

Fused satellite radar and radiometer data were integrated with comprehensive field studies of the gold occurrences in the Wadi Beitan–Wadi Rahaba area, supplemented by petrography of the quartz veins, highlighting the gold mineralization controlled by kilometer-scale shear zones. Field and microscopic investigations revealed that most of the mineralization occurs in the central-shear quartz veins within the main shear zone and are associated with strong stretching and mylonitic lineation in the wallrock. These features and the S-folded quartz lenses suggest that formation of the mineralized quartz veins took place under a sinistral transpression regime.

The multispectral and SAR data promoted a robust and fast automated mapping technique of the study area and promoted a better understanding of factors controlling the distribution of gold occurrences. The band-rationing technique, principal component and independent component analyses, spectral mineralogical indices, directional filtering, and automated lineament extraction algorithms were found to be powerful in highlighting and extracting lineaments and hydrothermal alteration zones. A density lineament map and the hydrothermal alteration images indicate that gold occurrences are mostly confined to chlorite–epidote alteration zones coinciding with high lineament intersections. This study concluded with a priority map with zones defined as exploration targets with high potential (Figure 17). These zones and the extensive ~WNW–ESE displacement shear/fault zones associated with epidote–chlorite and silicified hydrothermal breccia have to be sampled and assayed.

Author Contributions: Conceptualization, B.Z. and A.E.; methodology and flowchart set-up, A.E., B.Z., and N.S.; field work and validation, B.Z. and M.A.-W.; writing—review and editing, B.Z. and A.E. Publication fees were waived by Remote Sensing as B.Z. is a guest editor.

Funding: This research received no external funding.

Acknowledgments: Basem Zoheir wishes to acknowledge the Alexander von Humboldt foundation (AvH) for promoting this work. The four anonymous expert reviewers and Remote Sensing editors are thanked for providing constructive comments on the first draft of this paper.

Conflicts of Interest: The authors declare no conflict of interest.

References

1. Abdelsalam, M.G.; Stern, R.J. Sutures and shear zones in the Arabian-Nubian Shield. *J. Afr. Earth Sci.* **1996**, *23*, 289–310. [CrossRef]

2. Abdelsalam, M.G.; Abdeen, M.M.; Dowidar, H.M.; Stern, R.J.; Abdelghaffar, A.A. Structural evolution of the Neoproterozoic western Allaqi-Heiani suture zone, Southern Egypt. *Precambrian Res.* **2003**, *124*, 87–104. [CrossRef]
3. Johnson, P.R.; Andresen, A.; Collins, A.S.; Fowler, A.R.; Fritz, H.; Ghebreab, W.; Kusky, T.; Stern, R.J. Late Cryogenian–Ediacaran history of the Arabian–Nubian Shield: A review of depositional, plutonic, structural, and tectonic events in the closing stages of the northern East African Orogen. *J. Afr. Earth Sci.* **2011**, *61*, 167–232. [CrossRef]
4. Stern, R.J.; Hedge, C.E. Geochronologic and isotopic constraints on late Precambrian crustal evolution in the Eastern Desert of Egypt. *Am. J. Sci.* **1985**, *285*, 97–127. [CrossRef]
5. Noweir, A.M.; El-Amawy, M.A.; Rashwan, A.A.; Abdel-Aziz, A.M. Geology and structural evolution of the Pan-African basement rocks around Wadi Umm Araka, Northeast Wadi Allaqi, South Eastern Desert, Egypt. *Egypt. J. Geol.* **1996**, *40*, 477–512.
6. Miller, M.M.; Dixon, T.H. Late Proterozoic evolution of the northern part of the Hamisana zone, northeast Sudan: Constraints on Pan-African accretionary tectonics. *J. Geol. Soc.* **1992**, *149*, 743–750. [CrossRef]
7. Greiling, R.O.; Abdeen, M.M.; Dardir, A.A.; El Akhal, H.; El Ramly, M.F.; Kamal El Din, G.M.; Osman, A.F.; Rashwan, A.A.; Rice, A.H.N.; Sadek, M.F. A structural synthesis of the Proterozoic Arabian–Nubian Shield in Egypt. *Geologische Rundschau* **1994**, *83*, 484–501. [CrossRef]
8. Sultan, M.; Arvidson, R.E.; Duncan, I.J.; Stern, R.J.; El Kaliouby, B. Extension of the Najd shear system from Saudi Arabia to the central Eastern Desert of Egypt based on integrated field and Landsat observations. *Tectonics* **1988**, *7*, 1291–1306. [CrossRef]
9. Kusky, T.M.; Ramadan, T. Structural controls on Neoproterozoic mineralization in the south Eastern Desert, Egypt: An integrated field, Landsat TM, and SIR C/X approach. *J. Afr. Earth Sci.* **2002**, *35*, 107–121. [CrossRef]
10. Zoheir, B.; Emam, A.; El-Amawy, M.; Abu-Alam, T. Auriferous shear zones in the central Allaqi-Heiani belt: Orogenic gold in post-accretionary structures, SE Egypt. *J. Afr. Earth Sci.* **2018**, *146*, 118–131. [CrossRef]
11. Zoheir, B.; Emam, A. Integrating geologic and satellite imagery data for high-resolution mapping and gold exploration targets in the South Eastern Desert, Egypt. *J. Afr. Earth Sci.* **2012**, *66*, 22–34. [CrossRef]
12. Zoheir, B.; Emam, A. Field and ASTER imagery data for the setting of gold mineralization in Western Allaqi–Heiani belt, Egypt: A case study from the Haimur deposit. *J. Afr. Earth Sci.* **2014**, *99*, 150–164. [CrossRef]
13. Pour, A.B.; Hashim, M. Structural geology mapping using PALSAR data in the Bau gold mining district, Sarawak, Malaysia. *Adv. Space Res.* **2014**, *54*, 644–654. [CrossRef]
14. Gabr, S.S.; Hassan, S.M.; Sadek, M.F. Prospecting for new gold-bearing alteration zones at El-Hoteib area, South Eastern Desert, Egypt, using remote sensing data analysis. *Ore Geol. Rev.* **2015**, *71*, 1–13. [CrossRef]
15. Bannari, A.; El-Battay, A.; Saquaque, A.; Miri, A. PALSAR-FBS L-HH Mode and Landsat-TM Data Fusion for Geological Mapping. *Adv. Remote Sens.* **2016**, *5*, 246. [CrossRef]
16. Amer, R.; El Mezayen, A.; Hasanein, M. ASTER spectral analysis for alteration minerals associated with gold mineralization. *Ore Geol. Rev.* **2016**, *75*, 239–251. [CrossRef]
17. Emam, A.; Zoheir, B.; Johnson, P. ASTER-based mapping of ophiolitic rocks: Examples from the Allaqi–Heiani suture, SE Egypt. *Int. Geol. Rev.* **2016**, *58*, 525–539. [CrossRef]
18. Emam, A.; Hamimi, Z.; El-Fakharani, A.; Abdel-Rahman, E.; Barreiro, J.G.; Abo-Soliman, M.Y. Utilization of ASTER and OLI data for lithological mapping of Nugrus-Hafafit area, South Eastern Desert of Egypt. *Arabian J. Geosci.* **2018**, *11*, 756. [CrossRef]
19. Hassan, S.M.; Taha, M.M.; Mohammad, A.T. Late Neoproterozoic basement rocks of Meatiq area, Central Eastern Desert, Egypt: Petrography and remote sensing characterizations. *J. Afr. Earth Sci.* **2017**, *131*, 14–31. [CrossRef]
20. Pour, A.B.; Hashim, M.; Makoundi, C.; Zaw, K. Structural Mapping of the Bentong-Raub Suture Zone Using PALSAR Remote Sensing Data, Peninsular Malaysia: Implications for Sediment-hosted/Orogenic Gold Mineral Systems Exploration. *Res. Geol.* **2016**, *66*, 368–385. [CrossRef]
21. Pour, A.B.; Park, Y.; Park, T.Y.S.; Hong, J.K.; Hashim, M.; Woo, J.; Ayoobi, I. Regional geology mapping using satellite-based remote sensing approach in Northern Victoria Land, Antarctica. *Polar Sci.* **2018**, *16*, 23–46. [CrossRef]
22. Amer, R.; Kusky, T.; Ghulam, A. Lithological mapping in the Central Eastern Desert of Egypt using ASTER data. *J. Afr. Earth Sci.* **2010**, *56*, 75–82. [CrossRef]

23. Amer, R.; Kusky, T.; El Mezayen, A. Remote sensing detection of gold related alteration zones in Um Rus area, Central Eastern Desert of Egypt. *Adv. Space Res.* **2012**, *49*, 121–134. [CrossRef]
24. Mars, J.C.; Rowan, L.C. ASTER spectral analysis and lithologic mapping of the Khanneshin carbonatite volcano, Afghanistan. *Geosphere* **2011**, *7*, 276–289. [CrossRef]
25. Madani, A.A.; Emam, A.A. SWIR ASTER band ratios for lithological mapping and mineral exploration: A case study from El Hudi area, southeastern desert, Egypt. *Arabian J. Geosci.* **2011**, *4*, 45–52. [CrossRef]
26. Sadek, M.F.; Ali-Bik, M.W.; Hassan, S.M. Late Neoproterozoic basement rocks of Kadabora-Suwayqat area, Central Eastern Desert, Egypt: Geochemical and remote sensing characterization. *Arabian J. Geosci.* **2015**, *8*, 10459–10479. [CrossRef]
27. Asran, A.M.; Emam, A.; El-Fakharani, A. Geology, structure, geochemistry and ASTER-based mapping of Neoproterozoic Gebel El-Delihimmi granites, Central Eastern Desert of Egypt. *Lithos* **2017**, *282*, 358–372. [CrossRef]
28. Abrams, M.J.; Brown, D.; Lepley, L.; Sadowski, R. Remote sensing for porphyry copper deposits in Southern Arizona. *Econ. Geol.* **1983**, *78*, 591–604. [CrossRef]
29. Roy, D.P.; Wulder, M.A.; Loveland, T.R.; Woodcock, C.E.; Allen, R.G.; Anderson, M.C.; Helder, D.; Irons, J.R.; Johnson, D.M.; Kennedy, R.; et al. Landsat-8: Science and product vision for terrestrial global change research. *Remote Sens. Environ.* **2014**, *145*, 154–172. [CrossRef]
30. Zhang, X.; Pazner, M.; Duke, N. Lithologic and mineral information extraction for gold exploration using ASTER data in the south Chocolate Mountains (California). *ISPRS J. Photogramm. Remote Sens.* **2007**, *62*, 271–282. [CrossRef]
31. Crowley, J.K.; Brickey, D.W.; Rowan, L.C. Airborne imaging spectrometer data of the Ruby Mountains, Montana: Mineral discrimination using relative absorption band-depth images. *Remote Sens. Environ.* **1989**, *29*, 121–134. [CrossRef]
32. Karnieli, A.; Meisels, A.; Fisher, L.; Arkin, Y. Automatic extraction and evaluation of geological linear features from digital remote sensing data using a Hough transform. *Photogramm. Eng. Remote Sens.* **1996**, *62*, 525–531.
33. Kim, G.B.; Lee, J.Y.; Lee, K.K. Construction of lineament maps related to groundwater occurrence with ArcView and Avenue™ scripts. *Comput. Geosci.* **2004**, *30*, 1117–1126. [CrossRef]
34. Arlegui, L.E.; Soriano, M.A. Characterizing lineaments from satellite images and field studies in the central Ebro basin (NE Spain). *Int. J. Remote Sens.* **1998**, *19*, 3169–3185. [CrossRef]
35. Suzen, M.L.; Toprak, V. Filtering of satellite images in geological lineament analyses: An application to a fault zone in Central Turkey. *Int. J. Remote Sens.* **1998**, *19*, 1101–1114. [CrossRef]
36. Hung, L.Q.; Dinh, N.Q.; Batelaan, O.; Tam, V.T.; Lagrou, D. Remote sensing and GIS-based analysis of cave development in the Suoimuoi catchment (Son La-NW Vietnam). *J. Cave Karst Stud.* **2002**, *64*, 23–33.
37. Conoco Coral. *Geological Map of Egypt, Scale 1:500,000, Bernice Sheet, NF 36 NE.*; The Egyptian General Petroleum Corporation (EGPC): Cairo, Egypt, 1987.
38. Nano, L.; Kontny, A.; Sadek, M.F.; Greiling, R.O. Structural evolution of metavolcanics in the surrounding of the gold mineralization at El Beida, South Eastern Desert, Egypt. *Ann. Geol. Surv. Egypt* **2002**, *25*, 11–22.
39. Sadek, M.F. Geological and Structural Setting of Wadi Hodein Area Southeast Egypt with Remote Sensing Applications. In Proceedings of the International Archives of the Photogrammetry, Remote Sensing and Spatial Information Sciences, Beijing, China, 3–11 July 2008; Volume XXXVII-B8.
40. Zoheir, B.A. Controls on lode gold mineralization, Romite deposit, South Eastern Desert, Egypt. *Geosci. Front.* **2012**, *3*, 571–585. [CrossRef]
41. Sadek, M.F. Discrimination of basement rocks and alteration zones in Shalatein area, Southeastern Egypt using Landsat TM Imagery data. *Egypt. J. Remote Sens. Space Sci.* **2004**, *7*, 89–98.
42. Abdeen, M.M.; Sadek, M.F.; Greiling, R.O. Thrusting and multiple folding in the Neoproterozoic Pan-African basement of Wadi Hodein area, south Eastern Desert, Egypt. *J. Afr. Earth Sci.* **2008**, *52*, 21–29. [CrossRef]
43. Kontny, A.; Sadek, M.F.; Abdallah, M.; Marioth, R.; Greiling, R.O. First investigation on shear zone related gold mineralization at El Beida area, South Eastern Desert, Egypt. In *Aspects of Pan-African Tectonics. International Cooperation, Bilateral Seminars International Bureau*; De Wall, H., Greiling, R.O., Eds.; Forschungszentrum Julich: Jülich, Germany, 1999; Volume 32, pp. 91–97.
44. Obeid, M.; Ali, M.; Mohamed, N. Geochemical exploration on the stream sediments of Gabal El Mueilha area, central Eastern Desert, Egypt: An overview on the rare metals. *Res. Geol.* **2001**, *51*, 217–227. [CrossRef]

45. Kokaly, R.F.; Clark, R.N.; Swayze, G.A.; Livo, K.E.; Hoefen, T.M.; Pearson, N.C.; Klein, A.J. *USGS Spectral Library Version 7 (No. 1035)*; US Geological Survey: Reston, VA, USA, 2017.
46. Henderson, F.M.; Lewis, A.J. *Principles and Applications of Imaging Radar. (Manual of Remote Sensing, Volume 2)*, 3rd ed.; Wiley: Hoboken, NJ, USA, 1998.
47. J-Spacesystems. *PALSAR User's Guide*, 2nd ed.; 2012; 69p. Available online: http://gds.palsar.ersdac.jspacesystems.or.jp/e/guide/ (accessed on 21 March 2019).
48. Ma, J.; Xiao, X.; Qin, Y.; Chen, B.; Hu, Y.; Li, X.; Zhao, B. Estimating aboveground biomass of broadleaf, needleleaf, and mixed forests in northeastern China through analysis of 25-m ALOS/PALSAR mosaic data. *For. Ecol. Manag.* **2017**, *389*, 199–210. [CrossRef]
49. ESA. *Sentinel-1: ESA's Radar Observatory Mission for GMES Operational Services*; ESA SP-1322/1; European Space Agency: Paris, France, March 2012.
50. Marghany, M.; Hashim, M. Lineament mapping using multispectral remote sensing satellite data. *Int. J. Phys. Sci.* **2010**, *5*, 1501–1507. [CrossRef]
51. Bishta, A.Z.; Sonbul, A.R.; Kashghari, W. Utilization of supervised classification in structural and lithological mapping of Wadi Al-Marwah Area, NW Arabian Shield, Saudi Arabia. *Arabian J. Geosci.* **2014**, *7*, 3855–3869. [CrossRef]
52. Flaash, U.S.G. *Atmospheric Correction Module: QUAC and Flaash User Guide v. 4.7*; ITT Visual Information Solutions Inc.: Boulder, CO, USA, 2009.
53. Smith, M.J. A Comparison of DG AComp, FLAASH and QUAC Atmospheric Compensation Algorithms Using WorldView-2 Imagery. Master's Dissertation, University of Colorado, Boulder, CO, USA, 2015.
54. Cooley, T.; Anderson, G.P.; Felde, G.W.; Hoke, M.L.; Ratkowski, A.J.; Chetwynd, J.H.; Gardner, J.A.; Adler-Golden, S.M.; Matthew, M.W.; Berk, A.; et al. FLAASH, a MODTRAN4-based atmospheric correction algorithm, its application and validation. In Proceedings of the IEEE International Geoscience and Remote Sensing Symposium, Toronto, ON, USA, 24–28 June 2002; Volume 3, pp. 1414–1418.
55. Gelautz, M.; Frick, H.; Raggam, J.; Burgstaller, J.; Leberl, F. SAR image simulation and analysis of alpine terrain. *ISPRS J. Photogramm. Remote Sens.* **1998**, *53*, 17–38. [CrossRef]
56. Franceschetti, G.; Lanari, R. *Synthetic Aperture Radar Processing*; Electronic Engineering Systems Series; CRC Press: Boca Raton, FL, USA, 1999.
57. Lee, J.S.; Jurkevich, L.; Dewaele, P.; Wambacq, P.; Osterlinck, A. Speckle filtering of synthetic aperture radar images: A review. *Remote Sens. Rev.* **1994**, *8*, 313–340. [CrossRef]
58. Sveinsson, J.R.; Benediktsson, J.A. Speckle reduction and enhancement of SAR images in the wavelet domain. In Proceedings of the IGARSS'96, 1996 International Geoscience and Remote Sensing Symposium, Lincoln, NE, USA, 31 May 1996; Volume 1, pp. 63–66.
59. Jensen, J.R. *Introductory Image Processing: A Remote Sensing Perspective*; Prentice Hall: New York, NY, USA, 1996.
60. Glikson, A.Y. Mineral-Mapping in the North Pilbara Craton. In *A Directed Principal Components of the Band Ratios Method for Correlating Landsat-5 Thematic Mapper Spectral Data with Geology: AGSO-Australian Geol. Survey Organisation Research Newsletter*; AGSO: Canberra City, ACT, Australia, 1997; Volume 26, pp. 1–4.
61. Sabins, F.F. Remote sensing for mineral exploration. *Ore Geol. Rev.* **1999**, *14*, 157–183. [CrossRef]
62. Inzana, J.; Kusky, T.; Higgs, G.; Tucker, R. Supervised classifications of Landsat TM band ratio images and Landsat TM band ratio image with radar for geological interpretations of central Madagascar. *J. Afr. Earth Sci.* **2003**, *37*, 59–72. [CrossRef]
63. Zumsprekel, H.; Prinz, T. Computer-enhanced multispectral remote sensing data: A useful tool for the geological mapping of Archean terrains in (semi) arid environments. *Comput. Geosci.* **2000**, *26*, 87–100. [CrossRef]
64. Ninomiya, Y. A stabilized vegetation index and several mineralogic indices defined for ASTER VNIR and SWIR data. In Proceedings of the IGARSS 2003, 2003 IEEE International Geoscience and Remote Sensing Symposium. Proceedings (IEEE Cat. No. 03CH37477), Toulouse, France, 21–25 July 2003; Volume 3, pp. 1552–1554.
65. Ninomiya, Y.; Fu, B.; Cudahy, T.J. Detecting lithology with Advanced Spaceborne Thermal Emission and Reflection Radiometer (ASTER) multispectral thermal infrared "radiance-at-sensor" data. *Remote Sens. Environ.* **2005**, *99*, 127–139. [CrossRef]
66. Bonn, F.; Rochon, G. *Précis de télédétection, vol. 1 Principes et méthodes*; AUPELF-UREF, Collection Presses Universitaires du Quebec: Montreal, QC, Canada, 1992.

67. Klemm, D.; Klemm, R.; Murr, A. Gold of the Pharaohs–6000 years of gold mining in Egypt and Nubia. *J. Afr. Earth Sci.* **2001**, *33*, 643–659. [CrossRef]
68. Klemm, R.; Klemm, D. *Gold and Gold Mining in Ancient Egypt and Nubia, Geoarchaeology of the Ancient Gold Mining Sites in the Egyptian and Sudanese Eastern Deserts*; Springer: Berlin/Heidelberg, Germany, 2013; 663p.
69. Hassan, O.A.; El-Manakhly, M.M. Gold deposits in the southern Eastern Desert, Egypt. In *A Commodity Package*; Egyptian Geological Survey and Mining Authority: Cairo, Egypt, 1986.
70. Gabra, S.Z. *Gold in Egypt: A Commodity Package*; 86 Pages, 11 Figures, 8 Maps; Geological Survey of Egypt: Cairo, Egypt, 1986.
71. Zoheir, B.A.; Mehanna, A.M.; Qaoud, N.N. Geochemistry and geothermobarometry of the Um Eleiga Neoproterozoic island arc intrusive complex, SE Egypt: Genesis of a potential gold-hosting intrusion. *Appl. Earth Sci.* **2008**, *117*, 89–111. [CrossRef]
72. Takla, M.A.; El Dougdoug, A.A.; Gad, M.A.; Rasmay, A.H.; El Tabbal, H.K. Gold-bearing quartz veins in mafic and ultramafic rocks, Hutite and Um Tenedba, south Eastern Desert, Egypt. *Ann. Geol. Surv. Egypt* **1995**, *20*, 411–432.

© 2019 by the authors. Licensee MDPI, Basel, Switzerland. This article is an open access article distributed under the terms and conditions of the Creative Commons Attribution (CC BY) license (http://creativecommons.org/licenses/by/4.0/).

Article

Integrated Hyperspectral and Geochemical Study of Sediment-Hosted Disseminated Gold at the Goldstrike District, Utah

Lei Sun [1,*], Shuhab Khan [1] and Peter Shabestari [2]

1. Department of Earth and Atmospheric Sciences, University of Houston, Houston, TX 77204, USA
2. Pilot Goldstrike Inc., 1031 Railroad St Suite 110, Elko, NV 89801, USA
* Correspondence: lsun10@uh.edu; Tel.: +1-806-392-6708

Received: 30 July 2019; Accepted: 20 August 2019; Published: 23 August 2019

Abstract: The Goldstrike district in southwest Utah is similar to Carlin-type gold deposits in Nevada that are characterized by sediment-hosted disseminated gold. Suitable structural and stratigraphic conditions facilitated precipitation of gold in arsenian pyrite grains from ascending gold-bearing fluids. This study used ground-based hyperspectral imaging to study a core drilled in the Goldstrike district covering the basal Claron Formation and Callville Limestone. Spectral modeling of absorptions at 2340, 2200, and 500 nm allowed the extraction of calcite, clay minerals, and ferric iron abundances and identification of lithology. This study integrated remote sensing and geochemistry data and identified an optimum stratigraphic combination of limestone above and siliciclastic rocks below in the basal Claron Formation, as well as decarbonatization, argillization, and pyrite oxidation in the Callville Limestone, that are related with gold mineralization. This study shows an example of utilizing ground-based hyperspectral imaging in geological characterization, which can be broadly applied in the determination of mining interests and classification of ore grades. The utilization of this new terrestrial remote sensing technique has great potentials in resource exploration and exploitation.

Keywords: hyperspectral; Goldstrike; gold mineralization; Carlin-type; decarbonatization; argillization

1. Introduction

The Great Basin of western North America has produced a significant amount of gold, making the United States one of the largest gold producers in the world [1]. Among the gold mines, the most famous ones are of the Carlin-type, carbonate rock-hosted disseminated gold deposits that formed in the Eocene Epoch [1,2]. Since the discoveries of Nevada Carlin-type gold deposits, similar sediment-hosted gold has been searched for around the world [3–6] and in nearby states [7–9]. The Goldstrike district is a gold deposit in southwest Utah similar to Carlin-type deposits [10,11]. The modern production of disseminated gold in Goldstrike was active from 1988 to 1996 producing 209,835 ounces of gold and 197,654 ounces of silver, which ceased because of falling gold price, increasing strip ratios, production costs, and safety concerns [12]. The remaining gold is currently being explored by Pilot Goldstrike Inc., Elko, USA [11].

The exploration and mining of precious metals have been a challenge because of the high expense of drilling, geochemical analyses, and metallurgy tests. On the other hand, hyperspectral imaging as a non-destructive, low cost, and large areal coverage, remote sensing technique, can provide high-resolution mineralogical analyses and is becoming popular in geologic studies [13–16]. To test the applicability of hyperspectral imaging in mineral exploration, this study combines hyperspectral imaging with fire assay metallurgy and inductively coupled plasma mass spectrometry (ICP-MS) geochemistry data in the study of gold mineralization in a drilled core from the Goldstrike district. The identification of an optimum stratigraphic combination and the rock alterations related to gold precipitation using hyperspectral imaging demonstrate applications of this state-of-art technique in the mining industry.

Geological Settings

The Goldstrike district locates in the Bull Valley Mountains, Washington County, southwest Utah (Figure 1). This area is on the eastern edge of the Great Basin [17,18]. The oldest structures found in the Goldstrike region are the southeastward thrust faults related with the Late Cretaceous Sevier orogenic event, which emplaced Paleozoic strata over the Mesozoic rocks on the Colorado Plateau, as well as several coeval asymmetrical folds [17]. Unconformably above the Paleozoic and Mesozoic rocks lies Tertiary siliciclastic and volcanic ash-flow tuff rocks. A major basin and range faulting event and localized drag folding trending east-northeast and west-northwest most likely formed in the Miocene following the tuff deposits [17]. These created high-angle faults with normal and strike-slip displacements, which then created the Goldstrike graben trending east–west and the Arsenic Gulch graben trending northwest–southeast [17]. Gold deposits were mostly found near the high-angle faults bounding the Goldstrike graben (Figure 1) and along the Covington Hill fault [11,12].

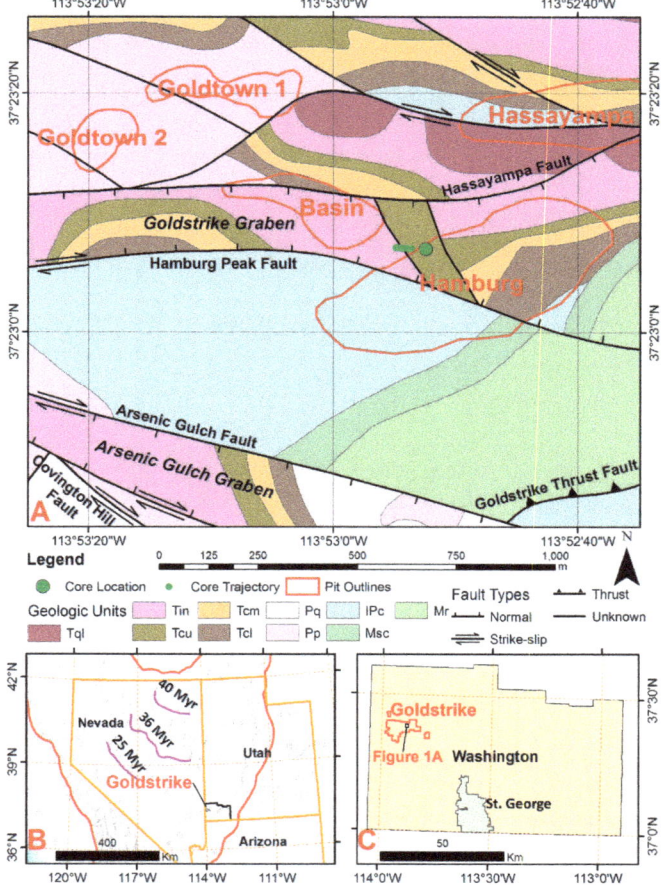

Figure 1. (**A**) Geologic map of the Hamburg pit and nearby area; (**B**) location of the Goldstrike district and Washington County, Utah in the basin and range setting of eastern Great Basin (red); and (**C**) location

of the Goldstrike district (red) in Washington County, Utah. The core location is shown with a large green circle in (**A**), and the map view trajectory of the retrieved core samples is shown with small green dots, pit outlines and names are shown in red. Mapped geologic units in (**A**) are: Tql = Leach Canyon Formation; Tin = Isom Formation and Wah Wah Springs Formation; Tcu = Claron Formation upper unit; Tcm = Claron Formation middle unit; Tcl = Claron Formation lower unit; Pq = Queantoweap Sandstone; Pp = Pakoon Formation; IPc = Callville Limestone; Msc = Scotty Wash Quartzite and Chainman Shale; Mr = Redwall Limestone. Original geologic map in (**A**) retrieved from Rowley et al. [19], the original scale of the map is 1:24,000. The magenta lines in (**B**) represent the mid-Tertiary magmatic fronts that swept the Great Basin from northeast to southwest (from Muntean et al. [2]), the ages of the magmatic fronts are also labeled. The black rectangle in (**C**) shows the extent of the map area in (**A**) within the Goldstrike District.

The stratigraphy of the Goldstrike district is shown in Figure 2. This district is underlain by a series of Tertiary ash-flow tuffs, limestone, sandstone, and conglomerate, and Mississippian through Permian carbonate and clastic sediments interbeds [10,11]. Late Cenozoic rocks include an undifferentiated tuff and andesite on top and the Quichapa ash-flow sheets. Below them are the early Cenozoic Isom limestones and tuff, the Needles Range Tuff, and the Claron Formation, which consists of an upper limestone, a middle Red Beds member of shale, siltstone, mudstone, sandstone, conglomerate, and limestone, and then the basal Claron sandstone and conglomeratic sandstone [20]. Triassic and Jurassic sediments are not widely exposed in the area, which consist of the Grapevine Wash Conglomerate and Navajo Sandstone. Paleozoic rocks include the Queantoweap—Coconino Sandstone, Pakoon Dolomite, Callville Limestone, Chainman Shale and Scotty Wash Quartzite, Redwall Limestone, and Muddy Peak Dolomite [10,11,19]. Gold mineralization in the Goldstrike district is hosted primarily in sandstone and conglomeratic sandstone of the basal Claron Formation, and in favorable carbonate rocks that underlie the unconformity including the Callville Limestone and the middle unit of the Pakoon Dolomite [10,11]. Karst cavities, collapse breccias, high-angle faults, and anticlinal folds are the main structural controls of mineralization [11,17].

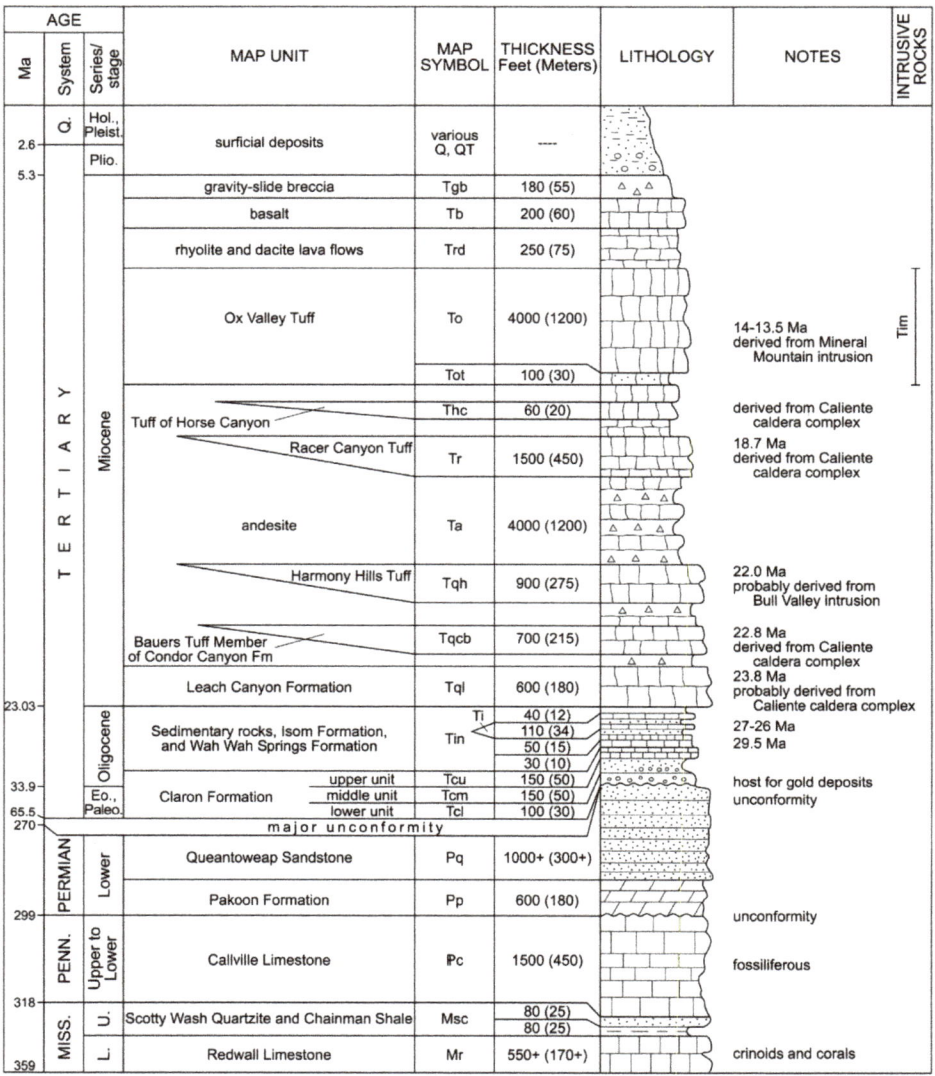

Figure 2. Stratigraphic column of the Goldstrike area, modified from Rowley et al. [19].

2. Materials and Methods

This study focused on a core drilled by Pilot Goldstrike Inc. at easting 244,910.51 m, northing 4,141,510.69 m (UTM Zone 12N), on an azimuth of 275° and a dip of −65°. Drilling retrieved core from apparent depths of 24 to 448 ft. (7.3 to 136.6 m), which was 8.5 cm in diameter. The core was split into halves, and one half was again split into two $\frac{1}{4}$ samples. Half of the core was sent for metallurgical test, one $\frac{1}{4}$ cut was sent for inductively coupled plasma mass spectrometry (ICP-MS) measurements, and the other $\frac{1}{4}$ cut was segmented into mostly 5 ft. (1.5 m) long sections and imaged by ground-based hyperspectral cameras (Figure 3A). The imaged core sections spanned 273 to 448 ft. (83.2 to 136.6 m), in which the core sections from 288 to 293 and 418 to 423 ft. (87.8 to 89.3 and 127.4 to 128.9 m) were duplicated for quality control of metallurgy and geochemical tests and were not scanned.

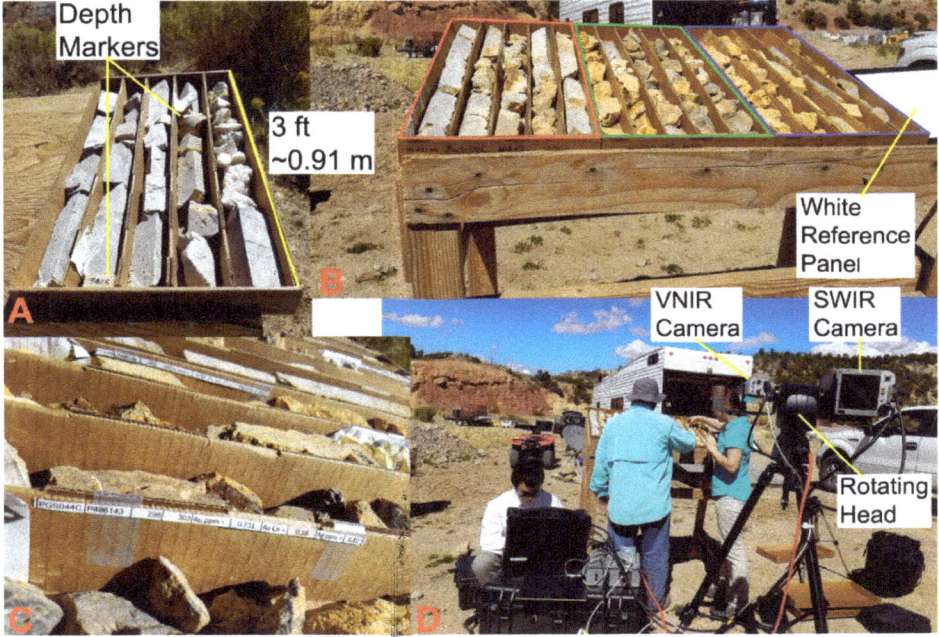

Figure 3. (**A**) A core box with five 3-ft-long columns containing core samples, with depth markers separating core sections; (**B**) each hyperspectral scan covers three core boxes, a white reference panel is included for calibration; (**C**) metallurgy and geochemical results are labeled for each core section; and (**D**) field setup of hyperspectral cameras mounted on a pan and tilt rotating head (FLIR Systems, Wilsonville, USA) to scan core samples.

2.1. Hyperspectral Imaging

Hyperspectral imaging is a remote sensing technique that collects the reflected light spectrum from material surfaces. The reflectance curve contains physical and chemical properties of the material since chemical bonds absorb light at specific wavelengths [21]. Ground-based hyperspectral imaging has been widely used in geologic characterizations [15,22–28], in which variations of the sub-centimeter or sub-millimeter scale can be resolved. This study used hyperspectral imaging to identify mineralogy as well as to extract relative abundances of the minerals.

The core samples were placed in fifteen core boxes and scanned with a Specim dual-camera system (Spectral Imaging Ltd., Oulu, Finland). Each scan imaged three core boxes that were placed on a table, and the table was held at around 20° towards the cameras by two people while scanning (Figure 3B). The hyperspectral camera system consisted of a visible and near-infrared (VNIR) camera over the spectral range of 394–1008 nm at a spectral resolution of 2.8 nm, and a short wave infrared (SWIR) camera over the spectral range of 896–2504 nm at a spectral resolution of 10 nm, and both cameras were push-broom scanners (Figure 3D). The cameras were mounted with roughly a −10° tilt towards samples on the two arms of a pan and tilt rotating head (FLIR Systems, USA) on top of a tripod, and the rotating head rotated on a horizontal plane so that the push broom cameras swept the core samples. The cameras were about 1.3 to 2.0 m away from the samples, and the spatial resolutions were 1.7 to 2.6 mm for VNIR (with four times of spatial binning) and 1.7 to 2.7 mm for SWIR. Dark current images were taken with lens caps covering lenses, these represented random background noises from the electronics and internal temperature. Due to the inevitable shaking by people holding the inclined table, the hyperspectral imagery is distorted in the across-track direction. The two cameras have different scanning angles on the two sides of the rotating head, thus produces different geometries in imagery (Figure 4). No efforts were made to correct the geometric distortions since geometry was not the focus of the study.

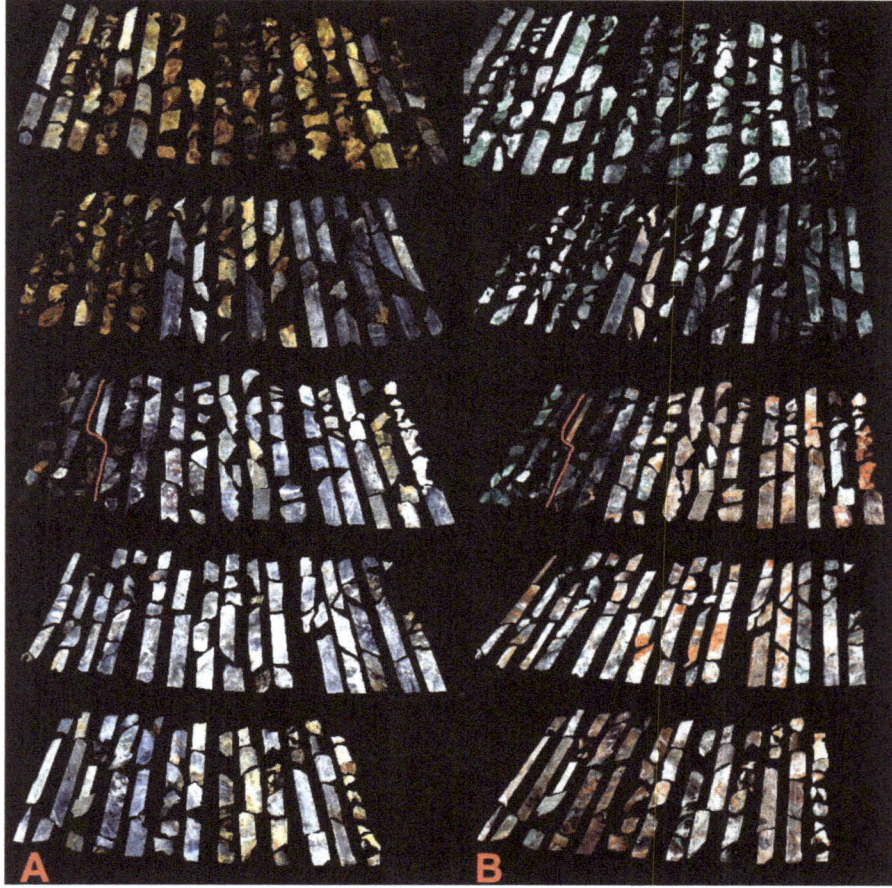

Figure 4. (**A**) True-color image of the core samples; and (**B**) false-color composite of bands 206-217-228 in the short wave infrared (SWIR) data of the core samples. Pixels other than the core samples have been masked out. The RGB bands in (**B**) have wavelengths of 2203, 2271, and 2340 nm, respectively. The red lines show the interpreted unconformity between the basal Claron Formation and the underlying Callville Limestone.

The dark current images were subtracted from the image spectra, and the results were converted into at-sensor reflectance using a flat field calibration with a white diffuse reflectance standard. Image stripes were corrected by comparing reflectance values of bad pixels with adjacent pixels. Assuming the noise were spatially related, the images were processed with forward and inverse minimum noise fraction [29] to smooth the spectra and to maximize the signal-to-noise ratio. After smoothing, the five scans were mosaicked together for more straightforward data processing. Regions of interest were manually created on the images for core samples, the other pixels, including core boxes, depth markers, and the white reference panel were masked out (Figure 4). Minerals were identified by comparing pixel spectra with the U.S. Geological Survey spectral library [30]. The spectra of identified mineral absorption features were continuum removed to isolate the non-selective scattering and spectrally inactive mineral effects [31,32]. These absorption features were then modeled with the modified Gaussian model [33] to extract the absorption depth as well as the absorption wavelength. This model is a deconvolution method that models electronic transition bands in reflectance spectra, enables the isolation of absorptions from the continuum and distinct absorption from overlapping wavelengths [33,34]. The absorption depth is a proxy of the mineral abundance [31,35], and the absorption wavelength helps to differentiate similar minerals and to imply mineral chemistry in solid solutions [36,37]. All these image processing steps were performed by Matlab 2016a (MathWorks, Natick, USA) and ENVI 5.5 (Harris Geospatial, Boulder, USA).

The reflected spectrum hosts several distinctive absorption features, including the overtones of C–O stretch and O–H stretch, combinations of O–H stretch and metal-OH bend vibrations, as well as crystal fields transitions of metal elements, these absorption features enables identification of many minerals [38–40]. Common carbonate and phyllosilicate minerals in sedimentary rocks (Figure 5A) show an Al–OH absorption near 2.2 µm, and a CO_3^{2-} absorption near 2.34 µm [39,40]. Common ferric iron oxide and hydroxide minerals (Figure 5B) show absorption features near 0.5 and 0.66 µm [41]. There is another ferric iron absorption near 0.9 µm [41]. However, the bands with long wavelengths (>800 nm) of the VNIR camera had low signal-to-noise ratios, so the absorption feature near 900 nm was not studied. In this study, we examined 50 spectral bands (2109 to 2416 nm) in the SWIR spectrum to look for two absorption bands near 2200 nm and 2340 nm, and examined 118 spectral bands (400 to 736 nm) in the VNIR spectrum to look for the absorption bands near 500 nm and 660 nm. For each core sample section, an average value of absorption depth was calculated for all pixels of the section, and this average absorption depth was compared with geochemical measurements.

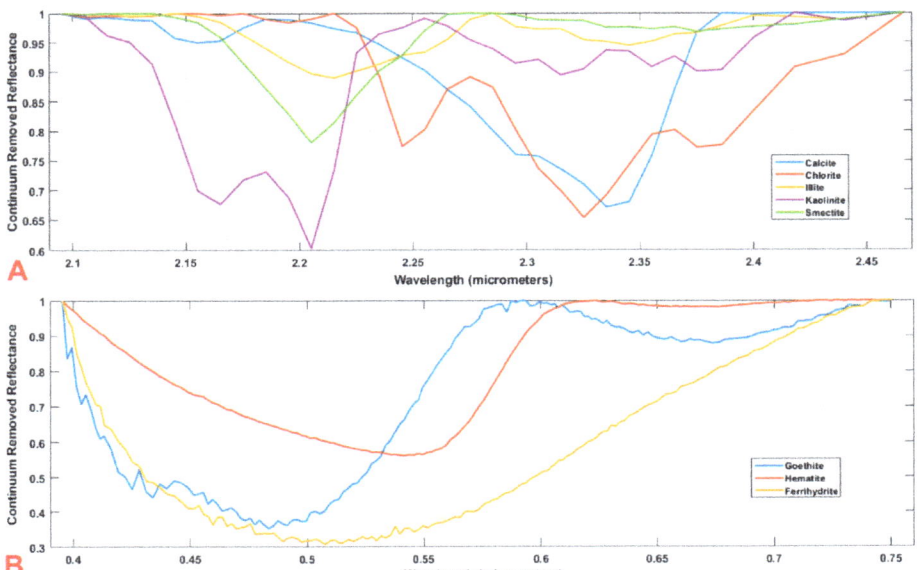

Figure 5. (**A**) Continuum removed spectra of common carbonate and phyllosilicate minerals in sedimentary rocks, and (**B**) continuum removed spectra of common ferric iron oxide and hydroxides, data from Clark et al. [30].

2.2. Geochemistry

The dry samples were crushed to 70% less than 2 mm particle size and then riffle-split. A 250 g sub-sample was pulverized to 85% less than 75 μm with ring-mill. Of the pulps 30 g was analyzed for gold by fire assay atomic absorption. After aqua regia digestion, another 1 g sub-sample of the pulps was analyzed by ICP-MS.

Multivariate principal component analyses (PCA) [42,43] were performed on the element concentration data and hyperspectral data to show the variability of different elements and affinity of elements. Eigenvectors of element concentrations (vectors showing correlation coefficients) and principal component scores of samples (points showing the linear combinations of eigenvectors) are plotted in bi-plots; the closer the vectors or dots are to each other, the closer affinity or similarity they have.

3. Results

3.1. Hyperspectral Imaging

An average was calculated for all the spectral bands and all the core sample pixels with the same lithology (Figure 6; for lithology classification see the text in this section); the mean spectral reflectance curves show the major spectral characteristics of core samples. The mean spectral reflectance curves in the SWIR spectrum (Figure 6A,C) show two strong absorption features at 2340 nm and 2200 nm. The depths of these absorption features were automatically modeled, which represent the relative abundances of calcite and clay minerals, respectively. Possible interference of chlorite with calcite because of chlorite's absorption near 2340 nm was ruled out due to the lack of absorption from chlorite near 2250 nm (Figure 6A,C). The distinctive duplet absorptions of kaolinite at 2165 nm and 2200 nm were also not observed (Figure 6A,C). As a result, the clay minerals should be illitic or smectitic. Illite and smectite have similar overlapping Al–OH absorption with each other; this study does not attempt to distinguish between the two species. With the presence of abundant calcite, detection of possible interference of the Mg–OH absorption near 2300 nm was not possible because this weak absorption would be masked out by the strong asymmetric absorption of CO_3^{2-} represented by two Gaussians at 2340 and 2300 nm in the modified Gaussian model. The Mg–OH feature only interferes with the Gaussian at 2300 nm and does not affect the abundance quantification based on the Gaussian at 2340 nm. All three mean spectral reflectance curves in the VNIR spectrum (Figure 6B,D) displayed the strong ferric iron absorption near 500 nm without an obvious absorption near 670 nm. The lack of the absorption feature near 670 nm indicated the presence of secondary oxidized ferric iron in ferrihydrite and the lack of primary ferric iron in cementing goethite and hematite [44]. The absorption depth at 500 nm was modeled to represent the relative abundance of the ferric iron.

It is shown that the upper sections were mostly deficient in calcite and the lower sections were rich in calcite (Figure 7A). Compared with the stratigraphic column and the drilling records, these sections were most probably the siliciclastic sediments of the basal Claron Formation on the top and the underlying Callville Limestone on the bottom. The unconformity separating them was probably near the third core column from the left side of Scan3 (Figure 7A).

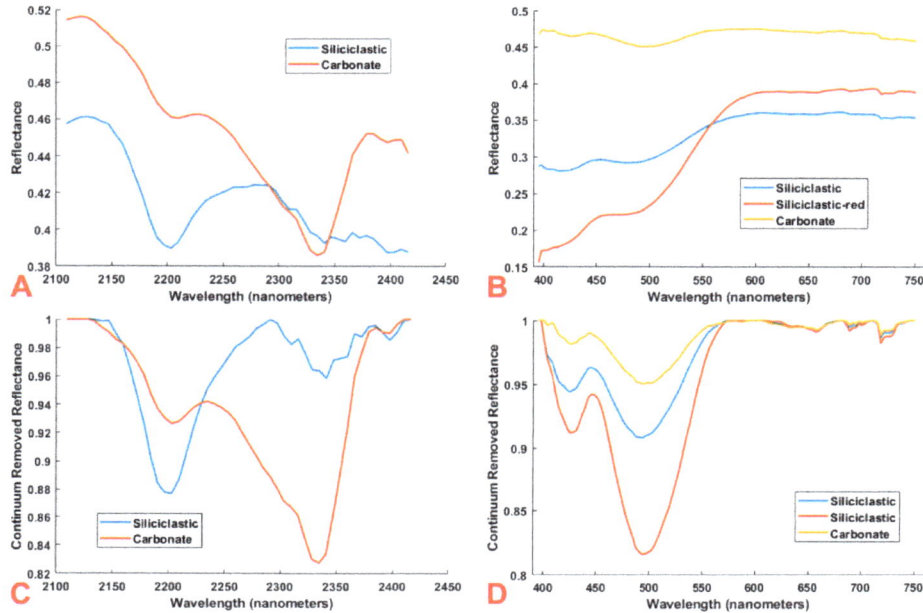

Figure 6. (**A**) Mean spectral reflectance curves of scanned core samples in the SWIR spectrum; (**B**) mean spectral reflectance curves of scanned core samples in the visible and near-infrared (VNIR) spectrum; (**C**) mean continuum removed reflectance curves of scanned core samples in the SWIR spectrum; and (**D**) mean continuum removed reflectance curves of scanned core samples in the VNIR spectrum. Some siliciclastic rocks showed strong red-yellow colors with abundant ferric iron, these rocks were plotted separately in the VNIR spectra (**B**,**D**).

Above the unconformity, the calcite abundance in the basal Claron Formation is generally very low, except for a section of high calcite content in the fifth and sixth core column from the left side of Scan2 (Figure 7A). Calcite abundance is variable in Callville Limestone, including some spots of high calcite concentrations in sections of mostly low calcite content. Clay mineral abundances are significantly higher and display more variability in the basal Claron Formation than in the Callville Limestone (Figure 7B). Within the siliciclastic sections of the basal Claron Formation, the samples with medium calcite content usually are higher in clay contents. In contrast, within the sections of higher calcite abundance in the Callville Limestone, the clay content is lower. The carbonate-rich section in the basal Claron Formation also shows lower clay content than the siliciclastic sections. Ferric iron abundance is generally higher in the basal Claron Formation than in the Callville Limestone (Figure 7C). Most of the core samples above the section of high calcite content in Scan 2 (Figure 7B) display much higher ferric iron content than other samples (Figure 7C), which is consistent with the red-yellow colors in the VNIR true-color image (Figure 4A). Some core samples in the Callville Limestone display higher ferric iron content.

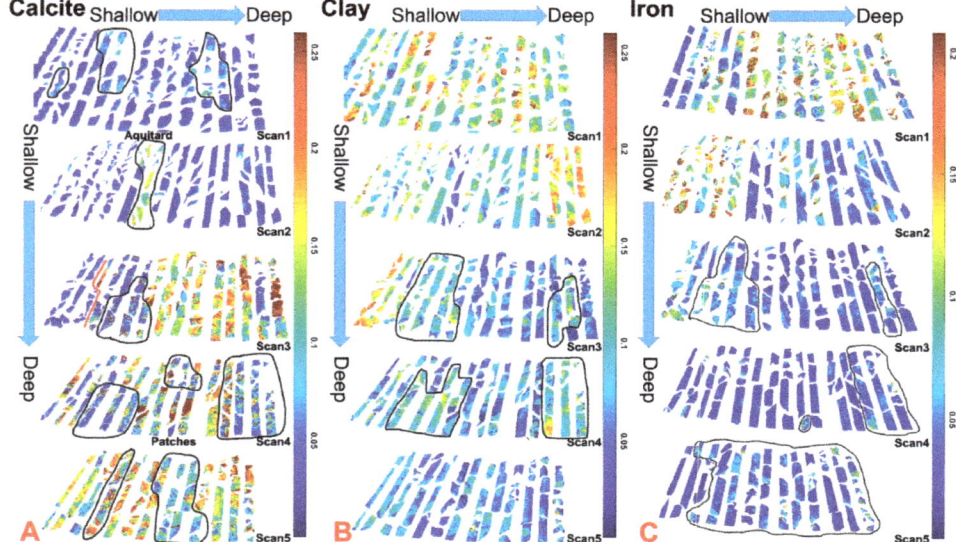

Figure 7. (**A**) Depths of continuum removed absorption features at 2340 nm representing calcite abundance; (**B**) depths of continuum removed absorption feature at 2200 nm representing clay mineral abundances; and (**C**) depths of continuum removed absorption feature at 500 nm representing ferric iron abundances. Red to yellow colors mean higher abundances, and blue to green colors mean lower abundances. Five scans were mosaicked together, core samples were shallower on the left and up, and deeper on the right and down. The red line in (**A**) shows the interpreted unconformity between the basal Claron Formation and the underlying Callville Limestone. The identified aquitard and samples with medium calcite content in the basal Claron Formation, and patches of decarbonatization in the Callville Limestone are labeled in (**A**); some argillization samples with medium clay contents are circled in (**B**); and some carbonate samples with relatively higher contents of pyrite oxidized into ferric iron are circled in (**C**).

3.2. Geochemistry

Fire assay metallurgy measured gold concentrations for the core sections, and fifty-one (51) element concentrations were measured with ICP-MS (see Table S1 in supplemental data). Gold (Au) concentrations reported from ICP-MS had fewer significant digits than measurements by the fire assay, so the values from the fire assay were used in the analyses. Concentrations of boron (B), germanium (Ge), indium (In), niobium (Nb), rhenium (Re), tantalum (Ta), and titanium (Ti) were often below the detection limit and were therefore not reported.

Forty-four (44) element measurements from fifty core sections from 213 to 448 ft. (64.9 to 136.6 m) were analyzed by PCA (Figure 8). Silver (Ag), tellurium (Te), and lead (Pb) showed closest affinities to Au; sulfur (S), mercury (Hg), thallium (Tl), arsenic (As), bismuth (Bi), antimony (Sb), selenium (Se), and iron (Fe) were also close to Au. Major elements like calcium (Ca), magnesium (Mg), aluminum (Al), and phosphorus (P) showed little or no affinity to Au; transition metals like copper (Cu), molybdenum (Mo), and zinc (Zn) that often co-occurred with Au in porphyry mines did not show characteristic affinity with Au.

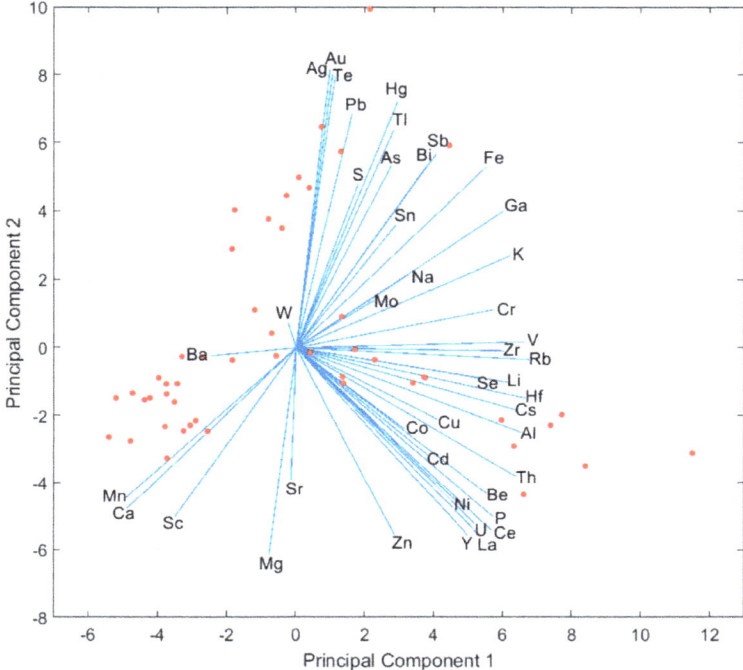

Figure 8. Bi-plot of principal components from the geochemical analyses. Blue vectors show loadings (correlation coefficients) of element concentrations, and red dots show sample scores (linear combinations of eigenvectors). The lengths of element vectors were magnified 30 times to increase legibility.

4. Discussion

4.1. Gold Mineralization in Carlin-Type Deposits

Carlin-type gold deposits form several spatial trends spanning the central Great Basin, northern Nevada, and central Nevada, and are characterized by concentrations of very finely micrometer to sub-micrometer sized disseminated arsenian pyrite grains found in structurally controlled, silty, carbonaceous, and calcareous rocks [1]. Regionally, after the flattening of the subducting Farallon plate under the North America plate around 65 Ma [45], extension prevailed in the Great Basin, and the rollback or delamination of the slab renewed magmatism in Eocene and Oligocene that swept the Great Basin southwestward from northern Nevada and Idaho to southern Nevada, southwest Utah, and north Arizona [2,46]. The dehydration of the slab, upwelling of asthenosphere, and the extensional environment enabled formation of hydrous, S- and Au-bearing, high-K, calc-alkaline magma with elevated Au/Cu ratio [47,48], which then released CO_2-, H_2S-, and Au-bearing aqueous fluid at a significantly deeper depth (about 10 km) than typical porphyry Cu–Au and associated deposits [2,49]. The fluid ascended along high-angle faults, and released vapor with high S/Fe ratios, which was trapped in permeable and reactive rocks by impermeable rocks, and precipitated gold in micro-sized arsenian pyrite grains, forming typical Carlin-type gold deposits. The Goldstrike district in southwest Utah is near the southeastward extension of the 25 million years magmatic front [2] with an active tectonic setting and high angle faults (Figure 1). Therefore gold mineralization may result from similar geologic processes to the Carlin-type deposits.

The host rock alteration of Carlin-type deposits is typically manifested by decarbonatization, argillization, silicification and/or jasperoid, fine-grained disseminated sulfide minerals and oxidation of sulfide minerals, remobilization and/or addition of carbon, and late-stage barite and/or calcite veining, with incipient collapse brecciation that enhances the migration of mineralization fluids [1,11]. Carlin-type deposits are typically stratiform, with mineralization localized with specific favorable stratigraphic units.

4.2. Stratigraphic Control

Gold concentrations measured with fire assay metallurgy were used to colorize outlines of scanned core sections and then compared with the mapped calcite and clay abundances (Figure 7) to infer the influences or patterns of mineralogy on gold mineralization. Gold concentrations are not homogeneous throughout whole core sections. Therefore necessary signals may be missed from sampling. However, geochemical analyses cannot sample small enough areas to be comparable to the resolution of hyperspectral imaging. Nevertheless, gold concentration data can show the general variations of the Au-bearing fluid flow patterns.

Gold mineralization is significantly influenced by stratigraphic and structural control in Carlin-type deposits. As shown in Figure 1 the oblique core in this study cut across a fault, which may have facilitated fluid flow. Rocks of low porosity and permeability act as aquitards to prevent fluid from ascending and the Au-bearing fluids react with the permeable rocks below aquitards to precipitate disseminated gold. The calcite-rich section in Scan2, the basal Claron Formation is such an aquitard due to its lower permeability compared with siliciclastic rocks below. Gold concentrations in the calcite-rich section were very low, then very high in the section below that, and decreasing downward (Figure 9B). Those samples with medium calcite content also showed relatively low gold concentrations. The combination of low permeability carbonate aquitard and high permeability siliciclastic rocks below in Carlin-type deposits were similar to the seal and reservoir rocks in conventional oil and gas industry. Exploration can be focused on such stratigraphic combinations near high-angle faults.

4.3. Mineralogical Alterations

Macroscopic rock alterations that are reported to be related to gold mineralization in the Goldstrike district as well as in the Carlin-type deposits include silicification, decarbonatization, argillization, and pyrite oxidation [1,11]. Quartz, chert, and amorphous silica are all not spectrally active in the visible to the short-wave infrared spectrum, only in the thermal infrared, so could not be studied by the hyperspectral cameras used in this study. Decarbonatization stands for the removal of carbonate, so the decline or diminishment of CO_3^{2-} absorption near 2340 nm represents decarbonatization. Argillization stands for the addition of argillic minerals (most commonly illite and kaolinite), so the increase or appearance of Al–OH absorption near 2200 nm represents argillic alteration. These argillic minerals may come from alterations of felsic minerals or the ore fluids. The basal Claron Formation is sandstone or conglomerate sandstone, which may have argillaceous material, but without a lateral comparison with unaltered strata, it is hard to confirm the argillic alteration in a single core. More cores or outcrops may be helpful. However, the authors only had limited access. On the other hand, Callville Limestone rarely has argillaceous components [50,51], and the detection of Al–OH absorption is interpreted to represent argillic alteration. Pyrite and arsenian pyrite in the strata can be oxidized into ferric iron minerals after mineralization, which is especially common in pyrite-rich silty limestones or limey siltstones [1], so the detection of ferric iron absorption near 500 nm may represent pyrite oxidation.

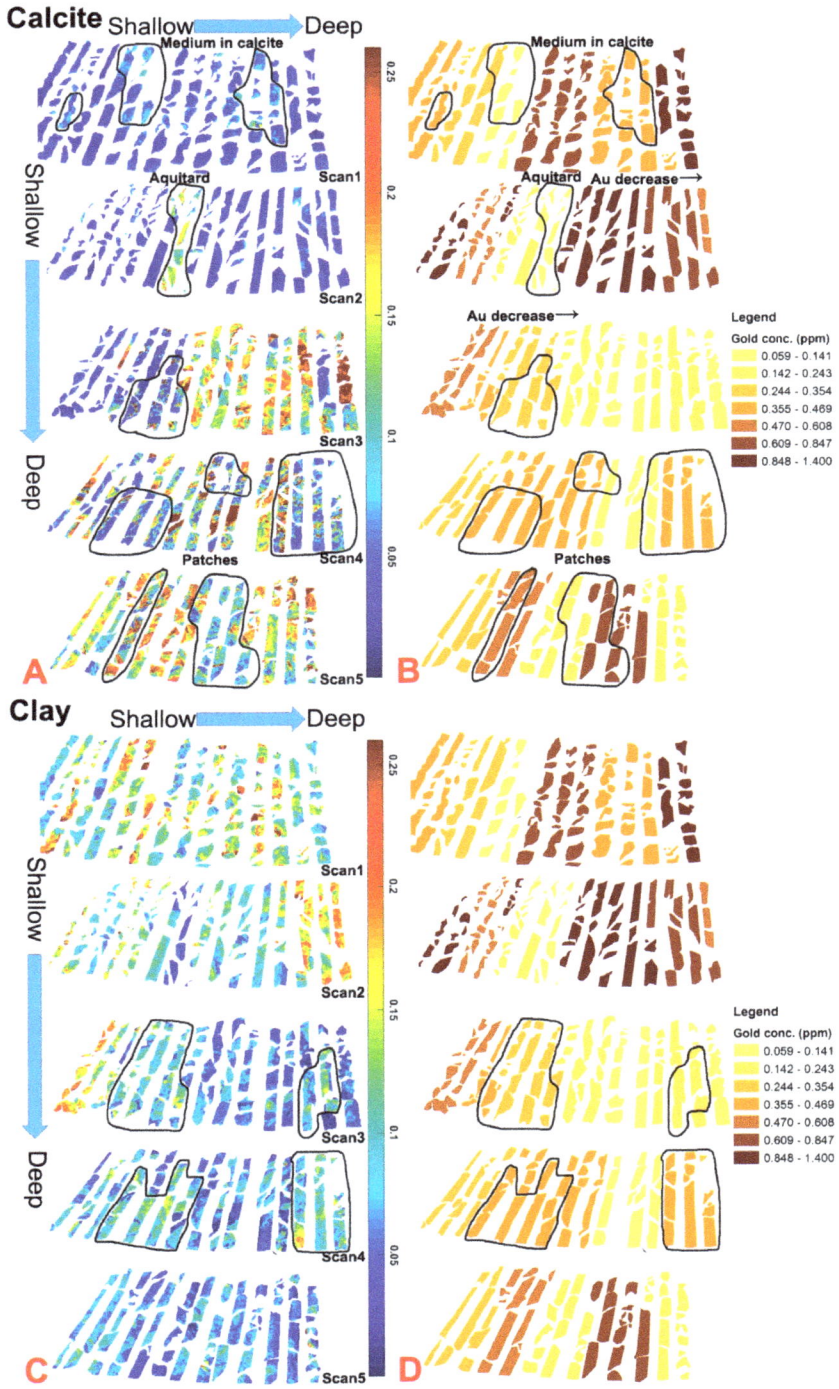

Figure 9. Cont.

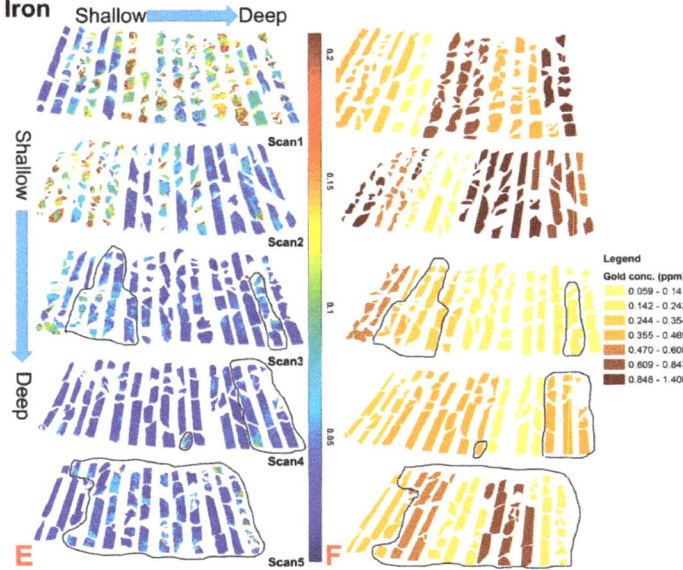

Figure 9. (**A**) Depths of continuum removed absorption features at 2340 nm representing calcite abundance; (**B**) core samples colorized by gold concentrations from fire assay metallurgy; (**C**) depths of continuum removed absorption features at 2200 nm representing clay mineral abundances; (**D**) core samples colorized by gold concentrations from fire assay metallurgy; (**E**) depths of continuum removed absorption feature at 500 nm representing ferric iron abundance; and (**F**) core samples colorized by gold concentrations from fire assay metallurgy. Identified aquitard, samples with medium calcite content, some argillization samples with medium clay contents, patches of decarbonatization, and some carbonate samples with relatively higher contents of pyrite oxidized into ferric iron are labeled.

Much of the scanned Callville Limestone has undergone various degrees of decarbonatization. Patches of high calcite content surrounded by areas of lower calcite content can be seen below the unconformity in Scan3, are prevalent in Scan4, and in some columns in Scan5. Some sections have gone through strong decarbonatization removing all the calcite. Decarbonatization may reflect more active gold mineralization, as the gold concentrations are relatively higher in these sections (Figure 9B). Some of the scanned Callville Limestone has also undergone various degrees of argillization. Samples of medium clay contents can be seen in Figure 9C, with argillization corresponding to more active gold mineralization and relatively higher gold concentrations (Figure 9D). Some samples in Callville Limestone showed higher ferric iron (Figure 9E), which might correlate with pyrite oxidation after mineralization with higher gold concentrations (Figure 9F).

These interpretations were supported by PCA of absorption depths together with element concentrations (Figure 10). Due to the significant differences in lithology and alteration patterns, measurements of seventeen (17) siliciclastic and seventeen carbonate rock sections (including the carbonate aquitard section in the basal Claron Formation) were processed separately. The average pixel value of absorption depths and 12 selected elements were analyzed by PCA since only seventeen sections were available. Analyses in siliciclastic rocks showed that calcite, clay, and ferric iron depths all had minimal affinities with gold (Figure 10A; See Table S2 in supplemental data). This confirmed that decarbonatization and argillization in siliciclastic rocks could not be easily mapped from a single core and correlate with gold mineralization. However, stratigraphic control by different lithologies played a major role (Figure 9). It was also shown that calcite and clay absorption depths were close to each other, supporting the observation that samples with medium calcite content usually were higher in clay contents. On the other hand, analyses in carbonate rocks showed that both clay and ferric iron depths had close affinities to silver and a little less affinity to gold, while calcite depth showed very limited affinity to gold, silver, and clay depth (Figure 10B) (See Table S4 in supplemental data). These facts confirmed that decarbonatization, argillization, and pyrite oxidation in carbonate rocks could be mapped from core samples, and correlated with gold mineralization. The identification of these mineralogical alterations could be used as a classifier for ore grades; intense alterations might correlate with higher grades, and weak alterations correlate with lower grades.

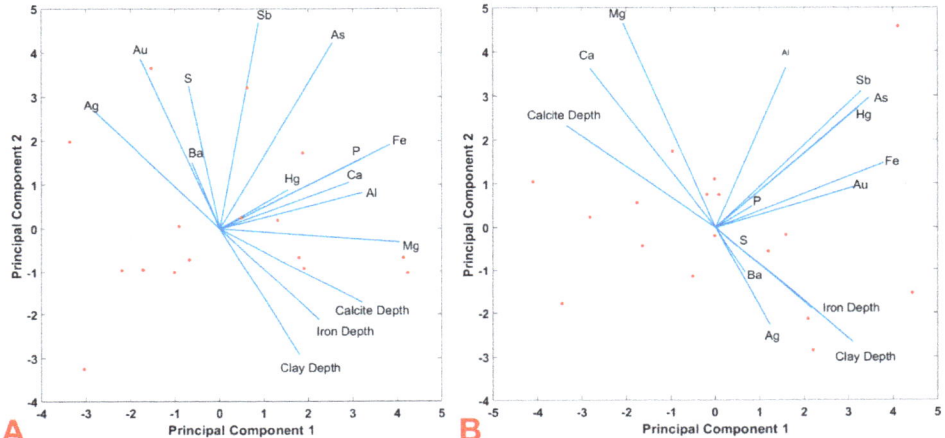

Figure 10. (**A**) Bi-plot of the principal components from average absorption depths and selected elements in the siliciclastic rocks of the basal Claron Formation; and (**B**) bi-plot of principal components from average absorption depths and selected elements in the carbonate rocks of the Callville Limestone as well as the carbonate section in the basal Claron Formation. Vectors show loadings of elements and absorption depths, and dots show sample scores. The lengths of element vectors were magnified ten times to increase legibility.

Besides the macroscopic alterations, some trace elements like As, Hg, and Sb are associated with gold mineralization, and they form sulfide minerals like orpiment, realgar, cinnabar, and stibnite. These minerals appeared in trace amount, and no investigation was performed on these minerals.

4.4. Implications

This study showed an example of using hyperspectral imaging to identify the optimum stratigraphic combination of limestone above and siliciclastic rocks below, as well as mineralogical alterations including decarbonatization, argillization, and pyrite oxidation that are related to gold mineralization. Such mineralogical information may shed light on the geologic mechanism of mineralization, and help in the determination of mining interests with similar stratigraphic and structural conditions, as well as help in the classification of ore grades based on the alteration patterns. Without expenses in chemical consumables and procedure, most of the costs of hyperspectral imaging are for personnel. As a result, hyperspectral imaging may be a cost-effective alternative or complementary method for geochemical methods [52,53]. Besides, the sub-centimeter spatial resolution and detailed mineralogical identification/semi-quantification from hyperspectral imaging are superior to lithologic logging that accompanies exploration drilling. Ground-based hyperspectral imaging, as a new direction in earth sciences, provides high spatial and spectral resolution measurements, fast data collection, sizeable areal coverage, and easy data processing. We envision more studies with hyperspectral imaging in various fields of earth sciences.

5. Conclusions

Hyperspectral imaging was used to scan a core covering the basal Claron Formation and Callville Limestone in the Goldstrike district, southwest Utah, which is believed to be similar to Carlin-type gold deposits in Nevada with sediment-hosted disseminated gold. This study used spectral modeling to identify and semi-quantify calcite, clay minerals, and ferric iron. Variations in mineralogy are used to identify lithology, as well as decarbonatization, argillization, and pyrite oxidation alterations within the core samples. Compared with metallurgy and ICP-MS geochemical data of the core, this study confirmed the correlation between stratigraphic control as well as mineralogical alterations with gold mineralization. Although the silicification and formation of jasperoids are essential indicators of gold mineralization, silica is not spectrally active in the reflected spectrum and cannot be studied by hyperspectral imaging. This state-of-art technology has excellent potentials in broader applications in the mining industry.

Supplementary Materials: The following are available online at http://www.mdpi.com/2072-4292/11/17/1987/s1, Table S1: Element concentrations from fire assay metallurgy and ICP-MS; Table S2: Correlation matrix of absorption depths and element concentrations in siliciclastic rocks; Table S3: Eigenvalues of the correlation matrix of absorption depths and element concentrations in siliciclastic rocks; Table S4: Correlation matrix of absorption depths and element concentrations in carbonate rocks; Table S5: Eigenvalues of the correlation matrix of absorption depths and element concentrations in carbonate rocks.

Author Contributions: Conceptualization, L.S. and S.K.; data curation, L.S.; formal analysis, L.S.; methodology, L.S.; resources, P.S.; writing—original draft preparation, L.S.; writing—review and editing, S.K.

Funding: This research received no external funding.

Acknowledgments: We are thankful to Pilot Goldstrike Inc. for their help and the permission to publish this paper. We thank three anonymous reviewers for their helpful suggestions and academic editor Amin Beiranvand Pour for his kind support. We also appreciate the help of Diana Krupnik and Macey Crockett in fieldwork.

Conflicts of Interest: The authors declare no conflict of interest.

References

1. Cline, J.S.; Hofstra, A.H.; Muntean, J.L.; Tosdal, R.M.; Hickey, K.A. Carlin-type gold deposits in Nevada: Critical geologic characteristics and viable models. *Econ. Geol. 100th Anniv. Vol.* **2005**, 451–484. [CrossRef]
2. Muntean, J.L.; Cline, J.S.; Simon, A.C.; Longo, A.A. Magmatic-hydrothermal origin of Nevada's Carlin-type gold deposits. *Nat. Geosci.* **2011**, *4*, 122–127. [CrossRef]
3. Gu, X. *Turbidite-Hosted Micro-Disseminated Gold Deposits*; Chengdu University of Science and Technology Press: Chengdu, China, 1996.
4. Ashley, R.P.; Cunningham, C.G.; Bostick, N.H.; Dean, W.E.; Chou, I.M. Geology and geochemistry of three sedimentary-rock-hosted disseminated gold deposits in Guizhou Province, People's Republic of China. *Ore Geol. Rev.* **1991**, *6*, 133–151. [CrossRef]
5. Xia, Y.; Su, W.; Zhang, X.; Liu, J. Geochemistry and metallogenic model of carlin-type gold deposits in southwest Guizhou province, China. In *Geochemistry-Earth's System Processes*; Panagiotaras, D., Ed.; InTech: Rijeka, Croatia, 2012; pp. 127–156.
6. Percival, T.J.; Radtke, A.S. Sedimentary-rock-hosted disseminated gold mineralization in the Alsar district, Macedonia. *Can. Mineral.* **1994**, *32*, 649–665.
7. Staude, J.M.G. Epithermal Mineralization in The Sierra Madre Occidental, and the Metallogeny of Northwestern Mexico. Ph.D. Thesis, University of Arizona, Tucson, AZ, USA, 1995.
8. Krahulec, K. Sedimentary rock-hosted gold and silver deposits of the Northeastern Basin and Range, Utah. *Gt. Basin Evolut. Metallog. Geol. Soc. Nev. Symp.* **2010**, *1*, 31–62.
9. Emsbo, P.; Hutchinson, R.W.; Hofstra, A.H.; Volk, J.A.; Bettles, K.H.; Baschuk, G.J.; Johnson, C.A. Syngenetic au on the carlin trend: Implications for carlin-type deposits. *Geology* **1999**, *27*, 59–62. [CrossRef]

10. Willden, R.; Adair, D.H. Gold deposits at Goldstrike, Utah. In *Thrusting and Extensional Structures and Mineralization in the Beaver Dam Mountains, Southwestern Utah*; Griffen, D.T., Phillips, W.R., Eds.; Utah Geological Association: Salt Lake City, UT, USA, 1986; pp. 137–147.
11. Gustin, M.M.; Smith, M.T. *Technical Report on the Goldstrike Project, Washington County, Utah, U.S.A.*; Technical Report No.; Mine Development Associates: Reno, NV, USA, 2016; pp. 1–106.
12. Willden, R. Goldstrike mining district, Washington County, Utah. In *Mining Districts of Utah*; Bon, R.L., Gloyn, R.W., Park, G.M., Eds.; Utah Geological Association: Salt Lake City, UT, USA, 2006; pp. 458–476.
13. Dalm, M.; Buxton, M.W.N.; van Ruitenbeek, F.J.A. Discriminating ore and waste in a porphyry copper deposit using short-wavelength infrared (SWIR) hyperspectral imagery. *Miner. Eng.* **2017**, *105*, 10–18. [CrossRef]
14. Gallie, E.A.; McArdle, S.; Rivard, B.; Francis, H. Estimating sulphide ore grade in broken rock using visible/infrared hyperspectral reflectance spectra. *Int. J. Remote Sens.* **2002**, *23*, 2229–2246. [CrossRef]
15. Kruse, F.A.; Bedell, R.L.; Taranik, J.V.; Peppin, W.A.; Weatherbee, O.; Calvin, W.M. Mapping alteration minerals at prospect, outcrop and drill core scales using imaging spectrometry. *Int. J. Remote Sens.* **2012**, *33*, 1780–1798. [CrossRef]
16. Riaza, A.; Müller, A. Hyperspectral remote sensing monitoring of pyrite mine wastes: A record of climate variability (Pyrite Belt, Spain). *Environ. Earth Sci.* **2010**, *61*, 575–594. [CrossRef]
17. Adair, D.H. Structural Setting of the Goldstrike District, Washington County, Utah. In *Thrusting and Extentional Structures and Mineralization in the Beaver Dam Mountains, Southwest Utah*; Griffen, D.T., Phillips, W.R., Eds.; 1986; Volume 15, pp. 129–135. Available online: http://archives.datapages.com/data/uga/data/057/057001/129_ugs570129.htm (accessed on 22 August 2019).
18. Christiansen, R.L.; Yeats, R.S.; Graham, S.A.; Niem, W.A.; Niem, A.R.; Snavely, P.D., Jr. Post-Laramide geology of the US Cordilleran region. In *The Cordilleran Orogen, Conterminous US: Geology of North America 3*; Burchfiel, B.C., Lipman, P.W., Zoback, M.L., Eds.; Geological Society of America: Boulder, CO, USA, 1992; pp. 261–406.
19. Rowley, P.D.; Anderson, R.E.; Hacker, D.B.; Boswell, J.T.; Maxwell, D.J.; Cox, D.P.; Willden, R.; Adair, D.H. *Interim Geologic Map of the Goldstrike Quadrangle and the East Part of the Docs Pass Quadrangle, Washington County, Utah*; Utah Geological Survey Open-File Report 510; Utah Geological Survey: Salt Lake City, Utah, 2007.
20. Ott, A.L. Detailed Stratigraphy and Stable Isotope Analysis of the Claron Formation, Bryce Canyon National Park, Southwestern Utah. Master's Thesis, Washington State University, Pullman, WA, USA, 1999.
21. Clark, R.N.; Swayze, G.A.; Livo, K.E.; Kokaly, R.F.; Sutley, S.J.; Dalton, J.B.; McDougal, R.R.; Gent, C.A. Imaging spectroscopy: Earth and planetary remote sensing with the USGS Tetracorder and expert systems. *J. Geophys. Res. Planets* **2003**, *108*, 5131. [CrossRef]
22. Alonso de Linaje, V.; Khan, S.D.; Bhattacharya, J. Study of carbonate concretions using imaging spectroscopy in the Frontier Formation, Wyoming. *Int. J. Appl. Earth Obs. Geoinf.* **2018**, *66*, 82–92. [CrossRef]
23. Entezari, I.; Rivard, B.; Geramian, M.; Lipsett, M.G. Predicting the abundance of clays and quartz in oil sands using hyperspectral measurements. *Int. J. Appl. Earth Obs. Geoinf.* **2017**, *59*, 1–8. [CrossRef]
24. Khan, S.D.; Okyay, U.; Ahmad, L.; Shah, M.T. Characterization of gold mineralization in northern Pakistan using imaging spectroscopy. *Photogramm. Eng. Remote Sens.* **2018**, *84*, 425–434. [CrossRef]
25. Krupnik, D.; Khan, S.; Okyay, U.; Hartzell, P.; Zhou, H.W. Study of Upper Albian rudist buildups in the Edwards Formation using ground-based hyperspectral imaging and terrestrial laser scanning. *Sediment. Geol.* **2016**, *345*, 154–167. [CrossRef]
26. Okyay, U.; Khan, S.D.; Lakshmikantha, M.R.; Sarmiento, S. Ground-based hyperspectral image analysis of the Lower Mississippian (Osagean) reeds spring formation rocks in southwestern Missouri. *Remote Sens.* **2016**, *8*, 1–21. [CrossRef]
27. Sun, L.; Khan, S.D.; Sarmiento, S.; Lakshmikantha, M.R.; Zhou, H. Ground-based hyperspectral imaging and terrestrial laser scanning for fracture characterization in the Mississippian Boone Formation. *Int. J. Appl. Earth Obs. Geoinf.* **2017**, *63*, 222–233. [CrossRef]
28. Sun, L.; Khan, S.; Godet, A. Integrated ground-based hyperspectral imaging and geochemical study of the Eagle Ford Group in West Texas. *Sediment. Geol.* **2018**, *363*, 34–47. [CrossRef]

29. Green, A.A.; Berman, M.; Switzer, P.; Graig, M.D. A transformation for ordering multispectral data in term of image quality with implications for noise removal. *IEEE Trans. Geosci. Remote Sens.* **1988**, *26*, 65–74. [CrossRef]
30. Clark, R.N.; Swayze, G.A.; Wise, R.; Livo, K.E.; Hoefen, T.M.; Kokaly, R.F.; Sutley, S.J. *USGS Digital Spectral Library Splib06a*; Digital Data Series 231; US Geological Survey: Reston, VA, USA, 2007.
31. Clark, R.N.; Roush, T.L. Reflectance spectroscopy: Quantitative analysis techniques for remote sensing applications. *J. Geophys. Res. Solid Earth* **1984**, *89*, 6329–6340. [CrossRef]
32. Mustard, J.F.; Sunshine, J. Spectral analysis for earth science: Investigations using remote sensing data. In *Remote Sensing for the Earth Sciences: Manual of Remote Sensing*; Rencz, A.N., Ed.; John Wiley & Sons, Inc.: New York, NY, USA, 1999; pp. 251–307.
33. Sunshine, J.M.; Pieters, C.M.; Pratt, S.F. Deconvolution of minerals absorption bands: An improved approach. *J. Geophys. Res* **1990**, *95*, 6955–6966. [CrossRef]
34. Asadzadeh, S.; de Souza Filho, C.R. Spectral remote sensing for onshore seepage characterization: A critical overview. *Earth Sci. Rev.* **2017**, *168*, 48–72. [CrossRef]
35. Hunt, G.R. Near-infrared (1.3–2.4 µm) spectra of alteration minerals–Potential for use in remote sensing. *Geophysics* **1979**, *44*, 1974–1986. [CrossRef]
36. Cloutis, E.A.; Gaffey, M.J.; Jackowski, T.L.; Reed, K.L. Calibrations of phase abundance, composition, and particle size distribution for olivine-orthopyroxene mixtures from reflectance spectra. *J. Geophys. Res.* **1986**, *91*, 11641–11653. [CrossRef]
37. Gaffey, S.J. Spectral reflectance of carbonate minerals in the visible and near infrared (0.35–2.55 µm): Anhydrous carbonate minerals. *J. Geophys.* **1987**, *92*, 1429–1440. [CrossRef]
38. Hunt, G.R.; Salisbury, J.W. Visible and near infrared spectra of minerals and rocks. II. Carbonates. *Mod. Geol.* **1971**, *2*, 23–30.
39. Hunt, G.R.; Salisbury, J.W. Visible and near infrared spectra of minerals and rocks: I silicate minerals. *Mod. Geol.* **1970**, *1*, 283–300.
40. Clark, R.N.; King, T.V.V.; Klejwa, M.; Swayze, G.; Vergo, N. High spectral resolution reflectance spectroscopy of minerals. *J. Geophys. Res* **1990**, *95*, 12653–12680. [CrossRef]
41. Morris, R.V.; Lauer, H.V.; Lawson, C.A.; Gibson, E.K.; Nace, G.A.; Stewart, C. Spectral and other physicochemical properties of submicron powders of hematite (alpha-Fe2o3), maghemite (gamma-Fe2o3), magnetite (Fe3o4), goethite (alpha-Feooh) and lepidocrocite (gamma-Feooh). *J. Geophys. Res* **1985**, *90*, 3126–3144. [CrossRef]
42. Pearson, K. Principal components analysis. *Lond. Edinb. Dublin Philos. Mag. J. Sci.* **1901**, *6*, 559. [CrossRef]
43. Wold, S.; Esbensen, K.; Geladi, P. Principal component analysis. *Chemom. Intell. Lab. Syst.* **1987**, *2*, 37–52. [CrossRef]
44. Sun, L.; Khan, S. Ground-based hyperspectral remote sensing of hydrocarbon-induced rock alterations at cement, Oklahoma. *Mar. Pet. Geol.* **2016**, *77*, 1243–1253. [CrossRef]
45. Coney, P.J.; Reynolds, S.J. Cordilleran Benioff zones. *Nature* **1977**, *270*, 403–406. [CrossRef]
46. Humphreys, E.D. Post-Laramide removal of the Farallon slab, western United States. *Geology* **1995**, *23*, 987–990. [CrossRef]
47. Richards, J.P. Postsubduction porphyry Cu-Au and epithermal Au deposits: Products of remelting of subduction-modified lithosphere. *Geology* **2009**, *37*, 247–250. [CrossRef]
48. Gans, P.B.; Mahood, G.A.; Schermer, E. *Synextensional Magmatism in the Basin and Range Province: A Case Study from the Eastern Great Basin*; Geological Society of America: Boulder, CO, USA, 1989.
49. Williams-Jones, A.E.; Heinrich, C.A. Vapor transport of metals and the formation of magmatic-hydrothermal ore deposits. *Econ. Geol.* **2005**, *100*, 1287–1312. [CrossRef]
50. Rice, J.A. Stratigraphy, Diagenesis, and Provenance of Upper Paleozoic Eolian Limestones, Western Grand Canyon and Southern Nevada. Ph.D. Thesis, University of Nebraska-Lincoln, Lincoln, NE, USA, May, 1990.
51. Wardlaw, B.R. The Pennsylvanian Callville Limestone of Beaver County, southwestern Utah. In *Paleozoic Paleogeography of the West-Central United States: Rocky Mountain Paleogeography Symposium 1*; Fouch, T.D., Magathan, E.R., Eds.; Rocky Mountain Section SEPM: Denver, CO, USA, 1980; pp. 175–179.

52. Choe, E.; van der Meer, F.; van Ruitenbeek, F.; van der Werff, H.; de Smith, B.; Kim, K.W. Mapping of heavy metal pollution in stream sediments using combined geochemistry, field spectroscopy, and hyperspectral remote sensing: A case study of the Rodalquilar mining area, SE Spain. *Remote Sens. Environ.* **2008**, *112*, 3222–3233. [CrossRef]
53. Hong-yan, R.E.N.; Da-fang, Z.; Singh, A.N.; Jian-jun, P.A.N.; Dong-sheng, Q.I.U. Estimation of As and Cu contamination in agricultural soils around a mining area by reflectance spectroscopy: A case study. *Pedosphere* **2009**, *19*, 719–726.

 © 2019 by the authors. Licensee MDPI, Basel, Switzerland. This article is an open access article distributed under the terms and conditions of the Creative Commons Attribution (CC BY) license (http://creativecommons.org/licenses/by/4.0/).

Article

Orogenic Gold in Transpression and Transtension Zones: Field and Remote Sensing Studies of the Barramiya–Mueilha Sector, Egypt

Basem Zoheir [1,2,*], Mohamed Abd El-Wahed [3], Amin Beiranvand Pour [4,5] and Amr Abdelnasser [1]

1. Geology Department, Faculty of Science, Benha University, Benha 13518, Egypt; amr.khalil@fsc.bu.edu.eg
2. Institute of Geosciences, University of Kiel, Ludewig-Meyn Str. 10, 24118 Kiel, Germany
3. Tanta University, Geology Department, Faculty of Science, Tanta 31527, Egypt; mohamed.abdelwahad@science.tanta.edu.eg
4. Korea Polar Research Institute (KOPRI), Songdomirae-ro, Yeonsu-gu, Incheon 21990, Korea; beiranvand.amin80@gmail.com
5. Institute of Oceanography and Environment (INOS), Universiti Malaysia Terengganu (UMT), 21030 Kuala Nerus, Terengganu, Malaysia
* Correspondence: basem.zoheir@ifg.uni-kiel.de or basem.zoheir@fsc.bu.edu.eg; Tel.: +49-177-210-4494

Received: 17 July 2019; Accepted: 10 September 2019; Published: 12 September 2019

Abstract: Multi-sensor satellite imagery data promote fast, cost-efficient regional geological mapping that constantly forms a criterion for successful gold exploration programs in harsh and inaccessible regions. The Barramiya–Mueilha sector in the Central Eastern Desert of Egypt contains several occurrences of shear/fault-associated gold-bearing quartz veins with consistently simple mineralogy and narrow hydrothermal alteration haloes. Gold-quartz veins and zones of carbonate alteration and listvenitization are widespread along the ENE–WSW Barramiya–Um Salatit and Dungash–Mueilha shear belts. These belts are characterized by heterogeneous shear fabrics and asymmetrical or overturned folds. Sentinel-1, Phased Array type L-band Synthetic Aperture Radar (PALSAR), Advanced Space borne Thermal Emission and Reflection Radiometer (ASTER), and Sentinel-2 are used herein to explicate the regional structural control of gold mineralization in the Barramiya–Mueilha sector. Feature-oriented Principal Components Selection (FPCS) applied to polarized backscatter ratio images of Sentinel-1 and PALSAR datasets show appreciable capability in tracing along the strike of regional structures and identification of potential dilation loci. The principal component analysis (PCA), band combination and band ratioing techniques are applied to the multispectral ASTER and Sentinel-2 datasets for lithological and hydrothermal alteration mapping. Ophiolites, island arc rocks, and Fe-oxides/hydroxides (ferrugination) and carbonate alteration zones are discriminated by using the PCA technique. Results of the band ratioing technique showed gossan, carbonate, and hydroxyl mineral assemblages in ductile shear zones, whereas irregular ferrugination zones are locally identified in the brittle shear zones. Gold occurrences are confined to major zones of fold superimposition and transpression along flexural planes in the foliated ophiolite-island arc belts. In the granitoid-gabbroid terranes, gold-quartz veins are rather controlled by fault and brittle shear zones. The uneven distribution of gold occurrences coupled with the variable recrystallization of the auriferous quartz veins suggests multistage gold mineralization in the area. Analysis of the host structures assessed by the remote sensing results denotes vein formation spanning the time–space from early transpression to late orogen collapse during the protracted tectonic evolution of the belt.

Keywords: Advanced Space borne Thermal Emission and Reflection Radiometer (ASTER); Sentinel 2; Synthetic Aperture Radar (SAR) data; Egyptian Eastern Desert; gold mineralization; structural control; transpression and transtension zones

1. Introduction

Remote sensing satellite imagery has a high capability of providing a synoptic view of geological structures, alteration zones and lithological units in metallogenic provinces. Typically, application of multi-sensor satellite imagery can be considered as a cost-efficient exploration strategy for prospecting orogenic gold mineralization in transpression and transtension zones, which are located in harsh regions around the world [1–13]. Synthetic Aperture Radar (SAR) is an active microwave remote sensing sensor that transmits and detects radiation with wavelengths between 2.0 and 100 cm, typically at 2.5–3.8 cm (X-band), 4.0–7.5 cm (C-band), and 15.0–30.0 cm (L-band) [7]. Longer wavelengths (L-band) can enhance the depth of penetration of radar signals through the Earth's surface and therefore provide valuable information for structural geology mapping related to orogenic gold mineralization [8,9]. The C-band and L-band SAR data, i.e., the Sentinel-1 and Phased Array type L-band Synthetic Aperture Radar (PALSAR) data, have successfully promoted mapping of structural lineaments that are associated with hydrothermal gold mineralization in tropical, arid, and semi-arid environments [4,14–18].

The hydrothermal alteration zones are normally impregnated with iron oxides, clay and carbonate ± sulfate, which have diagnostic spectral signatures in the visible, near infrared, and shortwave infrared radiation regions [19]. The electronic processes caused by the transitional elements in these minerals, such as Fe^{2+}, Fe^{3+}, Mn, Cr, Co and Ni, produces absorption features in the visible near infrared region (VNIR) (0.4 to 1.1 µm) [19]. The hydrous mineral phases with the OH groups (Mg–O–H, Al–O–H, Si–O–H) and CO_3 acid group have diagnostic absorption features in short wave infrared region (SWIR) (2.0–2.50 µm) [20,21]. Using data of multispectral and hyperspectral remote sensing sensors for lithological and mineralogical mapping in metallogenic provinces have been continually demonstrated around the world [6,22–24]. The Advanced Space-borne Thermal Emission and Reflection Radiometer (ASTER) and Sentinel-2 data have high capabilities in discriminating lithological units and alteration zones associated with hydrothermal ore deposits using the VNIR and SWIR spectral data [11–13,25–27].

The Central Eastern Desert (CED) of Egypt is built up mainly of tectonized ophiolites, metasedimentary rock successions, granitoid intrusions, and subordinate volcanic rocks and molasse sediments generally of Neoproterozoic age [28]. Gold–quartz veins cutting mainly through the metavolcanic–metasedimentary rock successions or in small granitic intrusions have been intensely mined out and produced gold during ancient times [29]. Gold mineralization is thought to have occurred during episodes of calc-alkaline granite magmatism in the evolution of the Eastern Desert shield [30–32]. Shear-related gold lodes have been described in several occurrences in the South and Central Eastern Desert (SED, CED) [33–38]. Zoheir [39] discussed the possibility of temporal and spatial relationships between discrete gold occurrences and regional transpression shear zones particularly between the ophiolite and island arc terranes. The Au-quartz veins are typically hosted by brittle–ductile ductile and fault zones attributed to post-accretionary, wrench-dominated defromation [40,41].

In this study, Sentinel-1, PALSAR, ASTER, and Sentinel-2 data are analyzed to decode the distribution of geological structures and hydrothermal alteration zones associated with gold–quartz veins in the Barramiya–Mueilha Sector of the CED (Figure 1). This contribution comes in response to the present-day surge in gold exploration in the Eastern Desert of Egypt and other parts of the Nubian Shield (i.e., Sudan, Arabia, Eritrea, and Ethiopia). The main objectives of this study are: (i) To map the major lineaments, curvilinear structures, and intersections in the study area using Sentinel-1 and PALSAR datasets by employing the Feature-oriented Principal Components Selection (FPCS) technique; (ii) to identify the alteration zones and lithological units in association with brittle and ductile shear zones by applying principal component analysis (PCA), band combination and specialized band ratioing techniques to ASTER and Sentinel-2 datasets; (iii) to integrate field, structural analysis, and multi-sensor satellite imagery for an ample understanding of the setting and structural controls of gold occurrences in the study area; and (iv) to inaugurate a cost-effective multi-sensor satellite imagery approach for orogenic gold in transpression and transtension zones in the Egyptian Eastern Desert and analogous areas.

2. Geologic Setting

The Neoproterozoic shield rocks of the western part of the CED is dominated by dismembered ophiolites, tectonic mélange of allochthonous blocks of serpentinite incorporated and intermixed with an intensively deformed pelitic and calcareous schists, locally with intercalations of quartzite and black marble bands [42]. In the Barramiya–Mueilha sector (Figure 1), serpentinite represents greenschist facies metamorphosed cumulus ultramafic rocks of an ophiolite sequence [43–46]. The island arc metavolcanic rocks in the study area (Figure 1) comprise metabasalt, basaltic meta-andesite, interbedded with dactic tuffs and agglomerate, with accidental carbonate fragments. In the Wadi Dungash area, the island arc assemblages comprises mainly medium- to fine-grained, massive or foliated metabasalt and meta-andesite and epidote-chlorite schist. Pillowed morphologies are commonly observed in exposures of the less tectonized metabasalt in the area between Wadi Dungash and Wadi Barramiya [47]. Metagabbro-diorite complexes are weakly deformed rocks underlying extensive areas around Wadi Beizah. Large sub-rounded masses and discrete elongate masses of tonalite and granodiorite cut the ophiolitic mélange, island arc rocks, and the syn-orogenic granitoids in the northern part of the study area. The late- or post-orogenic intrusions are mainly alkali-feldspar granites, granite porphyries, and less commonly albitite, i.e., the G. Mueilha (Figure 1).

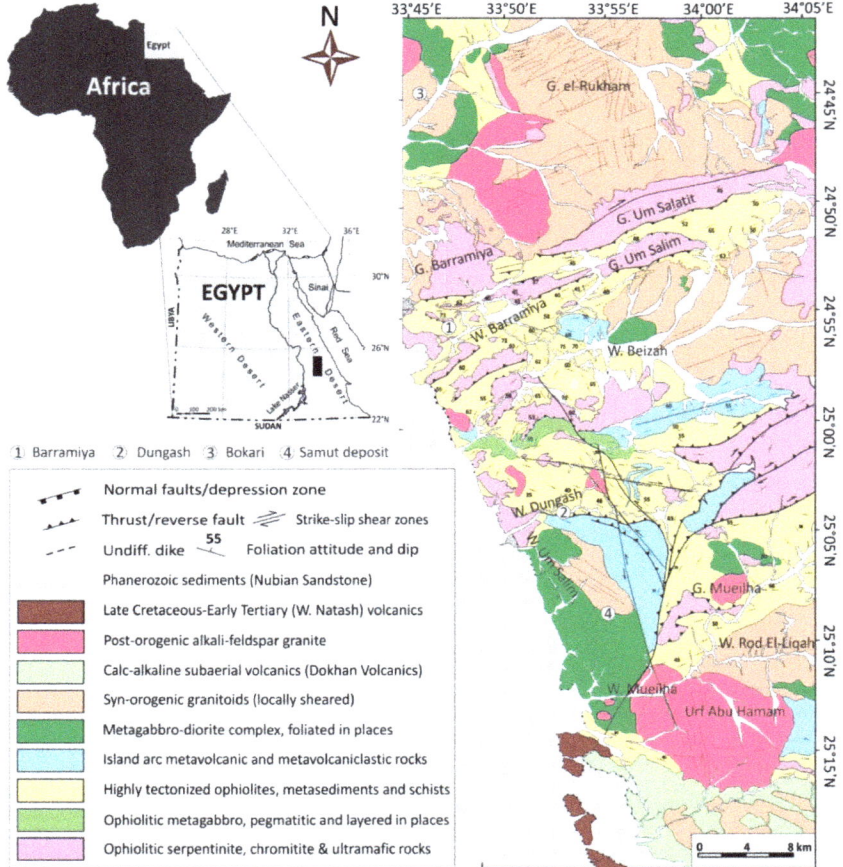

Figure 1. Simplified geological map of the Barramiya–Mueilha sector in the western part of the Central Eastern Desert of Egypt. Compiled from [35,36,44–47] and modifications based on new field verifications. Insets show the location of Egypt in Africa and the study area in Egypt.

Oval shaped and elongate masses of late- or post-tectonic granites cut the ophiolitic mélange, arc metavolcanics, and metagabbro-diorite rocks in the western and northern parts of the study area. These intrusions are cut by a system of NW- and NE-trending faults and shear zones. The Gabal Mueilha albite granite intrusion is known for historical economic Sn resources [48,49]. It exhibits sharp contacts with the country rocks and is intersected by NE and NW-trending faults/fracture sets filled by dikes and pegmatite veins. Rhyodacite–rhyolite rocks and their subvolcanic equivalents form a small exposure in the extreme southern part of the study area. The subvolcanic rocks are less deformed relative to the surrounding rocks and occur at the fault intersection zones [50,51]. Numerous sets of basaltic and dacitic dikes cut the granitoid rock terranes in different directions. The dominance of the ENE-trending dikes in the northern part of the study area adjacent to the ENE–WSW Barramiya–Um Salatit belt. The ~NNW-trending dikes in the south may imply different timing and geometry of the tensile deformation in the tectonic evolution of the area. The Barramiya–Mueilha sector contains several gold deposits that show ostensible features of structural control and association with hydrothermal carbonate zones.

3. Materials and Methods: SAR and Multispectral Satellite Data

3.1. Data Characteristics

Sentinel-1, Phased Array type L-band Synthetic Aperture Radar (PALSAR), Advanced Space borne Thermal Emission and Reflection Radiometer (ASTER), and Sentinel-2 datasets were used for this study. Sentinel-1 is a C-band synthetic aperture radar (SAR) instrument (frequency = 5.40 gigahertz) with a spatial resolution of down to 5 m and an up to 400 km-wide swath. Sentinel-1 was launched on 3 April 2014 (www.esa.int/copernicus). It has four operational modes, namely (i) Strip Map (SM) mode with 5 × 5 m spatial resolution and a 80 km swath, (ii) Interferometric Wide Swath (IW) mode with 5 × 20 m spatial resolution and a 250 km swath, (iii) Extra Wide Swath (EW) mode with 25 × 100 m spatial resolution and a 400 km swath, and (iv) Wave (WV) mode with 5 × 20 m resolution and a low data rate (20 × 20 km images along the orbit every 100 km) [52,53]. Sentinel-1 contains single polarization (VV or HH) for the Wave mode and dual polarization (VV + VH or HH + HV) for all other modes [54]. In this study, Sentinel-1 Wave Mode data (Granule ID: S1B_IW_GRDH_1SDV_20181130T154622_20181130T154647_013836_019A55, acquired on November 30, 2018) were obtained from the European Space Agency (ESA).

PALSAR sensor is L-band (1.27 gigahertz) synthetic aperture radar, and has a multi-mode observation function, which enables Fine, Direct Downlink, ScanSar, and Polarimetric modes. It contains multi-polarization configurations, namely HH, VV, HV, and VH, with variable off-nadir angles between 9.9 to 50.8 degrees, and different spatial resolutions of 10 m (Fine mode), 30 m (Polarimetric), and 100 m (ScanSar mode). The swath width is 30 km for the Polarimetric mode, 70 km for the Fine mode, and 250–350 km for the ScanSar mode [55–57]. A PALSAR scene covering the study area was acquired from the Earth and Remote Sensing Data Analysis Center (ERSDAC) Japan (http://gds.palsar.ersdac.jspacesystems.or.jp/e/). It was a Fine Mode Dual polarization (FBD) of HH + HV Level 1.5 product (ALPSRP080050480) acquired on 26 July 2007. The scene was of good quality (12.5 m pixel spacing), with an off-nadir angle of 34.3 and an incident angle of 38.8 degree, and was already geo-referenced to the UTM Zone 36 North projection with the WGS-84 datum.

ASTER is a multispectral sensor with 14 spectral bands, including three visible and near infrared radiation bands (VNIR; 0.52 to 0.86 µm) with a spatial resolution of 15 m, six shortwave infrared radiation bands (SWIR; 1.6 to 2.43 µm) with a spatial resolution of 30 m, and five thermal infrared radiation bands (TIR; 8.125 to 11.65 µm) with a spatial resolution of 90 m. A cloud-free, level 1T ASTER scene (AST_L1T_00303112003083059) of the study area was obtained from the U.S. Geological Survey Earth Resources Observation and Science Center (EROS) (https://earthexplorer.usgs.gov/). The scene was acquired on 11 March 2003 and was georeferenced to the UTM zone 36 North projection using the WGS-84 datum.

Sentinel-2 is multispectral sensor, launched on 23 June 2015, and provides 13 spectral bands, comprising of four bands in the VNIR region (0.45 to 0.66 μm), three narrow red edge bands (0.70 to 0.78 μm), two narrow NIR bands (0.84 to 0.86 μm) and two bands in SWIR region (1.6 to 2.20 μm), with spatial resolution ranging from 10, 20–60 m) with large swath width of 290 km [58,59]. In this study, a cloud-free, level-1 C Sentinel-2 (Granule ID: S2B_MSIL1C_20181207T082329_ N0207_R121_ T36RWN_20181207T120140, acquired on 7 December 2018) was obtained from the European Space Agency (ESA) (https://scihub.copernicus.eu/).

Processing of the different multispectral and radar data was made by using various software including the ENVI®(version 5.2, developed by L3Harris.com) and ArcGIS (version 10.3, developed by esri.com) packages.

3.2. Pre-Processing Methods

In this study, the Enhanced Lee filter was used to reduce speckle in Sentinel-1 and PALSAR radar imagery though concurrently conserving the texture evidence [60]. The Enhanced Lee filter is a revision of the Lee filter and likewise uses coefficient of variation within separate filter spaces [61,62]. The Enhanced Lee filter parameters used in this analysis are 7×7 m filter size, 1.00 damping factor, and the coefficient of variation in cutoffs for homogenous and heterogonous areas were arranged as 0.5230 and 1.7320, respectively. Every pixel is placed by applying the Enhanced Lee filter in one of the following three classes: (i) Homogeneous class—pixel value is replaced by the average of the filter window; (ii) heterogeneous treatment—pixel value is substituted by a weighted mean; and finally (iii) point target class—pixel value remains unchanged.

The atmospheric correction is used to minimize the influences of atmospheric factors in multispectral data. The Internal Average Relative Reflection (IARR) method was applied to the ASTER and Sentinel-2 data. The IARR technique for mineral mapping requires no prior knowledge of the geological features [63]. The IARR normalizes the images to a scene with an average spectrum. The 30 m-resolution ASTER SWIR bands were re-sampled to match with the VNIR 15-m. Sentinel-2 bands were geo-referenced to the zone 36 North UTM projection using the WGS-84 datum and the spectral bands were stacked on the 10-meter resolution bands via the nearest neighbor resampling method to preserve the original pixel values.

3.3. Image Processing Methods

The inverse relationship of HH or VH and HV or VV polarizations of the SAR data can optimize the geological features with different orientations. This is principally useful for the topographic applications and mapping structural pattern [4]. The combination of the different polarizations (HH or VH and HV or VV) results in numerous cross-polarized backscatter ratio pictures. In this study, ratio images of VV, VH, VH/VV and VH + VV of Sentinel-1 and HV, HH, HH/HV and HH + HV of PALSAR were produced. We used the Feature-oriented Principal Components Selection (FPCS) technique to analyze the backscatter ratio polarization images of Sentinel-1 and PALSAR datasets [64,65]. We assessed the PCA eigenvector loadings on the basis of correlation matrix in order to choose the most suitable PC which ascertains the important backscatter signatures and variability. A correlation matrix is normally used if the variances of individual variates are high, or if the units of measurement of the individual variates differ. The factor model is based on summarizing the total variances. Unities are used in the diagonal of the correlation matrix for PCA to imply that all variance is common or shared. Whether a surface feature appears as dark or bright pixels, it is based on the sign of the eigenvector loadings, coupled with the effect of topographic perception on the radar backscatter response [65]. Consistent improvements in image enhancement and signal to noise ratio (SNR) are possible by using a correlation matrix in the principal component analysis [4]. The computed correlation eigenvector values for Sentinel-1 and PALSAR datasets are shown Tables 1 and 2, respectively.

Table 1. Eigenvector matrix of copolarized and cross-polarized backscattering for Sentinel-1 Feature-oriented Principal Components Selection (FPCS) images.

Eigenvector	VH	VV	VH/VV	VH + VV
FPCS 1	0.230369	0.563194	0.563194	0.563194
FPCS 2	0.783316	−0.591162	0.591162	0.591162
FPCS 3	−0.003523	−0.003523	0.999990	0.999990
FPCS 4	−0.999990	−0.577350	−0.000000	0.577350

VH = verical horizontal, VV = vertical veritcal.

Table 2. Eigenvector matrix of copolarized and cross-polarized backscattering for PALSAR FPCS images.

Eigenvector	HH	HV	HH/HV	HH + HV
FPCS 1	0.589359	0.194688	0.004823	0.784047
FPCS 2	−0.564816	0.792703	−0.025923	0.227887
FPCS 3	0.017490	−0.019617	−0.999652	−0.002127
FPCS 4	−0.577350	−0.577350	0.000000	0.577350

HH = horizontal horizontal, HV = horizontal vertical.

The principal component analysis (PCA), band combination and specialized band ratios techniques were applied to ASTER and Sentinel-2 data to extract information related to lithological and hydrothermal alteration mapping. The PCA is a standard statistical method applied to minimize the independent principal components and highlight most of the variability inherited from the numerous combined band images [66,67]. In this study, the PCs are calculated using the covariance matrix (scaled sums of squares and cross products) of ASTER VNIR + SWIR and Sentinel-2 spectral bands. Sentinel-2 bands 1, 9, and 10 were excluded from this analysis as they do not contain mineralogical/geological information. These bands contain information related to atmospheric issues (e.g., aerosol scattering, water vapor absorption, and detection of thin cirrus) irrelevant to this study [64]. The eigenvector matrix for the ASTER and Sentinel-2 data derived from the PCA are given in Tables 3 and 4.

The PCA is capable of determining the direction of space containing the highest sample variance, and moving on to the orthogonal subspace in this direction to find the next highest variance. The result is iteratively discovering an ordered orthogonal basis of the highest variance. The subspace defined by the first n PCA vectors can explain a given percentage of the variance. The subspace of dimension n explains the largest possible fraction of the total variance [66]. The PCA1 contains the albedo is related largely to the topographic features. Three first PCA images (PCA1, PCA2, and PCA3), containing the highest topographical and spectral information, are suitable for lithological discrimination. The PCA images (except PCA 1) may have information related to alteration minerals, which could be reproduced in the eigenvector loading of the absorption and reflection bands. A PC image with moderate to high eigenvector loadings for the indicative bands (reflection and/or absorption bands) and opposite signs promotes an efficient discrimination of a given mineral. If the loading in the reflection band of a given mineral is positive, the enhanced pixels related to the mineral will appear as bright pixels. On the contrary, the enhanced related pixels will appear as dark if the loading to a given mineral in the reflection band is negative [67]. For inverting the dark pixels to bright pixels, negation can be accomplished by multiplication by −1.

Table 3. Eigenvector matrix for Advanced Space-borne Thermal Emission and Reflection Radiometer (ASTER) visible near infrared region (VNIR) + short wave infrared region (SWIR) principal component analysis (PCA) images.

Eigenvector	Band 1	Band 2	Band 3	Band 4	Band 5	Band 6	Band 7	Band 8	Band 9
PCA 1	−0.286530	−0.343729	−0.329190	−0.329130	−0.309398	−0.316583	−0.363650	−0.381071	−0.331134
PCA 2	0.549949	0.503725	0.370167	−0.145513	−0.182937	−0.175862	−0.266600	−0.302618	−0.242021
PCA 3	−0.226565	0.135415	0.271478	−0.498784	−0.406829	−0.358883	0.207509	0.453128	0.255251
PCA 4	−0.608617	0.126442	0.533556	0.496999	−0.100609	−0.083579	−0.132278	−0.102609	−0.191767
PCA 5	−0.296023	0.110156	0.237883	−0.545338	0.291567	0.563536	−0.300481	−0.167608	0.159023
PCA 6	−0.023765	−0.166820	0.260418	−0.267115	0.557757	−0.291015	0.480996	−0.115227	−0.438202
PCA 7	0.126345	−0.329776	0.249486	0.076552	0.288319	−0.444288	−0.282746	−0.222119	0.630381
PCA 8	−0.297947	0.666275	−0.453618	0.008140	0.296406	−0.318394	0.056334	−0.184583	0.187068
PCA 9	0.027941	0.033903	−0.061731	0.005964	0.357891	−0.178985	−0.575640	0.650058	−0.283134

Table 4. Eigenvectors and eigenvalues of correlation matrix of Sentinel-2 PCA images.

Eigenvector	Band 2	Band 3	Band 4	Band 5	Band 6	Band 7	Band 8	Band 8a	Band 11	Band 12
PCA 1	−0.130767	−0.195965	−0.294453	−0.305036	−0.323206	−0.340531	−0.332815	−0.346959	−0.421500	−0.368126
PCA 2	0.219830	0.283259	0.314141	0.182452	0.160849	0.138366	0.241307	0.086667	−0.547330	−0.573705
PCA 3	−0.330657	−0.392816	−0.335883	0.295142	0.344046	0.361373	−0.324495	0.361036	−0.123340	−0.191365
PCA 4	−0.504592	−0.418581	0.185161	−0.229417	−0.109091	−0.024353	0.653589	0.124095	−0.162823	0.040942
PCA 5	−0.104777	−0.066455	0.061494	−0.025064	0.018281	−0.049037	0.088132	−0.110870	0.690943	−0.692817
PCA 6	0.250524	0.098325	−0.259140	−0.241083	−0.368827	−0.148165	0.049876	0.789284	0.055270	−0.125685
PCA 7	−0.074844	−0.061944	0.038226	0.476730	0.238402	−0.824566	0.033654	0.150443	−0.015442	0.032866
PCA 8	0.306221	−0.016038	−0.648371	−0.181613	0.458534	−0.044668	0.453038	−0.186550	−0.015476	−0.008441
PCA 9	−0.054521	0.038123	0.324790	−0.643651	0.579072	−0.165281	−0.282445	0.182228	−0.008348	0.010264
PCA 10	0.628125	−0.729531	0.263138	0.025161	−0.024824	0.011445	−0.046073	−0.021424	0.006267	−0.000219

Several ASTER VNIR + SWIR and Sentinal-2 band combinations and band ratio images are tested and adopted for mapping lithological units associated with gold occurrences in the study area. Band ratio technique is used for reducing the effects of topography and enhancing the spectral differences between bands. It is a technique where the digital number value of one band is divided by the digital number value of another band. Band ratios are very useful for emphasizing hydrothermal alteration minerals and lithological units. Dividing one spectral band by another produces an image that provides relative band intensities. The final goal is to minimize the illumination differences due to topography [68]. Ratio images can be meaningfully interpreted because they can be directly related to the spectral properties of minerals. Ratioing can enhance minor differences between minerals by defining the slope of spectral curve between two bands. The band ratio technique is, therefore, specifically applicable to highly exposed areas and rugged terrains in arid regions [6]. The ASTER band combination images (R:4, G:3, B:1) and (R:4, G:7, B:3) are processed to map Fe_2O_3/MgO and Al_2O_3-rich rocks in the study area. The Sentinel-2 band combination (R:11, G:8, B:2) and (R:11, G:12, B:7) images are generated to map Fe_2O_3/MgO and Al_2O_3 -rich rocks [25]. The band ratio of ASTER bands, 4/2 (gossan), 4/3 (ferric oxide), (7+9)/8 (chlorite/epidote/clcite) [68], are used here to highlight the distribution of ophiolitic and island arc rocks in the study area. The hydrothermal alteration zones are characterized by substantial contents of hydrous ferromagnesian silicate and iron oxides. The Sentinel-2 band ratios, 11/4 (gossan), 11/8a (ferric oxide), 12/11 (ferrous silicates) and 11/12 (hydroxyl alteration) are used to map the different rock units [59] and to emphasize on the hydrothermal alteration zones.

4. Results

4.1. Remote Sensing Data Analysis

Analyzing the eigenvector loadings for Sentinel-1 cross-polarized backscatter ratio images indicates that the FPCS2 has strong loadings of VH (0.783316) and VV (−0.591162) with opposite signs (Table 1). Cross polarization (VH) is extremely sensitive to geologic structures, while surface roughness is reflected in robust reflection and high VV backscatter [4]. The inverse relationship between the VH and VV loadings in the FPCS2 emphasizes the topographic features with different orientations, contrast, and textural signatures. Figure 2A,B shows the FPCS2 Sentinal-1 image-map of the selected spatial subset covering the study area. Structural elements related to the transtension and transpression zones are traceable in Figure 2A, and are annotated in Figure 2B. Well-developed foliations and shear cleavages are dominated in the transpression zone (ductile zone), while zones accommodating displacements and form discontinuities are considered as transtension zones. The ENE-striking foliation and related close and overturned folds extend for several kilometers and enfold the huge ophiolite blocks. A considerable difference in the topography and foliation trajectories in terranes of the ophiolite nappes relative to the island arc metavolcanic rocks emphasizes distinctive deformation histories, where shortening should have brought together tectonically different terranes now are juxtaposed.

The depth of penetration of radar signal is improved by using the L-band radar data, therefore, PALSAR data have the capability to map geological structures that may be covered by sand in arid regions. The FPCS technique is applied to cross-polarized backscatter ratio images of PALSAR, including HV, HH, HH/HV and HH + HV (Table 2). Different polarizations are sensitive to ground surface features of different dimensions; they collectively bring out greater geological–geomorphological–structural details [4]. Considering the eigenvector loadings for PALSAR cross-polarized backscatter ratio images, it is evident that the FPCS2 contains information for structural mapping and topographic enhancement due to the strong contribution of HH (−0.564816) and HV (0.792703) with opposite signs (Table 2). Figure 3A,B shows the FPCS2 PALSAR image-map of the selected spatial subset covering the study area. The FPCS2 image-map clearly discriminates the relatively high topographic, intensely sheared ophiolitic nappes from the tectonically underlying, low topographic, island arc metavolcanic and metavolcaniclastic rocks. Folding of the ophiolitic rocks appears to be more recognizable in the southern part than in the northern part of the study area. Typically, high brightness contrast, textural variability, and tonal variation are recorded in the FPCS2 PALSAR image-map compared to the Sentinel-1 images.

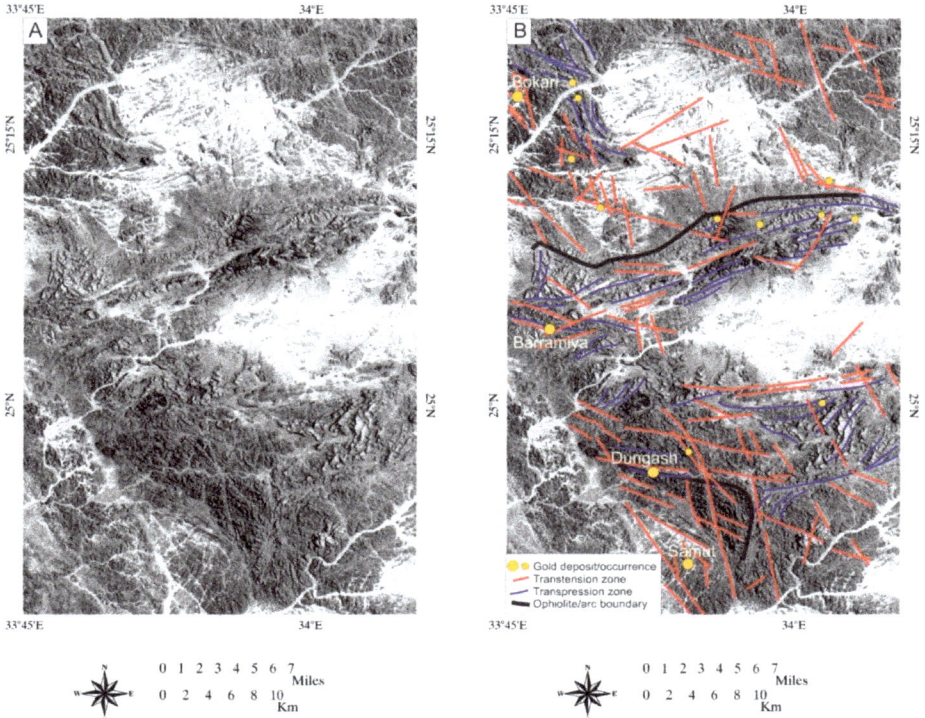

Figure 2. Processed Sentinel-1 data of the Barramiya–Mueilha area showing: (**A**) Feature-oriented Principal Components Selection (FPCS2) highlighting the variable styles of deformation in the study area. (**B**) Same as in A with interpreted structural elements and locations of gold deposits/occurrences. For more details please refer to the next sections.

A 9-band PCA is constructed from the original 9-band (VNIR + SWIR) of ASTER image covering the study area (Table 3). The PC1 band contains the largest percentage of data variance and albedo related to topographical features. The second and third PC bands contain the second largest data variance and spectral information. Thus, the first three high order PCs (1, 2, and 3) contain approximately 99% of spectral information, which can be used for lithological mapping rather than the subsequent low order principal components (4, 5, 6, etc.) which usually contain < 1% of spectral information and low signal-to-noise ratios. The last PCA bands may contain information related to alteration minerals or can appear noisy because they contain very little variance [4,11,12]. Accordingly, PC1, PC2, and PC3 can be considered in lithological mapping as Red–Green–Blue color composite, especially for arid region such study area. Figure 4A shows RGB color composite of the PC1, PC2, and PC3 for ASTER VNIR + SWIR bands covering the study area.

Most of the lithological units contain distinguishable spectral characteristics compliant with the geological map (Figure 1). However, the lithologies with analogous spectral features are hard to discriminate. The highly tectonized ophiolites, metasediments, and schists are expressed by yellow to orange pixels, demarcating the ductile shear zones in the central part of the study area (Figure 4A). Post-orogenic granites in the northwestern part of the study area appear as bluish green pixels, locally with pinkish yellow pixels. Spectral signature comparable to the granitic rocks are explained as alteration products of feldspar to clay minerals in brittle (transtension) shear zones. Metagabbro–diorite complex and the Nubian Sandstone appear as magenta to purple pixels. In Figure 4A, the ophiolitic rock terranes are notably separated from the island arc rocks by the pale green color, whereas the island arc metavolcaniclastic rocks, with a reddish image signature, form an autochthonous block in the southwestern part of the area and is surrounded by major thrusts.

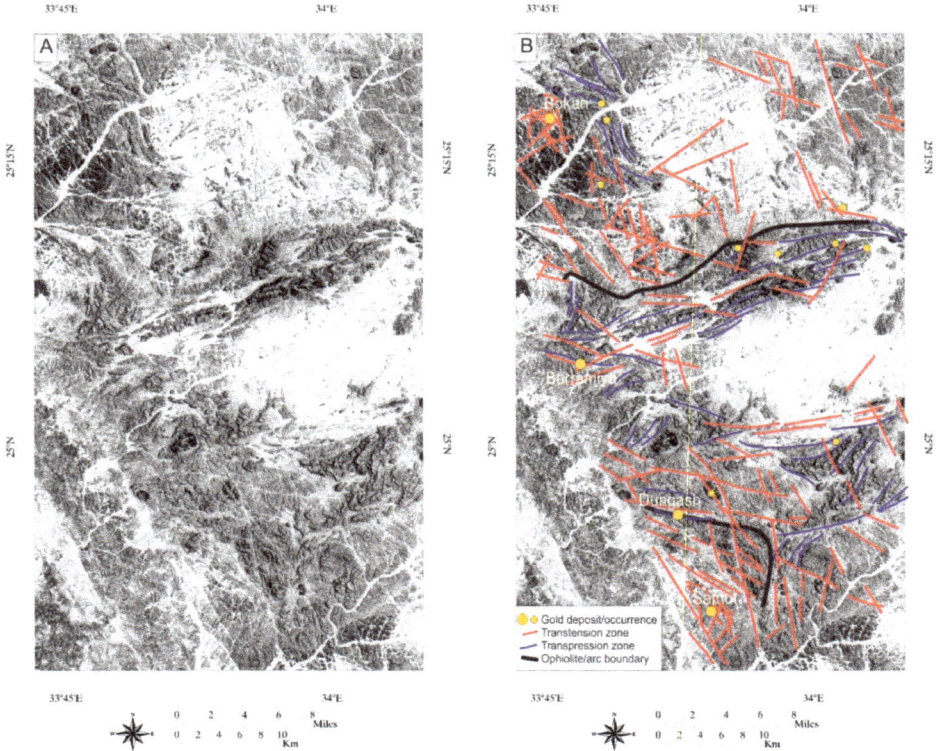

Figure 3. Processed Phased Array type L-band Synthetic Aperture Radar (PALSAR) data of the Barramiya–Mueilha area: (**A**) Feature-oriented Principal Components Selection (FPCS2) image showing fold traces and foliation trajectories most prominent in the central part of the area, (**B**) Same as A with interpreted structural elements and locations of gold deposits/occurrences. The different structural elements are classified based on fieldwork as detailed in the text.

Analysis of the eigenvector loadings for the ASTER VNIR + SWIR dataset shows that the PC4 has strong loadings of band 1 (−0.608617) and band 4 (0.496999) with opposite signs (Table 3). Hematite, jarosite, goethite, and limonite tend to have strong absorption features in 0.4–1.1 µm, coinciding with bands 1, 2, and 3 of ASTER, and high reflectance in 1.56–1.70 µm, coinciding with band 4 of ASTER [68,69]. Therefore, this PC image contains spectral information for mapping iron oxide/hydroxide minerals. The PC5 has high loadings of band 4 (−0.545338) and band 6 (0.563536) with opposed signs (Table 3). Al–OH mineral groups contain spectral absorption features in 2.1–2.2 µm and reflectance in 1.55–1.75 µm, which coincide with bands 5 and 6 (2.145 to 2.225 µm) and 4 (1.600 to 1.700 µm) of ASTER [69]. Accordingly, the PC5 can enhance kaolinite, alunite, and sericite (muscovite) mineral groups. It should be noted that the dark pixels in PC5 need to be inverted to bright pixels by negation. The PC6 shows strong positive loading in band 5 (0.557757) and strong negative loading in band 9 (−0.438202) (Table 3). Fe, Mg-OH-bearing alteration minerals and CO_3 mineral groups show high absorption characteristics in the position defined by bands 8 and 9 (2.295–2.430 µm) and low absorption features in band 5 (2.145–2.185 µm) of ASTER [69]. So, the PC6 may has spectral information for mapping epidote, chlorite, and calcite. Figure 4B shows an RGB color composite image of PC4 (R), PC5 (G), and PC6 (B) covering the study area. The red and magenta pixels have high surface abundance of iron oxide/hydroxide minerals, which are mostly associated with ophiolitic serpentinite, chromitite, and ultramafic rocks, highly tectonized ophiolites, metasediments and schists, and Nubian sandstone. The yellow pixels (admixture of iron oxide/hydroxide and Al–OH mineral groups) are typically occurred in alkali-feldspar granite and granitoids background (transtension zones). The green pixels are associated with Nubian sandstone, recent alluvium, ophiolitic serpentinite, chromitite and ultramafic rocks, syn-orogenic granitoids, and post-orogenic alkali-feldspar granites. The blue pixels strongly develop in the mafic and ultramafic rocks such as serpentinite, chromitite, and metagabbro-diorite complex (Figure 4B).

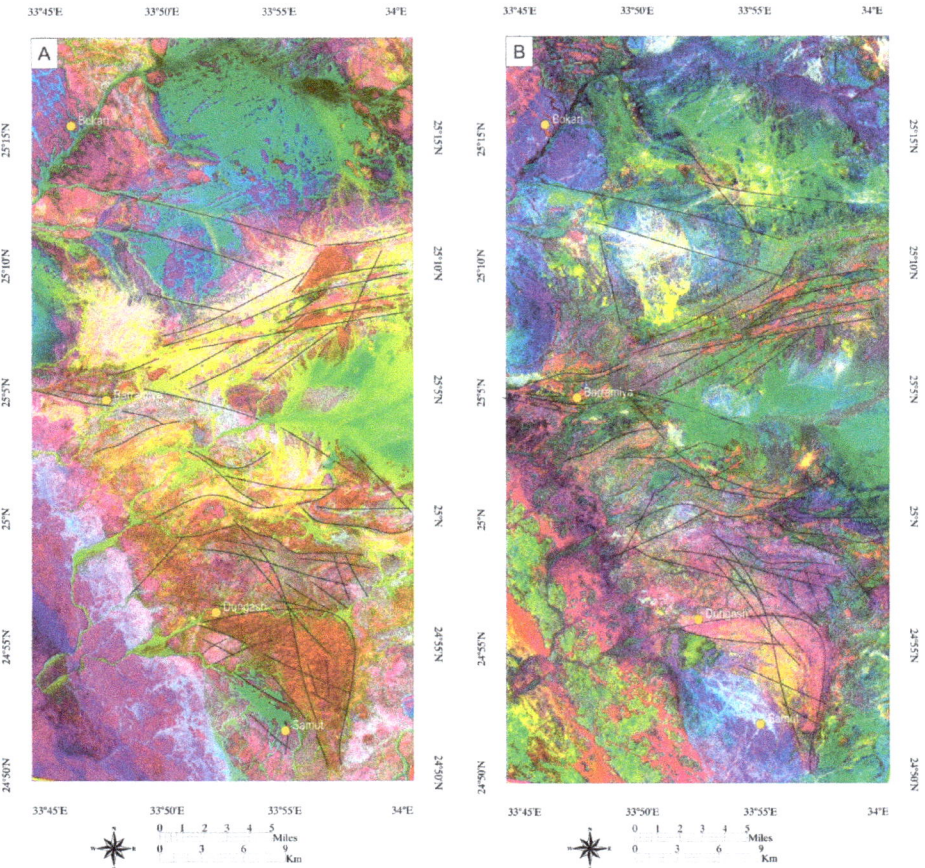

Figure 4. False-color composite, ASTER PCA images for the study area: (**A**) PC1, PC2, and PC3 (in RGB channels) shows the island arc metavolcanic/metavolcaniclastic rocks as red pixels, whereas the ophiolitic rocks and ophiolitic mélange rocks exhibit a greenish image signature. The unconformity boundary is marked by the transition to pale blue areas (of the Nubian Sandstone) in the southwestern part of the area, (**B**) PC4, PC5, and PC6 (in RGB channels) differentiates the iron oxide/hydroxide minerals (red pixels), Al–OH mineral groups (green pixels) and Fe, Mg–OH–bearing alteration minerals and CO_3 mineral groups (blue pixels). Solid black lines refer to the main lineaments (i.e., fault and shear zones) in the area. Notice that the curvilinear morphologies of the lineaments (transpression zones) in the central part, whereas straight lineaments (herein classified as transtension zones) prevail in the northern and extreme sourthern parts of the area.

Figure 5A (PC1, PC2, and PC3) shows the island arc metavolcanic/metavolcaniclastic rocks as reddish pixels. The light green domains in the ophiolitic mélange terranes are those which experienced high strain and demarcating transpression, which is highly tectonized ophiolites and schists as in the geological map. The boundary between ophiolites and island arc rocks appears prominent, and fold trajectories are easy to trace. The boundary between the Nubian sandstone and adjacent lithological units appear as light blue lines and changes into red and blue color towards the western part due to irregular abundance of ferrous silicates in this Phanerozoic sedimentary rocks. Post-orogenic alkali-feldspar granite and syn-orogenic granitoids represent in greenish-blue to blue color in northwestern part of the study area, while metagabbro–diorite complex depicts as magenta to light brown color (Figure 5A).

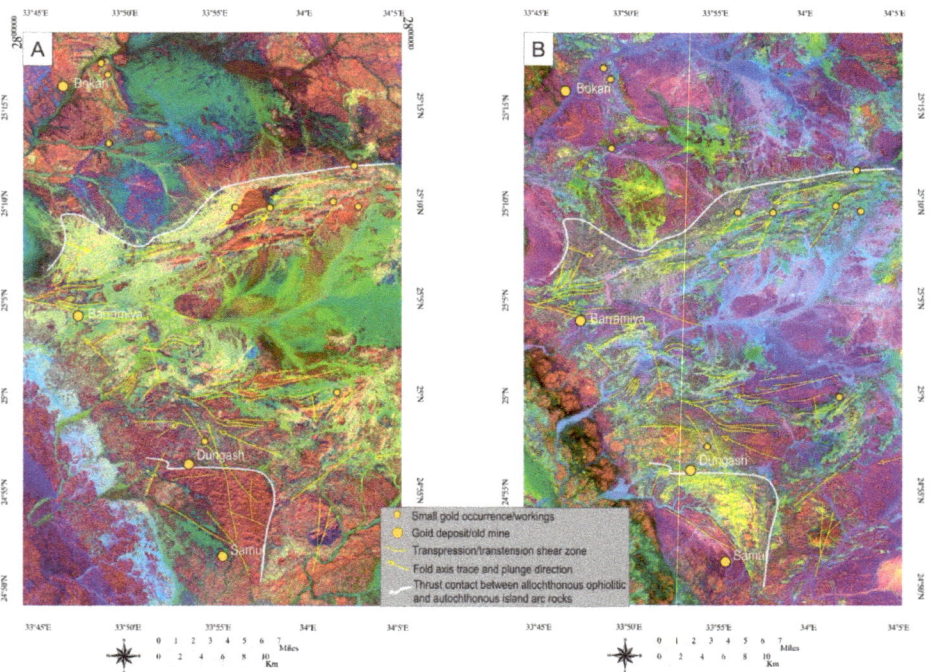

Figure 5. False-color composite, Sentinel-2 PCA images for the study area: (**A**) PC1, PC2, and PC3 in R, G, and B channels, respectively. Notice the island arc metavolcanic/metavolcaniclastic rocks are conspicuously distinctive as reddish color pixels. The light color domains within the ophiolitic mélange terranes are those experienced high strain and demarcating transpression. The boundary between ophiolites and island arc rocks appears prominent and fold trajectories are easy to trace, (**B**) PC6, PC5, and PC4 in R, G, and B channels, respectively. Iron oxide/hydroxide minerals (red pixels), Al–OH and Fe, Mg–OH-bearing and CO_3 alteration minerals (green pixels) and ferrous silicates mineral groups (blue pixels). The yellow lines refer to the main structural elements (faults, thrusts, and shear zones), while the yellow arrows are traces of the fold axial planes and the arrowhead point towards the plunge direction.

The magnitude and sign of eigenvector loadings for Sentinel-2 (Table 4) reveal that the PC4 contains a strong negative loading for band 2 (−0.504592), a strong positive loading for band 8 (0.653589) and a moderate negative loading for band 11 (−0.162823). Ferrous silicates such as biotite, chloride, and amphibole can be detected using band 2 (0.450–0.550 µm), band 8 (0.800–0.910 µm) and band 11 (1.520–1.850 µm) of Sentinel-2 due to their spectral features [25,27]. The PCA5 shows a strong loading for band 11 (0.690943) and band 12 (−0.692817) with opposite signs (Table 4). A PCA5 image can, therefore, discriminate between the hydrous and intensely deformed ophiolitic mélange rocks from the island arc rocks. Compressional structures are conspicuous in the central part of the study area, where hydroxyl mineral zones prevail. The PCA6 has great contribution for band 6 (−0.368827) and band 8a (0.789284) with opposite signs and a moderate contribution for band 2 (0.250524) (Table 4). The PCA6 image contains important information about Fe-oxides/hydroxides that are mainly enriched in the ultramafic rocks, i.e., ophiolites. A RGB (PC6, PC5 and PC4) composite image (Figure 5B) shows the different lithological units and alteration zones with distinct image signatures. The areas with high abundance of hydrous minerals and Fe-oxides/hydroxides appear as yellow pixels exemplify the highly tectonized ophiolites and schists. The yellow pixels also represent the post-orogenic alkali-feldspar granite, syn-orogenic granitoids and island arc metavolcanic and metavolcaniclastic rocks. Intense foliation and shear cleavages by aligned hydrous minerals typify the ductile shear (transpression) zones. Carbonate alteration is also associated with the transpression zones. Ferrugination zones that are associated with faults/brittle shear (transtension) zones are commonly irregular in shape and scattered.

When comparing the Sentinel-2 image (Figure 5B) with the ASTER image (Figure 4B), the hydroxyl minerals and iron oxide/hydroxide zones are found better discriminated on the ASTER image. The ASTER spectral bands are particularly designed to depict the Al–OH, Fe, Mg–OH–bearing alteration and CO_3 mineral groups and iron oxide/hydroxides. Bands 11 and 12 of Sentinel-2 do not have enough spectral width for distinguishing specific Al–OH, Fe, Mg–OH and CO_3 mineral assemblages. Bands 2–9 (0.450–1.20 µm) of Sentinel-2 contain enough spectral and spatial data for mapping iron absorption feature parameters [25], and can therefore be used for detecting goethite, jarosite, and hematite. In the Sentinel-2 image (Figure 5B), mafic and ultramafic lithological units such as metagabbro–diorite complex appear as magenta pixels because of the high content of iron in their composition.

Figure 6A shows an ASTER RGB combination image of band 4 (R), band 3 (G), and band 1(B) for mapping Fe_2O_3/MgO-rich rocks in the study area. It demarcates ophiolitic serpentinite as dark pixels, whereas rocks with lesser contents of ferromagnesian minerals exhibit lighter image signatures. Post-orogenic alkali-feldspar granite and syn-orogenic granitoids appear in light pink to white due to the low contents of Fe_2O_3 and MgO, while the mafic to ultramafic lithological units are expressed by dark green pixels (ophiolitic serpentinite, chromitite, and ultramafic rocks), green pixels (metagabbro–diorite complex) and brown to magenta pixels (the highly tectonized ophiolites and island arc metavolcanic and metavolcaniclastic rocks). Figure 6B displays an ASTER RGB combination image of band 4 (R), band 7 (G), and band 3 (B) for enhancing Al_2O_3-rich rocks. This image highlights post-orogenic alkali-feldspar granite, syn-orogenic granitoids and Nubian sandstone as green areas and carbonate alteration zones as greenish rafts in the high strain zones in the eastern part of the belt. The lithological units such as metagabbro–diorite complex and schists and metavolcaniclastic rocks contain moderate content of aluminosilicate minerals appear as dark green pixels. The ophiolitic serpentinite, chromitite, and ultramafic rocks and the highly tectonized ophiolites represent in black pixels attributed to very low content of Al_2O_3 in their composition.

Figure 7A,B are Sentinel-2 RGB combination images of band 11 (R), band 8 (G), and band 2 (B) for mapping Fe_2O_3/MgO-rich rocks and band 11 (R), band 12 (G) and band 7 (B) for identifying Al_2O_3-rich rocks. Mafic mineral-rich rocks are clearly delineated in Figure 7A as dark green in color, including the ophiolitic serpentinite, chromitite, and ultramafic rocks and highly tectonized ophiolites. The metagabbro–diorite complex appears as green color because of a high to moderate contribution of ferromagnesian minerals. Moreover, the different content of ferromagnesian minerals in ophiolitic and island arc rocks promotes the ease differentiation based on RGB color composite (Figure 7A). The folded ophiolitic rocks are noticeably distributed adjacent to the variably deformed island arc rocks in the southern part of the study area. The Al_2O_3-rich rocks are enhanced in Figure 7B as green color. The post-orogenic alkali-feldspar granite and syn-orogenic granitoids, Nubian sandstone, metagabbro–diorite complex and island arc metavolcanic and metavolcaniclastic rocks depict as green color. However, ophiolitic serpentinite, chromitite and ultramafic rocks and highly tectonized ophiolites only represent as black pixels. Compared to the Sentinel-2 image (Figure 7A,B), the ASTER RGB images (Figure 6A,B) are strongly capable of discriminating lithological units with different content of Fe$_2$O$_3$/MgO and Al$_2$O$_3$.

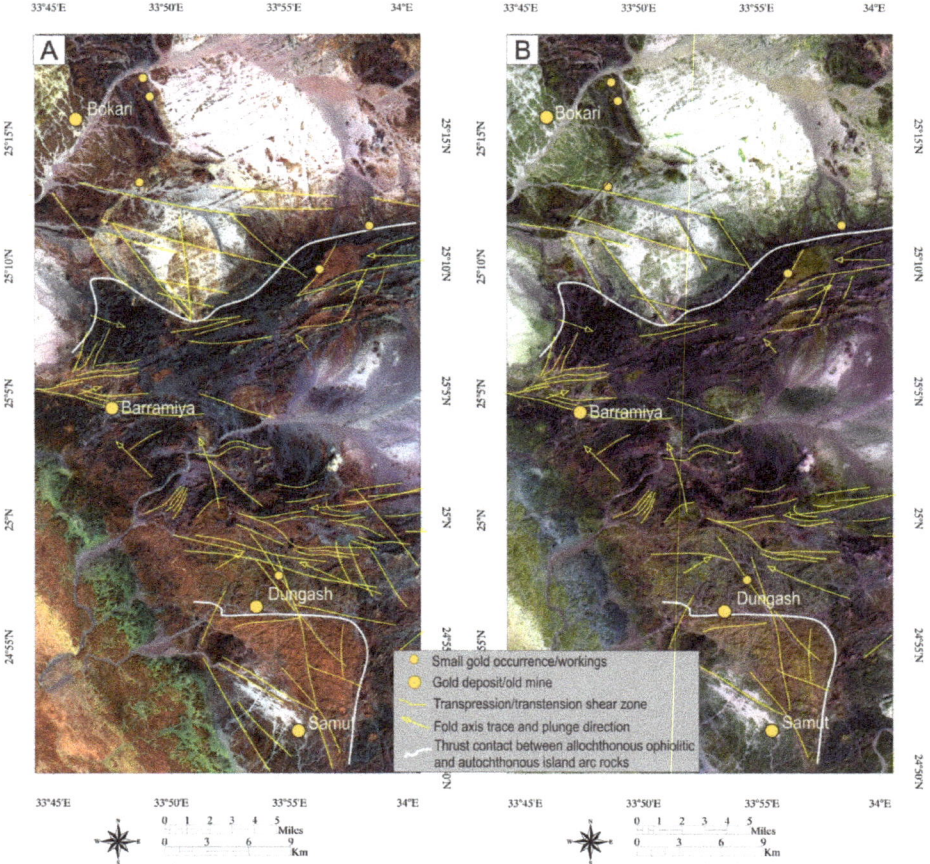

Figure 6. False-color composite, ASTER band combination images for the study area: (**A**) (R: Band 4, G: Band 3, B: Band 1) used to map Fe$_2$O$_3$/MgO-rich rocks, (**B**) (R: Band 4, G: Band 7, B: Band 3) used to enhance the signature of Al$_2$O$_3$-rich rocks to promote detailed mapping of the folded ophiolitic rocks.

Figure 8A–C are ASTER band ratio images for mapping gossan (4/2), ferric oxide (4/3), and the chlorite/epidote/calcite mineral group (7+9)/8. High digital number (DN) value pixels that appear as bright pixels indicate the spectral signatures of particular mineral or mineral groups [68,69]. Figure 8A shows the gossan zone as bright pixels, mostly associated with the highly tectonized ophiolites and schists, island arc metavolcanic and metavolcaniclastic rocks. The ferric oxide-rich zones are typically expressed by bright pixels in Figure 8B. The island arc rocks are seen as bright pixels in Figure 8C, most likely because of the abundant hydrous minerals such as mica, amphiboles, chlorite, and epidote. This observation ascertains that these rocks were tectonically overlain by thrusted ophiolitic nappes. Talc carbonate and listvenite zones in the ophiolitic domains appear as bright zones, and are commonly confined to the transpression zones. The bright pixels in the Nubian sandstone background can be attributed to the abundant detrital clay minerals.

Hydrothermal alteration zones also appear as bright pixels on the band ratios images of Sentinel-2 (Figure 9A–D). Band ratio 11/4 image shows gossan zones associated with tectonized ophiolites and island arc metavolcanic rocks (Figure 9A). Ferric oxides detected by band ratio 12/11 are mostly associated with the granitic rocks (Figure 9B). Ferrous silicates (biotite, chloride, and amphibole) mapped by band ratio 12/11 are distinguished in island arc metavolcanic and metavolcaniclastic rocks and highly tectonized ophiolites (Figure 9C). Band ratio 11/12 enhances highly carbonated and deformed ophiolitic rocks in the ductile deformation zone, which are associated with highly tectonized ophiolites, metasediments and schists and island arc metavolcanic and metavolcaniclastic rocks. The results of alteration mapping derived from ASTER and Sentinel-2 datasets match and show the altered zones in both ductile and brittle deformation zones. Gossan, the chlorite/epidote/calcite mineral group, ferrous silicates and hydroxyl alteration zones are mapped in ductile (transpression) deformation zones, whereas ferric oxides are identified in the brittle (transtension) deformation zones.

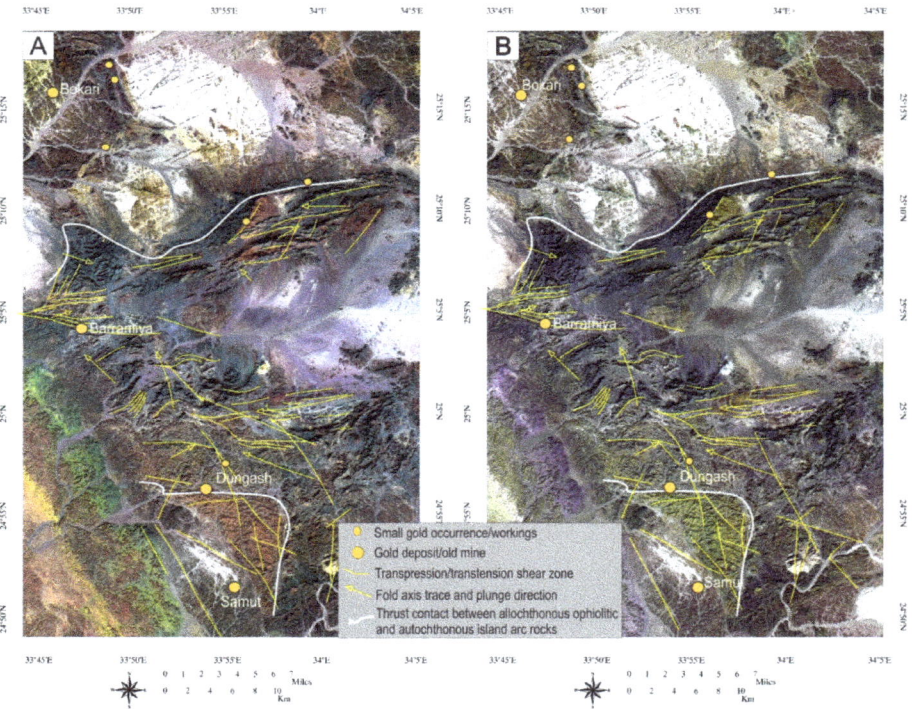

Figure 7. False-color composite, Sentinel-2 band combination images for the study area: (**A**) (R: Band 11, G: Band 8, B: Band 2), and (**B**) (R: Band 11, G: Band 12, B: Band 7) used to map Fe_2O_3/MgO and Al_2O_3-rich rocks, respectively. These images helped in differentiating the mafic and ultramafic ophiolitic from felsic rocks.

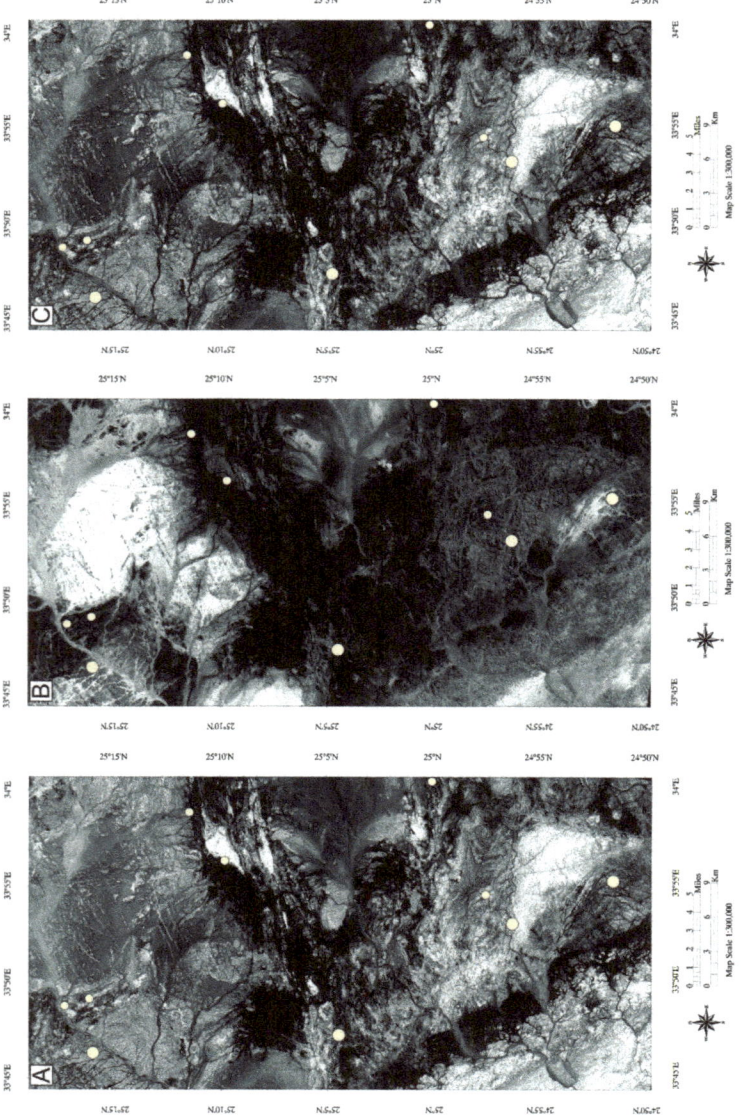

Figure 8. ASTER band ratio indices used to detect the hydrothermal alteration zones in the study area: (**A**) Band4/band2 image for mapping gossan zones, (**B**) Band4/band3 image for mapping ferric oxide-rich rocks, and (**C**) Band7+band9/band8 image for mapping chlorite/epidote/clacite-rich rocks. White circles are locations of gold deposits/occurrences.

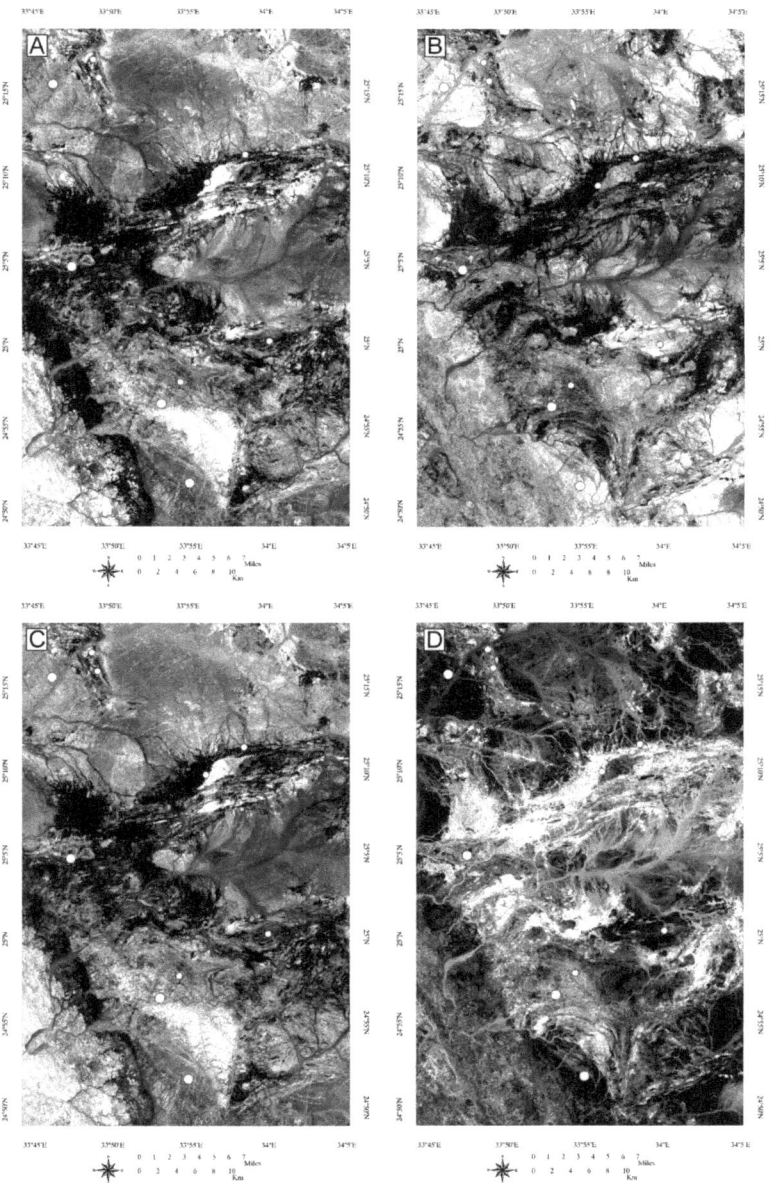

Figure 9. Band ratios of Sentinel-2 data used to map hydrothermal alteration zones in the study area: (**A**) band11/band4 image (for gossan mapping), (**B**) band11/band8a image (for ferric oxide-rich rocks), (**C**) band12/band11 image (for ferrous silicate-rich rocks), and (**D**) band11/band12 image (for hydroxyl alteration mapping).

4.2. Fieldwork Data Analysis

4.2.1. Structural Evolution of the Barramiya–Mueilha Area

Superimposed structural elements in the Barramiya–Mueilha area (Figure 10) are considered as manifestations of three phases of ductile deformations (D1, D2, D3). Extensional structures, e.g., fault/fracture zones and dike swarms cutting most of the ductile structures in different directions are signs of late brittle deformation (D4), most likely related to terrane cooling and exhumation during the orogen collapse [44,70–77].

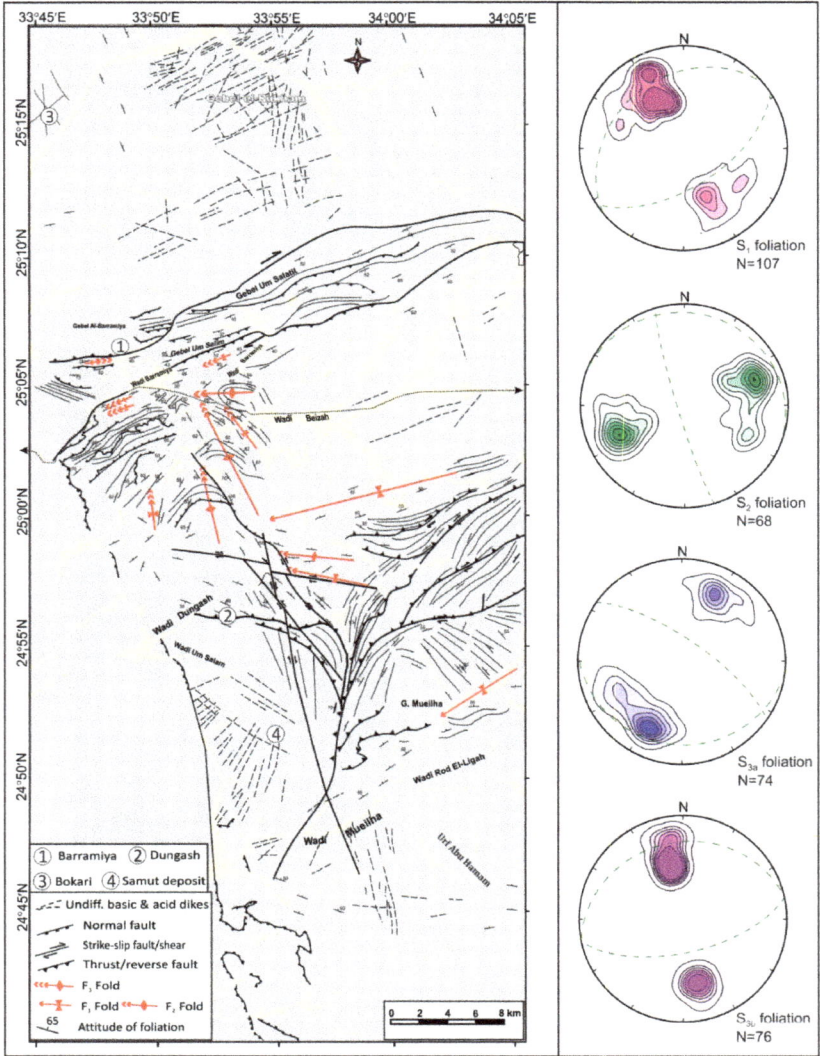

Figure 10. Simplified structural map of the Barramiya–Mueilha sector compiled from field measurements and satellite data interpretations. Several occurrences of gold–quartz veins are located in the area, but only significant gold occurrences/deposits are shown in the key. Insets are lower hemisphere stereographic projection for poles of the different generations of foliations.

4.2.2. NNW–SSE Shortening and Southward Tectonic Transport

This event in characterized by stacking and imbrication of large ophiolitic nappes and best exemplified by S_1 foliation (Figure 11A). Serpentinite in Barramiya and Dungash areas occurs as steeply dipping WNW- to NE-trending belts emplaced along thrust planes and stretched parallel to the penetrative foliation in the ophiolitic mélange. S_1 is ENE–striking shear foliation developed around the thrust-bound ophiolites. ENE-trending mineral stretching lineation (L_1) plunge gently (10–35°) mainly towards the ENE (Figure 11B). Schists, talc–magnesite rocks, listvenite and chromite pockets mark the tectonic contacts between the allochthonous serpentinite masses and the metasedimentary mélange matrix.

4.2.3. NNE-SSW Shortening Structures

Steeply dipping NW–SE to WNW–ESE mineral foliation (S_2), i.e., hornblende, epidote–chlorite, chlorite, tremolite, and graphite, are best developed in schistose rocks as well as sheared metagabbro, meta–agglomerate and metatuffs. The NW–SE Dungash–Beizah shear zone is marked by an intensification of S_2 schistosity (Figure 11C) and locally by NE-dipping mylonitic foliation. Feldspar-rich and amphibole-rich bands define S_2 foliation in the ophiolitic metagabbro. Mineral and stretching lineations are oriented close to fold axial plane.

4.2.4. E–W Oblique Convergence

This phase of deformation was characterized by the main stress component (sigma 1) swinging between NE and SE, but the main shortening is considered E–W coinciding with the Arabian–Nubian Shield collision with the Nile Craton in the west. During D3a deformation, the Wadi Beizah imbricated thrust sheets of ophiolitic serpentinte and metagabbro were openly folded into a series of kilometre-scale NNW-trending asymmetric and plunging synclinal and anticlinal folds (F_{3a}). These folds progressively open up westward, where fold hinges (L_{3a}) plunge gently (~25°) to the NNW. Minor F_{3a} folds around the Dungash–Beizah shear zone display complex interference geometry with the development of both S and Z patterns commonly seen in scattered exposures (Figure 11E). Three types of NE- to ENE-striking cleavage (S_{3b}) are recognized in the Barramiya–Mueilha area: Slaty, scaly and rhombohedral. Scaly and rhombohedral cleavage prevail in moderate to low strain sectors and in limbs of Z type and upright folds especially around Wadi Dungash and Mueilha (Figure 11F). A major NNE-trending shear zone occurs between metavolcanic and metavolcaniclastic rocks in the area between Wadi Dungash and Wadi Mueilha. This shear zone resembles a palm-tree structure, with the main segment striking NNE–SSW and dips steeply to WNW.

4.2.5. Exhumation Tectonics

This phase is marked by sub vertical fractures including major brittle strike slip faults and micro-faults (Figure 11G,H) and dykes associated with intrusion of late to post tectonic granites. Structures assigned to this deformation stage comprise WNW–ESE dextral and NNW–SSE sinistral strike–slip faults that locally controlling felsic and mafic dikes in Gebel el–Rukham syn-tectonic granites as well as Um Salam and Urf Abu Hamam post tectonic granite.

4.2.6. Gold Occurrences in the Barramiya–Mueilha Area

Several gold occurrences in the study area are mainly associated with high strain zones along the segmented thrust-bound ophiolitic belts in the central part of the map area. In the following we present some features of four important gold occurrences/deposits in the area, namely, Barramiya, Dungash, Bokari, and Samut (Figure 12).

Figure 11. Field relationships of (**A**) NE-striking fold and S_1 foliation preserved in the island arc metavolcanic rocks in north of Dungash mine, (**B**) Boudinage and shallowly-plunging stretching lineations (L_1) in strongly foliated chlorite schist north of Dungash mine, (**C**) Steeply-dipping thrust contact between island arc and ophiolitic rocks and best preserved ENE-striking S_2 foliation east of Dungash mine, (**D**) Carbonated ophiolitic serpentinite tectonically overlying chlorite schist and separated by a WNW-striking thrust fault north of the Barramiya mine, (**E**) F_{3a} fold with a sinisterly asymmetry developed in metasedimentary rocks south of the Barramiya mine, (**F**) F_{3b} asymmetrical fold indicating a dextral shear sense and is associated with dilation-controlled quartz pods, (**G**) ~E–W fracture system in talc carbonate north of the Barramiya mine, (**H**) Dissecting joint/fracture sets with quartz-hosted ones displaced by later sinistral exposure-scale faults south of the Dungash mine.

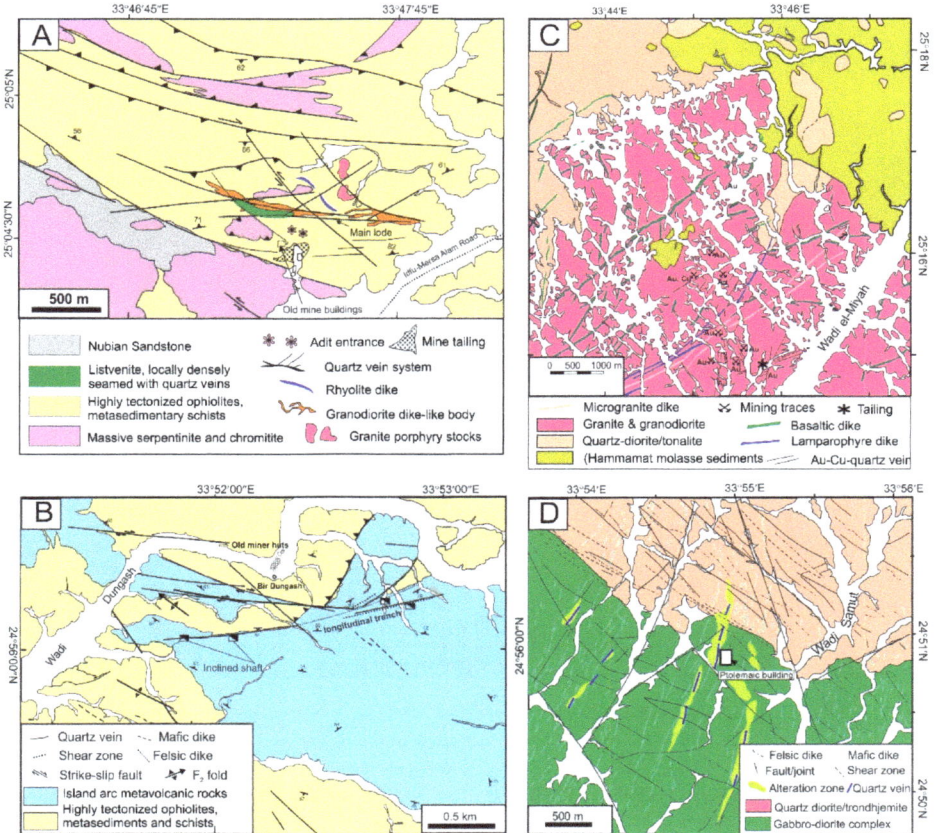

Figure 12. Detailed geological maps of the selected gold occurrences/deposits in the study area, (**A**) Barrmaiya deposit area, after [36], (**B**) Dungash deposit area, after [37], (**C**) Bokari deposit area, after [78], and (**D**) Samut gold deposit area, after [46].

Barramiya Gold Deposit

The Barramiya gold deposit (Figure 12A) is expressed in auriferous quartz and quartz–carbonate veins cutting altered and carbonaceous ophiolitic mélange serpentinite and metasiltstone (Figure 13A). The main ore body or the main lode extends for 900–1,100 m along the strike and exhibits common pinch and swell morphologies [79,80]. The internal structure of the mineralized veins comprises massive and laminated quartz veins with abundant slivers of pervasively carbonated wallrock. Zoheir and Lehmann [35] proposed a genetic relationship between gold mineralization and listvenite and listvenitized ophiolitic rocks in the Barramiya mine area (Figure 13A–D). Cr-chlorite and Cr-sericite are manifestations of K-metasomatism of the ophiolitic serpentinite, and reflect the crucial role played by a small granitic intrusion (exposed north of the ophiolitic belt) in formation of the Barramiya gold deposit. Fire assay analysis of the highly ferruginated/silicified listvenite samples gave high gold contents (up to 11 ppm) in several cases [81]. Gold-associated sulfides include arsenopyrite, pyrite and subordinate sphalerite, chalcopyrite, pyrrhotite, tetrahedrite, galena, and gersdorffite. Marcasite and covellite occur as supergene replacements of pyrite and chalcopyrite, respectively. Free gold and electrum blebs and scattered spikes are seen in the micro-fractured sulfides and recrystallized quartz veins.

Figure 13. Field relationships at the Barramiya mine area and its surroundings, including: (**A**) Sheared ophiolitic serpentinite and talc carbonate exposures north of the Barramiya mine where scattered old mining activities are observed, (**B**) ~E–W trending quartz vein significantly worked out by old miners and still preserves the trend of the main lode of the Barramiya deposit, (**C**) Listvenite sheet embedded within shear hornblende and actinolite schist north of the Barramiya mine, and (**D**) Typical listvenite exposure in the underground mine levels dissected by milky quartz veins.

Dungash Gold Deposit

The Dungash gold deposit is related to ~E-trending dilation zones in variably sheared island arc metavolcanic and metavolcaniclastic rocks (Figure 12B). The vein morphology and structures suggesting gold–quartz vein formation synchronous with dextral transpression and flexural displacement of heterogeneously folded greenstone belts (Figure 14A). In the eastern part of the mine area, the main quartz vein occurs along an undulating shear zone between the schistose rocks and heterogeneously foliated trachyandesite and extends further E into andesite and plunges into SE and is associated with pervasive epidote–chlorite alteration (Figure 14B). The hydrothermal alteration assemblages related to the gold–quartz veins include sulfide, carbonate, epidote, chlorite, iron oxides, and sericite (Figure 14B–D) replacing feldspar and ferromagnesian minerals in rhyodacitic metavolcanic rocks in the western mine [82,83]. The mineralized quartz veins are sulfide-rich and are associated with mylonite zones along reverse faults (Figure 14E). In places, quartz veins are bound by carbonate pockets with well crystalline calcite apparently late than ankerite disseminations (Figure 14F). Arsenopyrite, As-pyrite, gersdorffite and less abundant pyrrhotite are replaced in part by a late-paragenetic sulfide assemblage comprising tetrahedrite, chalcopyrite, sphalerite, galena, and free gold.

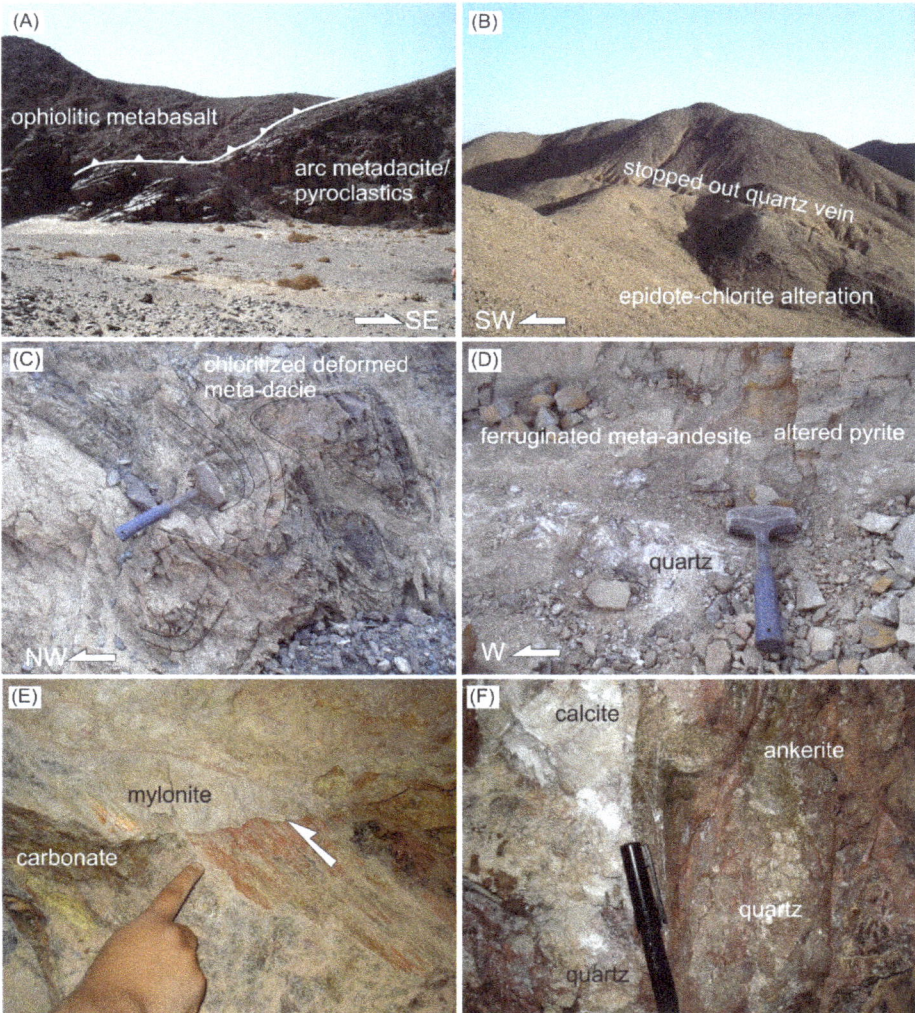

Figure 14. Field relationships at the Dungash mine area and its surroundings, including: (**A**) Ophiolitic metabasalt tectonically overlie the idland arc metavolcaniclastic rocks north of the Dungash mine, (**B**) Traces of ~ E–W trending gold-bearing quartz vein (main lode) and associated epitore–chlorite alteration in Dungash mine, (**C**) Deformed felsic metavolcanic rocks embedded in pervasively carbonated and chloritized material, (**D**) Disseminated altered sulfides and iron oxides (ferrugination) in the vicinity of mineralized quartz–ankerite vein, (**E**) Carbonate alteration along slicken planes on a reverse fault and associated mylonite zone in the underground levels of Dungash mine, (**F**) Ankerite and calcite associated with quartz veins.

Bokari Gold Occurrence

The Bokari gold occurrence (also known as Bakriya occurrence; 25°15′30″N, 33°45′15″E) occurs along Wadi Bokari to the north of Wadi el-Miyah, some 20 km north of the Barramiya mine. Gold mineralization is related mostly to milky quartz veins along fault and fracture zones cutting across variably deformed quartz–diorite to granodiorite intrusion and country gabbro–diorite complex

(Figure 12C). Shearing in the area is dominantly brittle, with alteration zones and quartz veins are generally associated with mylonite zones (commonly <0.5 m wide; Figure 15A). Scattered mine ruins of quartz vein and Wadi alluvium workings dating back to the Pharaohs and Early Arab times, but also from the Roman-Byzantine era, have been reported [78]. In places, sericite alteration is pervasive and kaolinite associates basic dikes and quartz veins (Figure 15B). The main feature in the Bokari mine is that gold-bearing quartz veins and ruins of old mining are confined to zones where the host granodioritic rocks are intensely brecciated (Figure 15C). The main quartz veins trend consistently to N (Figure 15D).

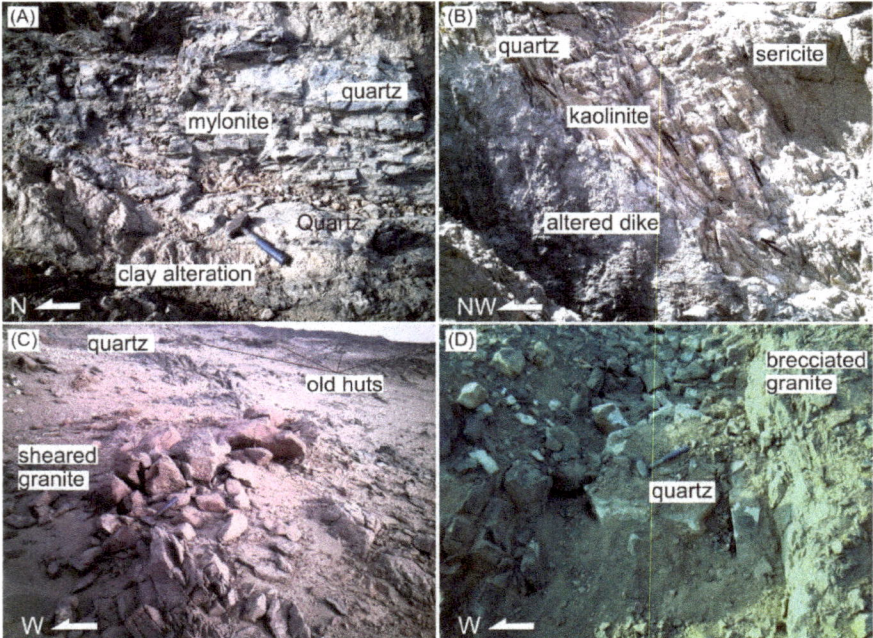

Figure 15. Field relationships of the mineralized quartz veins in the Bokari mine area: (**A**) Mylonite zone and associated quartz vein on one side and clay mineral alteration on the other side, (**B**) Kaolinite and quartz veins bordered by sericitized granodiorite and altered basic dike at the Bokari mine, (**C**) Scattered quartz vein segments and pockets in brecciated granodiorite near old miner huts at the Bokari mine area, (**D**) Thick (~1 m-thick) quartz vein trending ~N–S and is mined out selectively in areas where hydrothermal alteration selvages are abundant, whereas zones lacking the altered wallrock are not mined (barren?).

Samut Gold Deposit

The Samut gold deposit is related to milky and reddish quartz that cut through terranes of gabbro–diorite complex adjacent to its lithological contact with a quartz diorite/trondhjemite intrusion (Figure 12D). The main quartz vein (~2 m-thick) extends for more than 400 m along the ~ NNE–SSW strike direction. Other minor quartz veins are parallel to ~N or NNE-trending mafic dikes within and external of the Samut mine area [84]. The internal structures of the auriferous lodes comprise laminated quartz–chalcedony bands and vuggy comb, well-crystalline quartz. Hematite, sericite, kaolinite, ankerite, and calcite are common alteration phases in the marginal zones of quartz veins (Figure 16A–C). Quartz–carbonate veins (Figure 16D) are reported as the high grade ore bodies in the mine area. Gold-associated sulfides include abundant pyrite and minor sphalerite, chalcopyrite, galena, arsenopyrite, and marcasite scattered in the auriferous quartz veins

Figure 16. Field relationships of the orebodies at the Samut mine area showing: (**A**), (**B**), (**C**) Mineralized quartz veins at the Samut mine area, commonly associated with sericite, kaolinite, and sulfide alteration. Notice that the alteration zones are of limited thickness (commonly <0.5 m-wide), (**D**) Brecciated quartz vein exposed in the underground mine levels with abundant wall rock material and disseminated pyrite and chalcopyrite, locally altered into malachite and limonite, respectively.

5. Discussion

Orogenic gold deposits are commonly associated with large-scale, terrane-bounding fault systems and deformation belts, commonly described as orogenic belts [85,86]. Transpression and transtension tectonics occur where deformation is supplemented by a significant volume change, i.e., in oblique subduction margins in the arc, forearc, and back-arc environments [87–90]. These zones can also develop in restraining and releasing bends of major transform and dislocation zones, particularly throughout the early stages of continental rifting or during the late orogenic extension [87–95]. In such high strain zones, mineral deposits other than orogenic gold include different associations of elements such as Sn, W, U, Th, Mo, Cu, Au, Pb, Zn, Ag, Nb, Ta, Be, Sc, Li, Y, Zr, Sb, F, Bi, As, Hg, Fe, Ga, REEs [96–100]. Space-borne multispectral and radar imagery data provide great information for detailed structural, lithological, and alteration mapping to prospect a variety of these mineral resources in regional transpression and transtension zones [90].

Multi-sensor satellite imagery data, including Sentinel-1, PALSAR, ASTER, and Sentinel-2 are integrated here with field data are found efficient in mapping the different geological structures and alteration zones in the Barramiya–Mueilha sector of the Nubian Shield in Egypt. The belt is characterized by highly sheared ophiolite blocks, oppositely dipping, ENE–WSW thrusts, and less abundant NW-trending fault/shear zones [78,101,102]. Ductile shear zones developed in the highly sheared, refolded ophiolitic and island arc rocks in the central part of the study area host several occurrences of orogenic gold in the ophiolitic mélange (e.g., the Barramiya deposit) or along the decollement boundaries between ophiolites and island arc domains (e.g., the Dungash deposit) (Figure 17).

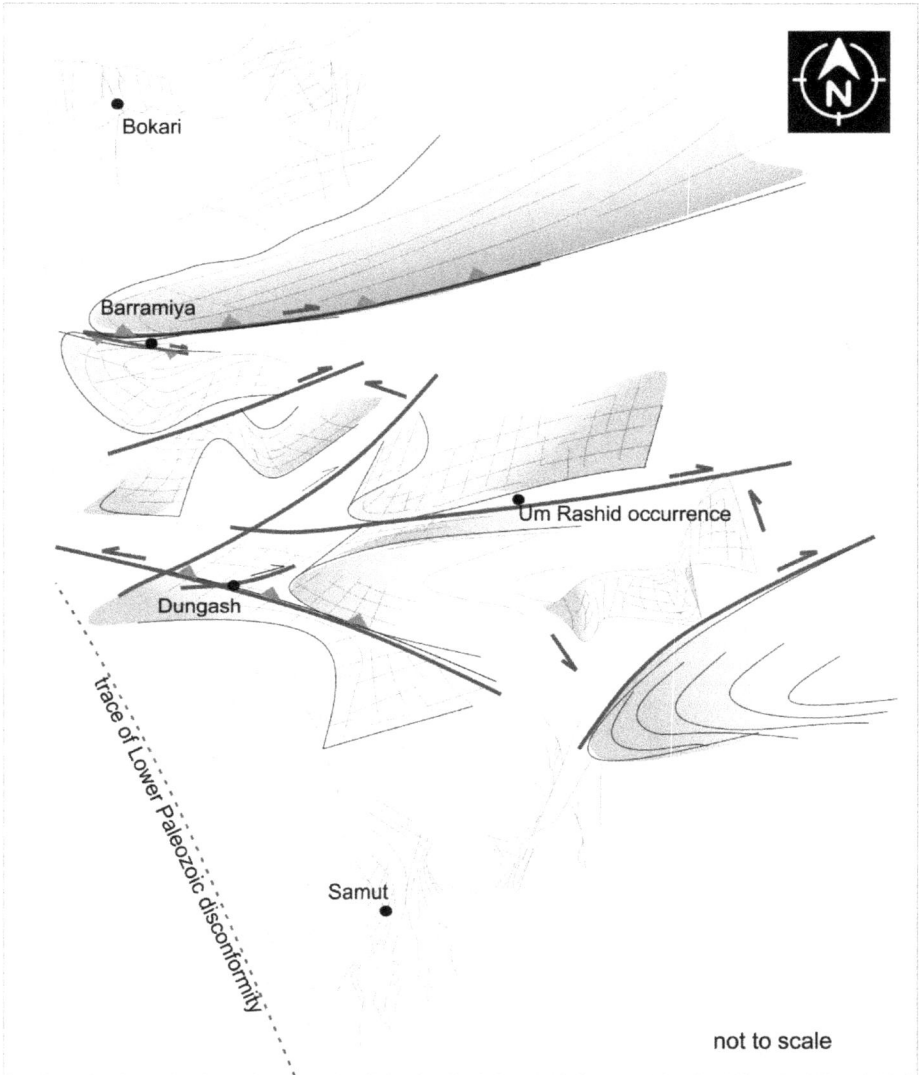

Figure 17. Sketch drawing explains the fold interference and location of transpression and transtension structures in the Barramiya–Mueilha sector. Notice that the Barrramiya and Dungash deposits (and many small occurrences from which only Um Rashid is shown) are confined to transpression zones within the highly strained rocks or at their margins. The Samut and Bokari gold deposits, in contrast, are hosted by transtension zones within the granitoid/gabbroid terranes.

Gold–quartz veins in these deposits show characteristically abundant plastic deformation textures, consistently in ~E–W direction conformable with the variably carbonated, talcous ophiolitic belts of serpentinite, chromitites, and listvenite. On the other hand, gold occurrences in the northern and southern parts of the study area are restricted to zones brittle (transtension) deformation in gabbro–diorite complex where cut by quartz-diorite/trondhjemite and alkali feldspar granite intrusions.

Gold–quartz veins in these occurrences are controlled mainly by rather narrow NNE–SSW or N–S fault/fracture zones (Figure 17) with abundant brecciation and mylonitization textures.

The FPCS2 image-maps (Figures 2 and 3) of Sentinel-1 and PALSAR data derived from cross-polarized backscatter ratio images detect structural elements related to transtension and transpression zones in the study area. The transtension zone is identified based on lack of intensity in foliation or any development of shear cleavage. However, intense foliation and shear cleavage are detected in transpression zones. PALSAR data hold more proficiency to map the geological structures covered by desert sand, hence, detailed textural variability and tonal variation are mapped for intensely sheared ophiolitic nappes and island arc metavolcanic and metavolcaniclastic rocks in FPCS2 PALSAR image-maps. The RGB color composite of the PC1, PC2, and PC3 of ASTER (VNIR + SWIR) and Sentinel-2 bands (Figures 4A and 5A) map variety of the lithological units contain discernible spectral features in the study area, while the lithologies with identical spectral characteristics exhibit similar hue in the image-maps. RGB color composite image of PC4, PC5, and PC6 of ASTER VNIR + SWIR bands highlights alteration mineral assemblages associated with the highly tectonized ophiolites and schists, island arc metavolcanic rocks, alkali-feldspar granite and granitoids background, ophiolitic serpentinite, chromitite and ultramafic rocks and metagabbro–diorite complex (Figure 4B). The RGB color composite image-map derived from PC6, PC5, and PC4 of Sentinel-2 datasets (Figure 5B) is capable of mapping hydrous minerals and Fe-oxides/hydroxides associated with the highly tectonized ophiolites, metasediments and schists, post-orogenic alkali-feldspar granite, syn-orogenic granitoids and island arc metavolcanic and metavolcaniclastic rocks in both ductile and brittle zones. Compressional structures in the central part of the study area accommodate hydrothermal extensive zones. Hydroxyl minerals and iron oxide/hydroxide zones are specifically mapped and discriminated by ASTER, despite the fact that they are mixed with each other in the Sentinel-2 results. Iron oxide/hydroxide minerals and mafic lithologies (metagabbro–diorite complex) are strongly detected by Sentinel-2 bands.

Fe_2O_3/MgO-rich rocks (ophiolitic serpentinite and metagabbro–diorite complex) and Al_2O_3–rich rocks (alkali-feldspar granite and granitoids) are easily mapped by ASTER and Sentinel-2 data band combination (Figures 6 and 7). A series of band ratio indices images for ASTER and Sentinel-2 data enabled the delineation of the hydrothermal alteration zones, which most associate with the major shear ductile zones. However, the resultant image-maps of alteration minerals derived from ASTER and Sentinel-2 datasets are almost identical. The alteration minerals zones are found in both ductile and brittle deformation zones. Gossan, the chlorite/epidote/calcite mineral group, ferrous silicates and hydroxyl alteration zones are mostly mapped in the transpression zones, whereas ferric oxides are typically detected in the transtension zones. Fieldwork and GPS surveying verified the scatter of alteration minerals and zones and their association with geological structures in the study area.

6. Conclusions

The space-borne radar (Sentinel-1 and PALSAR) and multispectral (ASTER and Sentinel-2) imagery data were coupled with comprehensive field observations and structural measurements to unravel the structural control of gold occurrences and alteration zones in the Barramiya–Mueilha sector. Gold mineralization in particular locations is constrained by the combined effect of transpression and transtension tectonics throughout the deformational history of the area. Application of the FPCS technique to the backscatter ratio images of Sentinel-1 and PALSAR datasets promoted detailed mapping of the major lineaments, curvilinear structures, and intersections associated with transpression and transtension zones in the study area. The results revealed that foliation and shear cleavage superimposition and shear-controlled gossan and carbonate alteration can be used as a criterion to distinguish the transpression zones. The transtension zones, on the other hand, are characterized by heterogeneous strain and irregular haloes of ferrugination and hydroxyl alteration. The PCA of the ASTER and Sentinel-2 bands, band combination and band ratioing techniques promoted the spectral discrimination between the lithological units and alteration zones. The combination of the remote sensing results and field data designates that gold-bearing quartz veins are restricted to high strain (transpression) zones in the ophiolite-island arc belts, particularly where carbonated or listvenized. Gold occurrences in granitoid–gabbroid domains are controlled by fault and brittle shear zones assigned to as antithetic sets in a rejuvenated transpression–transtension regime operated intermittently from the orogenic collision to orogen collapse. The radar and multispectral satellite data abetted a better understanding of the structural framework and unraveled settings of the scattered gold occurrences in the study area. An outlook study will need to consider the GIS-based morphometric analysis of the different geomorphometric features (i.e., regional ductile and brittle structures) based on high spatial resolution DEM data.

Author Contributions: Conceptualization & Research Initiation, B.Z.; Methodology, B.Z., M.A., A.B.P. and A.A.; Remote Sensing Data Acquisition and Analysis, B.Z. and A.B.P.; Field Investigation & Validation, B.Z., M.A. and A.A.; Writing-Original Draft Preparation, B.Z., M.A, A.B.P. and A.A.; Writing-Review & Editing, B.Z., A.B.P., M.A., and A.A. Publication fees are waived by Remote Sensing as A.B.P. and B.Z. are guest editors.

Funding: This research received no external funding

Acknowledgments: Basem Zoheir thanks the Alexander von Humboldt foundation for making this work possible. A.B. Pour is indebted to the Korea Polar Research Institute (KOPRI) for assigned time, providing the computer and software facilities.

Conflicts of Interest: The authors declare no conflict of interest.

References

1. Zoheir, B.A.; Emam, A. Field and ASTER imagery data for the setting of gold mineralization in Western Allaqi-Heiani belt, Egypt: A case study from the Haimur. *J. Afr. Earth Sci.* **2014**, *66–67*, 22–34. [CrossRef]
2. Zoheir, B.A.; Emam, A.; El-Amawy, M.; Abu-Alam, T. Auriferous shear zones in the central Allaqi-Heiani belt: Orogenic gold in post-accretionary structures, SE Egypt. *J. Afr. Earth Sci.* **2018**, *146*, 118–131. [CrossRef]
3. Testa, F.J.; Villanueva, C.; Cooke, D.R.; Zhang, L. Lithological and hydrothermal alteration mapping of epithermal, porphyry and tourmaline breccia districts in the Argentine Andes using ASTER imagery. *Remote Sens.* **2018**, *10*, 203. [CrossRef]
4. Pour, A.B.; Park, T.S.; Park, Y.; Hong, J.K.; Zoheir, B.; Pradhan, B.; Ayoobi, I.; Hashim, M. Application of multi-sensor satellite data for exploration of Zn-Pb sulfide mineralization in the Franklinian Basin, North Greenland. *Remote Sens.* **2018**, *10*, 1186. [CrossRef]
5. Noori, L.; Pour, B.A.; Askari, G.; Taghipour, N.; Pradhan, B.; Lee, C.-W.; Honarmand, M. Comparison of Different Algorithms to Map Hydrothermal Alteration Zones Using ASTER Remote Sensing Data for Polymetallic Vein-Type Ore Exploration: Toroud–Chahshirin Magmatic Belt (TCMB), North Iran. *Remote Sens.* **2019**, *11*, 495. [CrossRef]
6. Sheikhrahimi, A.; Pour, B.A.; Pradhan, B.; Zoheir, B. Mapping hydrothermal alteration zones and lineaments associated with orogenic gold mineralization using ASTER remote sensing data: A case study from the Sanandaj-Sirjan Zone, Iran. *Adv. Space Res.* **2019**, *63*, 3315–3332. [CrossRef]
7. Woodhouse, I.H. *Introduction to Microwave Remote Sensing*; CRC Press: Boca Raton, FL, USA; Taylor & Francis Group: Boca Raton, FL, USA, 2006.
8. Abdelsalam, M.G.; Robinson, C.; El-Baz, F.; Stern, R. Application of orbital imaging radar for geologic studies in arid regions: The Saharan Testimony. *Photogramm. Eng. Remote Sens.* **2000**, *66*, 717–726.
9. Kusky, T.M.; Ramadan, T.M. Structural controls on Neoproterozoic mineralization in the South Eastern Desert, Egypt: An integrated field, Landsat TM, and SIR-C/X SAR approach. *J. Afr. Earth Sci.* **2002**, *35*, 107–121. [CrossRef]
10. Pour, A.B.; Hashim, M. Structural geology mapping using PALSAR data in the Bau gold mining district, Sarawak, Malaysia. *Adv. Space Res.* **2014**, *54*, 644–654. [CrossRef]
11. Pour, A.B.; Park, Y.; Crispini, L.; Läufer, A.; Kuk Hong, J.; Park, T.-Y.S.; Zoheir, B.; Pradhan, B.; Muslim, A.M.; Hossain, M.S.; et al. Mapping Listvenite Occurrences in the Damage Zones of Northern Victoria Land, Antarctica Using ASTER Satellite Remote Sensing Data. *Remote Sens.* **2019**, *11*, 1408. [CrossRef]

12. Pour, A.B.; Hashim, M.; Park, Y.; Hong, J.K. Mapping alteration mineral zones and lithological units in Antarctic regions using spectral bands of ASTER remote sensing data. *Geocarto Int.* **2018**, *33*, 1281–1306. [CrossRef]
13. Pour, A.B.; Hashim, M.; Hong, J.K.; Park, Y. Lithological and alteration mineral mapping in poorly exposed lithologies using Landsat-8 and ASTER satellite data: North-eastern Graham Land, Antarctic Peninsula. *Ore Geol. Rev.* **2019**, *108*, 112–133. [CrossRef]
14. Pour, B.A.; Hashim, M. Structural mapping using PALSAR data in the Central Gold Belt, Peninsular Malaysia. *Ore Geol. Rev.* **2015**, *64*, 13–22. [CrossRef]
15. Pour, A.B.; Hashim, M.; Makoundi, C.; Zaw, K. Structural Mapping of the Bentong-Raub Suture Zone Using PALSAR Remote Sensing Data, Peninsular Malaysia: Implications for Sediment-hosted/Orogenic Gold Mineral Systems Exploration. *Resour. Geol.* **2016**, *66*, 368–385. [CrossRef]
16. Adiri, Z.; Harti, A.; Jellouli, A.; Lhissou, R.; Maacha, L.; Azmi, M.; Zouhair, M.; Bachaoui, M. Comparison of Landsat-8, ASTER and Sentinel 1 satellite remote sensing data in automatic lineaments extraction: A case study of Sidi Flah-Bouskour inlier, Moroccan Anti Atlas. *Adv. Space Res.* **2017**, *60*, 2355–2367. [CrossRef]
17. Hamimi, Z.; El-Fakharani, A.; Emam, A.; Barreiro, J.G.; Abdelrahman, E.; Abo-Soliman, M.Y. Reappraisal of the kinematic history of Nugrus shear zone using PALSAR and microstructural data: Implications for the tectonic evolution of the Eastern Desert tectonic terrane, northern Nubian Shield. *Arab. J. Geosci.* **2018**, *11*, 494. [CrossRef]
18. Javhar, A.; Chen, X.; Bao, A.; Jamshed, A.; Yunus, M.; Jovid, A.; Latipa, T. Comparison of Multi-Resolution Optical Landsat-8, Sentinel-2 and Radar Sentinel-1 Data for Automatic Lineament Extraction: A Case Study of Alichur Area, SE Pamir. *Remote Sens.* **2019**, *11*, 778. [CrossRef]
19. Hunt, G.R.; Ashley, P. Spectra of altered rocks in the visible and near infrared. *Econ. Geol.* **1979**, *74*, 1613–1629. [CrossRef]
20. Clark, R.N.; King, T.V.V.; Klejwa, M. and Swayze, G.A. High spectral resolution reflectance spectroscopy of minerals. *J. Geophys. Res.* **1990**, *95*, 12653–12680. [CrossRef]
21. Huntington, J.F. The role of remote sensing in finding hydrothermal mineral deposits on Earth. *Evol. Hydrother. Ecosyst. Earth Mars* **1996**, *21*, 214–234.
22. Kruse, F.A.; Perry, S.L. Mineral mapping using simulated Worldview-3 short-wave-infrared imagery. *Remote Sens.* **2013**, *5*, 2688–2703. [CrossRef]
23. Askari, G.; Pour, A.B.; Pradhan, B.; Sarfi, M.; Nazemnejad, F. Band Ratios Matrix Transformation (BRMT): A Sedimentary Lithology Mapping Approach Using ASTER Satellite Sensor. *Sensors* **2018**, *18*, 3213. [CrossRef]
24. Haselwimmer, C.E.; Riley, T.R.; Liu, J.G. Assessing the potential of multispectral remote sensing for lithological mapping on the Antarctic Peninsula: Case study from eastern Adelaide Island, Graham Land. *Antarct. Sci.* **2010**, *22*, 299–318. [CrossRef]
25. van der Werff, H.; van der Meer, F. Sentinel-2 for Mapping Iron Absorption Feature Parameters. *Remote Sens.* **2015**, *7*, 12635–12653. [CrossRef]
26. van der Werff, H.; van der Meer, F. Sentinel-2A MSI and Landsat 8 OLI Provide Data Continuity for Geological Remote Sensing. *Remote Sens.* **2016**, *8*, 883. [CrossRef]
27. Ibrahim, E.; Barnabé, P.; Ramanaidou, E.; Pirarda, E. Mapping mineral chemistry of a lateritic outcrop in new Caledonia through generalized regression using Sentinel-2 and field reflectance spectra. *Int. J. Appl. Earth Obs. Geoinf.* **2018**, *73*, 653–665. [CrossRef]
28. El Ramly, M.F.; Greiling, R.O.; Rashwan, A.A.; Rasmy, A.H. Explanatory note to accompany the geological and structural maps of Wadi Hafafit area, Eastern Desert of Egypt. *Geol. Surv. Egypt* **1993**, *9*, 1–53.
29. Gabra, S.Z. Gold in Egypt. A commodity package: Minerals, petroleum and groundwater assessment program. USAID project 363-0105. *Geol. Surv. Egypt* **1986**.
30. El-Gaby, S.; List, F.K.; Tehrani, R. Geology, evolution and metallogenesis of thePan-African Belt in Egypt. In *The Pan-African Belt of Northeast Africa and Adjacent Areas*; El Gaby, S., Greiling, R.O., Eds.; Friedrich Vieweg Sohn: Braunschweig, Germany, 1988; pp. 17–68.
31. Hussein, A.A. Mineral deposits. In *The Geology of Egypt: Rotterdam*; Said, R., Ed.; A.A. Balkema: Avereest, The Netherlands, 1990; pp. 511–566.
32. Botros, N.S. A new classification of the gold deposits of Egypt. *Ore Geol. Rev.* **2004**, *25*, 1–37. [CrossRef]
33. Zoheir, B.A. Characteristics and genesis of shear zone-related gold mineralization in Egypt: A case study from the Um El Tuyor mine, south Eastern Desert. *Ore Geol. Rev.* **2008**, *34*, 445–470. [CrossRef]

34. Zoheir, B.A. Structural controls, temperature-pressure conditions and fluid evolution of orogenic gold mineralisation in Egypt: A case study from the Betam gold mine, south Eastern Desert. *Miner. Depos.* **2008**, *43*, 79–95. [CrossRef]
35. Zoheir, B.A.; Lehmann, B. Listvenite–lode association at the Barramiya gold mine, Eastern Desert, Egypt. *Ore Geol. Rev.* **2011**, *39*, 101–115. [CrossRef]
36. Zoheir, B.A.; Weihed, P. Greenstone-hosted lode-gold mineralization at Dungash mine, Eastern Desert, Egypt. *J. Afr. Earth Sci.* **2014**, *99*, 165–187. [CrossRef]
37. Abd El-Wahed, M.A.; Harraz, H.Z.; El-Behairy, M.H. Transpressional imbricate thrust zones controlling gold mineralization in the Central Eastern Desert of Egypt. *Ore Geol. Rev.* **2016**, *78*, 424–446. [CrossRef]
38. Zoheir, B.A.; Emam, A.; Abd El-Wahed, M.; Soliman, N. Gold endowment in the evolution of the Allaqi-Heiani suture, Egypt: A synthesis of geological, structural, and space-borne imagery data. *Ore Geol. Rev.* **2019**, in press. [CrossRef]
39. Zoheir, B.A. Transpression zones in ophiolitic mélange terranes: Potential exploration targets for gold in South Eastern Desert of Egypt. *J. Geochem. Explor.* **2011**, *111*, 23–38. [CrossRef]
40. Moore, J.M.M.; Shanti, A.M. The use of stress trajectory analysis in the elucidation of part of the Najd Fault System, Saudi Arabia. *Proc. Geol. Assoc.* **1973**, *84*, 383–403. [CrossRef]
41. Sultan, M.; Arvidson, R.E.; Duncan, I.J.; Stern, R.; El Kaliouby, B. Extension of the Najd Fault System from Saudi Arabia to the central Eastern Desert of Egypt based on integrated field and Landsat observations. *Tectonics* **1988**, *7*, 1291–1306. [CrossRef]
42. El Gaby, S.; List, F.K.; Tehrani, R. The basement complex of the Eastern Desert and Sinai. In *The Geology of Egypt*; Rushdi, S., Ed.; Balkema: Rotterdam, The Netherlands, 1990; pp. 175–184.
43. Ali-Bik, M.W.; Taman, Z.; El Kalioubi, B.; Abdel Wahab, W. Serpentinite-hosted talc–magnesite deposits of Wadi Barramiya area, Eastern Desert, Egypt: Characteristics, petrogenesis and evolution. *J. Afr. Earth Sci.* **2012**, *64*, 77–89. [CrossRef]
44. Abd El-Wahed, M.A. Oppositely dipping thrusts and transpressional imbricate zone in the Central Eastern Desert of Egypt. *J. Afr. Earth Sci.* **2014**, *100*, 42–59. [CrossRef]
45. Abu El-Ela, A.M. Geology of Wadi Mubarak district, Eastern Desert, Egypt. Ph.D. Thesis, Tanta University, Tanta Chicago, IL, USA, 1985.
46. Zoheir, B.; Steele-MacInnis, M.; Garbe-Schönberg, D. Orogenic gold formation in an evolving, decompressing hydrothermal system: Genesis of the Samut gold deposit, Eastern Desert, Egypt. *Ore Geol. Rev.* **2019**, *105*, 236–257. [CrossRef]
47. Abdel-Karim, A.M.; El-Mahallawi, M.M.; Fringer, F. The Ophiolite Mélange of Wadi Dunqash and Wadi Arayis, Eastern Desert of Egypt: Petrogenesis and tectonic evolution. *Acta Mineral. Petrogr. (SzegedHung.)* **1996**, *47*, 129–141.
48. Aboelkhair, H.; Ninomiya, Y.; Watanabe, Y.; Sato, I. Processing and interpretation of ASTER TIR data for mapping of rare-metal-enriched albite granitoids in the Central Eastern Desert of Egypt. *J. Afr. Earth Sci.* **2010**, *58*, 141e151. [CrossRef]
49. Abu El-Rus, M.A.; Mohamed, M.A.; Lindh, A. Mueilha rare metals granite, Eastern Desert of Egypt: An example of a magmatic-hydrothermal system in the Arabian-Nubian Shield. *Lithos* **2017**, *294–295*, 362–382. [CrossRef]
50. Crawford, W.A.; Coulter, D.H.; Hubbard, H.B. The areal distribution, stratigraphy and major element chemistry of the Wadi Natash volcanic series, Eastern Desert, Egypt. *J. Afr. Earth Sci.* **1984**, *2*, 119–128. [CrossRef]
51. Mohamed, F.I.-I. The Natash alkaline volcanic field, Egypt: Geological and mineralogical inference on the evolution of a basalt to rhyolite eruptive suit. *J. Volcan Geotherm. Res.* **2001**, *105*, 291–322. [CrossRef]
52. Torres, R.; Snoeij, P.; Davidson, M.; Bibby, D.; Lokas, S. The Sentinel-1 mission and its application capabilities. *IEEE Int. Geosci. Remote Sens. Symp.* **2012**. [CrossRef]
53. Balzter, H.; Cole, B.; Thiel, C.; Schmullius, C. Mapping CORINE Land Cover from Sentinel-1A SAR and SRTM Digital Elevation Model Data using Random Forests. *Remote Sens.* **2015**, *7*, 14876–14898. [CrossRef]
54. Attema, E.; Bargellini, P.; Edwards, P.; Levrini, G.; Lokas, S.; Moeller, L.; Rosich-Tell, B.; Secchi, P.; Torres, R.; Davidson, M.; et al. Sentinel-1: The Radar Mission for GMES Operational Land and Sea Services. *Bulletin* **2007**, *131*, 10–17.
55. Igarashi, T. ALOS Mission requirement and sensor specification. *Adv. Space Res.* **2001**, *28*, 127–131. [CrossRef]

56. Rosenqvist, A.; Shimada, M.; Watanabe, M.; Tadono, T.; Yamauchi, K. Implementation of systematic data observation strategies for ALOS PALSAR, PRISM and AVNIR-2. In Proceedings of the 2004 IEEE International IEEE International Geoscience and Remote Sensing Symposium, Anchorage, AK, USA, 20–24 September 2004.
57. Earth Remote Sensing Data Analysis Center (ERSDAC). *PALSAR User's Guide*, 1st ed. 2019. Available online: http://www.eorc.jaxa.jp/ALOS/en/doc/alos_userhb_en.pdf (accessed on 25 April 2019).
58. Drusch, M.; Del Bello, U.; Carlier, S.; Colin, O.; Fernandez, V.; Gascon, F.; Hoersch, B.; Isola, C.; Laberinti, P.; Martimort, P.; et al. Sentinel-2: ESA's optical high-resolution mission for GMES operational services. *Remote Sens. Environ.* **2012**, *120*, 25–36. [CrossRef]
59. van der Meer, F.D.; van der Werff, H.M.A.; van Ruitenbeek, F.J.A. Potential of ESA's Sentinel-2 for geological applications. *Remote Sens. Environ.* **2014**, *148*, 124–133. [CrossRef]
60. Sheng, Y.; Xia, Z.G. A comprehensive evaluation of filters for radar speckle suppression. In Proceedings of the 1996 International Geoscience and Remote Sensing Symposium, Lincoln, NE, USA, 31 May 1996; pp. 1559–1561.
61. Lee, J.-S. Digital Image Enhancement and Noise Filtering by Use of Local Statistics. *IEEE Trans. Pattern Anal. Mach. Intell.* **1980**, *2*, 165–168. [CrossRef] [PubMed]
62. Lopes, A.; Touzi, R.; Nezry, E. Adaptive speckle filter and scene heterogeneity. *IEEE Trans. Geosci. Remote Sens.* **1990**, *28*, 992–1000. [CrossRef]
63. Ben-Dor, E.; Kruse, F.A.; Lefkoff, A.B.; Banin, A. Comparison of three calibration techniques for the utilization of GER 63 channel scanner data of Makhtesh Ramon, Negev, Israel. *Photogramm. Eng. Remote Sens.* **1994**, *60*, 1339–1354.
64. Paganelli, F.; Grunsky, E.C.; Richards, J.P.; Pryde, R. Use of RADARSAT-1 principal component imagery for structural mapping: A case study in the Buffalo Head Hills area, northern central Alberta, Canada. *Can. J. Remote Sens.* **2003**, *29*, 111–140. [CrossRef]
65. Pal, S.K.; Majumdar, T.J.; Bhattacharya, A.K. ERS-2 SAR and IRS-1C LISS III data fusion: A PCA approach to improve remote sensing based geological interpretation. *ISPRS J. Photogramm. Remote Sens.* **2007**, *61*, 281–297. [CrossRef]
66. Cheng, Q.; Jing, L.; Panahi, A. Principal component analysis with optimum order sample correlation coefficient for image enhancement. *Int. J. Remote Sens.* **2006**, *27*, 3387–3401.
67. Gupta, R.P.; Tiwari, R.K.; Saini, V.; Srivastava, N. A simplified approach for interpreting principal component images. *Adv. Remote Sens.* **2013**, *2*, 111–119. [CrossRef]
68. Kalinowski, A.; Oliver, S. ASTER Mineral Index Processing Manual. Technical Report; 2004; Geoscience Australia. Available online: http://www.ga.gov.au/image_cache/GA7833.pdf (accessed on 1 June 2019).
69. Mars, J.C.; Rowan, L.C. Regional mapping of phyllic- and argillic-altered rocks in the Zagros magmatic arc, Iran, using Advanced Spaceborne Thermal Emission and Reflection Radiometer (ASTER) data and logical operator algorithms. *Geosphere* **2006**, *2*, 161–186. [CrossRef]
70. Abdelsalam, M.G.; Stern, R.J. Sutures and shear zones in the Arabian–Nubian Shield. *J. Afr. Earth Sci.* **1996**, *23*, 289–310. [CrossRef]
71. Abdelsalam, M.G.; Stern, R.J.; Copeland, P.; Elfaki, E.M.; Elhur, B.; Ibrahim, F.M. The Neoproterozoic Keraf Suture in Ne Sudan: Sinistral Transpression along the Eastern Margin of West Gondwana. *J. Geol.* **1998**, *106*, 133–148. [CrossRef]
72. Abdelsalam, M.G.; Abdeen, M.M.; Dwaidar, H.M.; Stern, R.J. Structural evolution of the Neoproterozoic western Allaqi-Heiani Suture, southeastern Egypt. *Precambrian Res.* **2003**, *124*, 87–104. [CrossRef]
73. Loizenbauer, J.; Wallbrecher, E.; Fntz, H.; Neumayr, P.; Khudeir, A.A.; Kloetzil, U. Structural geology, single zircon ages and fluid inclusion studies of the Meatiq metamorphic core complex. Implications for Neoproterozoic tectonics in the Eastern Desert of Egypt. *Precambrian Res.* **2001**, *110*, 357–383. [CrossRef]
74. Fritz, H.; Dalmeyer, D.R.; Wallbrecher, E.; Loizenbauer, J.; Hoinkes, G.; Neumayr, P.; Khudeir, A.A. Neoproterozoic tectonothermal evolution of the Central Eastern Desert, Egypt: A slow velocity tectonic process of core complex exhumation. *J. Afr. Earth Sci.* **2002**, *34*, 543–576. [CrossRef]
75. Shalaby, A.; Stüwe, K.; Makroum, F.; Fritz, H.; Kebede, T.; Klotzli, U. The Wadi Mubarak belt, Eastern Desert of Egypt: A Neoproterozoic conjugate shear system in the Arabian-Nubian Shield. *Precambrian Res.* **2005**, *136*, 27–50. [CrossRef]

76. Abd El-Wahed, M.A. Thrusting and transpressional shearing in the Pan-African nappe southwest El-Sibai core complex, Central Eastern Desert, Egypt. *J. Afr. Earth Sci.* **2008**, *50*, 16–36. [CrossRef]
77. Bailo, T.; Schandelmeier, H.; Franz, G.; Sun, C.-H.; Stern, R. Plutonic and metamorphic rocks from the Keraf Suture (NE Sudan): A glimpse of Neoproterozoic tectonic evolution on the NE margin of W. Gondwana. *Precambrian Res.* **2003**, *123*, 67–80. [CrossRef]
78. Klemm, R.; Klemm, D. *Gold and Gold Mining in Ancient Egypt and Nubia. Geoarchaeology of the Ancient Gold Mining in the Egyptian and Sudanese Eastern Deserts*; Springer Science & Business Media: Berlin/Heidelberg, Germany, 2013; 649p.
79. Sabet, A.H.; Tscogoev, V.B.; Bordonosov, V.P.; Badourin, L.M.; Zalata, A.A.; Francis, M.H. On Gold mineralization in the Eastern Desert of Egypt Annals of the Geological. *Surv. Egypt* **1976**, *6*, 201–212.
80. Osman, A. The mode of occurrence of gold-bearing listvenite at El Barramiya gold mine, Eastern desert, Egypt. Middle East Research Centre. Ain Shams University. *Earth Sci. Ser.* **1995**, *9*, 93–103.
81. Osman, A. The gold metallotect in the Eastern Desert of Egypt. In *Mineral Deposits at the Beginning of the 21st Century*; Piestrzyński, A., Pieczonka, J., A. Głuszek, A., Eds.; Swets & Zeitinger: Lisse, Belgium, 2001; pp. 795–798.
82. Helba, H.A.; Khalil, K.I.; Abdou, N.M. Alteration patterns related to hydrothermal gold mineralization in meta-andesites at Dungash area, Eastern Desert Egypt. *Resour. Geol.* **2001**, *51*, 19–30. [CrossRef]
83. Khalil, I.K.; Helba, H.A.; Mücke, A. Genesis of the gold mineralization at the Dungash gold mine area, Eastern Desert, Egypt: A mineralogical–microchemical study. *J. Afr. Earth Sci.* **2003**, *37*, 111–122. [CrossRef]
84. Hassan, M.M.; Soliman, M.M.; Azzaz, S.A.; Attawiya, M.Y. Geological studies on gold mineralization at Sukkari, Um Ud, and Samut, Eastern Desert, Egypt. *Ann. Geol. Surv. Egypt* **1990**, *16*, 89–95.
85. Dewey, J.E.; Holdsworth, R.E.; Strachan, R.A. Transpression and transtension zones. In *Continental Transpressional and Transtensional Tectonics. Geological Society*; Holdsworth, R.E., Strachan, R.A., Dewey, J.E., Eds.; Special Publications: London, UK, 1998; Volume 135, pp. 1–14.
86. Goldfarb, R.J.; Taylor, R.D.; Collins, G.S.D.; Goryachev, N.A.; Orlandini, O.F. Phanerozoic continental growth and gold metallogeny of Asia. *Gondwana Res.* **2014**, *25*, 48–102. [CrossRef]
87. Harland, W.B. Tectonic transpression in Caledonian Spitsbergen. *Geol. Mag.* **1971**, *108*, 27–42. [CrossRef]
88. Fossen, H.; Tikoff, B.; Teyssier, C.T. Strain modeling of transpressional and transtensional deformation. *Nor. Geol. Tidsskrifl* **1994**, *74*, 134–145.
89. Jones, R.R.; Holdsworth, R.E.; Clegg, P.; McCaffrey, K.; Travarnelli, E. Inclined transpression. *J. Struct. Geol.* **2004**, *30*, 1531–1548. [CrossRef]
90. Sarkarinejad, K.; Faghih, A.; Grasemann, B. Transpressional deformations within the Sanandaj–Sirjan metamorphic belt (Zagros Mountains, Iran). *J. Struct. Geol.* **2008**, *30*, 818–826. [CrossRef]
91. Abd El-Wahed, M.A.; Ashmawy, M.H.; Tawfik, H.A. Structural setting of Cretaceous pull-apart basins and Miocene extensional folds in Quseir-Umm Gheig region, northwestern Red Sea, Egypt. *Lithosphere* **2010**, *2*, 13–32. [CrossRef]
92. Abd El-Wahed, M.A.; Kamh, S.; Ashmawy, M.; Shebl, A. Transpressive Structures in the Ghadir Shear Belt, Eastern Desert, Egypt: Evidence for Partitioning of Oblique Convergence in the Arabian-Nubian Shield during Gondwana Agglutination. *Acta Geol. Sin.-Engl. Ed.* **2019**. [CrossRef]
93. Abd El-Wahed, M.A.; Kamh, S.Z. Pan African dextral transpressive duplex and flower structure in the Central Eastern Desert of Egypt. *Gondwana Res.* **2010**, *18*, 315–336. [CrossRef]
94. Abd El-Wahed, M.A.; Kamh, S.Z. Evolution of conjugate strike-slip duplexes and wrench-related folding in the Central part of Al Jabal Al Ahkdar, NE Libya. *J. Geol.* **2013**, *121*, 173–195. [CrossRef]
95. Li, S.; Wilde, S.A.; Wanga, T.; Guo, Q. Latest Early Permian granitic magmatism in southern Inner Mongolia, China: Implications for the tectonic evolution of the southeastern Central Asian Orogenic Belt. *Gondwana Res.* **2016**, *29*, 168–180. [CrossRef]
96. Zaw, K.; Meffre, S.; Lai, C.K.; Burrett, C.; Santosh, M.; Graham, I.; Manaka, T.; Salam, A.; Kamvong, T.; Cromie, P. Tectonics and metallogeny of mainland Southeast Asia -A reviewand contribution. *Gondwana Res.* **2014**, *26*, 5–30.
97. Hedenquist, J.W.; Thompson, J.F.H.; Goldfarb, R.J.; Richards, J.P. Distribution, character, and genesis of gold deposits in metamorphic terranes. In *Economic geology one hundredth anniversary volume 1905–2005*; Hedenquist, J.W.; Thompson, J.F.H.; Goldfarb, R.J.; Richards, J.P. Society of Economic Geologists: Littleton, CO, USA, 2005; pp. 407–450.

98. Reinhardt, M.C.; Davison, I. Structural and lithological controls on gold deposition in the shear zone-hosted Fazenda Brasileiro Mine, Bahia state, Northeast Brazil. *Econ. Geol.* **1990**, *85*, 952–967. [CrossRef]
99. Almond, D.C.; Shaddad, M.Z. Setting of gold mineralization in the northern Red Sea Hills of Sudan. *Econ. Geol.* **1984**, *79*, 389–392. [CrossRef]
100. El-Samani, Y.; Al-Muslem, A.; El Tokhi, M. Geology and geotectonic classification of Pan-African gold mineralizations in the Red Sea Hills, Sudan. *Int. Geol. Rev.* **2001**, *43*, 1117–1128. [CrossRef]
101. Zoheir, B.A.; El-Shazly, A.K.; Helba, H.; Khalil, K.I.; Bodnar, R.J. Origin and evolution of the Um Egat and Dungash orogenic gold deposits, Egyptian eastern desert: Evidence from fluid inclusions in quartz. *Econ. Geol.* **2008**, *103*, 405–424. [CrossRef]
102. Murr, A. Geologische Kartierung der Goldlagerstättenbezirke Fatira und Dungash in der Ostwüste Ägyptens. unpubl. Master thesis, LMUUniversity, Munich, Germany, 1994.

© 2019 by the authors. Licensee MDPI, Basel, Switzerland. This article is an open access article distributed under the terms and conditions of the Creative Commons Attribution (CC BY) license (http://creativecommons.org/licenses/by/4.0/).

Article

Landsat-8, Advanced Spaceborne Thermal Emission and Reflection Radiometer, and WorldView-3 Multispectral Satellite Imagery for Prospecting Copper-Gold Mineralization in the Northeastern Inglefield Mobile Belt (IMB), Northwest Greenland

Amin Beiranvand Pour [1,2,*], Tae-Yoon S. Park [1], Yongcheol Park [1], Jong Kuk Hong [1], Aidy M Muslim [2], Andreas Läufer [3], Laura Crispini [4], Biswajeet Pradhan [5,6], Basem Zoheir [7,8], Omeid Rahmani [9], Mazlan Hashim [10] and Mohammad Shawkat Hossain [2]

1. Korea Polar Research Institute (KOPRI), Songdomirae-ro, Yeonsu-gu, Incheon 21990, Korea; typark@kopri.re.kr (T.-Y.S.P.); ypark@kopri.re.kr (Y.P.); jkhong@kopri.re.kr (J.K.H.)
2. Institute of Oceanography and Environment (INOS), University Malaysia Terengganu (UMT), Kuala Nerus 21030, Terengganu, Malaysia; aidy@umt.edu.my (A.M.M.); shawkat@umt.edu.my (M.S.H.)
3. Federal Institute for Geosciences and Natural Resources (BGR), Stilleweg 2, 30655 Hannover, Germany; andreas.laeufer@bgr.de
4. DISTAV– University of Genova – Corso Europa 26, 16132 Genova, Italy; laura.crispini@unige.it
5. Centre for Advanced Modelling and Geospatial Information Systems (CAMGIS), Faculty of Engineering and Information Technology, University of Technology Sydney, Ultimo 2007, New South Wales, Australia; Biswajeet.Pradhan@uts.edu.au
6. Department of Energy and Mineral Resources Engineering, Choongmu-gwan, Sejong University, 209 Neungdong-ro Gwangjin-gu, Seoul 05006, Korea
7. Department of Geology, Faculty of Science, Benha University, Benha 13518, Egypt; basem.zoheir@fsc.bu.edu.eg
8. Institute of Geosciences, University of Kiel, Ludewig-Meyn Str. 10, 24118 Kiel, Germany
9. Department of Natural Resources Engineering and Management, School of Science and Engineering, University of Kurdistan Hewlêr (UKH), Erbil 44001, Kurdistan Region, Iraq; omeid.rahmani@ukh.edu.krd
10. Geoscience and Digital Earth Centre (INSTeG), Research Institute for Sustainable Environment, Universiti Teknologi Malaysia, Johor Bahru, Skudai 81310, Malaysia; mazlanhashim@utm.my
* Correspondence: beiranvand.amin80@gmail.com; Tel.: +82-3-2760-5472

Received: 23 September 2019; Accepted: 17 October 2019; Published: 19 October 2019

Abstract: Several regions in the High Arctic still lingered poorly explored for a variety of mineralization types because of harsh climate environments and remoteness. Inglefield Land is an ice-free region in northwest Greenland that contains copper-gold mineralization associated with hydrothermal alteration mineral assemblages. In this study, Landsat-8, Advanced Spaceborne Thermal Emission and Reflection Radiometer (ASTER), and WorldView-3 multispectral remote sensing data were used for hydrothermal alteration mapping and mineral prospecting in the Inglefield Land at regional, local, and district scales. Directed principal components analysis (DPCA) technique was applied to map iron oxide/hydroxide, Al/Fe-OH, Mg-Fe-OH minerals, silicification (Si-OH), and SiO2 mineral groups using specialized band ratios of the multispectral datasets. For extracting reference spectra directly from the Landsat-8, ASTER, and WorldView-3 (WV-3) images to generate fraction images of end-member minerals, the automated spectral hourglass (ASH) approach was implemented. Linear spectral unmixing (LSU) algorithm was thereafter used to produce a mineral map of fractional images. Furthermore, adaptive coherence estimator (ACE) algorithm was applied to visible and near-infrared and shortwave infrared (VINR + SWIR) bands of ASTER using laboratory reflectance spectra extracted from the USGS spectral library for verifying the presence of mineral spectral signatures. Results indicate that the boundaries between the Franklinian sedimentary successions and the Etah metamorphic and meta-igneous complex, the orthogneiss in the northeastern part of the Cu-Au mineralization belt adjacent to Dallas Bugt, and the southern part of the Cu-Au mineralization belt nearby Marshall Bugt show high content of iron oxides/hydroxides and Si-OH/SiO2 mineral groups, which warrant high potential for Cu-Au prospecting. A high spatial distribution of hematite/jarosite, chalcedony/opal, and chlorite/epidote/biotite were identified with the documented Cu-Au occurrences in central and southwestern sectors of the Cu-Au mineralization belt. The calculation of confusion matrix and Kappa Coefficient proved appropriate overall accuracy and good rate of agreement for alteration mineral mapping. This investigation accomplished the application of multispectral/multi-sensor satellite imagery as a valuable and economical tool for reconnaissance stages of systematic mineral exploration projects in remote and inaccessible metallogenic provinces around the world, particularly in the High Arctic regions.

Keywords: Landsat-8; ASTER; WorldView-3; the Inglefield Mobile Belt (IMB); copper-gold mineralization; High Arctic regions

1. Introduction

The application of multispectral satellite imagery for mineral prospecting in remote and inaccessible metallogenic provinces is noteworthy for mining companies and the mineral exploration community for reconnaissance stages of systematic exploration projects. Many regions in the High Arctic remain poorly investigated for mineral exploration due to cold climate environments and remoteness, especially the northern part of Greenland containing Zn-Pb and Cu-Au mineralization [1–3]. The visible and near-infrared (VNIR), shortwave infrared (SWIR) and thermal infrared (TIR) bands of multispectral remote sensing data contain unprecedented spectral and spatial capabilities for detecting hydrothermal alteration minerals and lithological units associated with a variety of ore mineralization [4–22]. Numerous investigations successfully used Landsat data series, Advanced Spaceborne Thermal Emission and Reflection Radiometer (ASTER), and the Advanced Land Imager (ALI) multispectral data with moderate spatial resolution for the reconnaissance stages of mineral exploration around the world [23–29].

Landsat-8 carries two-sensors, including the Operational Land Imager (OLI) and the Thermal Infrared Sensor (TIRS). These two instruments collect data for nine visible, near-infrared, shortwave-infrared bands (from 0.433 to 2.290 µm) and two thermal-infrared bands (from 10.60 to 12.51 µm). The OLI bands have a 30 m spatial resolution, while the TIRS have a 100 m spatial resolution, which acquire in 185 km swaths and segmented into 185 × 180 km scenes. The data have a high signal to noise (SNR) radiometer performance, and 12-bit quantization of the data permits measurement of subtle variability in surface conditions [30,31]. High radiometric sensitivity in the TIR bands shows great potential for mapping exposed lithological units in polar regions through variation in temperature as felsic to mafic rocks show a modified response to solar heating due to different mineral compositions [31–33]. ASTER contains three VNIR bands from 0.52 to 0.86 µm with 15-m spatial resolution, six SWIR bands from 1.6 to 2.43 µm with 30-m spatial resolution, and five TIR bands from 8.0 to 14.0 µm with 90-m spatial resolution. Each scene of ASTER cuts 60 × 60 km^2 [34]. Iron oxide/hydroxide, hydroxyl-bearing, and carbonate mineral groups can be detected using VNIR and SWIR bands of ASTER due to diagnostic spectral absorption features of transition elements (Fe^{2+}, Fe^{3+} and REE) in the VNIR region and Al-OH, Mg-OH, Fe-OH, Si-OH, CO3, NH4, and SO4 groups in the SWIR region [35–37]. Discrimination of silicate lithological groups is feasible using TIR bands of ASTER due to different characteristics of the emissivity spectra derived from Si–O–Si stretching vibrations in the TIR region [18,38–41].

The multispectral commercial WorldView-3 (WV-3) sensor contains the highest spatial, spectral and radiation in the VNIR (eight bands with 1.2 m spatial resolution) and SWIR (eight bands with 3.7 m spatial resolution) portions among the multispectral satellite sensors, presently. WV-3 swath width is 13.2 km [42–45]. The VNIR and SWIR bands of WV-3 are worthy of particular attention for inclusive research related to detailed mineral exploration at district scale, particularly for remote and inaccessible regions in the High Arctic where availability of field data is limited. Recently, some investigations successfully used the VNIR and SWIR bands of WV-3 for mineral exploration and mapping of hydrothermal alteration zones and lithologies [14,19,32,46–49]. These studies established the efficiency of spatial resolution of the WV-3 dataset and emphasized the high capability of the VNIR and SWIR spectral bands as a valuable multispectral remote sensing data for detailed geological mapping and hydrothermal alteration mineral detection at district scale (1:10,000). The integration of multispectral/multi-sensor satellite imagery contains great applicability as a cost-effective tool compared to geophysical and geochemical techniques for mapping hydrothermal alteration minerals and lithological units at regional, local, and district scales in remote and inaccessible metallogenic provinces around the world.

Inglefield Land is an ice-free region (78°N–79°N and 72°30′W–66°W) in northwest Greenland (Figure 1), which contains copper-gold mineralization hosted by garnet-sillimanite paragneiss, orthogneiss, and mafic-ultramafic rocks [1–3,50–52]. A few geological investigations were carried out in Inglefield Land by the Geological Survey of Denmark and Greenland (GEUS) during years 1994 (an airborne geophysical survey) and 1995 (fieldwork geological mapping, mineralization studies, and a regional stream-sediment geochemical survey). A set of thematic maps with digital data in geographic information system (GIS) format were generated using the data acquired from these two field seasons [53,54]. From July to August 1999, fieldwork conducted in Inglefield Land by the GEUS (as part of a multidisciplinary Kane Basin 1999 project) was directed to the exploration of several remarkable gold mineralizations in the northeastern part of the Inglefield Mobile Belt (IMB) [55,56]. Since there is no remote sensing study available for hydrothermal alteration mineral and lithological mapping in the northeastern IMB, this study represents the first investigation on multispectral/multi-sensor satellite imagery for copper-gold prospecting in this region.

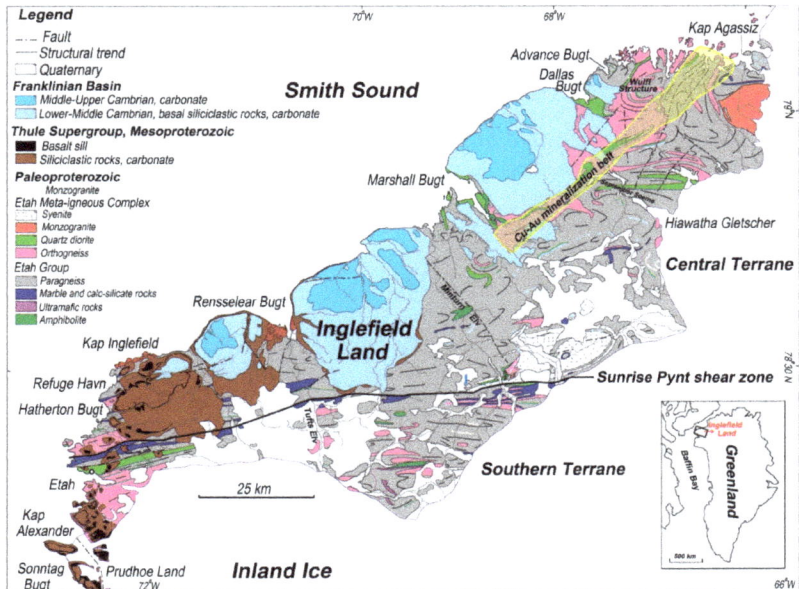

Figure 1. Geological map of the Inglefield Land. Cu-Au mineralized belt in the northeastern part of Inglefield Land shown as a yellow color semi-transparent polygon (modified after [3,42]).

In this research, Landsat-8, ASTER, and WV-3 data were used for hydrothermal alteration and lithological mapping at regional, local, and district scales in the northeastern Inglefield Mobile

Belt (IMB), Northwest Greenland (Figure 1). Mineralization in Inglefield Land is characterized by copper-gold ore associated with hydrothermal alteration assemblages such as hematite, jarosite, biotite, sericite, chlorite, epidote, and quartz (silicification), which overprint the altered areas (rust zones) and wall-rocks [2,3]. Typical landscape in the Cu-Au mineralization belt (rust zones) is extensive gossan in hilly terrain and meter-sized pyrrhotite mounds covered by gossan [3]. Consequently, this research has three main purposes: (1) to map hydrothermal alteration minerals associated with copper-gold mineralization in the northeastern IMB using Landsat-8, ASTER, and WV-3 satellite imagery at regional, local, and district scales; (2) to implement specialized/standardized image processing algorithms to VNIR/SWIR/TIR bands of multispectral/multi-sensor satellite imagery that are amendable for mineral detection and analysis; and (3) to establish the applicability of multispectral/multi-sensor satellite imagery as a valuable and cost-effective approach compared to costly geophysical and geochemical techniques for mining companies and the mineral exploration community for reconnaissance stages of systematic exploration projects in remote and inaccessible metallogenic provinces, specifically in the High Arctic regions.

2. Geological Setting of Inglefield Mobile Belt (IMB)

The IMB in northwest Greenland (approximately 7000 km^2) (Figure 1) forms the northern boundary of the Rae Craton and continues to the west across the Smith Sound into the Ellesmerian Belt in Canada [57,58]. It consists of quartzo-feldspathic gneisses, meta-igneous, and supracrustal rocks of the Palaeoproterozoic age [59–61]. The IMB is subdivided into two terranes by the E-W striking Sunrise Pynt Shear Zone, including (i) the Central Terrane and (ii) the Southern Terrane (Figure 1) [58]. The Central Terrane comprises of the Etah Group and Etah Meta-igneous Complex [57]. The Etah Group is characterized by paragneiss, marble, calc-silicate rocks, ultramafic rocks, amphiboloite, and quartzite [57–59]. The Etah Meta-igneous Complex consists of orthogneiss, tonalite, diorite, granodiorite and minor gabbro, monzogranite, and syenite [58].

The Southern Terrane is interpreted as the margin of the Rae Craton, where Paleoproterozoic sedimentation occurred probably in a passive margin setting [58]. In the Southern Terrane in Prudhoe Land, Paleoproterozoic rocks overly and intrude to Neoarchean rocks of the Rae Craton [58]. The Prudhoe Land Supracrustal Complex consists of garnet-mica schist, quartzite, marble, mafic granulite, and ultramafic rocks [55]. The IMB is unconformably overlain by an unmetamorphosed cover containing the successions of two sedimentary basins (Figure 1), including (i) the sedimentary–igneous rocks of the Mesoproterozoic Thule Basin that also includes basaltic sills and (ii) the Lower Palaeozoic sedimentary rocks of the Franklinian Basin [56,62]. The Cambrian rocks of the Franklinian Basin only remained in the IMB [60,61].

The copper-gold mineralization is delimited within an NE-trending structural belt (~70 × 4 km) in the northeastern part of Inglefield Land (Figure 1). This crustal-scale structural belt consists of sulphide + graphite-bearing bands, hydrothermal alteration zones (including hematite, jarosite, biotite, chlorite, epidote, sericite assemblages, and silicification) and quartzo-feldspathic gneiss that named rust zones [1,2,63,64]. Sulfide mineralization typically comprises of pyrrhotite, pyrite, chalcopyrite, graphite, and cubanite that endured intense supergene alteration. Mylonitic or cataclastic textures were also reported locally in the rust zones. Gossans strike for several meters to up to 5 km in the mylonite and cataclasite [2,3]. Gold in several rock samples was assayed up to 12.5 ppm Au and was characteristically associated with copper (up to 4 wt%) and enriched in Zn, Mo, Ni, Co, Ba, La, and Th [2,3].

3. Materials and Methods

3.1. Satellite Remote Sensing Data and Characteristics

Landsat-8, ASTER, and WV-3 data were used in this research for mapping and detection of hydrothermal alteration minerals and lithological units associated with copper-gold mineralization in the northeastern IMB at regional, local, and district scales. Technical characteristics of the Landsat-8, ASTER, and WV-3 sensors are shown in Table 1. Landsat-8 and ASTER data are successfully used in numerous mineral exploration projects around the world [6–12,15,16,27]. WV-3 is a high-spatial resolution commercial multispectral satellite sensor with eight VNIR (0.42 to 1.04 µm) and eight SWIR bands (1.2 to 2.33 µm), which was launched on 13 August 2014, by DigitalGlobe Incorporated from Vandenberg Air Force Base [43]. It provides high spatial resolution in panchromatic, VNIR, and SWIR with a nominal ground sample distance of 0.31 m, 1.24 m and 3.7 m, respectively (Table 1) (www.digitalglobe.com). Comparison between the spectral bands of WV-3 with Landsat-8 and ASTER emphasizes their priority and high potential for detailed mapping of alteration minerals in the VNIR and SWIR regions (Figure 2). Iron oxides/hydroxide minerals can be comprehensively mapped and discriminated by VNIR bands of WV-3 [14,19,47]. Additionally, SWIR bands of WV-3 contain excellent capability for detailed mapping of Al-OH, Mg-Fe-OH, CO3, and Si-OH key hydrothermal alteration minerals [44,45,47,49].

Table 1. Technical characteristics of the Landsat-8, Advanced Spaceborne Thermal Emission and Reflection Radiometer (ASTER), and WorldView-3 (WV-3) sensors [31,43,65].

Sensors	Subsystem	Band Number	Spectral Range (µm)	Ground Resolution (m)	Swath Width(m)
Landsat-8	VNIR	1	0.433–0.453	30	185
		2	0.450–0.515		
		3	0.525–0.600		
		4	0.630–0.680		
		5	0.845–0.885		
	SWIR	6	1.560–1.660	15	
		7	2.100–2.300		
		Pan	0.500–0.680		
	TIR	9	1.360–1.390	100	
		10	10.30–11.30		
		11	11.50–12.50		
ASTER	VNIR	1	0.520–0.600	15	60
		2	0.630–0.690		
		3	0.780–0.860		
	SWIR	4	1.600–1.700	30	
		5	2.145–2.185		
		6	2.185–2.225		
		7	2.235–2.285		
		8	2.295–2.365		
		9	2.360–2430		
	TIR	10	8.125–8.475	90	
		11	8.475–8.825		
		12	8.925–9.275		
		13	10.25–10.95		
		14	10.95–11.65		

Table 1. *Cont.*

Sensors	Subsystem	Band Number	Spectral Range (μm)	Ground Resolution (m)	Swath Width(m)
WV3	VNIR	Costal (1)	0.400–0.450	1.24	13.1
		Blue (2)	0.450–0.510		
		Green (3)	0.510–0.580		
		Yellow (4)	0.585–0.625		
		Red (5)	0.630–0.690		
		Red edge (6)	0.705–0.745		
		Near-IR1 (7)	0.770–0.895		
		Near-IR2 (8)	0.860–1.040		
	SWIR	SWIR-1 (9)	1.195–1.225	3.70	
		SWIR-1 (10)	1.550–1590		
		SWIR-1 (11)	1.640–1.680		
		SWIR-1 (12)	1.710–1.750		
		SWIR-1 (13)	2.145–2.185		
		SWIR-1 (14)	2.185–2.225		
		SWIR-1 (15)	2.235–2.285		
		SWIR-1 (16)	2.295–2.365		

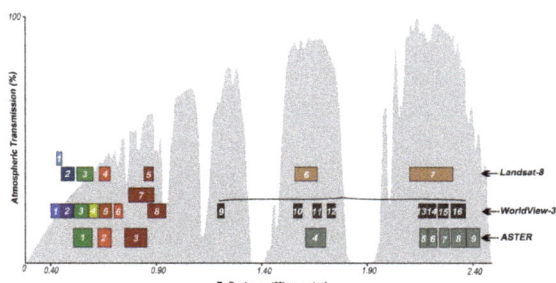

Figure 2. Comparison of the spectral bands of WV-3 with Landsat-8 and ASTER in the visible and near-infrared (VNIR) and shortwave infrared (SWIR) regions [46].

In this study, two Landsat-8 scenes (LC80350032018233LGN00 and LC80350042018233LGN00) covering Inglefield Land were acquired from the U.S. Geological Survey Earth Resources Observation and Science Center (EROS) (https://earthexplorer.usgs.gov/). The data set attributes of these images are summarized as follows: acquisition date: 21 August 2018, collection category: T1 (terrain corrected), Path/Raw: 035/003 and 035/004, scene cloud cover: 11.97% and 2.18%, sun elevation: 22.115 and 23.400 and sun azimuth: −158.241 and −163.695. An ASTER scene (AST_L1T_00307022003234340) covering the northeastern IMB was obtained from the EROS, USGS Global Visualization Viewer (GloVis) (https://glovis.usgs.gov/). It is a level 1T product which is cloud-free and it was acquired on 3 July 2003. The ASTER Level 1 Precision Terrain Corrected Registered At-Sensor Radiance (AST_L1T) data contains calibrated at-sensor radiance, which corresponds with the ASTER Level 1B (AST_L1B), that has been geometrically corrected, and rotated to a north up UTM projection (https://lpdaac.usgs.gov). Some WV-3 scenes were obtained by courtesy of the DigitalGlobe Foundation (www.digitalglobefoundation.org). The VNIR imagery (M2AS-059185278010_01_P001) of the northeastern IMB was granted by the DigitalGlobe Foundation (Copyright 2019 DigitalGlobe, Inc., Longmont CO USA 80503-6493), which was cloud-free, standard level 2 A and acquired on 25 August 2018. The Level 2A standard WV-3 imagery product contains a uniform Ground Sample Distance (GSD), which is radiometrically corrected, sensor corrected, and geometrically projected to the Universal Transverse Mercator (UTM) with the World Geodetic System 84 (WGS-84) datum [66,67]. The Environment for Visualizing Images (ENVI) (http://www.exelisvis.com) version 5.2 and ArcGIS version 10.3 (Esri, Redlands, CA, USA) software packages were utilized for processing Landsat-8, ASTER, and WV-3 datasets.

3.2. Pre-Processing of the Datasets

The Landsat-8 images were pre-georeferenced to the UTM zone 19 and 20 North projection using the WGS84 datum. The ASTER and WV-3 images were also pre-georeferenced to UTM zone 19 North projection using the WGS-84 datum. Atmospheric correction is required to eradicate the impact of atmospheric attenuation from remote sensing imagery and to re-scale the radiance at the sensor data to the surface reflectance data. The absolute radiometric correction and conversion to the top-of-atmosphere (TOA) spectral radiance are required for the WV-3 relative radiometrically corrected images [66]. Hence, these corrections were applied to WV-3 VNIR data used in this study. Crosstalk correction [68] was applied to ASTER data and layer staked of VNIR + SWIR bands with 15-meter spatial dimensions was generated. The Fast Line-of-sight Atmospheric Analysis of Hypercubes (FLAASH) algorithm [69] were applied to the remote sensing datasets used in this research by implementing the sub-arctic summer (SAS) atmospheric and the Maritime aerosol models [70]. ASTER TIR (radiance at the sensor) data without atmospheric corrections were used in this analysis for retaining the original radiance signature.

3.3. Image Processing Algorithms

3.3.1. Directed Principal Components Analysis (DPCA) Technique

The DPCA is a direct information extraction technique to analyze the principal component (PC) eigenvector loadings for selecting the most appropriate PC that focuses the most noteworthy information of interest [71–73]. The magnitude and sign of eigenvector loadings specify whether interesting information is characterized by a bright (positive loading) or a dark pixel (negative loading) in the DPCA image [74]. To map hydrothermal alteration mineral assemblages, including (i) hematite and jarosite (iron oxide/hydroxide group), (ii) biotite and sericite (Al/Fe-OH group), (iii) chlorite and epidote (Mg-Fe-OH group), and (iv) silicification (Si-OH group) (opal/chalcedony) and/or SiO2 group) in the study area, some specialized band ratios were defined to be used as input datasets for running the DPCA. The variance due to similarities in the spectral responses of the interfering component and the component of interest appear in eigenvector loadings of similar signs on input band ratio images. The DPCA contains strong eigenvector loadings of different signs on the input band ratio images, showing a specific contribution of the component [73,74].

For mapping hydrothermal alteration minerals associated with rust zones in the copper-gold mineralization belt, spectral characteristics of hematite, jarosite, biotite, muscovite, chlorite, epidote, chalcedony (hydrous-silica), and opal (hyalite) were considered to identify using the DPCA technique. Figure 3A–C shows laboratory reflectance spectra of hematite, jarosite, biotite, muscovite, chlorite, epidote, chalcedony (hydrous-silica), and opal (hyalite) resampled to response functions of VNIR + SWIR bands of Landsat-8, ASTER, and WV-3, that were extracted from the USGS spectral library version 7.0 [75]. For mapping the alteration mineral groups using Landsat-8 spectral bands, several band ratio indices were adopted and developed [7,8,76]. Band ratio indices of 4/2 (all iron oxides), 6/4 (ferrous iron oxides), 6/5 (ferric oxides), and 6/7 (hydroxyl bearing alteration) can be allotted as significant indicators of Fe^{3+}, Fe^{2+}, Al/Fe-OH, Mg-Fe-OH, and Si-OH groups using Landsat-8 spectral bands (see Figure 3A). Additionally, the normalized difference snow index (NDSI), Al-OH-bearing alteration minerals index (Al-OH-MI) and thermal radiance lithology index (TRLI) were used for mapping snow/ice, cloud, water, alteration OH minerals, and land and lithologies [7]. For mapping iron oxide/hydroxide mineral groups using Landsat-8 bands, three band ratios were developed on the basis of the laboratory spectra of the minerals [77,78]. Hematite, jarosite, goethite, and limonite tend to have strong absorption features in 0.4 to 1.1 μm (absorption features of Fe^{3+} near 0.45 to 0.90 μm and Fe^{2+} near 0.90 to 1.2 μm) [77,78] coincident with bands 2, 4, and 5 and high reflectance at 1.56 μm to 1.70 μm equivalent with band 6 (Figure 3A). As a result, bands 2, 4, 5, and 6 of Landsat-8 can be used for detecting Fe^{3+}/Fe^{2+} and Fe-OH iron oxides (4/2), ferrous iron oxides (6/4), and ferric oxides (6/5). Hydroxyl-bearing (Al-OH and Fe, Mg-OH) alteration has spectral absorption features in 2.1–2.4 μm and reflectance in 1.55–1.75 µm [35], corresponding band 7 (2.11–2.29 µm) and band 6 (1.57–1.65 µm) of Landsat-8 (Figure 3A), respectively. Therefore, band ratio of 6/7 can map hydroxyl bearing alteration. The DPCA was applied to the Landsat-8 band ratio indices (4/2, 6/4, 6/5, and 6/7) using a covariance matrix for obtaining the image eigenvectors and eigenvalues.

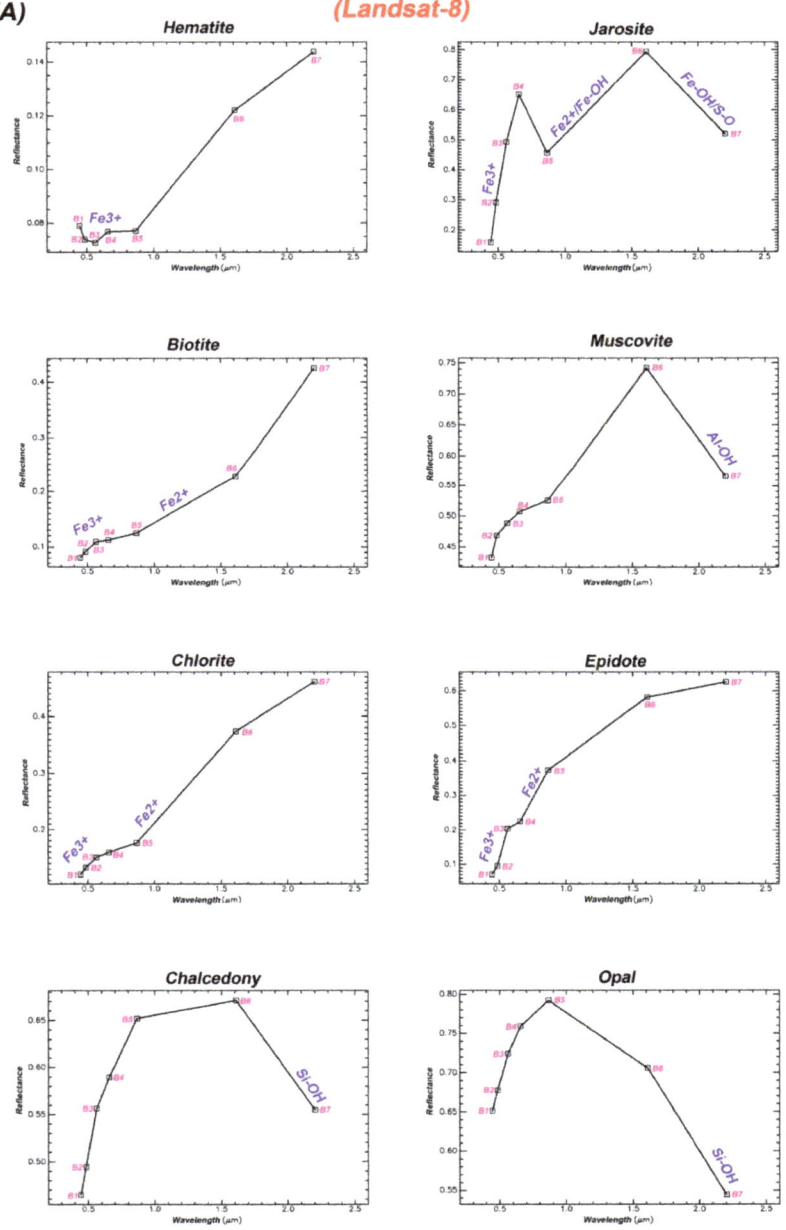

Figure 3. *Cont.*

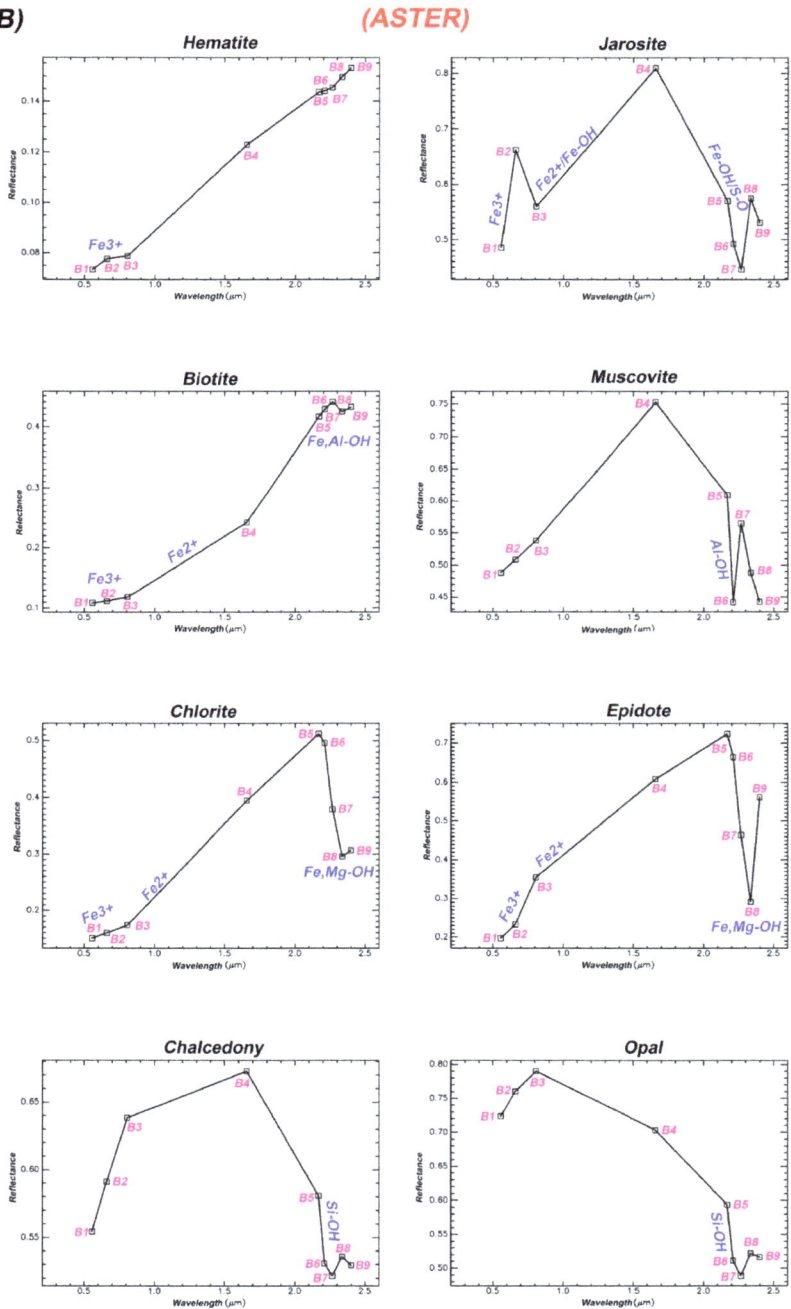

Figure 3. Cont.

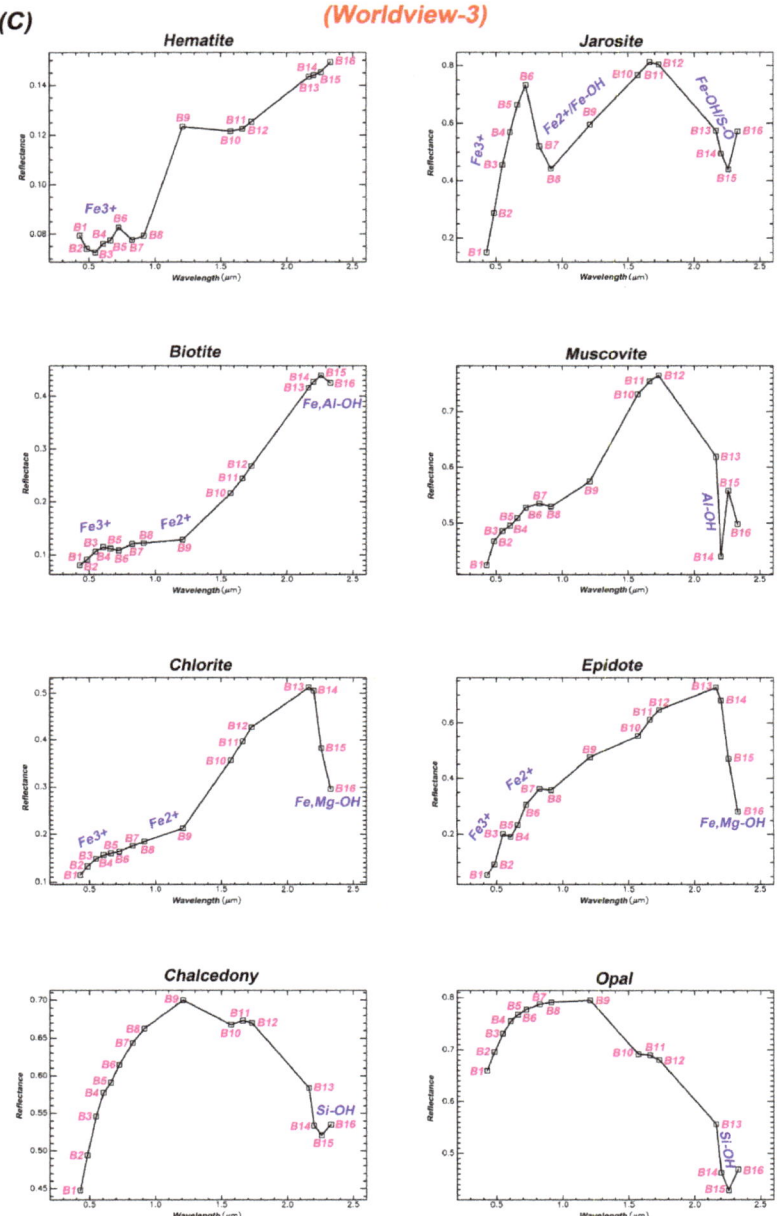

Figure 3. Laboratory reflectance spectra of hematite, jarosite, biotite, muscovite, chlorite, epidote, chalcedony (hydrous-silica), and opal (hyalite) resampled to response functions of VINR + SWIR bands of Landsat-8 (**A**), ASTER (**B**), and WV-3 (**C**) that were extracted from the USGS spectral library version 7.0 [75]. Cubes indicate the position of the VINR + SWIR bands of Landsat-8, ASTER, and WV-3 in the range of 0.4 µm to 2.5 µm.

Several band ratio indices were tested for mapping the alteration mineral groups using ASTER spectral bands (see Figure 3B). The band ratios of 2/1 and 4/2 were selected to map Fe^{3+}/Fe^{2+} iron oxides; (5 + 7)/6 was adopted to detect Al/Fe-OH minerals; (7 + 9)/8 was assigned to identify Mg-Fe-OH minerals [79]; and (6 + 8)/7 were developed to map Si-OH minerals, respectively. Bands 1 (0.520–0.600 μm), 2 (0.630–0.690 μm) and 4 (1.600–1.700 μm) of ASTER cover the spectral absorption and reflectance features of iron oxide/hydroxide minerals. Thus, band ratios of 2/1 and 4/2 can be utilized for detecting Fe^{3+}/Fe^{2+} iron oxides. Al-OH absorption features at 2.17 to 2.20 μm [35,77,78] are corresponded with bands 5 and 6, whereas Mg-Fe-OH absorption features are situated in 2.30 to 2.35 μm [35,77,78] that are equivalent with bands 7 and 8 of ASTER (Figure 3B). Si-OH absorption features are mostly concentrated at 2.20 to 2.30 μm, which are coincident with bands 6 and 7 of ASTER (Figure 3B). Subsequently, relative absorption band depth (RBD) [80] of these bands can be used to map Al/Fe-OH ((5 + 7)/6), Mg-Fe-OH ((7 + 9)/8), and Si-OH ((6 + 8)/7) minerals. The DPCA was implemented to the band ratio indices (2/1, 4/2, (5 + 7)/6 and (7 + 9)/8) using a covariance matrix for the spatial selected subset covering the Cu-Au mineralization belt and surrounding areas. Furthermore, for mapping silica-rich rocks containing SiO_2 group, Quartz Index (QI) = $11 \times 11/10 \times 12$, Carbonate Index (CI) = 13/14, and Mafic Index (MI) = 12/13 were selected [81] and applied to TIR bands of ASTER. These lithologic indices were defined by Ninomiya et al. [81] for discriminating quartz, carbonate, and mafic-ultramafic rocks, especially for mapping lithological units in arid and semi-arid regions. The DPCA was employed to these indices. Eigenvector matrix was calculated using a covariance matrix for the spatial selected subset covering the Cu-Au mineralization belt and surrounding areas.

The VNIR spectral bands of WV-3 contain the high capability to map Fe^{3+} and Fe^{2+} iron oxides (gossan), ferric, and ferrous silicates. Considering the laboratory reflectance spectra of selected minerals (see Figure 3C), the band ratio indices of 4 + 2/3 to map Fe^{3+} iron oxides, 6 + 8/7 for identifying Fe^{2+} iron oxides, 3 + 5/4 to detect ferric silicates (chlorite/epidote), and 5 + 7/6 for enhancing ferrous silicates (biotite) were developed. These indices were used to implement the DPCA using a covariance matrix for the spatial selected subset covering the southern part of the Cu-Au mineralization belt. The DPCA statistical results were also calculated for the WV-3 band ratio indices.

3.3.2. Linear Spectral Unmixing (LSU)

The LSU is a sub-pixel image processing algorithm, which is utilized to define the relative abundance of materials that can be diagnosed within optical imagery based on the materials' spectral properties [82–84]. The reflectance at each pixel of the image is presumed to be a linear combination of the reflectance of each material (or end-member) existing within the pixel. It is advocated in this algorithm that the pixel reflectance could be shown as a linear mixture of individual component reflectance multiplied by its relative fractions [85]. For extracting reference spectra directly from the Landsat-8, ASTER, and WV-3 images to generate fraction images of end-members using the LSU, the automated spectral hourglass (ASH) approach was implemented [86,87]. This approach contains the minimum noise fraction (MNF), the pixel purity index (PPI) and automatic end-member prediction from the n-Dimensional Visualizer to extract the most spectrally pure pixels (end-members) from the image [36,88]. Additionally, the continuum-removal process was performed to the extracted end-members for isolating their spectral features [89]. Then, the end-members were compared with the USGS library reflectance spectra of target minerals, including hematite, jarosite, biotite, muscovite, chlorite, epidote, chalcedony (hydrous-silica), and opal (hyalite) (see Figure 3A–C). Umix unit-sum constrained was adjusted 1.0 for running the LSU algorithm. This weighted unit-sum constraint is then added to the system of simultaneous equations in the unmixing inversion process. Larger weights in relation to the variance of the data cause the unmixing to honor the unit-sum constraint more closely. To strictly honor the constraint, the weight should be many times the spectral variance of the data. It also permits proper unmixing of MNF transform data, with zero-mean bands [70]. For interactive stretching histogram, auto apply option was selected to have stretching or histogram changes applied to the images automatically. Rule image classifier tool was used for post classification of the LSU rules images. Maximum value option was selected. Threshold value for classification of fraction images derived from the LSU algorithm was 0.750.

3.3.3. Adaptive Coherence Estimator (ACE)

The ACE is a target detection algorithm that carries out a partial unmixing approach to isolate feature of interest from the background and its input is a single score (abundance of the target) per pixel [90]. It is generated from the generalized likelihood ratio (GLR) approach, which is a homogenously most powerful invariant detection statistic [91,92]. The ACE is invariant to the relative scaling of input spectra and has a constant false alarm rate (CFAR) for such scaling [93]. Geometrically, it determines the squared cosine of the angle between a known target vector and a sample vector in a whitened coordinate space. The space is faded based on assessing the background statistics, which straightforwardly influences the presentation of the statistic as a target detector [94]. The standard formulation of the ACE detection statistic is defined as follows:

$$\text{ACE}(x) = \frac{\left[(t-\mu)^T \Sigma^{-1}(x-\mu)\right]^2}{\left[(t-\mu)^T \Sigma^{-1}(t-\mu)\right]\left[(x-\mu)^T \Sigma^{-1}(x-\mu)\right]} \tag{1}$$

where t is a known target signature (reference spectra from a spectral library signature) and x is a data sample. The background is assumed to be a Gaussian distribution parametrized by u and Σ which represent the mean and covariance, respectively. The ACE statistic is a number between zero and one, which can be interpreted as a measurement of the presence of t in x. The ACE can be estimated as the square of the cosine of the angle between x and t, in a coordinate space transformed by the background estimation. For example, if ACE produces 0.85, indicating a relatively strong presence of t in x. The key to effective ACE performance is accurate background estimation. Furthermore, the ACE does not need information about all the end-members within an image scene. In this study, the ACE algorithm was applied to VNIR + SWIR bands of ASTER covering the Cu-Au mineralization belt and surrounding areas. Laboratory reflectance spectra of hematite, jarosite, biotite, muscovite, chlorite, epidote, chalcedony (hydrous-silica), and opal (hyalite) extracted from USGS spectral library version 7.0 [75] were used for running the ACE algorithm. New covariance statistics were computed and subspace background was used. Background threshold was adjusted 0.900. The results of ACE appear as a series of grayscale images, one for each selected end-member.

4. Results

4.1. Regional Lithological-Mineralogical Mapping in Inglefield Land Using Lansat-8 Data

A regional view of the northwestern part of Greenland was generated using a mosaic of Landsat-8 images (Figure 4). The NDSI, Al-OH-MI, and TRLI [7] were used for mapping snow/ice, cloud, water, land, and lithologies. The NDSI (B3 − B6/B3 + B6), Al-OH-MI (B6/B7) × (B7), and TRLI (B10/B11) × (B11) were assigned to Red-Green-Blue false-color composite, respectively (Figure 4). The ice/snow zones appear in magenta, red, and orange shades that correspond to the different snow/ice facies. Stratocumulus cloud coverage is represented as golden yellow especially in the east and northeastern parts (inland ice) of the mosaic image-map. Water is depicted in a dark blue color. The Inglefield Land and Washington Land in the west and northwestern parts of the scene appear in light blue and cyan shades. The shelf-platform carbonate of the Franklinian Basin in the Washington Land and northwestern parts of the Inglefield Land (adjacent to Smith Sound) typically contains cyan shade. The exposed lithologies, including the complex metamorphic rocks of the Central Terrane and the Southern Terrane and Mesoproterozoic sedimentary–igneous rocks of the Thule Basin manifest in a light blue tone (Figure 4).

Band ratio indices of B4/B2, B4/B6, and B6/B7 were assigned to the RGB false-color composite for mapping iron oxides/hydroxides, ferrous iron oxides, and hydroxyl bearing alteration zones in the IMB at the regional scale, respectively. Figure 5A shows the resultant image-map. Regarding the geology map of the IMB (see Figure 1), the sedimentary successions of the Franklinian Basin and Thule Supergroup appear typically in cyan, pink, orange, and rose blush. Carbonate and siliciclastic rocks are dominant lithological units in these two sedimentary basins, which are mostly represented as cyan color. It could be due to the fact that most of Al-OH, Mg-Fe-OH, CO3, and Si-OH mineral groups show high reflectance at 1.55–1.75 µm and strong absorption at 2.1–2.4 µm coincident with bands 6 and 7 of Landsat-8, respectively [7,76]. Pink, orange, and rose blush zones may contain dolomite (Fe^{2+} absorption at 0.9–1.2 µm; the equivalent of band 5 of Landsat-8) or iron oxides/hydroxides minerals. Basaltic sills in the Thule Basin are depicted in purple color (western part of image-map) due to the high content of iron oxides/hydroxides minerals. Several golden yellow areas are recognizable at the boundaries between sedimentary successions and the Etah metamorphic complex rocks in the Central Terrane, which comprise Fe^{3+} and Fe^{2+} iron oxides/hydroxides. Paragneiss of the Etah Group manifests in magenta to tangerine tone in both the Southern and Central Terranes due to a strong amount of iron oxides/hydroxides, while Quaternary deposits appear as cyan color because of detrital clay minerals. Syenite of the Etah meta-igneous complex is characterized by brown color adjacent to the Sunrise Pynt Shear Zone. Orthogneiss in the western and northeastern parts of the IMB shows up in gray shade (Figure 5A). Syenite and orthogneiss probably contain a high amount of ferrous iron oxide minerals attributable to alteration products of primary mafic minerals such as biotite, hornblende, amphibole, and clinopyroxene (augite).

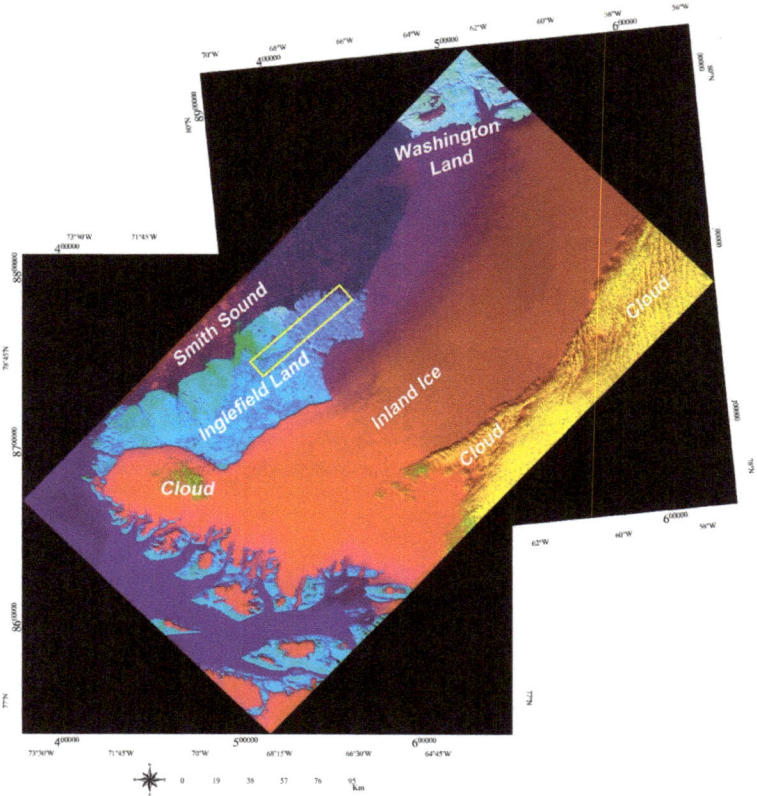

Figure 4. A regional view of the northwestern part of Greenland generated using a mosaic of Landsat-8 images as RGB false-color composite of the normalized difference snow index (NDSI), Al-OH-bearing alteration minerals index (Al-OH-MI), and thermal radiance lithology index (TRLI). Yellow rectangle shows the location of the Cu-Au mineralization belt.

Table 2 shows the eigenvector matrix of the Landsat-8 band ratio indices (4/2, 4/6, 6/5, and 6/7) derived from the DPCA for the selected subset covering the IMB. Analyzing the magnitude and sign of the eigenvector loadings derived from DPCA technique for the IMB selected subset scene (Table 2) indicates the DPCA1 contains positive eigenvector loadings for all input band ratio indices. Thus, it does not have any unique contribution of input band ratio indices and the discrimination of alteration mineral groups is impossible. The DPCA2 has a strong negative contribution (−0.770751) for ferric oxides (B6/B5). However, it contains moderate loadings of other alteration mineral groups with the opposite sign (Table 2). Ferric oxides manifest as dark pixel in the DPCA2 image due to negative loading. The DPCA3 contains strong positive loadings of B4/B2 (0.686248) and B4/B6 (0.714648) for iron oxides/hydroxides and ferrous iron oxides mineral groups, respectively (Table 2). However, the eigenvector loadings in the DPCA3 for ferric oxides (B6/B5) and hydroxyl bearing alteration (B6/B7) indices are weak and negative (−0.124892 and −0.052382). Therefore, the DPCA3 image shows desired information related to Fe^{3+} and Fe^{2+} iron oxides/hydroxides as bright pixel. Figure 5B shows a pseudocolor ramp of the DPCA3 rule image. The high concentration of Fe^{3+}/Fe^{2+} iron oxides/hydroxide minerals is observable in the boundaries between the Etah metamorphic complex rocks and sedimentary successions of the Franklinian Basin and Thule Supergroup in the Central Terrane. Moderate to low abundance of iron oxides/hydroxide minerals are associated with carbonate and siliciclastic rocks in both sedimentary basins. The southern part of the Cu-Au mineralization belt nearby Marshall Bugt contains high surface abundance of iron oxides/hydroxide minerals. The Etah group and meta-igneous complex rocks show moderate to low spatial distribution of iron oxides/hydroxide minerals. Some of the highly abundant iron oxides/hydroxide zones are located in Quaternary deposits and associated with Basaltic sills in the Thule Basin (Figure 5B).

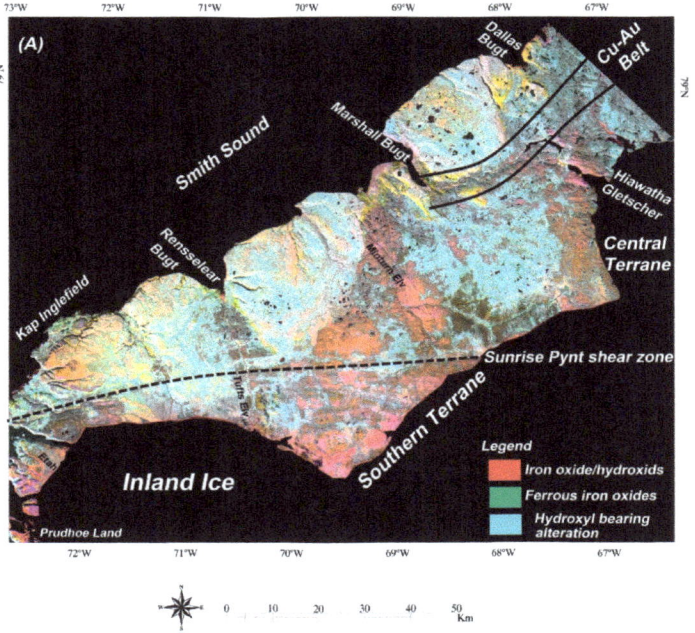

Figure 5. Cont.

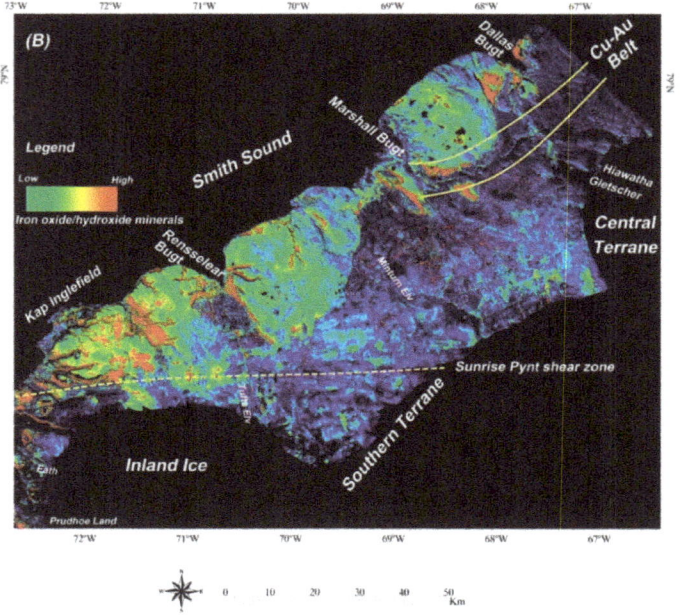

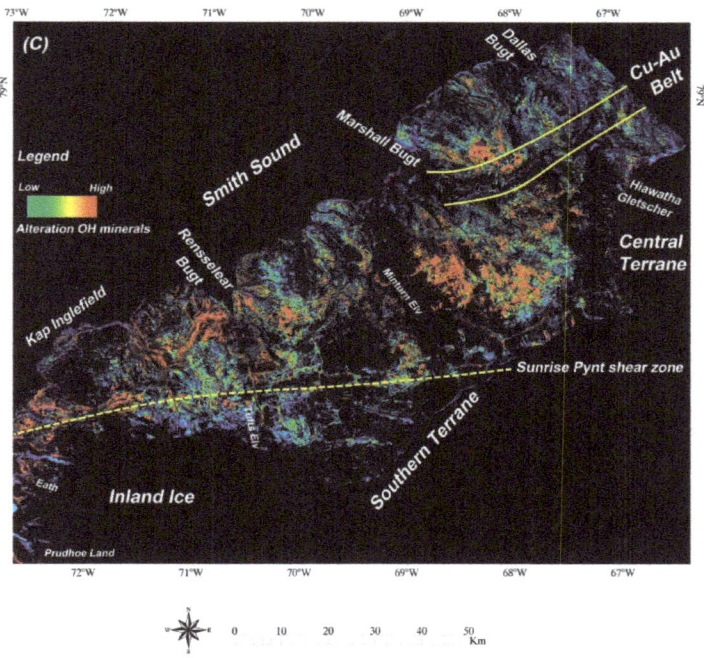

Figure 5. Landsat-8 image-maps of the IMB. (**A**) RGB false-color composite of B4/B2, B4/B6, and B6/B7 band ratio indices covering the IMB. (**B**) Pseudocolor ramp of the DPCA3 rule image covering the IMB. (**C**) Pseudocolor ramp of the DPCA4 rule image covering the IMB.

Table 2. Eigenvector matrix of the Landsat-8 band ratio indices derived from the directed principal components analysis (DPCA) for the Inglefield Mobile Belt (IMB) selected subset scene.

Eigenvector	B4/B2	B6/B4	B6/B5	B6/B7
DPCA 1	0.412529	0.470934	0.624086	0.467501
DPCA 2	0.459849	0.282028	−0.770751	0.339030
DPCA 3	0.686248	0.714648	−0.124892	−0.052382
DPCA 4	−0.383955	−0.433543	−0.029349	0.814713

The Al-OH, Mg-Fe-OH, CO3 and Si-OH alteration mineral groups are mapped in the DPCA4 image due to the great positive contribution of B6/B7 ratio index (0.814713) (Table 2). On the other hand, iron oxides/hydroxides (−0.383955), ferrous iron oxides (−0.433543), and ferric oxides (−0.029349) indices show moderate to weak eigenvector loadings with a negative sign in the DPCA4 (Table 2). It is evident that the DPCA4 image shows the alteration OH mineral groups as bright pixels. A pseudocolor ramp of the DPCA4 rule image was generated (Figure 5C). High spatial distribution of the alteration OH mineral groups is mostly associated with carbonate and siliciclastic units of the Franklinian Basin and Thule Supergroup as well as Quaternary deposits in the Central Terrane. Moreover, orthogneiss of the Etah meta-igneous complex and marble, amphibolite, and calc-silicate rocks of the Etah group show a high surface abundance of alteration OH mineral groups. The central part of the Cu-Au mineralization belt contains a remarkable concentration of the alteration OH mineral groups, which might be related to amphibolite or alteration products of quartz diorite units. Paragneiss of the Etah Group includes low to moderate surface distribution of the alteration OH minerals.

Figure 6A displays end-member spectra (n-D classes) extracted from the n-Dimensional analysis technique for a selected spatial subset of Landsat-8 covering the Cu-Au mineralization belt and surrounding areas. The n-D classes correspond to a set of unique pixels (a pure end-member), which are used to act as end-members for the LSU spectral mineral-mapping. Comparison of the extracted n-D classes with selected end-member reflectance spectra of the target minerals from the USGS spectral library (see Figure 3A) indicates that some of the n-D classes could be considered for the LSU spectral mineral-mapping. Some noticeable similarities between spectral signatures of the n-D classes and the target minerals could be utilized for identifying iron oxide/hydroxide, clay mineral groups and ferrous silicates (biotite, chlorite and epidote). The n-D class #1 and n-D class #6 typically represent Al-OH/Si-OH absorption characteristics (Figure 6A). Muscovite, chalcedony, and opal show high reflectance in band 6 (1.560–1.660 μm) and strong absorption in band 7 (2.100–2.300 μm) of Landsat-8 (see Figure 3A). The n-D class #2 and n-D class #4 can be considered as snow/ice/cloud group because these classes show high reflectance in the visible wavelengths from 0.40 μm to 0.75 μm (band 1 to band 4 of Landsat-8), lower reflectance in the near-infrared from 0.80 μm to 0.90 μm (band 5 of Landsat-8), and strong absorption in the short wave infrared from 1.57 μm to 1.78 μm (band 6 of Landsat-8) [95–97]. The n-D class #3 does not show any typical absorption features related to any geological materials and hydrothermal alteration minerals. The n-D class #5 contains some similar spectral signatures related to Mg-Fe-OH alteration minerals (ferrous silicates). Iron oxide (Fe^{+2}/Fe^{+3}) absorption features in bands 2 to 3 (0.50–0.60 μm) and bands 4 to 5 (0.70–0.90 μm) and Mg, Fe-OH absorption in bands 7 of Landsat-8 are recognizable for the n-D class #5 (Figure 6A). The n-D class #7 and n-D class #8 might be attributed to the iron oxide/hydroxide minerals because of Fe^{3+} and Fe-OH absorption features at 0.45 μm to 0.70 μm, 0.80-0.90 μm, and 2.20-2.30 μm coinciding with bands 2, 3, 4, 5, and 7 of Landsat-8.

Fraction images of the end-members resulted from the LSU algorithm manifest as a series of greyscale rule images (one for each extracted end-member). Considering the resultant fraction images and the n-D classes (extracted end-member spectra) for the Landsat-8 selected subset, it is evident that iron oxide/hydroxide minerals, clay minerals and ferrous silicates are main alteration mineral groups in the study area. For post-classification of the fraction images (excluding snow/ice/cloud group) the interactive density slicing tool was used to select colors for highlighting the high digital number (DN) value areas (bright pixels) in the grayscale rule images. The red color class was considered for iron oxide/hydroxide group, the green color class was selected for clay mineral groups, and the yellow color class was assigned for ferrous silicates, respectively. Figure 6B shows the LSU spectral mineral-map for the Landsat-8 selected subset covering the Cu-Au mineralization belt and surrounding areas. Iron oxide/hydroxide minerals (red pixels) are spectrally dominated in the image-map, whereas clay minerals and ferrous silicates show less spatial distribution in the selected subset. Comparison to the geological map of the study zone, suggests that an iron oxide/hydroxide group is typically concentrated in the southwestern part of the Cu-Au mineralization belt at the boundary between orthogneiss and paragneiss with the sedimentary succession of carbonate and basal siliciclastic rocks. However, an iron oxide/hydroxide group is also detected in the Franklinian Basin sedimentary succession (central north) and many other zones in orthogneiss and paragneiss of the Etah complex in the southwestern and southeastern parts of the scene (Figure 6B). The high surface abundance of clay minerals (green pixels) was mapped in orthogneiss, amphibolite, and quartz diorite units especially in the central part of the Cu-Au mineralization belt. Basal siliciclastic rocks of the Franklinian Basin show high concentrations of clay minerals in the central part of the scene. Ferrous silicates are lesser in the surface abundance and generally associated with an iron oxide/hydroxide mineral group (Figure 6B).

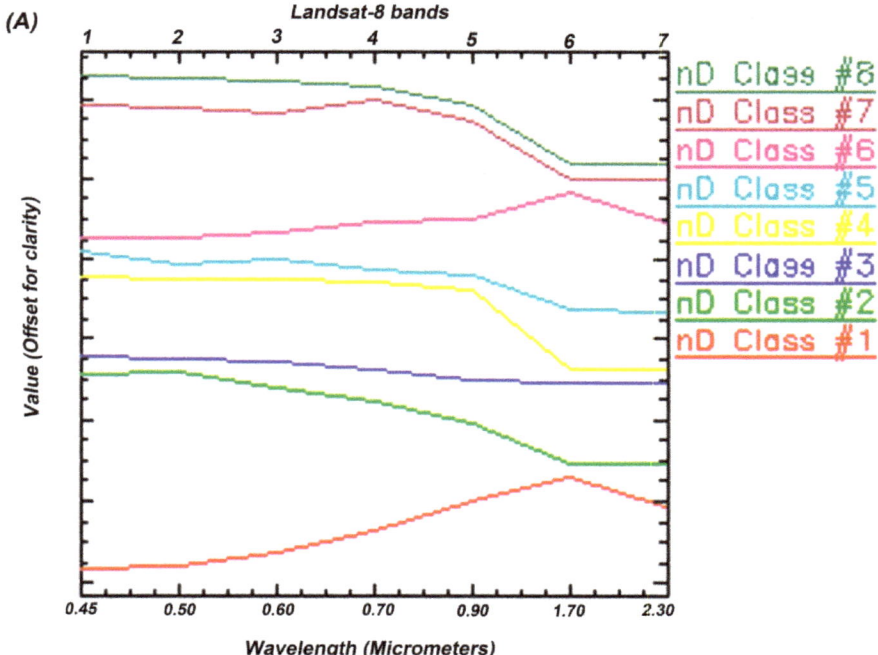

Figure 6. *Cont.*

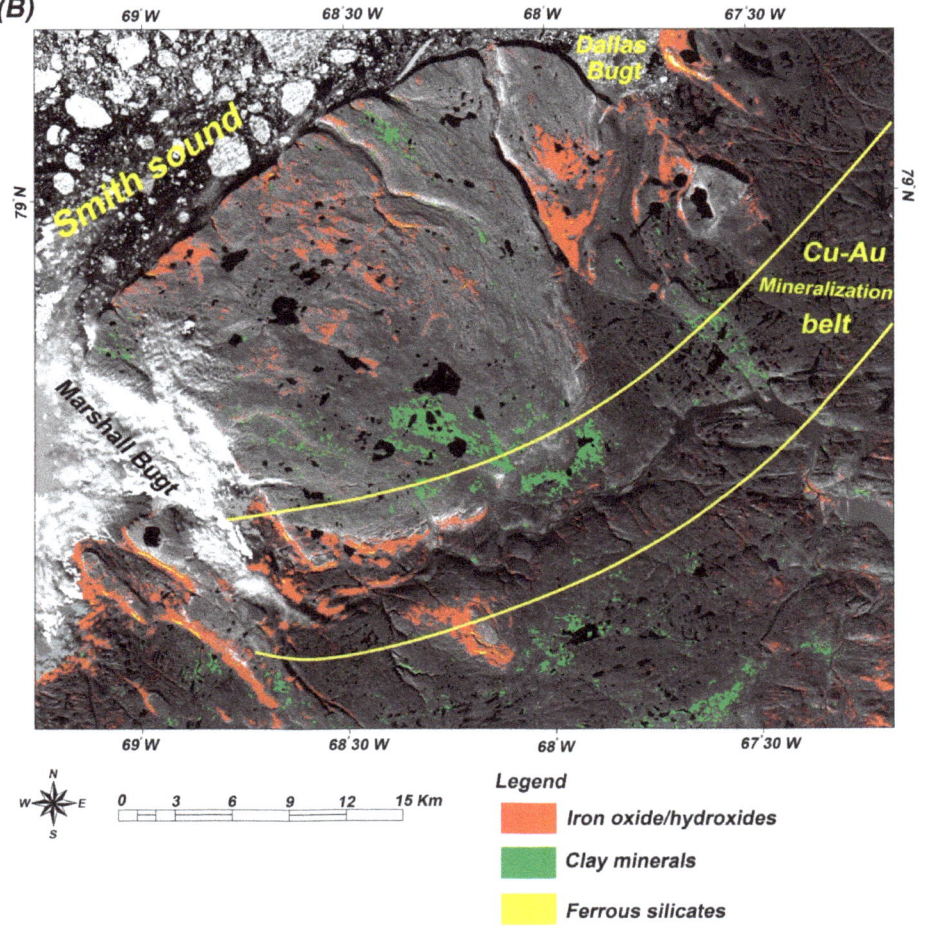

Figure 6. (**A**) The n-D classes (end-member spectra) extracted for a selected spatial subset (Landsat-8) covering Cu-Au mineralization belt and surrounding areas. Landsat-8 band center positions are shown. (**B**) LSU mineral map produced from fraction images overlaid on band 5 of Landsat-8 for the selected spatial subset covering the Cu-Au mineralization belt and surrounding areas.

4.2. Hydrothermal Alteration Mapping in the Northeastern IMB Using ASTER Data

Analyzing the eigenvector matrix of the band ratio indices for mapping hydrothermal alteration minerals using VNIR + SWIR bands of ASTER (Table 3) shows that the DPCA technique detected the surface distribution of Fe^{3+}/Fe^{2+} iron oxide/hydroxides, Al/Fe-OH, Mg-Fe-OH, and Si-OH minerals in some specific DPCA images with a strong contribution of the input band ratio components. Figure 7A–E shows the pseudocolor ramp of the DPCA rule images covering the selected spatial subset of the Cu-Au mineralization belt and surrounding areas (similar size as the Landsat-8 LSU image-map).

Table 3. Eigenvector matrix of the ASTER VNIR + SWIR band ratio indices derived from the DPCA for the selected subset covering the Cu-Au mineralization belt and surrounding areas.

Eigenvector	B2/B1	B4/B2	B5 + B7/B6	B7 + B9/B8	B6 + B8/B7
DPCA 1	−0.219557	−0.347765	−0.900627	−0.123124	−0.067567
DPCA 2	−0.547623	0.589087	−0.177962	−0.235635	−0.434915
DPCA 3	0.263209	0.141891	−0.759215	0.418035	0.399283
DPCA 4	−0.027870	−0.119071	0.531229	0.709489	−0.288482
DPCA 5	−0.568874	0.044535	0.108190	0.314737	−0.750756

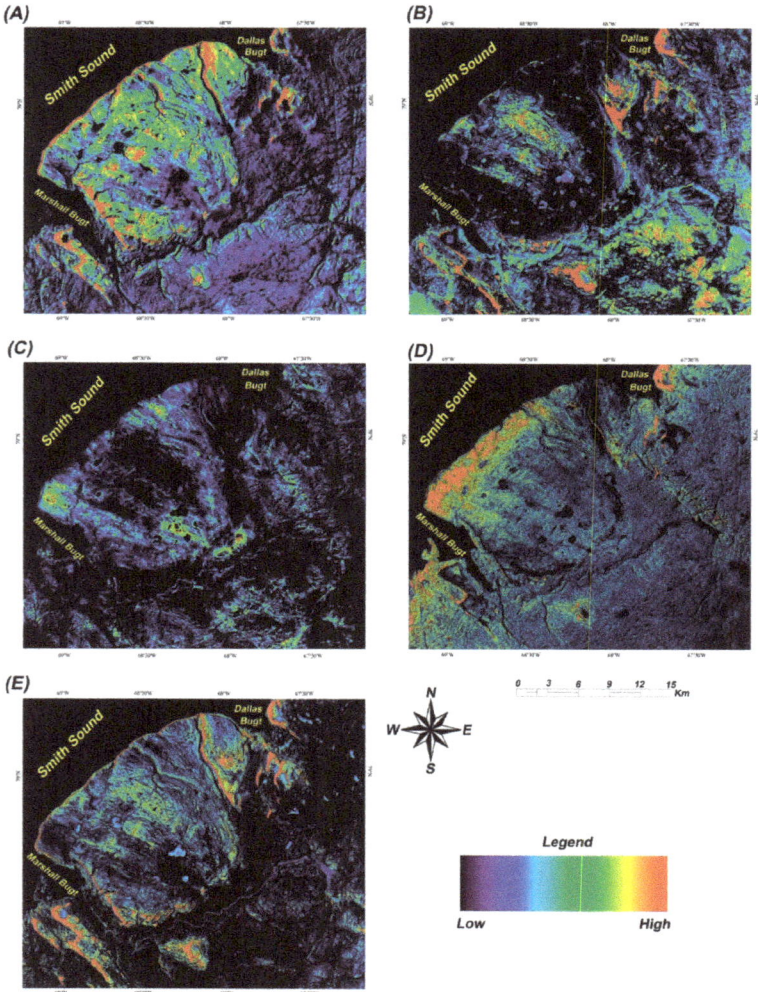

Figure 7. Pseudocolor ramp of the ASTER (VNIR + SWIR) DPCA rule images covering the selected spatial subset of the Cu-Au mineralization belt and surrounding areas. (**A**) Ferrous iron oxides (Fe^{+2})/Si-OH image-map; (**B**) ferric iron (Fe^{+3}) oxide/hydroxides image-map; (**C**) Al/Fe-OH minerals image-map; (**D**) Mg-Fe-OH minerals image-map; (**E**) Si-OH minerals image-map.

The DPCA2 contains a strong contribution of Fe^{3+}/Fe^{2+} iron oxide/hydroxides in the B2/B1 (−0.547623) and B4/B2 (0.589087), while the contributions of Al/Fe-OH, Mg-Fe-OH, Si-OH minerals are weak to moderate with negative signs (−0.177962, −0.235635, and −0.434915, respectively) (Table 3). Therefore, ferrous iron oxides (Fe^{+2}) can be characterized as dark pixels due to the strong magnitude and negative sign of eigenvector loadings (−0.547623) in the DPCA2. Considering the eigenvector loadings in this DPCA (see Table 3), the contribution of other mineral groups as dark pixels, especially Si-OH minerals, is also feasible. These dark pixels were converted to bright pixels by multiplication to −1, and then a pseudocolor ramp of greyscale rule image was generated for the DPCA2. Figure 7A shows the resultant image-map of ferrous iron oxides (Fe^{+2}) and silica-rich units. Referring to the geological map of the study area, high to moderate concentration of ferrous oxides/Si-OH was mostly mapped in the sedimentary successions of the Franklinian Basin, which can be attributed to dolomite and basal siliciclastic rocks. In the Cu-Au mineralization belt, some small zones show a high to moderate spatial distribution of ferrous oxides/Si-OH components.

In the DPCA2, ferric iron (Fe^{+3}) oxide/hydroxides can be mapped explicitly as bright pixels due to strong and positive loadings of the B4/B2 (0.589087) (Table 3). Figure 7B shows the pseudocolor ramp of the DPCA2 for ferric iron components. High to moderate surface abundance of ferric iron components is typically detected at the contact of orthogneiss and paragneiss with the Franklinian sedimentary successions. However, high concentration of ferric iron was also mapped in association with orthogneiss and quartz diorite in the northeastern part of the selected subset near Dallas Bugt. Carbonate successions of the Franklinian Basin and paragneiss of the Etah Group generally show a moderate to high surface abundance of ferric iron in some parts of the selected subset (Figure 7B). Several small zones of high to moderate concentration of ferric iron were identified within the Cu-Au mineralization belt, which can be considered as gossan zones (rust zones). The 4/2 band ratio of ASTER was documented as a reliable indicator for identifying gossan zones associated with massive sulfide mineralization in the Neoproterozoic Wadi Bidah shear zone, southwestern Saudi Arabia and many porphyry copper deposits around the world [98,99].

Al/Fe-OH minerals can be robustly detected in the DPCA3 image as dark pixels due to a high negative contribution of the B5+B7/B6 (−0.759215) (Table 3). For inverting the dark pixels to bright pixels, the DPCA3 image was negated. The pseudocolor ramp of the DPCA3 is shown in Figure 7C. The high concentration of Al/Fe-OH minerals was only mapped in some small sites in the carbonate/siliciclastic units of the Franklinian Basin, Quaternary deposits, quartz diorite, and amphibolite of the Etah Group. The orthogneiss and paragneiss units show low to moderate distribution of Al/Fe-OH minerals. The central part of the Cu-Au mineralization belt contains moderate to high spatial distribution of the mineral groups, which is related to the quartz diorite and amphibolite units (Figure 7C). The DPCA4 contains strong loadings of B7 + B9/B8 (0.709489) and B5 + B7/B6 (0.531229) with a positive sign (Table 3). Therefore, Mg-Fe-OH minerals can be mapped as bright pixels in the DPCA4 image. Although, this image might have some contribution of Al/Fe-OH minerals due to great and positive eigenvector loading of the B5 + B7/B6 component. Figure 7D shows a pseudocolor ramp of the DPCA4 image. High spatial distribution of Mg-Fe-OH minerals is typically concentrated in the Franklinian sedimentary successions and paragneiss units proximate to Marshall Bugt. However, the orthogneiss and quartz diorite units adjacent to Dallas Bugt also contain a strong surface abundance of the mineral groups. Few small locations inside the Cu-Au mineralization belt comprise high concentrations of Mg-Fe-OH minerals that are associated with rust zones (Figure 7D).

The B6 + B8/B7 component in the DPCA5 has strong weighting (−0.750756) with a negative sign, which can represent Si-OH minerals as dark pixels. Besides, the B2/B1 (ferrous iron oxides) shows high contribution (−0.568874) with a negative sign in the DPCA5 (Table 3). This image was negated for converting the dark pixels to bright pixels before applying pseudocolor ramp (Figure 7E). The resultant image-map shows spatial distribution of Si-OH minerals that may have some contribution of ferrous iron oxides. The high concentration of Si-OH minerals is characteristically mapped associated with quartz diorite and at the contact of orthogneiss and paragneiss with the Franklinian sedimentary successions. In the Cu-Au mineralization belt, the high concentration of Si-OH minerals was mapped in several localities associated with rust zones, especially in the southwestern part of the belt (Figure 7E).

Table 4 shows the eigenvector matrix of the ASTER TIR band ratio indices, including Quartz Index (QI) = 11 × 11/10 × 12, Carbonate Index (CI) = 13/14, and Mafic Index (MI) = 12/13 [81], for the selected subset covering the Cu-Au mineralization belt and surrounding areas. Considering eigenvector loadings for mapping altered silica-rich rocks (containing SiO2 group), it is evident that the DPCA2 is able to detect altered silica-rich rocks as bright pixels because of the strong contribution of QI (0.792423) with a positive sign. The CI (−0.302209) and MI (−0.097008) components contain weak contributions with a negative sign in the DPCA2 (Table 4). Figure 8A shows a pseudocolor ramp of the DPCA2 for the QI component. High to moderate concentration of quartz content was mostly mapped at the contact of orthogneiss with the Franklinian Basin successions, orthogneiss, and quartz diorite units. The low surface abundance of quartz was recorded for paragneiss and amphibolite. Several zones containing intense concentration of quartz content were identified in the Cu-Au mineralization belt (Figure 8A).

Table 4. Eigenvector matrix of the ASTER TIR band ratio indices derived from the DPCA for the selected subset covering the Cu-Au mineralization belt and surrounding areas.

Eigenvector	QI	CI	MI
DPCA 1	−0.596505	−0.527385	−0.605018
DPCA 2	0.792423	−0.302209	−0.097008
DPCA 3	−0.106481	0.790280	−0.530590

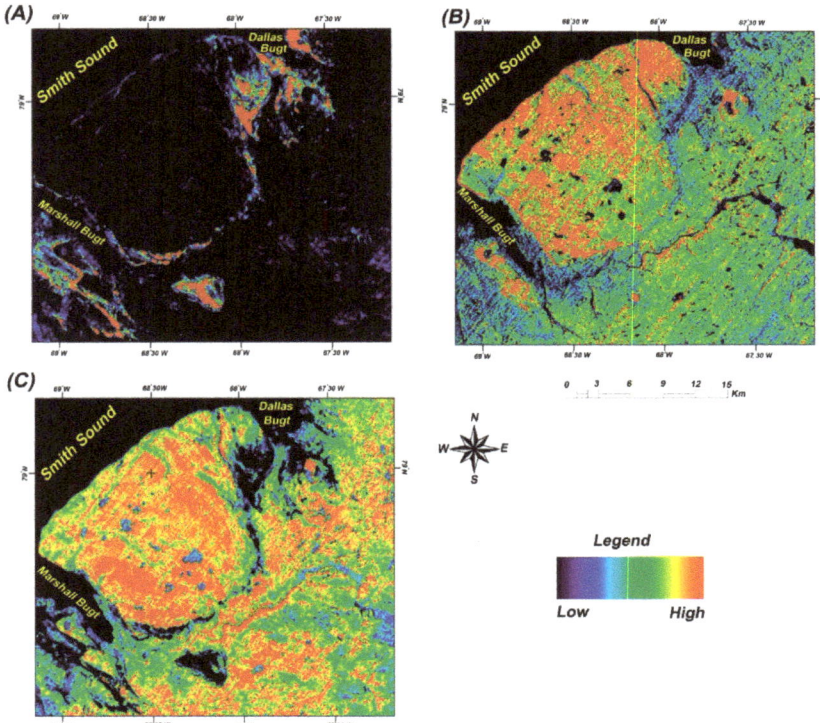

Figure 8. Pseudocolor ramp of the ASTER (TIR) DPCA rule images covering the selected spatial subset of the Cu-Au mineralization belt and surrounding areas. (**A**) Quartz Index (QI) image-map; (**B**) Carbonate Index (CI) image-map; (**C**) Mafic Index (MI) image-map.

The DPCA3 shows strong loadings for the CI (0.790280) with a positive sign and the MI (−0.530590) with a negative sign, respectively (Table 4). Therefore, carbonate minerals can be detected as bright pixels and mafic minerals as dark pixels in the DPCA3 rule image. Figure 8B displays a pseudocolor ramp of the DPCA3 for the CI component. High to moderate concentration of carbonate minerals were identified in carbonate successions of the Franklinian Basin. The Etah meta-igneous complex (orthogneiss and quartz diorite) and the Etah Group (paragneiss and amphibolite) generally show a low to moderate surface abundance of carbonate minerals. The Cu-Au mineralization belt mostly locates in a low to moderate range of carbonate content zone (Figure 8B). Moreover, a pseudocolor ramp of the MI was generated using the negation of the DPCA3 rule image (Figure 8C). Quartz-rich zones (contact boundaries of sedimentary successions with metamorphic units) appear in a very low range of mafic content in the MI image-map (Figure 8C). Mafic minerals show high to moderate ranges in the entire image-map, which are mostly concentrated in the Franklinian Basin, paragneiss, and orthogneiss units (Figure 8C).

The end-member spectra (n-D classes) extracted from the n-Dimensional analysis technique for the ASTER selected spatial subset covering the Cu-Au mineralization belt and surrounding areas are shown in Figure 9A. The n-D classes were compared with the end-member spectra of target minerals from the USGS spectral library (see Figure 3B). Results indicate that some of the n-D classes contain recognizable features similar to the target minerals. The n-D class #1 has an identical spectral signature with chalcedony and opal (see Figures 3B and 9A). Strong absorption features in bands 7, 8, and 9 could be attributed to Si-OH absorption characteristics. The n-D class #2 represents a combined spectral signature of jarosite and hematite due to Fe^{3+} (0.48 µm and 0.83–0.97 µm) and Fe-OH (2.27 µm) absorption features [89], coinciding with bands 1, 2, 3, and 7 of ASTER. The n-D class #3 and n-D class #5 do not contain any prominent spectral signatures related to the alteration minerals and can be considered as an unaltered/unknown mineral group. Snow/ice spectral signatures are recognizable in the n-D class #4 and n-D class #10 (Figure 9A). Strong reflectance in the VNIR portion (0.520–860 µm; bands 1, 2 and 3 of ASTER) and low reflectance in the SWIR portion (1.60–2.430 µm; bands 4 to 9 of ASTER) specify the snow/ice spectral properties [95]. The n-D class #6 contains spectral characteristics close to chlorite and epidote, which shows a dominant Mg, Fe-OH absorption at 2.30–2.35 µm [100] equivalent to bands 8 and 9 of ASTER. Biotite might be represented in the n-D class #7 because of slight iron absorption and a major Mg, Fe-OH absorption (Figure 9A). The n-D class #8 reveals mixed spectral features of hematite and jarosite. The n-D class #9 shows strong Al-OH spectral absorption features at 2.20 µm [89], which is related to muscovite/kaolinite spectral signatures coinciding with band 6 of ASTER.

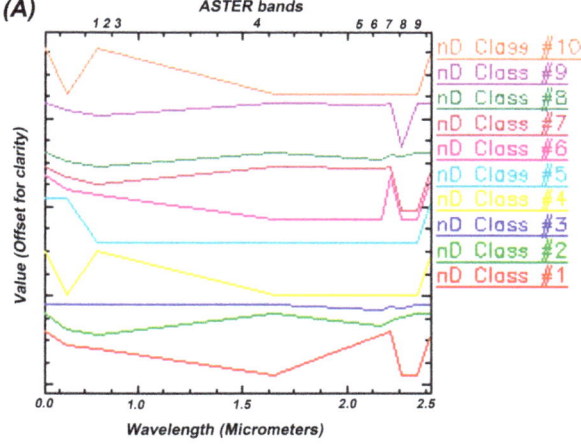

Figure 9. *Cont.*

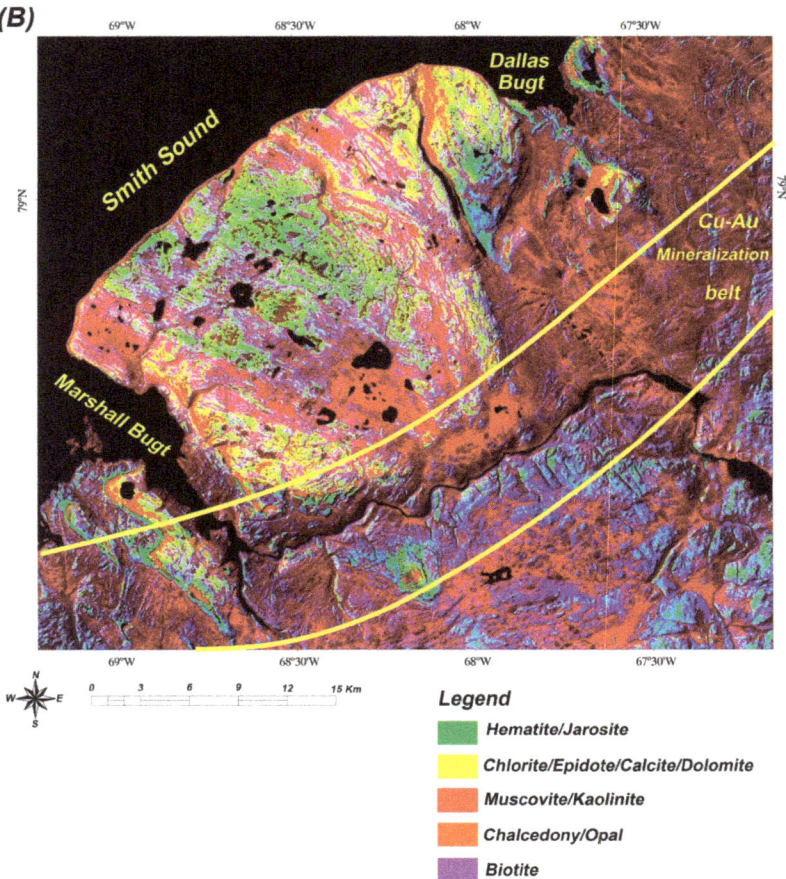

Figure 9. (**A**) The n-D classes (end-member spectra) extracted for a selected spatial subset (ASTER VNIR + SWIR) covering the Cu-Au mineralization belt and surrounding areas. ASTER band center positions are shown. (**B**) ASTER LSU classification mineral map for the selected spatial subset covering the Cu-Au mineralization belt and surrounding areas.

Figure 9B shows the LSU classification mineral map derived from fraction images of end-members (excluding snow/ice and unaltered/unknown groups) for the selected spatial subset covering the Cu-Au mineralization belt and surrounding areas. Results indicate that hematite/jarosite, muscovite/kaolinite, and biotite are spectrally strong, while chalcedony/opal and chlorite/epidote have a moderate contribution in total mixed spectral characteristics of the selected spatial subset. Comparison with the geological map of the study area (see Figure 1) suggests that muscovite/kaolinite is dominant in the Cu-Au mineralization belt, which is typically concentrated in the orthogneiss and amphibolite lithological units. In addition, a high surface abundance of biotite was mapped in both orthogneiss and paragneiss of the Etah meta-igneous complex and Etah group. The association of hematite/jarosite, chlorite/epidote, chalcedony/opal, and muscovite/kaolinite was identified in several parts of the central and southwestern sectors of the Cu-Au mineralization belt (Figure 9B), which are matched with the distribution of the main Cu-Au occurrences as documented by Pirajno et al. [2]. The Franklinian Basin sequences contain a high surface abundance of hematite/jarosite and chlorite/epidote and muscovite/kaolinite and a moderate to low surface abundance of chalcedony/opal and biotite. The high concentration of hematite/jarosite was mapped in the carbonate succession, while chlorite/epidote and muscovite/kaolinite were detected in the basal siliciclastic rocks. Chalcedony/opal is mostly concentrated at the contact between the Franklinian Basin sequences and Etah meta-igneous complex and Etah group. Low spatial distribution of biotite was detected in the basal siliciclastic rocks of the Franklinian Basin sequences (Figure 9B).

4.3. Mapping Iron Oxide/Hydroxide Minerals in the Southern Part of the Cu-Au Mineralization Belt Using WV-3 Data

A spatial selected subset of WV-3 imagery covering the southern part of the Cu-Au mineralization belt was considered (Figure 10) for mapping Fe^{3+} and Fe^{2+} iron oxides and ferric and ferrous silicates. Table 5 shows the eigenvector matrix of the WV-3 band ratio indices derived from the DPCA analysis, including B4 + B2/B3 (for mapping Fe^{3+} iron oxides), B6 + B8/B7 (for mapping Fe^{2+} iron oxides), B3 + B5/B4 (for mapping ferric silicates), and B5 + B7/B6 (for mapping ferrous silicates). The DPCA1 does not contain any specific contribution of band ratio indices with different signs (all of the eigenvector loadings are negative). Thus, this image-map contains spectral similarities and does not enhance any group of target minerals. The DPCA2 shows strong and positive eigenvector loading for mapping Fe^{3+} iron oxides (0.762743). However, the eigenvector loading for Fe^{2+} iron oxides (−0.369967) is weak and negative. The ferric (0.461865) and ferrous (0.262084) silicates have moderate to weak contribution with positive signs in the DPCA2 image. Therefore, the DPCA2 image-map represents the Fe^{3+} iron oxides as bright pixels, which might contain a very low contribution of ferric and ferrous silicates. Figure 10A shows a pseudocolor ramp of the DPCA2 covering the southern part of the Cu-Au mineralization belt, which includes two Cu-Au mineralization occurrences that have been already documented by Pirajno et al. [2]. High to moderate concentration of Fe^{3+} iron oxides is mapped in the vicinity of Cu-Au mineralization occurrences (Figure 10A). Moreover, many other parts inside the Cu-Au mineralization belt show strong to moderate spatial distribution of Fe^{3+} iron oxides (Figure 10A), which could be considered as high potential zones for Cu-Au mineralization.

The DPCA3 contains a significant contribution of Fe^{2+} iron oxides (−0.949469) and very low eigenvector loading of Fe^{3+} iron oxides (0.049338) and ferric silicates (0.090503), while a moderate contribution of ferrous silicates (0.396450) with a positive sign is present in this DPCA. Hence, the Fe^{2+} iron oxides will appear as dark pixels. The DPCA3 was negated (multiplication by −1) to generate the Fe^{2+} iron oxides as bright pixels. A pseudocolor ramp of the DPCA3 was generated to map Fe^{2+} iron oxides (Figure 10B). The high surface abundance of Fe^{2+} iron oxides was also detected proximate to the mineralization localities. For mapping ferrous silicates, a pseudocolor ramp was applied to the DPCA3 without negation (Figure 10C). Spatial distribution of ferrous silicates can be seen in many parts of the selected subset, especially in drainage systems and geological structures. However, a low concentration of the ferrous silicates is mapped close to the Cu-Au mineralization occurrences (Figure 10C).

The DPCA4 has a strong negative eigenvector loading of ferric silicates (−0.864801) and moderate positive contribution of Fe^{3+} iron oxides (0.496839), whereas eigenvector loadings for Fe^{2+} iron oxides (−0.004738) and ferrous silicates (0.072448) are meager. As a result, the ferric silicates will manifest as dark pixels in the DPCA4, which could be inverted to bright pixels by negation. The moderate contribution of Fe^{3+} iron oxides can affect the resultant map. Figure 10D shows a pseudocolor ramp for ferric silicates. In many parts, the surface abundance of ferric silicates is much stronger compared to ferrous silicates, especially adjacent to Cu-Au mineralization occurrences. The high concentration of ferric silicates shows a close spatial relationship with Fe^{3+} and Fe^{2+} iron oxides. The high to moderate surface abundance of ferric silicates was mapped nearby the Cu-Au mineralization localities in the selected subset (Figure 10D).

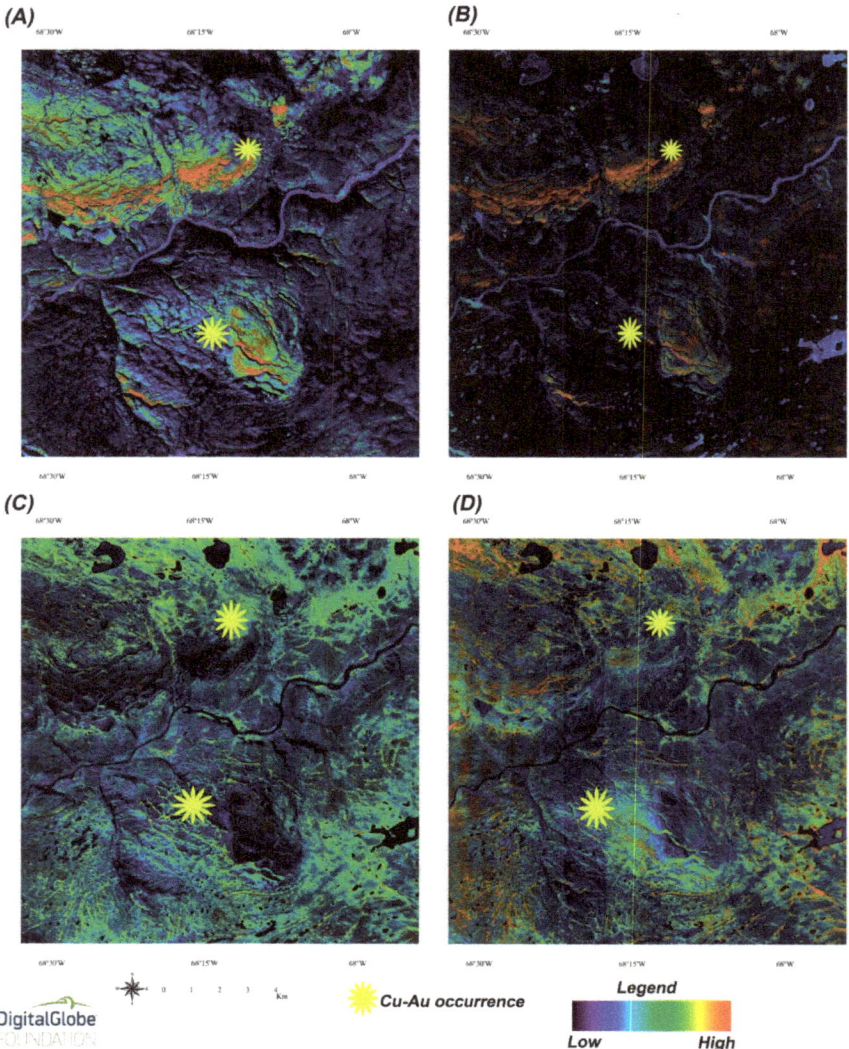

Figure 10. Pseudocolor ramp of the WV3 (VNIR) DPCA rule images covering the selected spatial subset of the southern part of the Cu-Au mineralization belt. (**A**) Fe^{3+} iron oxides image-map; (**B**) Fe^{2+} iron oxides image-map; (**C**) ferric silicates image-map; (**D**) ferrous silicates image-map (WV-3 image, courtesy of the DigitalGlobe Foundation (www.digitalglobefoundation.org)).

Table 5. Eigenvector matrix of the WV-3 band ratio indices derived from the DPCA for the selected subset covering the southern part of Cu-Au mineralization belt.

Eigenvector	B4 + B2/B3	B6 + B8/B7	B3 + B5/B4	B5 + B7/B6
DPCA 1	−0.155644	−0.927722	−0.174958	−0.290683
DPCA 2	0.762743	−0.369967	0.461865	0.262084
DPCA 3	0.049338	−0.949469	0.090503	0.396450
DPCA 4	0.496839	−0.004738	−0.864801	0.072448

End-member spectra (n-D classes) extracted from the n-Dimensional analysis technique for the WV-3 selected spatial subset of the southern part of the Cu-Au mineralization belt are presented in Figure 11A. Comparison with selected end-member reflectance spectra of the target minerals from the USGS spectral library (see Figure 3C) shows the presence of some n-D classes containing similar spectral characteristics with hematite, jarosite, ferric, and ferrous silicates. The n-D class #1, n-D class #3, n-D class #5, and n-D class #8 do not contain any particular spectral signature related to alteration minerals, which might be water/ice (snow/slush) or unknown geologic materials. The concentration of transition metal cations such as Fe^{3+} and Fe^{2+} can affect the intensities of absorption features [99]. Fe^{3+} produces absorption features near 0.45 to 0.90 μm, while broad absorption features near 0.90 to 1.2 μm are related to Fe^{2+} [100]. The n-D class #2 has absorption features related to ferric iron (Fe^{3+}), which corresponds with bands 5 (Red), 6 (Red edge), and 7 (Near-Infrared 1) of WV-3. It seems that this n-D class is related to ferric silicates. The n-D class #4 shows a similar spectral pattern with hematite (see Figures 3C and 11A). The n-D class #6 can be considered for jarosite. The n-D class #7 can be attributed to the admixture of hematite and jarosite. Charge transfer absorption features between 0.48 to 0.72 μm and crystal-field absorption properties between 0.63 to 0.72 μm are documented for iron oxide/hydroxide minerals such as hematite, limonite, goethite, and jarosite [101–103]. The n-D class #9 contains robust absorption features related to Fe^{2+}, coinciding with bands 7 (Near-Infrared 1) and 8 (Near-Infrared 2) of WV-3. Hence, it can be characterized by ferrous silicate.

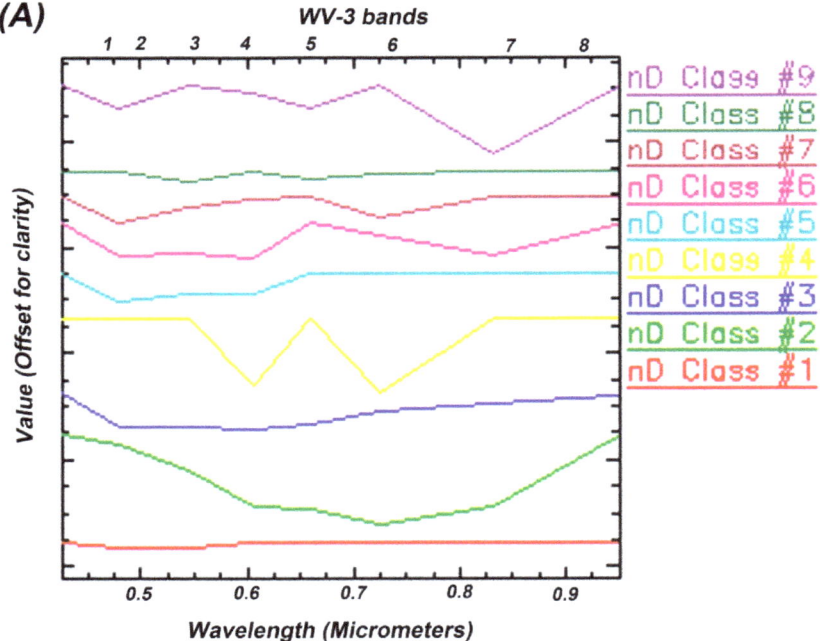

Figure 11. *Cont.*

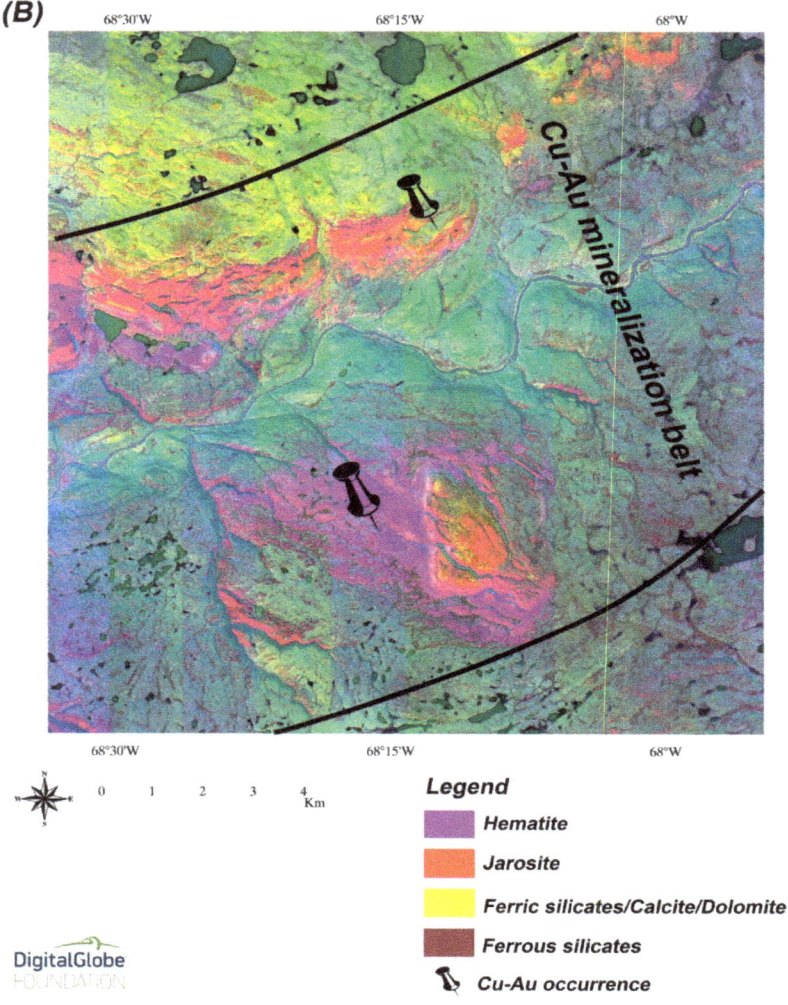

Figure 11. (**A**) The n-D classes (end-member spectra) extracted for the WV-3 selected spatial subset covering the southern part of the Cu-Au mineralization belt. WV-3 band center positions are shown. (**B**) LSU mineral map produced from fraction images for the selected spatial subset covering the southern part of the Cu-Au mineralization belt (WV-3 image, courtesy of the DigitalGlobe Foundation (www.digitalglobefoundation.org)).

The LSU spectral mineral-map of the WV-3 spatial selected subset covering the southern part of the Cu-Au mineralization belt was generated using fraction images derived from the n-D classes contain end-member reflectance spectra of the target minerals. Figure 11B shows the resultant image-map. In the vicinity of Cu-Au mineralization occurrences, high concentration of hematite, jarosite, and ferric silicates was identified. On the other hand, carbonates (calcite and dolomite) also appear in association with ferric silicate, especially in the central and northwestern parts of the selected subset. Most of the ferrous silicates are detected in the drainage systems and geological structures.

4.4. ACE Analysis for Detecting End-Member Minerals Using VINR + SWIR Bands of ASTER

For verifying the presence of mineral spectral signatures detected in the selected spatial subset covering the Cu-Au mineralization belt and surrounding areas, the ACE algorithm was applied to the VINR + SWIR bands of ASTER using laboratory reflectance spectra of hematite, jarosite, biotite, muscovite, chlorite, epidote, chalcedony (hydrous-silica), and opal (hyalite) extracted from the USGS spectral library [75]. Fraction images of the selected end-member were generated as a series of greyscale rule images using the ACE algorithm. To show the high fractional abundance (high DN value pixels) of the target minerals, a pseudo-color ramp of greyscale rule images was produced, one for each selected mineral (Figure 12). The ACE image-maps were visually compared with the LSU classification image-maps (see Figures 6B, 9B, 11B and 12). Results indicate that fractional abundances of hematite, chlorite, epidote, chalcedony, and opal are high, whereas jarosite and biotite are low in the detected altered zones. Spatial distribution of muscovite is typically different from other target minerals in the identified altered zones and selected subset (Figure 12). However, some of the high abundance zones contain jarosite, chalcedony, and opal that are spatially matched with muscovite. Comparison of the DPCA image-maps and LSU classification image-map of ASTER (see Figure 7 and Figure 9B) with the ACE fraction images indicates a little spatial dissimilarity between the DPCA4 image (Figure 7D) for mapping Mg-Fe-OH minerals and fraction images of biotite, chlorite, and epidote (Figure 12). However, the LSU classification image-map (Figure 9B) shows a high spatial similarity with fraction images of hematite, jarosite, biotite, muscovite, chlorite, epidote, chalcedony, and opal.

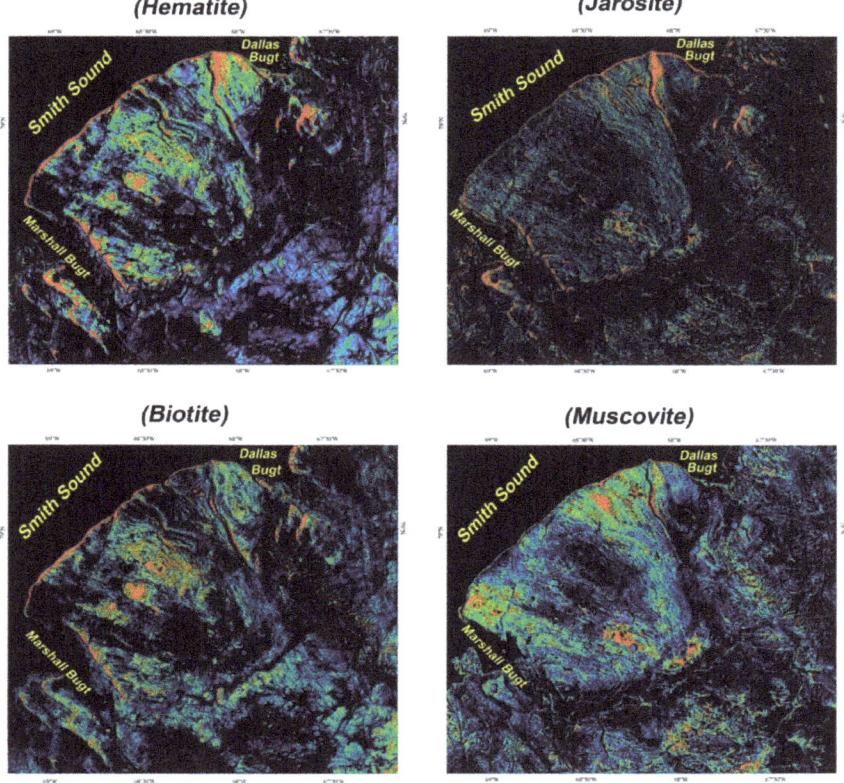

Figure 12. *Cont.*

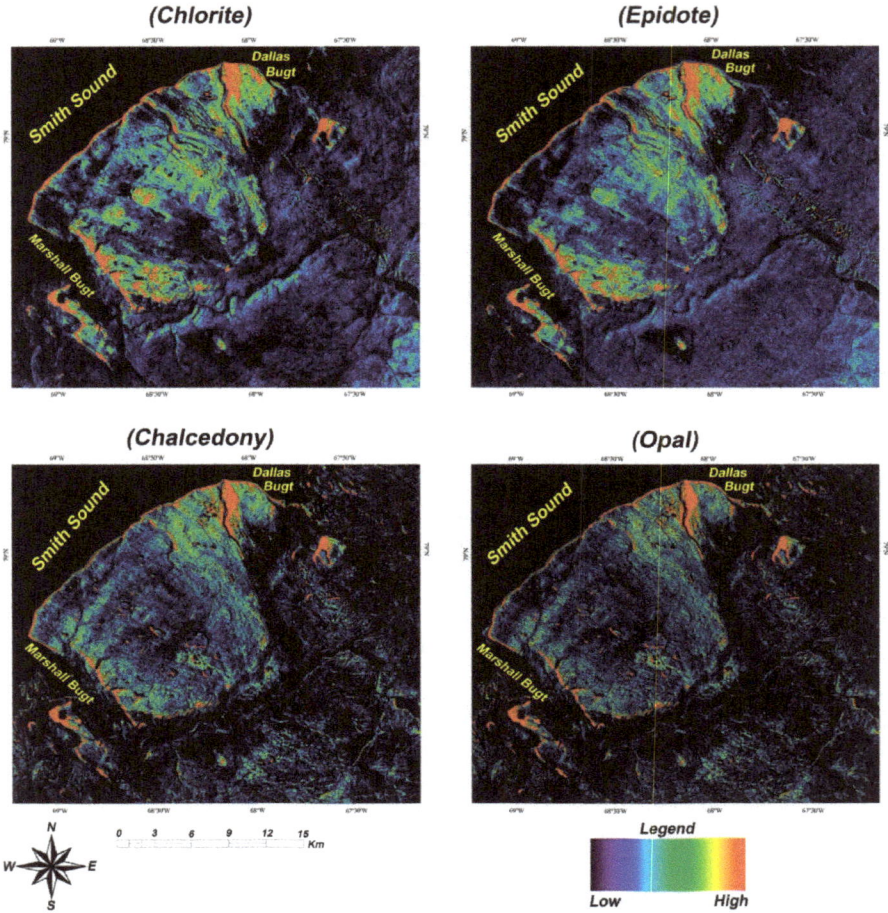

Figure 12. Fraction images of the selected end-member minerals derived from the adaptive coherence estimator (ACE) algorithm for the selected spatial subset covering the Cu-Au mineralization belt and surrounding areas. Pseudo-color ramp was applied to greyscale rule images.

4.5. Virtual Verification Assessment

Confusion matrix (error matrix) and Kappa Coefficient [101–103] were calculated for the LSU classification image-maps derived from Landsat-8, ASTER, and WV-3 versus the ACE fraction images derived from VINR + SWIR bands of ASTER (Tables 6–8). In this analysis, the confusion matrix was assumed based on one-class per pixel classifications. The pixels were selected inside the altered zones with high digital number values. The spatial resolutions of the pixels were considered and resampled to a similar size to the ACE fraction images using a pixel aggregation (neighborhood averaging). Furthermore, highly dissimilar pixels were excluded using a standard deviation threshold. Finally, 160 pixels of Landsat-8, 300 pixels of ASTER, and 200 pixels of WV-3 were selected and analyzed, respectively (Tables 6–8).

Table 6. Confusion matrix for the LSU classification image-maps derived from Landsat-8 versus the ACE fraction images derived from VINR + SWIR bands of ASTER.

LSU Classification Map Landsat-8	Detected Pixel Spectra by the ACE Algorithm				
	Iron Oxide/Hydroxides	Clay Minerals	Ferrous Silicates	Totals	User's Accuracy
Iron oxide/hydroxides	46	2	8	56	82%
Clay minerals	2	48	4	54	88%
Ferrous silicates	12	10	28	50	56%
Totals	60	60	40	160	
Producer's Accuracy	76%	80%	70%		
Overall accuracy = 76.25%			Kappa Coefficient = 0.64		

Table 7. Confusion matrix for the LSU classification image-maps derived from ASTER versus the ACE fraction images derived from VINR + SWIR bands of ASTER.

LSU Classification Map ASTER	Detected Pixel Spectra by the ACE Algorithm						
	Hematite/Jarosite	Chlorite/Epidote	Muscovite/Kaolinite	Chalcedony/Opal	Biotite	Totals	User's Accuracy
Hematite/jarosite	42	8	3	10	8	71	59%
Chlorite/epidote	6	39	1	7	6	59	66%
Muscovite/kaolinite	0	1	43	2	5	51	84%
Chalcedony/opal	7	8	8	38	6	67	56%
Biotite	5	4	5	3	35	52	67%
Totals	60	60	60	60	60	300	
Producer's Accuracy	70%	65%	71%	63%	58%		
Overall accuracy = 65.66%			Kappa Coefficient = 0.57				

Table 8. Confusion matrix for the LSU classification image-maps derived from WV-3 versus the ACE fraction images derived from VINR + SWIR bands of ASTER.

LSU Classification Map WV-3	Detected Pixel Spectra by the ACE Algorithm					
	Hematite	Jarosite	Ferric Silictes	Ferrous Silicates	Totals	User's Accuracy
Hematite	39	6	5	1	51	76%
Jarosite	7	40	4	3	54	74%
Ferric Silictes	3	4	38	9	54	70%
Ferrous Silicates	1	0	3	37	41	90%
Totals	50	50	50	50	200	
Producer's Accuracy	78%	80%	76%	74%		
Overall accuracy = 77%			Kappa Coefficient = 0.69			

Table 6 shows confusion matrix for the LSU classification image-maps derived from Landsat-8 versus the ACE fraction images derived from VINR + SWIR bands of ASTER. The overall accuracy and Kappa Coefficient are 76.25% and 0.64, respectively. Producer's accuracy (omission error) indicates the probability of a reference pixel being correctly classified and user's accuracy (commission error) shows the total number of correct pixels in a category, which is divided by a total number of pixels that were classified in the category [104,105]. The highest producer's accuracy (80%) and user's accuracy (88%) were achieved for the clay minerals class. However, the lowest producer's accuracy (70%) and user's accuracy (56%) were recorded for the ferrous silicates class. It shows that spectral mixing and confusion between the ferrous silicates and iron oxide/hydroxides classes is more feasible than the clay minerals class using Landsat-8 spectral bands.

The overall accuracy of 65.66% and Kappa Coefficient of 0.57 were assessed for the LSU classification image-maps versus the ACE fraction images derived from VINR + SWIR bands of ASTER (Table 7). The muscovite/kaolinite class has the highest producer's accuracy (71%) and user's accuracy (84%). The biotite class shows the lowest producer's accuracy (58%) and the chalcedony/opal class contains the lowest user's accuracy (56%). So, the muscovite/kaolinite class pixels were strongly mapped compared to other mineral classes in this study. Spectral mixing has been recorded for pixels contain hematite/jarosite, chlorite/epidote, chalcedony/opal, and biotite mineral assemblages. Consequently, detecting the subtle spectral differences between alteration mineral classes are challenging and might have some confusion using ASTER data. Calculation of confusion matrix for LSU classification image-maps derived from WV-3 versus the ACE fraction images derived from VINR + SWIR bands of ASTER indicates the overall accuracy of 77% and Kappa Coefficient of 0.69 (Table 8). Producer's accuracy and user's accuracy for ferrous silicates class pixels are 74% and 90%, respectively. The jarosite class pixels contain the highest producer's accuracy (80%). The ferric silicates class pixels show the lowest user's accuracy (70%). Results indicate some spectral mixing effects between mineral classes, but the overall accuracy (77%) and Kappa Coefficient (0.69) have a good potential for separating the classes using WV-3 VNIR spectral bands.

5. Discussion

Mineral exploration is very challenging in the Arctic regions due to cold and harsh environments and inaccessibility, especially in the northern part of Greenland that contains a variety of ore mineral resources [3,104]. Application of remote sensing satellite/airborne imagery for mineral identification, exploration, and prospecting in Greenland has been documented in the Sarfartoq carbonatite complex, southern West Greenland [105,106] and the Kap Simpson complex area, East Greenland [107] as well as the Franklinian Basin, North Greenland [8]. The Inglefield Mobile Belt (IMB), Northwest Greenland contains copper-gold mineralization hosted by garnet-sillimanite paragneiss, orthogneiss, and mafic-ultramafic rocks, which are confined in hydrothermal alteration zones (rust zones) [1–3]. In this study, the application of Landsat-8, ASTER, and WV-3 multispectral satellite remote sensing data were evaluated for mapping hydrothermal alteration minerals associated with Cu-Au mineralization in the IMB.

Using ratio indices of Landsat-8 spectral bands (B4/B2, B4/B6, and B6/B7) discriminate a variety of sedimentary, metamorphic, and igneous lithological units at the regional scale based on different content of iron oxides/hydroxides, ferrous iron oxides, and hydroxyl minerals (see Figure 5A). The sedimentary successions of the Franklinian Basin and Thule Supergroup were mapped due to high amounts of Al-OH, Mg-Fe-OH, CO_3, and Si-OH mineral groups related to carbonate and siliciclastic rocks and Fe^{2+} absorption that might be attributed to dolomitization. Basaltic sills of the Thule Basin, paragneiss of the Etah Group, and syenite and orthogneiss of the Etah meta-igneous complex were discriminated because of different surface abundance of Fe^{3+} and Fe^{2+} iron oxides/hydroxide minerals (see Figure 5A). Quaternary deposits were mapped owing to the high surface distribution of detrital clay minerals. The DPCA3 and DPCA4 images derived from Landsat-8 band ratio indices identified Fe^{3+}/Fe^{2+} iron oxides/hydroxide minerals and Al-OH, Mg-Fe-OH, CO_3, and Si-OH alteration mineral groups, respectively (see Figure 5B,C).

In the DPCA3 image-map (Figure 5B), the boundaries between the Etah metamorphic complex rocks and sedimentary successions of the Franklinian Basin and Thule Supergroup in the Central Terrane, as well as the southern part of the Cu-Au mineralization belt nearby Marshall Bugt, show high surface abundance of iron oxide/hydroxide minerals. These locations are typically matched with the documented rust zones, which are identified as Cu-Au sulfide mineralization areas [1–3]. Furthermore, a high concentration of the OH-alteration mineral groups was mapped in the DPCA4 image-map (see Figure 5C) that could be considered with some parts of the rust zones. The XRD analyses, as documented by Pirajno et al. [2] for mineralogy of rust zones indicate the presence of biotite, sericite, and chlorite. High spatial distribution of iron oxide/hydroxide minerals along the boundaries between the metamorphic complex rocks and sedimentary successions in the southwestern part of the Cu-Au mineralization belt was also detected in the LSU spectral mineral-map of the Landsat-8 (see Figure 6B). Ferrous silicates (biotite, chlorite, and epidote) were typically mapped with iron oxide/hydroxide minerals, while clay minerals, detected in the central part of the Cu-Au mineralization belt, are mostly adjacent to the amphibolite and quartz diorite lithological units (see Figures 1 and 6B).

Detailed maps of the spatial distribution of Fe^{3+}/Fe^{2+} iron oxide/hydroxides, Al/Fe-OH, Mg-Fe-OH, and Si-OH minerals in the Cu-Au mineralization belt and surrounding areas (see Figure 7) were generated by implementing the DPCA technique to ASTER band ratio components (B2/B1, B4/B2, B5 + B7/B6, B7 + B9/B8, and B6 + B8/B7). The DPCA2 mapped the Fe^{3+}/Fe^{2+} iron oxide/hydroxide minerals, which are highly concentrated at the contact of metamorphic complex rocks with the Franklinian sedimentary successions and orthogneiss in the northeastern and southern parts of the Cu-Au mineralization belt (see Figure 7A,B). Numerous zones of high to moderate concentration of iron oxide/hydroxide minerals were mapped inside the Cu-Au mineralization belt together with the rust zones. The occurrence of iron minerals such as rozenite, jarosite, cacoxenite, and jahnsite was reported in the Cu-Au mineralization belt as oxidation products of sulfide minerals associated with the rust zones [2]. The DPCA3 and DPCA4 images represented the spatial distribution of Al/Fe-OH and Mg-Fe-OH minerals, which show low abundances in the mineralization belt (Figure 7C,D). The DPCA5 detected Si-OH minerals and ferrous silicates, which are typically associated with quartz diorite and the contact between metamorphic complex rocks with the Franklinian sedimentary successions (Figure 7E). The high concentration of Si-OH minerals was mapped as associated with rust zones, particularly in the southwestern part of the mineralization belt. Pirajno et al. [2] documented the association of hydrolitic alteration assemblages (chlorite and biotite) and silicification that overprint the wallrocks and rust zones in the Cu-Au mineralization belt.

The implementation of the DPCA to ASTER TIR band ratio indices (QI, CI, and MI) provided complementary information for mapping of altered, silica-rich rocks (containing SiO2 group), carbonates, and mafic minerals in the DPCA2 and DPCA3 (Figure 8A–C). The boundaries of orthogneiss with the Franklinian Basin successions and quartz diorite units show a high to moderate concentration of quartz content, which is matched with the DPCA5 derived from the ASTER VNIR + SWIR ratio indices (see Figure 7E). In the Cu-Au mineralization belt, several zones containing high concentration of quartz content were identified (see Figure 8A). Carbonate minerals were clearly detected in carbonate successions of the Franklinian Basin, while mafic minerals were mostly mapped in the paragneiss and orthogneiss units (see Figure 8A,B). Boundaries of sedimentary successions with metamorphic rocks show a very low range of carbonates and mafic minerals. According to Pirajno et al. [1,2] and Kolb et al. [3] Cu–Au mineralization in rust zones is restricted to the NE-trending strip, which has a close spatial relationship with the contact of carbonate successions of the Franklinian Basin and the basement metamorphic rocks.

Hematite/jarosite, muscovite/kaolinite, and biotite are spectrally dominated in the ASTER LSU classification mineral map (Figure 9A), whereas chalcedony/opal and chlorite/epidote have a moderate contribution in the total mixed spectral properties. The assemblage of hematite/jarosite, chlorite/epidote, chalcedony/opal, and muscovite/kaolinite was detected in many parts of the Cu-Au mineralization belt (Figure 9B), especially in the central and southwestern parts, where the main occurrences of Cu-Au mineralization were reported by Pirajno et al. [2]. Muscovite/kaolinite has a high surface abundance in the Cu-Au mineralization belt, which is typically concentrated in the orthogneiss and amphibolite lithological units. Chalcedony/opal is generally concentrated at the contact between the Franklinian Basin sequences and basement metamorphic complex (Figure 9B). The spatial distribution of the alteration minerals in the ASTER LSU classification image-map was comparable with ASTER DPCA image-maps, however, a detailed surface abundance of alteration minerals was more apparent in the LSU classification image-map (see Figure 7A–E and Figure 9B).

Fe^{3+} and Fe^{2+} iron oxides and ferric and ferrous silicates were comprehensively mapped in the southern part of the Cu-Au mineralization belt by applying DPCA to WV-3 band ratio indices (see Figure 10 and Table 5). High to moderate surface abundance of Fe^{3+} iron oxides was mapped near to Cu-Au mineralization occurrences (reported by Pirajno et al. [2]) in the DPCA2 image-map (Figure 10A). Furthermore, ferric silicates and Fe^{2+} iron oxides were also mapped in the vicinity of Cu-Au mineralization occurrences, which are recorded in the DPCA3 and DPCA4 image-maps (Figure 10B–D). A number of zones containing high to moderate spatial distribution of Fe^{3+} and Fe^{2+} iron oxides and ferric silicates are recorded as feasible Cu-Au mineralization occurrences. The LSU spectral mineral-map of the WV-3 shows spatial distribution of hematite, jarosite, ferric silicates/calcite/dolomite, and ferrous silicates (see Figure 11B). The high concentration of hematite, jarosite, and ferric silicates was mapped in the vicinity of Cu-Au mineralization occurrences, which is coincident with the DPCA image-map (see Figure 10). As stated by Pirajno et al. [2], the whole-rock XRD analyses of the rust zones have shown hydrous Fe sulfate and phosphate, jarosite, biotite, sericite, and chlorite, which are paralleled with the remote sensing results derived from WV-3 VNIR data.

The presence of hematite, jarosite, biotite, muscovite, chlorite, epidote, chalcedony, and opal in the selected spatial subset covering the Cu-Au mineralization belt and surrounding areas was verified using the ACE fraction images (see Figure 12). Hematite, chlorite, epidote, chalcedony, and opal show high surface abundances in the altered zones, while jarosite, biotite, and muscovite are lesser in the altered zones and they are mostly associated with specific lithological units in the study area. The DPCA image-maps of ASTER dataset show a little spatial dissimilarity with the ACE fraction images, especially in the DPCA4 image (Figures 7 and 12). High spatial similarity with fraction images was recorded in the LSU classification image-map (see Figure 9B). The overall accuracy and Kappa Coefficient calculated for the LSU classification image-maps derived from Landsat-8 versus the ACE fraction images derived from VINR + SWIR bands of ASTER were 76.25% and 0.64, respectively (see Table 6). The overall accuracy of 65.66% and Kappa Coefficient of 0.57 were assessed for the ASTER LSU classification image-maps (see Table 7). Using ASTER datasets, muscovite/kaolinite was intensely mapped compared to hematite/jarosite, chlorite/epidote, chalcedony/opal, and biotite. On the other hand, spectral mixing for hematite/jarosite, chlorite/epidote, chalcedony/opal, and biotite was more feasible.

The overall accuracy of 77% and Kappa Coefficient of 0.69 were calculated for the WV-3 LSU classification image-maps (see Table 8), which show a good potential for separating iron mineral classes. Subsequently, the virtual verification indicates that the alteration zones mapped by the Landsat-8, ASTER, and WV-3 datasets reveal a good rate of agreement (Kappa Coefficient of 0.57 to 0.69) and reasonable accuracy (overall accuracy of 65.66% to 77%), which could be pondered for prospecting Cu-Au mineralization. As a result, the boundaries between the Etah metamorphic and meta-igneous complex and sedimentary successions of the Franklinian Basin in the Central Terrane, orthogneiss in the northeastern part of the Cu-Au mineralization belt adjacent to Dallas Bugt, as well as the southern part of the Cu-Au mineralization belt nearby Marshall Bugt, can be considered as high potential zones for Cu-Au prospecting in the IMB.

6. Conclusions

Landsat-8, ASTER, and WV-3 multispectral remote sensing datasets were processed, interpreted, and integrated for mapping hydrothermal alteration minerals and prospecting Cu-Au mineralization in the IMB, Northwest Greenland. Iron oxides/hydroxide minerals and Al-OH, Mg-Fe-OH, CO_3 and Si-OH/SiO$_2$ alteration mineral groups were mapped by executing the DPCA, LSU, and ACE image processing techniques to the Landsat-8, ASTER, and WV-3 datasets. The discrimination of lithological units and the zones contain high concentration of iron oxides/hydroxide and clay minerals in the IMB were achieved using Landsat-8 data at the regional scale. The information extracted from Landsat-8 provides a synoptic view of alteration mineral zones in the IMB metallogenic province. Iron oxides/hydroxide minerals typically concentrated at the contact between sedimentary successions of the Franklinian Basin and Thule Supergroup with the Etah metamorphic and meta-igneous complex rocks. ASTER datasets helped to map the spatial distribution of Fe^{3+}/Fe^{2+} iron oxide/hydroxides, Al/Fe-OH, Mg-Fe-OH, Si-OH/SiO2 mineral groups in the Cu-Au mineralization belt and surrounding areas, comprehensively. Fe^{3+}/Fe^{2+} iron oxide/hydroxides and Si-OH/SiO$_2$ were also detected in the contact between sedimentary successions and metamorphic and meta-igneous rocks, orthogneiss, and quartz diorite. Intense concentration of iron oxide/hydroxides and Si-OH/SiO2 was identified within documented rust zones (Cu-Au mineralization).

Furthermore, fraction abundance of hematite, jarosite, biotite, muscovite, chlorite, epidote, chalcedony, and opal was detected in the Cu-Au mineralization belt and surrounding areas using the VNIR + SWIR bands of ASTER. Hence, the rust zones contain the assemblage of hematite/jarosite, chalcedony/opal, and chlorite/epidote with little amount of muscovite/kaolinite. Using the WV-3 dataset, Fe^{3+} and Fe^{2+} iron oxides and ferric and ferrous silicates were broadly mapped and discriminated in the southern part of the Cu-Au mineralization belt. High to moderate spatial distribution of Fe^{3+} and Fe^{2+} iron oxides and ferric silicates were detected in the rust zones. Strong fraction abundance of hematite, jarosite, and ferric silicates was also mapped in the rust zones. The virtual verification shows an appropriate overall accuracy and reasonable rate of agreement for mapping alteration mineral zones using image processing techniques and remote sensing multispectral/multi-sensor satellite imagery. Consequently, high potential zones for Cu-Au prospecting were identified in the IMB, Northwest Greenland, including (i) the boundaries between the Etah metamorphic and meta-igneous complex rocks and sedimentary successions of the Franklinian Basin in the Central Terrane, (ii) orthogneiss in the northeastern part of the Cu-Au mineralization belt adjacent to Dallas Bugt, and (iii) the southern part of the Cu-Au mineralization belt nearby Marshall Bugt. It is recommended that these high prospective zones be considered for future comprehensive fieldwork and detailed geophysical and geochemical surveys in the IMB, Northwest Greenland. This investigation suggests the necessity of multispectral/multi-sensor satellite image processing analysis as a cost-effective tool for mining companies for reconnaissance stages of mineral prospecting before costly fieldwork, geophysical, and geochemical surveys in remote and inaccessible metallogenic provinces around the world.

Author Contributions: A.B.P. wrote the manuscript and analyzed the data; T.-Y.S.P. and Y.P. and J.K.H. supervision and funding acquisition; A.L. and L.C. and A.M.M. and B.Z. and B.P. writing, editing; and O.R. and M.H. and M.S.H. image processing and analysis.

Funding: This study was conducted as a part of KOPRI research grant PE19160.

Acknowledgments: We are thankful to the Korea Polar Research Institute (KOPRI) for providing all the facilities for this investigation. The DigitalGlobe Foundation was ethically acknowledged for providing and granting WV-3 data used in this study.

Conflicts of Interest: The authors declare no conflict of interest.

References

1. Pirajno, F.; Thomassen, B.; Iannelli, T.R.; Dawes, P.R. Copper—Gold mineralisation in Inglefield Land, NW Greenland. *Newsl. Int. Liaison Group Gold Miner.* **2000**, *30*, 49–53.
2. Pirajno, P.; Thomassen, B.; Dawes, P.R. Copper–gold occurrences in the Palaeoproterozoic Inglefield mobile belt, northwest Greenland: A new mineralisation style? *Ore Geol. Rev.* **2003**, *22*, 225–249. [CrossRef]
3. Kolb, J.; Keiding, J.K.; Steenfelt, A.; Secher, K.; Keulen, N.; Rosa, D.; Stengaard, B.M. Metallogeny of Greenland. *Ore Geol. Rev.* **2016**, *78*, 493–555. [CrossRef]
4. Gabr, S.S.; Hassan, S.M.; Sadek, M.F. Prospecting for new gold-bearing alteration zones at El-Hoteib area, South Eastern Desert, Egypt, using remote sensing data analysis. *Ore Geol. Rev.* **2015**, *71*, 1–13. [CrossRef]
5. Amer, R.; El Mezayen, A.; Hasanein, M. ASTER spectral analysis for alteration minerals associated with gold mineralization. *Ore Geol. Rev.* **2016**, *75*, 239–251. [CrossRef]
6. Pour, A.B.; Park, Y.; Park, T.S.; Hong, J.K.; Hashim, M.; Woo, J.; Ayoobi, I. Evaluation of ICA and CEM algorithms with Landsat-8/ASTER data for geological mapping in inaccessible regions. *Geocarto Int.* **2018**. [CrossRef]
7. Pour, A.B.; Park, Y.; Park, T.S.; Hong, J.K.; Hashim, M.; Woo, J.; Ayoobi, I. Regional geology mapping using satellite-based remote sensing approach in Northern Victoria Land, Antarctica. *Polar Sci.* **2018**, *16*, 23–46. [CrossRef]
8. Pour, A.B.; Park, T.S.; Park, Y.; Hong, J.K.; Zoheir, B.; Pradhan, B.; Ayoobi, I.; Hashim, M. Application of multi-sensor satellite data for exploration of Zn-Pb sulfide mineralization in the Franklinian Basin, North Greenland. *Remote Sens.* **2018**, *10*, 1186. [CrossRef]

9. Pour, A.B.; Hashim, M.; Park, Y.; Hong, J.K. Mapping alteration mineral zones and lithological units in Antarctic regions using spectral bands of ASTER remote sensing data. *Geocarto Int.* **2018**, *33*, 1281–1306. [CrossRef]
10. Testa, F.J.; Villanueva, C.; Cooke, D.R.; Zhang, L. Lithological and hydrothermal alteration mapping of epithermal, porphyry and tourmaline breccia districts in the Argentine Andes using ASTER imagery. *Remote Sens.* **2018**, *10*, 203. [CrossRef]
11. Sheikhrahimi, A.; Pour, B.A.; Pradhan, B.; Zoheir, B. Mapping hydrothermal alteration zones and lineaments associated with orogenic gold mineralization using ASTER remote sensing data: A case study from the Sanandaj-Sirjan Zone, Iran. *Adv. Space Res.* **2019**, *63*, 3315–3332. [CrossRef]
12. Noori, L.; Pour, B.A.; Askari, G.; Taghipour, N.; Pradhan, B.; Lee, C.-W.; Honarmand, M. Comparison of Different Algorithms to Map Hydrothermal Alteration Zones Using ASTER Remote Sensing Data for Polymetallic Vein-Type Ore Exploration: Toroud–Chahshirin Magmatic Belt (TCMB), North Iran. *Remote Sens.* **2019**, *11*, 495. [CrossRef]
13. Rajendran, S.; Nasir, S. Characterization of ASTER spectral bands for mapping of alteration zones of volcanogenic massive sulphide deposits. *Ore Geol. Rev.* **2017**, *88*, 317–335. [CrossRef]
14. Salehi, T.; Tangestani, M.H. Large-scale mapping of iron oxide and hydroxide minerals of Zefreh porphyry copper deposit, using Worldview-3 VNIR data in the Northeastern Isfahan, Iran. *Int. J. Appl. Earth Obs. Geoinf.* **2018**, *73*, 156–169. [CrossRef]
15. Pour, A.B.; Hashim, M.; Hong, J.K.; Park, Y. Lithological and alteration mineral mapping in poorly exposed lithologies using Landsat-8 and ASTER satellite data: North-eastern Graham Land, Antarctic Peninsula. *Ore Geol. Rev.* **2019**, *108*, 112–133. [CrossRef]
16. Pour, A.B.; Park, Y.; Crispini, L.; Läufer, A.; Hong, J.K.; Park, T.-Y.S.; Zoheir, B.; Pradhan, B.; Muslim, A.M.; Hossain, M.S.; et al. Mapping Listvenite Occurrences in the Damage Zones of Northern Victoria Land, Antarctica Using ASTER Satellite Remote Sensing Data. *Remote Sens.* **2019**, *11*, 1408. [CrossRef]
17. Safari, M.; Maghsodi, A.; Pour, A.B. Application of Landsat-8 and ASTER satellite remote sensing data for porphyry copper exploration: A case study from Shahr-e-Babak, Kerman, south of Iran. *Geocarto Int.* **2018**, *33*, 1186–1201. [CrossRef]
18. Ninomiya, Y.; Fu, B. Thermal infrared multispectral remote sensing of lithology and mineralogy based on spectral properties of materials. *Ore Geol. Rev.* **2019**, *108*, 54–72. [CrossRef]
19. Bedini, E. Application of WorldView-3 imagery and ASTER TIR data to map alteration minerals associated with the Rodalquilar gold deposits, southeast Spain. *Adv. Space Res.* **2019**, *63*, 3346–3357. [CrossRef]
20. Sun, L.; Khan, S.; Shabestari, P. Integrated Hyperspectral and Geochemical Study of Sediment-Hosted Disseminated Gold at the Goldstrike District, Utah. *Remote Sens.* **2019**, *11*, 1987. [CrossRef]
21. Zoheir, B.; Emam, A.; Abdel-Wahed, M.; Soliman, N. Multispectral and Radar Data for the Setting of Gold Mineralization in the South Eastern Desert, Egypt. *Remote Sens.* **2019**, *11*, 1450. [CrossRef]
22. Zoheir, B.; El-Wahed, M.A.; Pour, A.B.; Abdelnasser, A. Orogenic Gold in Transpression and Transtension Zones: Field and Remote Sensing Studies of the Barramiya–Mueilha Sector, Egypt. *Remote Sens.* **2019**, *11*, 2122. [CrossRef]
23. Leverington, D.W.; Moon, W.M. Landsat-TM-Based discrimination of Lithological units associated with the Purtuniq ophiolite, Quebec, Canada. *Remote Sens.* **2012**, *4*, 1208–1231. [CrossRef]
24. He, J.; Harris, J.R.; Sawada, M.; Behnia, P. A comparison of classification algorithms using Landsat-7 and Landsat-8 data for mapping lithology in Canada's Arctic. *Int. J. Remote Sens.* **2015**, *36*, 2252–2276. [CrossRef]
25. Pour, B.A.; Hashim, M.; Marghany, M. Exploration of gold mineralization in a tropical region using Earth Observing-1 (EO1) and JERS-1 SAR data: A case study from Bau gold field, Sarawak, Malaysia. *Arabian J. Geosci.* **2014**, *7*, 2393–2406. [CrossRef]
26. Pour, B.A.; Hashim, M.; van Genderen, J. Detection of hydrothermal alteration zones in a tropical region using satellite remote sensing data: Bau gold field, Sarawak, Malaysia. *Ore Geol. Rev.* **2013**, *54*, 181–196. [CrossRef]
27. Askari, G.; Pour, A.B.; Pradhan, B.; Sarfi, M.; Nazemnejad, F. Band Ratios Matrix Transformation (BRMT): A Sedimentary Lithology Mapping Approach Using ASTER Satellite Sensor. *Sensors* **2018**, *18*, 3213. [CrossRef]
28. Kurata, K.; Yamaguchi, Y. Integration and Visualization of Mineralogical and Topographical Information Derived from ASTER and DEM data. *Remote Sens.* **2019**, *11*, 162. [CrossRef]

29. Guha, A.; Yamaguchi, Y.; Chatterjee, S.; Rani, K.; Vinod Kumar, K. Emittance Spectroscopy and Broadband Thermal Remote Sensing Applied to Phosphorite and Its Utility in Geoexploration: A Study in the Parts of Rajasthan, India. *Remote Sens.* **2019**, *11*, 1003. [CrossRef]
30. Irons, J.R.; Dwyer, J.L.; Barsi, J.A. The next Landsat satellite: The Landsat Data Continuity Mission. *Remote Sens. Environ.* **2012**, *145*, 154–172. [CrossRef]
31. Roy, D.P.; Wulder, M.A.; Loveland, T.A.; Woodcock, C.E.; Allen, R.G.; Anderson, M.C.; Helder, D.; Irons, J.R.; Johnson, D.M.; Kennedy, R.; et al. Landsat-8: Science and product vision for terrestrial global change research. *Remote Sens. Environ.* **2014**, *145*, 154–172. [CrossRef]
32. Bhambri, R.; Bolch, T.; Chaujar, R.K. Mapping of debris-covered glaciers in the Garhwal Himalayas using ASTER DEMs and thermal data. *Int. J. Remote Sens.* **2011**, *32*, 8095–8119. [CrossRef]
33. Shukla, A.; Arora, M.K.; Gupta, R.P. Synergistic approach for mapping debris-covered glaciers using optical–thermal remote sensing data with inputs from geomorphometric parameters. *Remote Sens. Environ.* **2010**, *114*, 1378–1387. [CrossRef]
34. Abrams, M.; Hook, S.J. Simulated ASTER data for geologic studies. *IEEE Trans. Geosci. Remote Sens.* **1995**, *33*, 692–699. [CrossRef]
35. Hunt, G.R.; Ashley, R.P. Spectra of altered rocks in the visible and near-infrared. *Econ. Geol.* **1979**, *74*, 1613–1629. [CrossRef]
36. Clark, R.N. Spectroscopy of rocks and minerals, and principles of spectroscopy. In *Manual of Remote Sensing*; Rencz, A., Ed.; Wiley and Sons Inc.: New York, NY, USA, 1999; Volume 3, pp. 3–58.
37. Cloutis, E.A.; Hawthorne, F.C.; Mertzman, S.A.; Krenn, K.; Craig, M.A.; Marcino, D.; Methot, M.; Strong, J.; Mustard, J.F.; Blaney, D.L. Detection and discrimination of sulfate minerals using reflectance spectroscopy. *Icarus* **2006**, *184*, 121–157. [CrossRef]
38. Salisbury, J.W.; D'Aria, D.M. Emissivity of terrestrial material in the 8–14 μm atmospheric window. *Remote Sens. Environ.* **1992**, *42*, 83–106. [CrossRef]
39. Salisbury, J.W.; Walter, L.S. Thermal infrared (2.5–13.5 μm) spectroscopic remote sensing of igneous rock types on particulate planetary surfaces. *J. Geophys. Res.* **1989**, *94*, 9192–9202. [CrossRef]
40. Ninomiya, Y. Quantitative estimation of SiO_2 content in igneous rocks using thermal infrared spectra with a neural network approach. *IEEE TGRS* **1995**, *33*, 684–691.
41. Ninomiya, Y.; Fu, B. Regional lithological mapping using ASTER-TIR data: Case study for the Tibetan Plateau and the surrounding area. *Geosciences* **2016**, *6*, 39. [CrossRef]
42. DigitalGlobe. WorldView-3 Datasheet. 2014. Available online: https://www.digitalglobe.com/sites/default/files/DG_WorldView3_DS_forWeb_0.pdf (accessed on 7 September 2019).
43. Kruse, F.; Perry, S. Mineral mapping using simulated Worldview-3 short-wave infrared imagery. *Remote Sens.* **2013**, *5*, 2688–2703. [CrossRef]
44. Kruse, F.A.; Baugh, M.W.; Perry, S.L. Validation of DigitalGlobe Worldview-3 earth imaging satellite shortwave infrared bands for mineral mapping. *J. Appl. Remote Sens.* **2015**, *9*, 1–18. [CrossRef]
45. Asadzadeh, S.; Filho, C.R.S. Investigating the capability of WorldView-3 superspectral data for direct hydrocarbon detection. *Remote Sens. Environ.* **2016**, *173*, 162–173. [CrossRef]
46. Mars, J.C. Mineral and Lithologic Mapping Capability of WorldView 3 Data at Mountain Pass, California, Using True- and False-Color Composite Images, Band Ratios, and Logical Operator Algorithms. *Econ. Geol.* **2018**, *113*, 1587–1601. [CrossRef]
47. Sun, Y.; Tian, S.; Di, B. Extracting mineral alteration information using Worldview-3 data. *Geosci. Front.* **2017**, *8*, 1051–1062. [CrossRef]
48. Ye, B.; Tian, S.H.; Ge, J.; Sun, Y. Assessment of WorldView-3 data for lithological mapping. *Remote Sens.* **2017**, *9*, 1132. [CrossRef]
49. Dawes, P.R.; Frisch, T.; Garde, A.A.; Iannelli, T.R.; Ineson, J.R.; Pirajno, F.; Sønderholm, M.; Stemmerik, L.; Stouge, S.; Thomassen, B.; et al. Kane Basin 1999: Mapping, stratigraphic studies and economic assessment of Precambrian and Lower Palaeozoic provinces in North-West Greenland. *Geol. Greenl. Survey Bull.* **2000**, *186*, 11–28.
50. Thomassen, B.; Dawes, P.R.; Iannelli, T.R.; Pirajno, F. *Gold Indications in Northern Inglefield Land, North-West Greenland: A Preliminary Report from Project Kane Basin 1999*; the Geological Survey of Denmark and Greenland (GEUS): Copenhagen, Denmark, 2000; 14p.

51. Thomassen, B.; Pirajno, F.; Iannelli, T.R.; Dawes, P.R.; Jensen, S.M. *Economic Geology Investigations in Inglefield Land, North–West Greenland: Part of the Project Kane Basin 1999*; the Geological Survey of Denmark and Greenland (GEUS): Copenhagen, Denmark, 2000; 98p.
52. Schjøth, F.; Steenfelt, A.; Thorning, L. (Eds.) *Regional Compilations of Geoscience Data from Inglefield Land, North-West Greenland*; the Geological Survey of Denmark and Greenland (GEUS): Copenhagen, Denmark, 1996; 35p.
53. Schjøth, F.; Thorning, L. *GIS Compilation of Geoscience Data: An ArcView GIS Version of Previously Published Thematic Maps from Inglefield Land*; the Geological Survey of Denmark and Greenland (GEUS): Copenhagen, Denmark, 1998; 59p.
54. Dawes, P.R. *Explanatory Notes to the Geological Map of Greenland, 1:500,000, Thule, Sheet 5, Map Series 2*; Geological Survey of Denmark and Greenland: Copenhagen, Denmark, 2006.
55. Dawes, P.R. *Explanatory Notes to the Geological Map of Greenland, 1:500,000, Humboldt Gletscher, Sheet 6, Map Series 1*; Geological Survey of Denmark and Greenland: Copenhagen, Denmark, 2004.
56. Nutman, A.P.; Dawes, P.R.; Kalsbeek, F.; Hamilton, M.A. Palaeoproterozoic and Archaean gneiss complexes in northern Greenland: Palaeoproterozoic terrane assembly in the High Arctic. *Precambrian Res.* **2008**, *161*, 419–451. [CrossRef]
57. Dawes, P.R.; Larsen, O.; Kalsbeek, F. Archaean and Proterozoic crust in North-West Greenland: Evidence from Rb–Sr whole-rock age determinations. *Can. J. Earth Sci.* **1988**, *25*, 1365–1373. [CrossRef]
58. Dawes, P.R. *A Review of Geoscientific Exploration and Geology in the Kane Basin Region of Greenland, Central Nares Strait*; the Geological Survey of Denmark and Greenland (GEUS): Copenhagen, Denmark, 1999; 63p.
59. Henriksen, N.; Higgins, A.K.; Kalsbeek, F.; Pulvertaft, T.C.R. Greenland from Archaean to Quaternary: Descriptive text to the geological map of Greenland 1:2,500,000. *Geol. Greenl. Survey Bull.* **2000**, *185*, 93.
60. Higgins, A.K.; Ineson, J.R.; Peel, J.S.; Surlyk, F.; Sønderholm, M. Lower Palaeozoic Franklinian Basin of North Greenland. In *Sedimentary Basins of North Greenland*; Peel, J.S., Sønderholm, M., Eds.; Bulletin-Grønlands Geologiske Undersøgelse; the Geological Survey of Denmark and Greenland (GEUS): Copenhagen, Denmark, 1991; Volume 160, pp. 71–139.
61. Thomassen, B.; Appel, P.W. *Ground Check of Airborne Anomalies and Regional Rust Zones in Inglefield Land, North-West Greenland*; Rapport-Danmarks og Grønlands Geologiske Undersøgelse; the Geological Survey of Denmark and Greenland (GEUS): Copenhagen, Denmark, 1997; 43p.
62. Thompson, A.B. Some aspects of fluid motion during metamorphism. *J. Geol. Soc.* **1987**, *144*, 309–312. [CrossRef]
63. Abrams, M.; Hook, S.; Ramachandran, B. *ASTER User Handbook*; Version 2; Jet Propulsion Laboratory, California Institute of Technology: La Cañada Flintridge, CA, USA, 2004. Available online: http://asterweb.jpl.nasa.gov/content/03_data/04_Documents/aster_guide_v2.pdf (accessed on 7 September 2019).
64. Kuester, M. *Radiometric Use of WV-3 Imagery*; Technical Note; DigitalGlobe: Westminster, CO, USA, 2016; p. 12.
65. Kuester, M.A.; Ochoa, M.; Dayer, A.; Levin, J.; Aaron, D.; Helder, D.L.; Leigh, L.; Czapla-Meyers, J.; Anderson, N.; Bader, B.; et al. *Absolute Radiometric Calibration of the DigitalGlobe Fleet and Updates on the New WV-3 Sensor Suite*; Technical Note; DigitalGlobe: Westminster, CO, USA, 2015; p. 16.
66. Iwasaki, A.; Tonooka, H. Validation of a crosstalk correction algorithm for ASTER/SWIR. *IEEE Trans. Geosci. Remote Sens.* **2005**, *43*, 2747–2751. [CrossRef]
67. Cooley, T.; Anderson, G.P.; Felde, G.W.; Hoke, M.L.; Ratkowski, A.J.; Chetwynd, J.H.; Gardner, J.A.; Adler-Golden, S.M.; Matthew, M.W.; Berk, A.; et al. FLAASH, a MODTRAN4-based atmospheric correction algorithm, its application and validation. In Proceedings of the IEEE International on Geoscience and Remote Sensing Symposium, Toronto, ON, Canada, 24–28 June 2002; Volume 3, pp. 1414–1418.
68. Research Systems, Inc. *ENVI Tutorials*; Research Systems, Inc.: Boulder, CO, USA, 2008.
69. Fraser, S.J.; Green, A.A. A software defoliant for geological analysis of band ratios. *Int. J. Remote Sens.* **1987**, *8*, 525–532. [CrossRef]
70. Crosta, A.; Moore, J. Enhancement of Landsat Thematic Mapper imagery for residual soil mapping in SW Minais Gerais State, Brazil: A prospecting case history in Greenstone belt terrain. In Proceedings of the 7th ERIM Thematic Conference: Remote Sensing for Exploration Geology, Calgary, AB, Canada, 2–6 October 1989; pp. 1173–1187.

71. Crosta, A.P.; Souza Filho, C.R.; Azevedo, F.; Brodie, C. Targeting key alteration minerals in epithermal deposits in Patagonia, Argentina, Using ASTER imagery and principal component analysis. *Int. J. Remote Sens.* **2003**, *24*, 4233–4240. [CrossRef]
72. Loughlin, W.P. Principal components analysis for alteration mapping. *Photogramm. Eng. Remote Sens.* **1991**, *57*, 1163–1169.
73. Kokaly, R.F.; Clark, R.N.; Swayze, G.A.; Livo, K.E.; Hoefen, T.M.; Pearson, N.C.; Wise, R.A.; Benzel, W.M.; Lowers, H.A.; Driscoll, R.L.; et al. *USGS Spectral Library Version 7*; U.S. Geological Survey Data Series 1035; USGS Crustal Geophysics and Geochemistry Science Center: Denver, CO, USA, 2017; 61p. [CrossRef]
74. Van der Werff, H.; van der Meer, F. Sentinel-2A MSI and Landsat 8 OLI Provide Data Continuity for Geological Remote Sensing. *Remote Sens.* **2016**, *8*, 883. [CrossRef]
75. Clark, R.N.; Swayze, G.A.; Gallagher, A.; King, T.V.V.; Calvin, W.M. *The U.S. Geological Survey, Digital Spectral Library: Version 1: 0.2 to 3.0 Microns: U.S. Geological Survey Open File Report 93-592*; 1993; 1340p. Available online: http://speclab.cr.usgs.gov (accessed on 24 August 1999).
76. Clark, R.N.; Swayze, G.A. Mapping minerals, amorphous materials, environmental materials, vegetation, water, ice, and snow, and other materials. In Proceedings of the USGS Tricorder Algorithm, Summaries of the Fifth Annual JPL Airborne Earth Science Workshop, The United States Geological Survey, Reston, VA, USA, 3 August 1995.
77. Kalinowski, A.; Oliver, S. *ASTER Mineral Index Processing Manual*; Technical Report; Geoscience Australia: Canberra, Australia, 2004. Available online: http://www.ga.gov.au/image_cache/GA7833.pdf (accessed on 12 August 2018).
78. Crowley, J.K.; Brickey, D.W.; Rowan, L.C. Airborne imaging spectrometer data of the Ruby Mountains, Montana: Mineral discrimination using relative absorption band-depth images. *Remote Sens. Environ.* **1989**, *29*, 121–134. [CrossRef]
79. Ninomiya, Y.; Fu, B.; Cudahy, T.J. Detecting lithology with Advanced Spaceborne Thermal Emission and Reflection Radiometer (ASTER) multispectral thermal infrared radiance-at-sensor data. *Remote Sens. Environ.* **2005**, *99*, 127–139. [CrossRef]
80. Boardman, J.W. Inversion of imaging spectrometry data using singular value decomposition. In Proceedings of the IGARSS'89, 12th Canadian Symposium on Remote Sensing, Vancouver, BC, Canada, 10–14 July 1989; Volume 4, pp. 2069–2072.
81. Boardman, J.W. Sedimentary Facies Analysis Using Imaging Spectrometry: A Geophysical Inverse Problem. Ph. D. Thesis, University of Colorado, Boulder, CO, USA, 1992; p. 212.
82. Adams, J.B.; Smith, M.O.; Gillespie, A.R. Imaging spectroscopy: Interpretation based on spectral mixture analysis. In *Remote Geochemical Analysis: Elemental and Mineralogical Composition*; Pieters, C.M., Englert, P.A.J., Eds.; Cambridge University Press: New York, NY, USA, 1993; pp. 145–166.
83. Adams, J.B.; Sabol, D.E.; Kapos, V.; Filho, R.A.; Roberts, D.A.; Smith, M.O.; Gillespie, A.R. Classification of multispectral images based on fractions of endmembers: Application to land-cover change in the Brazilian Amazon. *Remote Sens. Environ.* **1995**, *52*, 137–154. [CrossRef]
84. Kruse, F.A.; Bordman, J.W.; Huntington, J.F. Comparison of airborne hyperspectral data and EO-1 Hyperion for mineral mapping. *IEEE Trans. Geosci. Remote Sens.* **2003**, *41*, 1388–1400. [CrossRef]
85. Kruse, F.A.; Perry, S.L. Regional mineral mapping by extending hyperspectral signatures using multispectral data. *IEEE Trans. Geosci. Remote Sens.* **2007**, *4*, 154–172.
86. Boardman, J.W.; Kruse, F.A. Automated spectral analysis: A geologic example using AVIRIS data, north Grapevine Mountains, Nevada. In Proceedings of the Tenth Thematic Conference on Geologic Remote Sensing, Environmental Research Institute of Michigan, Ann Arbor, MI, USA, 9 June 1994; pp. I-407–I-418.
87. Boardman, J.W.; Kruse, F.A.; Green, R.O. Mapping target signatures via partial unmixing of AVIRIS data. In Proceedings of the Fifth JPL Airborne Earth Science Workshop, Pasadena, CA, USA, 12 March 1995; Volume 1, pp. 23–26.
88. Manolakis, D.; Marden, D.; Shaw, G.A. Hyperspectral Image Processing for Automatic Target Detection Applications. *Linc. Lab. J.* **2003**, *14*, 79–116.
89. Kraut, S.; Scharf, L.L.; Butler, R.W. The adaptive coherence estimator: A uniformly most-powerful-invariant adaptive detection statistic. *IEEE Trans. Signal Process.* **2005**, *53*, 427–438. [CrossRef]
90. Bidon, S.; Besson, O.; Tourneret, J.Y. The Adaptive Coherence Estimator is the Generalized Likelihood Ratio Test for a Class of Heterogeneous Environments. *IEEE Signal Process. Lett.* **2008**, *15*, 281–284. [CrossRef]

91. Kraut, S.; Scharf, L.L.; McWhorter, L.T. Adaptive subspace detectors. *IEEE Trans. Signal Process.* **2001**, *49*, 1–16. [CrossRef]
92. Alvey, B.; Zare, A.; Cook, M.; Ho, D.K.C. Adaptive coherence estimator (ACE) for explosive hazard detection using wideband electromagnetic induction (WEMI). In Proceedings of the SPIE 9823, Detection and Sensing of Mines, Explosive Objects, and Obscured Targets XXI, Baltimore, MA, USA, 3 May 2016. [CrossRef]
93. Warren, S.G. Optical properties of snow. *Rev. Geophys. Space Phys.* **1982**, *20*, 67–89. [CrossRef]
94. Hall, D.K.; Riggs, G.A.; Salomonson, V.V.; DiGirolamo, N.E.; Bayr, K. MODIS snow-cover products. *Remote Sens. Environ.* **2002**, *83*, 181–1194. [CrossRef]
95. Gupta, R.P.; Haritashya, U.K.; Singh, P. Mapping dry/wet snow cover in the Indian Himalayas using IRS multispectral imagery. *Remote Sens. Environ.* **2005**, *97*, 458–469. [CrossRef]
96. Pour, B.A.; Hashim, M. Identification of hydrothermal alteration minerals for exploring of porphyry copper deposit using ASTER data, SE Iran. *J. Asian Earth Sci.* **2011**, *42*, 1309–1323. [CrossRef]
97. Velosky, J.C.; Stern, R.J.; Johnson, P.R. Geological control of massive sulfide mineralization in the Neoproterozoic Wadi Bidah shear zone, southwestern Saudi Arabia, inferences from orbital remote sensing and field studies. *Precambrian Res.* **2003**, *123*, 235–247. [CrossRef]
98. Bishop, J.L.; Lane, M.D.; Dyar, M.D.; Brwon, A.J. Reflectance and emission spectroscopy study of four groups of phyllosilicates: Smectites, kaolinite-serpentines, chlorites and micas. *Clay Min.* **2008**, *43*, 35–54. [CrossRef]
99. Sherman, D.M.; Waite, T.D. Electronic spectra of Fe^{3+} oxides and oxide-hydroxides in the near IR to near UV. *Am. Mineral.* **1985**, *70*, 1262–1269.
100. Morris, R.V.; Lauer, H.V.; Lawson, C.A.; Girson, E.K.; Nace, G.A.; Stewart, C. Spectral and other physicochemical properties of submicron powders of hematite (á-Fe_2O_3), maghemite (ã-Fe_2O_3), magnetite (Fe3O4), goethite (á-FeOOH), and lepidocrocite (ã-FeOOH). *J. Geophys. Res.* **1985**, *90*, 3126–3144. [CrossRef] [PubMed]
101. Story, M.; Congalton, R. Accuracy assessment: A user's perspective. *Photogramm. Eng. Remote Sens.* **1986**, *52*, 397–399.
102. Congalton, R.G. A review of assessing the accuracy of classification of remotely sensed data. *Remote Sens. Environ.* **1991**, *37*, 35–46. [CrossRef]
103. Lillesand, T.; Kiefer, R. *Remote Sensing and Image Interpretation*; John Wiley & Sons, Inc.: New York, NY, USA, 1994; Chapter 7.
104. Steenfelt, A.; Dam, E. *Reconnaissance Geochemical Mapping of Inglefield Land, North-West Greenland*; Rapport-Danmarks og Grønlands Geologiske Undersøgelse; the Geological Survey of Denmark and Greenland (GEUS): Copenhagen, Denmark, 1996; 27p.
105. Bedini, E. Mapping lithology of the Sarfartoq carbonatite complex, southern West Greenland, using HyMap imaging spectrometer data. *Remote Sens. Environ.* **2009**, *113*, 1208–1219. [CrossRef]
106. Bedini, E.; Rasmussen, T.M. Use of airborne hyperspectral and gamma-ray spectroscopy data for mineral exploration at the Sarfartoq carbonatite complex, southern West Greenland. *Geosci. J.* **2018**, *22*, 641–651. [CrossRef]
107. Bedini, E. Mineral mapping in the Kap Simpson complex, central East Greenland, using HyMap and ASTER remote sensing data. *Adv. Space Res.* **2011**, *47*, 60–73. [CrossRef]

© 2019 by the authors. Licensee MDPI, Basel, Switzerland. This article is an open access article distributed under the terms and conditions of the Creative Commons Attribution (CC BY) license (http://creativecommons.org/licenses/by/4.0/).

Article

A Remote Sensing-Based Application of Bayesian Networks for Epithermal Gold Potential Mapping in Ahar-Arasbaran Area, NW Iran

Seyed Mohammad Bolouki [1], Hamid Reza Ramazi [1], Abbas Maghsoudi [1,*], Amin Beiranvand Pour [2,3] and Ghahraman Sohrabi [4]

1. Department of Mining and Metallurgical Engineering, Amirkabir University of Technology (Tehran Polytechnic), Tehran 1591634311, Iran; mohammadbloki@aut.ac.ir (S.M.B.); ramazi@aut.ac.ir (H.R.R.)
2. Korea Polar Research Institute (KOPRI), Songdomirae-ro, Yeonsu-gu, Incheon 21990, Korea; Amin.Beiranvand@kopri.re.kr
3. Institute of Oceanography and Environment (INOS), University Malaysia Terengganu (UMT), 21030 Kuala Nerus, Terengganu, Malaysia
4. Department of Geology, University of Mohaghegh Ardabili, Ardabil 5619911367, Iran; q_sohrabi@uma.ac.ir
* Correspondence: a.maghsoudi@aut.ac.ir; Tel.: +98-21-64542929

Received: 19 November 2019; Accepted: 23 December 2019; Published: 27 December 2019

Abstract: Mapping hydrothermal alteration minerals using multispectral remote sensing satellite imagery provides vital information for the exploration of porphyry and epithermal ore mineralizations. The Ahar-Arasbaran region, NW Iran, contains a variety of porphyry, skarn and epithermal ore deposits. Gold mineralization occurs in the form of epithermal veins and veinlets, which is associated with hydrothermal alteration zones. Thus, the identification of hydrothermal alteration zones is one of the key indicators for targeting new prospective zones of epithermal gold mineralization in the Ahar-Arasbaran region. In this study, Landsat Enhanced Thematic Mapper+ (Landsat-7 ETM+), Landsat-8 and Advanced Spaceborne Thermal Emission and Reflection Radiometer (ASTER) multispectral remote sensing datasets were processed to detect hydrothermal alteration zones associated with epithermal gold mineralization in the Ahar-Arasbaran region. Band ratio techniques and principal component analysis (PCA) were applied on Landsat-7 ETM+ and Landsat-8 data to map hydrothermal alteration zones. Advanced argillic, argillic-phyllic, propylitic and hydrous silica alteration zones were detected and discriminated by implementing band ratio, relative absorption band depth (RBD) and selective PCA to ASTER data. Subsequently, the Bayesian network classifier was used to synthesize the thematic layers of hydrothermal alteration zones. A mineral potential map was generated by the Bayesian network classifier, which shows several new prospective zones of epithermal gold mineralization in the Ahar-Arasbaran region. Besides, comprehensive field surveying and laboratory analysis were conducted to verify the remote sensing results and mineral potential map produced by the Bayesian network classifier. A good rate of agreement with field and laboratory data is achieved for remote sensing results and consequential mineral potential map. It is recommended that the Bayesian network classifier can be broadly used as a valuable model for fusing multi-sensor remote sensing results to generate mineral potential map for reconnaissance stages of epithermal gold exploration in the Ahar-Arasbaran region and other analogous metallogenic provinces around the world.

Keywords: epithermal gold; hydrothermal alteration; Ahar-Arasbaran region; ASTER; Landsat-7 ETM+; Landsat-8; Bayesian Network Classifiers

1. Introduction

Hydrothermal alteration minerals such as iron oxide/hydroxides, Al-OH, Fe,Mg-OH, S-O, Si-OH and carbonate minerals show indicative spectral absorption signatures in the visible near-infrared (VNIR) and the shortwave infrared (SWIR) regions [1–5]. Multispectral and hyperspectral satellite imagery with appropriate spatial and spectral resolution is capable of recording the spectral absorption signatures of alteration minerals in the VNIR and SWIR spectral bands, which can be utilized to map and remotely detect hydrothermal alteration mineral zones associated with ore mineraliztions [6–9]. Recently, the identification of alteration mineral zones using remote sensing sensors is effectively and extensively used for prospecting porphyry copper, epithermal gold, uranium and massive sulfide deposits in metallogenic provinces around the world [10–20].

The Landsat-7 ETM+ imagery was used for mapping hydrothermal alteration zones related to epithermal gold and porphyry copper deposits in the reconnaissance stages of copper/gold exploration. The VNIR spectral bands of Landsat-7 ETM+ were utilized to map iron oxides/hydroxide minerals (gossan), while, SWIR spectral bands were used to detect hydroxyl-bearing minerals and carbonates [21–24]. Band ratio of 3/1 is able to identify iron oxides/hydroxide minerals (hematite, jarosite and limonite) due to strong reflectance in band 3 (0.63–0.69 μm) and absorption features in band 1 (0.45–0.52 μm) [23]. Band ratio of 5/7 is sensitive to hydroxyl-bearing minerals and carbonates because of reflectance features in band 5 (1.55–1.75 μm) and strong absorption in band 7 (2.09–2.35 μm) [23,25–27]. Equivalent bands of Landsat-8, bands 2 and 4 responsive to iron oxides/hydroxides and bands 6 and 7 sensitive to hydroxyl-bearing minerals and carbonates, were also extensively used for hydrothermal alteration mineral mapping in metallogenic provinces [12,16,18,19,28]. Discrimination of particular alteration zones and minerals (i.e., argillic, phyllic propylitic zones and muscovite, chlorite and kaolinite) using Landsat-7 ETM+ and Landsat-8 VNIR and SWIR spectral bands is challenging due to position, number and the broad extent of the bands [28,29].

Distinguishing hydrothermal alteration zones or specific mineral assemblages as an indicator of high-economic potential zones for exploring ore mineralizations is significant [30,31]. For instance, discriminating phyllic zone within the inner shell of mineralization for porphyry copper exploration is important and identification of advanced argillic zone situated near to hydrothermal mineralization system for epithermal gold exploration is essential [32–34]. ASTER multispectral satellite imagery is particularly useful for discriminating hydrothermal alteration zones associated with ore mineralizations [6,35–37]. Three VNIR spectral bands of ASTER (0.52 to 0.86 μm) are used for detecting iron oxide/hydroxide minerals [6,35]. Phyllic, argillic and propylitic zones are recognizable using six SWIR spectral bands of ASTER (1.6 to 2.43 μm) [35]. The phyllic zone containing illite/muscovite (sericite) and strong Al-OH absorption feature at 2.20 μm is detectable by band 6 of ASTER. The argillic zone (kaolinite/alunite) has Al-OH absorption feature at 2.17 μm, which is coincident with band 5 of ASTER. The propylitic zone comprising epidote, chlorite and calcite shows absorption features around 2.35 μm, which is corresponded with band 8 of ASTER [35–39].

Obtaining information from multi-sensor remote sensing satellite data can produce relevant results for detailed mapping of hydrothermal alteration zones [12]. The integration of the multi-sensor remote sensing results using geostatistical techniques can quickly produce a mineral potential map, which indicates the high potential zones of hydrothermal ore mineralizations [40]. Mineral potential map of a region is generally realized as the predictive classification of each spatial unit contains a particular combination of spatially coincident predictor patterns as mineralized or barren zones [41,42]. A Bayesian network is a type of statistical model (probabilistic graphical model), which represents a set of variables and their conditional dependencies through a Directed Acyclic Graph (DAG) [41,43,44]. It predicts the likelihood that anyone of several possible known causes was the contributing factor [45]. Therefore, the Bayesian network is a suitable model for fusing thematic layers derived from multi-sensor remote sensing satellite data to generate a mineral potential map.

In this study, Landsat-7 ETM+, Landsat-8 and ASTER multispectral remote sensing datasets were used to identify hydrothermal alteration zones associated with epithermal gold mineralization and producing thematic layers, which were afterward synthesized in the Bayesian networks for mineral potential mapping in the Ahar-Arasbaran region, NW Iran (Figure 1). This region is a well-endowed terrain hosting numerous known epithermal gold deposits, several porphyry and skarn Cu-Mo deposits, Fe skarn deposits, Cu-Au porphyry deposits and many other Cu-Mo-Au vein mineralizations [46–50]. The deposits are associated with extensive hydrothermal alteration mineral zones such as iron oxide/hydroxides, advanced argillic, argillic, phyllic and propylitic [48,51,52]. The Ahar-Arasbaran region has a high potential for exploring new prospective zones of epithermal gold and many other ore mineralizations. Pazand et al. [52] used ASTER satellite data for hydrothermal alteration mapping in the Ahar area, NW Iran. Some geo-referenced hydrothermal alteration maps were produced using RBD (relative absorption band depth), principal component analysis (PCA), minimum noise fraction (MNF) and matched filtering (MF) image processing techniques for reconnaissance stages of porphyry copper exploration in the Ahar area. Furthermore, Pazand and Hezarkhani [48] generated a favorability map for Cu porphyry mineralization using fuzzy modeling in the Ahar–Arasbaran zone, NW Iran. There is no comprehensive remote sensing research available for mapping hydrothermal alteration zones in the Ahar-Arasbaran region using multi-sensor satellite imagery at a regional scale. This study characterizes an extensive remote sensing analysis using Landsat-7 ETM+, Landsat-8 and ASTER datasets, detailed fieldwork and laboratory analysis for mineral potential mapping. Therefore, the primary purposes of the research are: (1) to map hydrothermal alteration mineral zones using Landsat-7 ETM+, Landsat-8 and ASTER datasets by implementing the band ratio, PCA, RBD and selective PCA image processing techniques; (2) to generate mineral potential map by fusing the alteration thematic layers using the Bayesian networks; and (3) to verify the high potential zones by checking the detailed global positioning system (GPS) surveying in the field and analyzing several microphotographs of hydrothermal alteration minerals and gold mineralization and X-ray diffraction (XRD) analysis of collected rock samples from alteration zones.

2. Geology of the Ahar-Arasbaran Region

The Ahar-Arasbaran region covers an area (approximately 5000 km^2), which is located between latitudes 38°07′N and 38°52′N and longitudes 46°15′E and 47°30′E (Figure 1). This zone is a part of Lesser Caucasus metallogenic zone and corresponding to tectono-magmatism activity from Jurassic to Quaternary [46,47,53,54]. The volcano-plutonic belt of Arasbaran-Lesser Caucasus is a mountainous and uplifted region that trending NW-SE from Georgia (Republic of Azerbaijan) to the Talesh region (Iran) [50]. Magmatic rocks in the Ahar-Arasbaran region containing tholeiitic, calc-alkaline, high calcium calc-alkaline, shoshonitic, adakitic, alkaline sodic and potassic rocks, which are formed in a continental margin of a subduction zone (subduction to post-collision stages) [50]. Cretaceous units (limestone and shale), flysch deposits, Paleocene and Eocene volcanic rocks are also exposed in the study area (Figure 1). Several intrusive bodies having different sizes are penetrated in the Eocene and Cretaceous volcanic-sedimentary rocks and caused folding, alteration and mineralization [49,51]. Structural trends of folds, faults, dykes and veins are mostly NW-SE, E-W and NE-SW, which show the main stresses that affected the study area [50,51].

The intrusion of the Oligo-Miocene batholiths into the Cretaceous to Eocene sedimentary and volcano-sedimentary deposits along with hydrothermal fluids is formed intensive alteration halos in the Eocene volcanic rocks [55]. The alteration zones such as argillic, silica and alunite are associated with Cu, Au, Mo, Ag, Pb and Zn mineralizations [49]. Moreover, several skarn zones are formed in the contact zone of intrusive masses with Cretaceous limestone [51]. A variety of ore mineralization zones were identified in the Ahar-Arasbaran region, including Fe, Cu, Pb-Zn, Cu-Au, Cu-Mo, Au-Ag, Fe-Au, which occurred in the form of sprains, veins, stokes and in relation to the skarn zones [49,51]. The gold mineralization in the study area is observed in the form of epithermal veins [55]. The Masjed Daghi (Siahrood) and AliJavad valley (Anjerd) are considered to be Au-Cu porphyry deposits. The Sharaf-abad, Hize-jan, Nabi-jan, Zailig, Miveh-roud, Safi-Khanloo, Noqdouz, Anniqh and Khoyneh-roud are known as epithermal gold deposits in the study area [55].

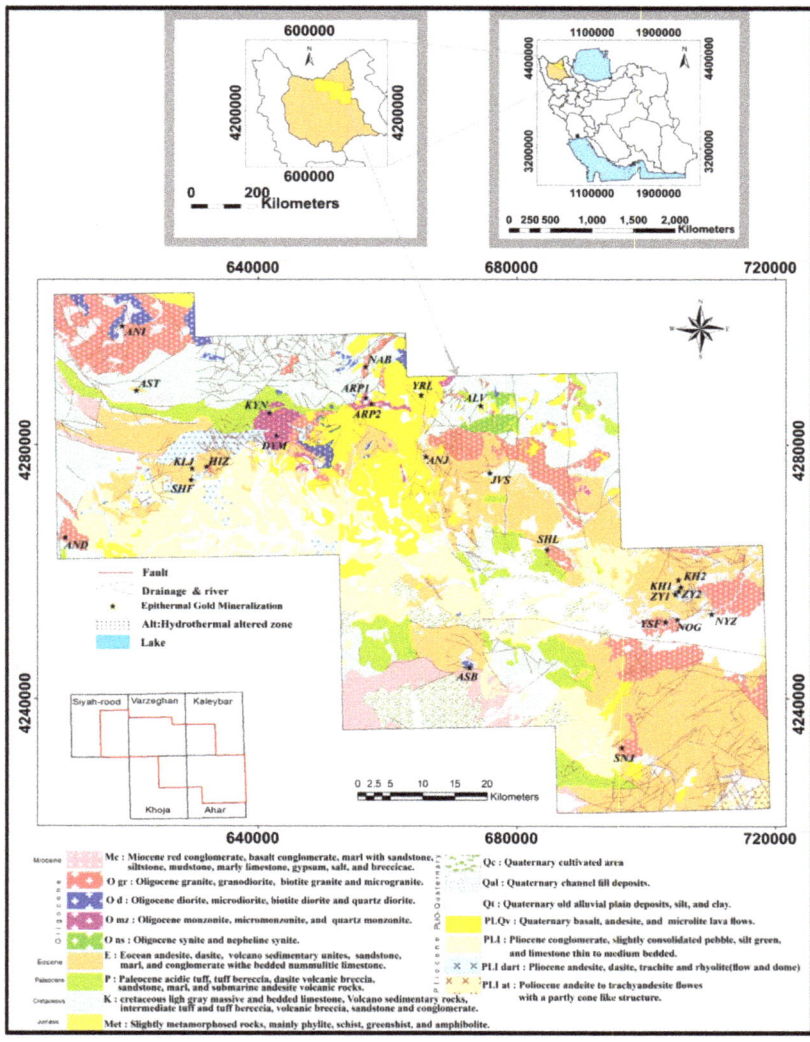

Figure 1. Geological map of the Ahar-Arasbaran region. Modified from five 1:100,000 geological map sheet provided by the Geological Survey of Iran [56]. Abbreviation to epithermal gold mineralization: ANI = Annigh; AST = Astamal; AND = Andiryan; KLJ = Kalijan; SHF = Sharaf-abad; HIZ = Hize-jan; KYN = Khoyneh-roud; DYM = Day-mamagh; NAB = Nabi-jan; ARP1 = Arpaligh1; ARP2 = Arpaligh2; YRL = Yaralojeh; ANJ = Anjerd; ALV = Alavigh; JVS = Javan-sheykh; SHL = Shaleh-boran; ASB = Asb-abad; SNJ = Sonajil; YSF = Yosoufloo; NOG = Noghdouz; NYZ = Niyaz; ZY1 = Zailigh1; ZY2 = Zailigh2; KH1 = Khiarloo1; KH2 = Khiarloo1.

3. Materials and Methods

3.1. Remote Sensing Data and Pre-Processing

The Landsat-7 ETM+, Landsat-8 and ASTER satellite remote sensing datasets were used in this study. Technical characteristics of Landsat-7 ETM+, Landsat-8 and ASTER remote sensing sensors are summarized in Table 1.

Table 1. Technical characteristics of the Landsat-7 ETM+, Landsat-8 and ASTER remote sensing sensors [23,57–59].

		Bands	Wavelength (μm)	Resolution (m)
Landsat 7 Enhanced Thematic Mapper Plus (Landsat-7 ETM+)		Band 1—Blue	0.45–0.52	30
		Band 2—Green	0.52–0.60	30
		Band 3—Red	0.63–0.69	30
		Band 4—Near Infrared (NIR)	0.77–0.90	30
		Band 5—Shortwave Infrared (SWIR) 1	1.55–1.75	30
		Band 6—Thermal	10.40–12.50	60 * (30)
		Band 7—Shortwave Infrared (SWIR) 2	2.09–2.35	30
		Band 8—Panchromatic	0.520–0.900	15
		Bands	**Wavelength (μm)**	**Resolution (m)**
Landsat 8 Operational Land Imager (OLI) and Thermal Infrared Sensor (TIRS)		Band 1—Ultra Blue (coastal/aerosol)	0.435–0.451	30
		Band 2—Blue	0.452–0.512	30
		Band 3—Green	0.533–0.590	30
		Band 4—Red	0.636–0.673	30
		Band 5—Near Infrared (NIR)	0.851–0.879	30
		Band 6—Shortwave Infrared (SWIR) 1	1.566–1.651	30
		Band 7—Shortwave Infrared (SWIR) 2	2.107–2.294	30
		Band 8—Panchromatic	0.503–0.676	15
		Band 9—Cirrus	1.363–1.384	30
		Band 10—Thermal Infrared (TIRS) 1	10.60–11.19	100 * (30)
		Band 11—Thermal Infrared (TIRS) 2	11.50–12.51	100 * (30)
	Band	**Label**	**Wavelength (μm)**	**Resolution (m)**	**Description**
ASTER Advanced Space borne Thermal Emission and Reflection Radiometer	B1	VNIR_Band1	0.520–0.60	15	Visible green/yellow
	B2	VNIR_Band2	0.630–0.690	15	Visible red
	B3N	VNIR_Band3N	0.760–0.860	15	Near infrared
	B3B	VNIR_Band3B	0.760–0.860	15	
	B4	SWIR_Band4	1.600–1.700	30	Short-wave infrared
	B5	SWIR_Band5	2.145–2.185	30	
	B6	SWIR_Band6	2.185–2.225	30	
	B7	SWIR_Band7	2.235–2.285	30	
	B8	SWIR_Band8	2.295–2.365	30	
	B9	SWIR_Band9	2.360–2.430	30	
	B10	TIR_Band10	8.125–8.475	90	Long-wave infrared or thermal IR
	B11	TIR_Band11	8.475–8.825	90	
	B12	TIR_Band12	8.925–9.275	90	
	B13	TIR_Band13	10.250–10.950	90	
	B14	TIR_Band14	10.950–11.650	90	

* The 60 m thermal band of Landsat-7 ETM+ is resampled and co-registered to the 30 m VNIR and SWIR bands. The 100 m TIRS bands are resampled and co-registered to the 30 m OLI bands.

A Landsat-7 ETM+ scene (Path/Raw: 168/33) covering the Ahar-Arasbaran region was acquired on 15 June 2001. A level 1T (terrain corrected) Landsat 8 scene (Path/Raw: 168/33) was also acquired on 10 June 2016 for the study area. Seven level 1B ASTER scenes covering the study area were acquired from 8 to 29 June 2002–2004. The data were obtained from the U.S. Geological Survey's Earth Resources Observation System (EROS) Data Center (EDC) (https://earthexploere.usgs.gov/ and https://glovis.usgs.gov). The scenes were cloud-free and have been already georeferenced to the UTM zone 38 North projection using the WGS-84 datum. For converting Landsat-7 ETM+ digital numbers to spectral radiance or exoatmospheric reflectance (reflectance above the atmosphere), the Landsat Calibration technique was adopted from Chander et al. [60]. This technique uses the published post-launch gain and offset values [61,62]. The mathematical details of the technical performance can be found in Chander et al. [60]. For Landsat 8 and ASTER datasets, Internal Average Relative Reflectance (IARR) was utilized. The IARR calibration method normalizes images to a scene average spectrum [61,63]. This is particularly effective for reducing imaging spectrometer data to relative reflectance in an area where no ground measurements exist and little is known about the scene [61,63].

It works best for arid areas with no vegetation. The IARR calibration is performed by calculating an average spectrum for the entire scene and using this as the reference spectrum. Apparent reflectance is calculated for each pixel of the image by dividing the reference spectrum into the spectrum for each pixel. The atmospheric correction was implemented to ASTER data after Crosstalk correction [64]. Moreover, the 15 m VNIR bands of ASTER were resampled to the 30 m SWIR bands using the cubic convolution technique. A masking procedure was applied to the remote sensing datasets for removing the effects of vegetation and Quaternary deposits. Normalized Difference Vegetation Index (NDVI) was calculated for the remote sensing datasets. As a result, a masking procedure was executed to the remote sensing datasets for eliminating the influences of sparse vegetation in the study area. For Quaternary deposits, we used geological map of the study area to identify the location of the Quaternary units, then a masking procedure was implemented to the remote sensing datasets. The ENVI (Environment for Visualizing Images, http://www.exelisvis.com) version 5.2 and ArcGIS version 10.3 (Esri, Redlands, CA, USA) software packages were employed for processing Landsat-7 ETM+, Landsat-8 and ASTER data.

3.2. Image Processing Techniques

The main objective of image processing techniques implemented in this analysis is to map hydrothermal alteration zones for generating thematic layers from multi-sensor remote sensing satellite datasets. Then, the thematic layers are fused using a Bayesian network model for producing a mineral potential map of the Ahar-Arasbaran region. Fieldwork and laboratory analysis are used to verify the results. A view of the methodological flowchart applied in this study is shown in Figure 2.

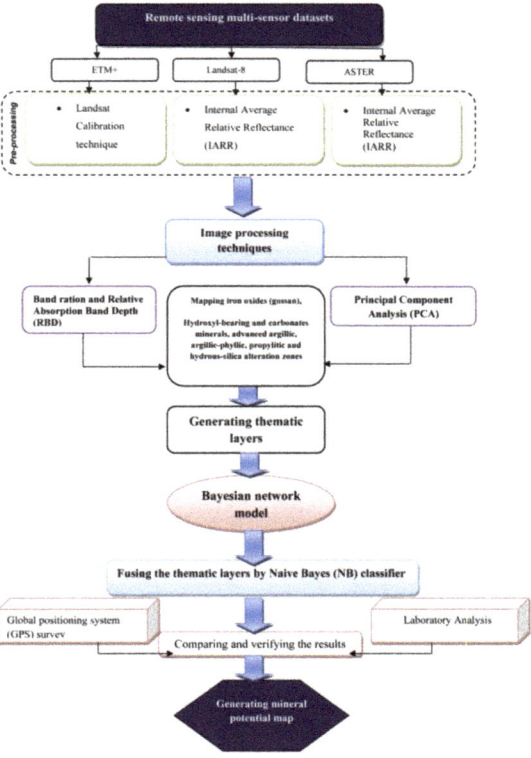

Figure 2. An overview of the methodological flowchart applied in this analysis.

3.2.1. Band Ratio

The band ratio technique is one of the most applicable image processing techniques for mapping hydrothermal alteration minerals and zones such muscovite, jarosite, gossan, advanced argillic, argillic-phyllic, propylitic and hydrous silica-affected zones [23,39,65,66]. The digital number (DN) value of a band is partitioned by the DN value of other band, which highlights particular spectral features related to minerals or materials that planned to map [23]. Relative Absorption Band Depth (RBD) uses three-point ratio formulation for detecting typical absorption features related to a specific mineral or alteration zone [67]. For a specific absorption characteristic, the numerator is the sum of the bands demonstrating the shoulders and the denominator is the band positioned adjoining the absorption feature minimum [67]. Therefore, the absorption intensities attributed to Al-OH, Fe,Mg-OH, Si-OH and CO_3 can be formulated for mapping advanced argillic, argillic-phyllic, propylitic and hydrous-silica alteration zones [35].

In this study, iron oxide-bearing minerals (gossan) were mapped using Landsat-7 ETM+ band ratio of band 3/band 1, Landsat-8 band ratio of band 4/band 2 and ASTER band ratio of band 2/band 1, respectively [23,39,68]. As mentioned before, iron oxide/hydroxide minerals contain diagnostic spectral characteristics coincident with selected bands of different sensors [19,23,28,35]. Hydroxyl-bearing (Al-OH and Fe,Mg-OH) and carbonates minerals were typically identified in the study region through Landsat-7 ETM+ band ratio of band 5/band 7, Landsat-8 band ratio of band 6/band 7 and ASTER band ratio of band 4/band 9, respectively [23,28,29,68]. The advanced argillic alteration (alunite and kaolinite) contain strong absorption about 2.17 µm (corresponding band 5 of ASTER) [35], thus, ASTER band ratio of band 4/band 6 was used to highlight the advanced argillic alteration zone [69] in the study area. The argillic-phyllic alteration zone is mostly dominated by sericite (muscovite/illite), which shows high absorption feature at 2.20 µm (equivalent to band 6 of ASTER) [35,36]. This alteration zone was detected by applying ASTER band ratio of band 5/band 6 [36]. Moreover, the propylitic alteration zone was mapped using ASTER band ratio of band 5/band 8 [35,36] in this analysis.

For detailed mapping of advanced argillic, argillic-phyllic, propylitic and hydrous silica-affected alteration zones, four RBDs were adopted using SWIR bands of ASTER (Table 2). The RDB1 = (band 4 + band 6)/band 5 for detecting advanced argillic zone, the RDB2 = (band 5 + band 7)/band 6 for identifying argillic-phyllic zone [38,39], the RDB3 = (band 6 + band 9)/(band 7+ band 8) for discriminating propylitic zone and RDB4 = (band 5 + band 8)/(band 6 + band 7) for mapping hydrous silica zone [70] were implemented.

Table 2. The RBD indices applied for hydrothermal alteration mapping in the study area using ASTER imagery.

Alteration Zone	Mineral Assemblages	RBD Band
Advanced Argillic	Alunite-Kaolinite-Pyrophyllite	(4 + 6)/5
Argillic-Phyllic	Sericitic-Illite-Smectite	(5 + 7)/6
Propylitic	Epidote-Chlorite-Amphibole-Biotite	(6 + 9)/(7 + 8)
Hydrous Silica	Hydrous Silica-Jarosite-Sericite	(5 + 8)/(6 + 7)

3.2.2. Principal Component Analysis

Principal Component Analysis (PCA) is a statistical approach that broadly and successfully used for decorrelation and enhancing the spectral contrast in remote sensing imagery [71]. This method transforms a number of correlated variables into several uncorrelated variables that termed PCs [72]. The eigenvector loadings (uncorrelated linear combinations) of variables were selected in a consistent way that each PC contains a smaller variance of extracted linear combination, sequentially [71,73]. The eigenvector loadings include key information linked to spectral features, which are anticipated from spectral bands of a remote sensing sensor [74]. For instance, a PC contains strong eigenvector loadings for indicative bands (reflection and absorption bands) of an alteration mineral with opposite signs enhances that mineral as bright pixels (if loading is positive in reflection band) or dark pixels (if loading is negative in reflection band) in the PC image [74,75].

In this study, the PCA method was implemented to some selected bands of Landsat-7 ETM+, Landsat-8 and ASTER using a covariance matrix for mapping hydrothermal alteration minerals. For identifying iron oxide-affected zones (gossan), bands 1, 3, 4 and 5 of Landsat-7 ETM+, bands 2, 4, 5 and 6 of Landsat-8 and bands 1, 2, 3 and 4 of ASTER were selected. The selected bands cover the iron oxide/hydroxide spectral properties in the VNIR region [3–5]. The eigenvector matrix for the selected bands and satellite sensors for mapping iron oxide/hydroxides are shown in Table 3A–C. Bands 1, 4, 5 and 7 of Landsat-7 ETM+, bands 2, 5, 6 and 7 of Landsat-8 and bands 1, 3, 4 and 6 of ASTER were used for detecting hydroxyl-bearing minerals. These bands cover the reflectance and absorption features of OH-minerals in the VNIR and SWIR regions [1,2]. Table 4A–C shows the eigenvector matrix for the selected bands and satellite sensors for mapping hydroxyl-bearing minerals. The reflectance properties and absorption intensities related to Al-OH, Fe,Mg-OH and CO_3 can be mapped by ASTER VNIR+SWIR bands [23,35,38]. Bands 1, 4, 6 and 7 of ASTER were utilized for mapping advanced argillic zone. Bands 1, 3, 5 and 6 of ASTER were executed to detect argillic-phyllic zone. Bands 1, 3, 5 and 8 of ASTER were performed for discriminating propylitic alteration zone. Table 5A–C shows eigenvector matrix for the selected bands of ASTER for mapping advanced argillic, argillic-phyllic and propylitic alteration zones. After implementing the algorithms for all band ratios and PCAs, firstly the obtained DN values were normalized, then the X+3S was used to obtain definite anomaly. It means all the DN values showing the number more than the X+3S have been considered as target alteration minerals and zones.

Table 3. The Eigenvector matrix values derived from principal component analysis (PCA) for mapping iron oxide/hydroxides. (**A**) Bands 1, 3, 4 and 5 of Landsat-7 ETM+; (**B**) Bands 2, 4, 5 and 6 of Landsat-8; and (**C**) Bands 1, 2, 3 and 4 of ASTER.

(A)				
Eigenvector	Band 1	Band 3	Band 4	Band 5
PCA 1	0.442	0.536	0.386	0.605
PCA 2	0.095	0.616	−0.771	−0.123
PCA 3	0.420	0.258	0.383	−0.780
PCA 4	0.786	−0.515	−0.328	0.091

(B)				
Eigenvector	Band 2	Band 4	Band 5	Band 6
PCA 1	0.399	0.444	0.544	0.587
PCA 2	−0.374	0.037	−0.555	0.741
PCA 3	−0.566	−0.510	0.614	0.201
PCA 4	0.616	−0.734	−0.126	0.253

(C)				
Eigenvector	Band 1	Band 2	Band 3	Band 4
PCA1	0.320	0.360	0.562	0.671
PCA2	0.265	0.506	−0.779	0.253
PCA3	−0.396	−0.539	−0.259	0.695
PCA4	0.817	−0.567	−0.092	−0.008

Table 4. The Eigenvector matrix values derived from PCA for mapping hydroxyl-bearing minerals. (**A**) Bands 1, 4, 5 and 7 of Landsat-7 ETM++; (**B**) Bands 2, 5, 6 and 7 of Landsat-8; and (**C**) Bands 1, 3, 4 and 6 of ASTER.

(A)				
Eigenvector	Band 1	Band 4	Band 5	Band 7
PCA 1	0.455	0.401	0.629	0.484
PCA 2	−0.002	−0.839	0.129	0.528
PCA 3	0.865	−0.130	−0.475	−0.086
PCA 4	−0.208	0.343	−0.599	0.692
(B)				
Eigenvector	Band 2	Band 5	Band 6	Band 7
PCA 1	0.388	0.529	0.573	0.490
PCA 2	0.294	0.698	−0.410	−0.506
PCA 3	0.841	−0.426	−0.298	0.142
PCA 4	−0.232	0.223	−0.643	0.694
(C)				
Eigenvector	Band 1	Band 3	Band 4	Band 6
PCA1	0.284	0.498	0.599	0.558
PCA2	0.062	−0.839	0.202	0.499
PCA3	0.839	0.011	−0.527	0.127
PCA4	0.457	−0.215	0.567	−0.649

Table 5. The Eigenvector matrix values derived from PCA for mapping advanced argillic, argillic-phyllic and propylitic alteration zones using ASTER VNIR+SWIR bands. (**A**) Bands 1, 4, 6 and 7 for advanced argillic zone mapping; (**B**) Bands 1, 3, 5 and 6 for argillic-phyllic zone mapping; and (**C**) Bands 1, 3, 5 and 8 for propylitic zone mapping.

(A)				
Eigenvector	Band 1	Band 4	Band 6	Band 7
PCA1	−0.28	−0.593	−0.553	−0.514
PCA2	−0.787	0.589	−0.08	−0.166
PCA3	−0.543	−0.549	0.509	0.382
PCA4	−0.089	0.003	−0.655	0.75
(B)				
Eigenvector	Band 1	Band 3	Band 5	Band 6
PCA1	0.297	0.512	0.551	0.586
PCA2	0.01	−0.845	0.366	0.389
PCA3	0.954	−0.15	−0.154	−0.207
PCA4	0.027	−0.004	−0.733	0.679
(C)				
Eigenvector	Band 1	Band 3	Band 5	Band 8
PCA1	0.307	0.528	0.568	0.552
PCA2	0.014	−0.832	0.341	0.438
PCA3	0.95	−0.154	−0.233	−0.142
PCA4	−0.059	0.073	−0.712	0.696

3.3. Bayesian Networks Model

A Bayesian network is an interpreted directed acyclic graph (DAG), which is able to model uncertain relationships between variables in a complex system [76–79]. The mathematical concepts of the Bayesian networks model can be summarized as follows [43,77]. The subclass x belongs to a class of a set of classes $\omega_1, \omega_2, \ldots, \omega_n$, if a class is defined by the highest conditional probability. The conditional probability is calculated using Equation (1):

$$P(\omega_i|x) = \frac{P(\omega_i|x)P(\omega_i)}{P(x)}, \qquad (1)$$

where P(x) is the non-conditional probability and $P(\omega_i)$ is the prior probability of each class. The prior probability is calculated by dividing the number of samples in each class by the total number of samples [43]. In this method, a probability distribution function (PDF) is assigned for each class. Then, the training data is exploited to estimate the parameters involved in the PDF. The covariance matrix and the mean vector are calculated as the parameters of a Gaussian probability function provided that the data is normally distributed [76]. In other words, it is mathematically formulated as follows:

$$g_i(x) = \frac{1}{(2\pi)^{\frac{m}{2}}|\Sigma_i|^{\frac{1}{2}}} \exp\left[\tfrac{1}{2}(x-\mu_i)^T \Sigma_i^{-1}(x-\mu_i)\right] \times P(\omega_i) \qquad (2)$$
$$i = 1, 2, \ldots, c$$

In this equation (Equation (2)), m is the number of variables, which is added to μ_i and Σ_i of the mean vector and an m*m covariance matrix of the ith class that calculated using Equations (3) and (4):

$$\mu_i = \frac{1}{n_i} \sum_{j=1}^{n_i} x_{ji} \qquad (3)$$

$$\sum_i = \frac{1}{n_i} \sum_{j=1}^{n_i} (x_{ji} - \mu_i)(x_{ji} - \mu_i)^T. \qquad (4)$$

Bayesian networks model uses a structural graph known as a DAG to represent the knowledge about different domains or random variables [41]. The DAG is defined by the nodes and the directed edges. The former and the latter represent random variables and the relationship among variables, respectively, as it is shown in Figure 3. As can be seen from the direction of the arrow in Figure 3, there is a direct relationship between x_i and x_j. The x_i (known as the parent node) is a dependent variable of the x_j (known as an offspring node) [43].

Figure 3. A schematic diagram depicting a general Bayesian network model [43].

There are different forms of Bayesian networks (See Reference [41] and references therein). One of the most popular forms of Bayesian networks is Naive Bayes (NB) classifier [80,81]. It is a simple structured algorithm with a single parent node and a number of offspring nodes [76,79,80]. A typical NB classifier diagram is shown in Figure 4. It is not only straightforward and easy to construct but also, no training procedure is required in the NB classifier [81]. The NB classifier undertakes comprehensive conditional independence between characteristics, which is impracticable for several predictor patterns utilized in mineral potential mapping [41]. In this study, the NB classifier was used for fusing the thematic layers derived from Landsat-7 ETM+, Landsat-8 and ASTER satellite sensors for generating a mineral potential map for the Ahar-Arasbaran region.

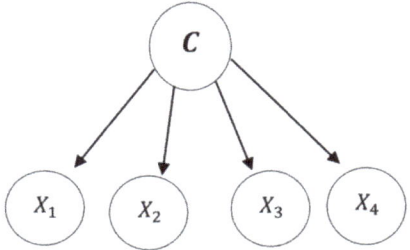

Figure 4. A typical Naive Bayes classifier diagram [79].

3.4. Fieldwork Data and Laboratory Analysis

The locations of hydrothermal alteration zones and their spatial relation with epithermal gold mineralization were systematically investigated using Global positioning system (GPS) survey in the study area (several field campaigns from June to August 2018). A handheld GPS (Garmin, Etrex Vista Hcx), with an average accuracy of 7 m, was used to record the hydrothermal alteration locations. Numerous field photographs and rock samples (120 samples) were collected from the alteration zones and ore mineralization. Rock samples were utilized for laboratory analysis to prepare thin and polished sections of altered rocks and ore mineralization as well as X-ray diffraction (XRD) analysis. Mineralogical compositions were analyzed using an Asenware AW-XDM 300 X-ray diffractometer (voltage: 40 Kv, current: 30 mA, step time: 1s and step size: 0.05° 2θ) at the Zarazma Mineral Studies Company, Tehran, Iran. Besides, the confusion matrix (error matrix) and Kappa Coefficient were calculated for hydrothermal alteration mineral mapping derived from remote sensing analysis versus field data.

4. Results

4.1. Generating Thematic Layers Using Multi-Sensor Remote Sensing Data

Figure 5A–C shows iron oxide/hydroxide zones (gossan) derived from 3/1 band ratio of Landsat-7 ETM+, 4/2 band ratio of Landsat-8 and 2/1 band ratio of ASTER, respectively. Figure 5A shows the spatial distribution of iron oxide/hydroxide minerals derived from the Landsat-7 ETM+ band ratio as red pixels. Most of the documented gold mineralizations are associated with iron oxide/hydroxide zones (gossan), especially in the northern and northeastern parts of the study area. The spatial distribution of iron oxide/hydroxide minerals in the Landsat-8 band ratio image (Figure 5B) is almost similar to Landsat-7 ETM+ resultant image. But, it is extensive in some locations in the northwestern and southeastern parts of the selected subset scene. Figure 5C shows the ASTER band ratio resultant image. The surface abundance of iron oxide/hydroxides in this image is lower compared to the Landsat-7 ETM+ and Landsat-8 results. However, the high concentration of iron oxide/hydroxides was mapped in the northwestern part of the study area using the ASTER band ratio (Figure 5C). Regarding the geological map of the Ahar-Arasbaran region (see Figure 1), iron oxide/hydroxide minerals were mapped along with geological lineament features and igneous rocks (granite, granodiorite, biotite granite, andesite, dasite and basalt), volcano sedimentary units and massive and bedded limestone.

Typically, hydroxyl-bearing (Al-OH and Fe,Mg-OH) minerals and carbonates zones were mapped in Figure 6A–C using the 5/7 band ratio of Landsat-7 ETM+ (A), 6/7 band ratio of Landsat-8 (B) and 4/9 band ratio of ASTER (C). The green pixels depict OH-alteration and carbonates, which are normally concentrated in igneous rock (granite, granodiorite, biotite granite, andesite and dasite), volcano sedimentary units and limestone. The OH-alteration minerals are more strongly mapped in the Landsat-8 and ASTER resultant images compared to Landsat-7 ETM+ image (Figure 6A–C). Almost all of the documented gold occurrences have an adjoining spatial relationship with hydroxyl-bearing alteration minerals; it is particularly observable in the Landsat-8 resultant image (Figure 6B). It may be due to the high signal to noise radiometer performance of Landsat-8 data, which allows detecting subtle variation in surface conditions [58].

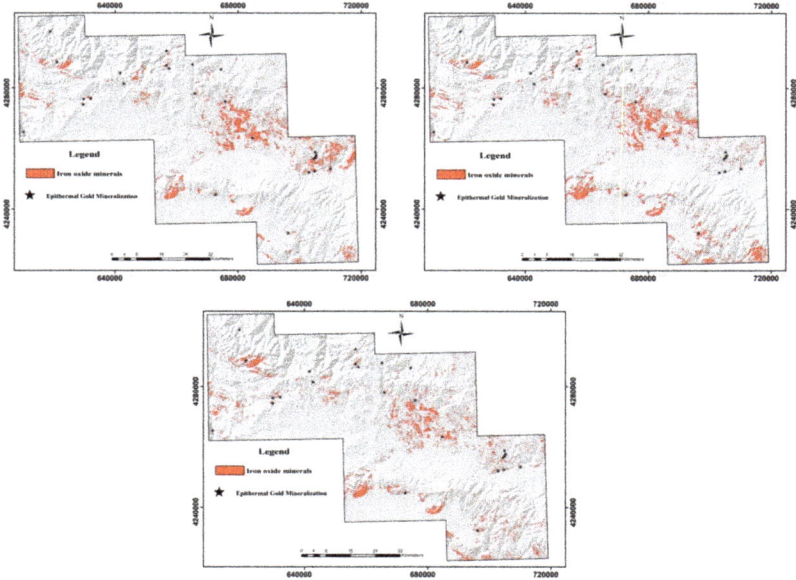

Figure 5. Spatial distribution of iron oxide/hydroxide zones (gossan) in the study area overlaid on hill shade. (**A**) The 3/1 band ratio image of Landsat-7 ETM+; (**B**) the 4/2 band ratio image of Landsat-8; (**C**) the 2/1 band ratio image of ASTER.

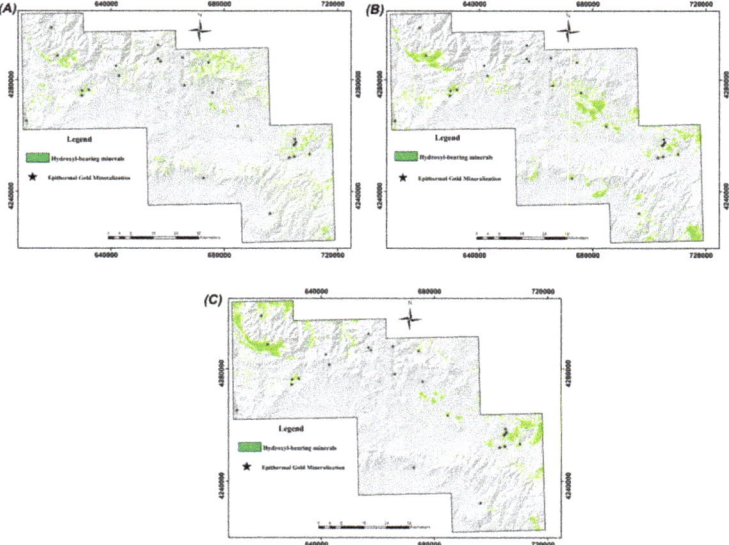

Figure 6. Spatial distribution of hydroxyl-bearing minerals and carbonates in the study area overlaid on hill shade. (**A**) The 5/7 band ratio image of Landsat-7 ETM+; (**B**) the 6/7 band ratio image of Landsat-8; (**C**) the 4/9 band ratio image of ASTER.

ASTER band ratios were used to specifically map the surface distribution of hydrothermal alteration zones in the study area. Figure 7A–C shows the advanced argillic alteration zone derived from 4/6 (A), the argillic-phyllic alteration zone derived from 5/6 (B) and the propylitic alteration zone derived from 5/8 (C), respectively. Concerning the geology map of the study area (see Figure 1), the advanced argillic alteration zone is corresponded to igneous, volcano sedimentary units and limestone; the argillic-phyllic alteration zone is associated with granite, granodiorite, andesite, dasite, rhyolite, trachyte, limestone units and sedimentary rocks; the propylitic alteration zone is typically concentrated with andesite, dasite, volcano sedimentary units and limestone (Figure 7A–C). The high surface abundance of argillic-phyllic and propylitic alteration zones was mainly mapped in the northwestern part of the study area. The spatial distribution of the advanced argillic alteration zone (Figure 7A) is intensely matched with hydroxyl-bearing mineral zones that mapped by Landsat-7 ETM+ and Landsat-8 band ratio images (Figure 6A,B). The documented gold mineralizations have closer spatial relationship with the advanced argillic alteration zone compared to the argillic-phyllic and propylitic alteration zones in the study area.

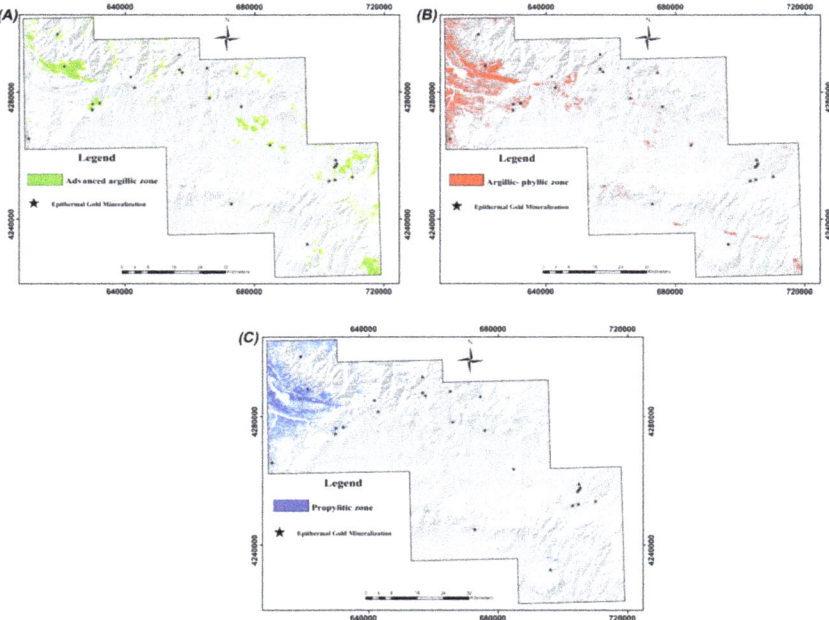

Figure 7. Spatial distribution of hydrothermal alteration zones in the study area overlaid on hill shade. (**A**) The advanced argillic alteration zone (4/6 band ratio image of ASTER); (**B**) the argillic-phyllic alteration zone (5/6 band ratio image of ASTER); (**C**) the propylitic alteration zone (5/8 band ratio image of ASTER).

Detailed mapping of advanced argillic, argillic-phyllic, propylitic and hydrous silica-affected alteration zones was obtained using the RDB1 (4 + 6/5), RDB2 (5 + 7/6), RDB3 (6 + 9/7 + 8) and RDB4 (5 + 8/6 + 7) of ASTER (Figure 8). Red pixels show advanced argillic zones, which are mostly distributed in the eastern and southeastern parts of the selected subset scene. Comparison to the geological map of the study area (see Figure 1), suggests that the advanced argillic zones are typically associated with granite and granodiorite rocks. Some of the documented gold mineralizations show close spatial relationship with the advanced argillic zones, especially in the eastern part of the study area (Figure 8). Argillic-phyllic alteration zone depict as green pixels. This alteration zone is distributed in many parts of the study area, which are normally associated with andesite, dasite, volcano sedimentary units and sedimentary rocks (e.g., sandstone, siltstone, marl and conglomerates). Due to high content of detrital clays (montmorillonite, illite and kaolinite) in the sedimentary units, argillic-phyllic alteration zone could also be mapped with exposures of sedimentary rocks [35]. The surface abundance of hydrous silica-affected alteration zone (blue pixels) is low and mostly detected in the southwestern and northwestern parts of the study zone (Figure 8). The hydrous silica zone was commonly identified with sedimentary units (conglomerates and sandstone), although this alteration zone is correspondingly adjacent to some of the gold mineralization zones in the northwestern part of the study area. Propylitic zone (yellow pixels) was strongly mapped in the selected subset scene (Figure 8). With regard to the geology map of the study area (see Figure 1), the spatial distribution of the propylitic zone typically corresponds with massive and bedded limestone, volcano sedimentary units and intermediate to mafic igneous rocks. It is because carbonates and alteration products of mafic minerals contain the strong contribution of CO_3 and Fe,Mg-OH mineral groups, which produce similar spectral features to propylitic alteration zone. However, this alteration zone is one of the dominant mineral assemblages that mapped near to the gold mineralization zones, especially in the northwestern and northern parts of the study area (Figure 8).

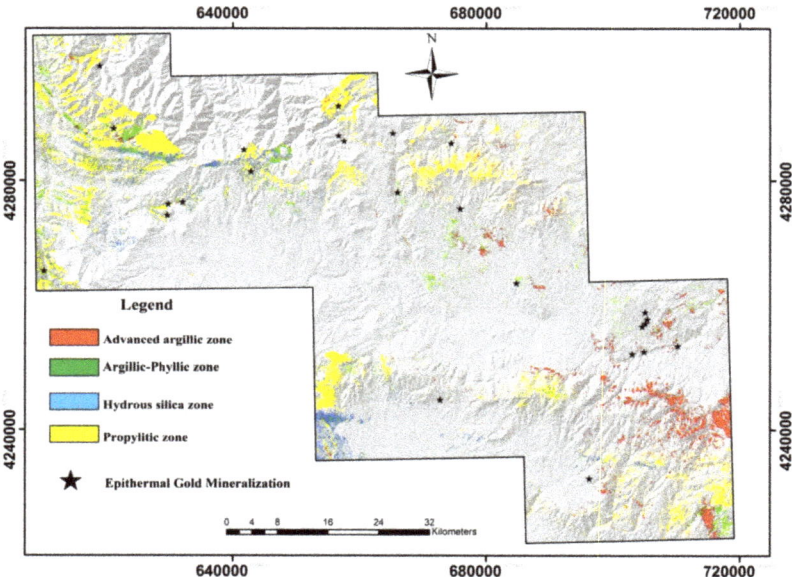

Figure 8. The RDB1(4 + 6/5), RDB2 (5 + 7/6), RDB3 (6 + 9/7 + 8) and RDB4 (5 + 8/6 + 7) of ASTER shows advanced argillic, argillic-phyllic, propylitic and hydrous silica-affected alteration zones in the study area overlaid on hill shade.

Table 3 shows the eigenvector loadings derived from PCA for mapping iron oxide/hydroxides (gossan) using bands 1, 3, 4 and 5 of Landsat-7 ETM+, bands 2, 4, 5 and 6 of Landsat-8 and bands 1, 2, 3 and 4 of ASTER. Analyzing the eigenvector loadings for Landsat-7 ETM+ selected bands (1, 3, 4 and 5) indicates that the PCA3 contains unique contribution (magnitude and sign of eigenvector loadings) of iron oxide/hydroxide minerals. The PCA3 has moderate loadings of band 1 (0.420) and strong loadings of band 5 (−0.780) with opposite signs (Table 3A). Band 1 (0.45–0.52 μm) of Landsat-7 ETM+ is positioned at absorption features of iron oxide/hydroxides (band 1 is considered an absorption band herein), while band 5 (1.55–1.75 μm) of Landsat-7 ETM+ is positioned at reflectance properties of iron oxide/hydroxides (band 5 is considered a reflection band herein). Thus, iron oxide/hydroxide minerals appear as dark pixels in the PCA3 due to negative sing in the reflection band (band 5), which were subsequently converted to bright pixels by negation. Figure 9A shows the resultant PCA3 image. Iron oxide/hydroxide minerals (red pixels) are mainly represented in the northern and northwestern parts of the study area, which are associated with some of the gold occurrences. However, a number of epithermal gold mineralizations do not show the spatial relationship with high abundance of iron oxide/hydroxide minerals, which are located in the southern and western parts of the study area.

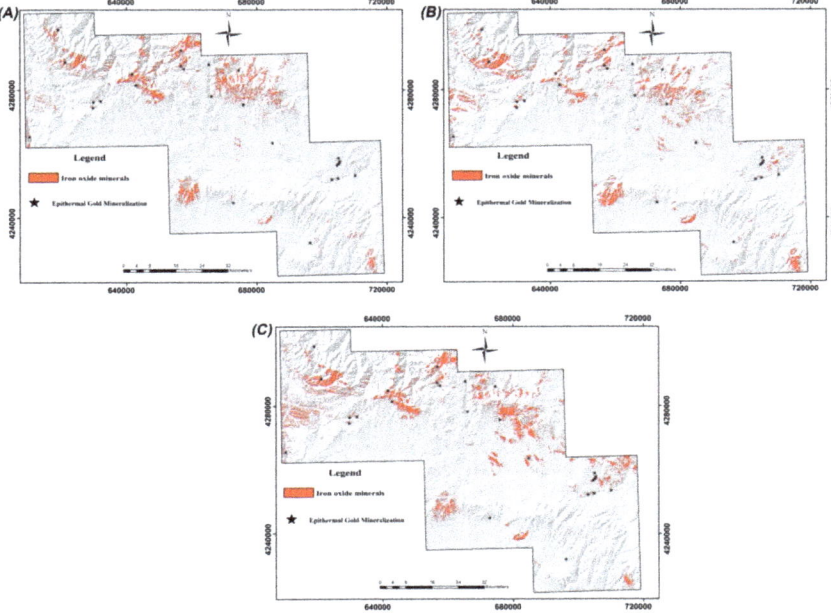

Figure 9. The PCA images derived from Landsat-7 ETM+, Landsat-8 and ASTER selected bands for mapping iron oxide/hydroxide zones (gossan) in the study area overlaid on hill shade. (**A**) The PCA3 image of Landsat-7 ETM+; (**B**) the PCA2 image of Landsat-8; (**C**) the PCA3 image of ASTER.

Analysis of the eigenvector loadings of Landsat-8 selected bands (2, 4, 5 and 6) shows that the PCA2 can be used for mapping oxide/hydroxide minerals. The PCA2 contains moderate to strong contribution of bands 2 (−0.374) and 5 (−0.555) as absorption bands and strong contribution of band 6 (0.741) with a positive sign as reflection band (Table 3B). As a result, iron oxide/hydroxide minerals manifest as bright pixels in the PCA2 image (Figure 9B). The spatial distribution of iron oxide/hydroxide minerals (red pixels) in Landsat-8 results is identical with Landsat-7 ETM+ PCA3 image but it is stronger in some parts, mainly in the southern and western sectors. Iron oxide/hydroxide minerals can be detected using the PCA3 derived from ASTER selected bands (1, 2, 3 and 4). The PCA3 shows moderate to strong loadings in absorption bands, including band 1 (−0.396), band 2 (−0.539) and band 3 (−0.259) with a negative sign and strong and positive loading in band 4 (0.695) as reflection band (Table 3C). Hence, iron oxide/hydroxide minerals represent bright pixels (Figure 9C). A higher abundance of iron oxide/hydroxide minerals was mapped in the ASTER PCA2 image compared to Landsat-7 ETM+ and Landsat-8 PCA images, which is typically matched with most of the gold mineralizations.

The pixels contain iron oxide/hydroxide minerals mapped by PCA images show a better spatial relationship with the gold mineralization zones compared to band ratio images (see Figure 5A–C). It indicates that the selective PCA can specially detect the alteration pixels in the spatial domain. Table 4 shows the eigenvector loadings derived from PCA for mapping hydroxyl-bearing minerals using bands 1, 4, 5 and 7 of Landsat-7 ETM+, bands 2, 5, 6 and 7 of Landsat-8 and bands 1, 3, 4 and

6 of ASTER. Considering the eigenvector loadings contain unique contribution of hydroxyl-bearing minerals in the absorption and reflection bands, it is discernible that the PCA4 includes the unique contribution of OH-minerals for all selected datasets (Table 4A–C). The PCA4 derived from Landsat-7 ETM+ selected bands (1, 4, 5 and 7) shows a strong negative loading in band 5 (−0.599) and a strong positive loading in band 7 (0.692) (Table 4A). Because of negative loading in the reflection band (band 5), the hydroxyl-bearing minerals are represented as dark pixels in the PCA4, which are inverted to bright pixels by multiplication to −1, subsequently (Figure 10A). Surface distribution of hydroxyl-bearing minerals (green pixels) depicts in the PCA4 image of Landsat-7 ETM+. The PCA4 derived from Landsat-8 selected bands (2, 5, 6 and 7) contains a strong negative loading of band 6 (−0.643) (the reflection band) and a strong positive loading of band 7 (0.694) (the absorption band) (Table 4B). Therefore, the PCA4 image was negated to depict the OH-minerals as bright pixels. Figure 10B shows the resultant image. For ASTER selected bands (1, 3, 4 and 6), the PCA4 has a strong positive contribution of band 4 (0.567) and a strong negative contribution of band 6 (−0.649) (Table 4C). Hence, the hydroxyl-bearing minerals appear as bright pixels in the PCA4 image. Figure 10C manifests the spatial distribution of the OH-minerals as green pixels in the PCA4 image of ASTER. Comparison of the PCA images to the band ratio images (see Figure 6A–C) suggests that the pixels detected in the selective PCA method show a closer spatial relationship to the gold mineralization zones and have a stronger manifestation in the image-maps.

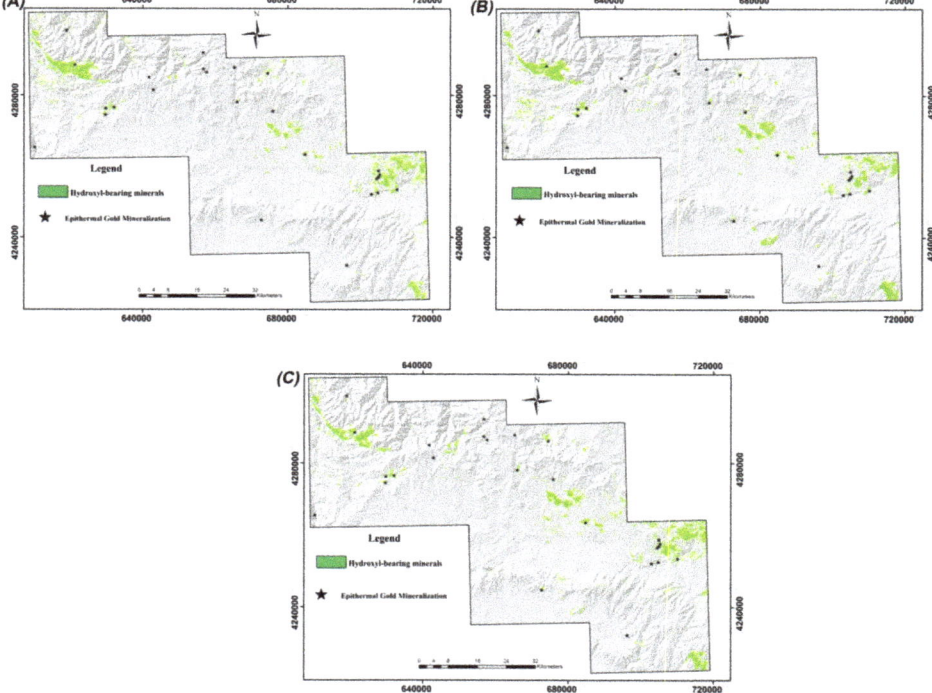

Figure 10. The PCA images derived from Landsat-7 ETM+, Landsat-8 and ASTER selected bands for mapping hydroxyl-bearing minerals in the study area overlaid on hill shade. (**A**) The PCA4 image of Landsat-7 ETM+; (**B**) the PCA4 image of Landsat-8; (**C**) the PCA4 image of ASTER.

Table 5 shows the eigenvector loadings for mapping advanced argillic, argillic-phyllic and propylitic alteration zones using ASTER bands such as bands 1, 4, 6 and 7 for the advanced argillic zone, bands 1, 3, 5 and 6 for the argillic-phyllic zone and bands 1, 3, 5 and 8 for the propylitic zone.

Considering the magnitude and sign of eigenvector loadings for mapping advanced argillic zone (Table 5A), it is evident that the PCA3 contains spectral information to map advanced argillic zone due to a strong negative loading in band 4 (−0.549) and a strong positive loading in band 6 (0.509). Dark pixels depict the alteration zone due to a negative sign in the reflection band (band 4), which are afterward converted to bright pixels. The analysis of eigenvector loadings for mapping argillic-phyllic zone indicates that the PCA4 can mainly detect argillic-phyllic zone because of the strong contribution of bands 5 (−0.733) and 6 (0.679) with inverse signs (Table 5B). Muscovite (as a typical and dominant mineral in the phyllic zone) shows strong absorption in band 6 of ASTER, while lower absorption in band 5 of ASTER [35,37]. Thus, band 5 is assumed to be a reflection band and band 6 is considered as a strong absorption band herein. As a result, argillic-phyllic zone manifests as dark pixels due to negative sign in the band 5 (reflection band). Then, dark pixels were inverted to bright pixels by negation. Propylitic alteration zone can be mapped in the PCA4 image because of strong eigenvector loadings in band 5 (−0.712) and band 8 (0.696) with opposed signs (Table 5C). Herein, band 5 is pondered as reflection band and band 8 is deliberated as absorption band. Fe,Mg-OH and CO_3 mineral groups (propylitic zone: chlorite, epidote and calcite) have high absorption properties in band 8 (2.295–2.365 µm) and reflection (very low absorption) features in band 5 (2.145–2.185 µm) of ASTER [35,39]. Accordingly, propylitic alteration zone appears as dark pixels that were negated to bright pixels in the PCA4 image.

Figure 11 shows PCA image-map derived from the PCA3 image for advanced argillic mapping, the PCA4 image for argillic-phyllic zone mapping and the PCA4 image for propylitic zone mapping.

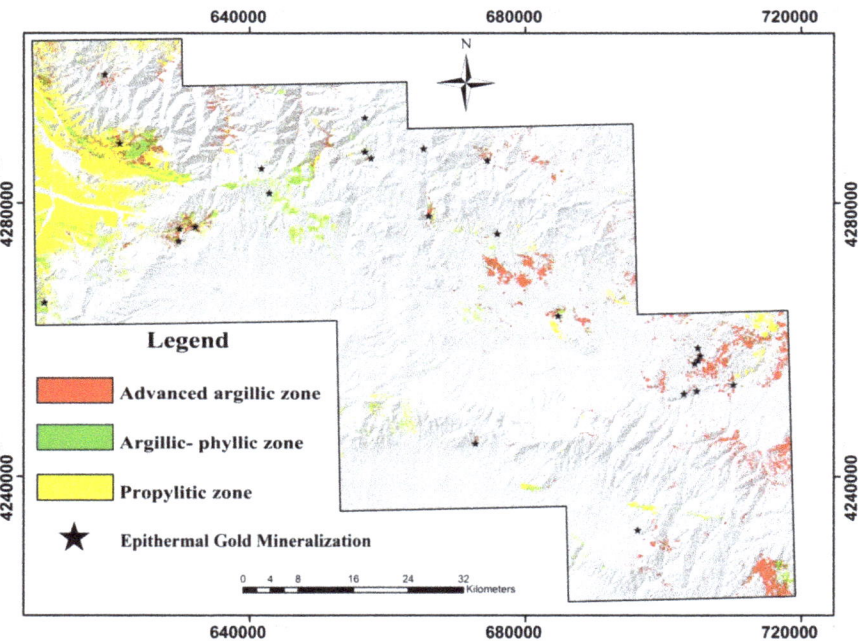

Figure 11. The PCA image-map derived from ASTER selected PCAs for mapping advanced argillic, argillic-phyllic and propylitic alteration zones in the study area that overlaid on hill shade.

The spatial distribution of advanced argillic zones is stronger in the northeastern parts and weaker in southeastern part of the study area compared to RDBs image-map (see Figure 8). The advanced argillic zone resulting from PCA shows remarkable vicinity to the gold mineralization (Figure 11). The argillic-phyllic zone shows nearly similar surface distribution to RDBs image-map. However, the high concentration of propylitic zone was mapped in the northwestern part of the study area in the PCA image-map compared to RDBs image-map (see Figure 8). On the other hand, the propylitic zone shows lower spatial distribution in the northeastern and southeastern parts of the PCA image-map (Figure 11).

4.2. Fusing Thematic Layers Using Naive Bayes (NB) Classifier

The thematic layers of hydrothermal alteration zones derived from Landsat-7 ETM+, Landsat-8 and ASTER datasets were fused using the NB classifier to generate a mineral potential map for the Ahar-Arasbaran region. A DAG was designed for the thematic layers produced by image processing techniques in this study (Figure 12). Eight distinct layers were employed as independent predictive layers, including iron oxide minerals derived from Landsat-7 ETM+, hydroxyl-bearing minerals derived from Landsat-7 ETM+, iron oxide minerals derived from Landsat-8, hydroxyl-bearing minerals derived from Landsat-8, iron oxide minerals derived from ASTER, advanced argillic alteration derived from ASTER, argillic-phyllic alteration derived from ASTER and propylitic alteration derived from ASTER.

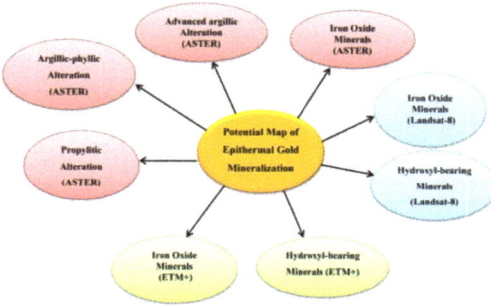

Figure 12. The DAG diagram used in this study for the fusing the thematic layers produced by image processing techniques.

The DAG was used to integrate the predictor variables. It yields a posterior probability map showing the probability of gold mineralization occurrences. Subsequently, the following steps were taken to generate the posterior probability map. To train the DAG, 25 known gold mineralizations in the study area were selected as positive sites and 26 non-mineralized locations were selected as non-deposit (negative) sites, which have already been verified by field survey. In the next stage, the thematic layers (alteration image-maps) were resampled to a cell size of 150 * 150 m and a buffer zone of 300 m was considered around the positive and negative sites. The training data, the pixels superimposed by the positive and negative sites, containing a total of 468 pixels. Each pixel was considered as a vector of 8 arrays, including the values of 8 thematic (alteration) layers. To train the model, 70% of these pixels were used, while 30% of the pixels were used to validate the model generated. The calculation of the confusion matrix shows a total accuracy of 85.1%, which indicates that the model has been hypothesized and established. Having the trained NB model, all the data were used as the input of the model to generate the posterior probability map. However, the map is also required subsequent classification; thus the natural breaks algorithm [81] was used for classification of the posterior probability map. Three threshold values (produced by the foregoing algorithm) were used to generate a four-class map showing the probability of epithermal gold occurrences. The classes are highly probable (red), probable (green), moderately probable (yellow) and improbable (gray). As a result, a mineral potential map for the Ahar-Arasbaran region was produced (Figure 13). Most of the known gold mineralizations are located in the highly probable (red) zone, although a small number of the gold occurrences can be seen in the probable (green) and moderately probable (yellow) zones. Many high probable zones in the northwestern, northern, northeastern, southeastern and southwestern parts of the study area contain high potential for undiscovered epithermal gold mineralizations (Figure 13).

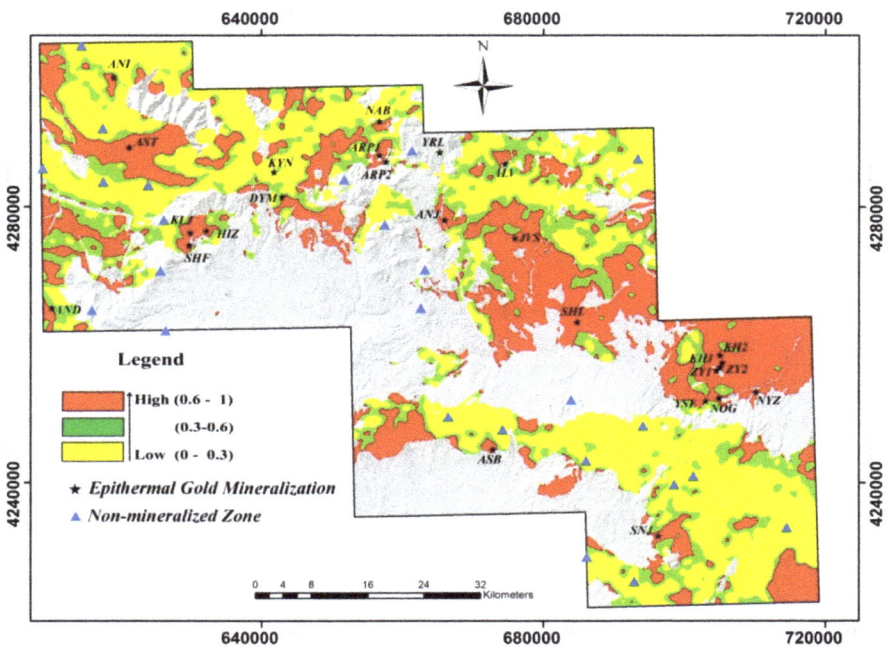

Figure 13. Mineral potential map of the Ahar-Arasbaran region produced using the NB classifier. For abbreviation to epithermal gold mineralizations, refer to Figure 1.

4.3. Verifying the Results Using Field Data and Laboratory Analysis

Several GPS surveys were carried out in different parts of the Ahar-Arasbaran region for verifying the mineral potential map and discovering new prospective zones of epithermal gold mineralizations, especially in highly probable zones. Numerous field photographs and rock samples were collected from different types of alteration zones related to gold mineralization such as advanced argillic, argillic-phyllic, propylitic and hydrous silica. In this investigation, some of the gold mineralization areas (highly probable zone), such as Zailig, Noghdouz, Javan-Sheikh, Nabi-Jan and Sonajil, were selected for a detailed field excursion, petrographic study and XRD analysis. The advanced argillic alteration, argillic-silica alteration, silica alteration and propylitic alteration were identified in the Zailig area (Figure 14A–D). The advanced argillic alteration is the most extensive alteration zone in the vicinity of gold mineralizations in the Zailig area (Figure 14A,B). The silica alteration is identified in the form of silica major clasts along with iron oxides (Figure 14C). The other type of alteration zones is argillic-silica alteration, which is placed around the silica veins associated with gold mineralization (Figure 14D). Figure 15A,B) shows microphotographs of argillic-silica alteration. Primary plagioclase replaced by sericite, clay minerals and jarosite (Figure 15A). Recrystallized quartz and relics of plagioclase are surrounded by clay minerals (Figure 15B). Propylitic alteration zone were also found as distal alteration zone in the Zailig area (Figure 16A–D). Secondary minerals for instance, chlorite, epidote and calcite replaced original mineralogy (feldspars) as vesicular and amygdaloidal textures in the propylitic zone (Figure 16B). Microphotographs of the propylitic zone show that the phenocrysts of plagioclase are replaced by chlorite, epidote and calcite (Figure 16C,D). Quartz is phenocrystalline and anhedral in the background, while plagioclase is euhedral and partially replaced by epidote (Figure 16C). The amygdaloidal texture is observable in Figure 16D, which amygdales are filled with calcite and quartz.

Figure 14. Field photographs of typical hydrothermal alteration zones in the Zailig area. (**A**) View of argillic alteration zone close to the quartz veins; (**B**) Regional view of advanced argillic alteration zone; (**C**) View of silicification alteration zone with iron oxides, (**D**) Close view of argillic-silica alteration and a sample of crustiform and colloform banded chalcedonic gold-quartz vein.

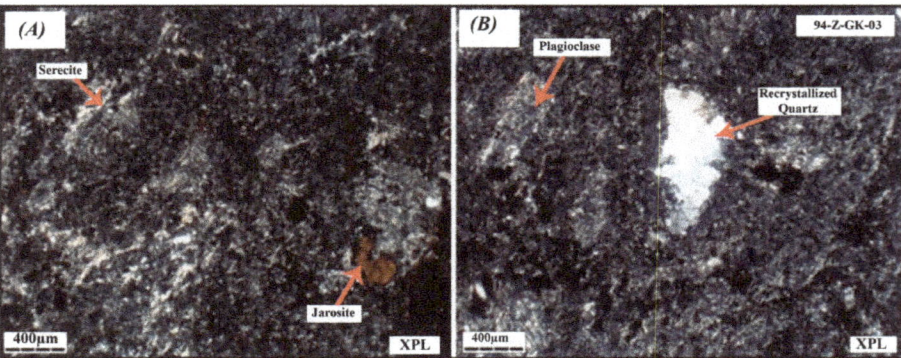

Figure 15. Microphotographs of argillic-silica alteration. (**A**) Original mineralogy (feldspars) replaced by sericite, jarosite and clay minerals; (**B**) Recrystallized large-grained quartz and relics of plagioclase in the background of clay minerals.

Figure 16. Propylitic alteration zone in the Zailig area. (**A**) View of Propylitic alteration zone; (**B**) Vesicular and amygdaloidal textures in a hand specimen of propylitic zone; (**C**) Microphotographs of plagioclase that is partially replaced by epidote and recrystallized large-grained quartz in the background; (**D**) Microphotographs of amygdaloidal texture in the propylitic zone that amygdales are filled with calcite, chlorite and quartz.

Some typical silicified and breccia (quartz veins) zones occur in altered granitic and andesitic rocks in the Noghdouz gold mineralization area (Figures 17 and 18). The specimens of silicified zone show breccia and clastic textures. The cement and major clasts of the breccia textures are composed of silicate minerals (Figure 17A–C). Epithermal gold mineralization occurs in the breccia zone (quartz veins) in the altered granitic host rocks. This mineralization is also associated with advanced argillic alteration (Figure 18A,B).

Figure 17. Silicified zone in the Noghdouz gold mineralization area. (**A**) Regional view of Silicified zone; (**B**) Close view of breccia textures in a hand specimen of the andesitic host rock; (**C**) View of silica clasts in a hand specimen of the breccia textures.

Figure 18. Breccia zone (quartz veins) in the granitic host rocks in the Noghdouz gold mineralization area. (**A**) View of breccia zone (quartz veins) in the granitic host rocks (Ahar-Meshkinshahr road); (**B**) Close view of quartz veins in the breccia zone and advanced argillic alteration.

Iron oxide alteration zone (limonitic-hematite rocks) and oxidized breccia with banded chalcedonic quartz are identified in Javan-Sheikh gold mineralization area (Figure 19A,B). The well-developed gossan covers (limonitic-hematite-silicic rocks) show rough and geologic relief features compared to surrounding altered rocks (Figure 19A). The size of gossan covers are around 200 to 300 m that are surrounded by more extensive zones of propylitic and phyllic-argillic alteration zones. Although, silicified zone is also associated with the gossan covers, partially. The epithermal gold mineralization of the Nabi-Jan area is located in quartz-sulfide veins and developed at the top of an intrusive body of granodiorite (Figure 20A,B). Gold mineralization is typically in the zones where intensely silicified and located in the advanced argillic alteration. In the Nabi-Jan area, the distal alteration zone is also propylitic alteration. The Sonajil gold mineralization occurs as a stock-work of thin quartz veins in granitoid rocks. The development of the argillic-phyllic alteration zone along with the siliceous zones and iron oxides were identified in the Kalijan area. Sphalerite, galena, chalcopyrite and pyrite are main sulfide mineralization associated with native gold mineralization (Figure 21A–E). Quartz, iron oxide/hydroxide and minor calcite are gangue minerals.

Mineralogical compositions of hydrothermal alteration zones were investigated by XRD analysis. Thirty samples from different hydrothermal alteration zones were analyzed for this study. Representative XRD analysis of samples collected from the iron oxide/hydroxide alteration (gossan covers), advanced argillic alteration, argillic-phyllic alteration, propylitic alteration and hydrous silica alteration (silicified zone) are shown in this paper (Figure 22A–E). Goethite, jarosite, gypsum and quartz are mineral phases that detected in the gossan cover (Figure 22A). In the advanced argillic alteration (Figure 22B), muscovite, illite, kaolinite, gypsum, orthoclase, albite and quartz are main mineralogical phases. The predominant minerals detected in the argillic-phyllic alteration are kaolinite, muscovite, illite, jarosite, albite and quartz (Figure 22C). Epidote, chlorite, calcite, albite and quartz are identified in the propylitic alteration (Figure 22D). Quartz, albite, jarosite, goethite, calcite, chlorite, gypsum and dolomite are observed in the silicified alteration zone (Figure 22E).

Figure 19. The iron oxides (limonitic-hematite rocks) in Javan-Sheikh area. (**A**) View of limonitic-hematite rocks; (**B**) View of oxidized breccia with banded chalcedonic quartz infill.

Figure 20. Gold mineralization in the Nabi-Jan area. (**A**) View of quartz-sulfide veins that developed at the top of an intrusive body of granodiorite; (**B**) Close view of quartz-sulfide gold mineralization in a hand specimen.

Figure 21. Microphotographs of sulfide mineralization in the Kalijan area (polished section). (**A**) Coarse anhedral sphalerite (Sp) in concordance with galena (Gn) and chalcopyrite (Chpy) (magnification: 10XPL); (**B**) Coarse anhedral sphalerite (Sp) intergrowth with galena (Gn) (white) and chalcopyrite (Chpy) (yellow) in quartz gangue (magnification: 10XPL); (**C**) Large anhedral form of chalcopyrite (Chpy), pyrite (Py) and sphalerite (Sp) (magnification: 20XPL); (**D**) Gold (Au) detected as an insulator inside the sphalerite (Sp) fracture (glossy yellow) (magnification: 20XPL); (**E**) Gold (Au) mineralization with a particle size of 10 microns along with an iron hydroxide (Lim) crystal surrounded by sphalerite (Sp) (magnification: 20XPL).

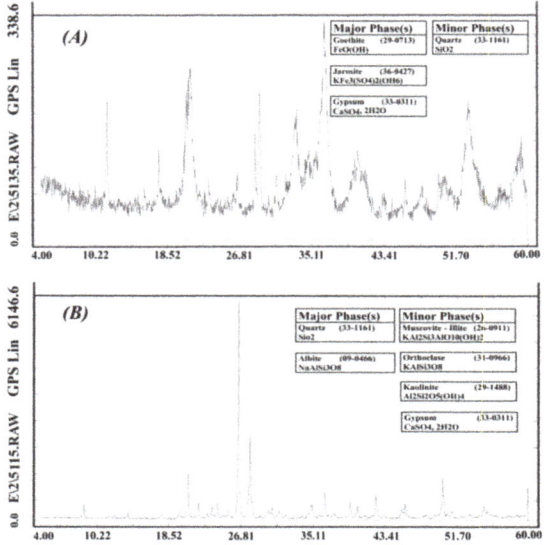

Figure 22. *Cont.*

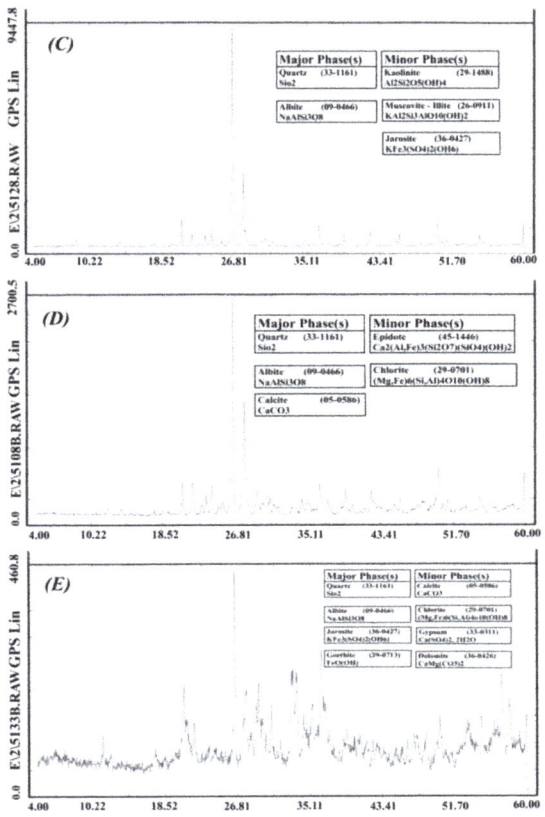

Figure 22. Representative XRD analysis of samples collected from hydrothermal alteration zones in the study area. (**A**) Iron oxide/hydroxide alteration (gossan covers); (**B**) Advanced argillic alteration; (**C**) Argillic-phyllic alteration; (**D**) Propylitic alteration; (**E**) Hydrous silica alteration (silicified zone).

In this analysis, confusion matrix and Kappa Coefficient [82–86] were used for assessing the accuracy of alteration mineral mapping derived from remote sensing analysis versus systematic GPS surveys collected from different alteration zones during fieldwork in the study area. Thirty representative GPS points were used for calculating the confusion matrix and Kappa Coefficient in this paper. Table 6 shows the locations of hydrothermal alteration zones recorded by a systematic GPS survey. Table 7 shows the confusion matrix for alteration mineral mapping versus field data. The results show the overall accuracy of 76.66% and Kappa Coefficient of 0.71. The advanced argillic alteration, argillic-phyllic and propylitic classes show the producer's accuracy of 83%, while the producer's accuracy for the iron oxide/hydroxide and hydrous silica classes is 67%. The highest user's accuracy is achieved for the argillic-phyllic and propylitic classes (100%), whereas the lowest user's accuracy is recorded for the iron oxide/hydroxide class (50%). The advanced argillic has the user's accuracy of 83% and hydrous silica class shows the user's accuracy of 67% (Table 7). Accordingly, the accuracy assessment results indicate that the alteration mineral mapping has appropriate match (overall accuracy: 76.66%) and very good degree of agreement (Kappa Coefficient: 0.71) with field data. However, some spectral mixing and confusion between alteration classes are also distinguishable. The iron oxide/hydroxide and hydrous silica classes show the highest feasibility for spectral mixing and confusion compared to other classes. The propylitic and argillic-phyllic classes contain the lowest spectral mixing and confusion. The advanced argillic class has some spectral mixing and confusion with the argillic-phyllic class.

Table 6. Locations of representative hydrothermal alteration zones recorded by systematic GPS survey during fieldwork in the study area.

	Alteration Zones	Coordinates
1	Advanced argillic	38°26.324′N–47°21.279′E
2	Advanced argillic	38°11.796′N–47°15.995′E
3	Advanced argillic	38°20.514′N–46°58.566′E
4	Advanced argillic	38°32.717′N–47°03.374′E
5	Advanced argillic	38°49.398′N–46°16.417′E
6	Advanced argillic	38°43.269′N–47°00.223′E
7	Iron oxide minerals	38°30.095′N–47°07.023′E
8	Iron oxide minerals	38°44.378′N–46°23.321′E
9	Iron oxide minerals	38°08.975′N–47°27.312′E
10	Iron oxide minerals	38°40.687′N–46°42.420′E
11	Iron oxide minerals	38°43.525′N–46°48.243′E
12	Iron oxide minerals	38°37.564′N–46°29.257′E
13	argillic-phyllic	38°10.102′N–47°28.384′E
14	argillic-phyllic	38°24.794′N–47°24.250′E
15	argillic-phyllic	38°36.792′N–46°43.712′E
16	argillic-phyllic	38°37.814′N–46°22.895′E
17	argillic-phyllic	38°21.812′N–46°51.621′E
18	argillic-phyllic	38°35.575′N–47°00.520′E
19	Hydrous silica	38°41.894′N–46°41.574′E
20	Hydrous silica	38°25.525′N–47°20.882′E
21	Hydrous silica	38°44.511′N–46°46.563′E
22	Hydrous silica	38°36.700′N–46°51.888′E
23	Hydrous silica	38°36.931′N–46°31.907′E
24	Hydrous silica	38°43.265′N–46°25.118′E
25	Propylitic	38°37.090′N–46°28.190′E
26	Propylitic	38°50.119′N–46°22.157′E
27	Propylitic	38°45.696′N–46°49.787′E
28	Propylitic	38°25.034′N–47°24.969′E
29	Propylitic	38°40.665′N–46°22.871′E
30	Propylitic	38°31.197′N–46°17.014′E

Table 7. Confusion matrix for alteration mineral mapping versus field data.

Class	Advanced Argillic	Iron Oxide/ Hydroxides	Argillic-Phyllic	Hydrous Silica	Propylitic	Totals (Field Data)	User's Accuracy
Advanced argillic	5	0	1	0	0	6	83%
Iron oxide/hydroxides	0	4	0	2	1	8	50%
Argillic-phyllic	1	0	5	0	0	5	100%
Hydrous silica	0	2	0	4	0	6	67%
Propylitic	0	0	0	0	5	5	100%
Totals (Remote sensing analysis)	6	6	6	6	6	30	
Producer's Accuracy	83%	67%	83%	67%	83%		
Overall accuracy = 76.66%				Kappa Coefficient = 0.71			

5. Discussion

Hydrothermal alteration mineral assemblages associated with gold mineralization that formed under low to medium temperatures (≤150 °C~300 °C) are deliberated as one of the most significant indicators for epithermal gold exploration [33,34,87–89]. Remote sensing satellite imagery is extensively and successfully used for mapping hydrothermal alteration zones for gold minerals exploration in many metallogenic provinces around the world [7,12,15,17–20,36,38,90–94]. In the Ahar-Arasbaran region, NW Iran, a variety of ore mineralizations such as Au, Cu-Au, Au-Ag, Fe-Au, Cu-Mo, Fe, Cu, Pb-Zn are identified, which are associated with widespread hydrothermal alteration minerals [47,49,51,95–97]. In this investigation, Landsat-7 ETM+, Landsat-8 and ASTER multi-sensor remote sensing satellite imagery was used to map hydrothermal alteration zones associated with epithermal gold mineralization in the Ahar-Arasbaran region. A Bayesian network model was subsequently used to fuse thematic layers of hydrothermal alteration zones derived from the multi-sensor satellite imagery for producing a mineral potential map of the study area.

Iron oxide/hydroxide zones (gossan cover), hydroxyl-bearing (Al-OH and Fe,Mg-OH) minerals and carbonates zones, advanced argillic, argillic-phyllic, propylitic and hydrous silica (silicified zone) alteration zones were mapped using band ratio, RBD and selective PCA image processing techniques. Using band ratios of 3/1 (Landsat-7 ETM+), 4/2 (Landsat-8) and 2/1 (ASTER) identify the spatial distribution of iron oxide/hydroxide zones, which are mainly associated with lineament features and igneous rocks, volcano sedimentary units and massive and bedded limestone (See Figure 5A–C). The documented epithermal gold occurrences mostly show close spatial locations with detected iron oxide/hydroxide zones. The PCA3 image of Landsat-7 ETM+ selected bands (1, 3, 4 and 5), the PCA2 image of Landsat-8 selected bands (2, 4, 5 and 6) and the PCA3 image of ASTER selected bands (1, 2, 3 and 4) were also represented iron oxide/hydroxide spatial distribution in the study area (see Figure 9A–C). The identified iron oxide/hydroxide zones using PCA are characteristically better matched with most of the gold mineralizations compared to band ratio images. Using band ratios of 5/7 (Landsat-7 ETM+), 6/7 (Landsat-8) and 4/9 (ASTER) detect the hydroxyl-bearing minerals and carbonates zones (see Figure 6A–C), which are generally matched with igneous rock, volcano sedimentary units and limestone. The gold mineralizations are typically located in the high abundance zones of hydroxyl-bearing/carbonate minerals. The advanced argillic, argillic-phyllic and propylitic alteration zones are mapped using ASTER band ratios of 4/6 (advanced argillic), 5/6 (argillic-phyllic), 5/8 (propylitic), respectively (see Figure 7A–C). The advanced argillic alteration shows closer spatial location with the gold mineralizations in comparison with the argillic-phyllic and propylitic alteration zones. The PCA4 image of Landsat-7 ETM+ selected bands (1, 4, 5 and 7), Landsat-8 selected bands (2, 5, 6 and 7) and ASTER selected bands (1, 3, 4 and 6) detects the surface distribution of hydroxyl-bearing minerals in the study area (see Figure 10A–C). The pixels detected in the PCA images show a stronger manifestation of OH-minerals compared to band ratio images and closer spatial relationship to the documented gold mineralization zones.

Implementing the RDB1 (4 + 6/5), RDB2 (5 + 7/6), RDB3 (6 + 9/7 + 8) and RDB4 (5 + 8/6 + 7) of ASTER reveal the advanced argillic, argillic-phyllic, propylitic and hydrous silica-affected alteration zones in the study area, comprehensively (see Figure 8). Some of the gold mineralizations in the eastern part of the study area are mainly situated in the advanced argillic zones. The hydrous silica zone was also mapped near some of the gold mineralization zones in the northwestern part of the study area. The propylitic zone is one of the main mineral assemblages associated with gold mineralization zones in the northern and northwestern parts of the study area. Only few gold occurrences were identified in the argillic-phyllic alteration zone. The PCA3 image derived from ASTER bands 1, 4, 6 and 7 for advanced argillic mapping, the PCA4 image derived from ASTER bands 1, 3, 5 and 6 for argillic-phyllic zone mapping and the PCA4 image derived from ASTER bands 1, 3, 5 and 8 for propylitic zone mapping show surface abundance of advanced argillic, argillic-phyllic and propylitic zone with some spatial discrepancies (see Figure 11) compared to RDBs image-map (see Figure 8). Notable vicinity to the documented gold mineralizations was mapped in the advanced argillic zone, which is detected with the PCA technique.

The produced thematic layers (see the DAG diagram in Figure 12) derived from band ratio and PCA image processing techniques are fused using the NB classifier. Consequently, a mineral potential map for the Ahar-Arasbaran region is produced (see Figure 13), which includes four classes such as highly probable, probable, moderately probable and improbable. Maximum numbers of the known gold occurrences are situated in highly probable class, while some of the gold mineralizations are located in the probable and moderately probable classes. Accordingly, several parts of the study area, such as the northwestern, northern, northeastern, southeastern and southwestern sectors, can be considered to be highly prospective zones for epithermal gold mineralizations and may contain undiscovered Au deposits (see Figure 13).

Detailed field expedition, petrographic study and XRD analysis in some of the prospective areas located in the highly probable zone show the presence of hydrothermal alteration zones associated with gold mineralizations. Extensive alteration mineral assemblages of the advanced argillic and argillic-silica alteration zones are found in the vicinity of gold mineralizations in the Zailig area (see Figure 14A–D). Microphotographs of argillic-silica alteration show that primary plagioclase replaced by sericite, clay minerals and jarosite and relics of plagioclase are surrounded by clay minerals. The distal alteration zone in the Zailig area is propylitic alteration zone, which contains chlorite, epidote and calcite that replaced original mineralogy (feldspars) as vesicular and amygdaloidal textures (see Figure 16A–D). In the Noghdouz area, gold mineralization is occurred in the breccia zone (quartz veins) in the altered granitic host rocks, which is associated with advanced argillic alteration (see Figure 17A–C and Figure 18A,B). Limonitic-hematite rocks and oxidized breccia with banded chalcedonic quartz are identified in Javan-Sheikh gold mineralization area (see Figure 19A,B), which are surrounded by propylitic and phyllic-argillic alteration zones. In the Nabi-Jan area, gold mineralization is associated with quartz-sulfide veins hosted by granodiorite (see Figure 20A,B), which strongly silicified and placed in advanced argillic alteration. Development of the argillic-phyllic alteration zone associated with the siliceous zones and iron oxides in granitoid rocks were identified with gold mineralization in the Sonajil area. Native gold mineralization is associated with sphalerite, galena, chalcopyrite and pyrite (see Figure 21A–E).

The XRD analysis of rock samples collected from different alteration zones is verified the presence of hydrothermal alteration minerals, including (i) goethite, jarosite, gypsum and quartz in the gossan cover; (ii) muscovite, illite, kaolinite, gypsum, orthoclase, albite and quartz in the advanced argillic alteration, (iii) kaolinite, muscovite, illite, jarosite, albite and quartz in the argillic-phyllic alteration; (iv) epidote, chlorite, calcite, albite and quartz in the propylitic alteration (see Figure 22A–E). The accuracy assessment results show the overall accuracy of 76.66% and Kappa Coefficient of 0.71 for hydrothermal alteration mapping using remote sensing datasets. It indicates that the alteration mineral mapping contains a suitable match and a very good degree of agreement with field data. Analyzing the producer's accuracy and user's accuracy shows that some spectral mixing and confusion between alteration classes are also feasible, especially for iron oxide/hydroxide and hydrous silica alteration groups and the advanced argillic and the argillic-phyllic alteration groups. Accordingly, the mineral potential map produced in this study using multi-sensor remote sensing imagery and Bayesian network model is viable and can be broadly applicable for epithermal gold exploration in the Ahar-Arasbaran region.

6. Conclusions

This investigation was accomplished to produce a mineral potential map for prospecting epithermal gold mineralization in the Ahar-Arasbaran region, NW Iran using multi-sensor remote sensing satellite imagery (e.g., Landsat-7 ETM+, Landsat-8 and ASTER) and the Bayesian network model. Iron oxide/hydroxide zones, hydroxyl-bearing minerals and carbonates zones, advanced argillic, argillic-phyllic, propylitic and silicified alteration zones were mapped in the Ahar-Arasbaran region using band ratio, RBD and selective PCA image processing techniques. The NB classifier was successfully implemented to fuse the thematic layers of hydrothermal alteration zones derived from the multi-sensor satellite imagery. As a result, a mineral potential map for the Ahar-Arasbaran region was produced, which highlighted the prospective zones as highly probable, probable and moderately probable zones. The northwestern, northern, northeastern, southeastern and southwestern parts of the study area were considered high potential zones for epithermal gold mineralizations, which might have undiscovered epithermal gold deposits. The high potential zones were verified by field and laboratory analysis such as systematic GPS surveying, analyzing several microphotographs of hydrothermal alteration minerals and ore mineralization and XRD analysis of collected rock samples from alteration zones. The advanced argillic and argillic-silica alteration zones were typically found in the vicinity of gold mineralizations. However, limonitic-hematite rocks, oxidized breccia and propylitic alteration zones were also documented as high potential zones in the study area. The field and laboratory results verified that the mineral potential map of the Ahar-Arasbaran region successfully indicates the known epithermal gold mineralizations and several new high prospective zones in the study area. The approach developed in this study is a cost-effective technique that can be used for epithermal gold exploration in metallogenic provinces before costly geophysical and geochemical studies. Briefly, this study suggests that geostatistical techniques (e.g., Bayesian network model, Fuzzy model, Artificial Neural Network Model etc.) are valuable approaches to fuse thematic layers of the multi-sensor imagery for generating the remote sensing-based mineral potential map for metallogenic provinces. The mineral exploration community and mining companies can consider the remote sensing-based mineral potential map as an economical and cost-effective tool for mineral prospecting before pricey geophysical and geochemical surveys in the metallogenic provinces.

Author Contributions: S.M.B. writing—Original draft preparation, software, analysis, validation; H.R.R. and A.M. supervision and conceptualization; A.B.P. writing, reconstructing—Review, editing and supervision; G.S. data curation and resources. All authors have read and agreed to the published version of the manuscript.

Funding: Publication fees are waived by Remote Sensing as A.B.P is guest editor of the Special Issue (Multispectral and Hyperspectral Remote Sensing Data for Mineral Exploration and Environmental Monitoring of Mined Areas).

Acknowledgments: Department of Mining and Metallurgical Engineering, Amirkabir University of Technology (Tehran Polytechnic) for providing all the facilities for this investigation. The Korea Polar Research Institute (KOPRI) for assigned time and providing the computer is also acknowledged.

Conflicts of Interest: The authors declare no conflict of interest.

References

1. Clark, R.N. Spectroscopy of rocks and minerals, and principles of spectroscopy. In *Manual of Remote Sensing*; Rencz, A., Ed.; Wiley and Sons Inc.: New York, NY, USA, 1999; Volume 3, pp. 3–58.
2. Bishop, J.L.; Lane, M.D.; Dyar, M.D.; Brwon, A.J. Reflectance and emission spectroscopy study of four groups of phyllosilicates: Smectites, kaolinite-serpentines, chlorites and micas. *Clay Miner.* **2008**, *43*, 35–54. [CrossRef]
3. Sherman, D.M.; Waite, T.D. Electronic spectra of Fe^{3+} oxides and oxide-hydroxides in the near IR to near UV. *Am. Mineral.* **1985**, *70*, 1262–1269.
4. Hunt, G.R. Spectral signatures of particulate minerals in the visible and near-infrared. *Geophysics* **1977**, *42*, 501–513. [CrossRef]
5. Hunt, G.R.; Ashley, R.P. Spectra of altered rocks in the visible and near-infrared. *Econ. Geol.* **1979**, *74*, 1613–1629. [CrossRef]
6. Abrams, M.; Hook, S.J. Simulated ASTER data for geologic studies. *IEEE Trans. Geosci. Remote Sens.* **1995**, *33*, 692–699. [CrossRef]
7. Kruse, F.A.; Perry, S.L. Regional mineral mapping by extending hyperspectral signatures using multispectral data. *IEEE Trans. Geosci. Remote Sens.* **2007**, *4*, 1–14.
8. Boardman, J.W.; Kruse, F.A. Automated spectral analysis: A geologic example using AVIRIS data, north Grapevine Mountains, Nevada. In Proceedings of the Tenth Thematic Conference on Geologic Remote Sensing, Environmental Research Institute of Michigan, Ann Arbor, MI, USA, 9–12 May 1994; pp. 407–418.
9. Kruse, F.A.; Bordman, J.W.; Huntington, J.F. Comparison of airborne hyperspectral data and EO-1 Hyperion for mineral mapping. *IEEE Trans. Geosci. Remote Sens.* **2003**, *41*, 1388–1400. [CrossRef]
10. Sheikhrahimi, A.; Pour, B.A.; Pradhan, B.; Zoheir, B. Mapping hydrothermal alteration zones and lineaments associated with orogenic gold mineralization using ASTER remote sensing data: A case study from the Sanandaj-Sirjan Zone, Iran. *Adv. Space Res.* **2019**, *63*, 3315–3332. [CrossRef]

11. Noori, L.; Pour, B.A.; Askari, G.; Taghipour, N.; Pradhan, B.; Lee, C.-W.; Honarmand, M. Comparison of different algorithms to map hydrothermal alteration zones using ASTER remote sensing data for polymetallic vein-type ore exploration: Toroud–Chahshirin Magmatic Belt (TCMB), North Iran. *Remote Sens.* **2019**, *11*, 495. [CrossRef]
12. Pour, A.B.; Park, T.S.; Park, Y.; Hong, J.K.; Zoheir, B.; Pradhan, B.; Ayoobi, I.; Hashim, M. Application of multi-sensor satellite data for exploration of Zn-Pb sulfide mineralization in the Franklinian Basin, North Greenland. *Remote Sens.* **2018**, *10*, 1186. [CrossRef]
13. Pour, A.B.; Hashim, M.; Park, Y.; Hong, J.K. Mapping alteration mineral zones and lithological units in Antarctic regions using spectral bands of ASTER remote sensing data. *Geocarto Int.* **2018**, *33*, 1281–1306. [CrossRef]
14. Pour, B.A.; Hashim, M. The application of ASTER remote sensing data to porphyry copper and epithermal gold deposits. *Ore Geol. Rev.* **2012**, *44*, 1–9. [CrossRef]
15. Pour, A.B.; Park, Y.; Crispini, L.; Läufer, A.; Hong, J.K.; Park, T.-Y.S.; Zoheir, B.; Pradhan, B.; Muslim, A.M.; Hossain, M.S.; et al. Mapping listvenite occurrences in the damage zones of northern victoria land, Antarctica using ASTER Satellite Remote Sensing Data. *Remote Sens.* **2019**, *11*, 1408. [CrossRef]
16. Pour, A.B.; Hashim, M.; Hong, J.K.; Park, Y. Lithological and alteration mineral mapping in poorly exposed lithologies using Landsat-8 and ASTER satellite data: North-eastern Graham Land, Antarctic Peninsula. *Ore Geol. Rev.* **2019**, *108*, 112–133. [CrossRef]
17. Sun, L.; Khan, S.; Shabestari, P. Integrated hyperspectral and geochemical study of sediment-hosted disseminated gold at the Goldstrike District, Utah. *Remote Sens.* **2019**, *11*, 1987. [CrossRef]
18. Zoheir, B.; Emam, A.; Abdel-Wahed, M.; Soliman, N. Multispectral and radar data for the setting of gold mineralization in the South Eastern Desert, Egypt. *Remote Sens.* **2019**, *11*, 1450. [CrossRef]
19. Zoheir, B.; El-Wahed, M.A.; Pour, A.B.; Abdelnasser, A. Orogenic gold in transpression and transtension zones: Field and remote sensing studies of the barramiya–mueilha sector, Egypt. *Remote Sens.* **2019**, *11*, 2122. [CrossRef]
20. Guha, A.; Yamaguchi, Y.; Chatterjee, S.; Rani, K.; Vinod Kumar, K. Emittance spectroscopy and broadband thermal remote sensing applied to phosphorite and its utility in geoexploration: A study in the parts of Rajasthan, India. *Remote Sens.* **2019**, *11*, 1003. [CrossRef]
21. Rowan, L.C.; Goetz, A.F.H.; Ashley, R.P. Discrimination of hydrothermally altered and unaltered rocks in visible and near infrared multispectral images. *Geophysics* **1977**, *42*, 522–535. [CrossRef]
22. Crosta, A.; Moore, J. Enhancement of Landsat Thematic Mapper imagery for residual soil mapping in SW Minais Gerais State, Brazil: A prospecting case history in Greenstone belt terrain. In Proceedings of the 7th ERIM Thematic Conference on Remote Sensing for Exploration Geology, Calgary, AL, Canada, 2–6 October 1989; pp. 1173–1187.
23. Sabins, F.F. Remote sensing for mineral exploration. *Ore Geol. Rev.* **1999**, *14*, 157–183. [CrossRef]
24. Abdelsalam, M.G.; Stern, R.J.; Woldegabriel, G.B. Mapping gossans in arid regions with Landsat TM and SIR-C images, the Beddaho Alteration Zone in northern Eritrea. *J. Afr. Earth Sci.* **2000**, *30*, 903–916. [CrossRef]
25. Kusky, T.M.; Ramadan, T.M. Structural controls on Neoproterozoic mineralization in the South Eastern Desert, Egypt: An integrated field, Landsat TM, and SIR-C/X SAR approach. *J. Afr. Earth Sci.* **2002**, *35*, 107–121. [CrossRef]
26. Aydal, D.; Ardal, E.; Dumanlilar, O. Application of the Crosta technique for alteration mapping of granitoidic rocks using ETM+ data: Case study from eastern Tauride belt (SE Turkey). *Int. J. Remote Sens.* **2007**, *28*, 3895–3913. [CrossRef]
27. Rajesh, H.M. Mapping Proterozoic unconformity-related uranium deposits in the Rockole area, Northern Territory, Australia using Landsat ETM+. *Ore Geol. Rev.* **2008**, *33*, 382–396. [CrossRef]
28. Pour, A.B.; Hashim, M. Hydrothermal alteration mapping from Landsat-8 data, Sar Cheshmeh copper mining district, south-eastern Islamic Republic of Iran. *J. Taibah Univ. Sci.* **2015**, *9*, 155–166. [CrossRef]
29. Pour, A.B.; Park, Y.; Park, T.S.; Hong, J.K.; Hashim, M.; Woo, J.; Ayoobi, I. Evaluation of ICA and CEM algorithms with Landsat-8/ASTER data for geological mapping in inaccessible regions. *Geocarto Int.* **2019**, *34*, 785–816. [CrossRef]
30. Lowell, J.D.; Guilbert, J.M. Lateral and vertical alteration-mineralization zoning in porphyry ore deposits. *Econ. Geol. Bull. Soc. Econ. Geol.* **1970**, *65*, 373–408. [CrossRef]

31. Dilles, J.H.; Einaudi, M.T. Wall-rock alteration and hydrothermal flow paths about the Ann-Mason porphyry copper deposit, Nevada-a 6-km vertical reconstruction. *Econ. Geol.* **1992**, *87*, 1963–2001. [CrossRef]
32. Sillitoe, R.H. Porphyry Copper Systems. *Econ. Geol.* **2010**, *105*, 3–41. [CrossRef]
33. Goldfarb, R.J.; Taylor, R.D.; Collins, G.S.D.; Goryachev, N.A.; Orlandini, O.F. Phanerozoic continental growth and gold metallogeny of Asia. *Gondwana Res.* **2014**, *25*, 48–102. [CrossRef]
34. Hedenquist, J.W.; Arribas, A.R.; Gonzalez-Urien, E. Exploration for epithermal gold deposits. *SEG Rev.* **2000**, *13*, 245–277.
35. Mars, J.C.; Rowan, L.C. Regional mapping of phyllic-and argillic-altered rocks in the Zagros magmatic arc, Iran, using Advanced Spaceborne Thermal Emission and Reflection Radiometer (ASTER) data and logical operator algorithms. *Geosphere* **2006**, *2*, 161–186. [CrossRef]
36. Mars, J.C.; Rowan, L.C. ASTER spectral analysis and lithologic mapping of the Khanneshin carbonate volcano, Afghanistan. *Geosphere* **2011**, *7*, 276–289. [CrossRef]
37. Rowan, L.C.; Hook, S.J.; Abrams, M.J.; Mars, J.C. Mapping hydrothermally altered rocks at Cuprite, Nevada, using the Advanced Spaceborne Thermal Emission and Reflection Radiometer (ASTER), a new satellite-imaging system. *Econ. Geol.* **2003**, *98*, 1019–1027. [CrossRef]
38. Pour, B.A.; Hashim, M. Identification of hydrothermal alteration minerals for exploring of porphyry copper deposit using ASTER data, SE Iran. *J. Asian Earth Sci.* **2011**, *42*, 1309–1323. [CrossRef]
39. Pour, A.B.; Park, Y.; Park, T.S.; Hong, J.K.; Hashim, M.; Woo, J.; Ayoobi, I. Regional geology mapping using satellite-based remote sensing approach in Northern Victoria Land, Antarctica. *Polar Sci.* **2018**, *16*, 23–46. [CrossRef]
40. Carranza, E.J.M. Geocomputation of mineral exploration targets. *Comput. Geosci.* **2011**, *37*, 1907–1916. [CrossRef]
41. Porwal, A.; Carranza, E.J.M.; Hale, M. Bayesian network classifiers for mineral potential mapping. *Comput. Geosci.* **2006**, *32*, 1–16. [CrossRef]
42. Skabar, A. Mineral potential mapping using Bayesian learning for multilayer perceptrons. *Math. Geol.* **2007**, *39*, 439–451. [CrossRef]
43. Jensen, F.V. *An Introduction to Bayesian Networks*; UCL Press: London, UK, 1996; Volume 210, pp. 1–178.
44. Friedman, N.; Geiger, D.; Goldszmidt, M. Bayesian network classifiers. *Mach. Learn.* **1997**, *29*, 131–163. [CrossRef]
45. Scanagatta, M.; de Campos, C.P.; Corani, G.; Zaffalon, M. Learning bayesian networks with thousands of variables. In Proceedings of the NIPS'15 28th International Conference on Neural Information Processing Systems, Montreal, QC, Canada, 7–12 December 2015; Volume 28, pp. 1855–1863.
46. Jamali, H.; Dilek, Y.; Daliran, F.; Yaghubpur, A.; Mehrabi, B. Metallogeny and tectonic evolution of the Cenozoic Ahar–Arasbaran volcanic belt, northern Iran. *Int. Geol. Rev.* **2010**, *52*, 608–630. [CrossRef]
47. Maghsoudi, A.; Yazdi, M.; Mehrpartou, M.; Vosoughi, M.; Younesi, S. Porphyry Cu–Au mineralization in the Mirkuh Ali Mirza magmatic complex, NW Iran. *J. Asian Earth Sci.* **2014**, *79*, 932–941. [CrossRef]
48. Pazand, K.; Hezarkhani, A. Predictive Cu porphyry potential mapping using fuzzy modelling in Ahar–Arasbaran zone, Iran. *Geol. Ecol. Landsc.* **2018**, *2*, 229–239. [CrossRef]
49. Kouhestani, H.; Mokhtari, M.A.A.; Chang, Z.; Stein, H.J.; Johnson, C.A. Timing and genesis of ore formation in the Qarachilar Cu-Mo-Au deposit, Ahar-Arasbaran metallogenic zone, NW Iran: Evidence from geology, fluid inclusions, O–S isotopes and Re–Os geochronology. *Ore Geol. Rev.* **2018**, *102*, 757–775. [CrossRef]
50. Sohrabi, G. Metallogenic and Geochemical Investigations of Molybdenum Reservoirs in Karad-e Ghad-e Sharidagh, East Azarbaijan, Northwest of Iran. Ph.D. Thesis, Tabriz University, Tabriz, Iran, 2015. (In Persian)
51. Parsa, M.; Maghsoudi, A.; Yousefi, M. Spatial analyses of exploration evidence data to model skarn-type copper prospectivity in the Varzaghan district, NW Iran. *Ore Geol. Rev.* **2018**, *92*, 97–112. [CrossRef]
52. Pazand, K.; Sarvestani, J.F.; Ravasan, M.R. Hydrothermal alteration mapping using ASTER data for reconnaissance porphyry copper mineralization in the Ahar area, NW Iran. *J. Indian Soc. Remote Sens.* **2013**, *41*, 379–389. [CrossRef]
53. Hassanpour, S. The alteration, mineralogy and geochronology (SHRIMP U–Pb and 40 Ar/39 Ar) of copper-bearing Anjerd skarn, north of the Shayvar Mountain, NW Iran. *Int. J. Earth Sci.* **2013**, *102*, 687–699. [CrossRef]
54. Aghazadeh, M.; Hou, Z.; Badrzadeh, Z.; Zhou, L. Temporal–spatial distribution and tectonic setting of porphyry copper deposits in Iran: Constraints from zircon U–Pb and molybdenite Re–Os geochronology. *Ore Geol. Rev.* **2015**, *70*, 385–406. [CrossRef]

55. Hassanpour, S. Metallogeny and Mineralization of Copper and Gold in Arasbaran Zone (Eastern Azerbaijan). Ph.D. Thesis, Shahid Beheshti University, Tehran, Iran, 2010. (In Persian with English abstract)
56. Maghsoudi, A.; Rahmani, M.; Rashidi, B. *Gold Deposits and Indications of Iran*; Pars Arain Zamin Geology Research Centre: Tehran, Iran, 2005.
57. Irons, J.R.; Dwyer, J.L.; Barsi, J.A. The next Landsat satellite: The Landsat Data Continuity Mission. *Remote Sens. Environ.* **2012**, *145*, 154–172. [CrossRef]
58. Roy, D.P.; Wulder, M.A.; Loveland, T.A.; Woodcock, C.E.; Allen, R.G.; Anderson, M.C.; Helder, D.; Irons, J.R.; Johnson, D.M.; Kennedy, R.; et al. Landsat-8: Science and product vision for terrestrial global change research. *Remote Sens. Environ.* **2014**, *145*, 154–172. [CrossRef]
59. Abrams, M.; Tsu, H.; Hulley, G.; Iwao, K.; Pieri, D.; Cudahy, T.; Kargel, J. The Advanced Spaceborne Thermal Emission and Reflection Radiometer (ASTER) after fifteen years: Review of global products. *Int. J. Appl. Earth Obs. Geoinf.* **2015**, *38*, 292–301. [CrossRef]
60. Chander, G.; Markham, B.; Helder, D. Summary of current radiometric calibration coefficients for Landsat MSS, TM, ETM+, and EO-1 ALI sensors. *Remote Sens. Environ.* **2009**, *113*, 893–903. [CrossRef]
61. Research Systems, Inc. *ENVI Tutorials*; Research Systems, Inc.: Boulder, CO, USA, 2008.
62. NASA Goddard Space Flight Centre. Landsat 7 Science Data Users Handbook. Updated March 2011. Available online: http://landsathandbook.gsfc.nasa.gov/inst_cal/prog_sect8_2.html (accessed on 12 October 2019).
63. Ben-Dor, E.; Kruse, F.A.; Lefkoff, A.B.; Banin, A. Comparison of three calibration techniques for utilization of GER 63-channel aircraft scanner data of Makhtesh Ramon, Nega, Israel. *Int. J. Rock Mech. Min. Sci. Geomech. Abstr.* **1995**, *32*, 164A.
64. Iwasaki, A.; Tonooka, H. Validation of a crosstalk correction algorithm for ASTER/SWIR. *IEEE Trans. Geosci. Remote Sens.* **2005**, *43*, 2747–2751. [CrossRef]
65. Goetz, A.F.; Rock, B.N.; Rowan, L.C. Application of remote sensing in exploration. *Econ. Geol.* **1975**, *79*, 644–711.
66. Inzana, J.; Kusky, T.; Higgs, G.; Tucker, R. Supervised classifications of Landsat TM band ratio images and Landsat TM band ratio image with radar for geological interpretations of central Madagascar. *J. Afr. Earth Sci.* **2003**, *37*, 59–72. [CrossRef]
67. Crowley, J.K.; Brickey, D.W.; Rowan, L.C. Airborne imaging spectrometer data of the Ruby Mountains, Montana: Mineral discrimination using relative absorption band-depth images. *Remote Sens. Environ.* **1989**, *29*, 121–134. [CrossRef]
68. Abubakar, A.J.A.; Hashim, M.; Pour, A.B. Remote sensing satellite imagery for prospecting geothermal systems in an aseismic geologic setting: Yankari Park, Nigeria. *Int. J. Appl. Earth Obs. Geoinf.* **2019**, *80*, 157–172. [CrossRef]
69. Van der Meer, F.D.; Van der Werff, H.M.; Van Ruitenbeek, F.J.; Hecker, C.A.; Bakker, W.H.; Noomen, M.F.; Van Der Meijde, M.; Carranza, E.J.M.; De Smeth, J.B.; Woldai, T. Multi-and hyperspectral geologic remote sensing: A review. *Int. J. Appl. Earth Obs. Geoinf.* **2012**, *14*, 112–128. [CrossRef]
70. Kalinowski, A.; Oliver, S. ASTER Mineral Index Processing Manual. Technical Report; Geoscience Australia. 2004. Available online: http://www.ga.gov.au/image_cache/GA7833.pdf (accessed on 1 June 2019).
71. Cheng, Q.; Jing, L.; Panahi, A. Principal component analysis with optimum order sample correlation coefficient for image enhancement. *Int. J. Remote Sens.* **2006**, *27*, 3387–3401.
72. Gupta, R.P.; Tiwari, R.K.; Saini, V.; Srivastava, N. A simplified approach for interpreting principal component images. *Adv. Remote Sens.* **2013**, *2*, 111–119. [CrossRef]
73. Singh, A.; Harrison, A. Standardized principal components. *Int. J. Remote Sens.* **1985**, *6*, 883–896. [CrossRef]
74. Crosta, A.P.; Souza Filho, C.R.; Azevedo, F.; Brodie, C. Targeting key alteration minerals in epithermal deposits in Patagonia, Argentina, Using ASTER imagery and principal component analysis. *Int. J. Remote Sens.* **2003**, *24*, 4233–4240. [CrossRef]
75. Crosta, A.P.; Filho, C.R. Searching for gold with ASTER. *Earth Obs. Mag.* **2003**, *12*, 38–41.
76. Gelman, A.; Carlin, J.B.; Stern, H.S.; Rubin, D.B. *Bayesian Data Analysis*; Chapman & Hall: London, UK, 1995.
77. Neal, R.M. *Bayesian Learning for Neural Networks*; Springer: Berlin, Germany, 1996.
78. Antal, P.; Fannes, G.; Timmerman, D.; Moreau, Y.; Moor, B.D. Bayesian applications of belief networks and multilayer perceptrons for ovarian tumor classification with rejection. *Artif. Intell. Med.* **2003**, *29*, 39–60. [CrossRef]

79. Duda, R.O.; Hart, P.E. *Pattern Classification and Scene Analysis*; John Wiley and Sons: New York, NY, UK, 1973; p. 482.
80. Langley, P.; Sage, S. Induction of selective Bayesian classifiers. In Proceedings of the Tenth Conference on Uncertainty in Artificial Intelligence (UAI '94), Seattle, WA, USA, 29–31 July 1994; Kaufmann, M., de Mantars, L., Poole, D., Eds.; Morgan Kaufmann: Burlington, MA, USA, 1994; pp. 399–406.
81. Jenks, G.F. The data model concept in statistical mapping. *Int. Yearb. Cartogr.* **1967**, *7*, 186–190.
82. Story, M.; Congalton, R. Accuracy assessment: A user's perspective. *Photogramm. Eng. Remote Sens.* **1986**, *52*, 397–399.
83. Congalton, R.G. A review of assessing the accuracy of classification of remotely sensed data. *Remote Sens. Environ.* **1991**, *37*, 35–46. [CrossRef]
84. Lillesand, T.; Kiefer, R. *Remote Sensing and Image Interpretation*; John Wiley & Sons, Inc.: New York, NY, USA, 1994; Chapter 7.
85. Kruse, F.; Perry, S. Mineral mapping using simulated Worldview-3 short-wave infrared imagery. *Remote Sens.* **2013**, *5*, 2688–2703. [CrossRef]
86. Monserud, R.A.; Leemans, R. Comparing global vegetation maps with the Kappa-statistic. *Ecol. Model.* **1992**, *62*, 275–293. [CrossRef]
87. Goldfarb, R.J.; Phillips, G.N.; Nokleberg, W.J. Tectonic setting of synorogenic gold deposits of the Pacific Rim. *Ore Geol. Rev.* **1998**, *13*, 185–218. [CrossRef]
88. Goldfarb, R.J.; Groves, D.I. Orogenic gold: Common or evolving fluid and metal sources through time. *Lithos* **2015**, *233*, 2–26. [CrossRef]
89. White, N.C.; Hedenquist, J.W. Epithermal gold deposits: Styles, characteristics and exploration. *SEG Newsl.* **1995**, *23*, 9–13.
90. Pour, B.A.; Hashim, M. Structural mapping using PALSAR data in the Central Gold Belt, Peninsular Malaysia. *Ore Geol. Rev.* **2015**, *64*, 13–22. [CrossRef]
91. Pour, B.A.; Hashim, M.; Makoundi, C.; Zaw, K. Structural mapping of the bentong-raub suture zone using PALSAR remote sensing data, Peninsular Malaysia: Implications for sediment-hosted/orogenic gold mineral systems exploration. *Resour. Geol.* **2016**, *66*, 368–385. [CrossRef]
92. Pour, B.A.; Hashim, M.; Marghany, M. Exploration of gold mineralization in a tropical region using Earth Observing-1 (EO1) and JERS-1 SAR data: A case study from Bau gold field, Sarawak, Malaysia. *Arab. J. Geosci.* **2014**, *7*, 2393–2406. [CrossRef]
93. Pour, B.A.; Hashim, M. Evaluation of Earth Observing-1 (EO1) Data for Lithological and Hydrothermal Alteration mapping: A case study from Urumieh-Dokhtar Volcanic Belt, SE Iran. *Indian Soc. Remote Sens.* **2015**, *43*, 583–597. [CrossRef]
94. Testa, F.J.; Villanueva, C.; Cooke, D.R.; Zhang, L. Lithological and hydrothermal alteration mapping of epithermal, porphyry and tourmaline breccia districts in the Argentine Andes using ASTER imagery. *Remote Sens.* **2018**, *10*, 203. [CrossRef]
95. Parsa, M.; Maghsoudi, A.; Yousefi, M.; Sadeghi, M. Recognition of significant multi-element geochemical signatures of porphyry Cu deposits in Noghdouz area, NW Iran. *J. Geochem. Explor.* **2016**, *165*, 111–124. [CrossRef]
96. Parsa, M.; Maghsoudi, A.; Yousefi, M. An improved data-driven fuzzy mineral prospectivity mapping procedure; cosine amplitude-based similarity approach to delineate exploration targets. *Int. J. Appl. Earth Obs. Geoinf.* **2017**, *58*, 157–167. [CrossRef]
97. Parsa, M.; Maghsoudi, A.; Yousefi, M. A receiver operating characteristics-based geochemical data fusion technique for targeting undiscovered mineral deposits. *Nat. Res. Res.* **2018**, *27*, 15–28. [CrossRef]

© 2019 by the authors. Licensee MDPI, Basel, Switzerland. This article is an open access article distributed under the terms and conditions of the Creative Commons Attribution (CC BY) license (http://creativecommons.org/licenses/by/4.0/).

Article

Drill-Core Mineral Abundance Estimation Using Hyperspectral and High-Resolution Mineralogical Data

Laura Tuşa *, Mahdi Khodadadzadeh, Cecilia Contreras, Kasra Rafiezadeh Shahi, Margret Fuchs, Richard Gloaguen and Jens Gutzmer

Helmholtz Institute Freiberg for Resource Technology, Helmholtz-Zentrum Dresden-Rossendorf, Chemnitzer Straße 40, 09599 Freiberg, Germany
* Correspondence: l.tusa@hzdr.de

Received: 16 March 2020; Accepted: 9 April 2020; Published: 9 April 2020

Abstract: Due to the extensive drilling performed every year in exploration campaigns for the discovery and evaluation of ore deposits, drill-core mapping is becoming an essential step. While valuable mineralogical information is extracted during core logging by on-site geologists, the process is time consuming and dependent on the observer and individual background. Hyperspectral short-wave infrared (SWIR) data is used in the mining industry as a tool to complement traditional logging techniques and to provide a rapid and non-invasive analytical method for mineralogical characterization. Additionally, Scanning Electron Microscopy-based image analyses using a Mineral Liberation Analyser (SEM-MLA) provide exhaustive high-resolution mineralogical maps, but can only be performed on small areas of the drill-cores. We propose to use machine learning algorithms to combine the two data types and upscale the quantitative SEM-MLA mineralogical data to drill-core scale. This way, quasi-quantitative maps over entire drill-core samples are obtained. Our upscaling approach increases result transparency and reproducibility by employing physical-based data acquisition (hyperspectral imaging) combined with mathematical models (machine learning). The procedure is tested on 5 drill-core samples with varying training data using random forests, support vector machines and neural network regression models. The obtained mineral abundance maps are further used for the extraction of mineralogical parameters such as mineral association.

Keywords: hyperspectral imaging; drill-core; SWIR; mineral abundance mapping; mineral association; machine learning

---

1. Introduction

Exploration campaigns are fundamental steps towards the discovery and evaluation of mineral deposits required to fulfil the global demand of raw materials. Drilling is an essential part of exploration surveys and consists of the extraction of long cylindrical core samples from underground areas associated with relevant exploration potential. Traditionally, drill-cores are visually analyzed by on-site geologists, who document characteristics such as mineralization type, lithology, structures and alteration types [1]. Subsequently, core samples are used for laboratory-based geochemical and mineralogical measurements to complement core logging results. While bulk geochemical analyses are often available for entire boreholes, quantitative mineralogical information is usually restricted to selected representative regions of interest. Standard quantitative analyses include X-Ray diffraction (XRD) applied on powder samples [2] or Scanning Electron Microscopy (SEM) based image analyses techniques [3] applied on polished thin sections prepared from areas of interest in the drill-cores. Additionally, qualitative mineralogical analyses are performed through optical microscopy on thin sections. These laboratory techniques provide valuable mineralogical information and derived mineralogical and metallurgical parameters, but they are of small scale, highly time-consuming, destructive, and rather expensive. This represents a challenge since thousands of meters of core are acquired during exploration campaigns.

Hyperspectral imaging is currently being used in the mining and exploration industries as an alternative tool to complement traditional logging techniques and to provide a rapid and non-invasive analytical method to obtain mineralogical information [4–7]. Typical hyperspectral core imaging systems can deliver data from a whole core tray (which holds approximately 5 m of core) in a matter of seconds. Available sensors cover a wide range of the electromagnetic spectrum and record data in several hundreds of contiguous spectral bands. Minerals have different spectral responses in specific portions of the electromagnetic spectrum. These responses are influenced by the vibrational and electronic absorption processes dependent on the bonds between atoms and electron orbitals [8]. Sensors covering the visible to near-infrared (VNIR) and short-wave infrared (SWIR) are commonly used to identify and estimate the relative abundance of minerals such as phyllosilicates, amphiboles, carbonates, iron oxides and hydroxides as well as sulphates [9].

Because of the increasing interest in hyperspectral data in the raw materials industry, with a wealth of hyperspectral data becoming available, the development of methods to effectively analyze these data is required. Traditional mapping methods include the use of spectral reference libraries (e.g., USGS spectral library) for mineral identification and mapping on hyperspectral imagery [10,11]. Slightly more automatic approaches, such as band ratios, or wavelength parameters such as position, depth and width of the absorption features are also used to map the distribution and relative abundance of specific minerals [12–14]. One of the most common procedures makes use of some of available tools in a software called Environment for Visualizing Images (ENVI, Exelis Visual Information Solutions, Boulder, Colorado). Such tools comprise endmember extraction, identification of the minerals using the Spectral Analysis or Material Identification by comparison to a specific library in the software (e.g., in ENVI) or online reference (e.g., USGS), and finally the mineral mapping task using similarity measure algorithms or determination of partial abundances using unmixing algorithms [5,15–17].

Although these approaches may produce good results, they require continuous expert input and thus, they tend to be time-consuming and difficult to automate for large dataset analysis. More importantly, the performance of available unmixing algorithms highly relies on the determination of the number of end-members and the selection of their representative spectra. In drill-core hyperspectral data, highly mixed pixels of hardly pure mineral associations represent a challenge. Methods such as unmixing, band ratios and minimum wavelength analysis can only provide mineral abundances for spectrally diagnostic phases. Additionally, due to the nature of the hyperspectral data and the spatial resolution allowed by commercially available sensors, the estimation of important mineralogical parameters in the characterization of complex ores (e.g., mineral association), is currently challenging.

We propose a novel machine learning approach to estimate mineral quantities in drill-core hyperspectral data. The procedure comprises four steps: 1) drill-core hyperspectral scanning (VNIR –SWIR), 2) computing mineral abundances in a small but representative area of a drill-core by using high-resolution mineralogical analyses (e.g., SEM-based image analyses using a Mineral Liberation Analyser), 3) linking the mineral abundances in this small area to their corresponding spectra by a multivariate regression model, and 4) estimating mineral abundances for the whole drill-core hyperspectral data by using the learned model. The multivariate regression problem in the proposed scheme is solved using three algorithms: random forest (RF), support vector machines (SVM) and feedforward artificial neural networks (FF-ANN). The proposed procedure allows the abundance estimation of the main mineral groups using their spectral characteristics (SWIR active) and using those SWIR active minerals additionally as proxies for the SWIR non-active minerals or mineral groups such as quartz, feldspar and sulphide. The obtained mineral abundance mapping results can be used for the calculation of additional mineralogical parameters, relevant to exploration and mining projects. As an example, the concept of mineral association at hyperspectral pixel scale based on relative abundances is introduced in the current study.

2. Data Acquisition

2.1. Hyperspectral Data

The hyperspectral data used in this study were acquired from unpolished halves of diamond drilling core samples with a SisuROCK drill-core scanner equipped with an AisaFENIX hyperspectral sensor (Spectral Imaging Ltd., Oulu, Finland). The scanner is a fully automatic hyperspectral imaging workstation which employs a tray table which carries the drill-core trays or samples under the field-of-view of the spectrometer. The AisaFENIX camera implements two sensors to cover the VNIR and SWIR regions of the electromagnetic spectrum. The sensor specifications and acquisition settings are presented in Table 1.

Table 1. AisaFENIX sensor specification and setup for hyperspectral data acquisition.

Wavelength Range	VNIR 380–970 nm SWIR 970–2500 nm	Integration Time	VNIR 15 ms SWIR 4 ms
Sampling Distance	VNIR 1.7 nm SWIR 5.7 nm	Spatial Binning	VNIR 2 SWIR 1
Number of Bands	450	Frame Rate	15 Hz
Samples	384	Scanning Speed	25.06 mm/s
Spatial Resolution	1.5 mm/pixel	Field of View (FOV)	32.3°
Detector	CMOS (VNIR) Stirling cooled MCT (SWIR)	Spectral Binning	VNIR 4 SWIR 1

The conversion from radiance to reflectance of the hyperspectral data was performed within the acquisition software (LUMO Scanner version 2018-5, Spectral Imaging Ltd., Oulu, Finland) using PTFE reference panels (>99% VNIR and >95% SWIR). To correct the sensor-specific optical distortions (i.e., fish-eye and slit-bending effects on the images) and the spatial shift between the VNIR and SWIR sensors, the toolbox MEPHySTo [18] was used. To avoid bands with little or no coherent information, the data were spectrally resampled to 480—2500 nm by removing the first 30 bands. The Savitzky–Golay filter was applied to decrease noise while preserving spectral features [19]. Principal component analysis (PCA) [20] was performed on the hyperspectral dataset for data dimensionality reduction and de-correlation while preserving 99.9% of the information.

2.2. Scanning Electron Microscopy-Based Mineral Liberation Analysis

Regions considered representative based on visual observations for the mineralogical variation within the drill-core samples were cut and prepared into polished thin sections. The preparation process consisted of grinding and polishing the sample surface followed by coating with a thin carbon layer to avoid surface charging during data acquisition. The grinding and polishing led to the removal of around 300 µm of material between the surface analyzed with the hyperspectral sensor and the surface subjected to the high-resolution mineralogical analysis. Considering the sample morphology and orientation of structural features the mineralogical variation is considered negligible for the encountered shift.

The quantitative mineralogical data were acquired from the thin sections using an automated approach. The analyses were carried out using Scanning Electron Microscope (SEM)-based Mineral Liberation Analysis (MLA) [3,21]. For this, a FEI Quanta 650 F field emission SEM instrument (FEI, Hillsboro, OR, USA), equipped with two Bruker Quantax X-Flash 5030 energy dispersive X-ray (EDX) detectors (Bruker, Billerica, MA, USA) and the MLA Suite software package (version 3.1.4.686, FEI, Hillsboro, OR, USA) were used. The grain-based X-ray mapping (GXMAP) mode was used to collect the mineralogical information as follows: the MLA software collects the back-scattered electron images (BSE) and uses them to effectively distinguish individual mineral grain boundaries based on the grey scale variations. The grey scale values of the BSE images are proportional to the average atomic density of the mineral grains and are used to provide a first mineralogical segmentation. The identification of minerals is performed based on X-ray analysis by placing a closely-spaced grid on a particle in the BSE image and collecting the X-ray data at the defined points of the grid. When dealing with fine grained material of lower size than the placed grid, the GXMAP mode allows us to collect additional spectra where variations in the BSE image are observed in between the measured grid points. Finally, the mineral is determined by matching the resultant spectrum of energy peaks with a reference library of X-ray spectra provided by the instrument company (FEI, Hillsboro, OR, USA), or from sample extracted spectra analyzed based on peak locations and intensities [22]. Specifications of the operating conditions used in this study are shown in Table 2.

Table 2. Operating conditions and parameters used for the acquisition of high-resolution SEM-MLA mineralogical data.

SEM Settings		MLA Settings	
Acceleration voltage (kV)	25	Pixel size (µm)	3
Probe current (nA)	10	Step size (pixels)	6 × 6
Frame width (pixels)	1500	Acquisition time (ms)	5
BSE calibration (Au)	254	Minimum grain size (pixels)	4

For classification, a mineral list was developed using the mineral reference editor in online mode. The resulting mineral list contained a total of 59 entries. However, for the integration of the HSI with SEM-MLA, further grouping was performed in this paper, such as considering all feldspars in one class, all white micas in another or, all sulphides, sulphosalts and gold in another. Accessory minerals were included in the final grouping labelled as "others". As a result, ten main mineral groups are considered: white mica (WM), biotite (Bt), chlorite (Chl), amphibole (Amp), carbonate (Cb), gypsum (Gp), feldspar (Fsp), quartz (Qz), sulphide (Sp) and other.

3. Data Description

For testing the proposed methodology, 5 samples, labelled DC-1 to DC-5, from different locations within the Bolcana porphyry copper-gold system [23–26] were analyzed. Hyperspectral images were acquired on the halves cores after which thin sections were prepared from selected regions of interest and analyzed by SEM-MLA. Each region is further labelled as a, b and/or c starting from the left-hand side of the drill-core sample as illustrated in Figure 1. The ore minerals in the studied system are chalcopyrite, bornite, covellite, chalcocite and gold. Gold is dominantly present as fine inclusions in pyrite and chalcopyrite. The main encountered alteration types are potassic, sodic—calcic, phyllic and argillic. In the studied samples the first three are present, some samples presenting a transitional character and are described in this section. Please see Sillitoe, 2010 [27] for details on the mineralogical characteristics of the alteration styles typically associated with porphyry Cu-Au systems.

While the summary of the results for each sample is presented in the results section, an emphasis is made on DC-1 in order to illustrate all the potential information that can be extracted using the proposed methodology. Therefore, a more detailed description of this sample is available in the current section. Sample DC-1 consists of a diorite porphyry. Hydrothermal alteration in this sample appears transitional between potassic, represented by the presence of biotite and potassic feldspar and sodic-calcic characterized by the plagioclase-chlorite assemblage. Chlorite is more abundant than biotite in the first two thin sections, "a" and "b". The third thin section, though, due to the lower vein density and implicit associated alteration presents significant amounts of biotite disseminated as well as in clusters in the matrix. Plagioclase feldspar is dominant in all three thin sections, near the veins however, an increase in potassic feldspar is observed.

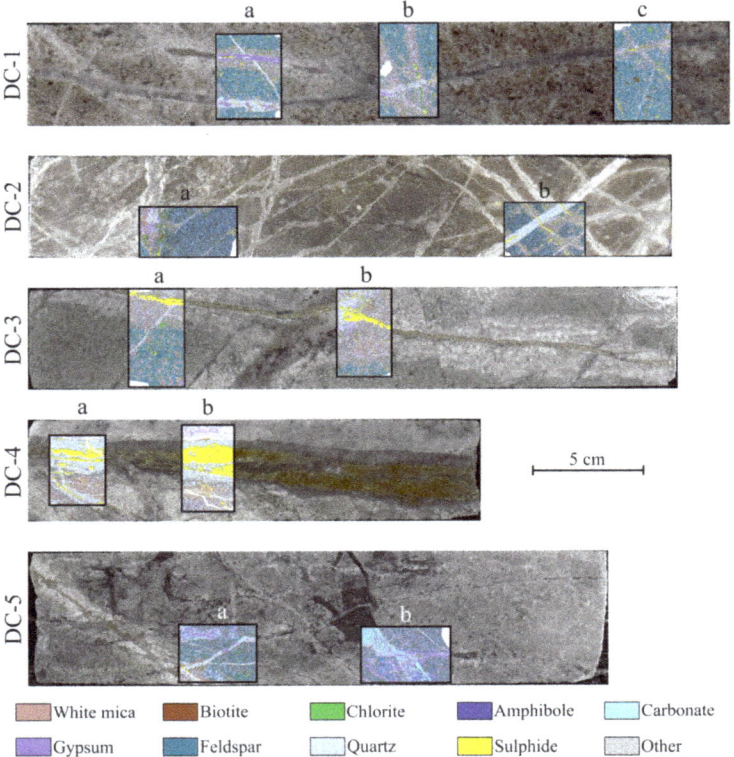

Figure 1. RGB photograph of the analyzed drill-cores (labelled on the left-hand side from DC-1 to DC-5) with overlain high-resolution mineral maps (labelled a, b and c) obtained by SEM-MLA.

Thin section "a" of sample DC-1 captures three main vein types: an oblique early quartz vein which exhibits a low intensity white mica alteration halo likely associated with a younger cross-cutting gypsum vein that has a sulphide centerline and a wide white mica-chlorite alteration halo (top). The alteration halo here is mica-dominant in the proximity of the vein and chlorite-dominant towards its edges. The third vein present in section "a" consists of quartz with a gypsum centerline and a spotty, low intensity white mica alteration halo (bottom). Thin section "b" captures three main vein types as well: two sub-vertical veins consist of variable ratios of gypsum and quartz and are surrounded by a strong white mica low-chlorite alteration halo. Compositionally, these veins appear to be a mixture between the first and third veins mentioned for thin section "a"; they have, however, a different morphology. In proximity to sub-horizontal veinlets in the lower half of the thin section, an increase in the pyrite and chlorite content is observed. The two sub-horizontal veinlets show strong similarity with the horizontal veins in the first thin section. The alteration intensity surrounding the sub-horizontal veinlets appears to be related to complex interactions with pre-existing veinlets in this area of section "b". Thin section "c" hosts several fine veinlets, of highest width, the two cross-cutting ones near the top of the thin section. The veinlets consist of variable amounts of quartz, gypsum, pyrite and white mica and present a white mica and chlorite alteration halo. Similar to the subvertical veins in thin section "b" these veins appear to have a composition intermediate between the horizontal veins in thin section "a". Unlike the two veins in thin section "b" however, the extent of the alteration halo is much lower.

Sample DC-2 is marked by pervasive potassic alteration characterized by the presence of K-feldspar, biotite and minor chlorite. Two main vein types are present in this sample: veins hosting dominantly sulphide which show a strong phyllic alteration halo caused by the late reaction of mineralizing hydrothermal fluids with the host rock. The second vein type comprises dominantly quartz with sulphide or with sulphide-calcium sulphate (gypsum or anhydrite) centerline. Additional veins of varying composition are present in the sample (left-hand side as illustrated in Figure 1). They appear to be the result of complex reopening and cross-cutting of the previously described veins. A sodic-phyllic rock matrix hosting two main vein-types characterizes sample DC-3. The first vein comprises of sulphide and presents a large white mica alteration halo. The second vein type consists predominantly of quartz, calcium sulphate and sulphide. The changing symmetry and mineral association in these latter veins indicate the reopening of an initially present quartz vein. Sample DC-4 is characterized by the presence of intense phyllic alteration in the matrix related to the thick pyrite-quartz-gypsum vein cross-cutting the sample. Additional fine veinlets comprising mostly quartz and pyrite are cutting the mica-rich matrix. The matrix in sample DC-5 consists of dominantly feldspar and subordinately white mica. Three main vein types can be observed in the samples: a sulphide dominant vein with a broad white mica alteration halo, quartz veinlets and carbonate iron-oxide veins which show low or absent alteration halos.

For the understanding of the modal composition of the available thin sections, the abundances of the minerals or mineral groups for all the analyzed thin sections are illustrated in the bar charts in Figure 2 (left). For most samples, quartz and feldspar represent the main rock-forming minerals. There is, however, a variation in the extent of alteration of feldspar to white mica ranging from low (DC-2a) to high (DC-4). In most of the analyzed samples, the amphibole is to a large extent altered to chlorite and/or biotite. Biotite is only present in significant amounts in sample DC-2 and DC-1 "c". The variation of the quartz, carbonate and gypsum contents is related to the surface abundance of the veins and veinlets filled mostly by these three minerals. While quartz and gypsum are present in significant amounts in all thin sections, carbonate is mainly represented in sample DC-5. The class "sulphide" comprises mainly pyrite, chalcopyrite, bornite, chalcocite and covellite but minor amounts of native gold hosted as inclusions in pyrite and chalcopyrite is also considered. While pyrite is not an ore mineral by itself, it frequently represents the host of micron-size native gold inclusions. The sulphide content in the thin sections ranges from around 1 area % in DC-5b to almost 30 area % in DC-4b. The main target being the quantification and understanding of the distribution of sulphide minerals within the presented samples, their mineral association is also analyzed and presented in the bar chart in Figure 2 (right).

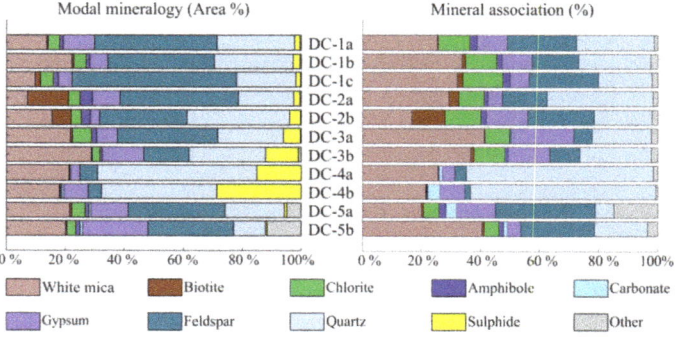

Figure 2. Modal mineralogy and mineral association of analyzed thin sections illustrated through mineral maps in Figure 1. The labels of each sample and thin section are illustrated between the two bar charts.

While an influence of the modal mineralogy can be observed on the mineral association, a strong increase in the white mica, chlorite, biotite, carbonate and gypsum can be seen. This is the result of the distribution of these minerals within or surrounding the veins also hosting the bulk of sulphides. The listed gangue minerals, unlike the sulphide, show distinct absorption features in the VNIR-SWIR region of the electromagnetic spectrum and may, therefore, be used as proxies for the distribution of the ore minerals.

4. Methodological Framework

4.1. HSI—SEM-MLA Data Integration

For the proposed approach, the SEM-MLA data is upscaled by adopting a re-sampling procedure. The two-dimensional SEM-MLA mineral map with high spatial resolution is transformed to a three-dimensional mineral abundance map with the lower spatial resolution of the hyperspectral data [7]. The third dimension consists of the relative abundance of each mineral present in each SEM-MLA map re-sampled to the hyperspectral pixel size (Figure 3). Note that a co-registration stage is needed after the re-sampling of the SEM-MLA data. Following Acosta et al., 2019, the structural features, such as veins, the mineral composition, and spectral responses are used to find suitable tie points. As a result of the co-registration each pixel where the SEM-MLA data is available is characterised by two vectors: the hyperspectral feature vector Xi of dimension d (i.e., the number of bands in the hyperspectral data) or r (number of extracted features) and a mineral abundance vector Yi containing the corresponding fractional abundances of the minerals identified by SEM-MLA.

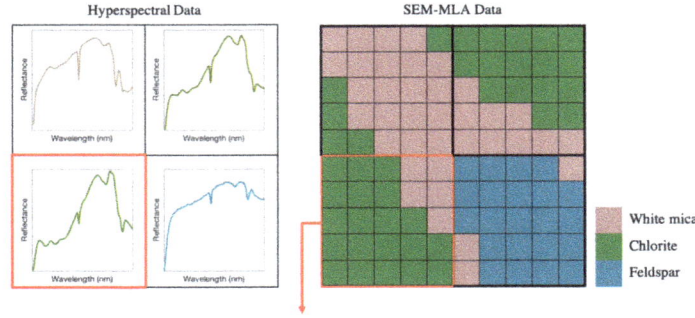

Figure 3. Graphical illustration of the co-registration and resampling process for the SEM-MLA to hyperspectral data. In red, the size of a hyperspectral pixel characterized by a mineral mixture in the SEM-MLA data and a spectrum in the hyperspectral data. The color of the spectra (left) is given by the mixture ratio of the minerals illustrated in the SEM-MLA simplified example (right).

Once the hyperspectral and SEM-MLA data are co-registered, they are divided into training and testing. For this procedure the following approach is adopted:

- Using 50% randomly selected pixels from all thin section regions within one drill-core sample for training, the remaining drill-core hyperspectral data for testing. The validation is performed using the remaining 50% data points from the MLA regions.
- Using 1 thin section for training and the second for testing and validation for all drill-core samples.
- For DC-1, where 3 thin sections are available, an additional test is performed using 2 thin sections for training and the last for testing and validation.

As can be seen from the main flowchart, shown in Figure 4, the proposed workflow is carried out in three main phases. In the training phase, different regression models (i.e., RF, SVM and FF-ANN) are trained following any of the three approaches mentioned before. In the prediction phase, the learned models are used to predict the mineral abundances on the entire drill-core samples. Finally, in the validation phase, the root mean square error (RMSE) [28] is calculated on the remaining SEM-MLA test data to assess the performance of the abundance mapping.

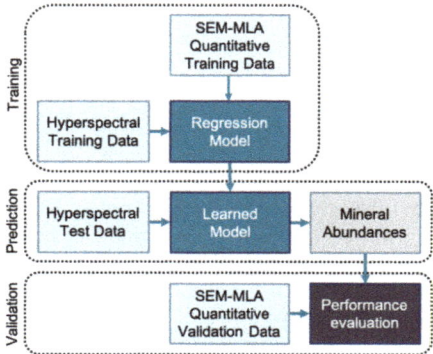

Figure 4. Flowchart illustrating the three main stages of the proposed workflow.

Two analysis types are further performed on the resulting mineral abundance data. For each validation set, the modal mineralogy is calculated based on the average abundance of each mineral phase in each pixel and compared to the modal mineralogy data obtained from SEM-MLA. Additionally, the concept of mineral association is adapted from the automated mineralogy field (Figure 2). There, the mineral association is calculated by counting the neighboring pixels to a specific target mineral. Slight changes in the approach have to be made when the spatial resolution of the hyperspectral data is used. The association of the main target group, i.e., sulphide, is a fundamental aspect in the present geological study. For each hyperspectral pixel the estimated mineral abundance of each mineral phase, except of the target, is normalized by the abundance of sulphide in the respective pixel. While this approach does not directly indicate the grain contact between the two minerals (or rather mineral groups) it can be seen as the probability of their association and occurrence at the scale of hyperspectral data resolution. The mineral association is calculated on the ground truth or validation data as well as on the estimated abundances calculated with the three proposed regression models.

4.2. Random Forest Regression

Random forests (RFs) are currently one of the most popular supervised learning techniques for classification and regression problems [29–31]. RFs are ensemble-based algorithms in which several models (trees) are running in parallel with randomized sampling. The individual results of these trees are then combined into the final prediction by an averaging process [32]. For regression purposes, the trees are given numerical values as predictors whereas in classification problems they are fed class labels. The RF technique is desirable in cases where only few training samples are available, as is usually the case in drill-core hyperspectral imaging.

4.3. Support Vector Regression

The aim of support vector machines (SVMs) is to search for hyperplane decision boundaries to define a linear prediction model [33,34]. To locate and orientate the hyperplane, only the samples that are close to the hyperplane, so-called support vectors, have an influence. Therefore, SVMs perform well when a limited number of well-chosen training samples are available [31,33,34]. This model can be used for classification or regression tasks. SVMs were originally proposed to solve linear problems. However, decision boundaries are often non-linear. To cope with the non-linearity problem, the kernel-based SVMs were introduced to project the data points into a higher dimensional feature space where the samples are linearly separable [31].

4.4. Artificial Neural Network Regression

Artificial neural networks have become some of the most popular methods in regression and classification because of their success in capturing the non-linearity relation between independent and dependent variables [35]. We chose a so-called "feedforward neural network" (FF-ANN) [36], as it fits the requirements of the problem at hand. In a feedforward network, each neuron in one layer is directly connected to neurons of the next layer with no cycle between layers. The applied neural network consists of an input layer, one hidden layer, and an output layer. Each neuron of a layer is computed by the product sum of the neurons of the previous layers plus a bias for the neuron [31]. A sigmoid function is applied for activation.

5. Experimental Results

In order to showcase the suitability of the proposed approach, the first drill-core sample presented in the data section (DC-1) is used. The remaining four samples have been analyzed following the same procedure. A summary of the results is presented in this section followed by a complete illustration of the results in Appendix A. Additionally, all numerical results are presented in the Electronic Supplementary Materials (Table S1).

From the entire drill-core sample (DC-1), the VNIR-SWIR hyperspectral data of size 33 by 189 pixels. The 420 spectral bands cover wavelengths from 480 nm to 2500 nm. The hyperspectral data is subjected to PCA leading to the reduction in dimensionality to 13 principal components in the third dimension. Moreover, the high-resolution mineralogical data obtained from representative regions (thin sections "a", "b" and "c") were used. In the thin section regions of the drill-core sample, each hyperspectral pixel covers an area of 1.5 by 1.5 mm^2, which is characterized by about 250,000 pixels in the SEM-MLA image. The fractional abundances were computed by considering the frequency of the identified minerals in the corresponding region of the SEM-MLA image for each hyperspectral pixel. To have more consistent results, we considered a threshold of 250,000 pixels (i.e., a hyperspectral pixel size) in each thin section region, for discarding minerals which have a very low frequency in the original SEM-MLA image. Taking this factor into consideration, the following six mineral classes remained: white mica (WM), biotite (Bt), chlorite (Chl), amphibole (Amp), gypsum (Gp), feldspar (Fsp), quartz (Qz), sulphide including sulphosalts and native gold (SP); less abundant minerals were grouped as "other". Because of the low abundance of biotite and accessory minerals in thin sections "a" and "b", the number of mineral classes considered was decreased accordingly. The test setups presented in the methodological framework section are used.

Cross-validation has been used to find the optimal parameters in order to train three models by internally resampling the training data. The main tested parameter ranges for each algorithm are presented in Table 3. The setups were chosen according to the lowest associated root-mean-square error (RMSE) based on cross-validation within 30 averaged iterations.

Table 3. Parameters and parameter ranges for the choice in optimum setup of the three tested algorithms.

RF	SVM	FF-ANN
Nb. of trees – 500 : 600	Kernel – Radial Basis Function Cost – 2 : 0.5 : 4 Sigma – 5 : 0.5 : 7	*Training function* – Scaled conjugate gradient backpropagation Nb. of hidden layers – 1 Nb. of neurons – 30 : 10 : 80

5.1. Mineral Abundance and Association Mapping

With the first experimental setup, presented in the methodological framework, 50% randomly distributed samples of the available thin section regions were used to train the regression models the mineral abundances estimation in the entire drill-core sample (Figure 5). Based on the visual analysis of the core and results analysis, RF and FF-ANN show better results in estimating the abundance of minerals with local distribution and small concentrations. With respect to matrix mineralogy, while biotite is well estimated by SVM in comparison with RF and FF-ANN, other major components of the matrix such as feldspar present a rather poor estimation. Similar performances of the algorithms can be observed for vein mineral components such as gypsum and sulphide.

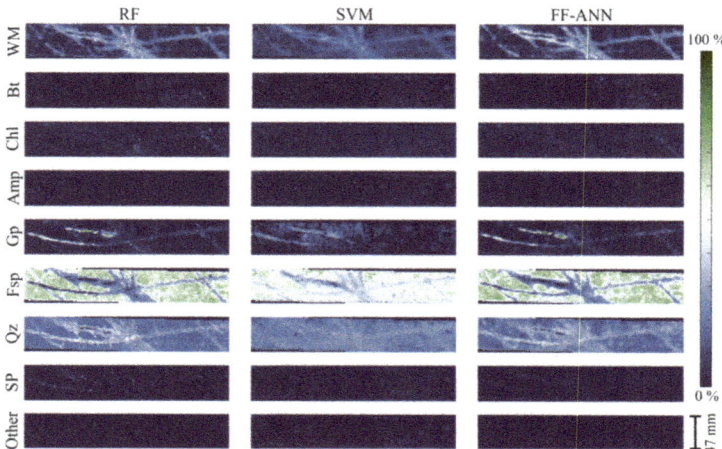

Figure 5. Drill-core mineral abundance maps of white mica (WM), biotite (Bt), chlorite (Chl), amphibole (Amp), gypsum (Gp), feldspar (Fsp), quartz (Qz), sulphide (SP) and accessory minerals (Other) using randomly distributed 50% of the available ground truth data for training for random forest (RF), support vector machine (SVM) and feed-forward neural network (FF-ANN) regressions.

With regards to the samples DC-2 (Figure A1), DC-3 (Figure A3), DC-4 (Figure A5) and DC-5 (Figure A7), using 50% of the available ground truth data for training, RF and FF-ANN show good, similar performances, while SVM shows limitations specifically in transitional areas between veins and matrix. Among the SWIR-diagnostic minerals, white mica, biotite and carbonate appear well mapped in all the samples, chlorite is slightly underestimated in samples DC-2 and DC-3 and gypsum is overestimated in sample DC-5. Among the SWIR non-diagnostic minerals, quartz shows the highest mapping inconsistencies between vein and matrix, particularly for samples DC-4 and DC-5. Sulphide, however, appears to be well mapped in most areas of the samples.

The quantitative evaluation of the mineral abundance mapping through the calculation of the RMSE supports the visual observations (Table 4). All three tested algorithms present low RMSEs and prove suitable to be used for mineral abundance mapping purposes. RF shows the lowest overall RMSE of 0.07, followed by FF-ANN with 0.08 and SVM with 0.1. Regarding the per class RMSE, RF and FF-ANN show similar results with the largest error associated with quartz, which can be the result of the lack of diagnostic absorption features in the VNIR-SWIR regions of the electromagnetic spectrum. SVM on the other hand shows larger per class errors for feldspar together with an increase in the error on white mica distribution. This can be explained by a misclassification between the two mineral groups. The mineral association of the sulphide in each pixel was calculated from the results of the mineral abundance mapping. Based on this calculation an equivalent overall performance of the methods was obtained (Table 5). For each of the methods, the error for the association of sulphide with feldspar is the largest.

Table 4. Evaluation of the three tested methods for the mineral abundance mapping of DC-1 through overall RMSE and per class RMSE values.

Method	RMSE	RMSE per Class								
		WM	Bt	Chl	Amp	Gp	Fsp	Qz	SP	Other
RF	0.07	0.06	0.06	0.06	0.06	0.06	0.05	0.08	0.06	0.05
SVM	0.10	0.12	0.03	0.05	0.01	0.12	0.21	0.12	0.04	0.02
NN	0.08	0.06	0.06	0.07	0.06	0.06	0.07	0.09	0.07	0.07

Table 5. Evaluation of the three tested methods for the mineral association mapping of DC-1 through overall RMSE and per class RMSE values.

Method	RMSE	RMSE per Class							
		WM	Bt	Chl	Amp	Gp	Fsp	Qz	Other
RF	0.05	0.05	0.00	0.01	0.00	0.02	0.13	0.05	0.00
SVM	0.06	0.06	0.01	0.02	0.00	0.02	0.15	0.06	0.00
NN	0.05	0.05	0.00	0.01	0.00	0.02	0.13	0.05	0.00

To assess the importance of sampling and representativeness of the SEM-MLA regions, thin sections "a", "b" (Figure 6) and "a + b" (Figure 7) of sample DC-1 were used for training the models in order to estimate the mineral abundance and association in thin section "c".

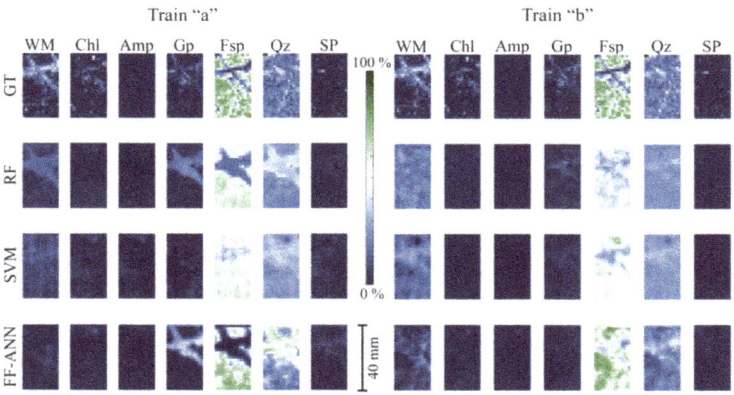

Figure 6. White mica (WM), chlorite (Chl), amphibole (Amp), gypsum (Gp), feldspar (Fsp), quartz (Qz) and sulphide (SP) abundance maps of TS-1c using TS-1a and TS-1b, respectively, for the training of random forest (RF), support vector machine (SVM) and feed-forward neural network (FF-ANN) regressions. The ground truth (GT) resized MLA data is presented for comparison.

For the three used methods, strong differences in the estimates of sample "c" mineralogy can be observed when using thin sections "a" and "b" for training (Table 6). The use of thin section "a" provides particularly better results for white mica and feldspar, which are confused using region "b" that hosts distinctly lower amounts of feldspar. On the other hand, using thin section "a" for training leads to an overestimation of the gypsum content. The use of both thin sections ("a" + "b") for training improves the classification leading to lower overall and per class RMSE values. As for the remaining drill-core samples, RF outperforms SVM and FF-ANN for most training scenarios, except when using thin section "b" for training. A similar effect of sampling on the RMSE evaluation can be seen for the mineral association mapping of DC-1 in all the scenarios (Table 7).

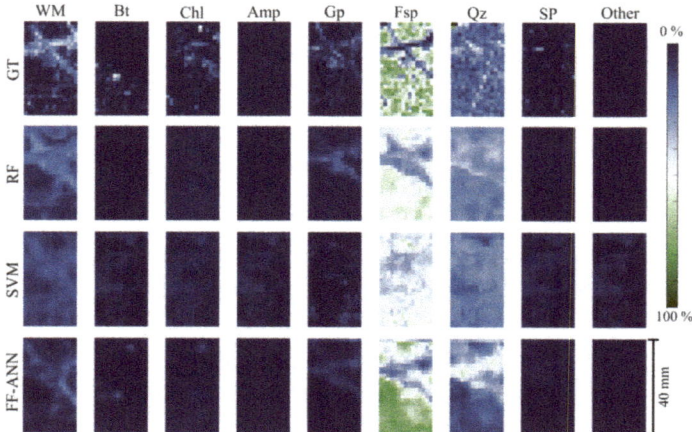

Figure 7. White mica (WM), biotite (Bt), chlorite (Chl), amphibole (Amp), gypsum (Gp), feldspar (Fsp), quartz (Qz), sulphide (SP) and accessory minerals (Other) abundance maps of TS-c using TS-a + TS-b for the training of random forest (RF), support vector machine (SVM) and feed-forward neural network (FF-ANN) regressions. The ground truth (GT) MLA data is presented for comparison.

Table 6. Evaluation of the three tested methods for the mineral abundance mapping of DC-1 thin section "c" through overall RMSE and per class RMSE values using different samples for training.

	Train and Valid. Data	Overall RMSE	RMSE per Class								
			WM	Bt	Chl	Amp	Gp	Fsp	Qz	SP	Other
RF	50%–50% rand. sel	0.07	0.06	0.06	0.06	0.06	0.06	0.05	0.08	0.06	0.05
RF	Train a—Test—c	0.10	0.08		0.06	0.01	0.06	0.18	0.13	0.03	
RF	Train b—Test—c	0.12	0.15		0.05	0.01	0.05	0.24	0.09	0.03	
RF	Train a + b—Test c	0.08	0.09	0.04	0.05	0.01	0.04	0.18	0.11	0.03	0.01
SVM	50%–50% rand. sel	0.10	0.12	0.03	0.05	0.01	0.12	0.21	0.12	0.04	0.02
SVM	Train a—Test—c	0.10	0.10		0.06	0.02	0.06	0.20	0.12	0.04	
SVM	Train b—Test—c	0.09	0.11		0.06	0.02	0.04	0.18	0.10	0.04	
SVM	Train a + b—Test c	0.09	0.09	0.05	0.06	0.03	0.07	0.21	0.09	0.05	0.03
FF-ANN	50"–50% rand. sel	0.08	0.06	0.06	0.07	0.06	0.06	0.07	0.09	0.07	0.07
FF-ANN	Train a—Test—c	0.17	0.12		0.07	0.02	0.14	0.28	0.27	0.06	
FF-ANN	Train b—Test—c	0.12	0.12		0.07	0.02	0.06	0.25	0.14	0.04	
FF-ANN	Train a + b—Test c	0.10	0.10	0.05	0.06	0.01	0.05	0.20	0.16	0.04	0.01

For the remaining samples, each having two regions analyzed by SEM-MLA, the mineral abundance estimations obtained using the second setup are illustrated in Figure A2 (DC-2), Figure A4 (DC-3), Figure A6 (DC-4) and Figure A8 (DC-5).

The tested methods show similar results for mineral abundance and association mapping on the remaining four drill-cores (Table 8). Overall, RF performs best, followed by FF-ANN and then SVM. For samples DC-1, DC-2, DC-3 and DC-5 each method results in comparable errors where similar amounts of training data are used. For sample DC-4 the overall RMSE values are higher, exceeding 0.2 depending on training data. For each sample the selection of the training data location plays an important role that is reflected into the RMSE evaluation.

Table 7. Evaluation of the three tested methods for the mineral association mapping of DC-1 thin section "c" through overall RMSE and per class RMSE values using different samples for training.

	Train and Validation Data	Overall RMSE	RMSE per Class							
			WM	Bt	Chl	Amp	Gp	Fsp	Qz	Other
RF	50%–50% rand. sel	0.05	0.05	0.00	0.01	0.00	0.02	0.13	0.05	0.00
	Train a—Test—c	0.06	0.05		0.04	0.01	0.05	0.05	0.10	
	Train b—Test—c	0.03	0.04		0.03	0.00	0.01	0.01	0.02	
	Train a + b—Test c	0.02	0.01	0.01	0.04	0.01	0.00	0.01	0.06	0.00
SVM	50%–50% rand. sel	0.06	0.06	0.01	0.02	0.00	0.02	0.15	0.06	0.00
	Train a—Test—c	0.05	0.05		0.02	0.02	0.07	0.06	0.06	
	Train b—Test—c	0.04	0.05		0.05	0.01	0.01	0.02	0.05	
	Train a + b—Test c	0.03	0.05	0.03	0.01	0.03	0.06	0.03	0.00	0.03
FF-ANN	50%–50% rand. sel	0.05	0.05	0.00	0.01	0.00	0.02	0.13	0.05	0.00
	Train a—Test—c	0.17	0.13		0.05	0.01	0.15	0.24	0.28	
	Train b—Test—c	0.07	0.07		0.04	0.00	0.04	0.15	0.00	
	Train a + b—Test c	0.05	0.09	0.00	0.05	0.01	0.02	0.06	0.10	0.00

Table 8. Methods evaluation for the mineral abundance and association mapping of the remaining four samples through overall RMSE and per class RMSE values using different data for training.

Sample ID	Train and Validation Data	Mineral Abundance Mapping			Mineral Association Mapping		
		RF	SVM	FF-ANN	RF	SVM	FF-ANN
DC-2	50%–50% rand. sel	0.07	0.09	0.08	0.07	0.07	0.07
	Train a—Test—b	0.11	0.18	0.10	0.09	0.17	0.09
	Train b—Test—a	0.14	0.14	0.19	0.13	0.16	0.13
DC-3	50%–50% rand. sel	0.08	0.11	0.09	0.12	0.12	0.12
	Train a—Test—b	0.14	0.14	0.17	0.09	0.18	0.07
	Train b—Test—a	0.11	0.14	0.14	0.10	0.11	0.09
DC-4	50%–50% rand. sel	0.12	0.20	0.14	0.12	0.12	0.12
	Train a—Test—b	0.24	0.29	0.24	0.08	0.10	0.05
	Train b—Test—a	0.16	0.20	0.19	0.04	0.16	0.07
DC-5	50%–50% rand. sel	0.07	0.10	0.08	0.03	0.03	0.03
	Train a—Test—b	0.11	0.13	0.11	0.05	0.18	0.05
	Train b—Test—a	0.13	0.13	0.15			

5.2. Modal Mineralogy

The modal mineralogy in area % is calculated by averaging the mineral abundances over the entire tested sample. To evaluate the modal mineralogy estimates sample DC-1 is used and the estimates are compared to the ground truth, using 50% of the available SEM-MLA data for training and 50% for testing (Table 9).

Table 9. Ground truth and estimated modal mineralogy of the SEM-MLA test regions of DC-1, using 50% randomly selected data for training.

Method	Modal Mineralogy (Area %)								
	WM	Bt	Chl	Amp	Gp	Fsp	Qz	SP	Other
GT	16.0%	1.1%	4.3%	1.8%	7.3%	42.3%	24.9%	1.8%	0.5%
RF	15.8%	1.1%	4.3%	1.7%	7.4%	42.2%	25.1%	2.0%	0.5%
SVM	14.2%	1.2%	3.8%	1.8%	7.8%	44.3%	24.4%	1.8%	0.6%
NN	15.8%	1.1%	4.3%	1.7%	7.3%	42.6%	24.8%	1.9%	0.5%

The estimates for all methods show good results with the highest RMSE value of 0.01 obtained with SVM. The complete modal mineralogy results are available in Table S1. The results for all the setups and all samples and methods are illustrated in Figure 8 by plotting the estimated values from RF (left), SVM (centre) and FF-ANN (right) against the ground truth values known from the re-sampled SEM-MLA data. The estimated and true values for RF and FF-ANN show overall a good correlation with local outliers related to mineral groups such as feldspar, as these do not have distinct spectral features in the VNIR-SWIR regions of the electromagnetic spectrum. Outliers can also be observed for white mica where the training and testing classes were unbalanced and confusions between mica and feldspar occurred. SVM, on the other hand, shows higher deviations from a linear correlation. Additionally, an important factor influencing the results is the data used for sampling. All test scenarios results are included in Figure 8 and as observed in the mineral abundance mapping results (Table 8), sampling plays a critical role in method performance.

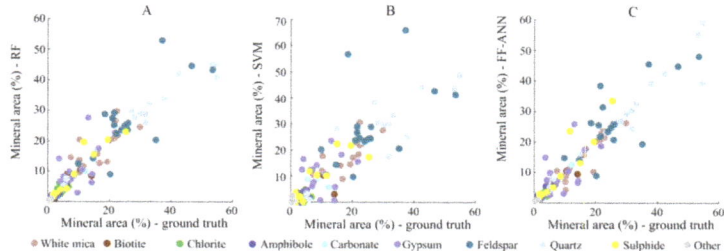

Figure 8. Scatter-plots of the ground truth vs. estimated mineral area % in all analyzed scenarios and samples using **A**. RF, **B**. SVM and **C**. FF-ANN.

5.3. Mineral Association

The overall mineral association is calculated by averaging the sulphide association in each classified pixel. The results for the setup consisting of 50% of the SEM-MLA regions of DC-1 for training and 50% for testing are presented in Table 10. For each regression method, the association of sulphide with white mica, chlorite, gypsum and quartz is underestimated, while the feldspar association is overestimated. The same tendency is observed for the rest of the calculated mineral associations in all samples and setups (Appendix A, Figures A1–A8). The relationship between ground truth and estimated data is illustrated in the scatter-plots in Figure 9. The results of the mineral association are strongly influenced by the estimation of the sulphide abundance as well as of the other mineral groups. Therefore, the highest errors in sulphide abundance mapping are consistent with the largest errors for sulphide association.

Table 10. Ground truth and estimated mineral association of the SEM-MLA test regions of DC-1, using 50% randomly selected data for training.

Method	Sulphide Association							
	WM	Bt	Chl	Amp	Gp	Fsp	Qz	Other
GT	21.0%	0.7%	5.6%	1.5%	9.9%	30.1%	30.5%	0.6%
RF	16.2%	1.1%	4.4%	1.8%	7.6%	42.8%	25.7%	0.5%
SVM	14.5%	1.2%	3.9%	1.9%	7.8%	45.1%	24.8%	0.6%
NN	16.2%	1.1%	4.4%	1.8%	7.5%	43.2%	25.3%	0.5%

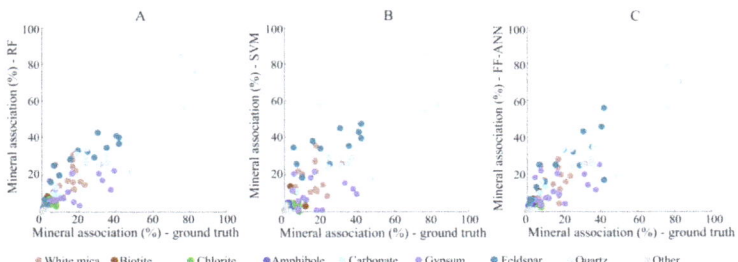

Figure 9. Scatter-plots of the ground truth vs. estimated mineral association in all analyzed scenarios and samples using **A**. RF, **B**. SVM and **C**. FF-ANN.

6. Discussion

The proposed approach for data preparation and analysis illustrates the potential to arrive at robust quantitative mineral abundance estimates from hyperspectral drill-core data—even for those minerals that do not have diagnostic absorption features in the VNIR-SWIR regions of the electromagnetic spectrum (e.g., feldspars, quartz, sulphides). Three regression methods were tested in this paper for mineral abundance estimation: random forest (RF), support vector machines (SVM) and feedforward artificial neural networks (FF-ANN). These methods were applied to quantify mineral abundances—also of minerals devoid of characteristic HS spectral features (here sulphide minerals). In addition, attempts were made to extract mineral association data from HS information at a lateral resolution far below the actual size of mineral grains in the studied ore. For this purpose, the abundance of each gangue mineral in each HS pixel is normalized to the content of ore minerals that are the main target in the currently studied porphyry system, thus constituting a rather simple proxy for the opportunity of two minerals or mineral groups to occur in direct contact with each other.

The abundance estimation of SWIR diagnostic mineral phases and groups is good overall, particularly for white mica, amphibole and chlorite. For the case of gypsum, however, due to its pervasive association with white mica in some training samples, errors in the abundance estimation occurred. Even though it is present in minor amounts in comparison to white mica, the estimation error can reach similar amplitudes as those of white mica. An additional reason for high errors associated with gypsum is related to its composition. The higher the degree of hydration of anhydrite towards gypsum the stronger and more distinct its absorption features. While SEM-MLA methods cannot measure the amount of water in the structure of the hydrated calcium sulphate, hyperspectral sensors are highly sensitive to these changes. Therefore, having training samples hosting mostly calcium sulphate with low amount of water can cause miss-estimation in test samples which may have low amounts of highly hydrated calcium sulphate. The local high errors in the estimation of biotite content can be assigned to the low amount of training samples containing relevant amounts of biotite. Sulphide is the main target in the current case study and this group comprises dominantly of pyrite, chalcopyrite, bornite, covellite, chalcocite, minor sulphosalts and native gold as an inclusion in the sulphides. While locally sulphide can be present as disseminations in the matrix, the highest fraction is present in veins. For all methods, the abundance estimation for SWIR non-diagnostic minerals is highly dependent on their association with the hydrothermal alteration minerals. To be able to estimate their abundance, representative sampling is required to avoid the erroneous estimation of these minerals based on local association with SWIR minerals that are not consistent at drill-core scale. For the analyzed samples the highest per-class errors are obtained for feldspar and quartz, both SWIR non-diagnostic minerals. In many cases feldspar was overestimated, particularly in samples where white mica abundance was underestimated. As white mica is present as an alteration product of feldspar in the proximity of veins, it can be assumed that the training samples consisted of lower alteration degrees of the feldspar to white mica while the test samples showed contrasting composition. As a result, feldspar particularly represented a bottleneck for the evaluation of the mineral association where their association with sulphide was in each case overestimated. Besides the fact that this mineral group does not show distinctive absorption features in the VNIR-SWIR regions of the electromagnetic spectrum, the spatial resolution of the used sensor can highly influence the misclassification and the overestimation in its association with sulphide. Feldspar is usually present in the host-rock matrix and is expected to have a low association with sulphide, usually being altered to white mica in the proximity of the sulphide-bearing veins. When the vein alteration halo is thinner than the spatial resolution of the sensor (here 1.5 mm), an increase in the apparent association of sulphide with feldspar is observed.

A potential limitation resides in the removal of the mineral fractions present in low concentrations (lower total surface abundance than the size of a hyperspectral pixel). Additionally, the compositional variation of minerals such as white mica and chlorites is not analyzed in the current work, but could be performed by auxiliary methods such and minimum wavelength analysis.

To evaluate the performance of the three regression methods employed in this paper, the RMSE was calculated. In general, for the mineral abundance estimation RF performed well and derived the lowest errors. The errors produced by FF-ANN tend to be higher than by SVMs in all the test scenarios, except in the case when 50% of the ground truth was randomly selected as the training data. This highlights the capabilities of SVM to perform well when a limited number of training samples are available and of FF-ANN to achieve good results when enough training data are available. The random selection of the training data allows for a more representative sampling per class than it is for the other two test scenarios where one thin section is used for training and the other thin section is used for the test. This is because certain minerals can be more abundant in one part of the core than in the other as it was previously stated for DC-1 in the results section. Although larger per class RMSE are obtained by minerals without diagnostic absorption features in the VNIR-SWIR, this is countered by random sampling and errors decrease considerably. From the analysis and evaluation of the results obtained by the utilized regression methods, the RF algorithm is the most suitable for the current dataset.

The proposed framework allows for fast evaluation of the modal mineralogy of analyzed samples and it shows potential for further upscaling. It proves that hyperspectral drill-core scanning provides a fast, non-invasive mineral identification and quantification if suitable training samples are available. Domaining of the hyperspectral data before the selection of representative samples for detailed analysis can minimize and focus the effort and amount of invasive measures related to sampling and high-resolution mineralogical analyses. The automated character of the approach can be later used on mine sites provided that hyperspectral drill-core scanning is available to support the geologists in the core-logging procedure, as well as training samples characterized by high resolution methods of mapping mineral distributions, such as SEM-based image analyses. The derived mineralogical parameters such as modal mineralogy and mineral association can additionally prove useful past exploration stages as they are essential in defining geometallurgical domains [37].

7. Conclusion and Remarks

Hyperspectral drill-core imaging provides fast, extensive and non-destructive mapping of certain minerals with spectral characteristic features in the VNIR-SWIR regions of the electromagnetic spectrum. SEM-MLA analyses allow a precise and exhaustive mineral mapping of selected small samples. We propose to combine both analytical techniques using machine learning in order to provide mineral abundance and association mapping over entire drill-cores. The proposed methodological framework is illustrated on samples collected from a porphyry type deposit, but the procedure is easily adaptable to other ore types. All tested ML algorithms deliver good results but RF is more robust to unbalanced and sparse training sets and is recommended for further work. As a result, quasi-quantitative maps are also produced and evaluated. The mineral abundance results can be further used to calculate parameters such as modal mineralogy, mineral association and other mineralogical indices. Therefore, this approach can be integrated in the standard core-logging procedure, complementing the on-site geologists, and can serve as background for the geometallurgical analysis of numerous ore types.

Supplementary Materials: The following are available online at http://www.mdpi.com/2072-4292/12/7/1218/s1, Table S1: Compilation of numerical results for mineral abundance and mineral association estimation.

Author Contributions: Conceptualization, L.T.; methodology, L.T. and M.K.; software, L.T., M.K. and C.C.; validation, L.T.; formal analysis, L.T.; investigation, L.T.; writing—original draft preparation, L.T., M.K., C.C. and K.R.S.; writing—all authors; visualization, L.T.; supervision, M.K., M.F., R.G. and J.G.; project administration, R.G. and J.G.; All authors have read and agreed to the published version of the manuscript.

Funding: This research received no external funding.

Acknowledgments: We would like to acknowledge Deva Gold S.A., subsidiary of Eldorado Gold for sample availability and on-site support. Robert Zimmerman (HZDR) is gratefully thanked for the support during data acquisition and Sandra Lorenz (HZDR) for the MEPHySTo toolbox for the hyperspectral data pre-processing. Sabine Gilbricht (TU Bergakademie Freiberg) is acknowledged for support during SEM-MLA data acquisition. The review by Timothy Baker (Eldorado Gold) helped improve the manuscript.

Conflicts of Interest: The authors declare no conflict of interest.

Appendix A

The results of mineral abundance mapping for DC-2 to DC-5 are shown in Figures A1–A8 using all test scenarios.

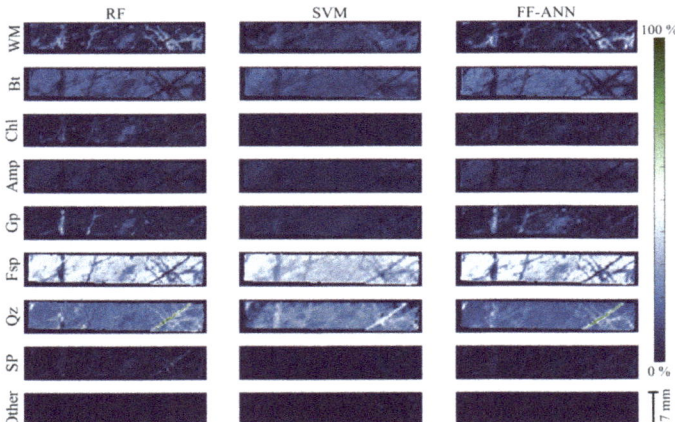

Figure A1. Drill-core abundance maps of white mica (WM), biotite (Bt), chlorite (Chl), amphibole (Amp), gypsum (Gp), feldspar (Fsp), quartz (Qz), sulphide (SP) and accessory minerals (Other) for DC-2 using randomly distributed 50% of the available ground truth data for training for random forest (RF), support vector machine (SVM) and feed-forward neural network (FF-ANN) regressions.

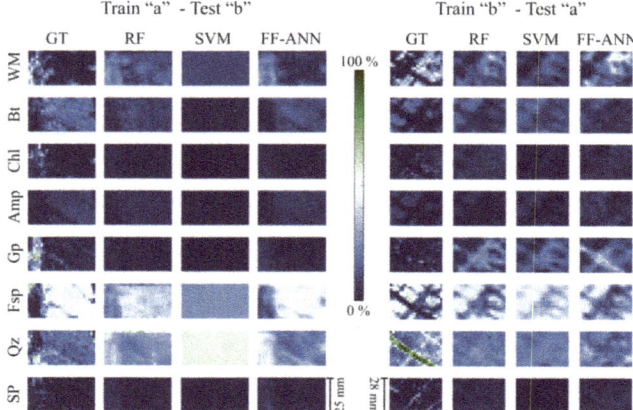

Figure A2. White mica (WM), biotite (Bt), chlorite (Chl), amphibole (Amp), gypsum (Gp), feldspar (Fsp), quartz (Qz) and sulphide (SP) abundance maps of TS-2a using TS-2b for training and of TS-2b using TS-2a respectively for training of random forest (RF), support vector machine (SVM) and feed-forward neural network (FF-ANN) regressions. The ground truth (GT) represented by resized MLA data is presented for comparison.

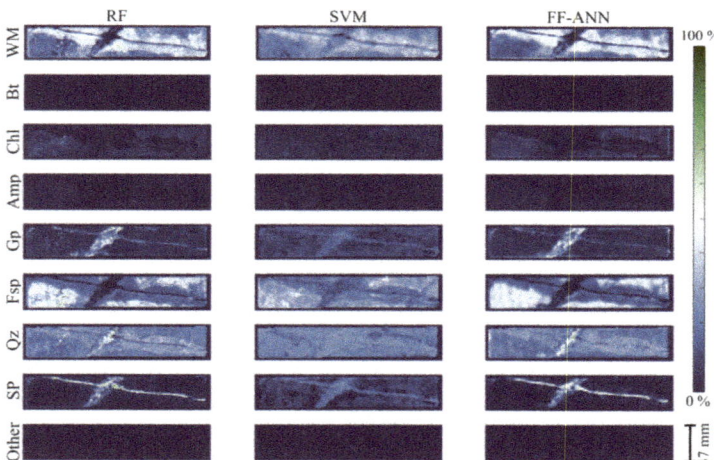

Figure A3. Drill-core abundance maps of white mica (WM), biotite (Bt), chlorite (Chl), amphibole (Amp), gypsum (Gp), feldspar (Fsp), quartz (Qz), sulphide (SP) and accessory minerals (Other) for DC-3 using randomly distributed 50% of the available ground truth data for training for random forest (RF), support vector machine (SVM) and feed-forward neural network (FF-ANN) regressions.

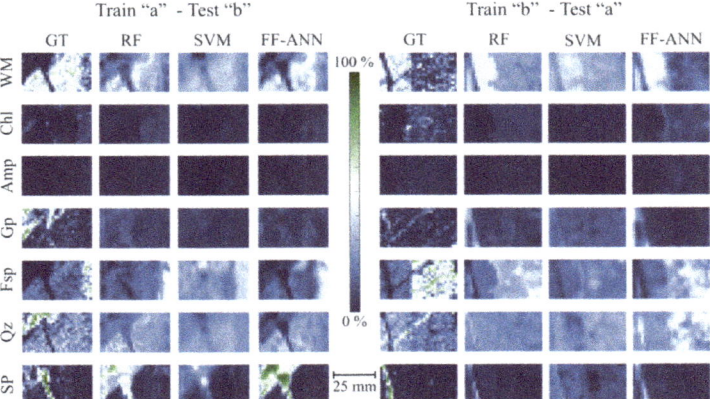

Figure A4. White mica (WM), chlorite (Chl), amphibole (Amp), gypsum (Gp), feldspar (Fsp), quartz (Qz) and sulphide (SP) abundance maps of TS-3a using TS-3b for training and of TS-3b using TS-3a respectively for training of random forest (RF), support vector machine (SVM) and feed-forward neural network (FF-ANN) regressions. The ground truth (GT) represented by resized MLA data is presented for comparison.

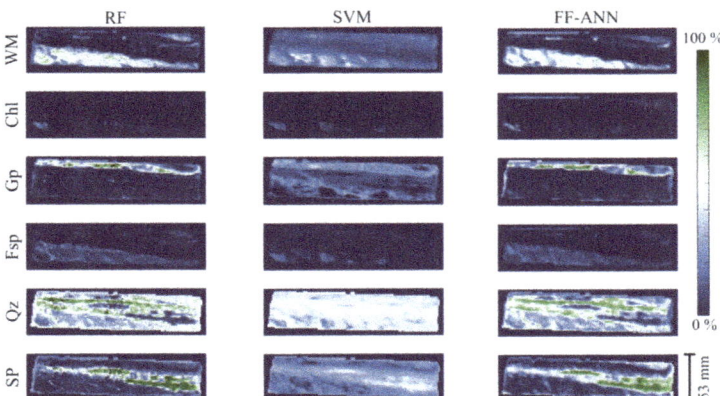

Figure A5. Drill-core abundance maps of white mica (WM), chlorite (Chl), gypsum (Gp), feldspar (Fsp), quartz (Qz) and sulphide (SP) for DC-4 using randomly distributed 50% of the available ground truth data for training for random forest (RF), support vector machine (SVM) and feed-forward neural network (FF-ANN) regressions.

Remote Sens. **2020**, *12*, 1218

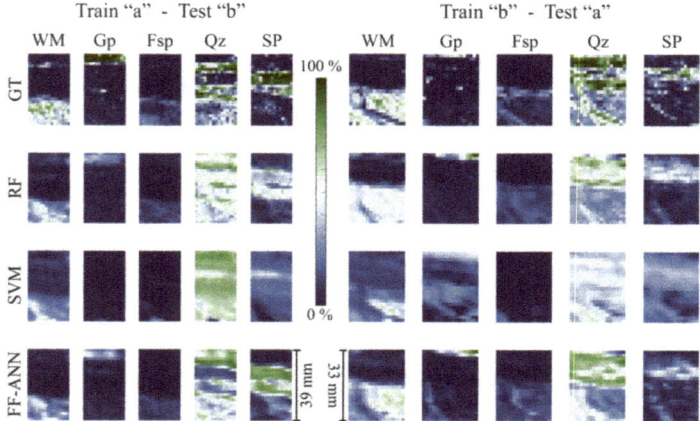

Figure A6. White mica (WM), gypsum (Gp), feldspar (Fsp), quartz (Qz) and sulphide (SP) abundance maps of TS-4a using TS-4b for training and of TS-4b using TS-4a respectively for training of random forest (RF), support vector machine (SVM) and feed-forward neural network (FF-ANN) regressions. The ground truth (GT) represented by resized MLA data is presented for comparison.

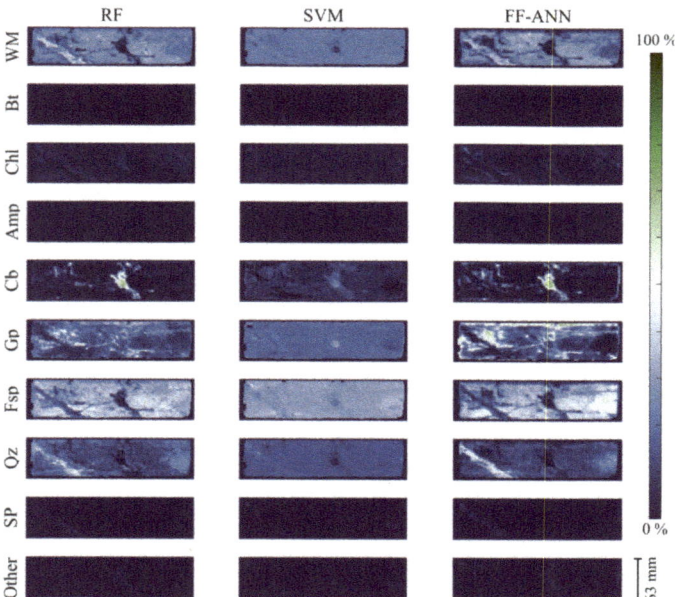

Figure A7. Drill-core abundance maps of white mica (WM), biotite (Bt), chlorite (Chl), amphibole (Amp), carbonate (Cb), gypsum (Gp), feldspar (Fsp), quartz (Qz), sulphide (SP) and accessory minerals (Other) for DC-5 using randomly distributed 50% of the available ground truth data for training for random forest (RF), support vector machine (SVM) and feed-forward neural network (FF-ANN) regressions.

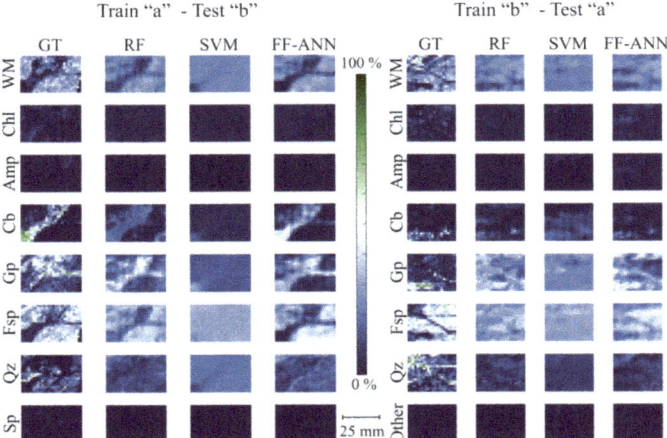

Figure A8. White mica (WM), chlorite (Chl), amphibole (Amp), carbonate (Cb), gypsum (Gp), feldspar (Fsp), quartz (Qz) and sulphide (SP) abundance maps of TS-5a using TS-5b for training and of TS-5b using TS-5a respectively for training of random forest (RF), support vector machine (SVM) and feed-forward neural network (FF-ANN) regressions. The ground truth (GT) represented by resized MLA data is presented for comparison.

References

1. Gandhi, S.M.; Sarkar, B.C.; Gandhi, S.M.; Sarkar, B.C. Chapter 8—Drilling. In *Essentials of Mineral Exploration and Evaluation*; Elsevier: Amsterdam, The Netherlands, 2016.
2. Lindholm, R.C. Mineral identification using X-ray diffraction. In *A Practical Approach to Sedimentology*; Springer: Dordrecht, The Netherlands, 1987; pp. 124–153.
3. Fandrich, R.; Gu, Y.; Burrows, D.; Moeller, K. Modern SEM-based mineral liberation analysis. *Int. J. Miner. Process.* **2007**, *84*, 310–320. [CrossRef]
4. Kirsch, M.; Lorenz, S.; Zimmermann, R.; Tusa, L.; Möckel, R.; Hödl, P.; Booysen, R.; Khodadadzadeh, M.; Gloaguen, R. Integration of terrestrial and drone-borne hyperspectral and photogrammetric sensing methods for exploration mapping and mining monitoring. *Remote Sens.* **2018**, *10*, 1366. [CrossRef]
5. Calvin, W.M.; Pace, E.L. Mapping alteration in geothermal drill core using a field portable spectroradiometer. *Geothermics* **2016**, *61*, 12–23. [CrossRef]
6. Kruse, F.A. Identification and mapping of minerals in drill core using hyperspectral image analysis of infrared reflectance spectra. *Int. J. Remote Sens.* **1996**, *17*, 1623–1632. [CrossRef]
7. Acosta, I.C.C.; Khodadadzadeh, M.; Tusa, L.; Ghamisi, P.; Gloaguen, R. A Machine Learning Framework for Drill-Core Mineral Mapping Using Hyperspectral and High-Resolution Mineralogical Data Fusion. *IEEE J. Sel. Top. Appl. Earth Obs. Remote Sens.* **2019**, *12*, 4829–4842. [CrossRef]
8. Clark, R.N. Spectroscopy of Rocks and Minerals and Principles of Spectroscopy. *Manual Remote Sens.* **1999**, *3*, 2–52.
9. Pontual, S.; Merry, N.; Gamson, P. *Spectral Interpretation Field Manual*; AusSpec International Ltd.: Queenstown, New Zealand, 1997; Volume I.
10. Mauger, A.J.; Keeling, J.L.; Huntington, J.F. Alteration mapping of the Tarcoola Goldfield (South Australia) using a suite of hyperspectral methods. *Appl. Earth Sci.* **2007**, *116*, 2–12. [CrossRef]
11. Huntington, J.F.; Mauger, A.J.; Skirrow, R.G.; Bastrakov, E.N.; Connor, P.; Mason, P.; Keeling, J.L.; Coward, D.; Berman, M.; Phillips, R.; et al. Automated mineralogical core logging at the Emmie Bluff iron oxide-copper-gold Prospect. *Mesa J.* **2006**, *41*, 38–44.
12. Roache, T.J.; Walshe, J.L.; Huntington, J.F.; Quigley, M.A.; Yang, K.; Bil, B.W.; Blake, K.L.; Hyvärinen, T. Epidote-clinozoisite as a hyperspectral tool in exploration for Archean gold. *Aust. J. Earth Sci.* **2011**, *58*, 813–822. [CrossRef]

13. Mathieu, M.; Roy, R.; Launeau, P.; Cathelineau, M.; Quirt, D. Alteration mapping on drill cores using a HySpex SWIR-320m hyperspectral camera: Application to the exploration of an unconformity-related uranium deposit (Saskatchewan, Canada). *J. Geochemical Explor.* **2017**, *172*, 71–88. [CrossRef]
14. Tappert, M.; Rivard, B.; Giles, D.; Tappert, R.; Mauger, A. Automated drill core logging using visible and near-infrared reflectance spectroscopy: A case study from the Olympic Dam Iocg deposit, South Australia. *Econ. Geol.* **2011**, *106*, 289–296. [CrossRef]
15. Littlefield, E.; Calvin, W.; Stelling, P.; Kent, T. Reflectance spectroscopy as a drill core logging technique: An example using core from the Akutan. *GRC Transaction* **2012**, *2011*, 1281–1283.
16. Bedini, E.; van der Meer, F.; van Ruitenbeek, F. Use of HyMap imaging spectrometer data to map mineralogy in the Rodalquilar caldera, southeast Spain. *Int. J. Remote Sens.* **2009**, *30*, 327–348. [CrossRef]
17. Kratt, C.; Calvin, W.M.; Coolbaugh, M.F. Mineral mapping in the Pyramid Lake basin: Hydrothermal alteration, chemical precipitates and geothermal energy potential. *Remote Sens. Environ.* **2010**, *114*, 2297–2304. [CrossRef]
18. Jakob, S.; Zimmermann, R.; Gloaguen, R. The Need for Accurate Geometric and Radiometric Corrections of Drone-Borne Hyperspectral Data for Mineral Exploration: MEPHySTo-A Toolbox for Pre-Processing Drone-Borne Hyperspectral Data. *Remote Sens.* **2017**, *9*, 88. [CrossRef]
19. Ruffin, C.; King, R.L. The analysis of hyperspectral data using Savitzky-Golay filtering-theoretical basis. In Proceedings of the IEEE 1999 International Geoscience and Remote Sensing Symposium IGARSS 99 (Cat. No.99CH36293), Hamburg, Germany, 28 June–2 July 1999; Volume 2, pp. 756–758.
20. Rodarmel, C.; Shan, J. Principal Component Analysis for Hyperspectral Image Classification. *Surv. L. Inf. Syst.* **2002**, *62*, 115–122.
21. Kern, M.; Möckel, R.; Krause, J.; Teichmann, J.; Gutzmer, J. Calculating the deportment of a fine-grained and compositionally complex Sn skarn with a modified approach for automated mineralogy. *Miner. Eng.* **2018**, *116*, 213–225. [CrossRef]
22. Gu, Y. Automated scanning electron microscope based mineral liberation analysis. *J. Miner. Mater. Charact. Eng.* **2003**, *2*, 33–41.
23. Milu, V.; Leroy, J.L.; Piantone, P. Le gisement de cuivre-or de Bolcana (monts Métallifères Roumanie) : Premières données sur les altérations et minéralisations associées. *Comptes Rendus Geosci.* **2003**, *335*, 671–680. [CrossRef]
24. Blannin, R.; Tusa, L.; Birtel, S.; Gutzmer, J.; Gilbricht, S.; Ivascanu, P. Metal deportment and ore variability of the Bolcana porphyry Au-Cu system (Apuseni Mts, Romania)-Implications for ore processing. In Proceedings of the 15th Biennial Meeting of the Society for Geology Applied to Mineral Deposits, Glasgow, UK, 27–30 August 2019.
25. Tusa, L.; Andreani, L.; Khodadadzadeh, M.; Contreras, C.; Ivascanu, P.; Gloaguen, R.; Gutzmer, J. Mineral Mapping and Vein Detection in Hyperspectral Drill-Core Scans: Application to Porphyry-Type Mineralization. *Minerals* **2019**, *9*, 122. [CrossRef]
26. Ivascanu, P.; Baker, T.; Lewis, P.; Kulcsar, Z.; Denes, R.; Tamas, C. *Bolcana, Romania: Geology and Discovery History of a Gold Rich Porphyry Deposit*; NewGenGold: Perth, Australia, 2019.
27. Sillitoe, R.H. Porphyry Copper Systems. *Econ. Geol.* **2010**, *105*, 3–41. [CrossRef]
28. Draper, C.; Reichle, R.; de Jeu, R.; Naeimi, V.; Parinussa, R.; Wagner, W. Estimating root mean square errors in remotely sensed soil moisture over continental scale domains. *Remote Sens. Environ.* **2013**, *137*, 288–298. [CrossRef]
29. Waske, B.; Benediktsson, J.A.; Arnason, K.; Sveinsson, J.R. Mapping of hyperspectral AVIRIS data using machine-learning algorithms. *Can. J. Remote Sens.* **2009**, *35*, S106–S116. [CrossRef]
30. Rodriguez-Galiano, V.; Sanchez-Castillo, M.; Chica-Olmo, M.; Chica-Rivas, M. Machine learning predictive models for mineral prospectivity: An evaluation of neural networks, random forest, regression trees and support vector machines. *Ore Geol. Rev.* **2015**, *71*, 804–818. [CrossRef]
31. Ghamisi, P.; Plaza, J.; Chen, Y.; Li, J.; Plaza, A.J. Advanced Spectral Classifiers for Hyperspectral Images: A review. *IEEE Geosci. Remote Sens. Mag.* **2017**, *5*, 8–32. [CrossRef]
32. Breiman, L. Random forests. *Mach. Learn.* **2001**, *45*, 5–32. [CrossRef]
33. Cortes, C.; Vapnik, V. Dispensing system boosts throughput 50 percent. *Assembly* **2001**, *44*, 97.
34. Vapnik, V.N. An overview of statistical learning theory. *IEEE Trans. Neural Netw.* **1999**, *10*, 988–999. [CrossRef]
35. Specht, D.F. A General Regression Neural Network. *IEEE Trans. Neural Netw.* **1991**, *2*, 568–576. [CrossRef]

36. Rumelhart, D.E.; Hintont, G.E. Learning representations by back-propagating errors. *Nature* **1986**, *323*, 533–536. [CrossRef]
37. Van den Boogaart, K.G.; Tolosana-Delgado, R. Predictive Geometallurgy: An Interdisciplinary Key Challenge for Mathematical Geosciences. In *Handbook of Mathematical Geosciences: Fifty Years of IAMG*; Daya Sagar, B.S., Cheng, Q., Agterberg, F., Eds.; Springer International Publishing: Cham, Switzerland, 2018; pp. 673–686.

© 2020 by the authors. Licensee MDPI, Basel, Switzerland. This article is an open access article distributed under the terms and conditions of the Creative Commons Attribution (CC BY) license (http://creativecommons.org/licenses/by/4.0/).

Article

Application of Landsat-8, Sentinel-2, ASTER and WorldView-3 Spectral Imagery for Exploration of Carbonate-Hosted Pb-Zn Deposits in the Central Iranian Terrane (CIT)

Milad Sekandari [1], Iman Masoumi [1,2], Amin Beiranvand Pour [3], Aidy M Muslim [3,*], Omeid Rahmani [4,5], Mazlan Hashim [6], Basem Zoheir [7,8], Biswajeet Pradhan [9,10], Ankita Misra [3] and Shahram M. Aminpour [11]

1. Department of Mining Engineering, Shahid Bahonar University of Kerman, Kerman 7616913439, Iran; m.sekandari@fadak-src.com (M.S.); imanmasoumi@eng.uk.ac.ir (I.M.)
2. The Iran Minerals Production and Supply Company (IMPASCO), 14155–3598, Vali-Asr Sq., Tehran 1594643118, Iran
3. Institute of Oceanography and Environment (INOS), Universiti Malaysia Terengganu (UMT), Kuala Nerus 21030, Malaysia; beiranvand.pour@umt.edu.my (A.B.P.); ankita.misra@umt.edu.my (A.M.)
4. Mahabad Branch, Islamic Azad University, Mahabad 59135-433, Iran; omeid.rahmani@iau-mahabad.ac.ir
5. Department of Natural Resources Engineering and Management, School of Science and Engineering, University of Kurdistan Hewlêr (UKH), Erbil 44001, Iraq
6. Geoscience and Digital Earth Centre (INSTeG), Research Institute for Sustainable Environment, Universiti Teknologi Malaysia, Johor Bahru, Skudai 81310, Malaysia; mazlanhashim@utm.my
7. Department of Geology, Faculty of Science, Benha University, Benha 13518, Egypt; basem.zoheir@ifg.uni-kiel.de
8. Institute of Geosciences, University of Kiel, Ludewig-Meyn Str. 10, 24118 Kiel, Germany
9. The Centre for Advanced Modelling and Geospatial Information Systems (CAMGIS), Faculty of Engineering and Information Technology, University of Technology Sydney, Sydney 2007, Australia; Biswajeet.Pradhan@uts.edu.au
10. Department of Energy and Mineral Resources Engineering, Sejong University, Choongmu-gwan, 209 Neungdong-ro Gwangjin-gu, Seoul 05006, Korea
11. Faculty of Chemical Engineering, Urmia University of Technology, Urmia 57155-419, Iran; shahram.aminpour@che.sharif.edu
* Correspondence: aidy@umt.edu.my; Tel.: +60-96683101; Fax: +60-96692166

Received: 18 March 2020; Accepted: 8 April 2020; Published: 13 April 2020

Abstract: The exploration of carbonate-hosted Pb-Zn mineralization is challenging due to the complex structural-geological settings and costly using geophysical and geochemical techniques. Hydrothermal alteration minerals and structural features are typically associated with this type of mineralization. Application of multi-sensor remote sensing satellite imagery as a fast and inexpensive tool for mapping alteration zones and lithological units associated with carbonate-hosted Pb-Zn deposits is worthwhile. Multiple sources of spectral data derived from different remote sensing sensors can be utilized for detailed mapping a variety of hydrothermal alteration minerals in the visible near infrared (VNIR) and the shortwave infrared (SWIR) regions. In this research, Landsat-8, Sentinel-2, Advanced Spaceborne Thermal Emission and Reflection Radiometer (ASTER) and WorldView-3 (WV-3) satellite remote sensing sensors were used for prospecting Zn-Pb mineralization in the central part of the Kashmar–Kerman Tectonic Zone (KKTZ), the Central Iranian Terrane (CIT). The KKTZ has high potential for hosting Pb-Zn mineralization due to its specific geodynamic conditions (folded and thrust belt) and the occurrence of large carbonate platforms. For the processing of the satellite remote sensing datasets, band ratios and principal component analysis (PCA) techniques were adopted and implemented. Fuzzy logic modeling was applied to integrate the thematic layers produced by image processing techniques for generating mineral prospectivity maps of the study area. The spatial distribution of iron oxide/hydroxides, hydroxyl-bearing and carbonate minerals and dolomite were mapped using specialized band ratios and analyzing eigenvector loadings of the PC images. Subsequently, mineral prospectivity maps of the study area were generated by fusing the selected PC thematic layers using fuzzy logic modeling. The most favorable/prospective zones for hydrothermal ore mineralizations and carbonate-hosted Pb-Zn mineralization in the study region were particularly mapped and indicated. Confusion matrix, field reconnaissance and laboratory analysis were carried out to verify the occurrence of alteration zones and highly prospective locations of carbonate-hosted Pb-Zn mineralization in the study area. Results indicate that the spectral data derived from multi-sensor remote sensing satellite datasets can be broadly used for generating remote sensing-based prospectivity maps for exploration of carbonate-hosted Pb-Zn mineralization in many metallogenic provinces around the world.

Keywords: band ratios; principal component analysis (PCA); fuzzy logic modeling; Kashmar–Kerman tectonic zone (KKTZ); carbonate-hosted Pb-Zn mineralization; Iran

1. Introduction

Remote sensing has provided tools for geological exploration for almost four decades. Nowadays, many satellite remote sensing datasets are accessible freely and can be extensively used for mineral exploration projects [1–15]. Pb-Zn sulfide mineralization is typically associated with hydrothermal alteration zones, the contact boundaries of lithological units and structural features such as faults and fractures [5,16–19]. According to World Bank Commodities Price Forecast (WBCPF), the price and consumption of Pb and Zn are increasing annually [20]. Pb and Zn are a necessity for the steady development of many countries around the world [21]. Accordingly, the exploration of Pb-Zn deposits using remote sensing satellite imagery as an available and inexpensive tool is of practical and economic interest.

Carbonate-hosted Pb-Zn deposits are some of the most significant sources of Pb and Zn [22]. The major hydrothermal alteration zones associated with carbonate-hosted Pb-Zn deposits are: (i) dissolution and hydrothermal brecciation, (ii) dolomite and calcite alteration, (iii) silicification and (iv) clay, mica, and feldspar diagenesis [22–24]. Besides, gossans as oxidation products of sulphide mineralized rocks are documented with carbonate-hosted Pb-Zn mineralization [25]. Particularly, studies on carbonate-hosted Pb-Zn deposits in Iran, India, China and Greenland have shown the possibility to identify hydrothermal alteration and iron oxides associated with Pb–Zn deposits using Landsat-8 and ASTER satellite imagery [5,16–19]. The application of multi-sensor remote sensing satellite imagery and fusing the most informative alteration thematic layers using geostatistical models can provide a low-cost exploration approach for generating remote sensing-based prospectivity maps [15]. Multiple sources of spectral data derived from different remote sensing sensors can be utilized for detailed mapping a variety of hydrothermal alteration minerals in the VNIR and the SWIR regions.

Landsat-8 imagery contains nine bands (0.433 to 2.290 µm; 30 m spatial resolution) in the VNIR and SWIR regions (Table 1). The VNIR spectral bands are particularly sensitive for mapping iron oxides/hydroxides, while SWIR spectral bands are responsive for detecting hydroxyl-bearing minerals and carbonates. These spectral bands have been broadly used for mapping hydrothermal alteration zones associated with hydrothermal ore mineralizations [3,5,12,15]. Sentinel-2 has thirteen spectral bands in the VNIR and the SWIR regions (0.433 to 2.280 µm; spatial resolutions from 10 to 60 m) (Table 1) which are useful to identify iron oxides/hydroxides and hydroxyl-bearing minerals [6]. Six spectral bands in the SWIR range (1.600 to 2.430 µm; 30 m spatial resolution) allow the ASTER sensor to map clay and carbonate minerals (Table 1). Detailed detection and discrimination of hydroxyl-bearing minerals and carbonates using ASTER SWIR bands is documented [1,4,8,13]. Moreover, ASTER VNIR bands

(0.52 to 0.86 µm; 15 m spatial resolution) can map iron oxides/hydroxides [2,9]. The VNIR spectral bands of WV-2 and WV-3 (0.400 to 1.040 µm; 1.24 m spatial resolution) were used to discriminate Fe^{2+} and Fe^{3+} mineral groups [10,14]. Al-OH, Mg-Fe-OH, CO3, and Si-OH alteration minerals were mapped in detail using SWIR bands of WV-3 (1.195 to 2.365 µm; 3.70 m spatial resolution) (Table 1) [10]. Therefore, multi-sensor satellite imagery can provide multiple sources of spectral data for mapping and discriminating hydrothermal alteration minerals to generate remote sensing-based prospectivity maps for metallogenic provinces.

The Central Iranian Terrane (CIT) of Iran has high potential for carbonate-hosted Pb-Zn deposits as a result of tectonic conditions related to its folded and thrust belt and the occurrence of large carbonate platforms [26]. The CIT consists of three N-S oriented crustal domains, namely the Lut, Tabas and Yazd blocks [27]. The Tabas and Yazd blocks are separated by a long, arcuate and structurally complex belt defined as the Kashmar–Kerman Tectonic Zone (KKTZ) (Figure 1), which has several occurrences of carbonate-hosted Pb-Zn deposits [28]. The KKTZ contains metamorphic rocks, limestones, pyroclastic and volcanic rocks, sandstone, dolomite and sandstone, slate and phyllite [28]. Ghanbari et al. [29] investigated the potential of rare earth element (REE) mineralization in the KKTZ using a fuzzy model. Geophysics, geochemistry, geology and remote sensing data were fused to indicate the prospective zones of REE mineralization. The favorability areas for REE mineralization sites were identified and prospectivity map for the study area was generated. However, lack of detailed geology map and comprehensive field surveying are the main issues that can be easily seen for the analysis.

Detailed mapping of hydrothermal alteration mineral zones associated with the carbonate-hosted Pb-Zn mineralization is one the essential factors for reconnaissance stages of Pb-Zn exploration in the CIT. However, there is no comprehensive remote sensing study for detailed identification of hydrothermal alteration mineral zones and lithological units for exploration Pb-Zn mineralization in the CIT, yet. In this research, Landsat-8, Sentinel-2, ASTER and WV-3 satellite remote sensing data were used for prospecting the carbonate-hosted Pb-Zn mineralization in the central part of the KKTZ, the CIT (Iran, Figure 1). The main objectives of this study are: (i) to map hydrothermal alteration minerals and lithological units by implementing band ratios and Principal Component Analysis (PCA) techniques to Landsat-8, Sentinel-2, ASTER and WV-3 datasets; (ii) to generate mineral prospectivity maps by fusing the most rational thematic layers using fuzzy logic modeling; and (iii) to verify the remote sensing results by field reconnaissance, laboratory analysis and confusion matrix.

2. Geologic Setting of the KKTZ

The formation of the CIT is attributed to the Late Precambrian Katangan/Pan-African orogenesis [30–32]. The KKTZ closely follows the trends of the predominant fault structures of the CIT. Three first-order fault systems are identified within the CIT, including (i) the N-trending system such as Nayband and Nehbandan faults, (ii) the NE system such as Poshteh-Badam and Kalmard faults and (iii) the NW system such as Kuhbanan and Rafsanjan faults. The NE and NW fault systems dominate the western part of the terrane, which also includes the KKTZ [31,33]. The KKTZ is part of the Poshte Badam-Bafgh basin [34]. Lithological outcrops in the KKTZ have an N-S trend. The folding and formation of these rocks is followed by a N-S faulting event [35]. Lithological units in this area are volcanic and schist units, dolomitic units and Quaternary deposits [33,36,37] (Figure 1).

The occurrence of magmatism in the Poshte Badam-Bafgh basin is associated with a back-arc extension zone during the Late Neoproterozoic to Early Cambrian [35,38,39]. Numerous ore mineralizations in the region have occurred in connection with the alkaline volcanic activity and extensional tectonics [36]. The carbonate-hosted Pb-Zn mineralization in the study area was formed during synchronous faulting activities with sedimentation, detrital sedimentation associated with faulting activities, emplacement of rhyolitic volcanic rocks and formation of rift sediments and subsidence [35,38]. In the KKTZ, Upper Precambrian series of volcano-sediments, detrital and carbonate rocks (especially dolomites) are overlying sandstone, conglomerate and tuff (the Tashk Formation) with a discontinuity [28,40,41].

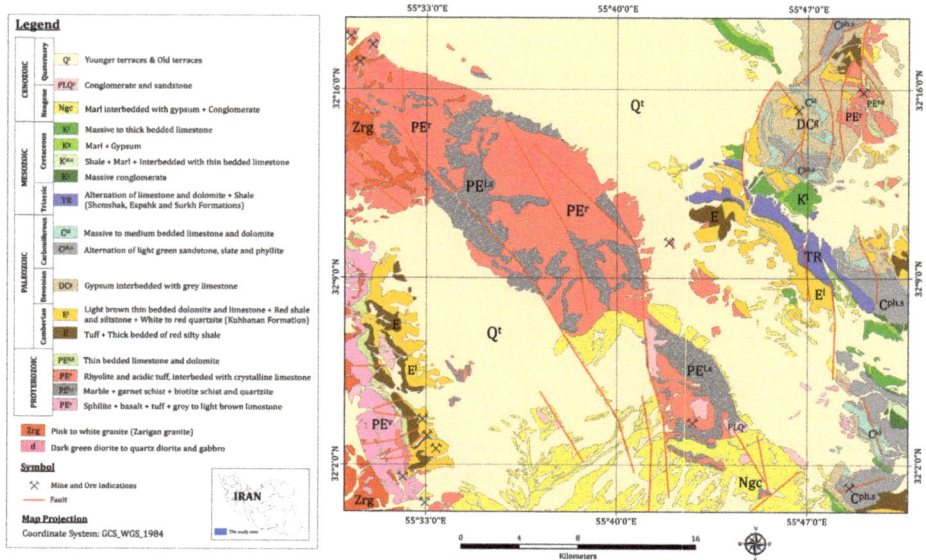

Figure 1. Geology map of the study area (modified from [36]).

3. Materials and Methods

3.1. Multi-Sensor Data Characteristics

In this investigation, multi-sensor satellite imagery, including Landsat-8, Sentinel-2, ASTER, and WV-3, was utilized for prospecting the carbonate-hosted Pb-Zn mineralization in the central part of the KKTZ region. Table 1 summarizes the technical performance and attributes of the Landsat-8, Sentinel-2, ASTER and WV-3 sensors. Broad spectral bands of Landsat-8 and Sentinel-2 were selected and used to discriminate lithological units and alteration zones for regional mapping of the study area. Narrow spectral bands of ASTER and WV-3 were used for detailed mapping of clay and carbonate minerals and Fe^{2+} and Fe^{3+} mineral groups.

Table 1. Technical performance and attributes of the Landsat-8, Sentinel-2, ASTER, and WV-3 sensors [42–47].

Sensor	Subsystem	Band Number	Spectral Range (µm)	Ground Resolution (m)	Swath Width (km)	Year of Launch
Landsat-8		PAN (8)	0.500–0.680	15		
	VNIR	Coastal aerosol (1) Blue (2) Green (3) Red (4) NIR (5)	0.433–0.453 0.450–0.515 0.525–0.600 0.630–0.680 0.845–0.885	30	185	2013
	SWIR	SWIR1 (6) SWIR2 (7) Cirrus (9)	1.560–1.660 2.100–2.300 1.360–1.390			
	TIR	TIRS1 (10) TIRS2 (11)	10.60–11.19 11.50–12.51	100		

Table 1. Cont.

Sensor	Subsystem	Band Number	Spectral Range (μm)	Ground Resolution (m)	Swath Width (km)	Year of Launch
Sentinel-2	VNIR	Coastal aerosol (1)	0.433–0.453	60	290	2015–2017
		Blue (2)	0.458–0.523	10		
		Green (3)	0.543–0.578			
		Red (4)	0.650–0.680			
		Vegetation Red Edge (5)	0.698–0.713	20		
		Vegetation Red Edge (6)	0.733–0.748			
		Vegetation Red Edge (7)	0.773–0.793			
		NIR (8)	0.785–0.900	10		
		Water-vapour (9)	0.935–0.955	60		
	SWIR	SWIR–Cirrus (10)	1.360–1.390			
		SWIR1 (11)	1.565–1.655	20		
		SWIR2 (12)	2.100–2.280			
ASTER	VNIR	1	0.52–0.60	15	60	1999
		2	0.63–0.69			
		3N	0.76–0.86			
		3B	0.76–0.86			
	SWIR	4	1.600–1.700	30		
		5	2.145–2.185			
		6	2.185–2.225			
		7	2.235–2.285			
		8	2.295–2.365			
		9	2.360–2.430			
	TIR	10	8.125–8.475	90		
		11	8.475–8.825			
		12	8.925–9.275			
		13	10.25–10.95			
		14	10.95–11.65			
WV3	VNIR	Coastal blue (1)	0.400–0.450	1.24	13.1	2014
		Blue (2)	0.450–0.510			
		Green (3)	0.510–0.580			
		Yellow (4)	0.585–0.625			
		Red (5)	0.630–0.690			
		Red-edge (6)	0.705–0.745			
		NIR1 (7)	0.770–0.895			
		NIR2 (8)	0.860–1.040			
	SWIR	SWIR-1 (9)	1.195–1.225	3.70		
		SWIR-1 (10)	1.550–1.590			
		SWIR-1 (11)	1.640–1.680			
		SWIR-1 (12)	1.710–1.750			
		SWIR-1 (13)	2.145–2.185			
		SWIR-1 (14)	2.185–2.225			
		SWIR-1 (15)	2.235–2.285			
		SWIR-1 (16)	2.295–2.365			

A cloud-free Landsat-8 scene (LC08_L1TP_161038_20170517, Path/Raw: 161/038) covering the central part of KKTZ was obtained from the U.S. Geological Survey Earth Resources Observation and Science Center (EROS) [48] for this analysis. It is level 1T (terrain corrected) data, which was acquired on 17 June 2017. A cloud-free Sentinel-2 (S2A_OPER_PRD_MSIL1C_PDMC_ 20160929T125040) scene covering the central part of KKTZ was obtained from the European Space Agency (Copernicus Open Access Hub), which was acquired on 29 September 2016. The Sentinel-2A utilized in this study is a Level-1C top-of-atmosphere (TOA) reflectance (100 km × 100 km tile) product, which includes radiometric and geometric corrections (UTM projection with WGS84 datum) along with orthorectification [49]. An ASTER scene covering the study region was acquired on 16 March 2003. It was cloud-free level 1T product that obtained from USGS EROS center [50]. A level 2 A WV-3 image (M2AS-056451539010_01_P001) covering the study area was purchased from the Arka Company (Tehran, Iran). Unfortunately, SWIR bands were not available for the study region. The WV-3 VNIR imagery was cloud-free and acquired on 20 April 2017. The level 2A WV-3 is a sensor and radiometrically corrected product, which is geometrically projected to the UTM with WGS84 datum [51,52]. The Landsat-8,

Sentinel-2, ASTER and WV-3 images used in this study have been already georeferenced to the UTM zone 40 North projection using the WGS84 datum. For processing the remote sensing datasets, the ENVI (Environment for Visualizing Images) [53], version 5.2 and ArcGIS version 10.3 (ESRI, Redlands, CA, USA) software packages were used.

3.2. PrePprocessing of the Remote Sensing Datasets

The Fast Line-of-sight Atmospheric Analysis of Hypercubes (FLAASH) algorithm [54] were applied to Landsat-8 (Operational Land Imager (OLI) bands) by implementing the Mid-Latitude Summer (MLS) and Rural aerosol models [55]. Sentinel-2 data layer stacked of VNIR+SWIR bands (bands 2, 3, 4, 8, 11 and 12) with 10 m spatial dimension was generated to obtain a six-band dataset. The QUick Atmospheric Correction (QUAC) was performed on this dataset by using mud filtering to eliminate highly structured materials such as shallow water, mud and vegetation [55]. ASTER data layer stacked of VNIR+SWIR bands with 30-meter spatial dimensions was generated by using Pan-sharpening method [56]. Internal Average Relative Reflectance (IARR) calibration [57] was applied to Crosstalk corrected [58] ASTER data for atmospheric correction. The conversion to the Top-of-Atmosphere (TOA) spectral radiance and absolute radiometric correction are required for the WV-3 relative radiometrically corrected images [59]. Therefore, the corrections were executed and FLAASH algorithm was applied to the WV-3 data.

3.3. Image Processing Techniques

Band Ratios and Principal Components Analysis (PCA) image processing procedures were executed to extract key information related to alteration minerals and lithological units from the pre-processed remote sensing datasets. Successively, the most rational thematic layers of the alteration zones were fused using fuzzy logic modeling to generate mineral prospectivity maps of the study area. Finally, field reconnaissance, laboratory analysis and confusion matrix were carried out for verifying the remote sensing results. An overview of the methodological flowchart used in this study is displayed in Figure 2.

3.3.1. Band Ratios

Band ratios method is broadly used for mapping hydrothermal alteration minerals and lithological units [2–4,60–63]. By ratioing bands that correspond to certain absorptions and reflectance, the pixels with particular mineral or mineral groups are highlighted [64–67]. Furthermore, this technique is proficient in reducing the topographic effects generated by slope orientations and solar illumination angles [68]. Several mathematical expressions were used for detecting alteration minerals or mineral groups, which are generally called Relative Absorption Band Depth (RBD) [69]. It includes three-point ratio formulation for revealing the mineral spectral intensities attributed to Fe^{2+}, Fe^{3+}, Fe-OH, Al-OH, Fe, Mg-OH, Si-OH, SO, CO_3 and SiO_2 [60,70–72]. For a particular absorption or emissivity distinction, the numerator is the sum of the bands indicating the shoulders and the denominator is the band placed nearby the absorption or emissivity feature minimum [69].

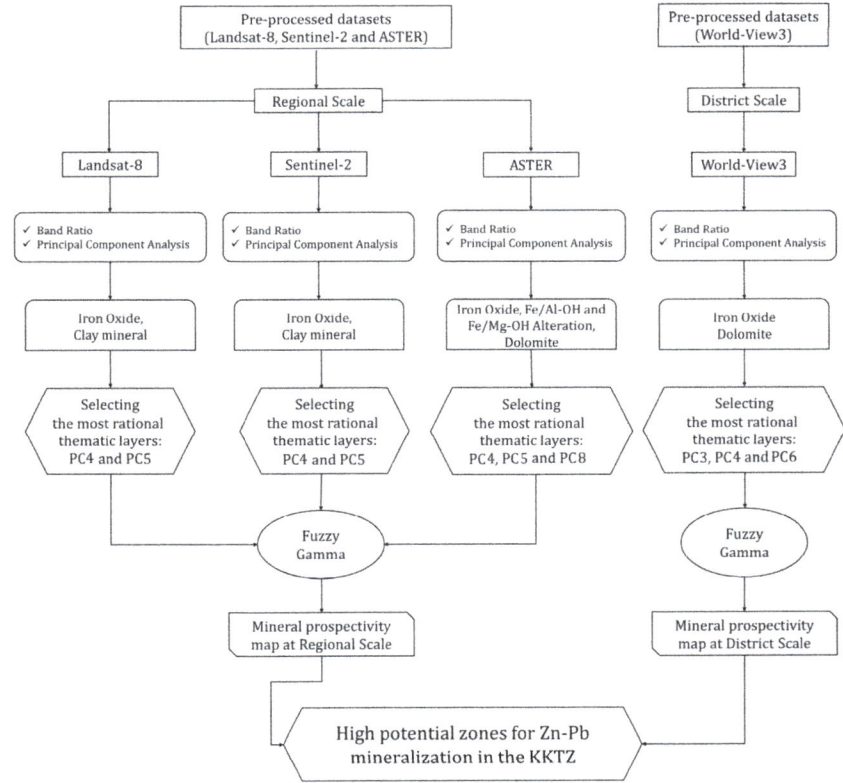

Figure 2. An overview of the methodological flowchart used in this study.

To characterize iron oxide/hydroxides (hematite, goethite and jarosite), the VNIR spectral bands contain the most important information due to electronic transitions of Fe^{3+}/Fe^{2+} in the VNIR region from 0.45 to 1.2 µm [73,74] (Figure 3). In this study for detecting iron oxide at a regional scale, 4/2 band ratio of Landsat-8, Sentinel-2 and ASTER was selected to highlight iron oxide/hydroxides. Hydroxyl-bearing (Al-OH) alteration and carbonates (muscovite, kaolinite, gypsum, calcite and dolomite) show spectral absorption features in the 2.1–2.5 µm region due to overtones and combinations of the fundamental vibrations [75], whereas their spectral reflectance typically occur in 1.55–1.75 µm in the SWIR regions (Figure 3). These characteristics are matched with band 7 (2.11–2.29 µm) and band 6 (1.57–1.65 µm) of Landsat-8, as well as band 12 (2.100–2.280 µm) and band 11 (1.565–1.655 µm) of Sentinel-2, respectively (Figure 3). Therefore, the 6/7 band ratio of Landsat-8 and 11/12 band ratio of Sentinel-2 were used to map hydroxyl bearing alteration minerals and carbonates in this study at the regional scale.

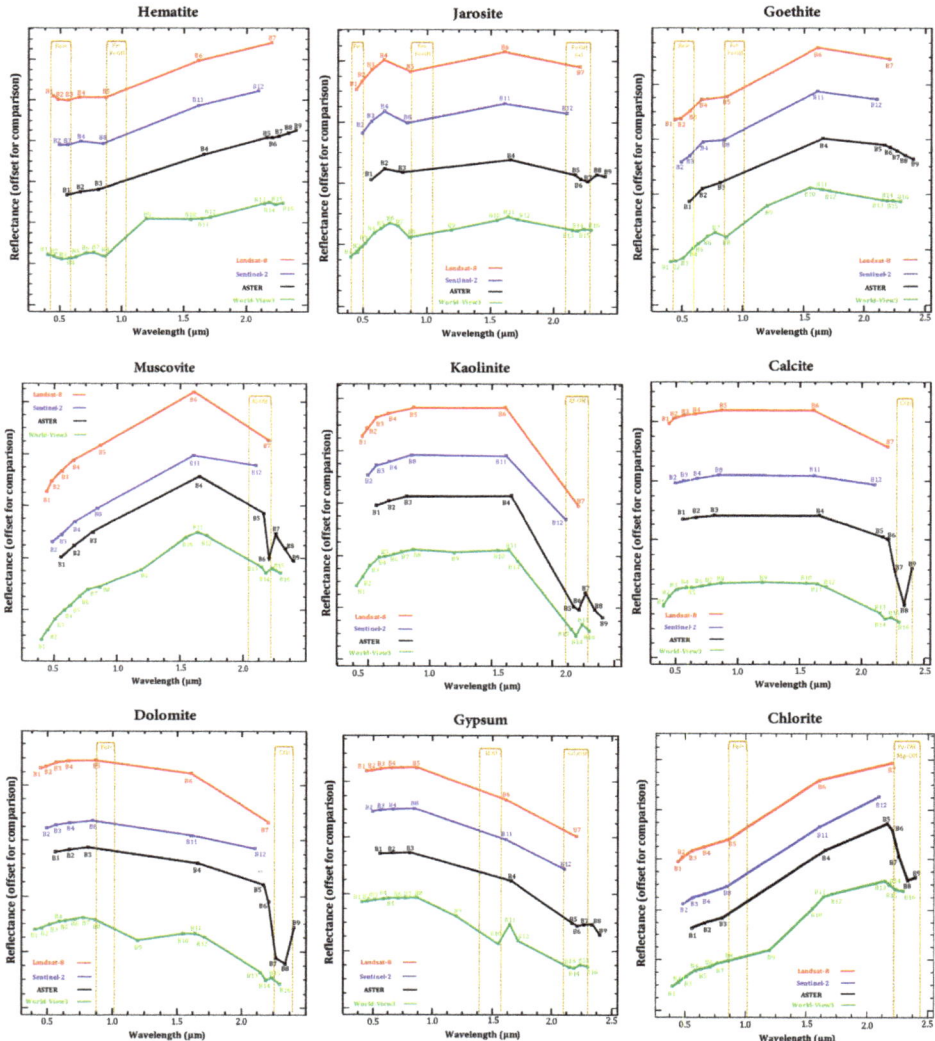

Figure 3. The laboratory reflectance spectra of hematite, jarosite, goethite, muscovite, kaolinite, calcite, dolomite, gypsum and chlorite resampled to response functions of VINR+SWIR bands of Landsat-8, Sentinel-2, ASTER and WV-3. Cubes indicate the position of the VINR+SWIR bands of Landsat-8, Sentinel-2, ASTER and WV-3 in the range of 0.4 µm to 2.5 µm. The main absorption feature spectra attributed to Fe^{+2}, Fe^{+3}, OH, H_2O, Fe-OH, S-O, Al-OH, Fe, Mg-OH, and CO_3 are delimited by dashed rectangles.

The Al-OH absorption features at 2.17 to 2.20 µm are corresponded with bands 5 and 6 of ASTER, whereas Mg-Fe-OH and CO3 absorption features are situated in 2.30 to 2.35 µm equivalent with bands 7 and 8 of ASTER [73,74,76] (Figure 3). Additionally, dolomite absorption features are mostly concentrated at 2.20 to 2.30 µm, which are coincident with bands 6 and 7 of ASTER [77] (Figure 3). Hence, calcite and dolomite minerals can be discriminated by different absorption features between 2.33 and 2.45 µm [76,78] (Figure 3). Subsequently in this study, (i) to map Al/Fe-OH minerals (muscovite, kaolinite and jarosite) the RBD1: (5 + 7)/6 was implemented; (ii) to identify Mg-Fe-OH/CO3 minerals (chlorite and calcite) the RBD2: (7 + 9)/8 was used; and (iii) for detecting dolomite the RBD3: (6 + 8)/7 was executed to ASTER (VNIR+SWIR) bands [67,76,77].

The VNIR spectral bands of WV-3 have high potential to map Fe^{3+}, gossan and dolomite/Fe^{2+}. Ferric/ferrous iron oxides contain a set of absorption features about 0.40 to 1.2 µm, which are matched with WV-3 VNIR bands 2, 3, 4, 6 and 8 [79–81] (Figure 3). The absorption features related to Fe^{3+} is typically at 0.49, 0.70 and 0.87 µm, while Fe^{2+} shows absorption properties at 0.51, 0.55 and 1.20 µm [73–75,82]. Considering the laboratory reflectance spectra of dolomite (Figure 3) indicate that band 7 (0.770–0.895 µm) of WV-3 can be assumed as a reflectance band, whereas band 4 (0.585–0.625 µm) can be considered as absorption features related to Fe^{2+} for detecting dolomite. The Fe^{2+} absorption in the dolomite spectrum normally can occur about 0.60 µm (in addition to 1.20 µm) [82] that is coincident with band 4 (yellow band) of WV-3. Accordingly, the band ratio of 5/3, 5/2 and 7/4 were adopted and developed for mapping Fe^{3+} and iron-stained alteration and dolomite/Fe^{2+}, respectively

3.3.2. Principal Components Analysis (PCA)

PCA is a mathematical technique that transforms a quantity of correlated variables into a number of uncorrelated linear variables called principal components (PCs) [83–85]. The PCA is usually implemented on a square symmetric matrix. It can be based on the covariance matrix (scaled sums of squares and cross products) or correlation matrix (sums of squares and cross products from standardized data) [83,86,87]. The PCA is broadly used to map hydrothermal alteration minerals and lithological units using spectral bands of remote sensing sensors [8,61,62,84,88–92]. The uncorrelated linear combinations (eigenvector loadings) contain indicative information allied to spectral characteristics of alteration minerals that can be expected from the specific spectral bands in the VNIR and SWIR regions [84,89]. Accordingly, a PC contains strong eigenvector loadings for indicative bands such as reflective and absorptive bands of an alteration mineral or mineral group with opposite signs enhances that mineral or mineral group as bright or dark pixels in the PC image. Positive loading in a reflective band enhances the alteration mineral as bright pixels, while negative loading is in a reflective band depicts the alteration mineral as dark pixels [84,89]. In this analysis, the PCA method was implemented based on covariance matrix to the selected bands of Landsat-8 (OLI bands), Sentinel-2 (bands 2, 3, 4, 8, 11 and 12) and ASTER (VNIR+SWIR bands) and WV-3 (VNIR bands) for identifying hydrothermal alteration mineral assemblages in the study area. Tables 2–5 show the eigenvector matrix for the selected bands of the remote sensing datasets.

Table 2. Eigenvector matrix derived from PCA for the selected bands of Landsat-8 bands (1 to 7) used in this study.

Eigenvector	Band 1	Band 2	Band 3	Band 4	Band 5	Band 6	Band 7
PC 1	0.098888	0.127588	0.222642	0.352134	0.470622	0.580154	0.49225
PC 2	0.322223	0.372444	0.433008	0.426627	0.230446	−0.374857	−0.440827
PC 3	−0.312408	−0.354582	−0.279999	0.084637	0.59951	0.168321	−0.550787
PC 4	0.247724	0.269361	0.117557	−0.327086	−0.247048	**0.667216**	**−0.48894**
PC 5	0.380773	**0.318038**	−0.297366	**−0.579504**	0.516661	−0.207254	0.140427
PC 6	−0.42878	−0.07607	0.726821	−0.486822	0.193733	−0.079913	0.034331
PC 7	0.631338	−0.734703	0.232424	−0.083794	0.022222	−0.007336	0.00581

Table 3. Eigenvector matrix derived from PCA for the selected bands of Sentinel-2 bands (2, 3, 4, 8, 11, 12) used in this study.

Eigenvector	Band 2	Band 3	Band 4	Band 8	Band 11	Band 12
PC 1	−0.15188	−0.235458	−0.383398	−0.466736	−0.570938	−0.480298
PC 2	0.355822	0.405278	0.438872	0.333419	−0.451093	−0.44931
PC 3	−0.486574	−0.409713	0.064894	0.557800	0.208124	−0.486532
PC 4	−0.310768	−0.246002	0.203541	0.259173	**−0.641949**	**0.567634**
PC 5	**−0.454829**	0.100882	**0.698634**	−0.523627	0.113057	−0.088865
PC 6	−0.556887	0.736037	−0.355904	0.135992	−0.048461	0.024823

Table 4. Eigenvector matrix derived from PCA for the selected bands of ASTER bands (VNIR+SWIR) used in this study.

Eigenvector	Band 1	Band 2	Band 3	Band 4	Band 5	Band 6	Band 7	Band 8	Band 9
PC 1	−0.996499	0.082351	−0.008442	0.008905	0.000780	0.005820	0.004598	−0.000774	−0.000742
PC 2	0.075083	0.954242	0.247544	−0.134085	−0.022691	−0.054521	−0.030563	0.009170	−0.004727
PC 3	−0.020004	−0.185786	−0.022226	**0.564435**	−0.157899	−0.054898	0.020914	**−0.896111**	−0.019443
PC 4	0.000139	**0.812776**	0.124735	**−0.429509**	−0.228138	−0.025162	0.314993	−0.251698	−0.002611
PC 5	−0.021063	−0.157428	0.221774	**−0.668838**	0.095360	**0.425904**	−0.014838	0.174277	**0.507113**
PC 6	0.013032	0.083318	−0.013620	0.319148	0.388530	0.009249	0.277101	−0.329300	−0.069032
PC 7	−0.006760	−0.037927	−0.034844	−0.164716	−0.322506	−0.352677	−0.157608	−0.042499	0.084469
PC 8	−0.016975	−0.119326	0.256410	**0.404433**	**0.558407**	0.158777	**−0.623420**	−0.103705	−0.140444
PC 9	0.002032	0.009398	−0.061213	0.025598	−0.153333	−0.047990	0.033235	−0.028976	−0.983741

Table 5. Eigenvector matrix derived from PCA for the selected bands of WV3 band (1 to 8 VNIR) used in this study.

Eigenvector	Band 1	Band 2	Band 3	Band 4	Band 5	Band 6	Band 7	Band 8
PC 1	−0.314986	−0.330951	−0.348156	−0.359256	−0.364601	−0.367182	−0.369097	−0.370119
PC 2	0.655926	0.454510	0.183457	−0.046042	−0.154854	−0.251952	−0.320189	−0.370709
PC 3	−0.331273	**−0.598506**	0.354295	−0.129646	**0.661001**	−0.220796	0.341420	0.108973
PC 4	−0.244961	**0.345377**	0.145561	**0.631659**	0.012267	0.368220	**−0.509311**	−0.142316
PC 5	−0.384633	0.279151	0.433976	−0.092808	0.081588	−0.370014	−0.142544	0.187618
PC 6	0.236442	**−0.427799**	**−0.515988**	−0.065670	**0.646312**	0.248715	0.043257	0.095274
PC 7	0.257771	−0.301701	−0.070317	−0.389055	0.471694	0.225588	−0.427691	0.035215
PC 8	0.174655	−0.560947	0.307690	−0.163685	−0.332755	0.108819	0.068151	−0.001993

3.3.3. Fuzzy Logic Modeling

Fuzzy logic modeling is based on the fuzzy set theory, which was proposed by Zadeh [93]. It is a form of many-valued logic in which the true values of variables may be any real number between 0 and 1 both inclusive [94]. A fuzzy set of A is a set of ordered pairs:

$$A = \{(x, \mu\_A(x) \mid x \in X\}, \qquad (1)$$

where $\mu\_A(x)$ is termed the membership function or membership grade of x in A. $\mu\_A(x)$ maps x to membership space (M), when M contains only the two points 0 and 1. The range of $\mu\_A(x)$ is [0, 1], where zero expresses non-membership and one expresses full membership [93]. Fuzzy logic modeling has been successfully applied for mineral prospectivity mapping in metallogenic provinces [29,95–98]. The application of fuzzy logic modeling for mineral prospectivity mapping normally includes three main feed-forward stages: (i) fuzzification of evidential data; (ii) logical integration of fuzzy evidential maps with the aid of an inference network and appropriate fuzzy set operations; and (iii) defuzzification of fuzzy mineral prospectivity output to aid its interpretation [96]. A set of fuzzy membership values is expressed in a continuous series from 0 to 1. This 0–1 scale, however, does not constitute a probability density function. Function-member values are established for each evidence map that will be integrated. In the fuzzy logic method, a total of sheet maps (fuzzy membership) based on the significance distance of features are weighted (for each pixel or spatial position, a particular weight between 0 and 1 is appointed) [96,99].

Five operators that are useful for combining mineral exploration datasets include the fuzzy AND, fuzzy OR, fuzzy algebraic product, fuzzy algebraic sum and fuzzy gamma [95,100–102]. In this analysis, the multiclass evidential image-maps were reclassified in 10 classes of equal interval and then were fuzzified using the linear membership function. The fuzzy gamma operator was used for mapping the prospective areas using Landsat-8, Sentinel-2, ASTER and WV-3 alteration thematic input layers. After testing several values for the γ parameter, it was adjusted 0.70 for ensuring a flexible compromise between the fuzzy algebraic sum and the fuzzy algebraic product [99]. Table 6 shows the fuzzification parameters for the input layers used in this analysis.

Table 6. Fuzzification parameters for the input layers used in this analysis.

Data Origin	Input Layer	Detection	Membership Type	Fuzzy Operator
Landsat-8 Dataset	PC4	OH-minerals and Carbonates	Linear	Gamma ($\gamma = 0.7$)
	PC5	Iron Oxide		
Sentinel-2 Dataset	PC4	OH-minerals and Carbonates		
	PC5	Iron Oxide		
ASTER Dataset	PC4	Iron oxide/hydroxides minerals	Linear	Gamma ($\gamma = 0.7$)
	PC5	OH/S-O/CO3-bearing minerals		
	PC8	Dolomite		
WV-3 Dataset	PC3	Iron-stained alteration	Linear	Gamma ($\gamma = 0.7$)
	PC4	Dolomite/Fe^{2+} oxides		
	PC6	Fe^{3+} oxides		

3.4. Fieldwork Data and Laboratory Analysis

Geological survey and laboratory analysis were carried out to confirm the image processing results and mineral prospectivity mapping for the central part of the KKTZ region. A global positioning system (GPS) survey was conducted in the study area for verifying the spatial distribution of alteration zones and lithological units using a handheld Monterra GPS (average accuracy of 3 m; Garmin, New Taipei City, Taiwan). Additionally, numerous photos were taken from alteration zones and lithological units during the field surveys. Thirty hand specimens were collected from alteration zones, ore mineralization and lithological units for laboratory analysis. Polish sections of ore mineralization and thin sections of alteration zones and lithological units were prepared. For a detailed mineralogical study of alteration zones, X-ray diffraction (XRD) analysis was implemented using an X'pert Pro XRD diffractometer (Philips, Amsterdam, The Netherlands) located at the Iran Mineral Processing Research Center (IMPRC, Tehran, Iran). Moreover, analytical spectral devices (ASD) spectroscopy was performed to the samples collected from the main lithological units exposed in the study area using a FieldSpec3® spectroradiometer (Malvern Panalytical Ltd., Malvern, UK, operating from 0.35 µm to 2.5 µm) located at the University of Kerman Institute of Science and High Technology (Kerman, Iran).

4. Results

4.1. Lithological and Alteration Mapping Using Landsat-8, Sentinel-2 and ASTER

For generating a regional view of lithological units in the study region, the Red-Green-Blue (RGB) false color composite of bands 2, 5 and 7 for Landsat-8 and bands 2, 8 and 12 for Sentinel-2 were considered, respectively. The resultant images show most of the lithological units having spectral features related to Fe^{3+} and Fe^{3+}/Fe^{2+} iron oxides and clay and carbonate minerals. Regarding the geological map of the study area (see Figure 1), the identification and lithological discrimination of the units in Landsat-8 and Sentinel-2 resultant images were almost similar. Therefore, the RGB false color composite image of the Sentinel-2 spectral bands was selected and presented herein (Figure 4).

The lithological units such as the rhyolite and acidic tuff interbedded with crystalline limestone (PE^r) unit, alternation of limestone, dolomite and shale (the Shemshak, Espahk and Surkh Formations) (TR), the alternation of light green sandstone, slate and phyllite ($C^{ph.s}$) and the Zarigan granite (Zrg) show a mixture of Fe^{3+} iron oxides and clay and carbonates (brown, blue and purple shades). The marble, garnet schist, biotite schist and quartzite ($PE^{l.s}$) unit, light brown thin bedded dolomite and limestone, red shale and siltstone and white to red quartzite (Kuhbanan Formation) (E^1), marl interbedded with gypsum and conglomerate (Ngc) unit and the massive to thick bedded limestone (K^1) unit mostly contain Fe^{3+}/Fe^{2+} iron oxides mixed with clay and carbonates (light green and cyan hues). The sphilite, basalt, tuff and grey to light brown limestone (PE^v) unit is characterized as dark brown color due to high content of Fe^{3+} iron oxides. The northeastern part of the scene (yellow polygon) shows a variety of colors related to absorption features of target alteration minerals including Fe^{3+} and Fe^{3+}/Fe^{2+} iron oxides and clay and carbonate minerals. The massive to medium bedded limestone and dolomite (C^{1d}) unit, gypsum interbedded with grey limestone (DCg), the $C^{ph.s}$ unit and the PE^r unit are the main lithological units in this zone. A high level of Fe^{3+} iron oxides (orange, pink to rose blush shades), mixture of Fe^{3+} and Fe^{3+}/Fe^{2+} iron oxides (yellow tone) and clay and carbonate minerals (light blue hue) in the zone is notable (Figure 4). Note that the occurrence of carbonate-hosted Pb-Zn mineralization is reported in this zone (yellow polygon) and detailed alteration mapping is presented by authors during this study.

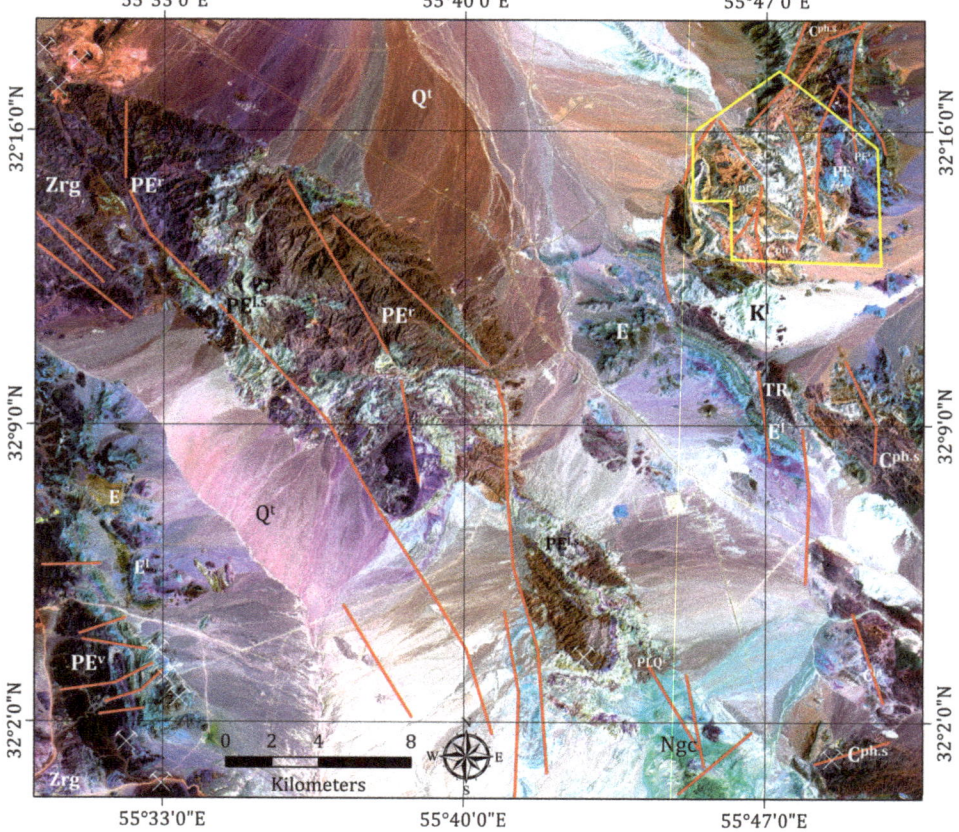

Figure 4. Regional view of lithological units and alteration zones in the study region. RGB false color composite of bands 2, 8 and 12 for Sentinel-2. Yellow polygon demarcates WV-3 imagery.

The RGB false color composite of bands 6, 2 and 8 of ASTER discriminate most of the lithological units containing Al/Fe-OH, Fe^{+2}/Fe^{3+} and Mg-Fe-OH/CO3 absorption properties (Figure 5). The PE^r unit, the Zarigan granite (Zrg), thin bedded limestone and dolomite (PE^{Ed}) unit and some parts of the Kuhbanan Formation (E^1) appear in purple due to Al/Fe-OH minerals that is slightly mixed with Mg-Fe-OH/CO$_3$ minerals. The DCg unit is depicted as a green tone because of Fe^{+2}/Fe^{3+} minerals. The $C^{ph.s}$ unit is characterized in cyan hue attributed to mixture of Fe^{+2}/Fe^{3+} and Mg-Fe-OH/CO$_3$ minerals. The Ngc unit and the K^1 unit are represented as bright yellow shade probably because of combination between Mg-Fe-OH/CO$_3$, Al/Fe-OH and Fe^{+2}/Fe^{3+} minerals. The PE^v unit is considered as dark shade due to high content of Fe^{3+} iron oxides. The TR unit having Al/Fe-OH and Mg-Fe-OH/CO$_3$ minerals is presented in a dark purple tone. The C^{1d} unit is manifested in brown-golden color due to Mg-Fe-OH/CO$_3$ minerals mixed with Fe^{+2}/Fe^{3+} minerals. The E unit, PE^r and some parts of the Kuhbanan Formation (E^1) appear as dark green hue because of Fe^{+2}/Fe^{3+} minerals (Figure 5).

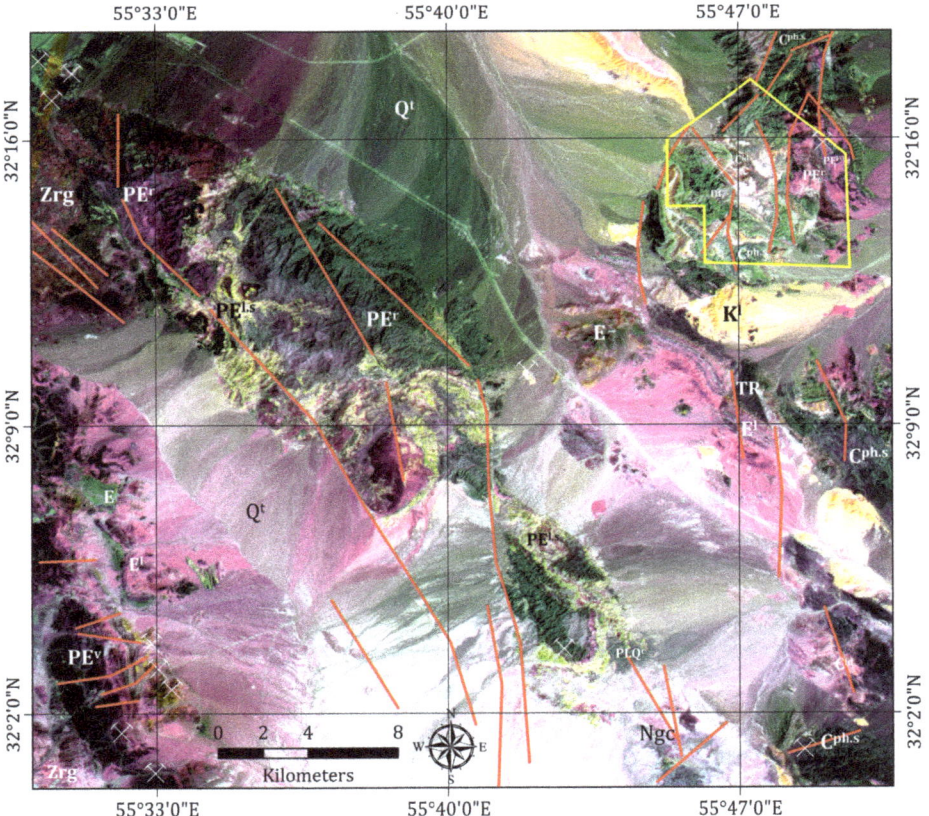

Figure 5. Regional view of lithological units and alteration zones in the study region. RGB false color composite of bands 6, 2 and 8 for ASTER. Yellow polygon demarcates WV-3 imagery.

Figure 6A–D shows the results of band ratios for mapping iron oxide/hydroxides and clay and carbonate minerals derived from Landsat-8 and Sentinel-2 spectral bands. The 4/2 band ratio of Landsat-8 and Sentinel-2 shows the spatial distribution of iron oxide/hydroxide minerals (red pixels), which are mostly mapped in the $PE^{1.s}$ unit, the Ngc unit, the K^1 unit, the PE^r unit, the DCg unit and Quaternary deposits (Q^t) (Figure 6A,B). The 6/7 band ratio of Landsat-8 and 11/12 band ratio of Sentinel-2 map the surface distribution of hydroxyl-bearing alteration minerals and carbonates (green pixels) (Figure 6C,D). Accordingly, the PE^{1,s} unit, tuff and thick bedded of red silty shale (E) unit, the K^1 unit, the PE^r unit, the pink to white granite (Zarigan granite) (Zrg), the C^{1d} unit and the Q^t deposits are mapped due to high content of clay and carbonate minerals in their composition. Clay and carbonate minerals show higher surface abundance in the Landsat-8 ratio-image (Figure 6C) compared to the Sentinel-2 ratio-image (Figure 6D). The K^1 unit is particularly mapped in the Landsat-8 ratio-image (Figure 6C), while it shows the smaller spatial distribution in the Sentinel-2 ratio-image (Figure 6D). It is because the band 12 (2.100–2.280 µm) of Sentinel-2 mostly covers the absorption features of hydroxyl-bearing minerals (2.10–2.20 µm), while the absorption features of carbonates typically concentrate around 2.350 to 2.450 µm [69,82]. Hence, carbonates could not be mapped using 11/12 band ratio of Sentinel-2, properly.

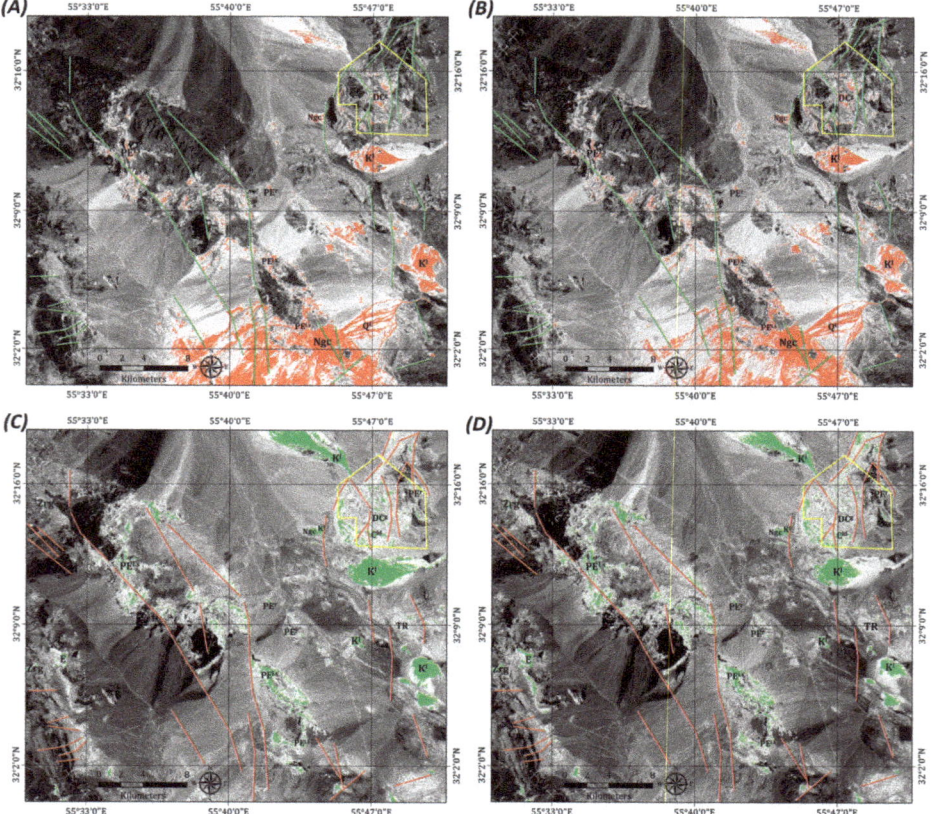

Figure 6. Band ratio image-maps showing spatial distribution of iron oxide/hydroxides and clay and carbonate minerals. (**A**) The 4/2 band ratio image-map of Landsat-8 showing iron oxide/hydroxides minerals as red pixels; (**B**) the 4/2 band ratio image-map of Sentinel-2 showing iron oxide/hydroxides minerals as red pixels; (**C**) the 6/7 band ratio image-map of Landsat-8 showing clay minerals and carbonates as green pixels; (**D**) the 11/12 band ratio image-map of Sentinel-2 showing hydroxyl-bearing minerals as green pixels. Yellow polygon demarcates WV-3 imagery.

The 4/2 band ratio of ASTER shows the iron oxide/hydroxide minerals in red pixels, which are typically mapped in the PE^{1,s} unit, the K^1 unit, the PE^r unit, the DCg unit, the PE^{Ed} unit, the Kuhbanan Formation (E^1) and the C^{1d} unit (Figure 7A). The Al/Fe-OH minerals are specifically mapped as mustard pixels in the RBD1 image of ASTER (Figure 7B). Comparison to the geology map of the study area, the high abundance of Al/Fe-OH minerals (mustard pixels) are associated with the lithological units of the $C^{ph.s}$, the PE^r, the Kuhbanan Formation (E^1), the PE^v unit, the PE^{Ed} unit, the TR and Quaternary deposits (Qt) (Figure 7B). The RBD2 image of ASTER shows the surface abundance of Mg-Fe-OH/CO3 minerals (green pixels; Figure 7C), which are typically associated with the K^1 unit and the C^{1d} unit. Although the Kuhbanan Formation (E^1), the $PE^{1.s}$ unit, the Ngc unit and the PE^r unit are also partially mapped in Figure 7C. Dolomitic units are specifically mapped in the RBD3 image of ASTER as yellow pixels (Figure 7D). The C^{1d} lithological unit is strongly represented in yellow pixels. Additionally, some parts of the Kuhbanan Formation (E^1), the TR unit, PE^{Ed} unit, the $PE^{1.s}$ unit and the PE^r unit are mapped in Figure 7D due to the high content of dolomite in their lithological composition.

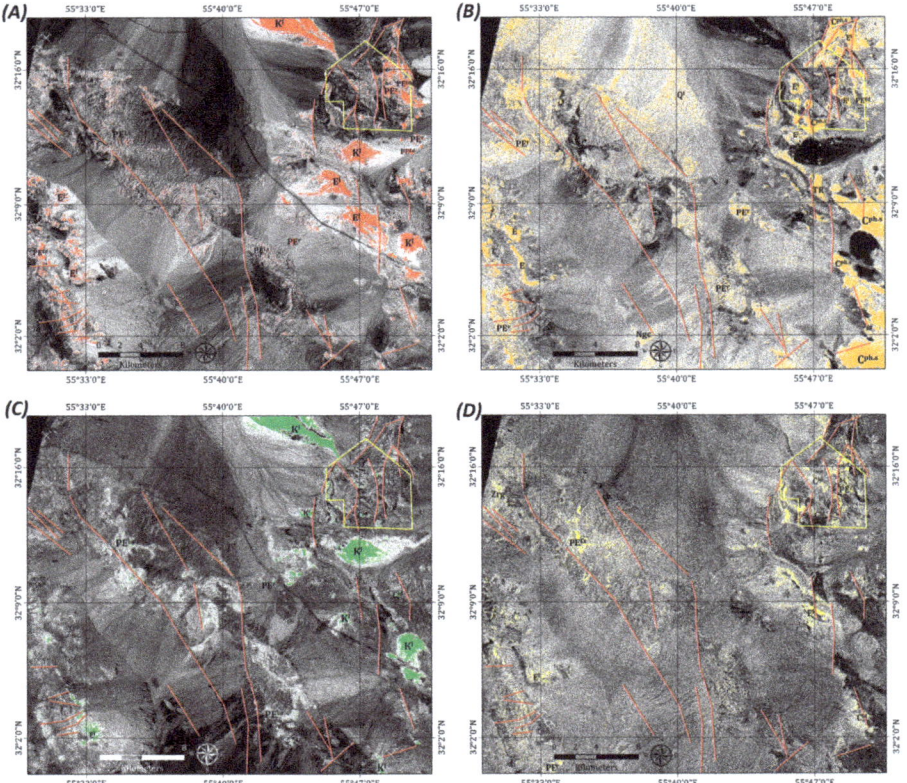

Figure 7. Band ratio image-maps showing spatial distribution of iron oxide/hydroxides Al/Fe-OH and Mg-Fe-OH/CO3 minerals and dolomite. (**A**) The 4/2 band ratio image-map of ASTER showing iron oxide/hydroxides minerals as red pixels; (**B**) the RBD1 image-map of ASTER showing Al/Fe-OH minerals as mustard pixels; (**C**) the RBD2 image-map of ASTER showing Mg-Fe-OH/CO3 minerals as green pixels; (**D**) the RBD3 image-map of ASTER showing dolomite as yellow pixels. Yellow polygon delineates WV-3 imagery.

Analyzing the eigenvector loadings derived from the PCA method for Landsat-8 and Sentinel-2 indicates that the PC4 and PC5 contain key information related to the hydroxyl-bearing and carbonate and iron oxide/hydroxide minerals. Table 2 shows the eigenvector matrix for Landsat-8 selected bands. The PC4 has strong loadings of bands 6 (0.667216) and 7 (−0.48894) with opposite signs. Thus, the PC4 image identify hydroxyl-bearing minerals and carbonates as bright pixels due to a positive sign in the reflection band (band 6). Figure 8A) shows the PC4 image of Landsat-8. The spatial distribution of hydroxyl-bearing minerals and carbonates (green pixels) is mostly identified with the units of the $PE^{1.s}$, the PE^r, the K^1, the $C^{ph.s}$, the DCg, the C^{1d} and Q^t deposits. The Zarigan granite (Zrg) and Kuhbanan Formation (E^1) also show some hydroxyl-bearing minerals in the PC4 image of Landsat-8 (Figure 8A). The PC5 of Landsat-8 has moderate positive contribution in band 2 (0.318038), negative strong loadings in band 4 (−0.579504) and strong positive contribution (0.516661) in band 5 (Table 2). Accordingly, iron oxide/hydroxide (Fe^{3+}/Fe^{2+}) minerals manifest as dark pixels in the PC5 image because of negative contribution of band 4 (reflection band). The dark pixels were inverted to bright pixels by negation (multiplication to −1). Figure 8B displays the spatial distribution of iron oxide/hydroxides (red pixels) in the study area. The units of the $PE^{1.s}$, the $C^{ph.s}$, the Ngc, the PE^r, the Zarigan granite (Zrg), the TR and the DCg show a high abundance of oxide/hydroxides in the PC5 image of Landsat-8 (Figure 8B). Comparison of the Landsat-8 PCA results to the band ratio indicates that the spatial distribution of iron oxide/hydroxides minerals and hydroxyl-bearing minerals and carbonates in the PC images is generally less widespread. Moreover, the alteration zones mapped by the PCA method show better spatial relationship with the documented ore mineral occurrences in the study area.

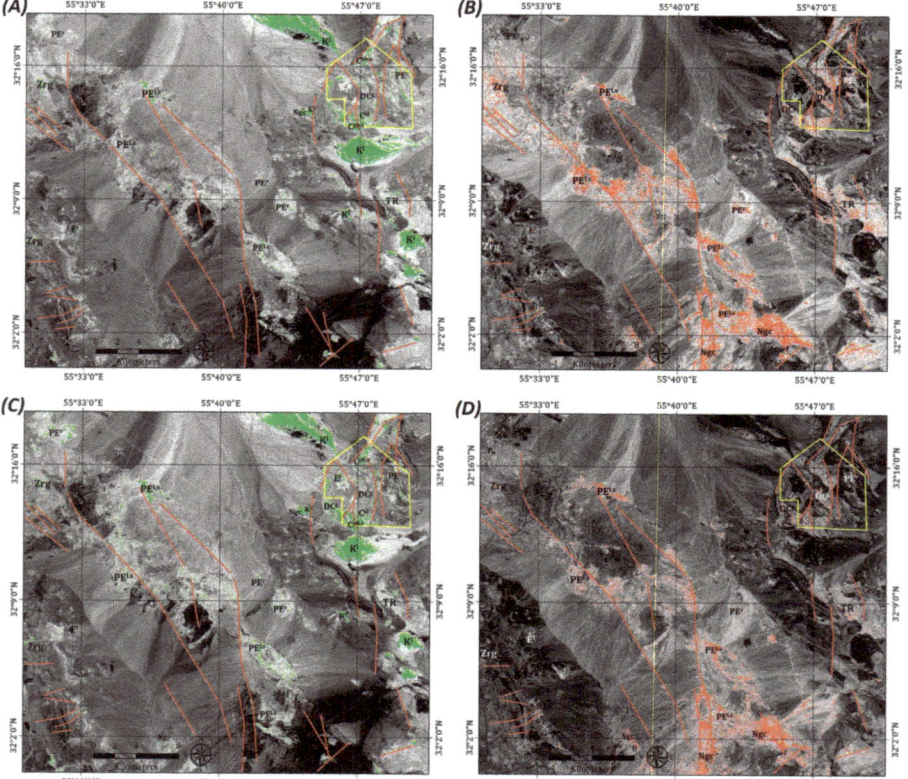

Figure 8. PC image-maps showing spatial distribution of iron oxide/hydroxides minerals and clay minerals and carbonates. (**A**) The PC4 image-map of Landsat-8 showing OH-bearing minerals and carbonates as green pixels; (**B**) the PC5 image-map of Landsat-8 showing iron oxide/hydroxides minerals as red pixels; (**C**) the PC4 image-map of Sentinel-2 showing OH-bearing minerals as green pixels; (**D**) the PC5 image-map of Sentinel-2 showing iron oxide/hydroxides minerals as red pixels. Yellow polygon demarcates WV-3 imagery.

Considering eigenvector loadings for selected bands of Sentinel-2 (see Table 3), the PCA4 includes a great contribution of band 11 (−0.641949) with a negative sign and band 12 (0.567634) with a positive sign. Hence, hydroxyl-bearing minerals represent in dark pixels of the PC4 image (Figure 8C). The dark pixels were converted to bright pixels by negation, subsequently. Results show that the spatial distribution of hydroxyl-bearing minerals is typically associated with the units of the $PE^{1.s}$, the $C^{ph.s}$, the Kuhbanan Formation (E^1), the PE^r, the K^1, the Zarigan granite (Zrg), the Ngc, the C^{1d} and Q^t deposits (Figure 8C). Iron oxide/hydroxide (Fe^{3+}/Fe^{2+}) minerals appear as bright pixels in the PCA5 image of Sentinel-2 due to a strong positive loading in band 4 (0.698634) and negative loadings of band 2 (−0.454829) and band 8 (−0.523627), respectively (see Table 3). Iron oxide/hydroxide minerals are represented in bright pixels as a result of a positive loading in the reflection band (band 4) (Figure 8D). The high surface abundance of iron oxide/hydroxide (red pixels) is highlighted with the $PE^{1.s}$ unit, the $C^{ph.s}$ unit, the Ngc unit, the PE^r unit, the DCg unit, the Zarigan granite (Zrg), the TR unit and Q^t deposits (Figure 8D). The spatial distribution of iron oxide/hydroxides minerals in the PCA5 image of Sentinel-2 is less prevalent compared to the PCA5 of Landsat-8 (see Figure 8B,D). Nevertheless, the PC4 image of Landsat-8 and Sentinel-2 are almost identical in many parts (see Figure 8A,C).

The PCA technique was also implemented on ASTER VNIR+SWIR bands for mapping the target alteration minerals. The eigenvector matrix for ASTER data is shown in Table 4. The PC3 contains strong positive loading in band 4 (0.564435) and strong negative loading in band 8 (−0.896111). The Mg-Fe-OH/CO3 has high reflectance about 1.6 µm that is coincident with band 4 of ASTER. The Fe-Mg-OH and CO3 minerals exhibit diagnostic absorption features near 2.350 µm, which is matched with bands 8 of ASTER [60,75]. So, the PC3 image maps Mg-Fe-OH/CO3 minerals as bright pixels because of positive loading in band 4 (reflection band). Figure 9A shows the PC3 image-map of the study area that overlain by green color. The green pixels show a high concentration of Mg-Fe-OH/CO_3 minerals. Referring to the local geology map of the study zone, the Mg-Fe-OH/CO_3 minerals are typically identified in the K^1 unit, the Ngc unit and the C^{1d}. The Kuhbanan Formation (E^1), the PE^r unit, the Zarigan granite (Zrg) and the $PE^{1.S}$ unit are also weakly mapped in the PC3 image (Figure 9A). These lithological units contain a high abundance of carbonates and Mg-Fe-OH minerals, which can be detected by band 8 (2.295–2.365 µm) of ASTER. The detected pixels in the PC3 image-map are comparable to the RBD2 image-map (see Figure 7C).

The PC4 has strong positive loadings in band 2 (0.812776), while it shows strong negative loadings in band 4 (−0.429509) (see Table 4). Iron oxide/hydroxides minerals characterize by high absorption features about 0.40 to 1.10 µm and high reflection around 1.60 µm [75]. Considering the spectral location of bands 2 (0.63–0.69 µm) and 4 (1.60–1.70 µm) of ASTER and the eigenvector loadings in PC4, it is discernable that iron oxide/hydroxides minerals depict as dark pixels in the PC4 image, which consequently negated to bright pixels and overlain by red color (Figure 9B). The Ngc unit, the Kuhbanan Formation (E^1) and the PE^{Ed} unit are strongly mapped in the image. However, some zones located in the K^1 unit, the $PE^{1.S}$ unit, the PE^r unit and Qt deposits are also distinguishable (Figure 9B). The PC5 shows a strong contribution in band 4 (-0.668838) and band 6 (0.425904) and band 9 (0.507113) with opposite signs (see Table 4). The OH/S-O/CO3-bearing minerals exhibit diagnostic absorption features at 2.20 to 2.50 µm, which are coincident with bands 6 to 9 of ASTER [74]. Therefore, OH/S-O/CO3-bearing minerals can be mapped as dark pixels in the PC5 image. Figure 9C shows the PC5 image that dark pixels are converted to bright and overlain by mustard color. Spatial distribution of OH/S-O/CO3-bearing minerals is clearly observable in the $C^{ph.s}$ unit, the DCg unit, the PE^{Ed} unit the TR unit the PE^r unit and Qt deposits. Some small parts of the Kuhbanan Formation (E^1) and the $PE^{1.S}$ unit are also appeared as mustard pixels (Figure 9C). The results of PC5 image is almost matched to the RBD1 image-map (see Figure 7).

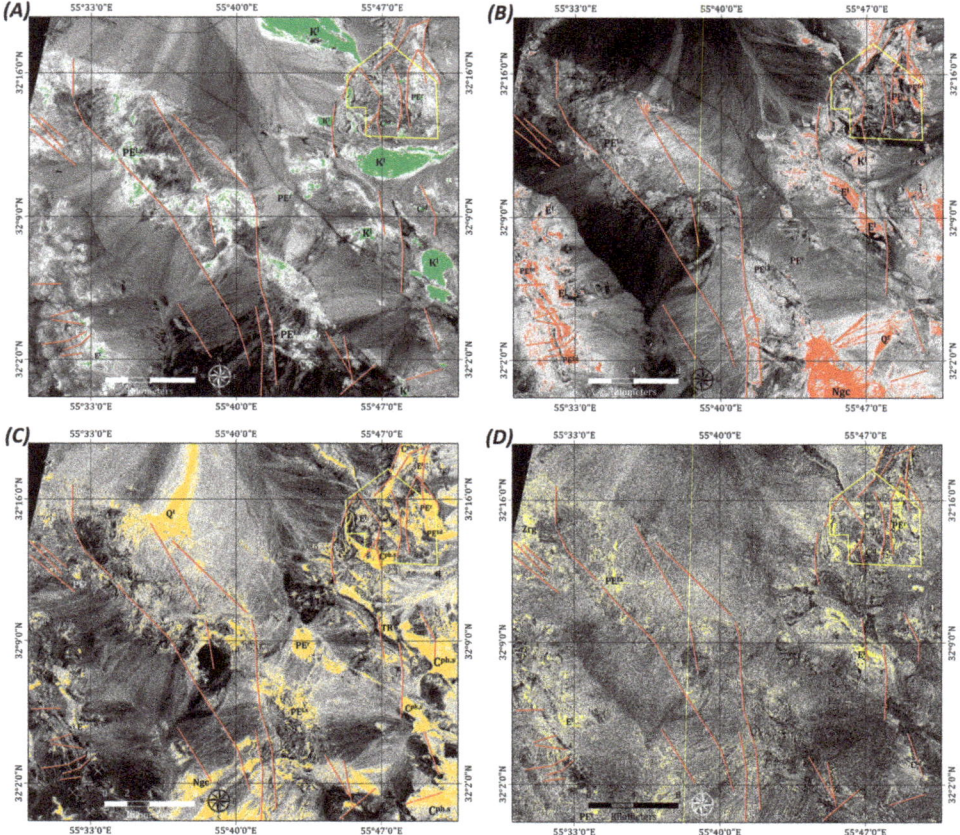

Figure 9. The PC image-maps of ASTER. (**A**) The PCA3 image-map showing Mg-Fe-OH/CO3 minerals as green pixels; (**B**) the PC4 image-map showing iron oxide/hydroxides minerals as red pixels; (**C**) the PC5 image-map showing OH/S-O/CO3-bearing minerals as mustard pixels; (**D**) the PC8 image-map showing dolomite as yellow pixels. Yellow polygon delineates WV-3 imagery.

Checking the PC6 and PC7 images indicate that they are mostly noisy and uninformative. Considering the laboratory reflectance spectra of dolomite resampled to response functions of VINR+SWIR bands of ASTER (see Figure 4) reveals that the PC8 image feasibly contains vital information related to the spatial distribution of dolomite. Bands 4 and 5 can be considered reflection bands, whereas band 7 can be deliberated absorption bands for detecting dolomite. The PCA8 has strong positive loadings in band 4 (0.404433) and band 5 (0.558407) and great contribution of band 7 (−0.623420) with a negative sign (see Table 4). The PC8 image-map of the study zone is shown in Figure 9D. Regarding the geology map of the study zone, the C^{1d} unit clearly appears as bright pixels in the PC8 image. Moreover, the Kuhbanan Formation (E^1), the TR, the PE^r unit, the $PE^{1.S}$ unit and the PE^{Ed} unit are partially characterized in yellow pixels. The resultant image-map of the PC8 is similar to the RBD3.

4.2. Detailed Detection of Iron Oxide/Hydroxides and Dolomit Using WV-3

Figure 10 shows the WV-3 scene covering the selected subset of the study area contains Zn–Pb mineralization. The band ratio image-map of 5/3 for mapping Fe^{3+} oxides (A), the band ratio image-map of 5/2 for identifying iron-stained alteration (B), the band ratio image-map of 7/4 for detecting dolomite/Fe^{2+} (C), and RGB false color composite image-map of 7/4, 5/3 and 5/2 for discriminating lithological units (D) are shown in Figure 10. The surface distribution of Fe^{3+} oxides (bright pixels) is generally associated with the DCg unit, the PE^r unit and PE^{Ed} unit (Figure 10A).

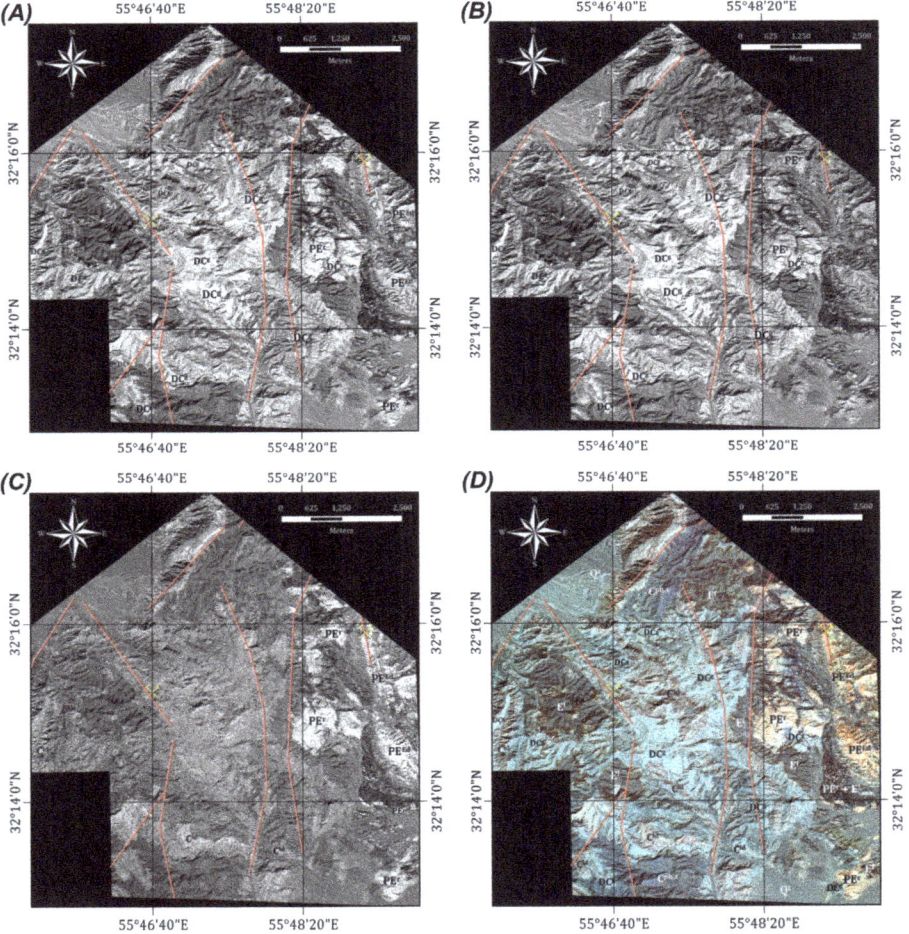

Figure 10. WV-3 scene covering the selected subset of the study area containing Zn–Pb mineralization. (**A**) The band ratio image-map of 5/3 for mapping Fe^{3+} oxides; (**B**) the band ratio image-map of 5/2 for identifying all iron oxides (iron-stained alteration); (**C**) the band ratio image-map of 7/4 for detecting dolomite/Fe^{2+}; (**D**) RGB false color composite image-map of 7/4, 5/3 and 5/2 for discriminating lithological units.

The band ratio image-map of 5/2 generates analogous results for mapping all iron oxides (iron-stained alteration) associated with the above-mentioned lithological units (Figure 10B). This similarity of iron oxide mapping results might be related to extensive iron-stained occurrences in the lithological units. The spatial distribution of dolomite/Fe^{2+} (bright pixels) is typically mapped in the eastern part of the selected subset (Figure 10C). The PE^r unit, PE^{Ed} unit and the C^{1d} unit are highlighted in the image-map of the 7/4 band ratio (Figure 10C). Figure 10D shows RGB false color composite of 7/4, 5/3 and 5/2 band ratio images for the selected subset. The discrimination of lithological units is characteristically discernable based on different composition of iron oxide/hydroxide minerals.

The lithological units with a high abundance of iron oxide/hydroxide minerals such as the DCg unit appear in cyan. The exposures of the PE^r, PE^{Ed} and C^{1d} units contain the high abundance of dolomite/Fe^{2+} represent as whitish-yellow, golden yellow and light brown. On the other hand, the Kuhbanan Formation (E^1), Q^t deposits and the $C^{ph.s}$ unit having a low abundance of iron oxides show recognizable colors (shades of gray) and boundaries with other lithologies (see Figure 10D).

Analyzing the PCA statistical results for the WV-3 selected subset shows the PC3, PC4 and PC6 contain essential information for mapping iron-stained alteration, dolomite/Fe^{2+} and Fe^{3+} oxides, respectively. The PC3 shows high negative loading in band 2 (−0.598506) and strong positive loading in band 5 (0.661001) (see Table 5). Thus, iron-stained alteration can be mapped as bright pixels because of band 5 that is assumed as a reflection band. Figure 11A shows the DCg unit and some small parts of the PE^r unit as bright pixels. The PC4 has moderate positive loading in band 2 (0.345377), strong positive loading in band 4 (0.631659) and high negative loading in band 7 (−0.509311) (see Table 5). Dolomite/Fe^{2+} can be mapped as dark pixels due to the negative sign in the reflection band (band 7). Figure 11B shows the negated image-map of the PC4. The PE^r unit, the PE^{Ed} unit and the C^{1d} unit are represented as bright pixels. The image-map of the PC4 is identical to the 7/4 band ratio resultant (see Figure 10C). The PC6 contains a strong contribution in band 2 (−0.427799) and band 3 (−0.515988) with a negative sign and strong loading in band 5 (0.646312) with a positive sign (see Table 5). Accordingly, Fe^{3+} oxides manifest in bright pixels because of positive sign in band 5, which is assumed as a reflection band (Figure 11C). The DCg unit, the PE^r unit and the PE^{Ed} unit are mapped in the PC6 image-map. Figure 11D shows RGB false color composite image-map of the PC4, PC6 and PC3, respectively. Lithological units are differentiated stronger than RGB false color composite of band ratios (see Figure 10D). The lithological boundaries of the PE^r unit, the PE^{Ed} unit and the C^{1d} unit (represented as golden to orange-yellow) with other lithological units such as the DCg, the $C^{ph.s}$ and the E^1 units (depicted in shades of gray and purple) are distinguishable (Figure 11D).

4.3. Generating Mineral Prospectivity Maps for the Study Area

The fuzzy-logic model was utilized to produce mineral prospectivity maps of the favorable areas for ore mineralizations in the study region using most rational alteration thematic layers derived from image processing techniques (see Table 6). In this analysis, the PCA output was considered more informative compared to band ratios output. It is due to the fact that the PCA is statistically based algorithm and uses uncorrelated linear combinations (eigenvector loadings) to map spectral characteristics of alteration minerals. Therefore, the PC4 and PC5 thematic layers of Landsat-8 and Sentinel-2 and PC4, PC5 and PC8 thematic layers of ASTER were selected to be integrated by application of the fuzzy gamma operator ($\gamma = 0.7$) for generating mineral prospectivity map at a regional scale. Figure 12 shows the regional mineral prospectivity map of the study area. Evaluating the fuzzy membership indicates that the high favorability index is associated with some of the lithological units in the study area. The PE1s unit, the PE^r unit, the Kuhbanan Formation (E^1), the TR unit, the $C^{ph.s}$ unit, the Ngc unit, the Zarigan granite (Zrg) and the DCg unit show high value (0.7 to 1.0) of the favorability index. Most of the mines and ore indications are located in the high value zones of the favorability index and associated with fault systems in the study area (Figure 12). Results demonstrate that the PE^{1s} unit, the PE^r unit and the Zarigan granite (Zrg) show the highest value (0.9 to 1.0) of the favorability index. The alteration zones associated with these lithological units are the most favorable/prospective areas for ore mineralizations at the regional scale. Some of the high prospective zone demarcated using dashed black ellipsoids and circles in Figure 12, which can be considered for future mineral exploration in the study region.

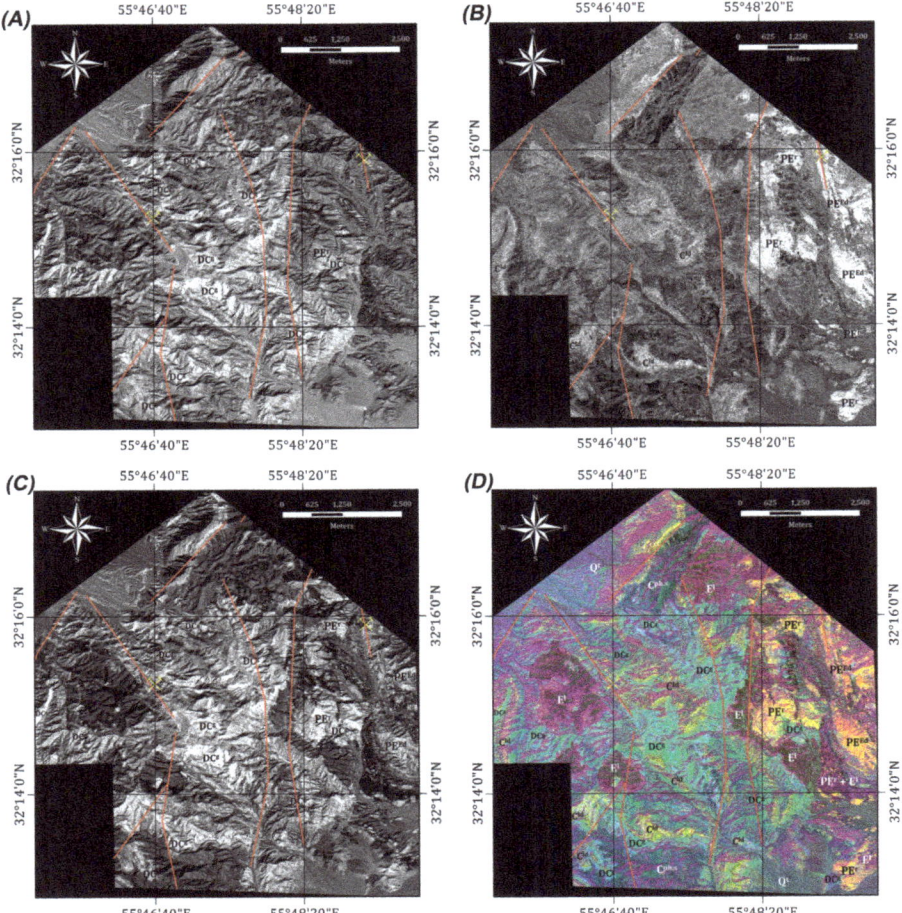

Figure 11. WV-3 scene covering the selected subset of the study area containing Zn–Pb mineralization. (**A**) The PC3 image-map showing iron-stained alteration as bright pixels; (**B**) the PC4 image-map showing dolomite/Fe^{2+} as bright pixels; (**C**) the PC6 image-map showing Fe^{3+} oxides as bright pixels; (**D**) RGB false color composite image-map of the PC4, PC6 and PC3 discriminates lithological units.

Figure 13 shows the local mineral prospectivity map of the study area derived from the PC3, PC4 and PC6 thematic layers (most rational alteration thematic layers) for WV-3 data. The fuzzy fuzzy gamma operator ($\gamma = 0.7$) was used to fuse the selected alteration thematic layers (see Table 3). The highest value of (0.8 to 1.0) the favorability index is obtained for the PE^r unit, the PE^{ED} unit and the C^{1d} unit. In addition, the DCg unit shows a high value (0.6 to 0.9) of the favorability index in some parts of the study area. The Pb-Zn mineralization zones have moderate to high favorability index value (0.6 to 1.0) and are also adjoining to fault systems (Figure 13). Accordingly, the most favorable/prospective zones for Pb-Zn mineralization in the study area are alterations (especially dolomitic zone) associated with the PE^r unit, the PE^{ED} unit and the C^{1d} unit, mainly in fault contact zones with impermeable lithological units. Black polygons, ellipsoids and circles show some of the high prospective zones for future mineral prospecting in the study region (Figure 13). These high prospective zones were selected to check during filed reconnaissance in this study. Locations of in situ observation are shown in Figure 13.

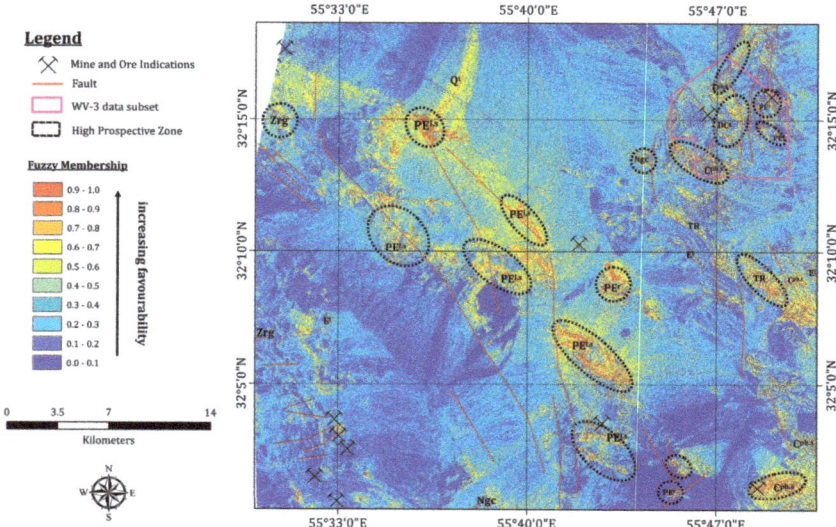

Figure 12. Mineral prospectivity map of the study area at regional scale derived from Landsat-8, Sentinal-2 and ASTER selected alteration thematic layers. Magenta polygon delineates WV-3 imagery.

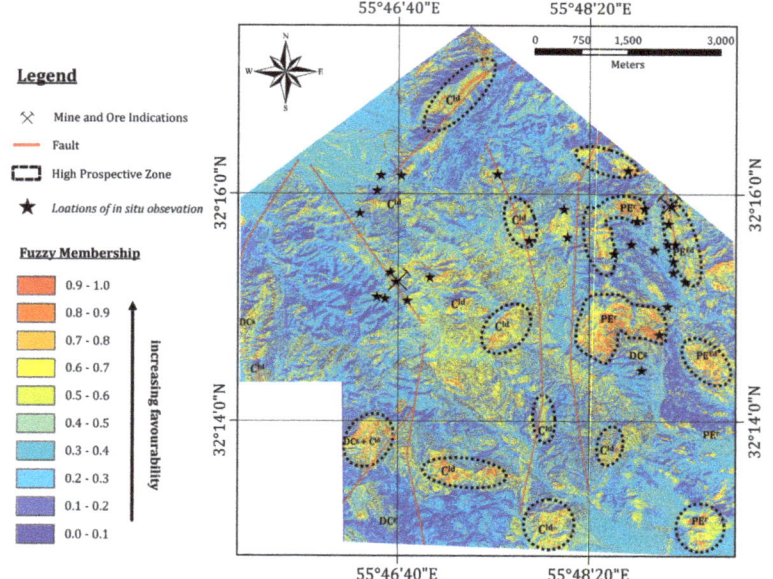

Figure 13. Mineral prospectivity map of the study area at local scale derived from WV-3 selected alteration thematic layers.

4.4. Verifying the Spatial Distribution of Alteration Zones and High Prospective Areas

Field reconnaissance was conducted to verify the occurrence of alteration zones and high prospective locations of Pb-Zn mineralization in the study area. The global positioning system (GPS) survey was undertaken for estimating the overall accuracy of the image processing techniques. Several field photos were taken to record the location, contact and characteristics of alteration zones and lithological units. Rock sample collection was carried out in several parts of the study regions and high prospective zones for laboratory analysis such as polish and thin sections and XRD analysis and the ASD spectroscopy. Surface manifestations of mineral assemblages such as iron oxide/hydroxides, dolomite, shale, calcite and gypsum are widespread in the study area. Some of the highly prospective zones, mineralogically interesting alteration zones and lithological units in the study area were particularly visited. The argillic alteration, sericitic zones, iron oxides and dolomitization were found in the PE^r unit, the PE^{ED} unit, the E^1 unit and the C^{1d} unit, which showed some surface expression of hematite, malachite, pyrite, galena and sphalerite. Some old open-pit quarries were also found in the alteration zones of the lithological units. In many parts of the study area, surface expression of Pb-Zn mineralization was typically observed in the fault contact of dolomite with other lithological units (Figure 14A–F).

Polish section study shows the presence of chalcopyrite, pyrite, malachite, smithsonite, sphalerite, galena, hematite and limonite. Thin section study typically displays the association of iron oxide/hydroxides with dolomite that mostly concentrated in the fractures. The XRD analysis of the samples collected from alteration zones inside old open-pit quarries and surrounding areas shows the presence of quartz, dolomite, calcite, muscovite, chlorite, gypsum, albite, illite, jarosite and malachite. The ASD spectroscopy analysis for main lithological units such as shale, gypsum, dolomite and calcite were measured in this study (Figure 15). The laboratory reflectance spectra from the shale sample display three distinguishable absorption features about 1.40 µm attributed to OH/H_2O stretches, 1.90 µm related to H_2O stretches and 2.20 µm due to combination of the OH-stretching fundamental with Al-OH bending mode, respectively (Figure 15). These absorption features occur in shale because of the high content of clay minerals (Al-rich phyllosilicates) [103]. The reflectance spectra derived from gypsum exhibits three absorption features, which are identical to the shale sample (Figure 15). But, the absorption feature at 1.90 µm (due to H_2O stretches) is stronger than the shale sample and the absorption feature near 2.20 µm is related to S-O bending mode [104]. The reflectance spectra of dolomite contain two main absorption features related to Fe^{2+} at 0.9 to 1.2 µm and CO_3 in 2.35 µm (Figure 15). Calcite absorption properties typically concentrated about 2.35 µm (Figure 15), which is attributed to the vibrational processes of the CO_3 bending mode [82]. Furthermore, the confusion matrix and Kappa Coefficient were calculated for alteration mapping results versus field GPS surveys (Table 7). The overall accuracy for Landsat-8 and Sentinel-2 datasets is 86.66% and 83.33%, respectively. The Kappa Coefficient of 0.83 for Landsat-8 and 0.81 for Sentinel-2 is also assessed. For the ASTER dataset, the overall accuracy and Kappa Coefficient are 70% and 0.68, respectively. The calculation of the confusion matrix for WV-3 shows the overall accuracy of 83.33% and Kappa Coefficient of 0.82 (Table 7).

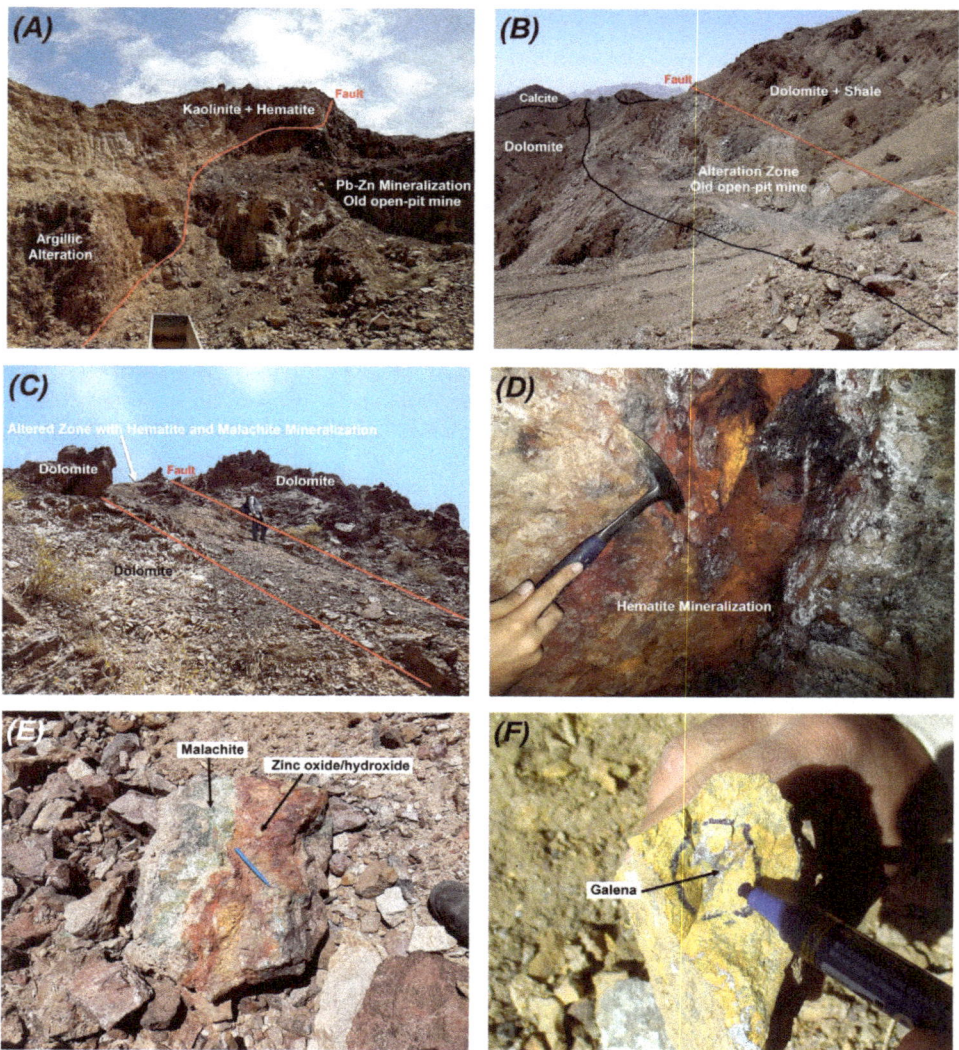

Figure 14. Field photographs of the alteration zones associated with Pb-Zn mineralization in the study area. (**A**) View old open-pit quarry located in the PE^r and E1 lithological units; (**B**) view of old open-pit quarry located in the PE^r, PE^{ED} and E1 lithological units; (**C**) view of ore mineralization associated with fault contact in the PE^r unit; (**D**) view of the hematite mineralization filling fault contact in the C^{1d} unit; (**E**) view of surface occurrence of malachite and zinc oxide/hydroxides in the PE^r unit; (**F**) view of galena mineralization in the dolomitic background in the C^{1d} unit.

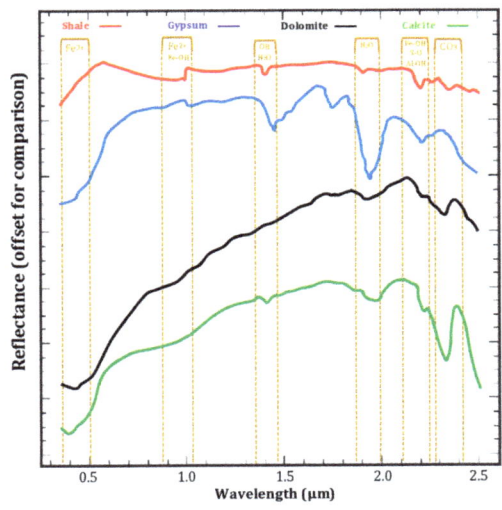

Figure 15. ASD spectroscopy results (0.35 to 2.5 µm spectral range) for rock samples collected from shale, gypsum, dolomite and calcite lithological units in the study area.

Table 7. Confusion matrix for alteration mapping derived from remote sensing datasets versus field GPS survey. (A) Landsat-8 data; (B) Sentinel-2 data; (C) ASTER data; (D) WV-3 data.

(A) Alteration Map Landsat-8	GPS Survey			
	Iron Oxide/Hydroxides	OH-Bearing and Carbonate Minerals	Totals	User's Accuracy
Iron oxide/hydroxides	4	2	6	67%
OH-bearing and carbonate minerals	2	22	24	91%
Totals	6	24	30	
Producer's accuracy	67%	91%		
Overall accuracy = 86.66%			Kappa Coefficient = 0.83	

(B) Alteration Map Sentinel-2	GPS Survey			
	Iron Oxide/Hydroxides	OH-Bearing Minerals	Totals	User's Accuracy
Iron oxide/hydroxides	5	2	7	71%
OH-bearing minerals	1	10	11	90%
Totals	6	12	18	
Producer's accuracy	83%	83%		
Overall accuracy = 83.33%			Kappa Coefficient = 0.81	

(C) Alteration Map ASTER	GPS Survey						
	Iron Oxide/ Hydroxides	Argillic Alteration+ Shale	Gypsum	Calcite	Dolomite	Totals	User's Accuracy
Iron oxide/hydroxides	4	0	0	0	1	5	80%
Argillic alteration + Shale	0	4	2	1	0	7	57%
Gypsum	0	2	4	0	0	6	67%
Calcite	0	0	0	5	1	6	83%
Dolomite	2	0	0	0	4	6	67%
Totals	6	6	6	6	6	30	
Producer's accuracy	67%	67%	67%	83%	67%		
Overall accuracy = 70%					Kappa Coefficient = 0.68		

Table 7. *Cont.*

(D) Alteration Map WV-3	GPS Survey			
	Iron Oxide/Hydroxides	Dolomite	Totals	User's Accuracy
Iron oxide/hydroxides	5	1	6	83%
Dolomite	1	5	6	83%
Totals	6	6	12	
Producer's accuracy	83%	83%		
Overall accuracy = 83.33%		Kappa Coefficient = 0.82		

5. Discussion

Remote sensing satellite imagery has been successfully utilized to detect the major hydrothermal alteration minerals associated with the carbonate-hosted Pb-Zn deposits in metallogenic provinces around the world [5,16–19]. In this investigation, multiple sources of spectral data derived from Landsat-8, Sentinel-2, ASTER and WV-3 sensors were utilized for detailed mapping a variety of hydrothermal alteration minerals in the central part of the KKTZ region, the CIT, Iran. Band ratios and PCA image processing techniques were used to produce thematic maps of hydrothermal alteration minerals for indicating the high prospective zones. Lithological units and alteration mineral zones were mapped based on spectral absorption characteristics of Fe^{3+}/Fe^{2+} and Al-OH/CO3 minerals by implementing band ratios and PCA techniques to spectral bands of Landsat-8 and Sentinel-2. The spatial distribution of iron oxide/hydroxide minerals was mapped in the lithological units of the PE^{1s}, the Ngc, the K^1, the PE^r, the DCg and Q^t deposits using the 4/2 band ratio and PC5 of Landsat-8 and Sentinel-2. Analyzing confusion matrix for mapping iron oxide/hydroxide minerals using these sensors (see Table 7) indicates that the highest the user's accuracy (71%) and producer's accuracy (83%) were obtained for Sentinel-2 dataset. It shows that the higher spatial resolution of Sentinel-2 (10 m) clearly impacts on mapping minerals. The 6/7 band ratio of Landsat-8 and 11/12 band ratio of Sentinel-2 and the PC4 images were used to identify the surface distribution of hydroxyl-bearing alteration minerals and carbonates in the study region. The alteration zones associated with lithological units of the PE^{1s}, the E, the K^1, the PE^r, the Zrg, the C^{1d} and the Q^t were highlighted. The Landsat-8 show user's accuracy (91%) and producer's accuracy (91%) for hydroxyl-bearing alteration minerals and carbonates, which are higher than user's accuracy (90%) and producer's accuracy (83%) of Sentinel-2 (see Table 7). It probably can be attributed to spectral coverage of band 12 (2.100–2.280 μm) of Sentinel-2, which is not able to map carbonates, properly. The overall accuracy of Landsat-8 for mapping target alteration minerals is 86.66% and higher than Sentinel-2 (83.33%) because of better spectral band placement in Landsat-8 for detection of hydroxyl-bearing alteration minerals and carbonates. Fuzzy logic modeling was used to fuse the most informative thematic alteration layers (the PC4 and PC5 images). Subsequently, mineral prospectivity map for the study area was generated. Several prospective zones were identified (see Figure 12), which are mostly associated with alteration zones in the PE^{1s} unit, the PE^r unit, the Zarigan granite (Zrg), the Ngc, the $C^{ph.s}$ and PE^{Ed} lithological units. Most of the prospective zones are located adjacent to the NW-SE and N-S fault systems, which likely acted as fluid pathways for hydrothermal ore mineralizations. Documented mineral occurrences show also close spatial relationship to the fault systems in the study area.

The ASTER dataset was used for detailed mapping iron oxide/hydroxides, Al/Fe-OH minerals, Mg-Fe-OH/CO3 minerals and dolomite associated with carbonate-hosted Pb-Zn deposits. The spatial distribution of these alteration minerals was comprehensively mapped. Results derived from the PCA for the study area show that calcite, gypsum, hematite and goethite are the main spectrally-spatially minerals. However, dolomite, jarosite, kaolinite and muscovite are minor spectrally-spatially minerals. The analysis of confusion matrix shows that calcite has the highest user's accuracy (83%) and producer's accuracy (83%). However, the lowest user's accuracy (57%) was obtained for argillic alteration + Shale (kaolinite and muscovite) (see Table 7). It means that the spectral mixing and confusion of calcite with other alteration minerals is low, while for argillic alteration + Shale is high. Iron oxide/hydroxides shows the user's accuracy of 80% and producer's accuracy of 67%. Gypsum and dolomitization have similar user's accuracy of 67% and producer's accuracy of 67%. Thus, spectral mixing with other alteration minerals especially dolomite could be feasible for iron oxide/hydroxides. The spectral mixing and confusion between gypsum and argillic alteration + shale can be expected.

Moreover, dolomite has some spectral mixing with calcite. ASTER dataset shows the overall accuracy of 70% and Kappa Coefficient of 0.68. Accordingly, ASTER can map and discriminate different alteration minerals appropriately, but some spectral mixing and confusion are also associated with alteration mapping. ASTER PC image-maps show iron oxide/hydroxides, OH/S-O/CO3-bearing minerals and dolomite associated with the C^{1d} unit, the PE^r unit, the PE^{ED} unit, the Kuhbanan Formation (E^1) and the $C^{ph.s}$ are high favorable/prospective zones for carbonate-hosted Pb-Zn mineralization (see Figure 12). The N-S faults are dominant structural features associated with the high potential zones; however, the NW-SE and NE-SW faults are also associated with few of the prospective zones.

More detailed surface distribution of Fe^{3+} oxides, iron-stained alteration and dolomite in the highly prospective zones were identified using WV-3 dataset. Dolomite is mostly identified associated with the PE^r unit, PE^{Ed} unit and the C^{1d} unit. Calculation of user's accuracy (83%) and producer's accuracy (83%) for iron oxide/hydroxides and dolomite shows a low rate of spectral mixing and confusion. The overall accuracy of 83.33% and Kappa Coefficient of 0.82 show a very good rate of accuracy and agreement for WV-3 dataset (see Table 7). WV-3 mineral prospectivity map (see Figure 13) shows the zones of most favorable/prospective zones for the Pb-Zn mineralization in the study area. The dolomitic zones in the PE^r unit, the PE^{ED} unit and the C^{1d} unit are typically indicated as the highest potential zone. In the vicinity of N-S, NW-SE and NE-SW fault systems, several of the most prospective zones are identified in the study area (see Figure 13).

Fieldwork was conducted comprehensively in the highly prospective zones for observing surface expression of Pb-Zn mineralization and related alteration zones and lithological units. Iron oxide/hydroxides and dolomite as well as weak argillic/sericitic zones was found with surface expression of hematite, malachite, pyrite, galena and sphalerite in the PE^r unit, the PE^{ED} unit, the E^1 unit and the C^{1d} unit. Surface manifestation of Pb-Zn mineralization was typically recorded in the fault contact of dolomite with impermeable lithological units. The presence of quartz, dolomite, calcite, muscovite, chlorite, gypsum, albite, illite, jarosite and malachite is verified by XRD analysis. These evidences emphasized that the fault systems provided fluid conduits for Pb-Zn mineralization that hydrothermally altered the host lithologies and afterwards oxidized during supergene processes. The ASD spectroscopy analysis of main lithological units showed distinct absorption features related to Fe^{2+}, OH/H_2O, H_2O, Al-OH, S-O and CO_3 for collected rock samples such as dolomite, shale, gypsum and calcite. Generally, confusion matrix and Kappa Coefficient calculated for alteration mapping results versus field GPS survey show reasonable overall accuracy (70% to 86.66%) and good rate of agreement (0.68 to 0.83). Landsat-8 and Sentinel-2 generally mapped iron oxide/hydroxides and hydroxyl-bearing and carbonate minerals and indicated favorable/prospective zones, while ASTER and WV-3 comprehensively detected and discriminated hematite, goethite, jarosite, gypsum, calcite, dolomite, kaolinite and muscovite and the highest prospective zones.

6. Conclusions

This investigation demonstrates the application of multi-source spectral data in the range of VNIR and SWIR wavelengths provided by Landsat-8, Sentinel-2, ASTER and WV-3 for indicating the highly prospective zones of carbonate-hosted Pb-Zn deposits in the KKTZ of the CIT region of Iran. Results show that significant information related to the iron oxide/hydroxide and hydroxyl-bearing and carbonate minerals can be easily obtained by implementing some specialized band ratio (e.g., 4/2, 6/7 and 11/12) and PCA technique to Landsat-8 and Sentinel-2 datasets. Fusing of the most informative alteration thematic layers by the fuzzy-logic model is a reliable approach for generating remote sensing-based mineral prospectivity map. Landsat-8/Sentinel-2/ASTER mineral prospectivity map for the KKTZ indicated the potential zones are mostly located in the altered zones of the PE^{1s} unit, the PE^r unit and the Zarigan granite (Zrg), which normally are near to the NW-SE and N-S fault systems. Detailed surface distribution of Al/Fe-OH minerals, Mg-Fe-OH/CO3 minerals and dolomite was detected using RBD and PCA methods to ASTER dataset. The alteration zones (especially gossan and dolomite) associated with C^{1d}, the PE^r, the PE^{ED}, the Kuhbanan Formation (E^1) and the $C^{ph.s}$ lithological units and close to N-S, NW-SE and NE-SW faults are considered highly favorable/prospective zones for carbonate-hosted Pb-Zn mineralization in the study area. Detailed identification of Fe^{3+} oxides, iron-stained alteration and dolomite in the highly prospective zones was obtained using WV-3 VNIR spectral bands processing (e.g., band ratios of 5/3, 5/2 and 7/4; PCA techniques). The most favorable/prospective zones for Pb-Zn mineralization in the study area are dolomite and gossan alteration zones located in the PE^r, the PE^{ED} and the C^{1d} lithological unit, which are exclusively placed in the fault contact zones of dolomitic occurrences with impermeable lithological units. The N-S, NW-SE and NE-SW trends fault systems provided fluid conduits for Pb-Zn mineralization and deposition and subsequent alteration zones in the study area. Therefore, it is recommended that detailed structural analysis of lineaments using Synthetic Aperture Radar (SAR) remote sensing data such as the Phased Array type L-band Synthetic Aperture Radar (PALSAR) is also required for future mineral prospection in the CIT. Additionally, SWIR bands of WV-3 can provide high spectral and spatial data for detailed alteration mapping. On the other hand, thermal infrared (TIR) data from Landsat-8 and ASTER are also valuable to map minerals like quartz, andradite, gypsum, calcite, dolomite, diopside and albite, comprehensively. In conclusion, the application of multi-sensor remote sensing satellite imagery and fusing the most informative alteration thematic layers using the fuzzy-logic model can provide a low-cost exploration approach for prospecting carbonate-hosted Pb-Zn mineralization in the KKTZ of CIT and other metllogenic provinces around the world. The results demonstrated in this investigation represent a significant contribution of space-borne multispectral systems to generate mineral prospectivity maps at various scales (regional, district and local scale). This approach could be very interesting for stakeholders and mining/exploration companies to use different types of space-borne multispectral data for distinct phases of mineral exploration.

Author Contributions: M.S. and I.M. writing—original draft preparation, software, analysis, validation; A.B.P. writing, reconstructing—review, editing and supervision; A.M.M. supervision; O.R. analysis; M.H. supervision; B.Z. supervision; B.P. supervision; A.M. and S.M.A. data processing. All authors have read and agreed to the published version of the manuscript.

Funding: Publication fees are not claim by Remote Sensing because A.B.P., M.H., B.Z., B.P. are guest editors for Special Issue of Multispectral and Hyperspectral Remote Sensing Data for Mineral Exploration and Environmental Monitoring of Mined Areas. They have right to publish feature papers free of charge due to serve for the Journal.

Acknowledgments: The Iran Minerals Production and Supply Company (IMPASCO) is acknowledged for providing and collecting the field and laboratory data for this investigation. Special thanks to Shahram Adibfard the exploration manager of IMPASCO. The Higher Institution Centre of Excellence (HICoE) Research Grant (Vote Number: 53209) awarded to the Institute of Oceanography and Environment (INOS), Universiti Malaysia Terengganu (UMT) is also acknowledged for providing required facilities for revising this manuscript.

Conflicts of Interest: The authors declare no conflict of interest.

References

1. Ahmadirouhani, R.; Karimpour, M.H.; Rahimi, B.; Malekzadeh-Shafaroudi, A.; Pour, A.B.; Pradhan, B. Integration of SPOT-5 and ASTER satellite data for structural tracing and hydrothermal alteration mineral mapping: Implications for Cu–Au prospecting. *Int. J. Image Data Fusion* **2018**, *9*, 237–262. [CrossRef]
2. Rajendran, S.; Sobhi, N. ASTER capability in mapping of mineral resources of arid region: A review on mapping of mineral resources of the Sultanate of Oman. *Ore Geol. Rev.* **2018**, *88*, 317–335. [CrossRef]
3. Pour, A.B.; Park, Y.; Park, T.S.; Hong, J.K.; Hashim, M.; Woo, J.; Ayoobi, I. Regional geology mapping using satellite-based remote sensing approach in Northern Victoria Land, Antarctica. *Polar Sci.* **2018**, *16*, 23–46. [CrossRef]

4. Pour, A.B.; Hashim, M.; Park, Y.; Hong, J.K. Mapping alteration mineral zones and lithological units in Antarctic regions using spectral bands of ASTER remote sensing data. *Geocarto Int.* **2018**, *33*, 1281–1306. [CrossRef]
5. Pour, A.B.; Park, T.S.; Park, Y.; Hong, J.K.; Zoheir, B.; Pradhan, B.; Ayoobi, I.; Hashim, M. Application of multi-sensor satellite data for exploration of Zn-Pb sulfide mineralization in the Franklinian Basin, North Greenland. *Remote Sens.* **2018**, *10*, 1186. [CrossRef]
6. Hu, B.; Xu, Y.; Wan, B.; Wu, X.; Yi, G. Hydrothermally altered mineral mapping using synthetic application of Sentinel-2A MSI, ASTER and Hyperion data in the Duolong area, Tibetan Plateau, China. *Ore Geol. Rev.* **2018**, *101*, 384–397. [CrossRef]
7. Abubakar, A.J.A.; Hashim, M.; Pour, A.B. Remote sensing satellite imagery for prospecting geothermal systems in an aseismic geologic setting: Yankari Park, Nigeria. *Int. J. Appl. Earth Obs. Geoinf.* **2019**, *80*, 157–172. [CrossRef]
8. Sheikhrahimi, A.; Pour, A.B.; Pradhan, B.; Zoheir, B. Mapping hydrothermal alteration zones and lineaments associated with orogenic gold mineralization using ASTER data: A case study from the Sanandaj-Sirjan Zone, Iran. *Adv. Space Res.* **2019**, *63*, 3315–3332. [CrossRef]
9. Feng, Y.; Xiao, B.; Li, R.; Deng, C.; Han, J.; Wu, C.; Lai, C. Alteration mapping with short wavelength infrared (SWIR) spectroscopy on Xiaokelehe porphyry Cu-Mo deposit in the Great Xing'an Range, NE China: Metallogenic and exploration implications. *Ore Geol. Rev.* **2019**, *112*, 103062. [CrossRef]
10. Xu, Y.; Meng, P.; Chen, J. Study on clues for gold prospecting in the Maizijing-Shulonggou area, Ningxia Hui autonomous region, China, using ALI, ASTER and WorldView-2 imagery. *J. Vis. Commun. Image Represent.* **2019**, *60*, 192–205. [CrossRef]
11. Hosseini, S.; Lashkaripour, G.R.; Moghadas, N.H.; Ghafoori, M.; Pour, A.B. Lineament mapping and fractal analysis using SPOT-ASTER satellite imagery for evaluating the severity of slope weathering process. *Adv. Space Res.* **2019**, *63*, 871–885. [CrossRef]
12. Pour, A.B.; Hashim, M.; Hong, J.K.; Park, Y. Lithological and alteration mineral mapping in poorly exposed lithologies using Landsat-8 and ASTER satellite data: North-eastern Graham Land, Antarctic Peninsula. *Ore Geol. Rev.* **2019**, *108*, 112–133. [CrossRef]
13. Pour, A.B.; Park, Y.; Crispini, L.; Läufer, A.; Kuk Hong, J.; Park, T.-Y.S.; Zoheir, B.; Pradhan, B.; Muslim, A.M.; Hossain, M.S.; et al. Mapping Listvenite Occurrences in the Damage Zones of Northern Victoria Land, Antarctica Using ASTER Satellite Remote Sensing Data. *Remote Sens.* **2019**, *11*, 1408. [CrossRef]
14. Pour, A.B.; Park, T.-Y.; Park, Y.; Hong, J.K.; Muslim, A.M.; Läufer, A.; Crispini, L.; Pradhan, B.; Zoheir, B.; Rahmani, O.; et al. Landsat-8, Advanced Spaceborne Thermal Emission and Reflection Radiometer, and WorldView-3 Multispectral Satellite Imagery for Prospecting Copper-Gold Mineralization in the Northeastern Inglefield Mobile Belt (IMB), Northwest Greenland. *Remote Sens.* **2019**, *11*, 2430. [CrossRef]
15. Bolouki, S.M.; Ramazi, H.R.; Maghsoudi, A.; Beiranvand Pour, A.; Sohrabi, G. A Remote Sensing-Based Application of Bayesian Networks for Epithermal Gold Potential Mapping in Ahar-Arasbaran Area, NW Iran. *Remote Sens.* **2020**, *12*, 105. [CrossRef]
16. Molan, Y.E.; Behnia, P. Prospectivity mapping of Pb–Zn SEDEX mineralization using remote-sensing data in the Behabad area, Central Iran. *Int. J. Remote Sens.* **2013**, *34*, 1164–1179. [CrossRef]
17. Niyeh, M.M.; Jafarirad, A.; Karami, J.; Bokani, S.J. Copper, Zinc, and Lead Mineral Prospectivity Mapping in the North of Tafresh, Markazi Province, Central Iran, Using the AHP-OWA Method. *Open J. Geol.* **2017**, *7*, 533. [CrossRef]
18. Govil, H.; Gill, N.; Rajendran, S.; Santosh, M.; Kumar, S. Identification of new base metal mineralization in Kumaon Himalaya, India, using hyperspectral remote sensing and hydrothermal alteration. *Ore Geol. Rev.* **2018**, *92*, 271–283. [CrossRef]
19. Yang, M.; Ren, G.; Han, L.; Yi, H.; Gao, T. Detection of Pb–Zn mineralization zones in west Kunlun using Landsat 8 and ASTER remote sensing data. *J. Appl. Remote Sens.* **2018**, *12*, 026018. [CrossRef]
20. World Bank Commodities Price Forecast (Nominal US Dollars). 2019. Available online: https://openknowledge.worldbank.org/bitstream/handle/10986/31549/CMO-April-19.pdf (accessed on 24 August 2019).
21. Bhavan, I.; Lines, C. *Indian Minerals Yearbook 2018 (Part-II: Metals and Alloys)*, 57th ed.; Lead & Zinc: Lisbon, Portugal, 2019; pp. 1–30.

22. Leach, D.L.; Bradley, D.C.; Huston, D.; Pisarevsky, S.A.; Taylor, R.D.; Gardoll, S.J. Sediment-hosted lead-zinc deposits in Earth history. *Econ. Geol.* **2010**, *105*, 593–625. [CrossRef]
23. Leach, D.L.; Taylor, R.D.; Fey, D.L.; Diehl, S.F.; Saltus, R.W. A deposit model for Mississippi Valley-Type lead-zinc ores. In *Mineral Deposit Models for Resource Assessment*; USGS Scientific Investigations Report 5070-A; U.S. Geological Survey: Reston, VA, USA, 2010; Chapter A; p. 52.
24. Taylor, R.D.; Leach, D.L.; Bradley, D.C.; Pisarevsky, S.A. *Compilation of Mineral Resource Data for Mississippi Valley-Type and Clastic-Dominated Sediment-Hosted Lead-Zinc Deposits*; USGS Open-File Report; U.S. Geological Survey: Reston, VA, USA, 2009; p. 42.
25. Parvaz, D.B. Oxidation Zones of Volcanogenic Massive Sulphide Deposits in the Troodos Ophiolite, Cyprus: Targeting Secondary Copper Deposits. Ph.D. Thesis, University of Exeter, Exeter, UK, 2014.
26. Rajabi, A.; Rastad, E.; Canet, C. Metallogeny of Cretaceous carbonate-hosted Zn–Pb deposits of Iran: Geotectonic setting and data integration for future mineral exploration. *Int. Geol. Rev.* **2012**, *54*, 1649–1672. [CrossRef]
27. Haghipour, A.; Pelissier, G. Geology of the Saghand Sector. In *Explanatory Text of the Ardekan Quadrangle Map*; Haghipour, A., Valeh, N., Pelissier, G., Davoudzadeh, M., Eds.; Geological Survey of Iran: Tehran, Iran, 1977; pp. 10–68.
28. Masoodi, M.; Yassaghi, A.; Nogole Sadat, M.A.A.; Neubauer, F.; Bernroider, M.; Friedl, G.; Houshmandzadeh, A. Cimmerian evolution of the Central Iranian basement: Evidence from metamorphic units of the Kashmar–Kerman Tectonic Zone. *Tectonophysics* **2013**, *588*, 189–208. [CrossRef]
29. Ghanbari, Y.; Hezarkhani, A.; Ataei, M.; Pazand, K. Mineral potential mapping with fuzzy models in the Kerman–Kashmar Tectonic Zone, Central Iran. *Appl. Geomat.* **2012**, *4*, 173–186. [CrossRef]
30. Huckriede, R.; Kürsten, M.; Venzlaff, H. Zur geologie des gebietes zwischen Kerman und Sagand (Iran): Beihefte zum. *Geol. Jahrb.* **1962**, *51*, 1–97.
31. Stöcklin, J. Structural history and tectonics of Iran: A review. *Bull. Am. Assoc. Pet. Geol.* **1968**, *52*, 1229–1258.
32. Davoudzadeh, M.; Lensch, G.; Weber-Diefenbach, K. Contribution to the paleogeography, stratigraphy and tectonics of the Infracambrian and lower Paleozoic of Iran. *Neues Jahrb. Geol. Paläontologie Abh.* **1986**, *172*, 245–269.
33. Ramezani, J.; Tucker, R.D. The Saghand region, central Iran: U-Pb geochronology, petrogenesis and implications for Gondwana tectonics. *Am. J. Sci.* **2003**, *303*, 622–665. [CrossRef]
34. Aghanabati, A. *Geology of Iran*; Geological Survey of Iran: Tehran, Iran, 2004; p. 587.
35. Berberian, M.; King, G.C.P. Towards a Paleogeography and tectonic evolution of Iran. *Can. J. Earth Sci.* **1981**, *18*, 210–265. [CrossRef]
36. Haghipour, A. *Geological Map of the Posht-e-Badam Area 1: 100 000*; Geological Survey of Iran: Tehran, Iran, 1977.
37. Haghipour, A.; Pelissier, G. Geology of the Posht-e-Badam-Saghand area (east central Iran). *Iran Geol. Surv. Note* **1968**, *48*, 144.
38. Samani, B.A. Metallogeny of the Precambrian in Iran. *Precambrian Res.* **1988**, *39*, 85–106. [CrossRef]
39. Husseini, M.I. Tectonic and deposition model of late Precambrian-Cambrian Arabian and adjoining plates. *AAPG Bull.* **1989**, *73*, 1117–1131.
40. Foerster, H.; Jafarzadeh, A. The Bafq mining district in central Iran; a highly mineralized Infracambrian volcanic field. *Econ. Geol.* **1994**, *89*, 1697–1721. [CrossRef]
41. Sennewald, S. Resurgent Cauldrons and Their mineralization between Narigan, Esfordi, Kushk, and Seh Chahoon, Central Iran. *Int. J. Eng.* **1988**, *1*, 149–161.
42. Drusch, M.; Del Bello, U.; Carlier, S.; Colin, O.; Fernandez, V.; Gascon, F.; Meygret, A.; Spoto, F.; Sy, O.; Marchese, F.; et al. Sentinel-2: ESA's optical high-resolution mission for GMES operational services. *Remote Sens. Environ.* **2012**, *120*, 25–36. [CrossRef]
43. Irons, J.R.; Dwyer, J.L.; Barsi, J.A. The next Landsat satellite: The Landsat data continuity mission. *Remote Sens. Environ.* **2012**, *122*, 11–21. [CrossRef]
44. Roy, D.P.; Wulder, M.A.; Loveland, T.A.; Woodcock, C.E.; Allen, R.G.; Anderson, M.C.; Scambos, T.A. Landsat-8: Science and product vision for terrestrial global change research. *Remote Sens. Environ.* **2014**, *145*, 154–172. [CrossRef]
45. DigitalGlobe. WorldView-3 Datasheet. 2014. Available online: https://www.digitalglobe.com/sites/default/files/DG_WorldView3_DS_forWeb_0.pdf (accessed on 24 September 2019).

46. Abrams, M.; Tsu, H.; Hulley, G.; Iwao, K.; Pieri, D.; Cudahy, T.; Kargel, J. The Advanced Spaceborne Thermal Emission and Reflection Radiometer (ASTER) after fifteen years: Review of global products. *Int. J. Appl. Earth Obs. Geoinf.* **2015**, *38*, 292–301. [CrossRef]
47. Lima, T.A.; Beuchle, R.; Langner, A.; Grecchi, R.C.; Griess, V.C.; Achard, F. Comparing Sentinel-2 MSI and Landsat 8 OLI Imagery for Monitoring Selective Logging in the Brazilian Amazon. *Remote Sens.* **2019**, *11*, 961. [CrossRef]
48. U.S. Geological Survey Earth Resources Observation and Science Center (EROS). Available online: https://earthexplorer.usgs.gov/ (accessed on 2 June 2019).
49. Pieschke, R.L. *US Geological Survey Distribution of European Space Agency's Sentinel-2 Data*; No. 2017-3026; US Geological Survey: Reston, VA, USA, 2017.
50. USGS EROS Center. Available online: https://earthexplorer.usgs.gov/ (accessed on 28 March 2019).
51. Kuester, M. *Radiometric Use of WV-3 Imagery*; Technical Note; DigitalGlobe: Westminster, CO, USA, 2016; p. 12.
52. Kuester, M.A.; Ochoa, M.; Dayer, A.; Levin, J.; Aaron, D.; Helder, D.L.; Leigh, L.; Czapla-Meyers, J.; Anderson, N.; Bader, B.; et al. *Absolute Radiometric Calibration of the DigitalGlobe Fleet and Updates on the New WV-3 Sensor Suite*; Technical Note; DigitalGlobe: Westminster, CO, USA, 2015; p. 16.
53. ENVI. Environment for Visualizing Images. Available online: http://www.exelisvis.com (accessed on 12 February 2019).
54. Cooley, T.; Anderson, G.P.; Felde, G.W.; Hoke, M.L.; Ratkowski, A.J.; Chetwynd, J.H.; Gardner, J.A.; Adler-Golden, S.M.; Matthew, M.W.; Berk, A.; et al. FLAASH, a MODTRAN4-based atmospheric correction algorithm, its application and validation. In Proceedings of the IEEE International Geoscience and Remote Sensing Symposium, Toronto, ON, Canada, 24–28 June 2002; Volume 3, pp. 1414–1418.
55. Research Systems, Inc. *ENVI Tutorials*; Research Systems, Inc.: Boulder, CO, USA, 2008.
56. Rahaman, K.; Hassan, Q.; Ahmed, M. Pan-sharpening of Landsat-8 images and its application in calculating vegetation greenness and canopy water contents. *ISPRS Int. J. Geo Inf.* **2017**, *6*, 168. [CrossRef]
57. Ben-Dor, E.; Kruse, F.A.; lefkoff, A.B.; Banin, A. Comparison of three calibration techniques for utilization of GER 63-channel aircraft scanner data of Makhtesh Ramon, Nega, Israel. *Int. J. Rock Mech. Min. Sci. Geomech. Abstr.* **1995**, *32*, 164A.
58. Iwasaki, A.; Tonooka, H. Validation of a crosstalk correction algorithm for ASTER/SWIR. *IEEE Trans. Geosci. Remote Sens.* **2005**, *43*, 2747–2751. [CrossRef]
59. Kalinowski, A.; Oliver, S. ASTER mineral index processing manual. *Remote Sens. Appl. Geosci. Aust.* **2004**, *37*, 36.
60. Mars, J.C.; Rowan, L.C. Regional mapping of phyllic- and argillic-altered rocks in the Zagros magmatic arc, Iran, using Advanced Spaceborne Thermal Emission and Reflection Radiometer (ASTER) data and logical operator algorithms. *Geosphere* **2006**, *7*, 276–289. [CrossRef]
61. Pour, B.A.; Hashim, M. Identification of hydrothermal alteration minerals for exploring of porphyry copper deposit using ASTER data, SE Iran. *J. Asian Earth Sci.* **2011**, *42*, 1309–1323. [CrossRef]
62. Pour, B.A.; Hashim, M. The application of ASTER remote sensing data to porphyry copper and epithermal gold deposits. *Ore Geol. Rev.* **2012**, *44*, 1–9. [CrossRef]
63. Eldosouky, A.M.; Sehsah, H.; Elkhateeb, S.O.; Pour, A.B. Integrating aeromagnetic data and Landsat-8 imagery for detection of post-accretionary shear zones controlling hydrothermal alterations: The Allaqi-Heiani Suture zone, South Eastern Desert, Egypt. *Adv. Space Res.* **2019**. [CrossRef]
64. Inzana, J.; Kusky, T.; Higgs, G.; Tucker, R. Supervised classifications of Landsat TM band ratio images and Landsat TM band ratio image with radar for geological interpretations of central Madagascar. *J. Afr. Earth Sci.* **2003**, *37*, 59–72. [CrossRef]
65. Di Tommaso, I.; Rubinstein, N. Hydrothermal alteration mapping using ASTER data in the Infiernillo porphyry deposit, Argentina. *Ore Geol. Rev.* **2007**, *32*, 275–290. [CrossRef]
66. Rockwell, B.W.; Hofstra, A.H. Identification of quartz and carbonate minerals across Northern Nevada using ASTER thermal infrared emissivity data, implications for geologic mapping and mineral resource investigations in well-studied and frontier areas. *Geosphere* **2008**, *4*, 218–246. [CrossRef]
67. Mars, J.C.; Rowan, L.C. ASTER spectral analysis and lithologic mapping of the Khanneshin carbonate volcano, Afghanistan. *Geosphere* **2011**, *7*, 276–289. [CrossRef]
68. Colby, J.D. Topographic normalization in rugged terrain. *Photogramm. Eng. Remote Sens.* **1991**, *57*, 531–537.

69. Crowley, J.K.; Brickey, D.W.; Rowan, L.C. Airborne imaging spectrometer data of the Ruby Mountains, Montana: Mineral discrimination using relative absorption band-depth images. *Remote Sens. Environ.* **1989**, *29*, 121–134. [CrossRef]
70. Ninomiya, Y.; Fu, B.; Cudahy, T.J. Detecting lithology with Advanced Spaceborne Thermal Emission and Reflection Radiometer (ASTER) multispectral thermal infrared radiance-at-sensor data. *Remote Sens. Environ.* **2005**, *99*, 127–139. [CrossRef]
71. Ninomiya, Y.; Fu, B. Regional lithological mapping using ASTER-TIR data: Case study for the Tibetan Plateau and the surrounding area. *Geosciences* **2016**, *6*, 39. [CrossRef]
72. Ninomiya, Y.; Fu, B. Thermal infrared multispectral remote sensing of lithology and mineralogy based on spectral properties of materials. *Ore Geol. Rev.* **2018**. [CrossRef]
73. Hunt, G.R. Spectral signatures of particulate minerals in the visible and near infrared. *Geophysics* **1977**, *42*, 501–513. [CrossRef]
74. Clark, R.N. Spectroscopy of rock and minerals and principles of spectroscopy. In *Remote Sensing for the Earth Sciences: Manual of Remote Sensing 3*; Rencz, A.N., Ed.; John Wiley Sons: New York, NY, USA, 1999; pp. 3–58.
75. Hunt, G.R.; Ashley, R.P. Spectra of altered rocks in the visible and near-infrared. *Econ. Geol.* **1979**, *74*, 1613–1629. [CrossRef]
76. Mars, J.C.; Rowan, L.C. Spectral assessment of new ASTER SWIR surface reflectance data products for spectroscopic mapping of rocks and minerals. *Remote Sens. Environ.* **2010**, *114*, 2011–2025. [CrossRef]
77. Rowan, L.C.; Mars, J.C. Lithologic mapping in the Mountain Pass area, California using Advanced Spaceborne Thermal Emission and Reflection Radiometer (ASTER) data. *Remote Sens. Environ.* **2003**, *84*, 350–366. [CrossRef]
78. Kuosmanen, V.; Laitinen, J.; Arkimaa, H.; Kuosmanen, E. *Hyperspectral Characterization of Selected Remote Detection Targets in the Mines of HYDO Partners*; Archive Report, RS/2000/02; Geological Survey of Finland: Espoo, Finland, 2000.
79. Sun, Y.; Tian, S.; Di, B. Extracting mineral alteration information using Worldview-3 data. *Geosci. Front.* **2017**, *8*, 1051–1062. [CrossRef]
80. Mars, J.C. Mineral and Lithologic Mapping Capability of WorldView 3 Data at Mountain Pass, California, Using True-and False-Color Composite Images, Band Ratios, and Logical Operator Algorithms. *Econ. Geol.* **2018**, *113*, 1587–1601. [CrossRef]
81. Bedini, E. Application of WorldView-3 imagery and ASTER TIR data to map alteration minerals associated with the Rodalquilar gold deposits, southeast Spain. *Adv. Space Res.* **2019**, *63*, 3346–3357. [CrossRef]
82. Gaffey, S.J. Spectral reflectance of carbonate minerals in the visible and near-infrared (0.35–2.55 microns): Calcite, aragonite, and dolomite. *Am. Mineral.* **1986**, *71*, 151–162.
83. Singh, A.; Harrison, A. Standardized principal components. *Int. J. Remote Sens.* **1985**, *6*, 883–896. [CrossRef]
84. Crosta, A.P.; Souza Filho, C.R.; Azevedo, F.; Brodie, C. Targeting key alteration minerals in epithermal deposits in Patagonia, Argentina, Using ASTER imagery and principal component analysis. *Int. J. Remote Sens.* **2003**, *24*, 4233–4240. [CrossRef]
85. Gupta, R.P.; Tiwari, R.K.; Saini, V.; Srivastava, N. A simplified approach for interpreting principal component images. *Adv. Remote Sens.* **2013**, *2*, 111–119. [CrossRef]
86. Eklundh, L.; Singh, A. A comparative analysis of standardized and unstandardized principal component analysis in remote sensing. *Int. J. Remote Sens.* **1993**, *14*, 1359–1370. [CrossRef]
87. Chang, Q.; Jing, L.; Panahi, A. Principal component analysis with optimum order sample correlation coefficient for image enhancement. *Int. J. Remote Sens.* **2006**, *27*, 3387–3401.
88. Crosta, A.; Moore, J. Enhancement of Landsat Thematic Mapper imagery for residual soil mapping in SW Minais Gerais State, Brazil: A prospecting case history in Greenstone belt terrain. In Proceedings of the 7th ERIM Thematic Conference: Remote Sensing for Exploration Geology, Calgary, AB, Canada, 2–6 October 1989; pp. 1173–1187.
89. Loughlin, W.P. Principal components analysis for alteration mapping. *Photogramm. Eng. Remote Sens.* **1991**, *57*, 1163–1169.
90. Pour, B.A.; Hashim, M.; Marghany, M. Exploration of gold mineralization in a tropical region using Earth Observing-1 (EO1) and JERS-1 SAR data: A case study from Bau gold field, Sarawak, Malaysia. *Arab. J. Geosci.* **2014**, *7*, 2393–2406. [CrossRef]

91. Noori, L.; Pour, B.A.; Askari, G.; Taghipour, N.; Pradhan, B.; Lee, C.-W.; Honarmand, M. Comparison of Different Algorithms to Map Hydrothermal Alteration Zones Using ASTER Remote Sensing Data for Polymetallic Vein-Type Ore Exploration: Toroud–Chahshirin Magmatic Belt (TCMB), North Iran. *Remote Sens.* **2019**, *11*, 495. [CrossRef]
92. Zoheir, B.; El-Wahed, M.A.; Pour, A.B.; Abdelnasser, A. Orogenic Gold in Transpression and Transtension Zones: Field and Remote Sensing Studies of the Barramiya–Mueilha Sector, Egypt. *Remote Sens.* **2019**, *11*, 2122. [CrossRef]
93. Zadeh, L.A. Fuzzy sets. *Inf. Control* **1965**, *8*, 338–353. [CrossRef]
94. Novák, V.; Perfilieva, I.; Močkoř, J. *Mathematical Principles of Fuzzy Logic*; Kluwer Academic: Dordrecht, The Netherlands, 1999; ISBN 978-0-7923-8595-0.
95. Nykänen, V.; Groves, D.I.; Ojala, V.J.; Eilu, P.; Gardoll, S.J. Reconnaissance-scale conceptual fuzzy-logic prospectivity modelling for iron oxide copper—Gold deposits in the northern Fennoscandian Shield, Finland. *Aust. J. Earth Sci.* **2008**, *55*, 25–38. [CrossRef]
96. Carranza, E.J.M. *Geochemical Anomaly and Mineral Prospectivity Mapping in GIS*; Elsevier: Amsterdam, The Netherlands, 2008; Volume 11.
97. Zhang, N.; Zhou, K.; Du, X. Application of fuzzy logic and fuzzy AHP to mineral prospectivity mapping of porphyry and hydrothermal vein copper deposits in the Dananhu-Tousuquan island arc, Xinjiang, NW China. *J. Afr. Earth Sci.* **2017**, *128*, 84–96. [CrossRef]
98. Kim, Y.H.; Choe, K.U.; Ri, R.K. Application of fuzzy logic and geometric average: A Cu sulfide deposits potential mapping case study from Kapsan Basin, DPR Korea. *Ore Geol. Rev.* **2019**, *107*, 239–247. [CrossRef]
99. Zimmermann, H.J.; Zysno, P. Latent connectives in human decision making. *Fuzzy Sets Syst.* **1980**, *4*, 37–51. [CrossRef]
100. An, P.; Moon, W.M.; Rencz, A. Application of fuzzy set theory to integrated mineral exploration. *Can. J. Explor. Geophys.* **1991**, *27*, 1–11.
101. Bonham-Carter, G.F. Geographic information systems for geoscientists-modeling with GIS. *Comput. Methods Geosci.* **1994**, *13*, 398.
102. Carranza, E.J.M.; Hale, M. Geologically constrained fuzzy mapping of gold mineralization potential, Bauio District, Philippines. *Nat. Resour. Res.* **2001**, *10*, 125–136. [CrossRef]
103. Bishop, J.L.; Lane, M.D.; Dyar, M.D.; Brwon, A.J. Reflectance and emission spectroscopy study of four groups of phyllosilicates: Smectites, kaolinite-serpentines, chlorites and micas. *Clay Miner.* **2008**, *43*, 35–54. [CrossRef]
104. Cloutis, E.A.; Hawthorne, F.C.; Mertzman, S.A.; Krenn, K.; Craig, M.A.; Marcino, D.; Methot, M.; Strong, J.; Mustard, J.F.; Blaney, D.L.; et al. Detection and discrimination of sulfate minerals using reflectance spectroscopy. *Icarus* **2006**, *184*, 121–157. [CrossRef]

© 2020 by the authors. Licensee MDPI, Basel, Switzerland. This article is an open access article distributed under the terms and conditions of the Creative Commons Attribution (CC BY) license (http://creativecommons.org/licenses/by/4.0/).

Article

Integration of Selective Dimensionality Reduction Techniques for Mineral Exploration Using ASTER Satellite Data

Hodjat Shirmard [1], Ehsan Farahbakhsh [2], Amin Beiranvand Pour [3,*], Aidy M Muslim [3], R. Dietmar Müller [4] and Rohitash Chandra [5]

[1] School of Mining Engineering, College of Engineering, University of Tehran, Tehran 143995-7131, Iran; hodjat.shirmard@ut.ac.ir
[2] Department of Mining Engineering, Amirkabir University of Technology (Tehran Polytechnic), Tehran 159163-4311, Iran; e.farahbakhsh@aut.ac.ir
[3] Institute of Oceanography and Environment (INOS), University Malaysia Terengganu (UMT), Kuala Nerus 21030, Terengganu, Malaysia; aidy@umt.edu.my
[4] EarthByte Group, School of Geosciences, University of Sydney, Sydney, NSW 2006, Australia; dietmar.muller@sydney.edu.au
[5] School of Mathematics and Statistics, University of New South Wales, Sydney, NSW 2052, Australia; rohitash.chandra@unsw.edu.au
* Correspondence: beiranvand.pour@umt.edu.my; Tel.: +60-96683824; Fax: +60-96692166

Received: 12 March 2020; Accepted: 14 April 2020; Published: 16 April 2020

Abstract: There are a significant number of image processing methods that have been developed during the past decades for detecting anomalous areas, such as hydrothermal alteration zones, using satellite images. Among these methods, dimensionality reduction or transformation techniques are known to be a robust type of methods, which are helpful, as they reduce the extent of a study area at the initial stage of mineral exploration. Principal component analysis (PCA), independent component analysis (ICA), and minimum noise fraction (MNF) are the dimensionality reduction techniques known as multivariate statistical methods that convert a set of observed and correlated input variables into uncorrelated or independent components. In this study, these techniques were comprehensively compared and integrated, to show how they could be jointly applied in remote sensing data analysis for mapping hydrothermal alteration zones associated with epithermal Cu–Au deposits in the Toroud-Chahshirin range, Central Iran. These techniques were applied on specific subsets of the advanced spaceborne thermal emission and reflection radiometer (ASTER) spectral bands for mapping gossans and hydrothermal alteration zones, such as argillic, propylitic, and phyllic zones. The fuzzy logic model was used for integrating the most rational thematic layers derived from the transformation techniques, which led to an efficient remote sensing evidential layer for mineral prospectivity mapping. The results showed that ICA was a more robust technique for generating hydrothermal alteration thematic layers, compared to the other dimensionality reduction techniques. The capabilities of this technique in separating source signals from noise led to improved enhancement of geological features, such as specific alteration zones. In this investigation, several previously unmapped prospective zones were detected using the integrated hydrothermal alteration map and most of the known hydrothermal mineral occurrences showed a high prospectivity value. Fieldwork and laboratory analysis were conducted to validate the results and to verify new prospective zones in the study area, which indicated a good consistency with the remote sensing output. This study demonstrated that the integration of remote sensing-based alteration thematic layers derived from the transformation techniques is a reliable and low-cost approach for mineral prospectivity mapping in metallogenic provinces, at the reconnaissance stage of mineral exploration.

Keywords: dimensionality reduction; principal component analysis; independent component analysis; minimum noise fraction; ASTER; hydrothermal alteration; fuzzy logic

1. Introduction

The interaction of hydrothermal fluids and wall rocks during the uprising process through conduits (e.g., faults and fractures), which results in the alteration of mineralogy and chemical composition of rocks, can lead to the generation of polymetallic epithermal and porphyry deposits [1–4]. The footprints of various types of hydrothermal alteration on the surface are key indicators through the exploration of outcropping or deep-seated deposits [5–8]. Each alteration type shows a specific spectral behavior due to different mineral assemblages. Exploration geologists use these spectral characteristics as diagnostic features for detecting and discriminating between different alteration types, using remote sensing data [9–11]. Detailed spectral information on the mineralogy and geochemistry of rock types comprising the Earth's surface are provided by multispectral and hyperspectral remote sensing instruments, and this technology has been used for decades to map rocks, mineral assemblages, and weathering characteristics in different regions [9,10,12–18]. Mapping prospective zones of various types of the hydrothermal alteration minerals is one of the most important applications of remote sensing in the field of mineral exploration [11,18–22].

The spectral and spatial resolution provided by the advanced spaceborne thermal emission and reflection radiometer (ASTER) sensor makes the identification of specific alteration assemblages feasible. The ASTER spectral subsets, including visible and near infrared, short-wave infrared, and thermal infrared wavelength regions provide complementary data for lithologic mapping and mineral exploration. The ASTER remote sensing data have been extensively used for alteration and lithological mapping [16,23–25]. Image processing approaches such as dimensionality reduction or transformation techniques are considered as efficient tools in identifying hydrothermal alteration zones in metallogenic provinces [10,15,16,26–29]. Transformation techniques such as principal component analysis (PCA), independent component analysis (ICA), and minimum noise fraction (MNF) are powerful statistical techniques that can be used for suppressing irradiance effects that dominate all bands, therefore, enhancing the spectral reflectance features of geological materials [30,31]. These techniques can be applied to multivariate data sets, such as multispectral satellite images, to extract specific spectral responses, as in the case of hydrothermal alteration minerals.

PCA has been used to transform remote sensing data in the form of image to uncover the most important features [32,33], by extracting a smaller set of variables with less redundancy from high-dimensional data sets [34,35]. This technique has been widely used for mapping lithological features and hydrothermal alteration zones, using different types of remote sensing data [23,36–38]. ICA has less been considered to be a common technique in image processing, although it has a wide range of applications in signal processing [39]. The lack of a comprehensive understanding of the underlying theory and foundations of ICA is one of the main reasons that ICA has not been applied commonly in geosciences, particularly, for multi- or hyper-spectral image-processing. There are only a few studies focused on the application of ICA in alteration mapping [40,41]. MNF is used to determine the inherent dimensionality of image data, segregate noise in the data, and reduce the computational requirements for subsequent processing [42,43]. This transformation can identify spectral signatures of spectral anomalies. MNF is of interest to exploration geologists because spectral anomalies are often indicative of hydrothermal alteration zones and has been applied on different data types for detecting such anomalies [44–46].

The integration and comparison of the dimensionality reduction techniques provide comprehensive information for creating the most informative thematic layers and generating a remote sensing evidential layer. In this study, we used the PCA, ICA, and MNF for mapping hydrothermal alteration zones, using ASTER remote sensing data in the Toroud-Chahshirin range, Central Iran (Figure 1). This region is mostly known for several epithermal polymetallic vein-type mineral occurrences, and anomalous Cu and Au concentration values have been reported to be associated with altered dacite, dacite-andesite, and volcaniclastics rocks. The presence of several mineral occurrences associated with widespread alteration zones suggests that the Toroud-Chahshirin range is a prospective zone for high-grade gold veins and base metal epithermal deposits [47]. There is no regional prospectivity map available for the study area. Accordingly, the main objectives of this study are: (1) to compare the PCA, ICA, and MNF techniques for mapping gossans and hydrothermal alteration zones, including argillic, phyllic, and propylitic, using selected spectral subsets of the ASTER data; (2) to select the most informative thematic layers for detecting gossans, argillic, phyllic, and propylitic zones, using statistical analyses; (3) to integrate the most informative thematic layers for generating a remote sensing evidential layer using fuzzy logic; and (4) to verify the prospective zones through detailed fieldwork and laboratory analysis.

2. Geological Setting

The magmatic arc of Toroud-Chahshirin located in Central Iran, lies between the Anjilow and Toroud faults (Figure 1) [48]. The rock outcrops of the study area are composed of Eocene volcano-pyroclastic rocks with an intermediate composition (andesite), which have been affected by Oligo-Miocene intrusive bodies. The magmatic activities commenced in the first and second geological eras along with tectonic events, gradually. The peak of magmatic activities occurred from middle to upper Eocene, which constitute the heights of the Toroud-Chahshirin region. Most of the magmatic products are made of andesite and basalt, which have an acidic or trachytic state. On the other hand, some magmatic products are basic in terms of composition and have changed into andesite lavas, breccias, and tuffs, at the end of Eocene. The volcanic rocks of the study area have been cut by multiple intrusive bodies aged Oligo-Miocene, which are known to be one of the key factors of mineralization. These rocks include granite, micro-granite, granodiorite, micro-granodiorite, micro-quartz monzonite, micro-monzonite, micro-monzodiorite, and micro-quartz monzodiorite. The major constituent minerals include quartz, alkali-feldspar, plagioclase, biotite, amphibole, pyroxene, apatite, titanite, zircon, tourmaline, magnetite, and ilmenite. The volcanic rocks are mainly from magmatic, subalkaline, and alkaline series [47]. The Toroud-Chahshirin range is the largest known gold and base metal province of Iran [47,49]. In this province, the Northern Iranian region hosts five gold and base metal deposits. Other types of deposits in this range include placer gold, an underground mine for turquoise at Baghu, skarn deposits, and Pb-Zn deposits in carbonate rocks.

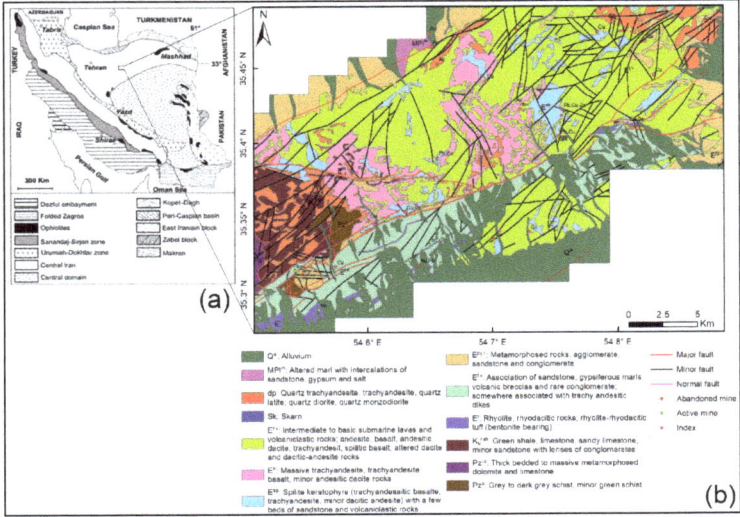

Figure 1. (**a**) Simplified tectonic scheme of Iran; and (**b**) geological map of the Toroud-Chahshirin range located in Central Iran [50].

3. Materials and Methods

3.1. ASTER Data Characteristics and Pre-Processing

The ASTER remote sensing data are the result of a joint plan between the United States and Japan, with a strong focus on geological and mineral exploration applications [51]. This sensor, which is aboard the Earth observing system (EOS) Terra platform, records solar radiation in 14 spectral bands [14,52]. It measures the reflected radiation in three subsets, including visible and near infrared (VNIR), short-wave infrared (SWIR), and thermal infrared (TIR). The VNIR consists of three bands between 0.52 and 0.86 micrometers (µm), the SWIR includes six bands from 1.6 to 2.43 µm, and emitted radiation in five bands in the 8.125–11.65 µm wavelength region constitute the TIR. The resolution of VNIR, SWIR, and TIR is 15 meters (m), 30 m, and 90 m, respectively [53]. Many clay and carbonate minerals show diagnostic spectral features in the short-wave infrared range, where the ASTER sensor provides six spectral bands [54]. According to the geological setting and metallogenetic characteristics of the study area, and different types of the hydrothermal alteration associated with epithermal mineral deposits [55–58], we used the ASTER data for mapping the hydrothermal alteration zones.

We used two cloud-free level 1 precision terrain corrected and registered at-sensor radiance (AST_L1T) ASTER scenes in this study. These scenes downloaded from the US Geological Survey Earth Explorer [59], were both acquired on October 3, 2004. The ASTER AST_L1T data was calibrated at-sensor radiance, which corresponded to ASTER Level 1B (AST_L1B); which was geometrically corrected, and rotated to a north-up universal transverse Mercator projection [60]. The ASTER scenes used in this study were pre-georeferenced to the UTM zone 40 North. The QUAC module within the ENVI software package [61], which works with the visible and near-infrared to short-wave infrared (VNIR–SWIR) wavelength range, was used to provide an atmospheric-corrected surface reflectance image of the study area. Moreover, this module was a quick solution for converting radiance-calibrated data to apparent reflectance. Eventually, the SWIR bands were resampled to the spatial resolution of VNIR using the nearest neighbor technique.

3.2. Image Processing

3.2.1. Principal Component Analysis

Principal component analysis aims at finding a set of linearly uncorrelated components called principal components, which can be considered to be projections from the original data [62–64]. In other words, the principal components are the projection of input data onto the principal axes or eigenvectors. The output components are arranged on the basis of the variance, in descending order. The first principal component has the largest variance and the next component has the next highest variance. There is a constraint that each component has to be orthogonal to the preceding components [65,66]. In PCA, the same number of output principal components as input spectral bands can be generated. Although, a small number of principal components often involve the majority of the variance in the data and provide most of the information about the structure of the data [67,68]. In this study, we assumed a normal data distribution and used the covariance matrix for calculating the principal components. The principal component with the loadings, which shows a similar trend to the spectral characteristics of the target alteration minerals, is considered to be the appropriate component for enhancing the target zones. The selected principal component image contains a unique contribution of eigenvector loadings in terms of magnitude and sign, for the absorption and reflection bands of an alteration mineral or mineral group. This feature helps by enhancing the target alteration zone.

3.2.2. Independent Component Analysis

Independent component analysis is known to be an efficient statistical signal processing technique for decomposing a set of multivariate signals into statistically independent streams, without losing much information [69]. ICA is able to reveal hidden features that underlie sets of random signals and attempts to make the separated signals as independent as possible. The independent components and the mixture signals are always assumed to have a zero mean and a unit variance, in order to simplify the model without a loss of generality. This assumption leads to no variance ranking of the independent components. There are many mature algorithms available for implementing ICA using various estimators of independence. In this study, we choose the fast ICA that uses a fixed-point algorithm for an approximation of negentropy as a measurement of independence, for data processing, due to its computing efficiency, flexible parameters, and robustness [70]. In information theory and statistics, negentropy is used as a measure of distance to normality [71]. Unlike the PCA, which is based on the assumptions of uncorrelation and normality, ICA is rooted in the assumption of statistical independence. PCA only requires the second-order statistics, while ICA looks for statistically independent components, a much stronger condition than being uncorrelated. In addition, ICA components are not necessarily geometrical orthogonal. The most important difference is that ICA needs a linear model to describe data while PCA does not. Therefore, ICA cannot be considered as a generalization of PCA [72].

3.2.3. Minimum Noise Fraction

Minimum noise fraction is known to be an efficient technique for reducing the dimensions of a large dataset into a smaller number of components that involve the majority of information [42]. This technique was similar to the PCA, but the resulting components were not necessarily orthogonal and were arranged according to the signal-to-noise ratio, in descending order. MNF is applied for discriminating between noise and signal in a dataset. Moreover, this technique is able to determine the inherent dimensionality of an image [28]. The MNF transform implemented in this study involved two cascaded PCA transformations. The first transformation is called noise-whitening and is based on an estimated noise covariance matrix that aims at decorrelating and rescaling the noise in the data. The second step is a standard PCA transformation of the noise-reduced data. The number of output components can be as many as the input bands, with a decreasing overall variance of the dataset from the first component to the last. Similar to other transformation techniques, only a small number of components were often required to describe most of the information for the entire dataset. The contribution of each component to the overall information in a multivariate dataset, such as multispectral or hyperspectral images, is measured by an eigenvalue. The output components can be divided into two parts, including the part associated with large eigenvalues and the other with near-unity eigenvalues and noise-dominated images. The part with large eigenvalues separates the noise from the data, and improves spectral results [42]. The contribution of each band to each component is measured by an eigenvector, which can be interpreted akin to a correlation coefficient [73]. The dimensionality reduction techniques used in this study, were executed using the ENVI software package [61].

3.2.4. Hydrothermal Alteration Mapping by the PCA, ICA, and MNF Techniques

In this study, the PCA, ICA, and MNF techniques were applied to specific subsets of the ASTER spectral bands. The subsets were selected according to the characteristic spectral features of key alteration minerals in the VNIR and SWIR ranges of the electromagnetic spectrum. The selected spectral bands involved absorption and reflection diagnostic features of the indicative minerals in each alteration zone. In this study, we targeted the detection of gossans and different types of hydrothermal alterations, including argillic, propylitic, and phyllic, which were mainly related to the epithermal ore deposits. The laboratory spectra of these types of alteration minerals are available in Figure 2, which were resampled to the ASTER spectral bands [74]. Gossans are important guides to buried metallic ore deposits and are usually found in the upper and exposed part of an ore deposit or mineral vein, which involves intensely oxidized and weathered rocks [75]. Iron oxide and hydroxide minerals such as goethite, hematite, jarosite, and limonite are known to be indicative minerals of a gossan [75]. According to Figure 2a, these minerals showed an absorption feature in bands 1–3, located in the VNIR portion of the electromagnetic spectrum due to electronic transitions, and a reflectance feature in band 4 (1.65 μm). Therefore, we selected bands 1–4 as the input to the PCA, ICA, and MNF techniques for mapping gossans.

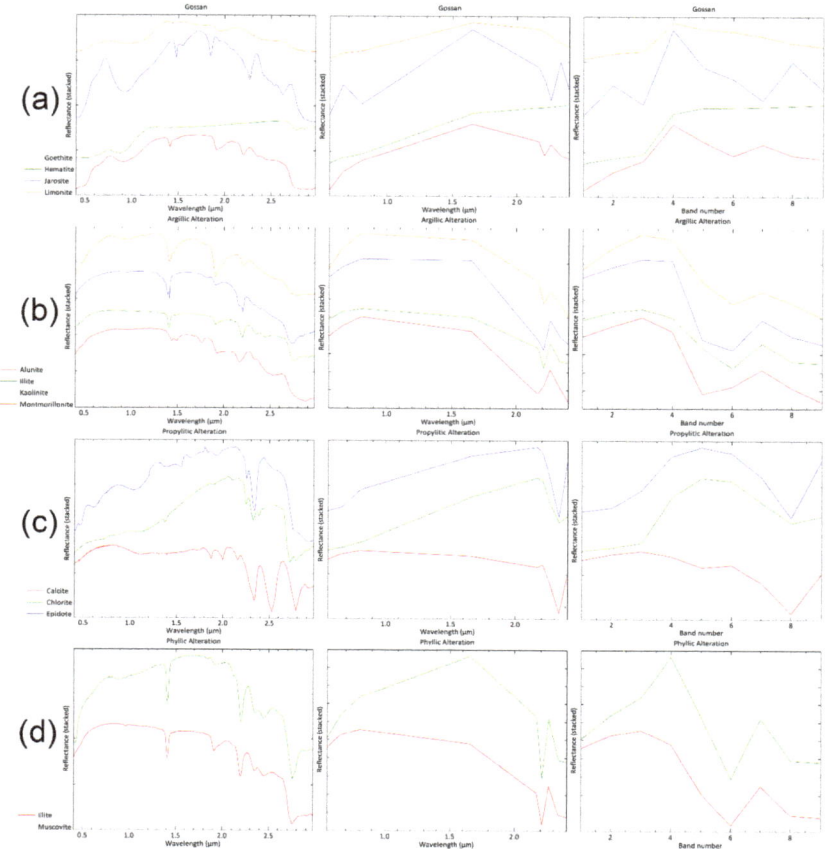

Figure 2. Laboratory spectra of the indicative minerals in (**a**) gossans, (**b**) argillic, (**c**) propylitic, and (**d**) phyllic hydrothermal alteration zones. These spectra were resampled to the advanced spaceborne thermal emission and reflection radiometer (ASTER) spectral bands and are shown against wavelength and band numbers in the second and third columns, respectively [74].

The indicative minerals that were considered for each hydrothermal alteration zone included alunite, illite, kaolinite, and montmorillonite for argillic; calcite, chlorite, and epidote for propylitic; and illite and muscovite for the phyllic alteration zones. As shown in Figure 2b, the clay minerals that constituted the major part of argillic alteration and usually exhibited aluminum hydroxide spectral features caused by vibrational processes, showed an absolute and relative reflectance feature in bands 4 (1.65 μm) and 7 (2.26 μm), respectively. Moreover, there was an absolute absorption in band 5 (2.165 μm). Therefore, we selected bands 1, 4, 5, and 7 as a spectral subset for enhancing the argillic alteration zones. According to Figure 2c, the indicative minerals of propylitic alteration, particularly chlorite and epidote showed an absolute and relative reflection in bands 5 (2.165 μm) and 4 (1.65 μm). Additionally, there was an absolute absorption in band 8 (2.33 μm) that was attributed to the vibrations of OH groups bound to the Fe and Mg cations. Therefore, we selected bands 1, 4, 5, and 8 as a spectral subset for enhancing propylitic alteration zones. According to the field observation, illite and muscovite minerals constituted most of the phyllic alteration zones in the study area. As shown in Figure 2d, there was an absolute and relative reflectance in bands 4 (1.65 µm) and 7 (2.26 µm), respectively. Additionally, there was an absolute absorption in band 6 (2.205 µm), due to the presence of aluminum hydroxide compound. Therefore, we selected bands 1, 4, 6, and 7 as a spectral subset for enhancing the phyllic alteration zones.

We used statistical analyses for selecting the meaningful component for enhancing each alteration type, derived from different transformation techniques. The concentration-area (C-A) fractal method was applied for determining an appropriate threshold for discriminating between the anomaly population and the background in each selected component [76]. The number of correctly classified rock samples was used to assess the accuracy of each selected component in terms of consistency with field observations, and to help us find the appropriate transformation technique for mapping each alteration type [77].

3.3. Integration of Hydrothermal Alteration Thematic Layers Using Fuzzy Logic

We used a logistic function for scaling input components to the integration process between 0 and 1 [78]. These components were integrated using a knowledge-driven approach based on fuzzy logic. The components were weighted from 1 to 10, using a subjective judgement based on the metallogenic models presented for hydrothermal mineralization and expert knowledge [79–81]. The more favorable the alteration type, the higher weight it took. The phyllic alteration is known to be highly associated with hydrothermal mineralization and is usually found close to the center of a mineralization system [82,83]. This alteration type is given the highest weight equal to 9. The argillic alteration and gossans are considered to be exploration guides and are usually not associated with target hydrothermal mineralization [84]. They are usually found in the surrounding regions of mineralization and were weighted 7. The propylitic alteration usually constitute the outermost ring of hydrothermal mineralization on the ground surface [85] and was given the lowest weight equal to 3. We applied the fuzzy gamma operator for integrating input components. The fuzzy gamma operator allowed a judicious choice of gamma, leading to an output that ensured a flexible compromise between the increasing trend of fuzzy algebraic sum and the decreasing effect of fuzzy algebraic product [86]. In this study, the fuzzy gamma operator was experimented with changing gamma values in the range of 0 and 1. The most satisfying map was obtained when the gamma equaled 0.9, which yielded the highest prediction rate based on the prediction-area (P-A) plots. We used the prediction-area plots in order to quantitatively validate the remote sensing evidential layers derived from the integration of transformation techniques, using different gamma values [87]. Moreover, we investigated the spatial association of anomalous zones and known hydrothermal mineral occurrences. The detailed methodology flowchart of this study is presented in Figure 3.

3.4. Field Survey

A field survey was planned for collecting samples from the detected hydrothermal alteration zones and verifying the results. We used a handheld global positioning system navigator (Garmin eTrex 10), with an accuracy of less than 15 m, for recording the coordinates of the samples. Overall, 55 rock samples were collected from different alteration zones and lithological units, for the microscopic studies and X-ray diffraction (XRD) analysis. The XRD analysis was carried out using Bruker AXS D8 Advance at the University of Tehran. The field data were used for selecting the appropriate transformation technique for enhancing each alteration type.

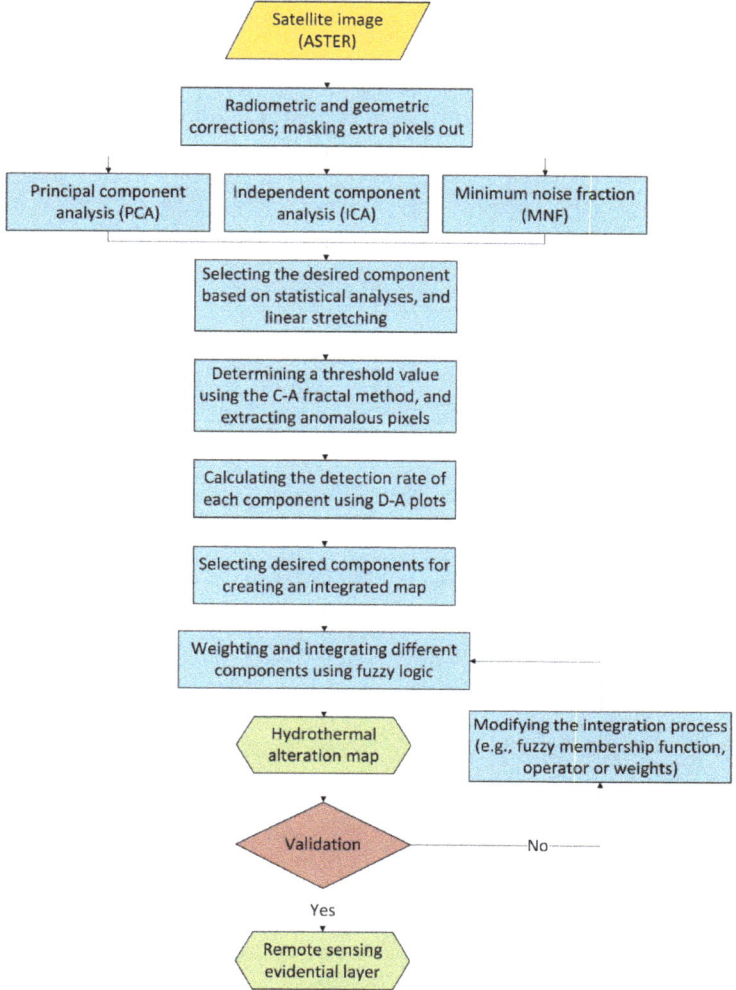

Figure 3. Methodology flowchart for generating a remote sensing evidential layer. We used ASTER remote sensing data as the input to this flowchart, which can be replaced with other types of satellite images.

4. Results

4.1. Hydrothermal Alteration Mapping Using PCA

The eigenvectors of each selective principal component analysis used for enhancing the gossans, argillic, propylitic, and phyllic alteration zones presented in Tables 1–4, respectively. The eigenvectors of each selective PCA are plotted in Figure 4. According to the spectral characteristics of the indicative minerals in the gossans shown in Figure 2a and the graphs presented in Figure 4a, negated PC 2 in PCA (1234) was considered for enhancing the gossans. This principal component showed a high negative loading in band 4 and an average constant and positive loading in bands 1–3. The relevant eigenvector showed a similar, but reverse trend, compared to the spectral graphs of the indicative minerals in gossans, such as jarosite. Therefore, the target areas appear in dark pixels in the original component, which had to be negated.

According to Table 2 and Figure 4b, the negated PC 4 in PCA (1457) was considered to be the component that could be used for enhancing the argillic alteration zones. In the relevant eigenvector, there was a large difference between the loadings in bands 5 and 7, and they showed opposite signs. This large difference was consistent with the spectral behavior of the indicative minerals in argillic alteration zones, such as alunite and kaolinite. The target areas appeared as dark pixels due to the reverse trend of the relevant eigenvector, compared to the target spectral behavior, thus the negated component was used for mapping the anomalous pixels. Based on the results presented in Table 3 and Figure 4c, the negated PC 4 in PCA (1458) enhanced the propylitic alteration zones. The relevant eigenvector to this PC showed a large difference between the loadings in bands 5 and 8, with opposite signs. This was similar to the spectral behavior of the indicative minerals in the propylitic alteration zones, such as chlorite and epidote. Similar to the reason mentioned above for the argillic alteration zones, the target zone in this component also appeared in dark pixels, and the negated component was used for enhancing the phyllic alteration zones.

According to Table 4 and Figure 4d, PC 3 in PCA (1467) was considered to be the component that could be used for enhancing the phyllic alteration zones. The relevant eigenvector showed a large difference between the loadings in bands 6 and 7, with opposite signs. Although PC 4 showed a larger difference, only the relevant eigenvector to PC 3 followed a similar trend to the spectral graphs of the indicative minerals in the phyllic alteration zones, such as muscovite. The anomalous areas in this component are displayed in bright pixels.

Table 1. Eigenvectors of the principal component analysis on a spectral subset of the ASTER data for detecting the gossans in the study area.

Eigenvectors	Band 1	Band 2	Band 3	Band 4
PC 1	0.341997	0.472648	0.499262	0.640608
PC 2	0.321681	0.413772	0.371346	−0.76643
PC 3	−0.680078	−0.19108	0.706279	−0.046394
PC 4	0.563075	−0.754244	0.337652	−0.007266

Table 2. Eigenvectors of the principal component analysis on a spectral subset of the ASTER data for detecting the argillic alteration zones in the study area.

Eigenvectors	Band 1	Band 4	Band 5	Band 7
PC 1	0.207235	0.505504	0.51344	0.66174
PC 2	0.947905	−0.240599	−0.203849	0.045107
PC 3	0.202174	0.78474	−0.113303	−0.574867
PC 4	0.132878	−0.266011	0.825825	−0.47916

Table 3. Eigenvectors of the principal component analysis on a spectral subset of the ASTER data for detecting the propylitic alteration zones in the study area.

Eigenvectors	Band 1	Band 4	Band 5	Band 8
PC 1	−0.264923	−0.649083	−0.657148	−0.276883
PC 2	−0.953077	0.22522	0.19047	−0.068121
PC 3	−0.050172	−0.725165	0.666135	0.166981
PC 4	−0.13763	−0.045865	−0.296886	0.943829

Table 4. Eigenvectors of the principal component analysis on a spectral subset of the ASTER data for detecting the phyllic alteration zones in the study area.

Eigenvectors	Band 1	Band 4	Band 6	Band 7
PC 1	0.18246	0.442498	0.657435	0.581965
PC 2	0.961565	−0.24002	−0.130325	0.028251
PC 3	0.120262	0.801479	−0.585627	0.014461
PC 4	−0.166245	−0.322827	−0.455889	0.812595

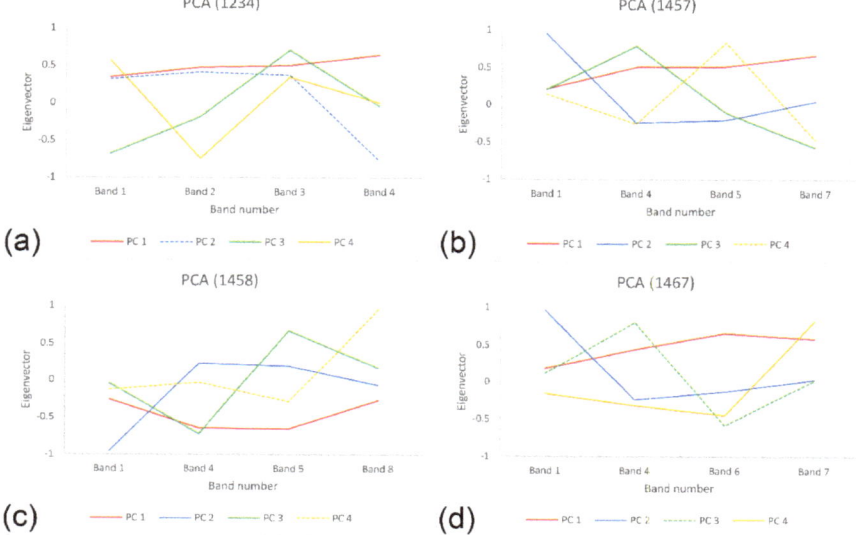

Figure 4. Trends of eigenvectors related to the selective principal component analysis on the ASTER spectral bands presented in Tables 1–4, are shown in (**a**–**d**), respectively.

The principal components obtained were in the form of grayscale images and needed to be converted into binary images, through mapping alteration zones. Based on the pixel values, C-A plots were generated on a logarithmic scale for each principal component selected, to enhance the different alteration zones. These plots are presented in Figure 5. The inflection points in these plots were considered to be the appropriate thresholds for separating the different populations, including background and anomaly. In Figure 6, we present the enhanced alteration zones, using PCA based on the C-A fractal method. The alteration zones were overlaid on the hillshade of the study area created by the ASTER digital elevation model.

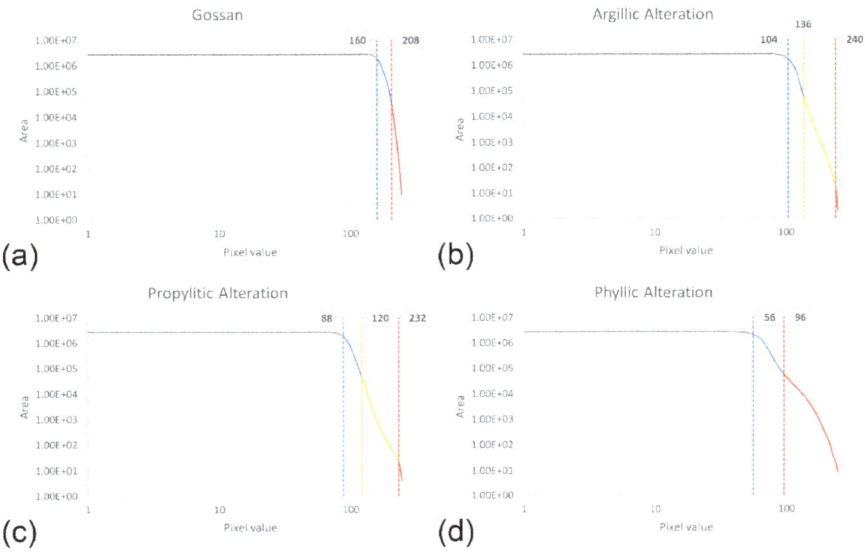

Figure 5. Logarithmic scaled plots of area versus the pixel values for the selected principal components for enhancing the (**a**) gossans, (**b**) argillic, (**c**) propylitic, and (**d**) the phyllic alteration zones.

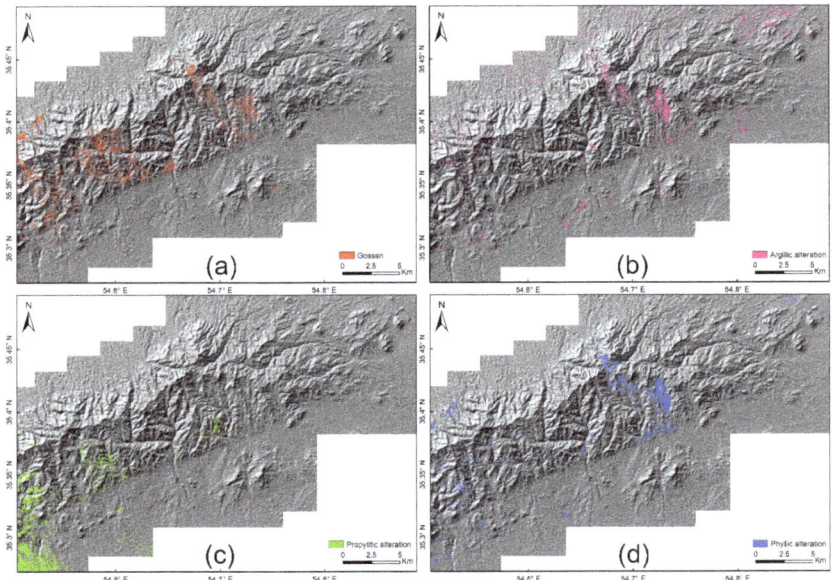

Figure 6. Anomalous pixels of different principal components determined using the C-A plots for enhancing (**a**) gossans, (**b**) argillic, (**c**) propylitic, and (**d**) phyllic alteration zones, overlaid on the hillshade of the study area.

4.2. Hydrothermal Alteration Mapping Using ICA

We applied the same spectral subsets used in PCA for independent component analysis and for enhancing the target alteration zones. Before performing the independent component forward calculation, we used the principal component rotation for data whitening with the same eigenvectors, as presented in Tables 1–4. Two main discrepancies for extracting components using ICA were: (i) There was no order of magnitude associated with each component in ICA. This meant that no better or worse component could be selected and other criteria such as two-dimensional (2D) spatial coherence might be considered by the user. (ii) The extracted components were invariant to the sign of the sources [88]. There are different ways to determine the most suitable IC for enhancing a target alteration zone. One way is to compare the spectral profile of both anomalous bright and dark pixels of each IC with the reference spectra. The other is to compare anomalous pixels of each IC with known criteria, such as the known color of each alteration zone, in specific false color composite images. For instance, argillic, propylitic, and phyllic alteration zones are displayed in pink, light green, and dark magenta, in the false color composite image created using bands 4, 6, and 8 of the ASTER data in red, green, and blue channels, respectively.

In this study, the independent components were sorted, based on the 2D spatial coherence, which is the average of the two correlation coefficients. One correlation coefficient was calculated between each spectral band and a version of itself, offset by one line. The other correlation coefficient was calculated between each spectral band and a version of itself, offset by one sample. Using the 2D spatial coherence sorting, independent components that contained the spatial structure and most of the information, appeared first, and those that contained little spatial structure and more noise appeared last. Based on these results, negated IC 2, IC 3, negated IC 2, and negated IC 3 were recognized as the most suitable components for enhancing gossans, argillic, propylitic, and phyllic alteration zones, respectively. As shown in Figure 7, the appropriate threshold for separating the anomalous pixels of the selected ICs were determined using the C-A fractal plots. The enhanced alteration zones using the ICA and based on the C-A fractal plots are presented in Figure 8.

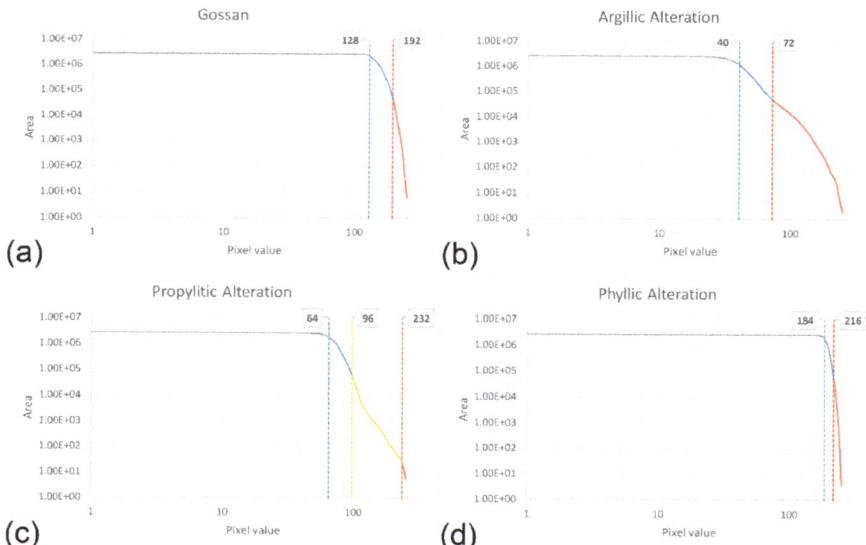

Figure 7. Logarithmic scaled plots of area versus pixel values for the selected independent components for enhancing the (**a**) gossans, (**b**) argillic, (**c**) propylitic, and (**d**) phyllic alteration zones.

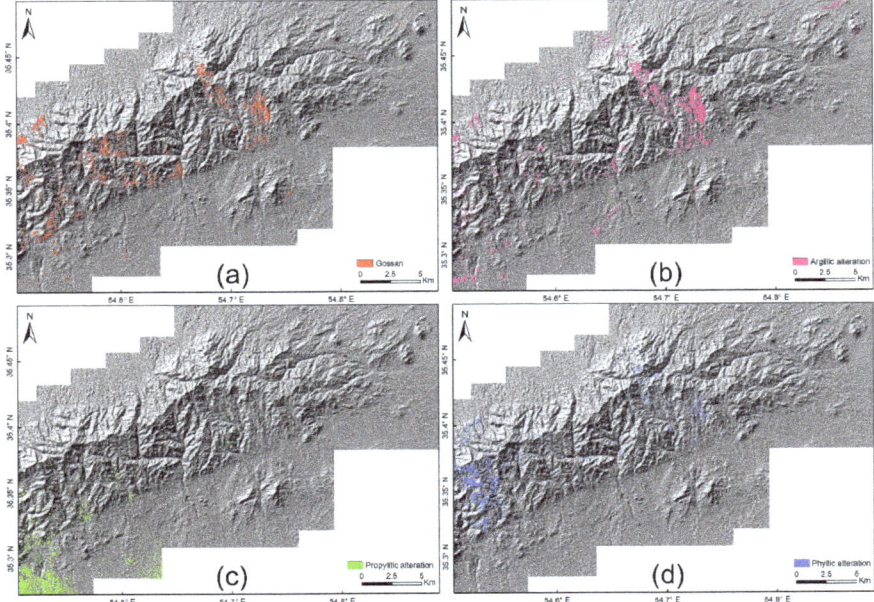

Figure 8. Anomalous pixels of different independent components determined using the C-A plots for enhancing the (**a**) gossans, (**b**) argillic, (**c**) propylitic, and (**d**) phyllic alteration zones, overlaid on the hillshade of the study area.

4.3. Hydrothermal Alteration Mapping Using MNF

We used identical spectral subsets to the PCA and ICA as input to this technique. We presented the transformation vectors of each selective MNF analysis used for enhancing the gossans, argillic, propylitic, and phyllic alteration zones in Tables 5–8, respectively. The transformation vectors of each analysis are shown in Figure 9. According to the spectral characteristics of the indicative minerals in gossans (shown in Figures 2a and 9a), the second component (C 2) in MNF (1234) was considered for enhancing gossans. This component showed a relatively similar trend to the spectral graphs of the indicative minerals in gossans. Therefore, the target areas appeared in bright pixels in this component.

Table 5. Transformation vectors of minimum noise fraction (MNF) analysis on a spectral subset of the ASTER data for detecting gossans in the study area.

Transformation Vectors	Band 1	Band 2	Band 3	Band 4
C 1	0.001754	−0.00002	0.000029	−0.01711
C 2	−0.01966	−0.00557	0.007738	0.005621
C 3	−0.024987	0.004426	0.018885	−0.00362
C 4	0.022636	−0.033569	0.017022	−0.000533

Table 6. Transformation vectors of MNF analysis on a spectral subset of the ASTER data for detecting the argillic alteration zones in the study area.

Transformation Vectors	Band 1	Band 2	Band 3	Band 4
C 1	0.001734	−0.013649	−0.011389	0.005532
C 2	−0.005871	0.034231	−0.025712	−0.005107
C 3	0.014524	0.007016	−0.019229	0.005719
C 4	−0.008015	−0.004828	−0.015279	0.018378

Table 7. Transformation vectors of MNF analysis on a spectral subset of the ASTER data for detecting the propylitic alteration zones in the study area.

Transformation Vectors	Band 1	Band 2	Band 3	Band 4
C 1	−0.001556	0.016958	0.013699	−0.028643
C 2	−0.006848	0.031257	−0.016662	−0.03121
C 3	−0.009486	−0.015755	0.028617	−0.024276
C 4	−0.013196	−0.00112	−0.012047	0.044362

Table 8. Transformation vectors of MNF analysis on a spectral subset of the ASTER data for detecting the phyllic alteration zones in the study area.

Transformation Vectors	Band 1	Band 2	Band 3	Band 4
C 1	0.001676	−0.01846	−0.003035	0.004273
C 2	0.004374	−0.032869	0.020412	0.001996
C 3	0.016131	0.003897	−0.010992	0.004757
C 4	0.005622	0.007379	0.010819	−0.019848

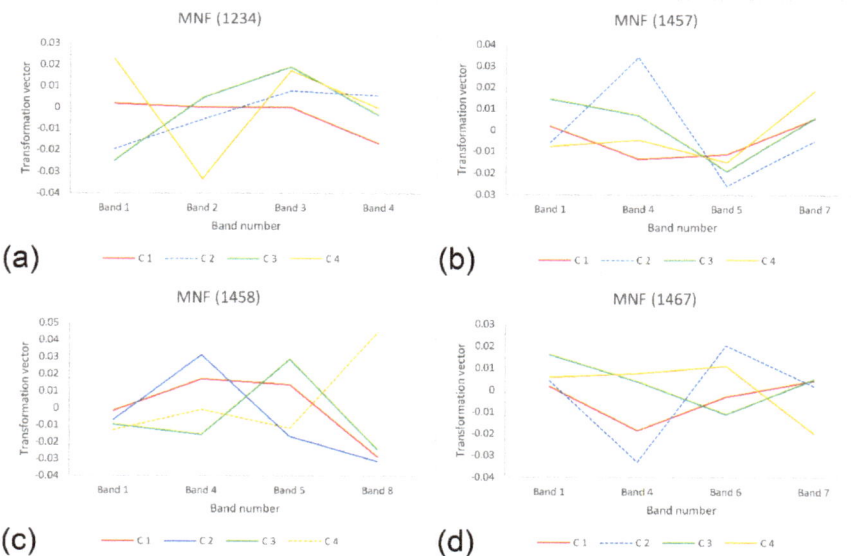

Figure 9. Trends of transformation vectors related to the selective MNF analysis on the ASTER spectral bands presented in Tables 5–8, are shown in (**a**–**d**), respectively.

According to Table 6 and Figure 9b, we considered C 2 in MNF (1457) as the appropriate component for enhancing argillic alteration zones. In the relevant transformation vector, there was a relatively high difference between the loadings in bands 5 and 7. This difference was consistent with the spectral behavior of the indicative minerals in the argillic alteration zones, such as alunite and kaolinite. The target areas appeared in bright pixels in this component. Based on the results in Table 7 and Figure 9c, negated C 4 in MNF (1458) enhanced the propylitic alteration zones. The relevant transformation vector to this component showed a relatively high difference between the loadings in bands 5 and 8, with opposite signs. This was similar to the spectral behavior of the indicative minerals in propylitic alteration zones, such as chlorite and epidote. The target zone in this component appeared in dark pixels, and the negated component was used for enhancing the phyllic alteration zones.

According to Table 8 and Figure 9d, we considered negated C 2 in MNF (1467) as the component that could be used for enhancing the phyllic alteration zones. The relevant transformation vector showed a relatively large difference between the loadings in bands 6 and 7, with opposite signs. Although C 4 showed a higher difference, only the relevant transformation vector to C 2 followed a similar trend to the spectral graphs of the indicative minerals of phyllic alteration zones, such as muscovite. The anomalous areas in this component are displayed in dark pixels.

As shown in Figure 10, we determined the appropriate thresholds for separating anomalous pixels of the selected MNF components, using the C-A fractal method. We present the enhanced alteration zones using the MNF analysis based on the C-A fractal method in Figure 11.

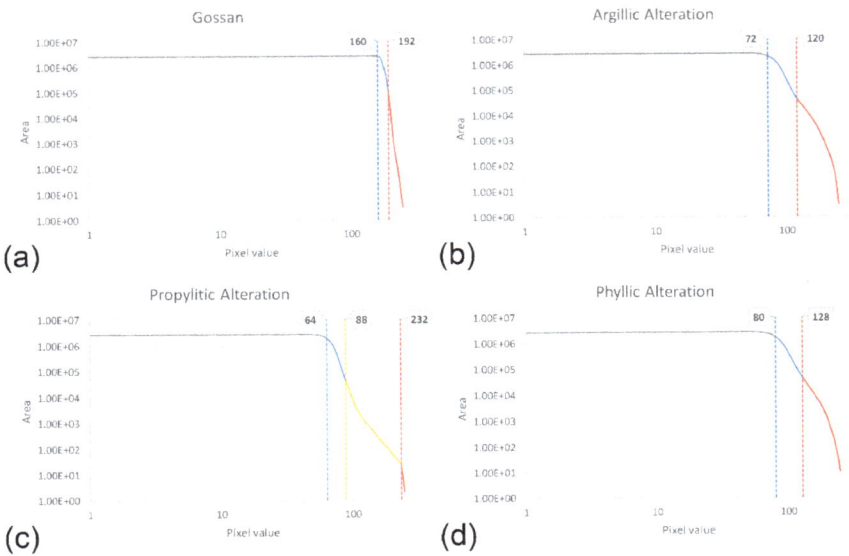

Figure 10. Logarithmic scaled plots of area versus pixel values for the selected MNF components for enhancing the (**a**) gossans, (**b**) argillic, (**c**) propylitic, and (**d**) phyllic alteration zones.

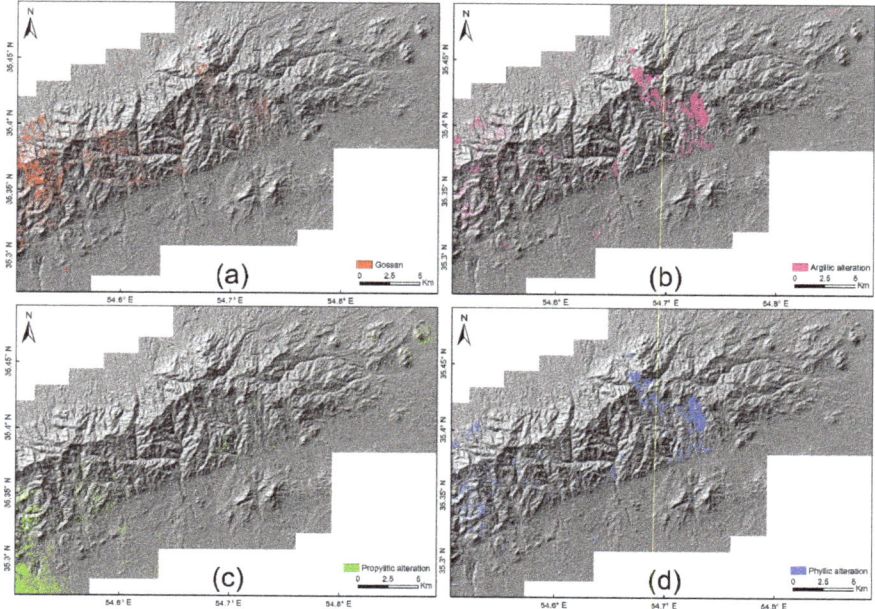

Figure 11. Anomalous pixels of different MNF components determined using the C-A plots for enhancing the (**a**) gossans, (**b**) argillic, (**c**) propylitic, and (**d**) phyllic alteration zones, overlaid on the hillshade of the study area.

4.4. Field Data and Laboratory Analysis

A comprehensive geological fieldwork was carried out in the study area, particularly in the alteration zones detected using the applied transformation techniques. The photos taken from different alteration types, such as phyllic, argillic, propylitic, and gossan are presented in Figure 12. We collected 55 rock samples from the prospects and used some of them for creating thin sections and the rest were sent for the XRD analysis (Figure 15). We carried out petrographic studies on the thin sections shown in Figure 13, which indicated the transformation of primary silicate minerals (feldspars) such as plagioclase to secondary altered minerals (calcite, clay minerals, epidote, and sericite). The opaque minerals constituted a notable part of the thin sections created using the rock samples collected from argillic and phyllic alteration zones. In the propylitic zone, the original minerals were fully replaced with secondary minerals (calcite and epidote). The minerals identified using the XRD analysis shown in Figure 14, included montmorillonite, illite, goethite, hematite, muscovite, albite, orthoclase, and quartz in the argillic zone; epidote, calcite, chlorite, albite, anorthite, and quartz in the propylitic zone; muscovite, illite, hematite, magnetite, albite, epidote, calcite, montmorillonite, and quartz in the phyllic zone; and goethite, hematite, kaolinite, muscovite, illite, and quartz in gossans.

The results using the XRD analysis indicated that most of the diagnostic spectral features in the argillic alteration zones were due to the presence of montmorillonite and illite; in the propylitic zone these were associated with chlorite, epidote, and calcite; in the phyllic zone these were derived from muscovite; and in gossans these were related to goethite and hematite. Moreover, we found that iron oxide or hydroxide minerals were associated with the alteration mineral assemblages in the argillic, propylitic, and phyllic alteration zones.

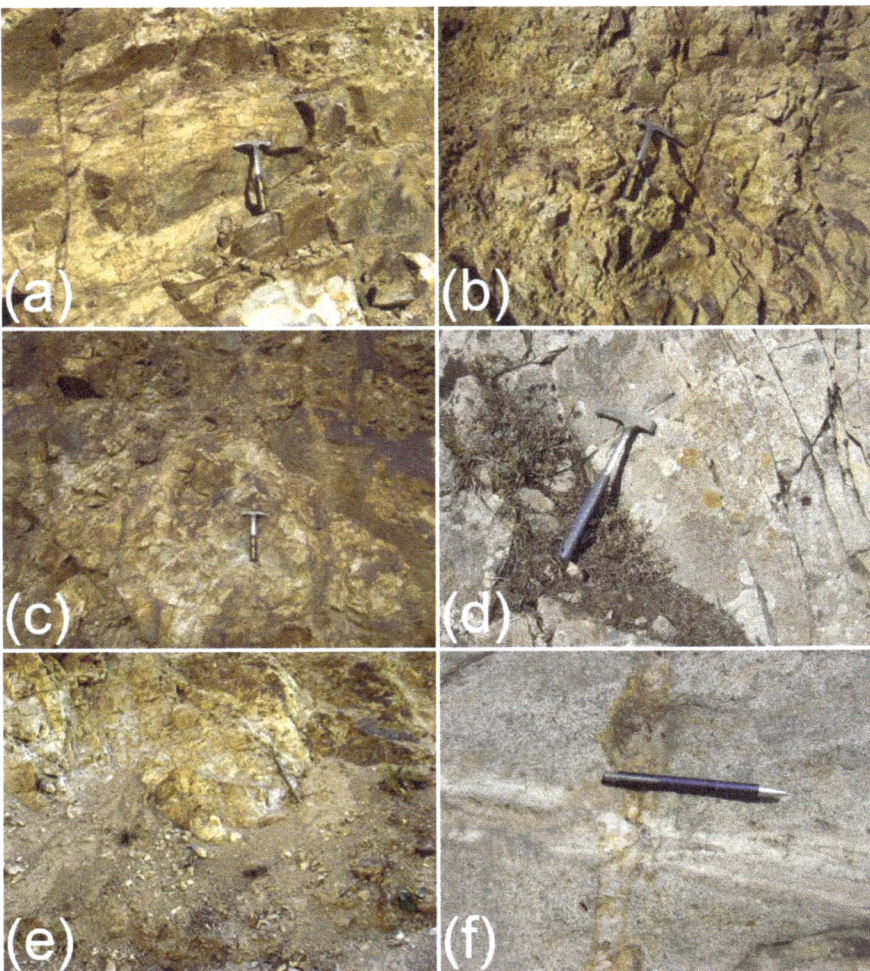

Figure 12. Photos taken from different alteration types including (**a**) phyllic alteration, (**b**) phyllic alteration associated with clay minerals known as the indicators of argillic alteration, (**c**) phyllic alteration associated with iron oxide and hydroxide minerals, (**d**) propylitic alteration, and (**e**) gossan. (**f**) Close view of a vein-type mineralization hosted by a silicified rock.

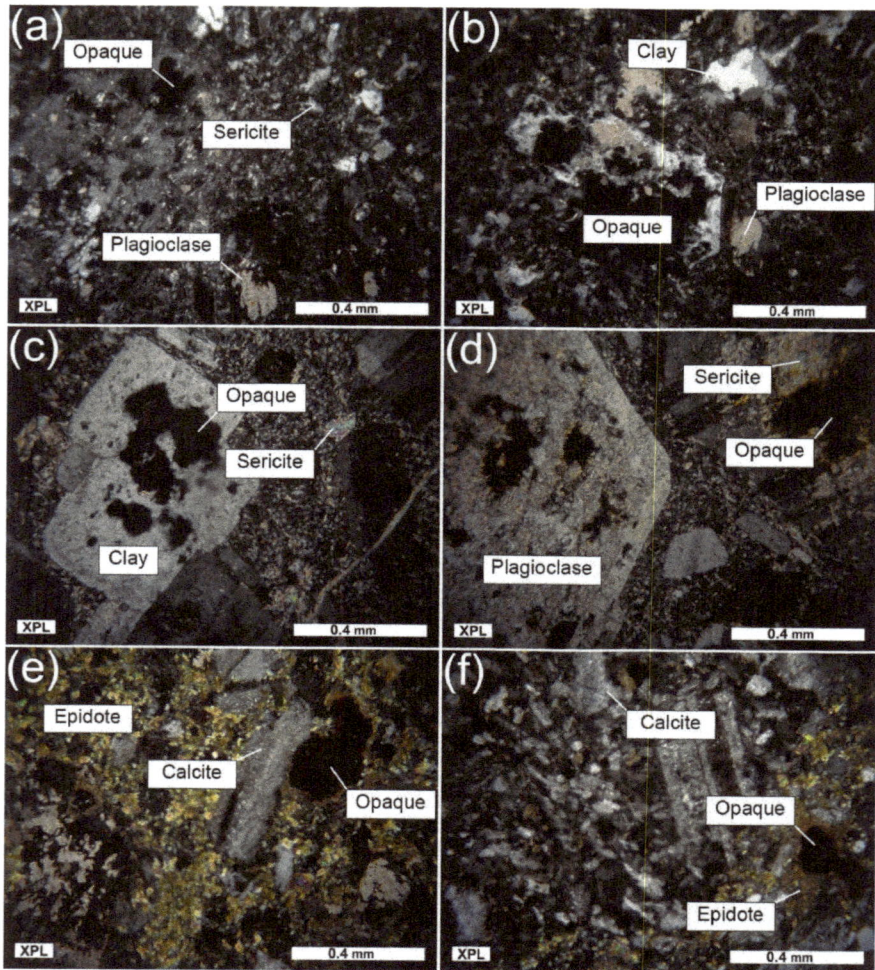

Figure 13. Thin sections of different types of alteration mineralogy. Microphotographs of (**a,b**) argillic alteration zone—plagioclase replaced with sericite and clay mineral groups; (**c,d**) phyllic zone—opaque minerals and plagioclase replaced with clay mineral groups and quartz; and (**e,f**) propylitic zone—completely replaced original mineralogy by calcite, epidote, chlorite, and quartz.

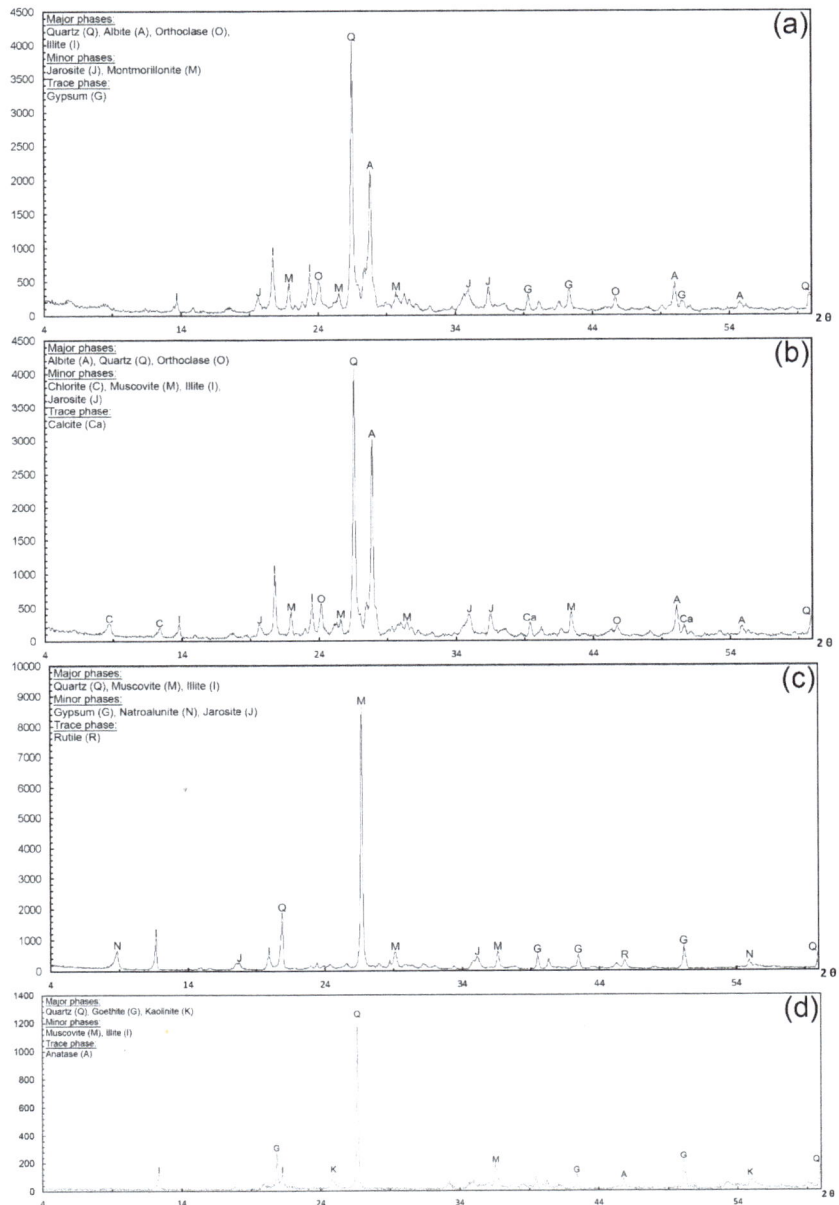

Figure 14. Results of the XRD analysis showing the indicative minerals of the representative samples collected from (**a**) argillic zone; (**b**) propylitic zone; (**c**) phyllic zone; and (**d**) gossans.

4.5. Integration of Selected Components

Fifty-five rock samples that were collected from the study area (shown in Figure 15) were used to generate the confusion matrix of each transformation technique (Table 9). The component with the highest number of correctly classified samples was selected for enhancing each alteration type. These components were integrated by generating the remote sensing evidential layer. Moreover, the overall accuracy of each transformation technique is presented in Table 9, which was the ratio of the total number of correctly classified samples in each confusion matrix to the total number of samples. The overall accuracy was used to assess the accuracy of each transformation technique. This ratio had a clear meaning and was simple to estimate [89].

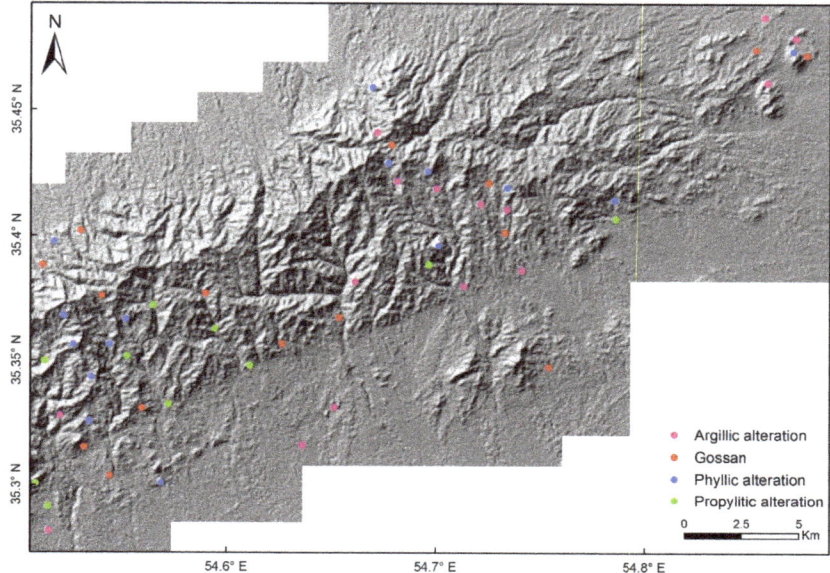

Figure 15. Samples collected from different alteration zones overlaid on the hillshade of the study area.

Table 9. Confusion matrices of the transformation techniques used for enhancing different alteration types.

PCA	Ground Truth			
	Gossan	Argillic	Propylitic	Phyllic
Gossan	14	1	0	0
Argillic	1	11	0	3
Propylitic	1	0	9	0
Phyllic	0	2	0	13
Overall accuracy		0.85		
ICA				
Gossan	13	1	0	1
Argillic	0	13	0	2
Propylitic	2	0	8	0
Phyllic	0	1	0	14
Overall accuracy		0.87		
MNF				
Gossan	12	2	0	1
Argillic	0	14	0	1
Propylitic	3	0	7	0
Phyllic	0	3	0	12
Overall accuracy		0.81		

We experimented with different gamma values for integrating the weighted components selected for enhancing different alteration types. Based on the prediction rates obtained by the P-A plots, gamma 0.9 was used for generating the remote sensing evidential layer. Using a P-A plot, we showed the cumulative percentage of predicted mineral occurrences and the corresponding cumulative occupied area, with respect to the total area, against the pixel values. Therefore, the prediction ability of the integrated map and its ability to delimit the exploration area for further investigation was evaluated in a scheme. The P-A plot showed a curve of the percentage (prediction rate) of known mineralization and a curve of the percentage of the occupied area corresponding to the classes of a map [90]. When an intersection point of the two curves was at a higher place, it portrayed a small area containing a large number of known mineral occurrences. The prediction rate in the P-A plots helped analyze the efficiency and association of a map in predicting target mineralization. We presented the P-A plot of the integrated map obtained using gamma 0.9 in Figure 16, which showed a higher prediction rate compared to other gamma values. According to the plot, the integrated map was able to predict 70% of the mineral occurrences in 30% of the study area. It is noteworthy that for assigning probabilistic values to the map, in terms of prospecting for the hydrothermal mineralization and distribution of the pixel values between 0 and 1, these were transformed to a fuzzy space using a linear function. We used the C-A fractal plot for classifying the integrated map and separating the anomaly population; shown in Figure 17. The classified integrated map which was suggested to be used as the remote sensing evidential layer, was presented in Figure 18a. The red-colored class could be considered to be the certain anomaly population. Moreover, we presented another map classified using the Jenks Natural Breaks [91] in Figure 18b, to provide a higher number of classes.

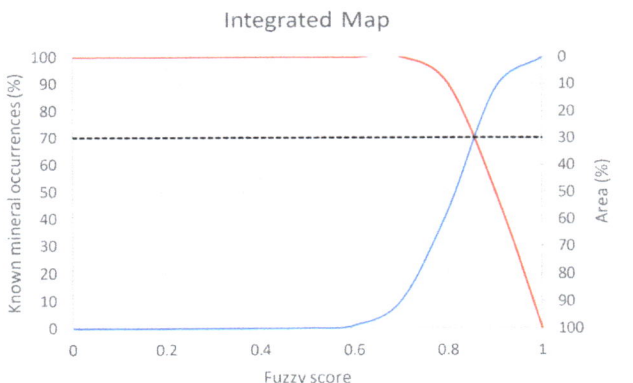

Figure 16. P-A plot of the integrated map obtained using fuzzy gamma operator (gamma = 0.9).

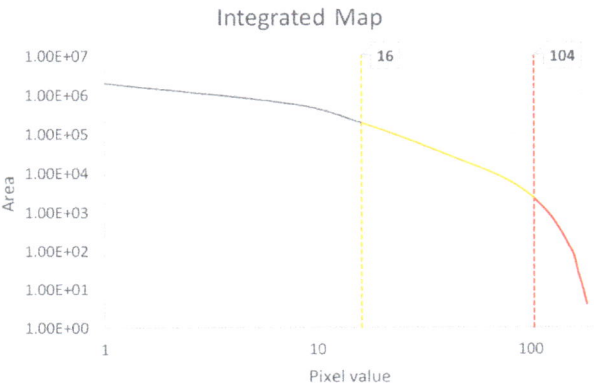

Figure 17. C-A plot of the integrated map used for classifying and separating the anomaly population.

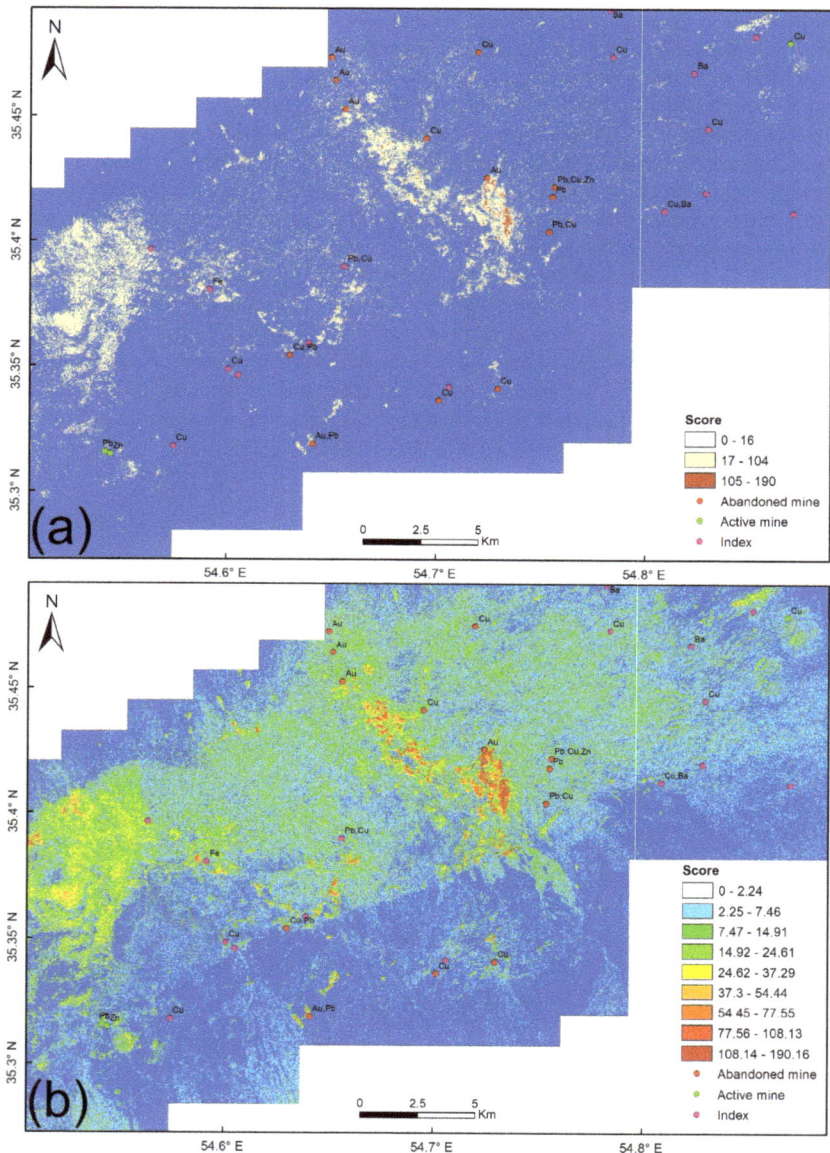

Figure 18. Known hydrothermal mineral occurrences overlaid on the classified integrated maps by (**a**) the C-A plot, and (**b**) Natural Jenks Breaks, generated by integrating the weighted-selected components.

5. Discussion

In this study, we applied different dimensionality reduction or transformation techniques on the ASTER data for mapping the gossans and hydrothermal alteration zones in a mineral-rich range located in Central Iran called Toroud-Chahshirin. It is thought that subpopulation samples, such as hydrothermal alteration zones, are more distinguishable using the components derived from the transformation techniques, such as PCA, ICA, and MNF. Thus, these components might provide clear geological meanings for interpretation. However, principal and independent components do not genetically correspond to distinct geological features, due to the intimate mixing of geological units or alteration types. The transformation techniques provided exploratory tools to view data from another perspective. We used this type of techniques to provide useful information, based on the knowledge of actual geological problems.

Principal component analysis is an orthogonal decomposition and we used it based on covariance matrix analysis and the Gaussian assumption, while independent component analysis is based on a non-Gaussian assumption of independent sources. The PCA and MNF use only second-order statistics, while the ICA uses higher-order statistics. Higher-order statistics is a stronger statistical assumption that reveals interesting features in the usually non-Gaussian datasets. If the feature of interest, such as a hydrothermal alteration zone only occupies a small portion of all pixels, it makes an insignificant contribution to the covariance matrix. In the PCA, the feature of interest would be probably buried in the noisy bands, while in ICA and MNF, the features are distinguished from the noisy bands. Based on the spectral characteristics of the indicative minerals of each hydrothermal alteration zone, we selected four spectral bands with diagnostic absorption and reflection characteristics as the input to the transformation techniques. The selective approach was preferred to the standard approach, due to the ease of interpreting the results based on the characteristic spectral behavior of each target zone. In PCA and MNF, we used the eigenvectors and transformation vectors for determining the appropriate component for detecting each alteration zone. In the ICA, we determined the target components based on the spectral behavior of known alteration zones.

We used the C-A plots for discriminating between the anomalous pixels and the background in each component were determined to be appropriate for enhancing the target zones. The maps created using the PCA and ICA for enhancing the gossans, showed a similar pattern, along with the number of correctly classified samples, whereas the ICA component showed a less noisy pattern. The MNF technique yielded a less accurate map and interference of vegetation and iron oxide/hydroxide-bearing pixels could be observed. The PCA failed to provide an acceptable map for enhancing the argillic alteration zones and the relevant component showed a noisy pattern and a low number of correctly classified samples. The ICA and MNF components showed a relatively similar pattern for the argillic alteration zones, whereas the ICA provided a less noisy pattern. The respective transformations applied in this study, showed a relatively similar number of correctly classified samples and pattern for the propylitic alteration zones. However, the MNF component was noisier compared to the other components. The ICA and MNF components provided a similar map for enhancing the phyllic alteration zones, while the PCA yielded a noisy map. Nevertheless, the number of correctly classified samples for all components was almost the same.

According to the field observations and Table 9, the applied techniques and to a greater extent, ICA, efficiently revealed the alteration halos. The components derived from the ICA showed the highest overall accuracy. The results indicated that the ICA has a great ability for providing comprehensive and significant exploration information at a regional scale. In general, the study area was dominantly covered by advanced argillic, argillic, and phyllic alteration zones. Additionally, gossans and propylitic alteration zones covered large parts of the study area. The argillic and phyllic alteration zones, along with gossans were mostly focused in the central and western portions of the study area. The propylitic alteration zone was located in the southwestern portion of the study area. Typically, gossans and phyllic alteration zones showed the closest spatial relationship in many parts of the study area with the hydrothermal mineralization. According to the results, there were a number of pixels which showed anomalous values for different alteration zones. This confirmed the presence of alteration minerals and their spatial distribution at the subpixel level in the study area.

According to the geological map presented in Figure 1, the gossans detected using the ICA were correlated with andesitic and dacitic units. The argillic alteration zones were associated with andesitic and dacitic units in central and northwestern, and the rhyolitic units in the southwestern portions of the study area. The propylitic alteration zones located mostly in the southwest of the study area were related to andesitic and rhyolitic units. The phyllic alteration zones in the central portion of the study area were associated with the andesitic and dacitic units, and were correlated with the dioritic units in the western portion.

Based on the number of correctly classified samples presented in Table 9, we selected the appropriate technique for enhancing each alteration type. The PCA components were selected for enhancing the gossans and propylitic alteration, and we considered the ICA and MNF components for enhancing phyllic and argillic alteration, respectively. The integrated map generated by combining the weighted selected components using the fuzzy gamma operator (see Figure 18) showed an acceptable prediction rate, based on the relevant P-A plot (see Figure 16). The results of fieldwork and XRD analysis verified the presence of gossan, argillic, phyllic, and propylitic zones in the study area. Distinctive spectral features related to montmorillonite, illite, chlorite, epidote, calcite, muscovite, goethite, and hematite were found in the alteration zones. Moreover, new prospective areas were detected in the western portions of the study area, along with the ring structures in northeast. The central portion of the study area was a high-altered zone associated with different metallic ore deposits. This map could be used as an efficient evidential layer through mineral prospectivity mapping of the study area. This methodology could be extrapolated to the unexplored regions for identifying new prospects of high-potential base metal mineralization zones in Central Iran and other arid or semi-arid regions on earth.

6. Conclusions

This study compared the efficiency of different dimensionality reduction or transformation techniques in terms of enhancing various types of the hydrothermal alteration in the Toroud-Chahshirin range located in Central Iran, using ASTER satellite data. Moreover, a framework was proposed for selecting and integrating the appropriate techniques through enhancing each alteration type, which led to generating a reliable remote sensing evidential layer. In this framework, a selective approach was used for implementing the transformation techniques based on the spectral characteristics of the indicative minerals in each alteration zone. Based on the field observations, we used the number of correctly classified rock samples for investigating the accuracy of each component in detecting different types of hydrothermal alteration zones. In parts of the study area, different alteration types were collocated and the results confirmed the presence of alteration minerals and their spatial distribution at the subpixel level. In general, ICA provided more accurate and less noisy maps, compared to the other techniques. We selected the components with the highest number of correctly classified samples for the integration process by providing a remote sensing evidential layer. The fuzzy gamma operator was used for generating an integrated map based on the components derived from the transformation techniques. The integrated map showed a high prediction rate that implied the efficiency of the proposed framework. Moreover, this map was consistent with the results of petrographic and XRD analysis. Using the integrated map, high potential zones of the hydrothermal mineralization were identified in the study area, particularly in the western and northeastern portions, which could be considered for future systematic exploration programs. The methodology used in this study could be applied for mapping hydrothermal alteration zones in other metallogenic provinces in the arid and semi-arid regions around the world.

Author Contributions: H.S. and E.F. writing—original draft preparation, software, analysis, validation; A.B.P. writing, reconstructing—review, R.D.M. and R.C. editing and supervision; A.M.M. supervision. All authors have read and agreed to the published version of the manuscript.

Funding: Publication fees are not claimed by Remote Sensing because A.B.P is one of the Editorial Board of Remote Sensing (Section of Environmental Remote Sensing). He has the right to publish one paper per year free of charge as he serves the Journal.

Acknowledgments: The authors would like to acknowledge the School of Mining Engineering at the University of Tehran for providing required field and laboratory data for this investigation. We appreciate the Department of Mining Engineering at the Amirkabir University of Technology for logistics support. Also, we are grateful to the Institute of Oceanography and Environment (INOS) and the University Malaysia Terengganu (UMT) for the assigned time and providing required facilities during writing–editing and reconstructing the manuscript.

Conflicts of Interest: The authors declare no conflict of interest.

References

1. Cooke, D.R.; Deyell, C.L.; Waters, P.J.; Gonzales, R.I.; Zaw, K. Evidence for magmatic-hydrothermal fluids and ore-forming processes in epithermal and porphyry deposits of the Baguio District, Philippines. *Econ. Geol.* **2011**, *106*, 1399–1424. [CrossRef]
2. Koděra, P.; Lexa, J.; Fallick, A.E.; Wälle, M.; Biroň, A. Hydrothermal fluids in epithermal and porphyry Au deposits in the Central Slovakia Volcanic Field. *Geol. Soc. London Spec. Publ.* **2014**, *402*, 177–206. [CrossRef]
3. Smith, D.J.; Naden, J.; Jenkin, G.R.T.; Keith, M. Hydrothermal alteration and fluid pH in alkaline-hosted epithermal systems. *Ore. Geol. Rev.* **2017**, *89*, 772–779. [CrossRef]
4. Takahashi, R.; Tagiri, R.; Blamey, N.J.F.; Imai, A.; Watanabe, Y.; Takeuchi, A. Characteristics and behavior of hydrothermal fluids for gold mineralization at the Hishikari Deposits, Kyushu, Japan. *Resour. Geol.* **2017**, *67*, 279–299. [CrossRef]
5. Bérubé, C.L.; Olivo, G.R.; Chouteau, M.; Perrouty, S.; Shamsipour, P.; Enkin, R.J.; Morris, W.A.; Feltrin, L.; Thiémonge, R. Predicting rock type and detecting hydrothermal alteration using machine learning and petrophysical properties of the Canadian Malartic ore and host rocks, Pontiac Subprovince, Québec, Canada. *Ore Geol. Rev.* **2018**, *96*, 130–145. [CrossRef]
6. Mathieu, L. Quantifying hydrothermal alteration: A review of methods. *Geosciences* **2018**, *8*, 245. [CrossRef]
7. Perrouty, S.; Linnen, R.L.; Lesher, C.M.; Olivo, G.R.; Piercey, S.J.; Gaillard, N.; Clark, J.R.; Enkin, R.J. Expanding the size of multi-parameter metasomatic footprints in gold exploration: Utilization of mafic dykes in the Canadian Malartic district, Québec, Canada. *Miner. Depos.* **2019**, *54*, 761–786. [CrossRef]
8. Simpson, M.P.; Christie, A.B. Hydrothermal alteration mineralogical footprints for New Zealand epithermal Au-Ag deposits. *N. Z. J. Geol. Geophys.* **2019**, *62*, 483–512. [CrossRef]
9. Rowan, L.C.; Hook, S.J.; Abrams, M.J.; Mars, J.C. Mapping hydrothermally altered rocks at Cuprite, Nevada, using the Advanced Spaceborne Thermal Emission and Reflection Radiometer (ASTER), a new satellite-imaging system. *Econ. Geol.* **2003**, *98*, 1019–1027. [CrossRef]
10. Van der Meer, F.D.; van der Werff, H.M.A.; van Ruitenbeek, F.J.A.; Hecker, C.A.; Bakker, W.H.; Noomen, M.F.; van der Meijde, M.; Carranza, E.J.M.; de Smeth, J.B.; Woldai, T. Multi- and hyperspectral geologic remote sensing: A review. *Int. J. Appl. Earth Obs. Geoinf.* **2012**, *14*, 112–128. [CrossRef]
11. Beiranvand Pour, A.; S Park, T.-Y.; Park, Y.; Hong, J.K.; M Muslim, A.; Läufer, A.; Crispini, L.; Pradhan, B.; Zoheir, B.; Rahmani, O. Landsat-8, advanced spaceborne thermal emission and reflection radiometer, and WorldView-3 multispectral satellite imagery for prospecting copper-gold mineralization in the Northeastern Inglefield Mobile Belt (IMB), Northwest Greenland. *Remote Sens.* **2019**, *11*, 2430. [CrossRef]
12. Abrams, M.J.; Ashley, R.P.; Rowan, L.C.; Goetz, A.F.H.; Kahle, A.B. Mapping of hydrothermal alteration in the Cuprite mining district, Nevada, using aircraft scanner images for the spectral region 0.46 to 2.36 µm. *Geology* **1977**, *5*, 713–718. [CrossRef]
13. Abrams, M.J.; Brown, D.; Lepley, L.; Sadowski, R. Remote sensing for porphyry copper deposits in southern Arizona. *Econ. Geol.* **1983**, *78*, 591–604. [CrossRef]
14. Rowan, L.C.; Mars, J.C. Lithologic mapping in the Mountain Pass, California area using Advanced Spaceborne Thermal Emission and Reflection Radiometer (ASTER) data. *Remote Sens. Environ.* **2003**, *84*, 350–366. [CrossRef]
15. Perry, S.L. Spaceborne and airborne remote sensing systems for mineral exploration-case histories using infrared spectroscopy. *Infrared Spectrosc. Geochem. Explor. Geochem. Remote Sens.* **2004**, *33*, 227–240.
16. Mars, J.C.; Rowan, L.C. Regional mapping of phyllic- and argillic-altered rocks in the Zagros magmatic arc, Iran, using advanced spaceborne thermal emission and reflection radiometer (ASTER) data and logical operator algorithms. *Geosphere* **2006**, *2*, 161–186. [CrossRef]
17. Di Tommaso, I.; Rubinstein, N. Hydrothermal alteration mapping using ASTER data in the Infiernillo porphyry deposit, Argentina. *Ore Geol. Rev.* **2007**, *32*, 275–290. [CrossRef]
18. Beiranvand Pour, A.; Hashim, M.; Hong, J.K.; Park, Y. Lithological and alteration mineral mapping in poorly exposed lithologies using Landsat-8 and ASTER satellite data: North-eastern Graham Land, Antarctic Peninsula. *Ore Geol. Rev.* **2019**, *108*, 112–133. [CrossRef]

19. Ahmadirouhani, R.; Karimpour, M.-H.; Rahimi, B.; Malekzadeh-Shafaroudi, A.; Pour, A.B.; Pradhan, B. Integration of SPOT-5 and ASTER satellite data for structural tracing and hydrothermal alteration mineral mapping: Implications for Cu–Au prospecting. *Int. J. Image Data Fusion* **2018**, *9*, 237–262. [CrossRef]
20. Sheikhrahimi, A.; Pour, A.B.; Pradhan, B.; Zoheir, B. Mapping hydrothermal alteration zones and lineaments associated with orogenic gold mineralization using ASTER data: A case study from the Sanandaj-Sirjan Zone, Iran. *Adv. Sp. Res.* **2019**, *63*, 3315–3332. [CrossRef]
21. Beiranvand Pour, A.; Park, Y.; Crispini, L.; Läufer, A.; Kuk Hong, J.; Park, T.-Y.S.; Zoheir, B.; Pradhan, B.; Muslim, A.M.; Hossain, M.S. Mapping listvenite occurrences in the damage zones of Northern Victoria Land, Antarctica using ASTER satellite remote sensing data. *Remote Sens.* **2019**, *11*, 1408. [CrossRef]
22. Bolouki, S.M.; Ramazi, H.R.; Maghsoudi, A.; Beiranvand Pour, A.; Sohrabi, G. A remote sensing-based application of Bayesian networks for epithermal gold potential mapping in Ahar-Arasbaran area, NW Iran. *Remote Sens.* **2020**, *12*, 105. [CrossRef]
23. Crósta, A.P.; De Souza Filho, C.R.; Azevedo, F.; Brodie, C. Targeting key alteration minerals in epithermal deposits in Patagonia, Argentina, using ASTER imagery and principal component analysis. *Int. J. Remote Sens.* **2003**, *24*, 4233–4240. [CrossRef]
24. Beiranvand Pour, A.; Hashim, M. Application of advanced spaceborne thermal emission and reflection radiometer (ASTER) data in geological mapping. *Int. J. Phys. Sci.* **2011**, *6*, 7657–7668.
25. Farahbakhsh, E.; Shirmard, H.; Bahroudi, A.; Eslamkish, T. Fusing ASTER and QuickBird-2 satellite data for detailed investigation of porphyry copper deposits using PCA; case study of Naysian deposit, Iran. *J. Indian Soc. Remote Sens.* **2016**, *44*, 525–537. [CrossRef]
26. Rowan, L.; Goetz, A.; Ashley, R. Discrimination of hydrothermally altered and unaltered rocks in visible and near infrared multispectral images. *Geophysics* **1977**, *42*, 522–535. [CrossRef]
27. Goetz, A.F.H.; Rock, B.N.; Rowan, L.C. Remote sensing for exploration; an overview. *Econ. Geol.* **1983**, *78*, 573–590. [CrossRef]
28. Boardman, J.W.; Kruse, F.A. Automated spectral analysis: A geological example using AVIRIS data, north Grapevine Mountains, Nevada. In *Proceedings of the 10th Thematic Conference on Geologic Remote Sensing*; Environmental Research Institute of Michigan: Ann Arbor, MI, USA, 1994; pp. 1407–1418.
29. Kruse, F.A.; Boardman, J.W.; Huntington, J.F. Comparison of airborne hyperspectral data and EO-1 Hyperion for mineral mapping. *IEEE Trans. Geosci. Remote Sens.* **2003**, *41*, 1388–1400. [CrossRef]
30. Serkan Öztan, N.; Lütfi Süzen, M. Mapping evaporate minerals by ASTER. *Int. J. Remote Sens.* **2011**, *32*, 1651–1673. [CrossRef]
31. Testa, F.J.; Villanueva, C.; Cooke, D.R.; Zhang, L. Lithological and hydrothermal alteration mapping of epithermal, porphyry and tourmaline breccia districts in the Argentine Andes using ASTER imagery. *Remote Sens.* **2018**, *10*, 203. [CrossRef]
32. Singh, A.; Harrison, A. Standardized principal components. *Int. J. Remote Sens.* **1985**, *6*, 883–896. [CrossRef]
33. Maćkiewicz, A.; Ratajczak, W. Principal components analysis (PCA). *Comput. Geosci.* **1993**, *19*, 303–342. [CrossRef]
34. Eklundh, L.; Singh, A. A comparative analysis of standardised and unstandardised Principal Components Analysis in remote sensing. *Int. J. Remote Sens.* **1993**, *14*, 1359–1370. [CrossRef]
35. Liu, J.G.; Mason, P.J. *Essential Image Processing and GIS for Remote Sensing*; John Wiley & Sons: Hoboken, NJ, USA, 2009; ISBN 1118724186.
36. Loughlin, W.P. Principal component analysis for alteration mapping. *Photogramm. Eng. Remote Sens.* **1991**, *57*, 1163–1169.
37. Zhang, X.; Pazner, M.; Duke, N. Lithologic and mineral information extraction for gold exploration using ASTER data in the south Chocolate Mountains (California). *ISPRS J. Photogramm. Remote Sens.* **2007**, *62*, 271–282. [CrossRef]
38. Pour, A.B.; Hashim, M. The application of ASTER remote sensing data to porphyry copper and epithermal gold deposits. *Ore Geol. Rev.* **2012**, *44*, 1–9. [CrossRef]
39. Benlin, X.; Fangfang, L.; Xigliang, M.; Huazhong, J. Study on independent component analysis' application in classification and change detection of multispectral images. *Int. Arch. Photogramm. Remote Sens. Spat. Inf. Sci.* **2008**, *37*, 871–876.

40. Beiranvand Pour, A.; Park, Y.; Park, T.-Y.S.; Hong, J.K.; Hashim, M.; Woo, J.; Ayoobi, I. Evaluation of ICA and CEM algorithms with Landsat-8/ASTER data for geological mapping in inaccessible regions. *Geocarto Int.* **2019**, *34*, 785–816. [CrossRef]
41. Beiranvand Pour, A.; Park, T.S.; Park, Y.; Hong, J.K.; Pradhan, B. Fusion of DPCA and ICA algorithms for mineral detection using Landsat-8 spectral bands. In Proceedings of the IEEE International Geoscience and Remote Sensing Symposium, Yokohama, Japan, 28 July–2 August 2019; pp. 6067–6070.
42. Green, A.A.; Berman, M.; Switzer, P.; Craig, M.D. A transformation for ordering multispectral data in terms of image quality with implications for noise removal. *IEEE Trans. Geosci. Remote Sens.* **1988**, *26*, 65–74. [CrossRef]
43. Boardman, J.W.; Kruse, F.A.; Green, R.O. Mapping target signatures via partial unmixing of AVIRIS data. *Summ. 5 Annu. JPL Airborne Geosci. Work.* **1995**, *1*, 23–26.
44. Ferrier, G.; White, K.; Griffiths, G.; Bryant, R.; Stefouli, M. The mapping of hydrothermal alteration zones on the island of Lesvos, Greece using an integrated remote sensing dataset. *Int. J. Remote Sens.* **2002**, *23*, 341–356. [CrossRef]
45. Poormirzaee, R.; Oskouei, M.M. Use of spectral analysis for detection of alterations in ETM data, Yazd, Iran. *Appl. Geomatics* **2010**, *2*, 147–154. [CrossRef]
46. Pazand, K.; Sarvestani, J.F.; Ravasan, M.R.S. Hydrothermal alteration mapping using ASTER data for reconnaissance porphyry copper mineralization in the Ahar area, NW Iran. *J. Indian Soc. Remote Sens.* **2013**, *41*, 379–389. [CrossRef]
47. Shamanian, G.H.; Hedenquist, J.W.; Hattori, K.H.; Hassanzadeh, J. The Gandy and Abolhassani epithermal prospects in the Alborz magmatic arc, Semnan province, Northern Iran. *Econ. Geol.* **2004**, *99*, 691–712. [CrossRef]
48. Alavi, M. Tectonostratigraphic synthesis and structural style of the Alborz mountain system in Northern Iran. *J. Geodyn.* **1996**, *21*, 1–33. [CrossRef]
49. Safonov, Y.G. Hydrothermal gold deposits: Distribution, geological-genetic types, and productivity of ore-forming systems. *Geol. Ore Depos.* **1997**, *39*, 20–32.
50. Eshraghi, S.A.; Jalali, A. *Moaleman (1:100,000) Geological Map*; Geological Survey of Iran: Tehran, Iran, 2006.
51. Ramachandran, B.; Justice, C.O.; Abrams, M.J. *Land Remote Sensing and Global Environmental Change: NASA's Earth Observing System and the Science of ASTER and MODIS*; Springer Science & Business Media: Berlin/Heidelberg, Germany, 2010; Volume 11, ISBN 1441967494.
52. Rowan, L.C.; Mars, J.C.; Simpson, C.J. Lithologic mapping of the Mordor, NT, Australia ultramafic complex by using the Advanced Spaceborne Thermal Emission and Reflection Radiometer (ASTER). *Remote Sens. Environ.* **2005**, *99*, 105–126. [CrossRef]
53. Fujisada, H. Design and performance of ASTER instrument. In Proceedings of the Advanced and Next-Generation Satellites, Paris, France, 15 December 1995; Volume 2583, pp. 16–25.
54. Abrams, M. The Advanced Spaceborne Thermal Emission and Reflection Radiometer (ASTER): Data products for the high spatial resolution imager on NASA's Terra platform. *Int. J. Remote Sens.* **2000**, *21*, 847–859. [CrossRef]
55. Schwartz, G.M. Hydrothermal alteration in the "porphyry copper" deposits. *Econ. Geol.* **1947**, *42*, 319–352. [CrossRef]
56. Gemmell, J.B. Exploration implications of hydrothermal alteration associated with epithermal Au-Ag deposits. *ASEG Ext. Abstr.* **2006**, *2006*, 1–5. [CrossRef]
57. Simpson, M.P.; Mauk, J.L. Hydrothermal alteration and veins at the epithermal Au-Ag deposits and prospects of the Waitekauri Area, Hauraki Goldfield, New Zealand. *Econ. Geol.* **2011**, *106*, 945–973. [CrossRef]
58. Alizadeh Sevari, B.; Hezarkhani, A. Fluid evolution of the magmatic hydrothermal porphyry copper deposit based on fluid inclusion and stable isotope studies at Darrehzar, Iran. *ISRN Geol.* **2014**. [CrossRef]
59. US Geological Survey EarthExplorer. Available online: https://earthexplorer.usgs.gov (accessed on 17 October 2019).
60. Duda, K.; Daucsavage, J. ASTER Level 1T User Guide. 2015. Available online: https://lpdaac.usgs.gov/documents/262/ASTER_User_Handbook_v2.pdf (accessed on 17 October 2019).
61. L3Harris Geospatial ENVI v5.3.1. Available online: https://www.harrisgeospatial.com/Software-Technology/ENVI (accessed on 9 November 2019).

62. Pearson, K. On lines and planes of closest fit to systems of points in space. *Philos. Mag. Ser. 6* **1901**, *2*, 559–572. [CrossRef]
63. Hotelling, H. Analysis of a complex of statistical variables into principal components. *J. Educ. Psychol.* **1933**, *24*, 417–441. [CrossRef]
64. Jolliffe, I.T. *Principal Component Analysis*, 2nd ed.; Springer: New York, NY, USA, 2002.
65. Yang, J.; Cheng, Q. A comparative study of independent component analysis with principal component analysis in geological objects identification, Part I: Simulations. *J. Geochemical Explor.* **2015**, *149*, 127–135. [CrossRef]
66. Farahbakhsh, E.; Chandra, R.; Olierook, H.K.H.; Scalzo, R.; Clark, C.; Reddy, S.M.; Müller, R.D. Computer vision-based framework for extracting tectonic lineaments from optical remote sensing data. *Int. J. Remote Sens.* **2020**, *41*, 1760–1787. [CrossRef]
67. Hall, D.L.; Llinas, J. An introduction to multisensor data fusion. In *Proceedings of the IEEE*; IEEE: Piscataway, NJ, USA, 1997; Volume 85, pp. 6–23.
68. Tobin, D.C.; Antonelli, P.; Revercomb, H.E.; Dutcher, S.; Turner, D.D.; Taylor, J.K.; Knuteson, R.O.; Vinson, K. Hyperspectral data noise characterization using principle component analysis: Application to the atmospheric infrared sounder. *J. Appl. Remote Sens.* **2007**, *1*, 013515. [CrossRef]
69. Fiori, S. Overview of independent component analysis technique with an application to synthetic aperture radar (SAR) imagery processing. *Neural Networks* **2003**, *16*, 453–467. [CrossRef]
70. Hyvärinen, A. Fast and robust fixed-point algorithms for independent component analysis. *IEEE Trans. Neural Networks* **1999**, *10*, 626–634. [CrossRef]
71. Novey, M.; Adali, T. Complex ICA by Negentropy Maximization. *IEEE Trans. Neural Networks* **2008**, *19*, 596–609. [CrossRef]
72. Chiang, S.S.; Chang, C.I.; Ginsberg, I.W. Unsupervised hyperspectral image analysis using independent component analysis. In Proceedings of the IEEE International Geoscience and Remote Sensing Symposium, Honolulu, HI, USA, 24–28 July 2000; pp. 3136–3138.
73. Harris, J.R.; Rogge, D.; Hitchcock, R.; Ijewliw, O.; Wright, D. Mapping lithology in Canada's Arctic: Application of hyperspectral data using the minimum noise fraction transformation and matched filtering. *Can. J. Earth Sci.* **2005**, *42*, 2173–2193. [CrossRef]
74. Clark, R.N.; Swayze, G.A.; Wise, R.; Livo, E.; Hoefen, T.; Kokaly, R.; Sutley, S.J. *USGS Digital Spectral Library splib06a*; US Geological Survey, Digital Data Series: Reston, VA, USA, 2007.
75. Ozdemir, A.; Sahinoglu, A. Important of gossans in mineral exploration: A case study in Northern Turkey. *Int. J. Earth Sci. Geophys.* **2018**, *4*, 1–20.
76. Cheng, Q.; Agterberg, F.P.; Ballantyne, S.B. The separation of geochemical anomalies from background by fractal methods. *J. Geochemical Explor.* **1994**, *51*, 109–130. [CrossRef]
77. Landis, J.R.; Koch, G.G. The measurement of observer agreement for categorical data. *Biometrics* **1977**, *33*, 159–174. [CrossRef]
78. Yousefi, M.; Nykänen, V. Data-driven logistic-based weighting of geochemical and geological evidence layers in mineral prospectivity mapping. *J. Geochem. Explor.* **2016**, *164*, 94–106. [CrossRef]
79. D'Ercole, C.; Groves, D.I.; Knox-Robinson, C.M. Using fuzzy logic in a geographic information system environment to enhance conceptually based prospectivity analysis of Mississippi Valley-type mineralisation. *Aust. J. Earth Sci.* **2000**, *47*, 913–927. [CrossRef]
80. Moradi, M.; Basiri, S.; Kananian, A.; Kabiri, K. Fuzzy logic modeling for hydrothermal gold mineralization mapping using geochemical, geological, ASTER imageries and other geo-data, a case study in Central Alborz, Iran. *Earth Sci. Informatics* **2015**, *8*, 197–205. [CrossRef]
81. Zhang, N.; Zhou, K.; Du, X. Application of fuzzy logic and fuzzy AHP to mineral prospectivity mapping of porphyry and hydrothermal vein copper deposits in the Dananhu-Tousuquan island arc, Xinjiang, NW China. *J. African Earth Sci.* **2017**, *128*, 84–96. [CrossRef]
82. Esmaeily, D.; Afshooni, S.Z.; Mirnejad, H.; Rashidnejad-e-Omran, N. Mass changes during hydrothermal alteration associated with gold mineralization in the Astaneh granitoid rocks, western Iran. *Geochem. Explor. Environ. Anal.* **2012**, *12*, 161–175. [CrossRef]
83. Zhang, Y.; Gao, J.-F.; Ma, D.; Pan, J. The role of hydrothermal alteration in tungsten mineralization at the Dahutang tungsten deposit, South China. *Ore Geol. Rev.* **2018**, *95*, 1008–1027. [CrossRef]

84. Pollard, P.J. An intrusion-related origin for Cu–Au mineralization in iron oxide–copper–gold (IOCG) provinces. *Miner. Depos.* **2006**, *41*, 179. [CrossRef]
85. Xiao, B.; Chen, H.; Hollings, P.; Wang, Y.; Yang, J.; Wang, F. Element transport and enrichment during propylitic alteration in Paleozoic porphyry Cu mineralization systems: Insights from chlorite chemistry. *Ore Geol. Rev.* **2018**, *102*, 437–448. [CrossRef]
86. Bonham-Carter, G.F. Geographic Information Systems for Geoscientists: Modeling with GIS. *Comput. Methods Geosci.* **1994**, *13*, 398.
87. Yousefi, M.; Carranza, E.J.M. Prediction-area (P-A) plot and C-A fractal analysis to classify and evaluate evidential maps for mineral prospectivity modeling. *Comput. Geosci.* **2015**, *79*, 69–81. [CrossRef]
88. Langlois, D.; Chartier, S.; Gosselin, D. An introduction to independent component analysis: InfoMax and FastICA algorithms. *Tutor. Quant. Methods Psychol.* **2010**, *6*, 31–38. [CrossRef]
89. Foody, G.M. Explaining the unsuitability of the kappa coefficient in the assessment and comparison of the accuracy of thematic maps obtained by image classification. *Remote Sens. Environ.* **2020**, *239*, 111630. [CrossRef]
90. Shokouh Saljoughi, B.; Hezarkhani, A.; Farahbakhsh, E. A comparison between knowledge-driven fuzzy and data-driven artificial neural network approaches for prospecting porphyry Cu mineralization; A case study of Shahr-e-Babak area, Kerman Province, SE Iran. *J. Min. Environ.* **2018**, *9*, 917–940.
91. De Smith, M.J.; Goodchild, M.F.; Longley, P. *Geospatial Analysis: A Comprehensive Guide to Principles, Techniques and Software Tools*; Troubador Publishing Ltd.: Leicester, UK, 2007; ISBN 1905886608.

© 2020 by the authors. Licensee MDPI, Basel, Switzerland. This article is an open access article distributed under the terms and conditions of the Creative Commons Attribution (CC BY) license (http://creativecommons.org/licenses/by/4.0/).

Article

Abandoned Mine Tailings Affecting Riverbed Sediments in the Cartagena–La Union District, Mediterranean Coastal Area (Spain)

Tomás Martín-Crespo [1,*], David Gómez-Ortiz [1], Silvia Martín-Velázquez [1], Pedro Martínez-Pagán [2], Cristina de Ignacio-San José [3], Javier Lillo [1,4] and Ángel Faz [5]

[1] Department Biología y Geología, Física y Química Inorgánica, ESCET, Universidad Rey Juan Carlos, C/Tulipán s/n, Móstoles, 28933 Madrid, Spain; david.gomez@urjc.es (D.G.-O.); silvia.martin@urjc.es (S.M.-V.); javier.lillo@urjc.es (J.L.)
[2] Department Ingeniería Minera y Civil, Universidad Politécnica de Cartagena, Paseo Alfonso XIII, 52, Cartagena, 30203 Madrid, Spain; p.martinez@upct.es
[3] Department Petrología y Geoquímica, Fac. CC. Geológicas, Universidad Complutense de Madrid, C/Antonio Nováis s/n, 28040 Madrid, Spain; cris@geo.ucm.es
[4] IMDEA Water Institute, Av/ Punto Com, 2, Parque Científico Tecnológico de la Universidad de Alcalá, Alcalá de Henares, 28805 Madrid, Spain
[5] Department Ciencia y Tecnología Agraria, Universidad Politécnica de Cartagena, Paseo Alfonso XIII 52, Cartagena, 30203 Madrid, Spain; angel.fazcano@upct.es
* Correspondence: tomas.martin@urjc.es; Tel.: +34-914-888-098; Fax: +34-916-647-490

Received: 28 May 2020; Accepted: 23 June 2020; Published: 25 June 2020

Abstract: This study presents the results of the geoenvironmental characterization of La Matildes riverbed, affected by mine tailings in the Cartagena–La Unión district, Murcia (southeast Spain). Soils and riverbeds in this area are highly polluted. Two Electrical Resistivity Imaging (ERI) profiles were carried out to obtain information about the thickness of the deposits and their internal structure. For the mine tailings deposits of La Murla, a tributary of the El Miedo riverbed, the geophysical method imaged two different units: the upper one characterized by low resistivity values and 5–8 m thickness, correlated with the mine tailings deposits; and the lower more resistive unit corresponding to the Paleozoic metasediments bedrock. The ERI profile transverse to the Las Matildes dry riverbed revealed the existence of three different units. The uppermost one has the lowest resistivity values and corresponds to the tailings deposits discharged to the riverbeds. An intermediate unit, with intermediate resistivity values, corresponds to the riverbed sediments before the mining operations. The lower unit is more resistive and corresponds to the bedrock. Significant amounts of pyrite, sphalerite, and galena were found both in tailings and riverbed sediments. The geochemical composition of borehole samples from the riverbed materials shows significantly high contents of As, Cd, Cu, Fe, Pb, and Zn being released to the environment. Mining works have modified the natural landscape near La Unión town. Surface extraction in three open-pit mines have changed the summits of Sierra de Cartagena–La Unión. Rock and metallurgical wastes have altered the drainage pattern and buried the headwaters of ephemeral channels. The environmental hazards require remediation to minimize the environmental impact on the Mar Menor coastal lagoon, one of the most touristic areas in SE Spain.

Keywords: riverbed; metals; electrical resistivity imaging; tailings; Mar Menor; Cartagena–La Unión

1. Introduction

The Province of Murcia, situated in the southeastern most part of the Iberian Peninsula, is one of those mining areas in Spain that today still suffer from serious environmental problems due to the presence of nearly eighty abandoned mineral waste structures, especially in the Sierra de Cartagena area and La Unión town [1]. A large amount of mine wastes was abandoned up until the 1980s, representing serious geochemical hazards. The composition and emplacement sites of these mine wastes have generated environmental hazards related to geochemical pollution (among others), that negatively affect soils, groundwater, flora, fauna, and humans [1,2]. It is therefore necessary to gain deeper knowledge of the current condition of these deposits with high potential risk. There are examples of studies on tailings ponds using (i) magnetometry, enabling us to deduce the variations of the tailings structures [3,4]; (ii) electromagnetics, used to investigate the structural and hydrogeological settings of oil sands tailings dykes [5]; (iii) seismic, to derive the internal pond boundaries by means of refraction and reflection waves [6,7]; and direct-current geoelectrical imaging surveys [8], where the electrical resistivity tomography (ERI) method is strongly affected by the variation of some important properties of tailings such as moisture, soil salinization, particle size distribution, acid mine drainage, etc. [9,10]. The ERI method has been the fundamental tool to support physical–chemical analysis in phytoremediation works on the abandoned mining ponds in the Sierra de Cartagena area. The application of geophysical, mineralogical, and geochemical techniques, together with landscape evolution studies, could allow the analysis and quantitative assessment of the pollution risk [11–15].

This work presents the results obtained by the joint application of geophysical, mineralogical, and geochemical techniques to both mine tailings and riverbed sediments from Las Matildes (Cartagena–La Unión district, Murcia, SE Spain). The shallow non-destructive geophysical technique applied in the study was electrical resistivity imaging. To gain better control of the geophysical results, aerial photographs of the area from different times have been used to evaluate the changes in relief that have progressively occurred due to the tailings accumulation. Geochemical and mineralogical characterization techniques include X-ray diffraction, Inductively Coupled Plasma Mass Spectrometry (ICP-MS), and Atomic Absorption Spectroscopy (AAS). Thus, a representative characterization of the riverbed affected by mine tailings has been established based on the geometry and internal structure of deposits, mineralogical and chemical composition, and thickness of tailings and alluvial sediments. Besides the high levels of contamination in abandoned mine sites, the land surface is intensely transformed. The transfer of huge amounts of earth causes the destruction of natural landscapes and the formation of new landforms. As these new environments are also modelled by earth surface processes (involving water, wind, mass movement), different studies have recently focused on geomorphic changes due to mining activity ([16,17] and references therein). In the area of 100 km^2 of the Sierra de Cartagena–La Unión mining district, 12 open-pits and 2351 waste deposits from ore-processing have been documented [1]. Some of these residues are in the headwaters of ephemeral channels to the east of La Unión town. To identify the geomorphic changes through time in this area, we have mapped the main mining wastes and excavations by means of aerial photograms, orthoimages, and anaglyphs from 1956 and 2016. The temporal evolution has also been used to evaluate landscape changes at the two survey sites and gain better control of the geophysical results.

Thus, the main objective of this study was to characterize the present conditions of tailings and riverbed sediments to identify potential environmental problems. This part of SE Spain is still highly polluted because the numerous mine wastes and mining structures (buildings, shafts, ...) remain abandoned. The goal is particularly relevant as this situation provokes a continual flow of Acid Mine Drainage (AMD) with associated metal pollution of soils and waters that are dispersed to the Mediterranean Sea.

2. Location and Features of the Mine Site

The mine tailings area studied is located around 1 km to the east of the town of La Unión (SE Spain) (Figure 1). The Descargador mine pond is located just on a tributary of the El Miedo ephemeral riverbed and shows maximum dimensions of 180 m × 30 m. The mine tailings are flotation deposits of medium-to-fine-grained, sand-size material. They were produced from grinding and metallurgical processing of pyrite, sphalerite, and galena from the Emilia mine, in works carried out between 1952 and 1981 [18,19]. The main ore minerals are pyrite, sphalerite, galena, and cassiterite. Other minor sulfides include chalcopyrite, minerals of the tetrahedrite-tenantite group, and arsenopyrite. Gangue minerals include chalcedony, quartz, siderite, and greenalite [20,21].

Figure 1. (a) Location of the Cartagena–La Unión mine district, the geophysical surveys, and the sampling points; (b) mine tailings and metallurgical slag, mixed with riverbed sediments in the Las Matildes riverbed; (c) remaining medium-to-fine-grained mine tailings deposits, showing embedded metamorphic bedrock fragments.

This mine pond was placed on Palaeozoic schists, gneisses, and metabasites from the so-called "Complejo Nevado-Filábride" of the Betic Cordillera, and the riverbed ran upon limestone and phyllites from the "Complejo Alpujárride" of the same Cordillera. In this area, significant amounts (tonnes) of metallurgical slag were dumped in ponds over the riverbed sediments. The Las Matildes riverbed is an ephemeral watercourse running from the Descargador area to the Mediterranean Sea. Both ephemeral rivers flow into the Mar Menor coastal lagoon. The Mar Menor is one of the largest hypersaline coastal lagoons of the Mediterranean Area, measuring about 170 km² with a mean depth of 2.5 m. It is separated from the Mediterranean Sea by a narrow sandy coastal barrier and is surrounded by the Campo de Cartagena, one of the most intensive agricultural zones of Spain. This fragile ecosystem suffers intense human pressure, such as excess nutrient and sediment inputs from agriculture and abandoned mining activities, respectively, as well as decreases in salinity.

The region is characterized by a semiarid Mediterranean climate with dry summers and mild winters [22,23]. According to data collected by the closest and most complete automatic weather station of the Agricultural Information System of Murcia (La Aljorra, 2000–2016 period; [24], the mean annual temperature was 18 °C, with a minimum in January (−2 °C) and a maximum in July (41 °C). Rainfall was extremely variable, with intense storms in October and September (maximum rainfall event in a 24 h period of 163 mm). Due to the low precipitation (annual average of 279 mm) and high evapotranspiration (annual average from the Penman–Monteith method: 1368 mm) only ephemeral streams that drain into the Mar Menor coastal lagoon during heavy rainfalls exist. The area is subjected to strong annual stormy episodes that can induce significant flash flooding. As an example of these episodes, in October 1972, an extremely intense rainfall event caused a flash flood of the tailings from the Brunita mine pond, located only 2 km to the west of the studied area, killing one person and causing serious material damage [14,25].

3. Methodology

3.1. Temporal Evolution of the Mining Landscape

This study focused on an area of ~4 km² to the east of La Unión town, where numerous mining works extend along the small headwaters of the El Miedo and Las Matildes dry riverbeds in the Sierra de Cartagena–La Unión region. The geophysical and geochemical survey was conducted in deposits located in both riverbeds. The mountain range (Sierra) extends ~26 km from east to west, parallel to the Mediterranean coast. The headwater tributaries transport sediments under torrential rainfall from the highest peak (Sancti Spiritus 3, 396 m) to the Mar Menor lagoon, ~8 km away to the North.

Main fluvial channels, mining wastes (rock and slag dumps, tailings ponds), and mining excavations (canals, open-pit mines) were mapped by using orthoimages from the years 1956 and 2016 using ArcMap 10.4.1 (GIS software, ESRI España Soluciones Geoespaciales, S.L.: Madrid, Spain). The orthoimages were added from the cartography service of the Murcia Region (https://geoportal.imida.es/gis/rest/services/2_03_ORTOIMAGENES). Aerial photograms from 1956 (USA Army Map Service flight-Geographic Service of the Spanish Army, scale of 1:32,000) were used to distinguish landscape changes by means of a stereoscope. The stereo pairs were downloaded from the IGN website (Instituto Geográfico Nacional, https://fototeca.cnig.es/). Regarding the 2016 orthoimage (IGN flight), an anaglyph was checked in the cartographic viewer of IGN (https://www.ign.es/iberpix2/visor/) with a 3D red cyan glass.

3.2. Non-Destructive Geophysical Method

Electrical resistivity imaging (ERI) is a shallow geophysical prospecting method designed to unravel complex geological structures with changes in both vertical and horizontal resistivity. The fundamentals of the resistivity methods can be found in different reviews [8,26–29]. The method implies the use of a variable number of electrodes, typically between 24 and 96, connected to a switching box via a multi-core cable [30]. The electrodes are switched to obtain measurements of subsurface resistivity at different depths. By means of a computer connected to the resistivity meter, different combinations of four electrodes are selected automatically for each measurement. By using different electrode spacing at different locations along the cable, a 2-D profile of the subsurface is finally obtained (e.g., [8]). As a rule of thumb, it is important to note that the measurements for each quadripole provide values of apparent resistivity at different depths, in such a manner that the greater the spacing between the electrodes, the deeper the depth of investigation. A Syscal Junior Switch 48 equipment was used in this work. The number of electrodes is 48 with a maximum electrode spacing of 5 m. This equipment uses an injection cycle of 4 s with an output voltage of up to 400 V and rejection filters for 50 and 60 Hz. Noise reduction is obtained by applying a continuous stacking selectable from 1 to 255 stacks. With these parameters, an average value (and the associated level of uncertainty) of the apparent resistivity is obtained for each quadripole. For the geolocation of the profiles, a MAGELLAN MobileMapper CX GPS with a submetrical accuracy has been used.

After the acquisition of the apparent resistivity values, the data must be processed to remove spurious data points. This is done by filtering raw data with low signal values (V/I < 10-6 Ω) or repeatability errors greater than 2%. In addition to this, topography data for each electrode of the profile must also be included at this stage. Once filtered, an iterative inversion process must be carried out to obtain a cross-section of true resistivity along the survey's profile. The code selected to iteratively convert the apparent resistivity values to true resistivity values is RES2DINV, which uses the L1 norm for the data misfit and the inversion is performed using the L1 norm (robust) for the model roughness filter [31]. A robust inversion has been chosen because this kind of inversion is more accurate when sharp discrete boundaries exist in the model. In the study area, sharp boundaries exist between the high resistive metamorphic rocks constituting the basement and the low resistive loose materials that define both the mine pond and dry riverbed infilling, so robust inversion is the most appropriate choice. The method uses a finite element scheme for solving the 2-D forward problem and blocky inversion method for inverting the ERI data [31]. The result of the process is a true resistivity image for each profile that is used to obtain the final interpretation of the variations of the subsurface lithology.

Regarding the application of electrical techniques for the characterization of mine waste deposits, good examples can be found in [32,33]. The ERI technique has mainly been used on waste piles [34] and tailings dams [35], whereas few studies have focused on the internal structure of mine tailings ponds (e.g., [2,11–14,36–39]). To combine a good penetration depth, a reasonable vertical and horizontal resolution, and a good signal-to-noise ratio, the Wenner-Schlumberger array was chosen for this study. This array has been successfully used in similar studies (e.g., [2,14]).

3.3. Description of Sampling Methods

In the Las Matildes dry riverbed, sequential sampling was carried out using a TP-50/400 rotary drilling machine, with a minimum core bit diameter of 100 mm (Figure 1). Non-disturbed rock drill core samples were collected with a constant vertical spacing of 50 cm, up to a sampling depth of 8 m. This sampling was carried out by digging down below the surface of each pond, casting aside the parts corresponding to surficial sealing to prevent wall material falling during drilling. In total, 17 unaltered samples (five from the borehole, 11 from the ponds and one from the bedrock) were collected, air-dried for 7 days, passed through a 2-mm sieve, homogenized, and stored in plastic bags at room temperature prior to laboratory analyses. Sampling was conducted in March.

3.4. Mineralogical and Geochemical Methods

Mineralogical characterization of samples was performed by X-ray diffraction (XRD) using a Philips X'Pert powder device with a Cu anticathode and standard conditions: speed 2° 2θ/min between 2° and 70° at 40 mA and 45 KV. The whole sample was studied by crystalline non-oriented powder diffraction on a side-loading sample holder. Semi-quantitative results were obtained by the normalized reference intensity ratio (RIR) method. The XRD analyses were performed at the Centro de Apoyo Tecnológico (CAT Universidad Rey Juan Carlos, Móstoles, Spain, http://www.urjc.es/cat). From the total list of major, minor and trace elements analyzed, 12 were chosen for the geochemical study (Ag, As, Cd, Cr, Cu, Fe, Ni, Pb, S, Sb, Sn and Zn) owing to their abundance in these types of sludge and also because most of them are included in the priority contaminant list of the environmental protection agencies [40]. The selected elements were analyzed by TD (Total Digestion) or FUS (lithium metaborate/tetraborate fusion) ICP-MS (Inductively Coupled Plasma-Mass Spectrometry) at Activation Laboratories Ltd. (1428 Sandhill Drive, Ancaster, Ontario, Canada; http://www.actlabs.com). Quality control at Actlabs is done by analyzing duplicate samples and blanks to check precision, whereas accuracy is ensured by using Certified Reference Materials (GXR series; see http://www.actlabs.com). Detection limits for the analyzed elements are (data in $\mu g \cdot g^{-1}$): Ag (0.3), As (5), Cd (0.5), Cr (20), Cu (1), Fe (100), Ni (1), Pb (5), S (10), Sb (0.5), Sn (1), and Zn (1). Concentrations of Pb > 5000 and Zn > 10000 $\mu g \cdot g^{-1}$ (above the ICP-MS maximum detection limits) were measured by an Atomic Absorption Spectrometer (AAnalyst 800, Perkin Elmer spectrometer) using flame or graphite-furnace technique in the Universidad Politécnica de Cartagena (Murcia, Spain) laboratories. pH was measured using an electronic pH meter (CRISON), calibrated at two points (pH 7 and pH 4) using standard buffer solutions. This parameter was determined in a slurry system with an air-dried sample (10 g) mixed with distilled water (25 mL). Before reading the pH values, the mixture was vigorously stirred in a mechanical shaker for 10 min and left to stand for 30 min.

Statistical data processing was done using Minitab 17 software (Minitab Ltd., Brandon Court, Unit E1-E2, Progress Way, Coventry CV3 2TE, United Kingdom). The multivariate analysis was based on clustering (group average linkage dendrograms, Euclidean distance) of the set of samples and significant metals (Ag, Sb, Fe, Cd, Cu, Pb, Zn, and Sn) plus As.

4. Results and Discussion

The landscape evolution in a sector of the La Union mining district, as well as the results of the geophysical study concerning the structure of the tailings and riverbed, and those obtained from mineralogical and geochemical characterization of the borehole samples, are presented and discussed here.

4.1. Temporal Evolution of Mining Landscape of La Union

In the 1956 orthoimage (Figure 2), the orography to the E of La Unión consisted of N–S to NNW–SSE V-shaped valleys with headwaters in Sancti Spiritus hill (431 m: [41]). The westernmost one, the La Murla riverbed (~900 m long), is a tributary of the El Miedo riverbed, whereas the rest (average length of ~600 m) are tributaries of the Las Matildes riverbed. Some works related to underground mines are recognized in the southern slopes of Sancti Spiritus. The first operations at the Emilia open pit mine are visible at the easternmost tributary headwater. However, the most important modifications occur on the northern slopes of Sancti Spiritus where many waste residues are scattered over valley bottoms and foothills. The mapped deposits originated to a certain extent by rock accumulations from open-pits and mines and mainly from waste accumulation during metallurgical processes (slag dumps from melting and tailings heaps from the post-flotation process); 35 rock and slag dumps and the five tailings dams occupy an area of ~0.4 km² and modified some hydrological pathways. Several slag deposits abruptly changed the courses of Las Señales, El Humo, and La Hoya del Agua riverbeds, and a tailings dam interrupted the flow of La Hoya del Agua tributary. Therefore, two canals had to be built to evacuate surface runoff.

Geomorphic changes mapped in the 2016 orthoimage (Figure 2) are mostly related to the movement of huge rock volumes from the open pit of Emilia, San Valentín, and Tomasa. The mineral extraction from San Valentín and Tomasa mines completely changed the summit orography in Sancti Spiritus hill (Santi Spiritus 3, 396 m: [42]). At present, the open-pit mines are partially filled with wastes from metallurgical, building, and industrial activities [1]. Another profound transformation is associated with the growth of open-pit spoils in the vicinity of mines, which buried the Las Señales and El Humo headwaters as well as the channel of La Hoya del Agua and its tributaries. Residues from mineral treatment accumulated in the previous tailings dams located in La Murla and Las Señales riverbeds, increasing their sizes, but also in Las Matildes riverbed. Furthermore, these younger materials buried previous polluted residues wastes (see, for example, the area of the open-pit spoils to the west of Emilia mine, or the tailings dams in Las Matildes riverbed). Regarding the slags, some dumps are smaller because they were excavated. The area of mine deposits (21 rock and slag dumps and 10 tailings ponds) on the 2016 image is ~0.8 km², two times larger than that in 1956.

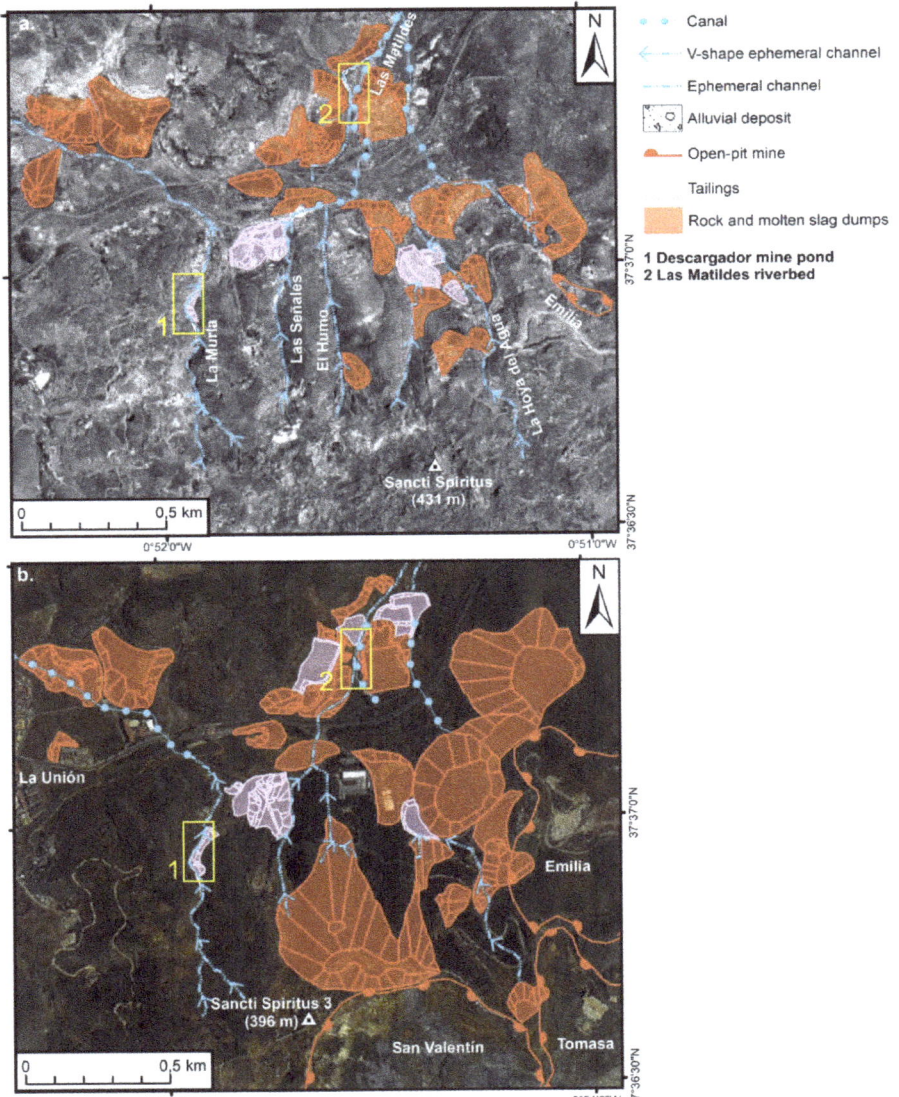

Figure 2. Mapping of the temporal evolution of the mining landscape to the east of La Unión town: (**a**) orthoimage from 1956 (Cartography Service of the Murcia Region); (**b**) orthoimage from 2016 (Cartography Service of the Murcia Region). Yellow rectangles: areas where geophysical prospection has been carried out: 1. Descargador mine pond, and 2. Las Matildes riverbed.

The temporal evolution of the mining works explains the geophysical results (Section 4.2) and the stratigraphic sequence of the core samples collected from Las Matildes riverbed (Section 4.3). Sludge in the Descargador mine pond in the La Murla valley was initially deposited in the riverbed (Figure 2a). However, due to the continued discharge, the tailings dam was raised over the metamorphic rocks of the foothill (Figure 2b). The uppermost material at the Las Matildes site, identified both in the borehole and the geophysical surveys, is related to the tailings deposit mapped in the riverbed in the

2016 orthoimage (Figure 2b). This mining waste overlapped the alluvial sediments that were identified in the 1956 orthoimage (Figure 2a). The intensive mining activities have deeply affected the landscape in this sector of the Sierra de Cartagena–La Unión. Open-pit mines and mining and metallurgical residues disrupted the natural fluvial network; furthermore, new canals were created to evacuate surface runoff. Polluted waste that accumulated in riverbeds and foothills as colluvium and alluvium sediments can be eroded and therefore mobilized by mass movements and surface runoff. The volume of tailings in the three largest dams has reached 623,000 m^3 [19].

Particulate and dissolved contaminants move from the anthropic deposits along the El Miedo and Las Matildes ephemeral riverbeds to the Mar Menor lagoon. Indeed, high concentrations of Pb and Zn have been measured in the sediments of the nearby El Beal riverbed (mean values of 39,000 and 2000 ppm, respectively: [1]). Therefore, the environmental impact of modified landscapes should also be considered, both in the estimation of potential risk, and in the proposal of management and reclamations solutions [17].

4.2. Structure of the Mine Pond and Riverbed Deposits

ERI has provided information about both the thickness and geometry of the waste deposits related to the mine tailings. Two ERI profiles have been carried out (one longitudinal to the mine pond and one transverse to the watercourse) (Figure 1). The borehole provided (from top to bottom) thicknesses of 2 m of mine tailings, 1.5 m of tailings mixed with riverbed sediments, and 4 m of riverbed sediments.

4.2.1. Mine Pond

The ERI profile (Figure 3) imaged two different units: the upper one is characterized by low resistivity (<20 ohm·m) values and a thickness ranging from 5–8 m. It can be correlated with the mine tailings deposits observed at the outcrop (Figure 1). In contrast, the lower unit is more resistive (>200 ohm·m) and corresponds to the Paleozoic metasediments that constitute the bedrock of the area.

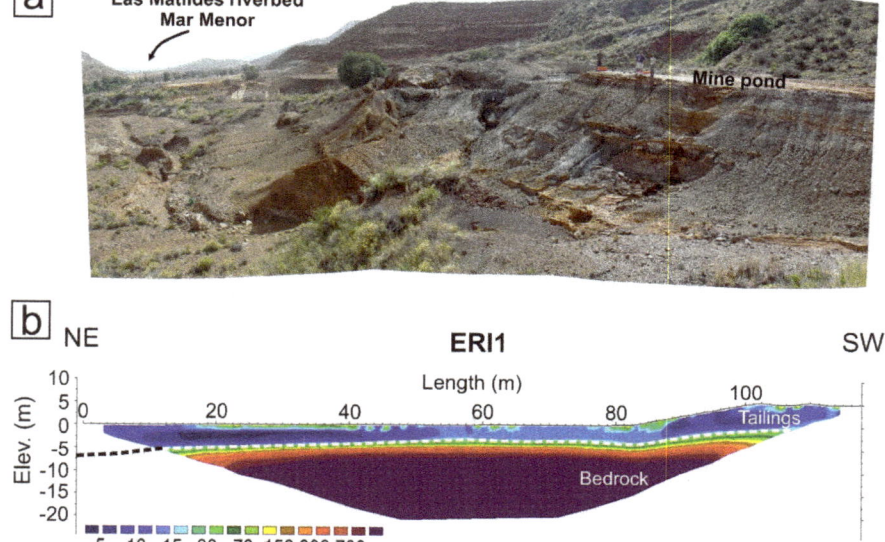

Figure 3. Descargador mine pond: (**a**) general view of the non-eroded remaining tailings; (**b**) ERI profile showing the two identified units.

As can be seen in the profile, the bedrock below the mine pond deposits is very homogeneous and consequently, the occurrence of faults or discontinuities can be disregarded. The ERI profile does not have enough lateral and vertical resolution to discriminate different units inside the mine pond deposits. Extremely low resistivity values characterize the occurrence of AMD inside the mine ponds, as demonstrated in nearby similar areas (e.g., Brunita mine pond, [14]), or even different mine districts (e.g., Mina Concepción, [15]). In Mina Concepción, values lower than 5 ohm·m delineate the preferential path of AMD (pH ranging from 2.5 to 3) flow inside the mine ponds. In the case of Descargador mine pond, the low (5–20 ohm·m) resistivity values measured in the mine pond infilling are compatible with the clayish to sandy-clayish texture and high water content of the deposits, as observed in the field, and thus internal AMD flow can be disregarded. A certain acidic character of the water (pH ~5–6) has been obtained for the same resistivity values in similar mine ponds (e.g., San Quintin mine pond, [15]) and this would be the case here. Moreover, the highly homogeneous and resistive character of the metamorphic bedrock imaged below the mine pond allows us to confirm that there is no AMD flow escaping from the bottom of the mine pond that could be affecting the groundwater of nearby fractured aquifers.

4.2.2. Dry Riverbed

The ERI profile transverse to the dry riverbed (Figure 4) has revealed the existence of three different units whose interpretation has been made by comparing them with the borehole data. The uppermost one, with a mean thickness of 2–3 m, extends from 21 to 100 m along the profile and has the lowest resistivity values (<15 ohm·m). It corresponds to the tailings deposits transported episodically by the riverbed during periods of strong rainfall. An intermediate unit, with resistivity values ranging from 15 to 40 ohm·m and varying thickness (1 to 4 m), corresponds to the watercourse sediments before the construction of the mine tailings pond located upstream. The lower unit is more resistive (>50 ohm·m) and corresponds to the bedrock.

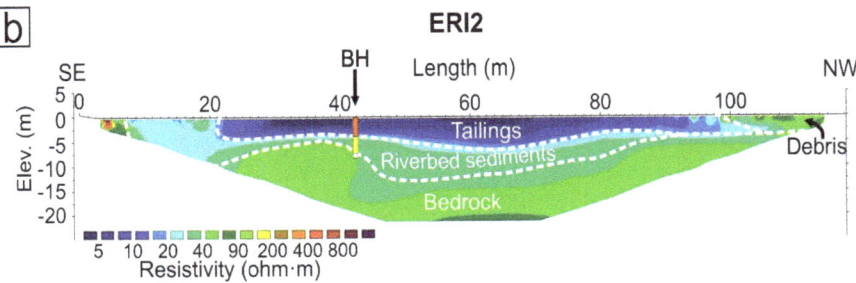

Figure 4. Las Matildes riverbed: (**a**) general view with borehole location; (**b**) ERI profile showing the three identified units; BH: borehole.

Two small heterogeneous highly resistive (>200 ohm·m) units located at both ends of the profile can be associated with debris deposits from the neighboring slag heap. As for the Descargador mine pond, the resistivity values observed at the uppermost unit (5–15 ohm·m) are not low enough to infer that AMD flow occurs in the transported tailings deposits. The mean pH value obtained from the borehole samples is close to 5, confirming the slightly acidic character of the deposits but far from the extremely low pH values of ~2.5 typical of AMD observed in similar deposits, as previously cited. The EC values obtained from the borehole samples, although greater than the intermediate and lower units (3–8 mS/cm vs. 0.2–3 mS/cm) are also very far from the extremely high EC values (>8000 mS/cm) obtained in the nearby Brunita mine pond where AMD flow has been detected [14]. For the intermediate unit, the measured pH values are higher than 6, in good agreement with the higher resistivity values and coarse texture of the deposit that characterizes this unit. No evidence of faulting or AMD can be observed in the lower resistive unit corresponding to the bedrock.

4.3. Mineralogical Characterization

The semi-quantitative data on the mineralogical composition of the Las Matildes samples from the borehole (BH) are displayed in Table 1.

Table 1. Semi-Quantitative Mineralogical Composition (wt %) of the Studied Samples from the Las Matildes Riverbed Borehole. Qtz: Quartz, Ill-kn: Illite-Kaolinite; Chl: Chlorite; Gre: Greenalite; Py: Pyrite; Sp: Sphalerite; Gn: Galena; Ang: Anglesite; St: Stannite; Ja: Pb-Jarosite; Sd: Siderite; Gp: Gypsum. Groups of Samples: Tailings, Tailings and Riverbed Sediments (T+S), Riverbed Sediments, and Bedrock (R).

Sample	Group	Depth (m)	Qtz	Ill-kn	Chl	Gre	Py	Sp	Gn	Ang	St	Ja	Sd	Gp
MAT-1	Tailings	0	30	10	5	1–5	10	10	5	5		5	5	15
MAT-2		0.5	30	10	10		5	10	5	5		5	5	15
MAT-3		1	30	10		1–5	5	5	5	5	5	15	5	15
MAT-4		1.5	25	5			5	10	5–10	5	5	15	5	15
MAT-5		2	25	5		5–10		5	5–10	1–5	5	15	5	15
MAT-6	T+S	2.5	40	15	5	10	5	5	5–10	10				5
MAT-7		3	45	10	5	1–5	5	5	5	10				10
MAT-8		3.5	45	10	5	1–5	5	1–5		10				15
MAT-9	Riverbed Sediments	4	50	20		5	5	5		10				5
MAT-10		4.5	55	20	1–5	5	5			10				
MAT-11		5	60	20		5			1–5	10				5
MAT-12		5.5	55	20		5	5			10				5
MAT-13		6	60	25					1–5	5			5	5
MAT-14		6.5	55	25		1–5			1–5	10				5
MAT-15		7	60	35	1–5					5				
MAT-16		7.5	60	20		10				10				
MAT-17	R	8	65	30	5									

Three levels showing different but nearly homogeneous mineralogical compositions can be inferred from the X-ray diffraction data (Table 1): an upper level (2 m thick) of reddish clay mine tailings, an intermediate 1.5 m of tailings mixed with riverbed sediments, and a lower, 4 m level of coarse to sandy riverbed sediments. Significant amounts of pyrite, sphalerite, and galena from the mined ore deposit have been identified in all three levels, although they are higher in the mine tailings upper levels (10%) and lower in the riverbed sediments (5%), probably reflecting less efficient ore benefiting processes. The upper level still includes a significant proportion of the original ore. Fe-carbonates (siderite), secondary sulfates (Pb-jarosite, gypsum), and ore-sulfides (stannite) have also been determined in the mine tailings upper levels. It can be inferred that silicates make up 30–50 wt% of the mine tailings upper level, 60–65 wt% of the intermediate level, and 70–95 wt% of the riverbed sediments' deeper level. Quartz, and Illite-kaolinite interstratified are the main minerals. In addition, a 5–10 wt% greenalite ($Fe_{2-3}Si_2O_5(OH)_4$) content has also been estimated in the three sampled levels. Greenalite is an ore mineral associated with the hydrothermal alteration caused by intense volcanic processes. Anglesite (5–10%) has also been identified in all samples from the three levels. The transition to the in-situ host rock was nearly sharp, with the only presence of silicates (quartz, illite-kaolinite, and chlorite). This assemblage may indicate the occurrence of hydrothermal alteration in the volcanic host rocks.

4.4. Geochemical Characterization

The chemical analyses confirmed the similar infilling structure and major and trace element contents for Las Matildes riverbed. The total Fe_2O_3 contents, the trace-element content, and pH at the different sampling depths are shown in Table 2.

The geochemistry of the sampled mine tailings and riverbed sediments matches the described mineral abundances and geophysical features well. Distribution with depth plots of these geochemical data (Figure 5) clearly shows the three-level structure previously described: tailings from the surface down to a 2.0 m depth, tailings mixed with riverbed deposits from 2.0 to 3.5 m depth, and riverbed deposits from 3.5 to 7.5 m depth. High contents of Fe, Pb, Zn, and other heavy and transitional metals characterize the composition of all borehole samples. The total iron content ranges from 9.1 to 45.3 wt%, Pb from 2340 to 8640 µg/g, and Zn from 897 to 12,310 µg/g. Other trace elements also show high contents: As (up to 1620 µg/g), Cd (up to 306 µg/g), Cu (up to 730 µg/g), and Sb (up to 236 µg/g). These significant amounts are due to the nature of the deposits, composed of pyrite, sphalerite, galena, and cassiterite. [43] reported similar trace element contents in other riverbeds of the district. The metamorphic host rock (8 m depth) in turn displays significantly lower amounts of metals, as is usually the case in mine deposits [2,14]. Variation in trace element trends and iron contents are related to the textural features and mineralogical composition previously described in Section 4.3. The highest values are generally observed to be associated with the surficial level (0–2 m), representing mine tailings directly stored over the riverbed sediments. The highest Ag, As, Fe, Pb, and Zn contents are clearly located at this uppermost level (Figure 5).

The lowest contents are clearly observed in the data from the deepest borehole samples (4–7.5 m). Textural (coarse to sandy), mineralogical (silicate content), and geochemical features clearly define the deepest level as riverbed sediments. Intermediate trace elements and iron contents are correspondingly shown in the intermediate level (2.5–3.5 m depth) and are related to mixing between tailings and the riverbed sediment level. Samples from the tailings level show lower pH values ranging from 2.5 to 5.3 compared to the almost neutral (6.0 to 7.7) pH in riverbed deposits and host rock. The lowest pH and highest electrical conductivity (EC) values from Table 2 are also associated with the upper level, confirming the three-level structure previously described.

Table 2. Fe_2O_3 Total and Trace Elements Content, pH, and Electrical Conductivity (EC) Values in the Las Matildes Riverbed Borehole (T+S: Tailings and Riverbed Sediments; R: Bedrock).

Sample		Depth (m)	Ag (µg/g)	As (µg/g)	Cd (µg/g)	Cr (µg/g)	Cu (µg/g)	Fe_2O_3 Total (wt%)	Ni (µg/g)	Pb (µg/g)	S (wt%)	Sb (µg/g)	Sn (µg/g)	Zn (µg/g)	pH	EC (mS/cm)
MAT-1	Tailings	0	33.1	452	19.1	40	350	39.17	28	8130	3.11	87.2	75	11220	4.42	3.20
MAT-2		0.5	4.4	456	14.1	30	201	36.04	24	4130	4.10	59	31	6930	5.30	4.61
MAT-3		1	5.7	262	11	b.d.	40	45.26	15	5320	1.82	130	38	7230	4.94	3.44
MAT-4		1.5	67.7	1620	306	b.d.	730	40.75	44	8640	16.80	211.5	122	12310	4.88	8.78
MAT-5		2	37	1270	9.3	40	215	41.68	14	7270	3.43	236	46	6350	2.50	5.19
MAT-6	T + S	2.5	2	27	5.5	90	34	12.88	47	7950	0.46	17.4	8	3010	3.83	1.55
MAT-7		3	3.8	61	38	60	76	20.47	46	7640	2.94	42.5	18	12010	4.51	3.38
MAT-8		3.5	4.3	26	133	60	69	9.11	86	2340	3.38	18.6	9	9870	6.38	2.53
MAT-9	Riverbed Sediments	4	1.7	42	8.4	70	169	15.38	50	3640	0.45	21	29	4680	6.08	1.85
MAT-10		4.5	1.7	48	3.6	70	110	11.80	42	2720	0.15	25	28	1140	6.28	0.63
MAT-11		5	2.1	38	3.2	60	89	9.78	32	2450	0.16	19.7	21	897	6.51	0.58
MAT-12		5.5	1.6	51	10.5	50	120	27.37	41	5870	0.34	39.7	19	3890	6.58	1.23
MAT-13		6	1.9	92	6.9	60	126	22.82	42	5960	0.23	44.8	20	1980	6.20	1.33
MAT-14		6.5	1.2	79	9.9	80	87	14.54	43	4210	0.17	32.6	44	1240	5.45	0.81
MAT-15		7	1.2	51	11.9	80	98	13.39	47	3460	0.14	21.5	234	1050	4.55	0.93
MAT-16		7.5	14.6	147	9.3	50	298	30.77	20	6770	0.27	37.5	1190	994	7.74	0.26
MAT-17	R	8	0.8	b.d.	3.8	80	1	9.08	40	888	0.02	8.3	6	764	7.75	0.15

b.d. below detection.

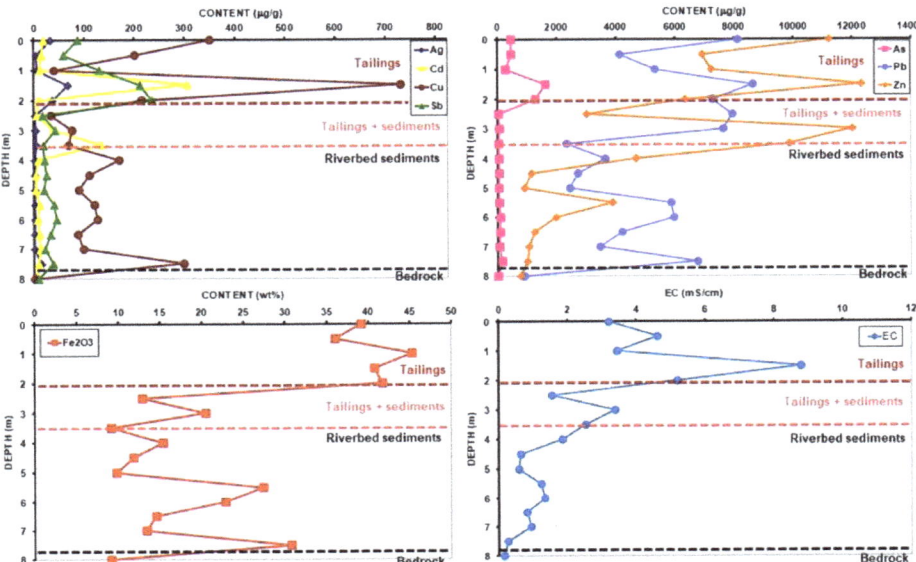

Figure 5. Distribution-with-depth profiles for total ferric iron and trace element concentrations from Las Matildes borehole samples.

The dendrogram (cluster membership) of metals (Ag, Sb, Fe, Cd, Cu, Pb, Zn, Sn) and As in the borehole samples (Figure 6a) is consistent with the metallic signature of the ore in the district (Pb-Zn-Cd-Cu-Sn, and Ag-As-Sb-Fe [20,21]), with As being mainly related to Sb (tetrahedrite-tenantite mineral group). The Ag-Pb-Cd-Zn signature is strongly defined due to the source of the minerals: the Emilia mine. [2,13,14] reported the same geochemical behavior in other works on similar mine ponds.

The cluster analysis of samples from the Las Matildes borehole shows the existence of two groups defining two main associations: mine tailings (MT) and mine tailings mixed with riverbed deposits (MTR) (Figure 6b). The MT association displays the uppermost 2 m thick level, characterized by the presence of ore minerals galena and sphalerite (with lower amounts of pyrite, stannite, jarosite, and siderite). The MTR association represents the mine tailings mixed with riverbed deposits and is characterized by higher amounts of silicates (quartz, illite-kaolinite, chlorite, and greenalite) and anglesite, with sulphide abundance decreasing at depth.

The highest trace element values observed in the upper mine tailings level are related to inefficient metallurgical processing of the benefited ore during the working years. High trace elements and iron contents in mine tailings have been obtained by the authors in similar deposits from Cartagena–La Unión [14], Mazarrón [2], Valle de Alcudia [13,44], and Iberian Pyrite Belt districts [11,12]. Significant metal amounts at the deepest level (riverbed sediments) is noticeable, probably related to the strong annual stormy episodes in this Mediterranean coastal area. Significant flash flood phenomena have also been described affecting abandoned mine deposits and structures [14,45].

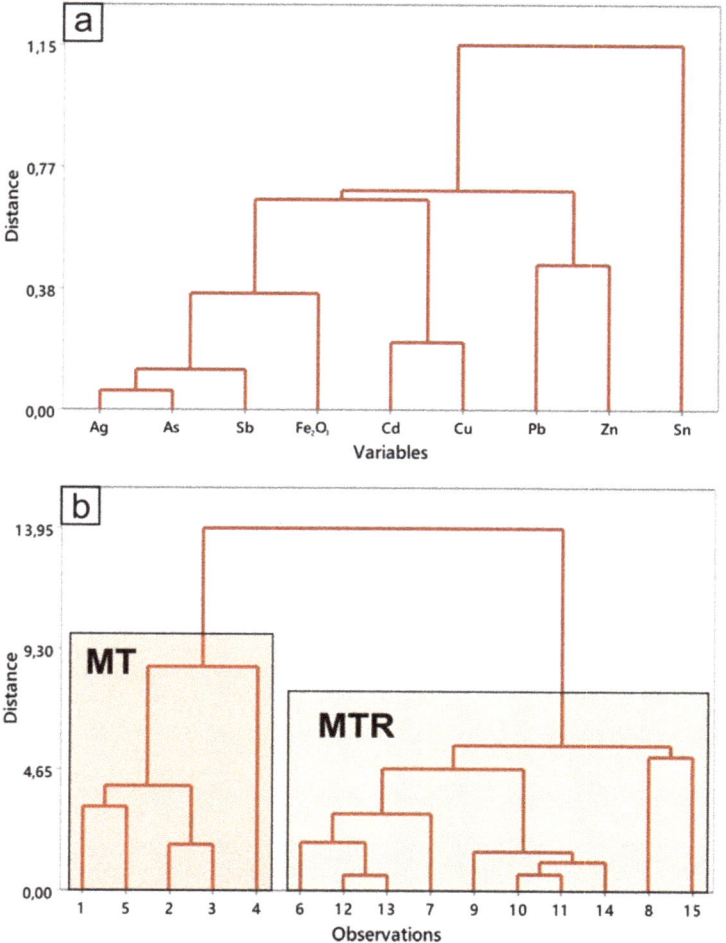

Figure 6. (**a**) Dendrogram (distance: simple) of metals from all borehole samples; (**b**) dendrogram (distance: Euclidean) defining two main associations of samples: mine tailings (MT) and mine tailings mixed with riverbed deposits (MTR).

4.5. Environmental Concerns

Potential environmental concerns are usually classified into three broad categories: (i) human health risks; (ii) ecosystem risks; and (iii) physical hazards. Physical hazards (open pits, open shafts, instable ponds) are broadly found in the Cartagena–La Unión district. Ecosystem and human risks have been assessed using the Geoaccumulation Index (I_{geo}). I_{geo} enables the assessment of contamination of sediments by comparing current and pre-industrial concentrations of heavy metals [46]. It is expressed as $I_{geo} = \log_2 Cn/1.5Bn$, where Cn is the concentration of an element in the sediment sample and Bn is the background concentration of that element in the Earth's crust, according to [47]. The factor 1.5 is usually used to address possible variations due to lithogenic effects. [46] defined six possible ranges: uncontaminated ($I_{geo} \leq 0$), uncontaminated to moderately contaminated ($0 < I_{geo} < 1$), moderately contaminated ($1 < I_{geo} < 2$), moderately to strongly contaminated ($2 < I_{geo} < 3$), strongly contaminated ($3 < I_{geo} < 4$), strongly to extremely contaminated ($4 < I_{geo} < 5$), and extremely contaminated ($I_{geo} > 5$). Metal and As I_{geo} values in borehole samples were calculated and are shown in Figure 7. As expected from contents in Table 2 and Figure 5, most trace elements (As, Cd, Pb, Sb, and Zn) show strong to extreme contamination in all borehole samples. Tailings samples (0–2 m depth) are plotted in the extremely contaminated range ($I_{geo} > 5$), whereas tailings mixed with riverbed sediments and riverbed sediments show strongly to extremely contaminated ranges. Cu and Sn show moderate to uncontaminated values, and Ag is classified as a non-pollutant due to low I_{geo} values. The present study reveals the significant contamination of riverbed sedimentary deposits from mine tailings, not only at the surface, but also at depth. Martínez-Martínez, et al [48] calculated similar I_{geo} values for tailings ponds and natural soils from the southern slope of the mine district. The same behavior has also been observed by the authors in similar abandoned Spanish mine sites: Iberian Pyrite Belt [11,12], San Quintín mining group [13], and Mazarrón district [2].

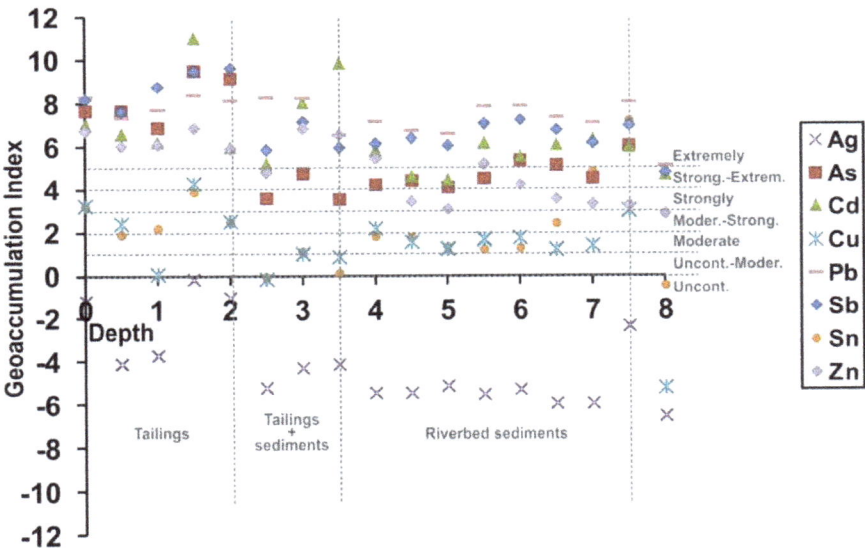

Figure 7. Variation at depth of the Geoaccumulation Index (I_{geo}) of metals and for the borehole samples.

Mineral wastes, intensive agricultural activities, and the population of the Campo de Cartagena area (400,000 people; 600,000 people in summertime) overlap within the district. Cultivated lands (50,000 hectares of horticultural irrigated crops) are located in the adjacent areas downstream of the Cartagena–La Unión mine district (Figure 1). Thus, dispersion of metal-rich particles to agricultural soils surrounding the mine tailings has revealed an important hazard for the environment. This is especially significant in the Mediterranean context, where river courses remain dry for a 5–10-year period due to the semi-arid climate of SE Spain, except when sporadic and torrential rainfall occurs. Since all mining activities ceased in 1991, potentially toxic elements from the mine tailings are being released by water and aeolian erosion. Thus, intense water erosion removes and transports particles from tailings during strong stormy episodes, and aeolian erosion induces the fine particle dispersion. In the study area, the aeolian dispersion of tailings could be considered negligible because of low to moderate wind velocity (2–4 m/s) [49]. The most important erosion mechanism in the Cartagena–La Unión mine district is hydraulic erosion by stormy episodes inducing erosive flash flood phenomena. As a result, very significant amounts of abandoned mine tailings from the Descargador mine pond have been eroded (Figure 2), transported, and re-deposited in El Miedo riverbed, and finally, in the Mar Menor.

There, 100,000 people usually spend the summer in the surrounding touristic areas. There is a perceptible environmental concern, as the sediments from the Mar Menor lagoon are highly polluted by heavy metals transferred from the Cartagena–La Unión mine district [50]. Reference [51] measured concentration values of Zn (>3500 µg/g) and Pb (≈3000 µg/g) 48 and 43 times higher, respectively, than those in the Mediterranean Sea reference sediment sample. Erena et al. [52] develops an operational system to monitor the evolution and variability of the turbidity and chlorophyll-a levels through the joint use of remote sensing techniques (Operational Land Imager and SPOT images) and in situ data.

Together with the use of geochemical data to infer environmental concerns, as previously stated, the geophysical characterization of abandoned mine deposits constitutes a useful, cheap, and fast acquisition technique to obtain complementary information. The use of ERI allows obtaining crucial information necessary for a proper evaluation, such as the thickness and internal structure of the deposits, the occurrence or not of AMD flows that would affect both the surface and groundwater, and the computation of the volume of potentially hazardous material stored in the mine deposits. The latter constitutes essential information in the case where reclamation and/or remediation of the deposits is envisaged. Urgent work must be clearly focused on (i) gaining in-depth knowledge of hydraulic erosion and dispersion, (ii) permanent monitoring of toxic element contents in waters flowing towards the Mar Menor, (iii) remediating resulting soils, and (iv) re-analyzing agricultural practices from an ecotoxicological point of view.

The inputs that can be highlighted compared to previous works are (i) a more detailed geochemical and mineralogical characterization of tailings and riverbed sediments from surface to bedrock samples (borehole), not only from surficial ones; (ii) the assessment of contamination of riverbed sediments estimated using the Geoaccumulation Index; (iii) application of the ERI method in this area to obtain more detailed information about the internal structure of the deposits and bedrock; (iv) the temporal evolution of the mining landscape of La Unión area by the use of orthoimages covering the last 70 years, the most active mining period.

5. Conclusions

The joint use of landscape evolution studies, and geophysical, mineralogical, and geochemical techniques has been confirmed as providing a complete environmental characterization of abandoned mine sites, also allowing estimations of the pollution grade and the extent of affected zones. The natural landscape to the E of La Unión town has been deeply modified by the mining activity. The former summits of Sierra de Cartagena–La Unión disappeared while huge depressions were created due to mineral extraction in open-pit mines. Rock, molten slags, and tailings deposits accumulated in ephemeral riverbeds and foothills, altering the drainage pattern and burying the fluvial channels. Nearly one fourth of the studied territory is overlaid by mining wastes. These transformations are evidenced in the geophysical, mineralogical, and geochemical results obtained in this work. At the Descargador mine site, geophysical surveys and mineralogical–geochemical data from borehole samples provide similar results. For the mine pond infilling, ERI surveys have allowed imaging a 5 to 8 m thick three-unit sequence with low resistivity values, in good agreement with the alternating layers of clay and sandy texture seen at the available outcrop in the area. The geometry of the mine pond boundary is clearly imaged at depth due to the high resistivity values and high dielectric permittivity contrast of the metasediments bounding the mine pond.

At the dry riverbed site, a ~4-m-thick upper unit of low resistivity values and a ~3.5–4-m-thick lower unit of higher resistivity values were distinguished and respectively correlated with surficial mine tailings and underlying riverbed sediments. The highest resistivity values correspond to the metasediments located below the two previous units. These results are in good agreement with borehole data. Significant amounts of pyrite, sphalerite, and galena were identified in the borehole samples, and are much more concentrated in the tailings upper levels than in the sediment's lower ones. The alluvial material from the riverbed showed significant toxic metal contents (As, Cd, Fe, Pb, and Zn) as well. Mineralogical and geochemical data have proved the important removal of pollutants by water erosion from mine waste deposits that finally reach the water courses. In fact, rock and metallurgical wastes alter the drainage pattern and bury the headwaters of ephemeral channels. The strongly seasonal character of the Mediterranean climate, which concentrates almost all the annual rainfall in heavy, episodic storms, increases the potential hazard in this area. Thus, strong water erosion and the transport of extremely contaminated tailings (I_{geo}) affect the agricultural soils and the Mar Menor, one of the most important touristic destinations in SE Spain. Both reclamation of the Cartagena–La Unión mining district and intensive monitoring of waters and soils are highly recommended to recover one of the most problematic yet tourist-popular areas of SE Spain from the environmental point of view.

Author Contributions: Data curation, T.M.-C. and D.G.-O.; Formal analysis, S.M.-V., C.d.I.-S.J. and J.L.; Investigation, T.M.-C. and P.M.-P.; Methodology, D.G.-O.; Writing—original draft, T.M.-C., D.G.-O. and S.M.-V.; Writing—review & editing, P.M.-P., C.d.I.-S.J., J.L. and Á.F. All authors have read and agreed to the published version of the manuscript.

Funding: This work has been accomplished within the framework of project URJC-CM-2006-CET-0636 funded by Comunidad de Madrid and Universidad Rey Juan Carlos.

Acknowledgments: We would like to thank the three anonymous reviewers and the editor for their suggestions and comments that considerably improved the quality of the paper.

Conflicts of Interest: The authors declare no conflict of interest.

References

1. Robles-Arenas, V.M.; Rodríguez, R.; García, C.; Manteca, J.I.; Candela, L. Sulphide-mining impacts in the physical environment: Sierra de Cartagena–La Unión (SE Spain) case study. *Environ. Geol.* **2006**, *51*, 47–64. [CrossRef]
2. Martín-Crespo, T.; Gómez-Ortiz, D.; Martínez-Pagán, P.; de Ignacio-San José, C.; Martín-Velázquez, S.; Lillo, J.; Faz, A. Geoenvironmental characterization of riverbeds affected by mine tailings in the Mazarrón district (Spain). *J. Geochem. Explor.* **2012**, *119–120*, 6–16. [CrossRef]
3. Morris, B.; Shang, J.; Howarth, P.; Witherly, K. Application of Remote Sensing and Airborne Geophysics to Mine Tailings Monitoring. In *Copper Cliff, Ontario. Symposium on the Application of Geophysics to Engineering and Environmental Problems*; Environmental & Engineering Geophysical Society: Denver, CO, USA, 2002. [CrossRef]
4. Smith, B.D.; McDougal, R.R.; McCafferty, A.E.; Deszcz-Pan, M.; Yager, D.B. Helicopter Electromagnetic and Magnetic Survey of the Upper Animas River Watershed; Application to Abandoned Mine Land Studies. In Proceedings of the 17th EEGS Symposium on the Application of Geophysics to Engineering and Environmental Problems, Colorado Springs, CO, USA, 22–26 February 2004; pp. 140–155.
5. Booterbaugh, A.P.; Bentley, L.R.; Mendoza, C.A. Geophysical Characterization of an Undrained Dyke Containing an Oil Sands Tailings Pond, Alberta, Canada. *J. Environ. Eng. Geoph.* **2015**, *20*, 303–317. [CrossRef]
6. Haegeman, W.; Van Impe, W.F. Characterization of Disposal Sites from Surface Wave Measurements. *J. Environ. Eng. Geophys.* **1999**, *4*, 27–33. [CrossRef]
7. Lghoul, M.; Teixidó, T.; Peña, J.A.; Hakkou, R.; Kchikach, A.; Guérin, R.; Jaffal, M.; Zouhri, L. Electrical and Seismic Tomography Used to Image the Structure of a Tailings Pond at the Abandoned Kettara Mine, Morocco. *Mine Water Environ.* **2012**, *31*, 53–61. [CrossRef]
8. Loke, M.H.; Chambers, J.E.; Rucker, D.F.; Kuras, O.; Wilkinson, P.B. Recent developments in the direct-current geoelectrical imaging method. *J. Appl. Geophys.* **2013**, *95*, 135–156. [CrossRef]
9. Martínez, J.; Rey, J.; Hidalgo, M.C.; Garrido, J.; Rojas, D. Influence of measurement conditions on the resolution of electrical resistivity imaging: The example of abandoned mining dams in the La Carolina District (Southern Spain). *Int. J. Min. Process* **2014**, *133*, 67–72. [CrossRef]
10. Acosta, J.A.; Martínez-Pagán, P.; Martínez-Martínez, S.; Faz, A.; Zornoza, R.; Carmona, D.M. Assessment of environmental risk of reclaimed mining ponds using geophysics and geochemical techniques. *J. Geochem. Explor.* **2014**, *147*, 80–90. [CrossRef]

11. Martín-Crespo, T.; de Ignacio-San José, C.; Gómez-Ortiz, D.; Martín-Velázquez, S.; Lillo, J. Monitoring study of the mine pond reclamation of Mina Concepción, Iberian Pyrite Belt (Spain). *Environ. Earth Sci.* **2010**, *54*, 1275–1284. [CrossRef]
12. Martín-Crespo, T.; Martín-Velázquez, S.; Gómez-Ortiz, D.; de Ignacio-San José, C.; Lillo, J. A geophysical and geochemical characterization of sulphide mine ponds at the Iberian Pyrite Belt (Spain). *Water Air Soil Pollut.* **2011**, *217*, 287–405. [CrossRef]
13. Martín-Crespo, T.; Gómez-Ortiz, D.; Martínez-Pagán, P.; Martín-Velázquez, S.; Esbrí, J.M.; de Ignacio-San José, C.; Sánchez-García, M.J.; Montoya-Montes, I.; Martín-González, F. Abandoned mine tailings in cultural itineraries: Don Quijote Route (Spain). *Eng. Geol.* **2015**, *197*, 82–93. [CrossRef]
14. Martín-Crespo, T.; Gómez-Ortiz, D.; Martín-Velázquez, S.; Martínez-Pagán, P.; de Ignacio-San José, C.; Lillo, J.; Faz, A. Geoenvironmental characterization of unstable abandoned mine tailings combining geophysical and geochemical methods (Cartagena–La Unión). *Eng. Geol.* **2018**, *232*, 135–146. [CrossRef]
15. Martín-Crespo, T.; Gómez-Ortiz, D.; Martín-Velázquez, S. Geoenvironmental Characterization of Sulfide Mine Tailings. In *Applied Geochemistry with Case Studies on Geological Formations, Exploration Techniques and Environmental Issues*; Mazadiego, F.L., Barrio-Parra, F., Izquierdo-Díaz, M., Eds.; IntechOpen: London, UK, 2020; pp. 1–26. [CrossRef]
16. Montoya-Montes, I.; Cano-Bermejo, I.; Sánchez-García, M.J.; de Ignacio-San José, C.; Martín-Velázquez, S.; Gómez-Ortiz, D.; Martín-Crespo, T.; Martín-González, F. Formación de cuerpos dunares a partir de lodos mineros: Mina de San Quintín (Ciudad Real, España). *Geotemas* **2012**, *13*, 1487–1490.
17. Martín Duque, J.F.; Zapico, I.; Oyarzun, R.; López García, J.A.; Cubas, P. A descriptive and quantitative approach regarding erosion and development of landforms on abandoned mine tailings: New insights and environmental implications from SE Spain. *Geomorphology* **2015**, *239*, 1–16. [CrossRef]
18. Villar, J.B.; Egea, P.; Fernández, J.C. *La Minería Murciana Contemporánea (1930–1985)*; IGME: Madrid, Spain, 1991; 256p.
19. IGME. *Inventario Nacional de Depósitos de Lodos*; Published database; IGME (Instituto Geológico y Minero de España): Madrid, Spain, 2002.
20. López García, J.A. Estudio Mineralógico, Textural y Geoquímico de Las Zonas de Oxidación de Los Yacimientos de Fe-Pb y Zn de la Sierra de Cartagena (Murcia). Ph.D. Dissertation, Universidad Complutense, Madrid, Spain, 1985.
21. Sanmartí, L.; Gaya, A.; Molina, C.; Amores, S.; Villanovade-Benavent, C.; Torró, L.; Melgarejo, J.C.; Manteca, J.I. The mineral association in the Emilia and Brunita Pb-Zn-Ag-(Sn) deposits, La Unión, Murcia (Spain). In Proceedings of the 2th SGA Biennial Meeting, Uppsala, Sweden, 12–15 August 2013.
22. Martín-Rosales, W.; Pulido-Bosch, A.; Vallejos, Á.; Gisbert, J.; Andreu, J.M.; Sánchez-Martos, F. Hydrological implications of desertification in southeastern Spain. *Hydrol. Sci. J.* **2007**, *52*, 1146–1161. [CrossRef]
23. Robles-Arenas, V.M.; Candela, L. Hydrogeological conceptual model characterisation of an abandoned mine site in semiarid climate. The Sierra de Cartagena–La Unión (SE Spain). *Geol. Acta* **2010**, *8*, 235–248.
24. SIAM. *Sistema de Información Agraria de Murcia*; SIAM: Murcia, Spain, 2017; Available online: http://siam.imida.es (accessed on 28 September 2017).
25. Mouzo Pagán, R. Crónicas Mineras de Rogelio Mouzo Pagán. 2012. Available online: http://cronicasmineras.blogspot.com.es/2012/10/inundaciones-en-la-sierra-minera-de.html (accessed on 14 April 2020).
26. Kearey, P.; Brooks, M.; Hill, I. *An Introduction to Geophysical Exploration*; Blackwell Science: Oxford, UK, 2002; 272p.
27. Revil, A.; Karaoulis, M.; Johnson, T.; Kemna, A. Review: Some low-frequency electrical methods for subsurface characterization and monitoring in hydrogeology. *Hydrogeol. J.* **2012**, *20*, 617–658. [CrossRef]
28. Reynolds, J.M. *An Introduction to Applied and Environmental Geophysics*, 2nd ed.; John Wiley & Sons: New York, NY, USA, 2011; 710p.
29. Telford, W.M.; Geldart, L.P.; Sheriff, R.E.; Keys, D.A. *Applied Geophysics*; University Press: Cambridge, UK, 1990; 770p.
30. Janik, M.; Krummel, H. Geoelectrical methods: 2D measurements. In *Groundwater Geophysics: A Tool for Hydrogeology*; Kirsch, R., Ed.; Springer: Heidelberg, Germany, 2009; pp. 109–117.
31. Loke, M.H.; Acworth, I.; Dahlin, T. A comparison of smooth and blocky inversion method in 2D electrical imaging surveys. *Explor. Geophys.* **2003**, *34*, 182–187. [CrossRef]

32. Buselli, G.; Hwang, H.S.; Lu, K. Minesite groundwater contamination mapping. *Explor. Geophys.* **1998**, *29*, 296–300. [CrossRef]
33. Campbell, D.L.; Fitterman, D.V. Geoelectrical methods for investigating mine dumps. In Proceedings of the Fifth International Conference on Acid Rock Drainage (ICARD2000), Denver, CO, USA, 21–24 May 2000; Society for Mining, Metallurgy, and Exploration, Inc.: Englewood, CO, USA, 2000; pp. 1513–1523.
34. Yuval, D.; Oldenburg, W. DC resistivity and IP methods in acid mine drainage problems: Results from the Copper Cliff mine tailings impoundments. *J. Appl. Geophys.* **1996**, *34*, 187–198. [CrossRef]
35. Niederleithinger, E.; Kruschwitz, S. Multi-channel Spectral Induced Polarization (SIP) measurements on tailings dams. In *Near Surface Meeting*; EAGE: Palermo, Italy, 2005.
36. Martínez-Pagán, P.; Faz, A.; Aracil, E.; Arocena, J.M. Electrical resistivity imaging revealed the spatial properties of mine tailings ponds in the Sierra Minera of Southeast Spain. *J. Eng. Environ. Geophys.* **2009**, *14*, 63–76. [CrossRef]
37. Gómez-Ortiz, D.; Martín-Velázquez, S.; Martín-Crespo, T.; De Ignacio-San José, C.; Lillo-Ramos, J. Application of electrical resistivity tomography to the environmental characterization of abandoned massive sulphide mine ponds (Iberian Pyrite Belt, SW Spain). *Near. Surf. Geophys.* **2010**, *8*, 65–74. [CrossRef]
38. Martínez, J.; Hidalgo, M.C.; Rey, J.; Garrido, J.; Kohfahl, C.; Benavente, J.; Rojas, D. A multidisciplinary characterization of a tailings pond in the Linares-La Carolina mining district, Spain. *J. Geochem. Explor.* **2016**, *162*, 62–71.
39. Cortada, U.; Martínez, J.; Rey, J.; Hidalgo, M.C.; Sandoval, S. Assessment of tailings pond seals using geophysical and hydrochemical techniques. *Eng. Geol.* **2017**, *227*, 59–70. [CrossRef]
40. US EPA. *Clean Water Act*; Section 503; No. 32; Environmental Protection Agency: Washington, DC, USA, 1993; Volume 58.
41. IGN. *Topographic Map of Cartagena, 977, 1:50.000*, 1st ed.; IGN: Madrid, Spain, 1945.
42. IGN. *Topographic Map of Cartagena, 977, 1:50.000*, 2nd ed.; IGN: Madrid, Spain, 2018.
43. Conesa-Alcaraz, H.; Jiménez-Cárceles, F.J.; María-Cervantes, A.; González-Alcaraz, M.N.; Egea-Nicolás, C.; Álvarez-Roger, J. Heavy metal contamination caused by mining activities in the Mar Menor lagoon (SE Spain). In *Heavy Metal Sediments*; Lucía, H.S., Ed.; Nova Science Publishers: New York, NY, USA, 2012; pp. 115–134.
44. Gómez-Ortiz, D.; Martín-Crespo, T.; Esbrí, J.M. Geoenvironmental characterization of the San Quintín mine tailings, Ciudad Real (Spain). *DYNA* **2010**, *161*, 131–140.
45. Oyarzun, R.; Lillo, J.; López-García, J.A.; Esbrí, J.M.; Cubas, P.; Llanos, W.; Higueras, P. The Mazarrón Pb–(Ag)–Zn mining district (SE Spain) as a source of heavy metal contamination in a semiarid realm: Geochemical data from mine wastes, soils, and stream sediments. *J. Geochem. Explor.* **2011**, *109*, 113–124. [CrossRef]
46. Müller, G. Index of geoaccumulation in sediments of the Rhine River. *Geojournal* **1969**, *2*, 108–118.
47. Taylor, S.R.; McLennan, S.M. The geochemical evolution of the continental crust. *Rev. Geophys.* **1995**, *33*, 241–265. [CrossRef]
48. Martínez-Martínez, S.; Acosta, J.A.; Faz Cano, A.; Carmona, D.M.; Zornoza, R.; Cerca, C. Assessment of the lead and zinc contents in natural soils and tailings ponds from the Cartagena–La Unión mining district, SE Spain. *J. Geochem. Explor.* **2013**, *124*, 166–175. [CrossRef]
49. Ministerio de Medio Ambiente. *Rosas de Viento (1971–2000)*; CD-Rom: Madrid, Spain, 2005.
50. Marín-Guirao, L.; Cesar, A.; Marín, A.; Vita, R. Assessment of sediment metal contamination in the Mar Menor coastal lagoon (SE Spain): Metal distribution, toxicity, bioaccumulation and benthic community structure. *Cienc. Mar.* **2005**, *31*, 413–428.
51. Dassenakis, M.; García, G.; Diamantopoulou, E.; Girona, J.D.; García-Marin, E.M.; Filippi, G.; Fioraki, V. The impact of mining activities on the hypersaline Mar Menor lagoon. *Desalin. Water Treat.* **2010**, *13*, 282–289. [CrossRef]
52. Erena, M.; Domínguez, J.A.; Aguado-Giménez, F.; Soria, J.; García-Galiano, S. Monitoring coastal lagoon water quality through remote sensing: The Mar Menor as a case study. *Water-Sui* **2019**, *11*, 1468. [CrossRef]

© 2020 by the authors. Licensee MDPI, Basel, Switzerland. This article is an open access article distributed under the terms and conditions of the Creative Commons Attribution (CC BY) license (http://creativecommons.org/licenses/by/4.0/).

Article

Integrated Geological and Geophysical Mapping of a Carbonatite-Hosting Outcrop in Siilinjärvi, Finland, Using Unmanned Aerial Systems

Robert Jackisch [1,*], Sandra Lorenz [1], Moritz Kirsch [1], Robert Zimmermann [1], Laura Tusa [1], Markku Pirttijärvi [2], Ari Saartenoja [2], Hernan Ugalde [3], Yuleika Madriz [1], Mikko Savolainen [4] and Richard Gloaguen [1]

[1] Helmholtz-Zentrum Dresden-Rossendorf, Helmholtz Institute Freiberg for Resource Technology, Division "Exploration Technology", Chemnitzer Str. 40, 09599 Freiberg, Germany; s.lorenz@hzdr.de (S.L.); m.kirsch@hzdr.de (M.K.); r.zimmermann@hzdr.de (R.Z.); l.tusa@hzdr.de (L.T.); y.madriz-diaz@hzdr.de (Y.M.); r.gloaguen@hzdr.de (R.G.)
[2] Radai Oy, Teknologiantie 18, 90590 Oulu, Finland; markku.pirttijarvi@radai.fi (M.P.); ari.saartenoja@radai.fi (A.S.)
[3] DIP Geosciences, 100 Burris Street, Hamilton, ON L8M 2J5, Canada; hernan.ugalde@dipgeosciences.com
[4] Yara Suomi Oy, Nilsiäntie 50, 71801 Siilinjärvi, Finland; mikko.savolainen@yara.com
* Correspondence: r.jackisch@hzdr.de; Tel.: +49-0351-260-4750

Received: 3 August 2020; Accepted: 10 September 2020; Published: 15 September 2020

Abstract: Mapping geological outcrops is a crucial part of mineral exploration, mine planning and ore extraction. With the advent of unmanned aerial systems (UASs) for rapid spatial and spectral mapping, opportunities arise in fields where traditional ground-based approaches are established and trusted, but fail to cover sufficient area or compromise personal safety. Multi-sensor UAS are a technology that change geoscientific research, but they are still not routinely used for geological mapping in exploration and mining due to lack of trust in their added value and missing expertise and guidance in the selection and combination of drones and sensors. To address these limitations and highlight the potential of using UAS in exploration settings, we present an UAS multi-sensor mapping approach based on the integration of drone-borne photography, multi- and hyperspectral imaging and magnetics. Data are processed with conventional methods as well as innovative machine learning algorithms and validated by geological field mapping, yielding a comprehensive and geologically interpretable product. As a case study, we chose the northern extension of the Siilinjärvi apatite mine in Finland, in a brownfield exploration setting with plenty of ground truth data available and a survey area that is partly covered by vegetation. We conducted rapid UAS surveys from which we created a multi-layered data set to investigate properties of the ore-bearing carbonatite-glimmerite body. Our resulting geologic map discriminates between the principal lithologic units and distinguishes ore-bearing from waste rocks. Structural orientations and lithological units are deduced based on high-resolution, hyperspectral image-enhanced point clouds. UAS-based magnetic data allow an insight into their subsurface geometry through modeling based on magnetic interpretation. We validate our results via ground survey including rock specimen sampling, geochemical and mineralogical analysis and spectroscopic point measurements. We are convinced that the presented non-invasive, data-driven mapping approach can complement traditional workflows in mineral exploration as a flexible tool. Mapping products based on UAS data increase efficiency and maximize safety of the resource extraction process, and reduce expenses and incidental wastes.

Keywords: unmanned aerial systems; hyperspectral; multispectral; magnetic; geologic mapping; drones; UAV

1. Introduction

Investigating the earth's surface using unmanned aerial systems (UASs) is becoming popular in the earth sciences, as they provide a tool for fast, flexible and high-resolution data acquisition. The integration of spectral and geophysical UAS-based information offers a refined scale between airborne and ground surveys. Numerous studies and reviews have investigated the potential of UASs for various applications, e.g., in the fields of agriculture and forestry, structural geology, and sedimentology [1–7].

UASs offer multiple potential applications in the exploration and mining industry. In mining environments, UASs are nowadays routinely used for topographical surveys, material volume calculation and post-mining environmental monitoring [8,9]. In the context of mineral exploration, UASs provide a non-invasive way to determine vectors towards ore occurrence at deposit scale.

Successful applications of UAS-based surveys in mineral exploration were used to explore rare earths using spectral imaging [10] and target uranium deposits using radiometric gamma survey [11]. UAS geophysical magnetic mapping was employed in exploration for iron, zinc, chromite, or gold deposits [12–15]. UAS-based photogrammetric surface models were used to explore structurally controlled gold deposits [16].

Within the development of an exploration project, drilling is the decisive step for validation and modeling. It represents one primary decision-making tool [17] and at the same time is the most cost-intensive part of mine planning [18]. Hence, UAS-based non-invasive and socially acceptable data acquisition (e.g., geophysical and hyperspectral) combined with robust data-processing methods can help decision-makers minimize investment risks and optimize the drilling program [19].

Most of the above-mentioned studies only employ single sensors to derive geoscientific data. A combination of information from different sensors allows for a more robust geological interpretation. The combination of spectral and magnetic data has long been recognized as a potent tool in airborne mineral exploration [20], because of their capability to provide both surface and subsurface information. Bridging the observation gap between airborne and ground surveying, UASs provide the possibility of carrying different sensors to acquire high-resolution spatial, spectral and temporal data [21,22] which contribute to the understanding of geologic settings [23].

UAS-based hyperspectral imaging and magnetics were identified as a promising sensor combination for direct targeting of iron ores [24], using surficial proxy iron-bearing minerals and high magnetic susceptibility. While there is ample scientific literature on using UAS for geological investigations, UAS are not established in the mineral exploration and mining industry. Arguably, that is due to a lack of case studies, processing and validation schemes, and dedicated software. This study showcases the value of multi-sensor UAS data and provides a guideline to maximize UAS potential in exploration scenarios in order to provide support to exploration geologists.

Here, multi- and hyperspectral drone-based imagery is used to delineate and classify surface lithologies using data fusion. Magnetic data are used to survey the extension of lithologic features and close observation gaps. The data provided by the different sensors are fused and supervised image classification is used to separate spectrally non-distinct rock types. Thus, we can link surface and subsurface information as indicators for mineral occurrences, relating surface classifications to magnetic minerals as lithologic proxies. Our final result is a UAS-borne digital geologic outcrop model, augmented by UAS data-based magnetic forward modeling and validated by a ground-truthing strategy for indirect exploration targeting. This study, to our knowledge at the time, is the first to attempt this integrated approach used for UAS data in geologic mapping and mineral exploration.

Our area of investigation is the Siilinjärvi apatite ore mine in Finland [25]. The site is an ideal testing ground due to the wealth of existing evaluation data, including geophysical [26–29] structural–geological [30–32], geochronological, and mineralogical information [33,34]. We used two on-site survey days to acquire high-resolution UAS data and ground validation in an area of about 1 km^2. We introduce our general and transferable workflow, which we adapt to the specifications of our survey site, show results and interpretation and finalize in five concluding statements.

2. Materials and Methods

In this section, we lay out the UAS survey approach. Our proposed workflow is based upon two fixed-wing UASs, one for magnetic and one for RGB and multispectral measurements, and one multicopter UAS for detailed hyperspectral data acquisition. Both fixed-wings cover the complete target area with high spatial resolution but in reduced spectral detail. The multicopter, on the other hand, provides high spectral resolution but reduced spatial coverage as it acquires data at a lower altitude and pace. This allows higher detail for selected areas of interest within the survey area. We show that the methodic combination of fixed-wings and multicopter complement each other. In the following subsection, we define the proposed workflow (Figure 1), introducing data processing routines and the used ground truthing methods that include spectroscopy, magnetic susceptibility, and structural measurements for a successful field campaign.

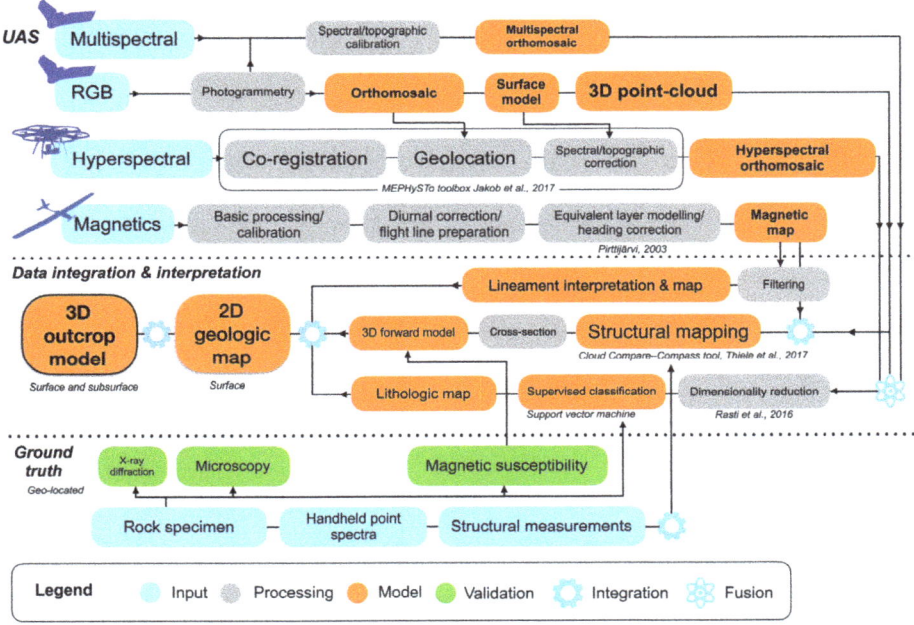

Figure 1. Detailed chart of proposed data-driven unmanned aerial system (UAS) based integration and modeling workflow.

2.1. UAS Data Acquisition Method

We collect RGB and multispectral images (MSI) with a fixed-wing UAS. Structure-from-motion multi view stereo (SfM-MVS) photogrammetric workflows allow us to construct a digital surface model and an orthomosaic from RGB and MSI orthophotos. RGB information, that provides the highest spatial resolution, is used to identify geological structures. MSIs provide additional spectral information compared to RGB images, and a much larger footprint than hyperspectral image (HSI) data in this acquisition setup. All images are geotagged from the drone's onboard GPS. Images are rectified using a number of ground control points.

The resulting SfM-MVS digital surface model (DSM) is used for topographic correction and referencing of the HSIs, and for structural analysis. By means of CloudCompare (www.danielgm.net/cc, vers. 2.11) and its Compass tool plugin [35], we semi-automatically trace and define best-fit planes for faults, foliation, and lithologic contacts directly on the point cloud. For ambiguous areas, supporting UAS data layers (e.g., HSIs, magnetics) are re-examined in the 3D environment.

We acquire UAS-based hyperspectral data frames with pre-coded flight paths in stop-and-go mode along the outcrop to maximize UAS surface coverage. We employ UAS-borne frame-based cameras because of their advantage in creating full image frames which, in our experience, are inherently less distorted than push-broom scanner data. For all HSI data, we manually crop water bodies and non-geologic structures such as roads and vegetated zones from the mosaics, or use semi-automatic masking with a spectral vegetation index.

We conduct UAS-based magnetic surveys using a fixed-wing drone to collect a high-resolution magnetic data set over the survey area, using predefined flight plans. Subsequently, we apply standard magnetic interpretation methods to inspect the shape and dimensions of the measured magnetic anomalies. The analytic signal or total gradient amplitude method [36] is utilized to estimate the location and depth of anomaly sources, as this function is independent of source magnetization direction [37]. Furthermore, we compute the first vertical derivative from total magnetic intensity (TMI) data to enhance the magnetic anomalies and reduce residual influences [38].

2.2. Data Products: Feature Extraction, Supervised Image Classification and Magnetic Forward Modeling

We perform data fusion on a "noisy" outcrop to reduce ambiguity of interpretation while increasing detection confidence and accuracy of classifications [39]. The feasibility of such a fusion approach was laid out for different lithologies at laboratory scale where multi-source hyperspectral and photogrammetric techniques were combined [40]. We apply spatially constrained feature extraction on the UAS-based optical imagery for a consistent classification as part of our multi-sensor data approach to enhance image classification results. The orthogonal total variation component analysis (OTVCA) is used to reduce data dimensionality [41]. It optimizes a cost function to obtain the best representation for multi-layer image data in lower-dimensional feature space, while giving a spatial smoothness over local neighboring pixels by minimizing the total variation of the image signal. OTVCA is robust towards non-systematic, random noise (e.g., salt-and-pepper noise) and has increased weight on neighboring pixels during the dimensionality reduction [42].

For supervised image classification, we choose the support vector machine (SVM) algorithm with Gaussian radial basis function (RBF) kernel, using the library for support vector machines (LibSVM) toolbox [43]. RBF-SVM is proven to perform well with heterogeneous classes and sparse training data, both of which are common cases in geological mapping [42]. Training and validation samples or pixels are defined by selecting pixel aggregates from the HSI data in a GIS environment from points with defined lithologies. The number of training/validation classes varies according to our field observations of the local lithologies.

For a 3D integration and interpretation of our UAS magnetic data, we use forward modeling. Model geometries are established by the UAS-based orthoimagery, hyperspectral mosaics and the DSM. The photogrammetric 3D outcrop model and ground measurements provide constraints on strike/dip and azimuth of the source bodies. Magnetic susceptibility values assigned to the modeled bodies are taken from published literature [27,28,44] and from additional measurements collected with a handheld susceptibility sensor over selected rock samples.

2.3. The Adapted Workflow Conducted for This Survey

We summarize the main characteristics of used sensors here (Table 1) and for specific technical details of our UAS workflow and data acquisition, we refer to Appendix A and [24].

Table 1. Sensors with technical specifications and platforms used for experimental data during this study.

Senor Type/Carrier Platform	Sensor	Resolution Spatial/Spectral	Bands/Sampling Range/Frequency	Data Product
Snapshot camera/Fixed-wing UAS	Parrot S.O.D.A.	5472 × 3648/ –	3/RGB/0.3 Hz	Orthomosaic-RGB, digital surface model
Snapshot camera/Fixed-wing UAS	Parrot Sequoia	1280 × 960/10–40 nm (FWHM)	4/550–790 nm/0.3 Hz	Orthomosaic multispectral
Frame-based camera/Multicopter UAS	Senop Rikola	1010 × 648/8 nm	50/504–900 nm/manual	Orthomosaic hyperspectral
Three-component fluxgate/Fixed-wing UAS	Radai magnetometer	– /0.5 nT	1/±100,000 nT/10 Hz	Magnetic raster grid

We used the senseFly eBee Plus fixed-wing (www.sensefly.com, senseFly, Cheseaux-sur-Lausanne, Switzerland) equipped with either a high-resolution RGB camera (www.parrot.com, Parrot S.O.D.A., Parrot SA, Paris, France), or a multispectral camera (Parrot Sequoia). Processing of RGB and multispectral drone-based data was conducted in Agisoft Photoscan (vers. 1.4, Agisoft Ltd., St.Petersburg, Russia) following recommended protocols [45,46].

Our used hyperspectral frame camera was the Senop Rikola hyperspectral imager (www.senop.fi, Senop, Oulu, Finland). The camera was stabilized by a gimbal (roll and pitch axes) and transported on board of the Aibotix Aibot X6v2 multicopter (www.leica-geosystems.com, Leica Geosystems, Heerbrugg, Switzerland). Automatic HSI georeferencing, mosaicking and application of topographic corrections (c-factor method) on each HSI scene based on the photogrammetric DSM was conducted after Jakob et al., 2017 [47]. We applied the empirical line method [48] to convert the images from radiance to reflectance units, using ground calibration targets.

Magnetics were flown with a composite material fixed-wing UAS Albatros VT2 from Radai Oy (www.radai.fi, Radai Ltd., Oulu, Finland). This UAS utilizes a three-component fluxgate magnetometer, a cost-reducing drone-based sensor [49], attached to the drone's tail boom. With 2.5 m of wingspan and a flight endurance of roughly 3 h, it can easily cover outcrops at square kilometer scales. The survey was flown with traverse lines at 30 m spacing, 99.4^0 azimuth and tie lines at 60 m spacing and 9.4^0 azimuth. The fixed-wing follows the topography along the flight plan based on any available high-resolution digital elevation model. In this case, we used publicly available data from the National Land Survey of Finland.

Magnetic data processing involved removal of spikes and duplicate points, compensation of the fluxgate magnetometer, computation of the total magnetic intensity from the compensated component magnetic data and removal of diurnal effects. Position coordinates, time stamps, barometric pressure and the three-component magnetic data were recorded simultaneously by data logging hardware. An equivalent source algorithm (equivalent layer model (ELM) after [50]) was utilized to prepare the final TMI grid for the survey with the minimum curvature gridding method of ELM data at 15 m cell size. The software Model Vision (vers. 16.0, Tensor Research Pty Ltd., Greenwich, Australia) was used for subsequent forward modeling. Five magnetic profiles crossing along the E–W direction on top and near the main trenches were used in the forward model. A number of simplified bodies with tabular geometries were modeled until a reasonable root mean square error (3–5%) between the measured and synthetic TMI response was achieved.

Covering the known lithologies, ground sampling locations of rock specimens (n = 23) and ground control points (n = 19) were localized with a Trimble global navigation satellite system (GNSS) kit (Trimble R5 base station, Trimble R10 rover; Trimble Inc., Sunnyvale, USA). An overview of the complete workflow is shown in Figure 1.

2.4. Ground Truthing and Laboratory Validation

Data integration at multiple scales, using local ground truth, airborne magnetics, and regional geology is an established method that can provide excellent results and meaningful geologic interpretations [51]. Our ground-truthing program involves rock sampling, as well as structural (n = 38) and spectral measurements (n = 336) and ground-based photogrammetry. All ground samples are geolocated using GNSS. All rock samples are cut and polished for optical investigation and some for analysis with selected geochemical and mineralogical methods.

We take several structural measurements (geological compass), which we incorporate in forward modeling of magnetic data. Main observations are made for contacts, orientation of dykes, and foliation.

During the outcrop studies, we record point representative spectra using a portable spectroradiometer in the available wavelength range of 400–2500 nm. We use selected scans as reference for the supervised image classifications (see Appendix A for point distribution and spectrometer specifications).

Laboratory validation methods, which represent traditional geological, mineralogical, and petrophysical verification methods, are selected to confirm our field observations, and to extract further geologic information from the study site itself. All measurements are conducted on collected rock specimens in the laboratory. Thin section samples are created from specimens covering all main lithologies of the outcrop and examined with optical and polarized light microscopy. Magnetic susceptibility and X-ray diffraction analysis is applied on selected samples (see Appendix D for additional information).

3. Case Study: The Siilinjärvi Carbonatite Complex

Here, we introduce the test area together with the geology. The Siilinjärvi carbonatite complex is situated 20 km north of the city of Kuopio in central Finland and extends for 16 km in N–S and 1.5 km in E–W directions (Figure 2a), with an estimated depth of 800 m [27]. It is one of the oldest known carbonatites with an Archean age of 2.6 Ga±10 Ma, according to U-Pb zircon dating [52]. The Siilinjärvi mine extracts carbonatite–glimmerite-hosted apatite ore for fertilizer production as one of the biggest producers in Europe.

3.1. Local Geology and Study Area

The carbonatite intrusion was emplaced into basement gneiss and deformed by the Svecofennian orogeny at 1.8 Ga [53]. Local rock types are fenite, gneiss, carbonatite–glimmerite, diabase, and other dykes (e.g., local diorites). The central carbonatite–glimmerite ore body has a tabular form, is up to 900 m in width, and is surrounded by a fenite margin created by carbonatite-derived alkali metasomatism of the granite–gneiss country rock and syenite [54].

Brittle and ductile deformation caused structural segmentation of the carbonatite complex and surrounding rock, expressed as sharp boundaries within some areas of intermixed diabase, fenite, tonalite and carbonatite–glimmerite. Fenites as metasomatic products of diorite and gneiss are found in the magmatic contact zones between country rock and carbonatite–glimmerite [25]. This halo of fenitized rocks contains microcline, orthoclase, amphibole, and pyroxene, as well as carbonate, zircon, and quartz.

Several generations of mafic dykes (dolerite) cut the Siilinjärvi intrusion in NW–SE and NNW–SSE directions, with widths ranging from centimeters to meters [54]. Most of the dykes are steeply dipping and, depending on their generation, were subjected to deformation [31]. Sheared feldspar-rich pegmatite dykes with widths varying from 1–50 m were recently discovered by a large-scale drilling program in the Jaakonlampi area [55] and are exposed on the surface. Structural emplacement of the dykes is still not fully understood, but given their size and increased magnetic susceptibilities, they could be an important component of forward modeling.

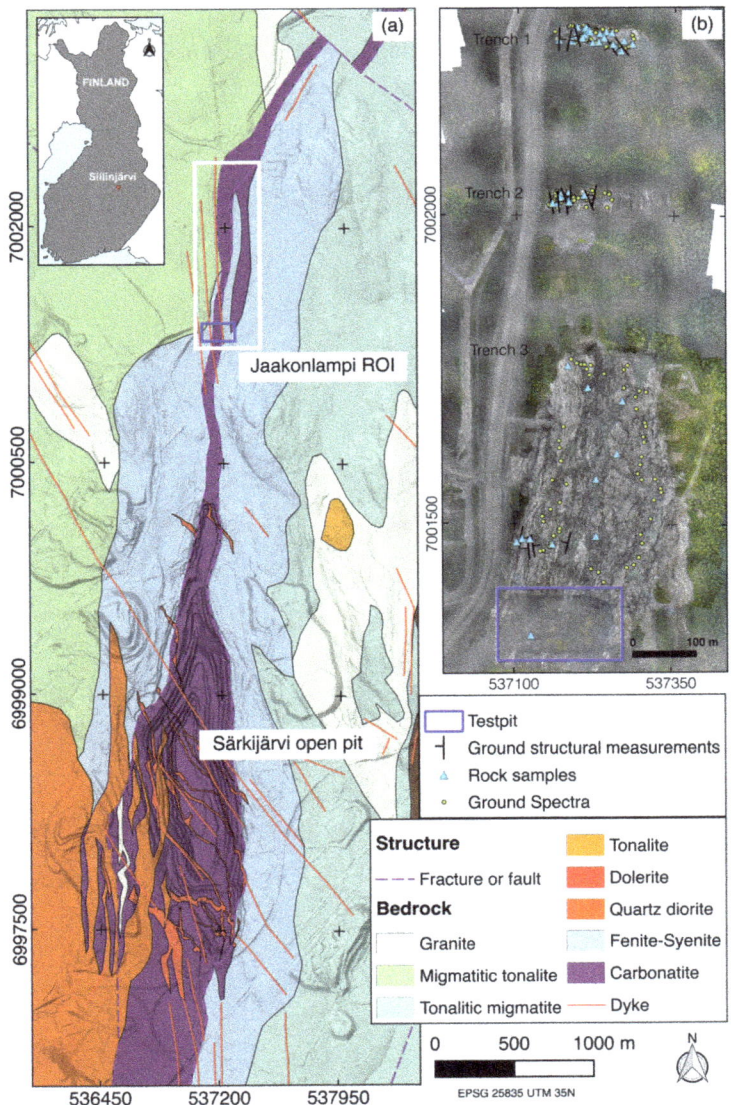

Figure 2. (**a**) Official geologic map (bedrock of Finland scale-free map © Geological Survey of Finland 2019, http://hakku.gtk.fi) that combines data of different map scales. The Jaakonlampi region of interest (ROI) includes our test area for UAS survey. (**b**) UAS-based orthophoto of the Jaakonlampi ROI, showing structural measurements, rock sample positions and ground spectroscopy.

3.2. The Jaakonlampi Test Area

Situated 1.2 km north of the Särkijärvi main pit, the Jaakonlampi area (Figure 2b) provided the test zone for our UAS survey. Jaakonlampi extends ~1 km in the northern direction and is characterized by three distinct exploration trenches, which from north to south, henceforth we refer to as trench 1, trench 2, and trench 3. The mine company expanded the exploration program for trench 3 in 2018 and removed significant soil overburden, uncovering a large exploration trench (Figure 3c).

However, the recent uncovering resulted in some remains of sand and clays on top of trench 3's surface, challenging subsequent image classifications.

Figure 3. Photographic illustrations of the applied field methods, data acquisition by UASs and ground truthing, and overviews for the visited outcrops in the Jaakonlampi area. (**a**) Hyperspectral survey using multicopter UAS. (**b**) Magnetic fixed-wing UAS. (**c**) Ground spectroscopy and geo-locating on trench 3. (**d**) Trench 1 during hyperspectral survey. (**e**) Ground sampling on trench 2 including structural measurements and spectral surface scans. (**f**) Contact between dolerite dyke and feldspar-rich pegmatite intrusions. (**g**) Photograph of the test pit wall that marks the southern survey end zone.

Within the glimmerite, the carbonatite is featured as thin, sub-vertical veins. The composition of carbonatite is mainly calcite, apatite (1.4–2.3 vol.%,) and magnetite (1 vol.%). On average, the ore contains 65% phlogopite, 19% carbonates, 10% apatite, 5% richterite, and 1% accessories that are mainly magnetite and zircon [54]. The composition of the three trenches (Figure 3c–f) is similar to the general configuration of the Siilinjärvi deposit. The southern-located trench 3 connects seamlessly to a so-called test pit (Figure 3g), an outcrop wall which presents a vertical geologic cross section of the lithological units further used in this study:

- Carbonatite–glimmerite (CGL) and carbonatite (CRB)
- Dolerite (DL)
- Felspar-rich pegmatite veins (FSP-PEG)
- Fenite (resp. syenitic fenite or fenite-syenite) (FEN-SYN)
- Glimmerite (GL)
- Granite–gneiss (GRGN)

4. Results

We present the mapping results sorted by method. Survey conditions, camera settings, and technical UAS-related data are found in Appendix A (Table A1). All trenches and the forested areas in between were surveyed by high-resolution RGB and multispectral UAS images and UAS magnetics. Additional hyperspectral imaging covers trench 1 completely, the western half of trench 2 (the other half was submerged by water), and the northern half of trench 3. Visual observation of the test pit wall showed dipping bodies between 70–90°, broadly striking along N–S.

4.1. Ground Spectroscopy and Principal Lithologic Representation

We measured the three trenches in situ with a representative dense spectral point sampling campaign (Figure 2b) at trench 1 and 2 (275 locations). For trench 3, we conducted a broader sampling sweep (61 locations, 37 of those covered by UAS-based HSIs and MSIs). While understanding the spectral differences of the lithologies, we selected training samples for the supervised classification (Figure 1, last row) guided by the ground spectra (representative spectra in Figure 4), the RGB mosaic, and the OTVCA layers.

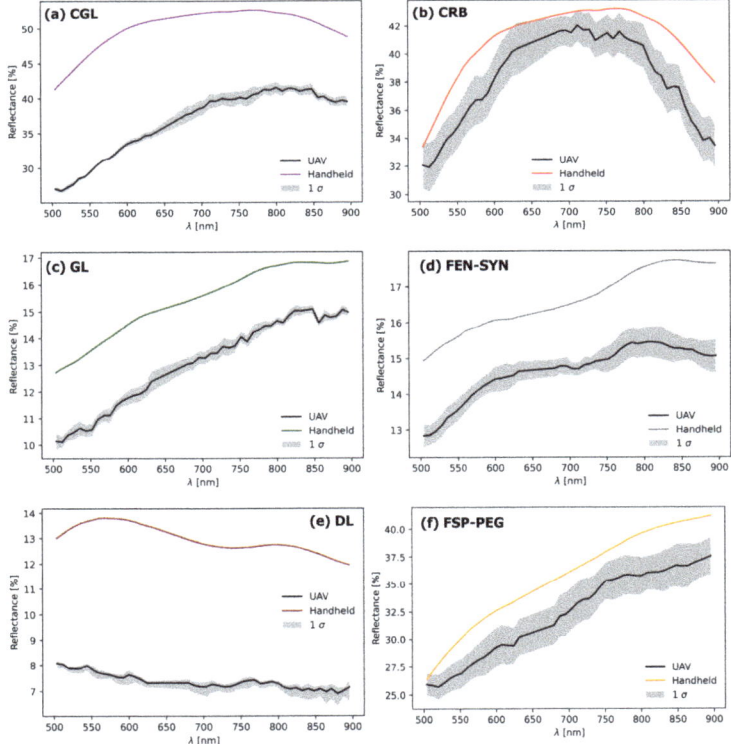

Figure 4. (a–f) Representative hyperspectral image (HSI) drone-based spectra compared to handheld point scans from the same lithologies and in direct spatial neighborhood, plotted between 504–900 nm. Spectra were manually extracted from representative spots of the main lithologies. GCL = Carbonatite–glimmerite; CRB = Carbonatite; GL = Glimmerite; FEN-SYN = Fenite–syenite; DL = Dolerite; FSP-PEG = Feldspar–pegmatite.

A relatively broad absorption between 900~1200 nm is attributed to the Fe^{2+} content in calcite and dolomite-rich carbonatite [56]. We detected rare earth element (REE) related absorptions at 580 ± 10 nm, 740 ± 10 nm, and 800 ± 10 nm (Figure 5b) [57]. A spectral shift from calcite-rich to dolomite-rich carbonatite is visible in our point scans, at the spectral minima transition from 2320 nm to 2340 nm (Figure 5c), related to vibrational processes of CO_3 combinations and overtones [58,59]. For glimmerite spectra, rich in phlogopite and biotite, we observe characteristic OH⁻ features at 1380 ± 10 nm and Mg-OH vibrational bands at 2320 ± 10 nm and 2380 ± 5 nm [58]. Carbonates are likely to influence the position of the absorption minima here. Hydroxyl group absorption features are seen for fenitized syenite spectra at 2315 nm and 2385 nm. Dolerite spectra show the lowest overall reflectance, weak Fe^{2+}/Fe^{3+} charge-transfer absorptions at 800 nm [60] due to iron alteration but a prominent absorption at 1920 nm (OH⁻ related). Feldspar-rich pegmatites, expressing a larger spectral variety and incorporating Fe^{2+} and pronounced OH⁻ features are found at 1410 nm, 2200 nm (Al-OH), and 2350 nm (Mg-OH). We observed apatite in carbonatite–glimmerite rock samples as a possible proxy for REE occurrence.

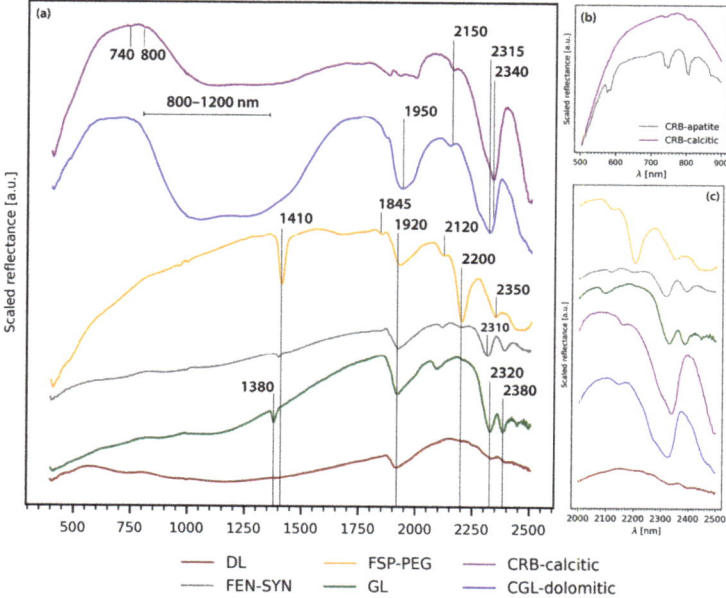

Figure 5. (a) Six selected handheld scans, representative for the mapped lithologies, plotted between 450–2500 nm and with indicated positions of spectral absorptions. (b) Zoom within the available UAS-based HSI wavelength window (504–900 nm) showing two carbonatites, where both apatite-rich carbonatites express some rare earth element-related absorption. (c) Enhanced view of the shortwave-infrared region between 2000–2500 nm, same color legend. DL = Dolerite; FSP-PEG = Feldspar–pegmatite; CRB = Carbonatite; CGL = Carbonatite–glimmerite; FEN-SYN = Fenite–syenite; GL = Glimmerite.

4.2. UAS-Based Optical Remote Sensing Observations

The RGB orthophoto (Figure 6a), the MSI mosaic (Figure 6b) and the HSI mosaics (Figure 6c) provide first-order information for subsequent interpretation. Low ceiling clouds were present during the RGB acquisition flight, producing horizontal gray stripes in the data. Occasional leftover dirt patches reduce the spectral quality in some HSI scans of trench 3. Topographic expressions are seen in the UAS-based DSM (ground sampling distance 10.6 cm; Figure 6d). The eBee RGB and MSI orthomosaics envelope the complete rock outcrop extension, which is covered by vegetation stripes between trenches 1 and 2 and between trenches 2 and 3. HSI mosaics were acquired completely for trenches 1 and 2. Trench 2 was partly covered with water on large surface portions. Low illumination conditions during the HSI acquisition of trench 3 reduced the spectral quality for all scans there. We augment the data set of trench 3 by using two additional data layers (DSM, MSIs) from the area. Those additional layers were resampled to the common lowest resolution (from the DSM) and fused with the HSI data set before applying the dimensionality reduction by OTVCA to improve supervised image classification.

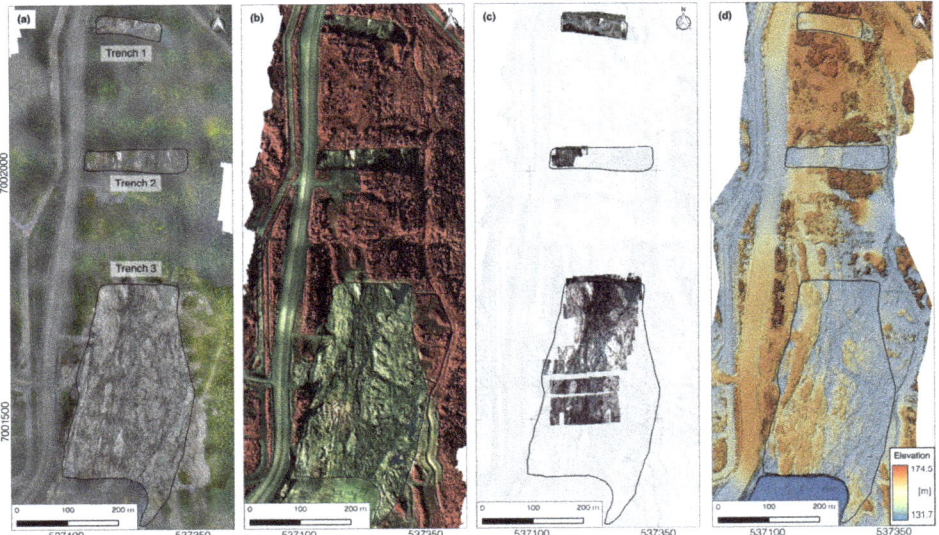

Figure 6. Overview of image-based data products showing the three trenches, with the test pit at the southern end of trench 3. (**a**) RGB orthomosaic from the eBee Plus UAS and S.O.D.A. camera. (**b**) Multispectral false-color infrared mosaic from the eBee Plus UAS and Sequoia camera (bands 735 nm, 660 nm, 550 nm). (**c**) Hyperspectral false-color RGB mosaic from Rikola camera images (bands 650 nm, 551 nm, 504 nm) flown on multicopter UAS. (**d**) Hillshaded digital surface model derived from SfM-MVS photogrammetry, based on eBee Plus orthophotos, elevation in meters above sea level.

The OTVCA-based false-color band combinations we selected for high variations are shown in Figure 7a,b. Only the merged multi-sensor OTVCA bands for trench 3 (Figure 7c) contain MS, RGB, and DSM data. Fusing those additional data layers for the classification of trench 3 helped to close some data coverage gaps of the hyperspectral survey (Figure 7c). The final classification produced by the SVM classifier and visual inspection was used to create the surface geology map. The resulting overall accuracy (OA) for all three trenches (>90% OA each) is acceptable. Overall supervised classification accuracies with used ground truth are as follows in mean accuracy (MA), OA, and kappa coefficient (κ): trench 1—MA 96.5, OA: 95.3, κ: 0.94; trench 2—MA 91.0, OA: 90.0, κ: 0.88; trench 3—MA 95.3, OA: 95.3, κ: 0.95. We refer to Appendix A for visualized training and validation samples, as well as confusion matrices per trench classification.

Although we achieved high classification accuracies, three falsely classified zones are identified (Figure 7f), i.e., a large block of carbonatite (25 m length) in the fenitized syenite and a stripe of dolerite extending into the feldspar–pegmatites and the mine road.

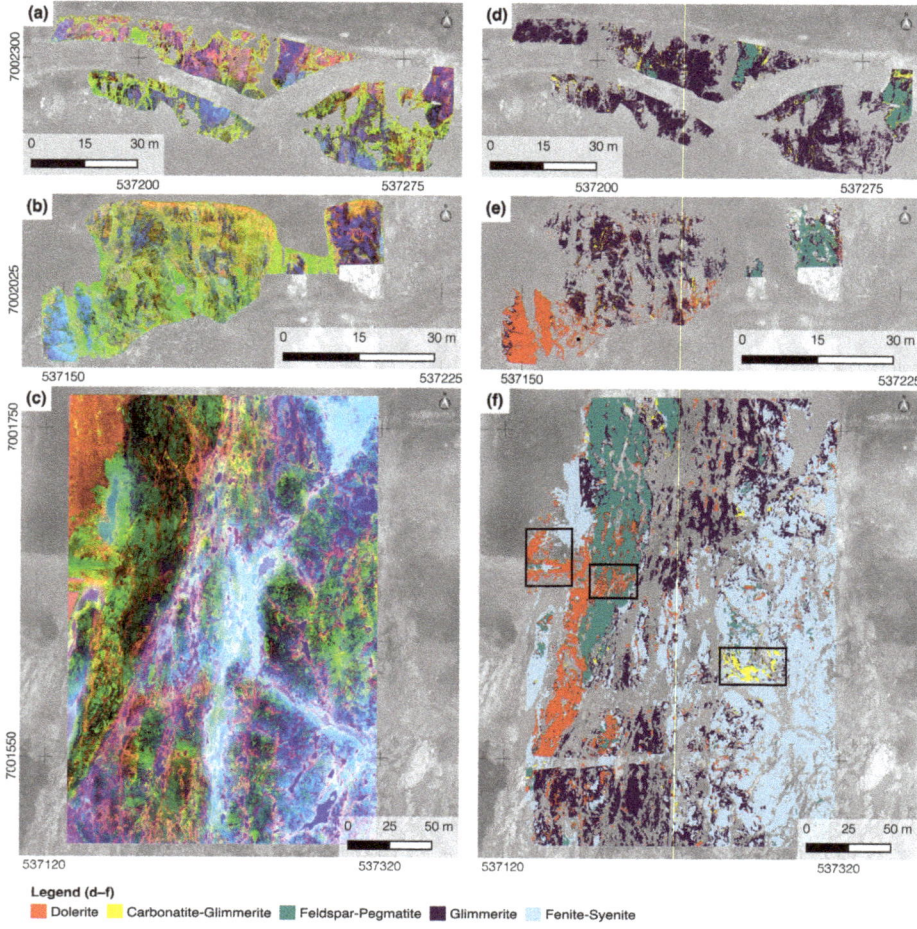

Figure 7. Display of feature extractions (**a**–**c**) and supervised classification maps, where only the geologically meaningful classes are shown for comparison (**d**–**f**), plotted on a grayscale UAS-RGB background orthophoto. (**a**) Trench 1—Orthogonal total variation component analysis (OTVCA) color combination bands 2,1,4. (**b**) Trench 2—OTVCA color combination bands 2,1,3. (**c**) Trench 3—OTVCA color combination bands 3,5,2. (**d**) Trench 1—Support vector machine (SVM) supervised image classification. (**e**) Trench 2—SVM supervised image classification. (**f**) Trench 3—SVM supervised image classification. Black frames highlight misclassified zones.

4.3. UAS-Based Magnetic Observations

Magnetic data interpretation is based on the processed TMI (Figure 8a) and filtered data products. The total survey length was ~ 39 km, with a mean flight height of 48 m above sea level (a.s.l.), a sampling line point distance of 2.1 m, and a mean velocity of 17.7 m/s. We show regional airborne magnetics ([61] modified after Geologic Survey of Finland © 2016) for comparison (Figure 8b). The regional field shows a decreasing tendency towards the west. A pronounced magnetic anomaly, with values reaching 400 nT, is heading in the north to south direction. At the center of trench 3, the TMI trend is decreasing. A TMI field strength reduction is visible at the southern end of trench 3 above the vertical wall of the test pit.

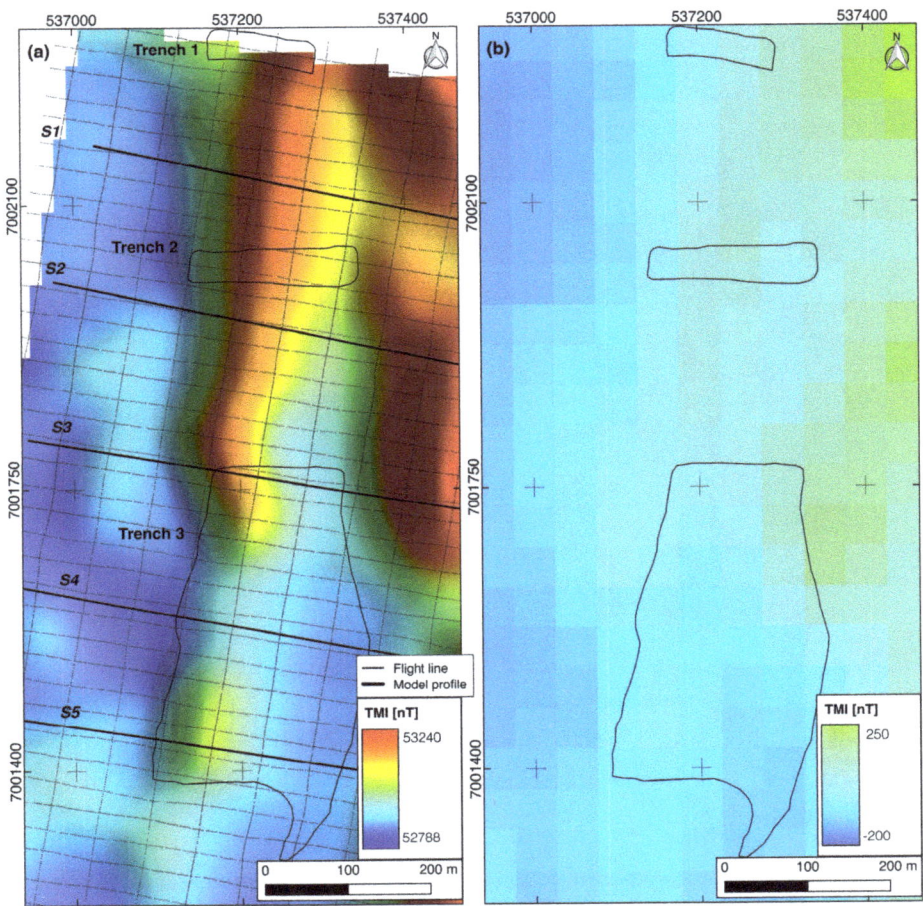

Figure 8. (**a**) Total magnetic intensity data plotted with shaded relief and UAS flight paths as stippled lines from fixed-wing magnetics. Recovered in-line sampling distance after processing varies between 1.5–2.2 m. Bold black profile lines are used in magnetic forward modeling. (**b**) Regional aeromagnetic data from the Geological Survey of Finland (40 m nominal flight altitude, 200 m line spacing; colors are hard-coded; definitive magnetic reference field version 1965 removed from the data).

The first vertical derivative (1VD, Figure 9a) sharpens the edge of the N–S trending anomaly and the 1VD outlines the distinct transition from low to high TMI values, which we interpreted as possible lithologic contact between country rock and fenite. By using the analytic signal (AS), which serves to minimize the impact of any magnetic remanence on the observed magnetic anomaly pattern, we enhance magnetic contacts, interpreted here as carbonatite–glimmerite and country rock (Figure 9b). Based on the aforementioned image classification (Figure 7f), the western border of the dolerite unit could be traced, which is running from N–S through the whole study area. A decrease in the vertical gradient magnitude is seen again in the center of trench 3, where the shear zone is located (Figures 2 and 9) [55]. The spatial width and field strength of the central anomaly could be related to the volume of material replaced by the non-magnetic feldspar-rich pegmatite dykes. The magnetic low at the center of trench 3, starting 50 m north from the test pit, is measured atop the observed fold and shear tectonics, where magnetic minerals are altered, displaced, or destroyed [62]. The two spatially large, oval-shaped anomalies cross above the eastern map border of Figure 8a.

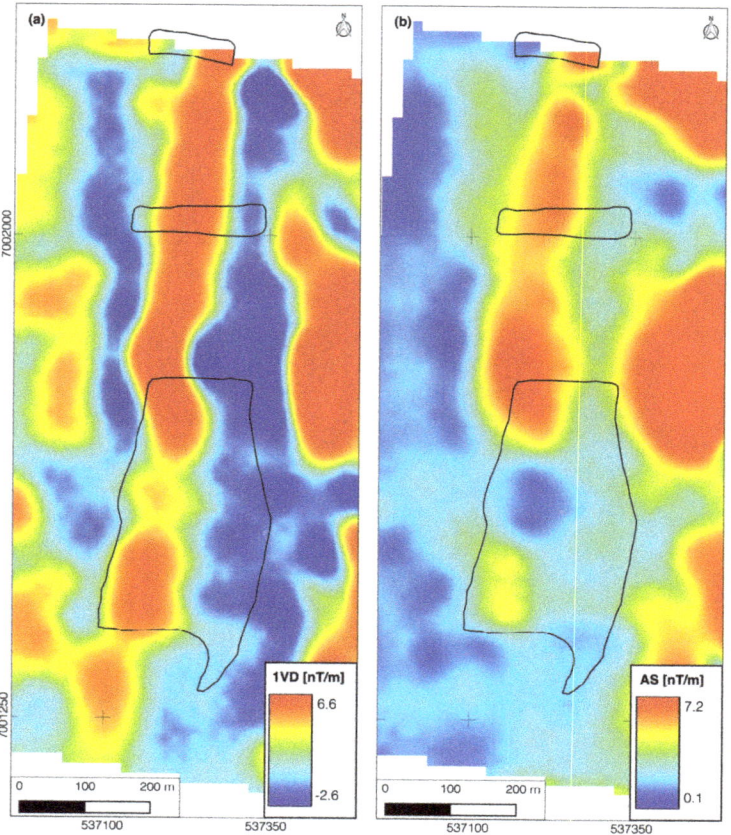

Figure 9. Comparison of magnetic data at different scales with black outlines representing the trenches. (**a**) Analytical signal from UAS total magnetic intensity (TMI) data. (**b**) First vertical derivative from UAS TMI data.

4.4. Geologic Modeling and Ground Magnetic Susceptibility

Magnetic susceptibility measurements are imperative for a supporting forward model as a secondary data derivative, based on UAS magnetics. The susceptibility ranges of our sampled lithologies are aligned with values presented in the literature and our own sampling. Table 2 lists susceptibility ranges for the relevant lithologies.

Table 2. Augmented value range for magnetic susceptibilities based on reference literature and own measurements, values given in SI units.

Lithology	Almqvist et al., 2017 [44]	V. Laakso, 2019 [28]	Measured	Used
Dolerite	$1.26 \times 10^{-4} - 1.29 \times 10^{-3}$	$1.0 \times 10^{-2} - 1.6 \times 10^{-1}$	$7.0 \times 10^{-4} - 1.35 \times 10^{-2}$	$1.0 \times 10^{-5} - 1.7 \times 10^{-2}$
Carbonatite–Glimmerite	$4.27 \times 10^{-4} - 2.09 \times 10^{-1}$	$1.3 \times 10^{-1} - 2.1 \times 10^{-1}$	$1.0 \times 10^{-4} - 1.1 \times 10^{-2}$	$3.2 \times 10^{-3} - 2.5 \times 10^{-2}$
Feldspar–Pegmatite	–	$0 - 5.0 \times 10^{-4}$	$7 \times 10^{-5} - 1.4 \times 10^{-4}$	$1.0 \times 10^{-5} - 5.0 \times 10^{-4}$
Fenite	–	$1.3 \times 10^{-1} - 1.5 \times 10^{-1}$	$1 \times 10^{-6} - 1 \times 10^{-5}$	–

We constructed a model, starting with simple cuboidal geometries, and advanced to polygonal tabular sheets, with their surface geometry constrained by our UAS-based surface geologic map (Figure 10). UAS-based DSM data were used to constrain the top surface of each polygon. An approximate maximum depth of 250 m meters was imposed, based here on available literature information for the study area. Body geometry (strikes and dip, width, azimuth) were taken from photogrammetric interpretation and compared with our own ground measurements. Initial susceptibility values were assigned to geological units on the basis of the literature and measured susceptibilities (Table 2). Optimization of the model was achieved using the inversion tool provided with the ModelVision software. After continuous reiterations, a root mean squared error between synthetic and modeled TMI response of 3–5% was reached per profile. In our model (cross section in Appendix B) one implication could be that the dolerites we measured can reach magnetic susceptibilities close to carbonatite–glimmerite. Yet, this could be an observation at only some depth or related to shearing. The dolerites are known to be low or non-magnetic in the mine area (personal communication, Yara chief mine geologist). The modeling results are integrated in Section 5.1 with the surface data for the final mapping. Extracted body boundaries are used to refine the surface map in a 2D cross section depth map over trench 3 (Figure 10c).

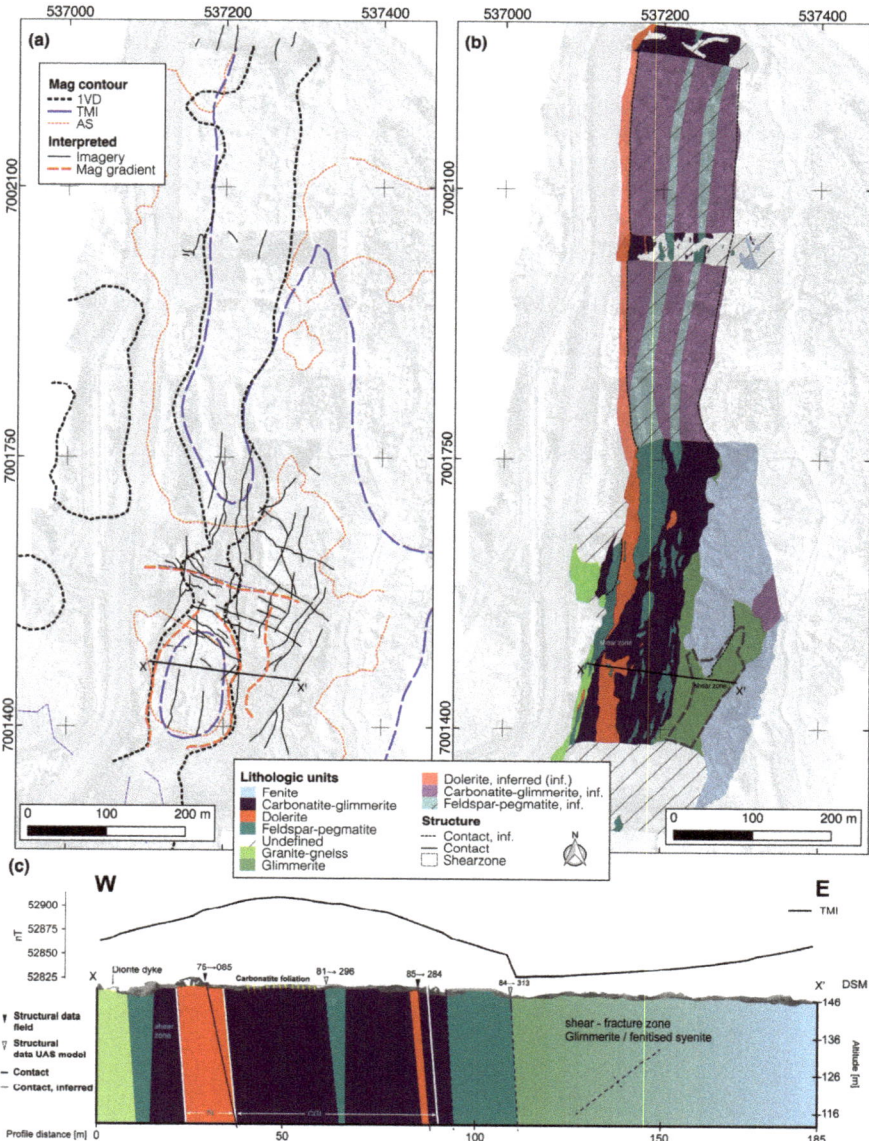

Figure 10. Structural and geological information is synthesized. A semi-transparent MSI mosaic is set as a background for referencing on both maps. (**a**) Extracted isolines from magnetics (Mag contour) are shown together with structural interpretations as observed discontinuities and lithologic contacts, based on UAS-based point clouds, digital surface model (DSM), and orthophotos. (**b**) Interpreted geologic map of surface lithologies. Color legend valid for (**b**) and (**c**), which shows an interpreted profile plot crossing trench 3. Main lithologies are drawn from surface mapping and extended in depth, based on forward modeling and structural measurements from photogrammetry; TMI response plotted above DSM. Orientation of planar features is indicated in dip→dip direction. In the shear zone, the magnetic anomaly is diminished, possibly caused by subsequent alteration and relocation of magnetite. A small diorite intrusion was observed during field mapping.

5. Data Integration and Validation

In this section, we present the integrated results of our UAS mapping approach, bringing together data acquired with UAS platforms and ground survey. All analyses and maps were conducted and created in Quantum GIS (vers. 3.4, QGIS development team). The inferred lithologies between the outcrop trenches are mapped using the UAS magnetic observations. The following link to the integrated 3D model is available online at https://skfb.ly/6U6Xo.

5.1. Geologic Mapping and Interpretation

Structural features (e.g., foliations, discontinuities, lineaments) and contours are interpreted visually in magnetic and DSM data, and with finer detail aided by the RGB orthophotos (Figure 8a). We produced magnetic contours from TMI, AS, and 1VD data. To do so, we calculated the contour lines from TMI and for filtered magnetics, to obtain magnetic isolines per data set in quartered data range steps and subsequently kept only each isoline representing the 50% data threshold. Thus, one isoline shows the arithmetic data threshold representing a mean. We observe that the TMI and 1VD isoline are superimposed along the western border of the main anomaly in the center of trench 3. This might reflect a well-expressed, deep contact of carbonatite–glimmerite and country rock. The 'mag gradient' outlines the observed field decrease (center of trench 3; Figure 9). The geologic surface interpretation (Figure 10b) brings together all data sources: RGB orthophoto, supervised classification of HSIs, and fused data. We extracted 66 discontinuities manually for the three trenches (sum of length: 4.46 km), with a mean length of 50 m per structure. We mapped a high density of features along trench 3, as a result of high contrast in both RGB and HSI mosaics. The visual overlap of RGB, HSIs, DSM and magnetics aided the extraction when contacts or boundaries were blurred or ambiguous. The shear zone in the south-east of trench 3 (Figure 10c) expresses visible lineament offsets and a dense fracture pattern in RGB data. We do not infer fenite as there are too few surface observations for reference, but the magnetics indicate a contact between carbonatite–glimmerites and fenites.

We infer that the lithologies carbonatite–glimmerite, dolerite, and feldspar–pegmatite continue their N–S trend and intersect with the surficial identified structures. A good example is the case for dolerite and feldspar–pegmatite, which we can observe for trenches 1 and 2 (Figure 10a,b compare observed vs. inferred lithologies). Additionally, we map the smaller carbonatite features based on HSI classifications and show them as overlaying foliation (Figure 11). A 3D representation of the pit wall is seen in Figure 12.

By applying the Cloud Compare compass tool [35], we could extract 21 contact planes between feldspar–pegmatite and glimmerite, 10 dolerite contacts, and 6 glimmerite–fenite contact planes, all of which were located in trench 3 (Figure 12). The largest dolerite dyke had a diameter of ~30 m. Trenches 1 and 2 expressed few topographic differences to extract meaningful contact planes.

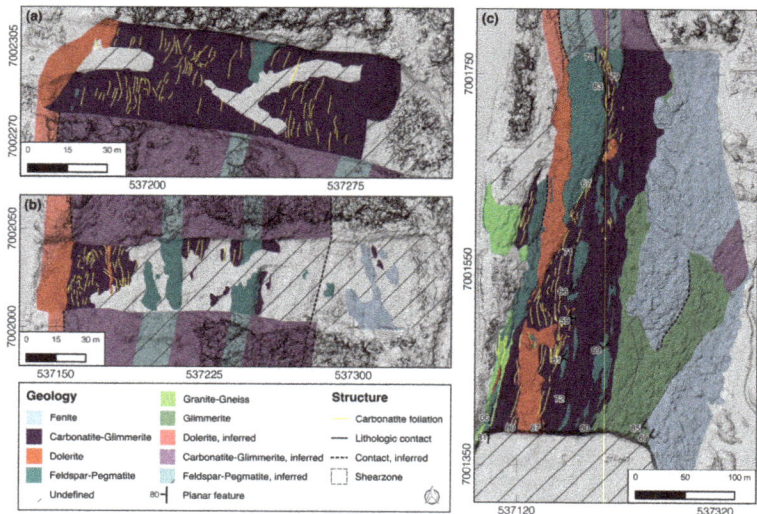

Figure 11. Enlarged maps of the interpreted geology from the three surveyed trenches. Gray background shows a hillshaded representation of UAS-based DSM to add topographic contrast. (**a**) Trench 1. (**b**) Trench 2. (**c**) Trench 3.

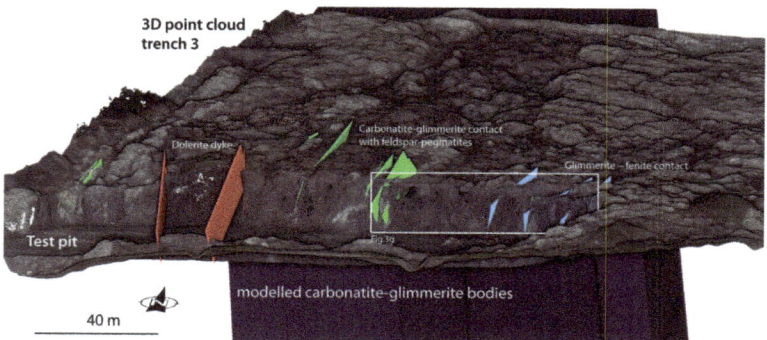

Figure 12. Enlarged view on an orthographically projected point cloud of trench 3 (see also Figures 1a and 2g), showing the test pit wall with 3D best-fit planes for digital structural measurements. The white box highlights the field photograph of Figure 3g. A 3D version is found online at https://skfb.ly/6U6Xo.

5.2. Mineralogic Validation and Additional Observation

We deployed optical microscopy (Appendix C) and X-ray diffraction (XRD) methods for mineralogical analysis. The microscopy of carbonatite–glimmerite shows calcite, a homogeneous distribution of magnetite grains ranging in size from microns to millimeters, and larger pyrite crystals. We observed idiomorph magnetite in rock thin sections of carbonatite–glimmerite, glimmerite, and dolerite. Magnetite seems to be in co-occurrence with pyrite. Combining microscopy and XRD, we detect some presence of magnetite in several mapped carbonatite–glimmerite and glimmerite units of this study. XRD of a bulk handheld specimen collected from carbonatite–glimmerite shows 1.8 wt.% of magnetite. Further evidence of magnetic minerals was only observed in one dolerite sample (2.4 wt.%). We did not identify magnetic minerals in the remaining lithologies from microscopy (fenitized syenite, feldspar–pegmatite). Moreover, XRD patterns detect calcite, apatite, biotite, pyrite, quartz, albite, ankerite, and actinolite (Appendix D).

5.3. Validation of Structural Observations

The results of the digitally extracted structural measurements are summarized (Figure 13) and compared with the ground measurements. High image contrast and geometric expression were found at the test pit of trench 3, and therefore used for extraction. Thirty-two contact points, 6 foliations, and 2 dykes (carbonatite, dolerite) were measured in situ during the field campaign. Digital point cloud measurements of apparent large units were extracted mainly on the test pit wall for dolerite, carbonatite–glimmerite and fenite features. Twenty contacts between carbonatite–glimmerite and feldspar–pegmatite, 10 dolerite dykes, and 6 glimmerite–syenite–fenite contacts were extracted digitally. Our structural observations of the Jaakonlampi area show an N–S trend, which is consistent with the formerly described N–S striking foliation trend of the host rock [25], and shearing along the contacts of intrusions with host rocks [54]. Structural orientations of contacts, dykes and foliations are comparable in their main trends (Figure 13a,b,d). Smaller feldspar–pegmatite units (Figure 13e,f) were measurable along the carbonatite–glimmerite in trench 3. The rather flat surfaces, low topography and reduced RGB image contrast of trenches 1 and 2 could not provide sufficient contrast for usable structural measurements. NW–SE-oriented shearing affects structural expressions in our study area (Figure 13c). Several shearing events were identified in the Jaakonlampi area (four deformation stages with D1 ∥ D3 identified in [55]). At the shear zone of trench 3, we observed contacts of carbonatite–glimmerite with granite–gneiss and an occasional absence of the fenite–syenite halo.

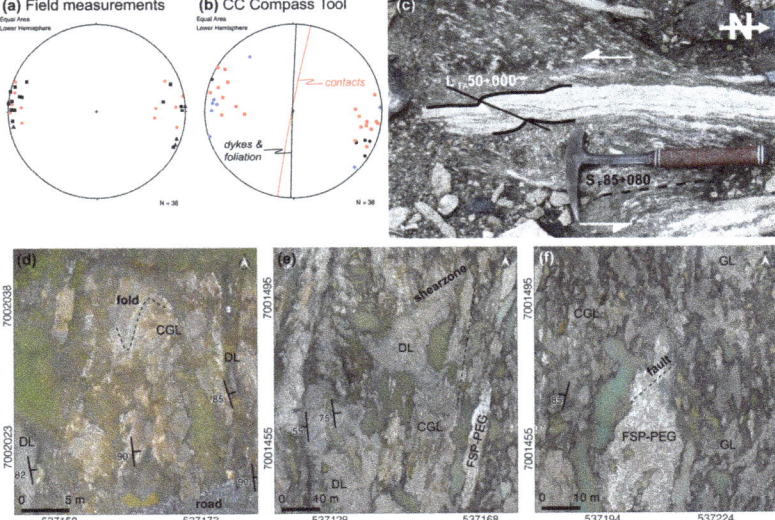

Figure 13. First row (**a–c**) shows a compilation of structural data from field work and point cloud analysis. Second row (**d–f**) presents UAS-based RGB orthophoto zooms with exemplary structural features. (**a**) Structural orientations obtained from field measurements. Triangle: foliation, circle: contacts, box: dykes. (**b**) Structural orientations resulting from point cloud analysis using the Cloud Compare Compass tool. Circle: contact FSP-GL, box: contact FSP-CGL; diamond: dolerite dykes. Large circles in (**b**) are the mean planes derived from weighted contouring (Kamb contours [63]) for the respective sub-groups. (**c**) Field photograph showing detail of the structures and relationship of carbonatite and glimmerite from trench 2. Hammer for scale (length 33 cm). Notation is "Plunge→Trend" for linear (L) and "Dip→Dip Direction" for planar (S) features. (**d**) Close-up of RGB UAS orthophoto of trench 2, with a folded carbonatite–glimmerite section. (**e–f**) UAS-RGB close-ups of trench 3's southern shear zone, showing a larger block of dolerite, relocated. Feldspar–pegmatite (pinch and swell and/or boudinage) dyke indicates horizontal displacement. Planar features measured with compass in the field.

6. Discussion

6.1. Assessing the General UAS Survey Workflow with Focus on Image Data

We tested a survey approach that is only limited by the external conditions for UAS operations, such as weather and legislation. Our multi-sensor UAS toolkit aids geologic ground mapping, i.e., at around 1 km^2 [64]. Our combination of different UAS-based sensors fills spatial gaps during the survey, and provides a wealth of interpretable data. Extracted spectroscopic and magnetic observations complement each other to capture surface and subsurface information, which allows an integrated geologic interpretation. Furthermore, we expand the coverage of the survey area by complementing missing areas with data from other sensors.

As expected from our lithologies at hand, a full class distinction based solely on HSI and RGB data was not feasible at first. Here, sensor integration substantially improved the UAS-based supervised image classifications. Some lithological boundaries seen in spectral data are expressed in the DSM topography. For example, classification accuracy for the feldspar–pegmatite intrusion and dolerite contacts was improved by including the DSM layer in the OTVCA feature extraction of trench 3, because those lithologies are more extruded. Particularly for trench 3, the occasional clay–soil patches smear larger surfaces and the cloudy weather during this data acquisition made it worthwhile to include additional information. OTVCA takes spatial relationships of multi-dimensional data (i.e., dozens of image channels) into consideration. By optical inspection, the selection of 13–20 bands of each extracted OTVCA data set of the three trenches (equaling 20–30% of the provided number of input bands) for the SVM classifier was feasible. Optical inspection means here that OTVCA bands with obvious noise content (stripes, artifacts, contrast gradients) are discarded. With a careful selection of training samples, we obtained a classification in good agreement with geologic ground mapping.

The multicopter-based hyperspectral data could identify spatially small (~5 cm), spectrally pronounced anomalies, i.e., fine carbonatite lenses and is effective at the given outcrop dimension. The same lenses are visible in RGB, but cannot be distinguished spectrally, e.g., from feldspar–pegmatite rubble. Some lithologies (feldspar–pegmatite, fenite–syenite, granite–gneiss) are hardly discernable due to their lack of characteristic spectral features in the VNIR range. For example, average reflectance of fenite–syenite was similar or higher than for feldspar–pegmatite and granite–gneiss. However, we could still discriminate those rocks by using the machine learning-based spatially constrained feature extraction. OTVCA allowed us to pass not only spectral information, but also slight spatial, textural, or overall reflectance changes to the classifier. With a set of representative, well-defined training points, the classifier is able to assign meaningful labels even to classes lacking any indicative spectral features. While delivering a good classification performance, this approach is highly dependent on good-quality training data. UAS short-wave infrared (SWIR) sensors would add more confidence to the classification and allow a direct, spectroscopic analysis of a much wider range of mineralogical features, however, their pricing and weight is still an obstacle. Light-weight VNIR sensors in combination with advanced, open-source machine learning techniques, have been shown to offer a cheaper, but still reliable, alternative for the discrimination of known lithological domains.

Furthermore, we see a high feasibility when UAS spectroscopy is used for, e.g., iron oxides and rare earth element identification. Neodymium and dysprosium are promising targets for remote sensing studies [57]. We observed specific rare earth element-related absorptions in VNIR regions of handheld spectra in local apatite (Figure 5b). For mapping, we are particularly interested in spectral absorption of Fe^{2+} bands in the range of 800–1200 nm as a target for the HSI camera. Further CO$_3$ related absorption around 2330 nm, indicative for carbonate mineralogy (i.e., carbonatite), is only detectable in the SWIR range of handheld spectroscopy [65,66]. To assist with UAS magnetic mapping, first-order results from UAS-based RGB orthophotos are available directly after each flight (Figure 6a). Orthomosaics could be further used to optimize and refine magnetic flight plans in the field, if important anomalies are identified. While atmospheric conditions influenced the data quality acquired from optical sensors, the magnetics could be flown with a low cloudy ceiling or over wet surfaces without any disturbance.

Line spacing, altitude, and sampling frequency of UAS magnetics define the features we can resolve physically, and therefore the size of targets we can model and interpret. We consider that the fixed-wing UAS probably created more valuable data for mapping with high surface coverage. Fixed-wing flight endurance was not exhausted with the current target area. In this case study, the following surface coverages were achieved per sensor:

- Magnetics: 0.695 km^2 (interpolated grid surface from 39-line km);
- MSIs: 0.649 km^2;
- RGB: 0.623 km^2;
- HSIs: 0.047 km^2 (sum of HSI flights).

The used UAS-fitted workflows are matured to a high user friendliness and could be flexibly adapted to all mining and exploration scenarios, where high resolution and spatial coverage is required. Safety concerns for detailed mapping along pit walls are mitigated by UAS mapping, when used for vertical outcrop scanning along unstable wall sections [67].

Our UAS mapping could improve the planning of material extraction processes in the mine. The volume of less profitable rock material can be reduced, which limits resource use and costs for additional drilling and curtails waste rock. Production schedules and mine layout planning could be improved. As example from UAS magnetics, we infer that the ore body cuts or continues below a mine road in the west on the outcrops, which could require a geotechnical repositioning of said infrastructure (Figures 2b and 9a, west of trench 1). Once regular UAS surveys become best practice for open-pit drilling, drill locations could be predefined in detailed orthophotos and subsurface drill orientations could be optimized by model-based interpretation of 3D data. In active mines, optical imagery is already implemented for explosive energy distribution optimization [68].

6.2. Further Implications of UAS Magnetic Surveys and Added Understanding of the Local Geology

UAS-based magnetics revealed the subsurface extension and trend of the glimmerite–carbonatite body between the trenches, and was validated on the trench surface. A high potential for ground- or UAS-based magnetic surveys to study lateral extension of those ore bodies was noted before [27], together with the recognition of the high magnetic susceptibility of Siilinjärvi carbonatite. The shape and direction of magnetic anomalies directly correlate with the extension of the lithologies at hand. For example, we interpret the pronounced trend (Figure 9a, eastern trench border) in the TMI-1VD as contact of the magnetic carbonatite with an intruded dolerite dyke. Furthermore, we interpret the TMI-AS as the estimated maximum width of glimmerite–carbonatite for this survey site. The two large anomalies crossing the eastern survey border (Figure 8) are likely part of much deeper granite–gneiss country rocks, however, neither hyperspectral data nor rock samples of those zones were acquired. We conclude that the abundant magnetite in the targeted lithologies is mostly responsible for the detected magnetic anomalies in UAS data, while fenite can be disregarded (Matias Carlsson, personal communication). The average magnetite content in the deposit is 1 wt.% [25], and is a highly abundant accessory mineral of both glimmerite and carbonatite [69]. Minor contents of pyrite, pyrrhotite, and some chalcopyrite occurrence form sulfide minerals in locally high abundance [54]. Sövite, a carbonatite variety, can carry 1–2% of magnetite, often together with apatite, biotite, and pyrochlore [70]. Although another source for high susceptibilities could be the mafic dykes, those are smaller in dimension as compared to the carbonatite–glimmerite and local fenite.

In a rock thin section of a dolerite sample, pyrite and magnetite were observed and confirmed by XRD measurements. For the glimmerite rocks, para- and ferrimagnetic effects can increase magnetic susceptibility in phlogopite due to magnetite domains in significant fractions [71].

Two-dimensional structural interpretation of the shear zones suggests an increasing mixture of ore and waste rocks in trench 3 (Figures 12 and 13e,f). Possibly, feldspar–pegmatites ascended near trench 3 and extruded laterally along the carbonatite–glimmerite contacts, following a path of least resistance.

To magnetically detect and model smaller dolerite dykes, a denser flight line pattern is recommended for higher spatial resolution. It was noted before [25] that aeromagnetic surveying cannot resolve the carbonatite–glimmerite, however, this is now possible with UAS-based magnetic surveying.

7. Conclusions

This study introduced a cohesive multi-sensor survey approach using optical and geophysical UAS sensors. We integrated UAS-based surface and sub-surface data to create a digital outcrop model for precise geology mapping. Detailed surface information from high-resolution orthophotos and structural trends from point clouds provided information to map geologic features at the centimeter scale. We measured structural constraints of carbonatite–glimmerite, mafic dykes, and feldspar-rich pegmatite on digital outcrop twins. Furthermore, we used a sensor fusion approach and machine learning methods for a supervised classification of outcropping rocks, partially covered by soil and captured during unfavorable atmospheric conditions. With hyperspectral data, we were able to identify and distinguish apatite-bearing lithologies from waste rock, i.e., feldspar-rich pegmatite intrusions and country rock. Based on UAS-borne magnetics, we created a surface-constrained forward model aided by measured and adapted magnetic susceptibilities to extract subsurface information, which revealed the extent of ore-bearing carbonatite-glimmerite. We observed this carbonatite structure at outcropping trenches, visible along the test pit wall, plunging into the subsurface and traced further based on magnetic data. The presumed high magnetic anomaly of carbonatite–glimmerite was measured in detail by a UAS. The scale and resolution of the magnetics covered all trenches in one UAS flight. Our survey lasted for two field work days, and included a spectral surface sampling campaign. All UAS flights were conducted in parallel to the sampling with a combined flight time of <6 hours in total.

The principal conclusions and highlights of this study are:

1. Rapid, flexible and automatized UAS-based surveying of lithologic surface and subsurface features, using light-weight multi-sensor technology, resulted in a 3D outcrop interpretation and provided material and structural information as a valuable alternative to time-consuming ground surveying.
2. Forward modeling of UAS-based magnetic data provided insight on orientation and depth of lithologies concealed from surface observation, here, UASs provided a link between 2D and 3D mapping.
3. Challenges arose in the integration of high-resolution HSI data at smaller scales and missing overlap between outcrops, together with spectrally inert rock types at the given spectral range.
4. Integration and fusion of topographic and spectral data using supervised surface classification of spectrally non-distinct targets with a support vector machine on dimensionality-reduced feature extraction data was successful in overcoming the challenges.
5. We recommend the use and combination of fixed-wing UASs for target-based surveying in the RGB, multispectral, and magnetic domains for advanced geologic mapping and interpretation, while using multicopter-borne HSI data for potential non-distinct lithology discrimination, sub-decimeter feature mapping and to identify features of narrow spectral range.

From this study, we observe that photo-based geology is transformed by UAS imaging techniques into automatic procedures, where magnetic and hyperspectral methods could become state of the art. MSIs and HSIs would stand next to the already implemented photogrammetric methods, to add potential for less invasive, data-driven mineral exploration and mining. UAS-based SWIR cameras will extend the range of identification for target lithologies, and future geophysical UAS sensors such as gravity, radiometric, and electromagnetic methods will extend the depth and resolution of observations.

Author Contributions: Conceptualization, R.J., M.K., S.L., R.Z., and R.G.; analysis: HSI, MSI, 3D: R.J.; analysis Mag: R.J. with support from H.U.; investigation: R.J., R.Z.; ground work: R.J., R.Z., M.K., Y.M., and L.T.; resources: R.G., A.S., and M.S.; software: S.L., M.P., R.J., and H.U.; validation: R.J., R.Z., and M.K.; visualization: R.J.; Writing—original draft: R.J.; writing—review and editing: M.K., R.Z., S.L., L.T., Y.M., R.G., and H.U.; supervision: R.G. All authors have read and agreed to the published version of the manuscript.

Funding: The research work was funded through the European Union and the EIT Raw Materials project "MULSEDRO" (grant id: 16193).

Acknowledgments: We thank Yara Oy and Aleksi Salo for allowing our research on the mine site and providing geological insights, and Martin Sonntag for magnetic susceptibility measurements at the petrophysical laboratory of TU Bergakademie Freiberg. We thank Björn H. Heincke (GEUS) and Heikki Salmirinne (GTK) for support and expertise during the work. Furthermore, we thank Robert Möckel and Doreen Ebert for conducting XRD measurements and Emer. William Morris for support in drafting the manuscript, and Lucas Pereira and Florian Rau for text improvements. We thank Louis Andreani, Gabriel Unger and Benjamin Melzer for supportive mapping during field work and Ziad Altoumh for aiding in laboratory preparation (HZDR-HIF). The work was funded through the European Union and the EIT Raw Materials project "MULSEDRO".

Conflicts of Interest: The authors declare no conflict of interest.

Appendix A

Table A1. Properties of test trenches, information for the UAS surveys, and further details of the HSI mapping, as we only surveyed the exposed trench rocks by HSIs. Altitude in m above sea level. The last column refers to the input layers used in the OTVCA for supervised image classification. GSD = ground sampling distance.

Outcrop/ Method	Coordinates	Dimension x-y	Survey Condition	Used Bands/ Integration Time	Flights/ Coverage	GSD	Altitude	OTVCA Layers
Method *(Hyperspectral only)*								
Trench 1	63.147N, 27.738E	130 × 36 m	sunny, windless	50/10 ms	1/5500 m^2	2.7 cm	40 m	HSI
Trench 2	63.145N, 27.738E	200 × 40 m	sunny, windless	50/10 ms	1/3050 m^2	2.3 cm	30 m	HSI
Trench 3	63.141N, 27.738E	220 × 400 m	low clouds, breeze	50/30 ms	3/38,200 m^2	3.4 cm	50 m	HSI, MSI, RGB
Multi-spectral	63.143N, 27.738E	450 × 1430 m	sunny, windless	4/automatic	1/0.649 km^2	10.5 cm	100 m	–
RGB	63.143N, 27.738E	540 × 1290 m	low clouds, breeze	3/automatic	2/0.623 km^2	2.7 (1.5) cm	100 m/70 m	–
Magnetic	63.143N, 27.738E	620 × 1100 m	sunny, windless	–	1/0.695 km^2	30 m *	40 m	–

* 15 m after interpolation.

Technical details for the used multi- and hyperspectral cameras are provided in Table A2.

Table A2. Technical specifications of used cameras.

Sensor	Senop Rikola	Parrot Sequoia	senseFly S.O.D.A.
Dynamic range	12 bits	10 bits	–
Horizontal field of view	36.5°	70.6°	90°
Vertical field of view	23.5°	52.6°	60°
Focal length	9 mm	4 mm	2.8–11
Mass	720 g	135 g (with sunshine sensor)	111 g
Frame rate	30 Hz	1 Hz	0.3 Hz
Spectral resolution	8 nm	40 nm (10 nm)	–

Training and validation samples used for the supervised image classification used a cross-referencing support vector machine algorithm. The final classification maps are used to approximate the geologic contacts which were indifferentiable in RGB orthophotos. Additionally, the carbonatite classification is possible, mainly for trenches 1 and 2, represented by the higher amount of training and validation pixels. The labels for the test and training points were determined with the handheld spectrometer. Each spectral signal was measured with a Spectral Evolution PSR-3500. A spectral resolution of 3.5 nm (1.5 nm sampling interval) in the visible and near-infrared (VNIR) range and 7 nm (2.5 nm sampling interval) in the SWIR range is provided, using a contact probe. Each spectral record consists of 10 individual measurements taken consecutively and averaged.

To convert radiance to reflectance, we use a PTFE panel (Zenith Polymer with >99% reflectance VNIR; >95% reflectance SWIR).

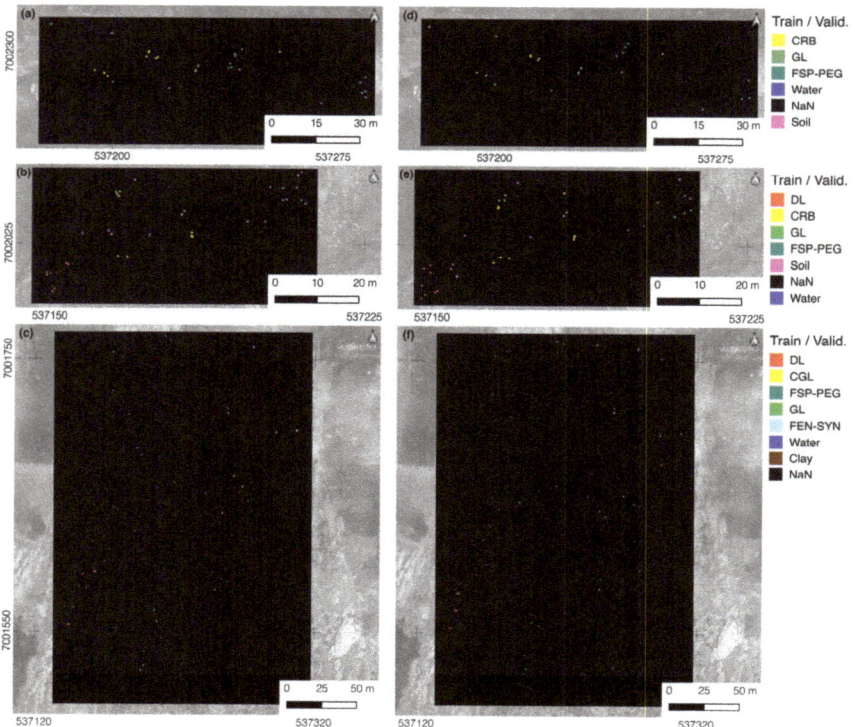

Figure A1. Training and validation for support vector machine classification in column-wise order. (**a**) Training samples trench 1. (**b**) Training samples trench 2. (**c**) Training samples trench 3. (**d**) Validation samples trench 1. (**e**) Validation samples trench 2. (**f**) Validation samples trench 3. CRB = Carbonatite; GL = Glimmerite; CGL = Carbonatite–glimmerite; FSP-PEG = Feldspar–pegmatite; NaN = Not a number; DL = Dolerite; FEN-SYN = Fenite–syenite.

Table A3. Confusion matrix trench 1. Indef./NaN = black pixel.

Predicted \ Truth	Carbonatite	Glimmerite	Feldspar–Pegmatite	Water	Indef.	Soil
Carbonatite	123	0	17	0	0	7
Glimmerite	0	120	0	0	0	0
Feldspar–Pegmatite	4	0	172	0	0	0
Water	0	0	0	63	0	0
Indef./Nan	0	0	0	0	42	0
Soil	0	0	2	0	0	87

Table A4. Confusion matrix trench 2. We observe that the differentiation between the water and soil pixels is ambiguous, however, both classes were rejected from the geological interpretation.

Predicted \ Truth	Dolerite	Carbonatite	Glimmerite	Feldspar–Pegmatite	Soil	Indef./Nan	Water
Dolerite	83	0	0	0	0	0	0
Carbonatite	0	147	0	6	0	0	0
Glimmerite	0	4	80	0	0	0	0
Feldspar–Pegmatite	0	8	0	124	0	0	0
Soil	0	0	1	0	50	0	0
Indef./Nan	0	0	0	0	0	32	0
Water	0	0	0	0	48	0	90

Table A5. Confusion matrix trench 2.

Predicted \ Truth	Dolerite	Glimmerite–Carbonatite	Feldspar–Pegmatite	Glimmerite	Fenite–Syenite	Water	Soil	Indef./Nan
Dolerite	649	0	0	15	0	0	0	0
Glimmerite–Carbonatite	0	34	11	0	0	0	0	0
Feldspar–Pegmatite	31	0	1141	0	80	0	0	0
Glimmerite	8	0	13	650	0	0	2	0
Fenite–Syenite	17	6	39	0	1296	0	0	0
Water	0	0	0	0	0	532	0	0
Soil	2	0	0	0	2	0	353	0
Indef./Nan	0	0	0	0	0	0	0	4

Appendix B

Profile plots across the DSM and the underlying modeled carbonatite–glimmerite bodies are shown. Note the increasing length scale. Corresponding magnetic profiles are shown in Figure 8 in the manuscript. Here, the calculated magnetic response per profile is plotted on the UAS-measured TMI signal. Due to the ambiguous nature of geophysical forward models, all available constraints were employed to create the model bodies. Starting parameters for each profile are given by the user. We iterated 20 sessions with various starting parameters for magnetic susceptibility, as well as position and depth of initial body geometry. We assumed tabular body shapes. Strike direction, dip, and length of each body were estimated based on UAS-RGB, hyperspectral and structural data. For example, the depth of the body for profile 4 (S4) seems to be overestimated, and constrained possible susceptibility. This corresponds with the magnetic low of profile 4, directly above a shear zone. Even with an apparent good model fit, an interpretation is complicated. As stated above, shear stress could have decreased the amount of magnetic minerals. For profile S1, a gap between two carbonatite bodies exists, caused by the absence of magnetic rock material, caused by an observed feldspar–pegmatite intrusion. Data of a comprehensive exploration drill campaign would solidify further interpretations.

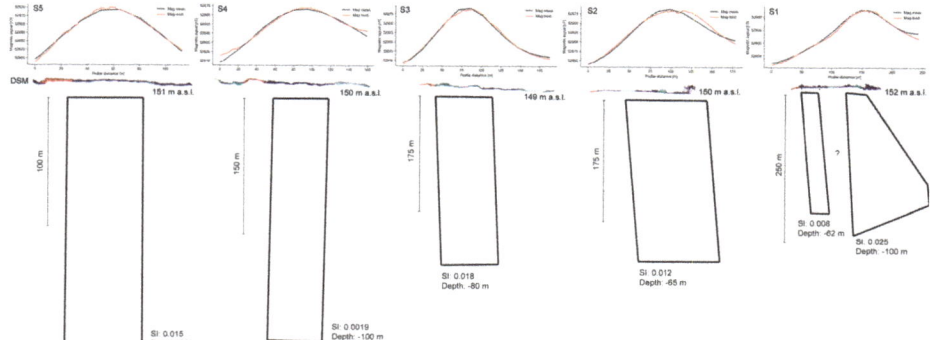

Figure A2. Cross-section profile plots across the DSM and the underlying, modeled tabular carbonatite–glimmerite bodies.

Appendix C

Figure A3. Optical microscopy (with the Zeiss Axio Imager M2m with Axiocam MRc 5 imaging module) conducted for thin sections of representative samples; Cal = calcite; Phl = phlogopite; Apt = apatite; Mag = magnetite; Py = pyrite. (**a**) Carbonatite–glimmerite, reflected light. (**b**) Carbonatite–glimmerite, transmitted light, crossed nicols. (**c**) Magnetite (subhedral–euhedral), reflected light. (**d**) Carbonatite–glimmerite, reflected light. (**e**) Carbonatite–glimmerite, transmitted light, parallel nicols. (**f**) Feldspar–pegmatite, transmitted light, crossed nicols.

Appendix D

Magnetic susceptibility, detecting magnetite signature, among others, is measured with a Bartington MS2 magnetic susceptibility system (Bartington Instruments, Witney, Oxon, United Kingdom). A mass fraction of material per sample was crushed to a fine powder (<0.1 mm grain size), weighed to 10.00 g and its susceptibility was measured with the sample tray holder of the MS2 system. The values are augmented with additional susceptibility values taken from the literature for those lithologies without available rock specimens.

XRD is conducted with the PANalytical Empyrean diffractometer with cobalt as the X-ray source and equipped with a PIXcel 3D Medipix detector. The main targets are mineral content, including detection and quantification of magnetic minerals. X-ray diffraction patterns for two selected samples are shown in Figures A4 and A5.

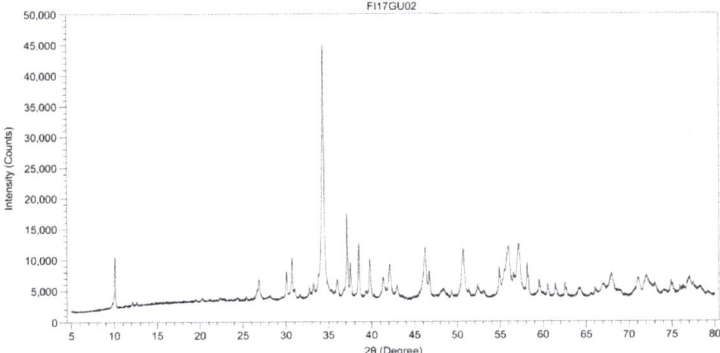

Figure A4. X-ray diffraction pattern for the carbonatite sample.

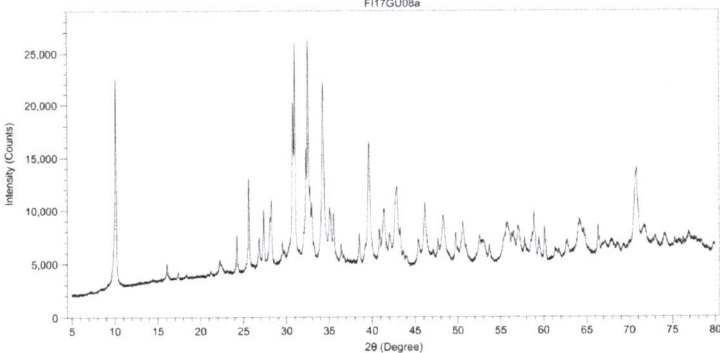

Figure A5. X-ray diffraction pattern for the dolerite sample.

Table A6. Mineral abundance from a carbonatite–glimmerite zone (GU02) and a dolerite dyke (GU08a) sample is listed below, with the mineral content in weight % (wt.%).

Mineral (wt.%)	Carbonatite (and Glimmerite)	Dolerite
Coordinates: UTM zone 35N	537156E, 7002020N	537124E, 7001475E
Calcite	59.6	16.6
Magnetite	1.8	2.4
Pyrite	–	2.0
Actinolite	3.7	–
Ankerite	4.1	–
Albite	–	37.4
Annite	9.8	–
Apatite	21.0	–
Biotite	–	29.7
K-Feldspar	–	4.8
Quartz	–	7.2

References

1. Kim, J.; Kim, S.; Ju, C.; Son, H. Il Unmanned Aerial Vehicles in Agriculture: A Review of Perspective of Platform, Control, and Applications. *IEEE Access* **2019**, *7*, 105100–105115. [CrossRef]
2. Adão, T.; Hruška, J.; Pádua, L.; Bessa, J.; Peres, E.; Morais, R.; Sousa, J.J. Hyperspectral imaging: A review on UAV-based sensors, data processing and applications for agriculture and forestry. *Remote Sens.* **2017**, *9*, 1110. [CrossRef]

3. Bemis, S.P.; Micklethwaite, S.; Turner, D.; James, M.R.; Akciz, S.; Thiele, S.T.; Bangash, H.A. Ground-based and UAV-Based photogrammetry: A multi-scale, high-resolution mapping tool for structural geology and paleoseismology. *J. Struct. Geol.* **2014**, *69*, 163–178. [CrossRef]
4. Dering, G.M.; Micklethwaite, S.; Thiele, S.T.; Vollgger, S.A.; Cruden, A.R. Review of drones, photogrammetry and emerging sensor technology for the study of dykes: Best practises and future potential. *J. Volcanol. Geotherm. Res.* **2019**, *373*, 148–166. [CrossRef]
5. Fairley, I.; Mendzil, A.; Togneri, M.; Reeve, D.E. The use of unmanned aerial systems to map intertidal sediment. *Remote Sens.* **2018**, *10*, 1918. [CrossRef]
6. Jackisch, R.; Lorenz, S.; Zimmermann, R.; Möckel, R.; Gloaguen, R. Drone-Borne Hyperspectral Monitoring of Acid Mine Drainage: An Example from the Sokolov Lignite. Hyperspectral Monitoring of Acid Mine Drainage: An Example from the Sokolov Lignite District. *Remote Sens.* **2018**, *10*, 385. [CrossRef]
7. Padró, J.; Carabassa, V.; Balagué, J.; Brotons, L.; Alcañiz, J.M.; Pons, X. Science of the Total Environment Monitoring opencast mine restorations using Unmanned Aerial System (UAS) imagery. *Sci. Total Environ.* **2019**, *657*, 1602–1614. [CrossRef]
8. Lee, S.; Choi, Y. Reviews of unmanned aerial vehicle (drone) technology trends and its applications in the mining industry. *Geosystem Eng.* **2016**, *19*, 197–204. [CrossRef]
9. Ren, H.; Zhao, Y.; Xiao, W.; Hu, Z. A review of UAV monitoring in mining areas: Current status and future perspectives. *Int. J. Coal Sci. Technol.* **2019**, *6*, 320–333. [CrossRef]
10. Booysen, R.; Zimmermann, R.; Lorenz, S.; Gloaguen, R.; Nex, P.A.M.; Andreani, L.; Möckel, R. Towards multiscale and multisource remote sensing mineral exploration using RPAS: A case study in the Lofdal Carbonatite-Hosted REE Deposit, Namibia. *Remote Sens.* **2019**, *11*, 2500. [CrossRef]
11. Parshin, A.; Grebenkin, N.; Morozov, V.; Shikalenko, F. Research Note: First results of a low-altitude unmanned aircraft system gamma survey by comparison with the terrestrial and aerial gamma survey data. *Geophys. Prospect.* **2018**, *66*, 1433–1438. [CrossRef]
12. Malehmir, A.; Dynesius, L.; Paulusson, K.; Paulusson, A.; Johansson, H.; Bastani, M.; Wedmark, M. The potential of rotary-wing UAV-based magnetic surveys for mineral exploration: A case study from central Sweden. *Leading Edge.* **2017**, *7*, 552–557. [CrossRef]
13. Cunningham, M.; Samson, C.; Wood, A.; Cook, I. Aeromagnetic Surveying with a Rotary-Wing Unmanned Aircraft System: A Case Study from a Zinc Deposit in Nash Creek, New Brunswick, Canada. *Pure Appl. Geophys.* **2018**, *175*, 3145–3158. [CrossRef]
14. Parvar, K.; Braun, A.; Layton-Matthews, D.; Burns, M. UAV magnetometry for chromite exploration in the Samail ophiolite sequence, Oman. *J. Unmanned Veh. Syst.* **2018**, *6*, 57–69. [CrossRef]
15. Walter, C.; Braun, A.; Fotopoulos, G. High-resolution unmanned aerial vehicle aeromagnetic surveys for mineral exploration targets. *Geophys. Prospect.* **2020**, *68*, 334–349. [CrossRef]
16. Sayab, M.; Aerden, D.; Paananen, M.; Saarela, P. Virtual structural analysis of Jokisivu open pit using "structure-from-motion" Unmanned Aerial Vehicles (UAV) photogrammetry: Implications for structurally-controlled gold deposits in Southwest Finland. *Remote Sens.* **2018**, *10*, 1296. [CrossRef]
17. Haldar, S. *Mineral Exploration Principles and Applications*, 2nd ed.; Elsevier: Amsterdam, The Netherlands, 2018; ISBN 978-0-12-814022-2.
18. Marjoribanks, R. *Geological Methods in Mineral Exploration and Mining*; Springer Science & Business Media: Berlin, Germany, 2010; ISBN 9783540743705.
19. Abedi, M.; Norouzi, G.H. Integration of various geophysical data with geological and geochemical data to determine additional drilling for copper exploration. *J. Appl. Geophys.* **2012**, *83*, 35–45. [CrossRef]
20. Slavinski, H.; Morris, B.; Ugalde, H.; Spicer, B.; Skulski, T.; Rogers, N. Integration of lithological, geophysical, and remote sensing information: A basis for remote predictive geological mapping of the Baie Verte Peninsula, Newfoundland. *Can. J. Remote Sens.* **2010**, *2*, 99–118. [CrossRef]
21. Beyer, F.; Jurasinski, G.; Couwenberg, J.; Grenzdörffer, G. Multisensor data to derive peatland vegetation communities using a fixed-wing unmanned aerial vehicle. *Int. J. Remote Sens.* **2019**, *40*, 9103–9125. [CrossRef]
22. Heincke, B.; Jackisch, R.; Saartenoja, A.; Salmirinne, H.; Rapp, S.; Zimmermann, R.; Pirttijärvi, M.; Vest Sörensen, E.; Gloaguen, R.; Ek, L.; et al. Developing multi-sensor drones for geological mapping and mineral exploration: Setup and first results from the MULSEDRO project. *Geol. Surv. Denmark Greenl. Bull.* **2019**, *43*, 2–6. [CrossRef]

23. Van der Meer, F.D.; van der Werff, H.M.A.; van Ruitenbeek, F.J.A. Multi- and hyperspectral geologic remote sensing: A review. *Int. J. Appl. Earth Obs. Geoinf.* **2012**, *14*, 112–128. [CrossRef]
24. Jackisch, R.; Madriz, Y.; Zimmermann, R.; Pirttijärvi, M.; Saartenoja, A.; Heincke, B.H.; Salmirinne, H.; Kujasalo, J.-P.; Andreani, L.; Gloaguen, R. Drone-borne hyperspectral and magnetic data integration: Otanmäki Fe-Ti-V deposit in Finland. *Remote Sens.* **2019**, *11*, 2084. [CrossRef]
25. Puustinen, K. Geology of the Siilinjärvi Carbonatite Complex, Eastern Finland. *Bull. la Commision Geol. Finlande* **1971**, *249*, 1–43.
26. Luoma, S.; Majaniemi, J.; Kaipainen, T.; Pasanen, A. GPR survey and field work summary in Siilinjärvi mine during July 2014. *Geol. Surv. Finland. Arch. Rep.* **2016**, *39*, 1–39.
27. Malehmir, A.; Heinonen, S.; Dehghannejad, M.; Heino, P.; Maries, G.; Karell, F.; Suikkanen, M.; Salo, A. Landstreamer seismics and physical property measurements in the siilinjärvi open-pit apatite (phosphate) mine, central Finland. *Geophysics* **2017**, *82*, B29–B48. [CrossRef]
28. Laakso, V. Testing of Reflection Seismic, GPR and Magnetic Methods for Mineral Exploration and Mine Planning at the Siilinjärvi Phosphate Mine Site in Finland. Master's Thesis, University of Helsinki, Helsinki, Finland, 2019.
29. Da Col, F.; Papadopoulou, M.; Koivisto, E.; Sito, Ł.; Savolainen, M.; Socco, L.V. Application of surface-wave tomography to mineral exploration: A case study from Siilinjärvi, Finland. *Geophys. Prospect.* **2020**, *68*, 254–269. [CrossRef]
30. Pajunen, M.; Salo, A.; Suikkanen, M.; Ullgren, A.-K.; Oy, Y.S. Brittle structures in the south-western corner of the Särkijärvi open pit, Siilinjärvi carbonatite occurrence. *Geol. Surv. Finland. Arch. Rep.* **2017**, *38*, 1–38.
31. Mattsson, H.B.; Högdahl, K.; Carlsson, M.; Malehmir, A. The role of mafic dykes in the petrogenesis of the Archean Siilinjärvi carbonatite complex, east-central Finland. *Lithos* **2019**, *342–343*, 468–479. [CrossRef]
32. Tuomas, K.; Pietari, S.; Emilia, K.; Savolainen, M. 3D modelling of the dolerite dyke network within the Siilinjärvi phosphate deposit. In Proceedings of the Visual3D Conference—Visualization of 3D/4D Models in Geosciences, Exploration and Mining, Luleå, Sweden, 1–2 October 2019; p. 33.
33. Tichomirowa, M.; Grosche, G.; Götze, J.; Belyatsky, B.V.; Savva, E.V.; Keller, J.; Todt, W. The mineral isotope composition of two Precambrian carbonatite complexes from the Kola Alkaline Province—Alteration versus primary magmatic signatures. *Lithos* **2006**, *91*, 229–249. [CrossRef]
34. Carlsson, M.; Eklund, O.; Fröjdö, S.; Savolainen, M. Petrographic and geochemical characterization of fenites in the northern part of the Siilinjärvi carbonatite-glimmerite complex, Central Finland. In Proceedings of the Geological Society of Finland, Abstracts of the 5th Finnish National Colloquium of Geosciences, Helsinki, Finland, 6–7 March 2019; p. 29.
35. Thiele, S.T.; Grose, L.; Samsu, A.; Micklethwaite, S.; Vollgger, S.A.; Cruden, A.R. Rapid, semi-automatic fracture and contact mapping for point clouds, images and geophysical data. *Solid Earth* **2017**, *8*, 1241–1253. [CrossRef]
36. Nabighian, M.N. The Analytic Signal Of Two-Dimensional Magnetic Bodies With Polygonal Cross-Section: Its Properties And Use For Automated Anomaly Interpretation. *Geophysics* **1972**, *37*, 507–517. [CrossRef]
37. Hinze, W.J.; von Frese, R.R.B.; Saad, A.H. *Gravity and Magnetic Exploration*; Cambridge University Press: Cambridge, UK, 2013; ISBN 9780511843129.
38. Vacquier, V.; Steenland, N.C.; Henderson, R.G.; Zietz, I. *Interpretation of Aeromagnetic Maps*; Geological Society of America: Boulder, CO, USA, 1951; ISBN 9780813710471.
39. Khaleghi, B.; Khamis, A.; Karray, F.O.; Razavi, S.N. Multisensor data fusion: A review of the state-of-the-art. *Inf. Fusion* **2013**, *14*, 28–44. [CrossRef]
40. Lorenz, S.; Seidel, P.; Ghamisi, P.; Zimmermann, R.; Tusa, L.; Khodadadzadeh, M.; Contreras, I.C.; Gloaguen, R. Multi-sensor spectral imaging of geological samples: A data fusion approach using spatio-spectral feature extraction. *Sensors* **2019**, *19*, 2787. [CrossRef] [PubMed]
41. Rasti, B.; Ulfarsson, M.O.; Sveinsson, J.R. Hyperspectral Feature Extraction Using Total Variation Component Analysis. *IEEE Trans. Geosci. Remote Sens.* **2016**, *54*, 6976–6985. [CrossRef]
42. Ghamisi, P.; Yokoya, N.; Li, J.; Liao, W.; Liu, S.; Plaza, J.; Rasti, B.; Plaza, A. Advances in Hyperspectral Image and Signal Processing: A Comprehensive Overview of the State of the Art. *IEEE Geosci. Remote Sens. Mag.* **2017**, *5*, 37–78. [CrossRef]
43. Chang, C.C.; Lin, C.J. LIBSVM: A Library for support vector machines. *ACM Trans. Intell. Syst. Technol.* **2011**, *2*, 27. [CrossRef]

44. Almqvist, B.; Högdah, K.; Karell, F.; Malehmir, A. Anisotropy of magnetic susceptibility (AMS) in the Siilinjärvi carbonatite complex, eastern Finland. In Proceedings of the Geophysical Research Abstracts, EGU General Assembly, Vienna, Austria, 23–28 April 2017; p. 9887.
45. James, M.R.; Robson, S.; D'Oleire-Oltmanns, S.; Niethammer, U. Optimising UAV topographic surveys processed with structure-from-motion: Ground control quality, quantity and bundle adjustment. *Geomorphology* **2016**, *280*, 51–66. [CrossRef]
46. James, M.R.; Chandler, J.H.; Eltner, A.; Fraser, C.; Miller, P.E.; Mills, J.P.; Noble, T.; Robson, S.; Lane, S.N. Guidelines on the use of structure-from-motion photogrammetry in geomorphic research. *Earth Surf. Process. Landforms* **2019**, *2084*, 2081–2084. [CrossRef]
47. Jakob, S.; Zimmermann, R.; Gloaguen, R. The Need for Accurate Geometric and Radiometric Corrections of Drone-Borne Hyperspectral Data for Mineral Exploration: MEPHySTo-A Toolbox for Pre-Processing Drone-Borne Hyperspectral Data. *Remote Sens.* **2017**, *9*, 88. [CrossRef]
48. Karpouzli, E.; Malthus, T. The empirical line method for the atmospheric correction of IKONOS imagery. *Int. J. Remote Sens.* **2003**, *5*, 1143–1150. [CrossRef]
49. Gavazzi, B.; Le Maire, P.; Mercier de Lépinay, J.; Calou, P.; Munschy, M. Fluxgate three-component magnetometers for cost-effective ground, UAV and airborne magnetic surveys for industrial and academic geoscience applications and comparison with current industrial standards through case studies. *Geomech. Energy Environ.* **2019**, *20*, 100117. [CrossRef]
50. Pirttijärvi, M. Numerical Modeling and Inversion of Geophysical Electromagnetic Measurements Using a Thin Plate Model. Ph.D. Dissertation, University of Oulu, Oulu, Finland, 2003.
51. Austin, J.R.; Schmidt, P.W.; Foss, C.A. Magnetic modeling of iron oxide copper-gold mineralization constrained by 3D multiscale integration of petrophysical and geochemical data: Cloncurry District, Australia. *Interpretation* **2013**, *1*, T63–T84. [CrossRef]
52. Tichomirowa, M.; Whitehouse, M.J.; Gerdes, A.; Götze, J.; Schulz, B.; Belyatsky, B.V. Different zircon recrystallization types in carbonatites caused by magma mixing: Evidence from U-Pb dating, trace element and isotope composition (Hf and O) of zircons from two Precambrian carbonatites from Fennoscandia. *Chem. Geol.* **2013**, *353*, 173–198. [CrossRef]
53. Poutiainen, M. Fluids in the Siilinjarvi carbonatite complex, eastern Finland: Fluid inclusion evidence for the formation conditions of zircon and apatite. *Bull. Geol. Soc. Finl.* **1995**, *67*, 3–18. [CrossRef]
54. O'Brien, H.; Heilimo, E.; Heino, P. The Archean Siilinjärvi Carbonatite Complex. *Miner. Depos. Finl.* **2015**, *1*, 327–343.
55. Salo, A. Geology of the Jaakonlampi Area in the Siilinjärvi Carbonatite Complex. Bachelor's Thesis, University of Oulu, Oulu, Finland, 2016.
56. Gaffey, S.J. Reflectance spectroscopy in the visible and near- infrared (0.35–2.55 micrometers): Applications in carbonate petrology. *Geology* **1985**, *4*, 270–273. [CrossRef]
57. Neave, D.A.; Black, M.; Riley, T.R.; Gibson, S.A.; Ferrier, G.; Wall, F.; Broom-Fendley, S. On the feasibility of imaging carbonatite-hosted rare earth element deposits using remote sensing. *Econ. Geol.* **2016**, *111*, 641–665. [CrossRef]
58. Hunt, G.R. Spectral signatures of particulate minerals in the visible and near infrared. *Geophysics* **1977**, *42*, 501–513. [CrossRef]
59. Clark, R.N. Spectroscopy of rocks and minerals, and principles of spectroscopy. *Man. Remote Sens.* **1999**, *3*, 2.
60. Hunt, G.R.; Ashley, R.P. Spectra of altered rocks in the visible and near infrared. *Econ. Geol.* **1979**, *74*, 1613–1629. [CrossRef]
61. Airo, M.-L. Aerogeophysics in Finland 1972–2004: Methods, System Characteristics and Applications. *Spec. Pap. Geol. Surv. Finl.* **2005**, *39*, 197.
62. Burkin, J.N.; Lindsay, M.D.; Occhipinti, S.A.; Holden, E.J. Incorporating conceptual and interpretation uncertainty to mineral prospectivity modelling. *Geosci. Front.* **2019**, *10*, 1383–1396. [CrossRef]
63. Cardozo, N.; Allmendinger, R.W. Spherical projections with OSXStereonet. *Comput. Geosci.* **2013**, *51*, 193–205. [CrossRef]
64. Jackisch, R. Drone-based surveys of mineral deposits. *Nat. Rev. Earth Environ.* **2020**, *1*, 187. [CrossRef]
65. Rowan, L.C.; Kingston, M.J.; Crowley, J.K. Spectral reflectance of carbonatites and related alkalic igneous rocks: Selected samples from four North American localities. *Econ. Geol.* **1986**, *81*, 857–871. [CrossRef]

66. Rowan, L.C.; Mars, J.C. Lithologic mapping in the Mountain Pass, California area using Advanced Spaceborne Thermal Emission and Reflection Radiometer (ASTER) data. *Remote Sens. Environ.* **2003**, *84*, 350–366. [CrossRef]
67. Kirsch, M.; Lorenz, S.; Zimmermann, R.; Andreani, L.; Tusa, L.; Pospiech, S.; Jackisch, R.; Khodadadzadeh, M.; Ghamisi, P.; Unger, G.; et al. Hyperspectral outcrop models for palaeoseismic studies. *Photogramm. Rec.* **2019**, *34*, 385–407. [CrossRef]
68. Valencia, J.; Battulwar, R.; Naghadehi, M.Z.; Sattarvand, J. Enhancement of explosive energy distribution using uavs and machine learning. In Proceedings of the Mining Goes Digital 39th International Symposium on Application of Computers and Operations Research in the Mineral Industry, Leiden, The Netherlands, 4–6 June 2019.
69. Heilimo, E.; Brien, H.O.; Heino, P. Constraints on the Formation of the Archean Siilinjärvi Carbonatite-Glimmerite Complex, Fennoscandian Shield. 2015. Available online: https://bit.ly/339EGyI (accessed on 2 June 2020).
70. Le Bas, M.J. Nephelinites and carbonatites. *Geol. Soc. Spec. Publ.* **1987**, *30*, 53–83. [CrossRef]
71. Borradaile, G.J.; Werner, T. Magnetic anisotropy of some phyllosilicates. *Tectonophysics* **1994**, *235*, 223–248. [CrossRef]

© 2020 by the authors. Licensee MDPI, Basel, Switzerland. This article is an open access article distributed under the terms and conditions of the Creative Commons Attribution (CC BY) license (http://creativecommons.org/licenses/by/4.0/).

Letter

Dust Dispersion and Its Effect on Vegetation Spectra at Canopy and Pixel Scales in an Open-Pit Mining Area

Baodong Ma [1,2,*], Xuexin Li [1,2], Ziwei Jiang [1,2], Ruiliang Pu [3], Aiman Liang [1,2] and Defu Che [1,2]

1. Key Laboratory of Ministry of Education on Safe Mining of Deep Metal Mines, Northeastern University, Shenyang 110819, China; 1701015@stu.neu.edu.cn (X.L.); 1900996@stu.neu.edu.cn (Z.J.); 1800993@stu.neu.edu.cn (A.L.); chedefu@mail.neu.edu.cn (D.C.)
2. Institute for Geoinformatics & Digital Mine Research, Northeastern University, Shenyang 110819, China
3. School of Geosciences, University of South Florida, Tampa, FL 33620, USA; rpu@usf.edu
* Correspondence: mabaodong@mail.neu.edu.cn; Tel.: +86-24-83691628

Received: 24 September 2020; Accepted: 12 November 2020; Published: 16 November 2020

Abstract: Dust pollution is severe in some mining areas in China due to rapid industrial development. Dust deposited on the vegetation canopy may change its spectra. However, a relationship between canopy spectra and dust amount has not been quantitatively studied, and a pixel-scale condition for remote sensing application has not been considered yet. In this study, the dust dispersion characteristics in an iron mining area were investigated using the American Meteorological Society (AMS) and the U.S. Environmental Protection Agency (EPA) regulatory model (AERMOD). Further, based on the three-dimensional discrete anisotropic radiative transfer (DART) model, the spectral characteristics of vegetation canopy under the dusty condition were simulated, and the influence of dustfall on vegetation canopy spectra was studied. Finally, the dust effect on vegetation spectra at the canopy scale was extended to a pixel scale, and the response of dust effect on vegetation spectra at the pixel scale was determined under different fractional vegetation covers (FVCs). The experimental results show that the dust pollution along a haul road was more severe and extensive than that in a stope. Taking dust dispersion along the road as an example, the variation of vegetation canopy spectra increased with the height of dust deposited on the vegetation canopy. At the pixel scale, a lower vegetation FVC would weaken the influence of dust on the spectra. The results derived from simulation spectral data were tested using satellite remote sensing images. The tested result indicates that the influence of dust retention on the pixel spectra with different FVCs was consistent with that created with the simulated data. The finding could be beneficial for those making decisions on monitoring vegetation under dusty conditions and reducing dust pollution in mining areas using remote sensing technology.

Keywords: dust dispersion; spectra; canopy scale; pixel scale; mining area

1. Introduction

Dust is a dominant feature of the global aerosol system [1]. Dust can affect air quality, climate, biosphere and atmospheric chemistry [2]. There are two main dust sources: natural sources and human activity [3]. Natural sources include wind erosion, rock weathering, dust storms, etc. [4,5], while human activity-induced dust is mainly caused by construction [6], road transportation [7], fuel combustion [8], open-pit mining [9], etc. In China, mining industry has developed rapidly in recent years [10]. A large-scale and high-intensity mining in open-pit mines may lead to serious dust pollution by drilling, overburden loading and unloading, mineral processing, vehicular movement on the haul roads [11].

There may be different dispersion characteristics of dust between excavation, transportation and mineral processing in mining areas. These dispersion differences could be simulated using models. Several models have been developed for air dispersion simulation, such as the American Meteorological Society (AMS) and the U.S. Environmental Protection Agency (EPA) regulatory model (AERMOD) [12], industrial source complex-short term (ISCST) [13], California puff model (CALPUFF) [14], etc. These models are suitable for different spatial scales, processes and particle sources [15]. AERMOD is one of the most commonly used models worldwide based on Gaussian dispersion [16]. It is a near-field, steady-state Gaussian plume model based on planetary boundary layer turbulence structure and scaling concepts, which can model multiple sources of different types including point, area and volume sources. The distribution is assumed to be Gaussian in both the horizontal and vertical directions in the stable boundary layer [15]. In other words, the model has a clear physical concept, can use the measured data to determine the required parameters, and is easy to modify for different situations. This model can be used in short-range (up to 50 km) dispersion from various polluting sources. Thus, in this study, the AERMOD model would be used to predict dust diffusion characteristics of different pollution sources and to obtain the spatial distribution of the dust in the mining area.

After dispersion, a large amount of dust falls on the land surface around the mining area. For vegetation, dust can deposit on the leaf [17]. From the perspective of remote sensing, the dust retention on vegetation canopy will change vegetation's spectral characteristics because vegetation spectra are mixed with dust spectra. If the measured spectra of dusty leaves were used to retrieve physiological parameters of vegetation directly, it would reduce the retrieval accuracy due to the mixed spectral information [18]. Therefore, it is necessary to study the dust effect on the vegetation spectra. For example, the effect of foliar dust (atmospheric pollution and limestone dust) on spectra of pear and *Fagus sylvatica* leaves has been studied, respectively [19,20]. However, most studies have studied the influence of dust retention on leaf spectra, and the related result at a leaf scale cannot be applied to the canopy scale directly due to variable canopy structure [21]. Moreover, due to the limitation of spatial resolution, pixels in a remote sensing image are usually mixed with other features [22]. Therefore, the dust effect on the spectra of pixels with different fractional vegetation covers (FVCs) should be considered for the remote sensing application.

Spectral data acquisition is time-consuming and laborious by field or laboratory measurement due to the complexity of the field environment. Moreover, collected spectral data are often difficult to meet the needs of research due to the various measuring conditions [23]. Therefore, simulation methods can frequently be used to obtain spectra. Various radiative transfer models (RTMs) are valuable tools for spectra simulation. In the case of heterogeneous canopies with complex architectures, three-dimensional (3D) RTMs are more appropriate by describing canopy structures explicitly [24]. The discrete anisotropic radiative transfer (DART) model is one of the most used 3D RTMs to simulate the spectra of crops [25,26] and forests [27,28]. DART could be used to simulate and compute radiation propagation through the entire earth-atmosphere system in the electromagnetic spectrum from visible to thermal infrared parts [24]. Thus, the DART was selected as the basic model to simulate spectra of dust at canopy and pixel scales in this study.

Therefore, in this study, we propose to study the dust effect on vegetation spectra at both canopy and pixel scales in a mining area through spectral modeling and remote sensing technology. More specific research objectives are to (1) quantify the dust dispersion in a mining area, and (2) assess the dust effect on vegetation spectra at canopy and pixel scales. To achieve the two research objectives, AERMOD would be used to study the characteristics of dust dispersion and the DART model to study the dust effect on vegetation spectra.

2. Models and Methods

2.1. Study Area

The study area was Kuancheng Mining Area, which is located in Hebei Province, North China (Figure 1). This area is covered with dense natural vegetation, such as *Populus cathayana*, *Pinus tabuliformis*, *Armeniaca sibirica*, and *Castanea mollissima*, and maize is the main crop. The annual output of ore is about 100 million tons. It has a continental monsoon climate with a mean annual precipitation of 662.5 mm and a mean annual temperature of 8.7 °C [29]. There are many open-pit iron mines in the mining area. Mining development is the main economic activity in this area, which includes mining, ore transportation and beneficiation. Dust pollution is an environmental problem that cannot be ignored in this area due to the high-intensity mining development.

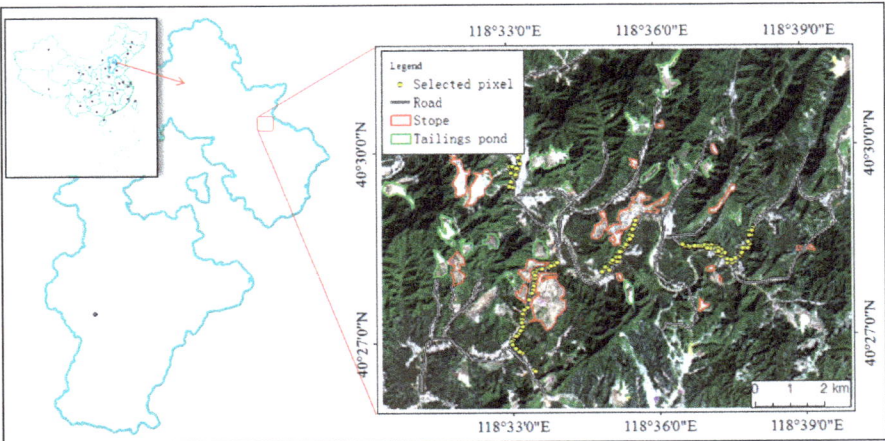

Figure 1. Location of Kuancheng Mining Area in a false-color composite image of Landsat 8 image (RGB432).

2.2. A General Work Flowchart

In the mining area, there are different dust sources that show dispersion characteristics. To quantify and assess the dust characteristics and effects on vegetation spectra, three major research components were included in this study. (1) The dust source was classified into point and line types. The spatial distribution of dust was studied by using AERMOD. (2) The dust effect on vegetation spectral reflectance at a canopy scale was studied using the DART model, and the dust effect on vegetation reflectance at a pixel scale was investigated by considering FVC in pixels with simulated spectra. (3) The analysis result with the simulated spectra was finally verified with a satellite remote sensing image. The detailed flowchart is presented in Figure 2.

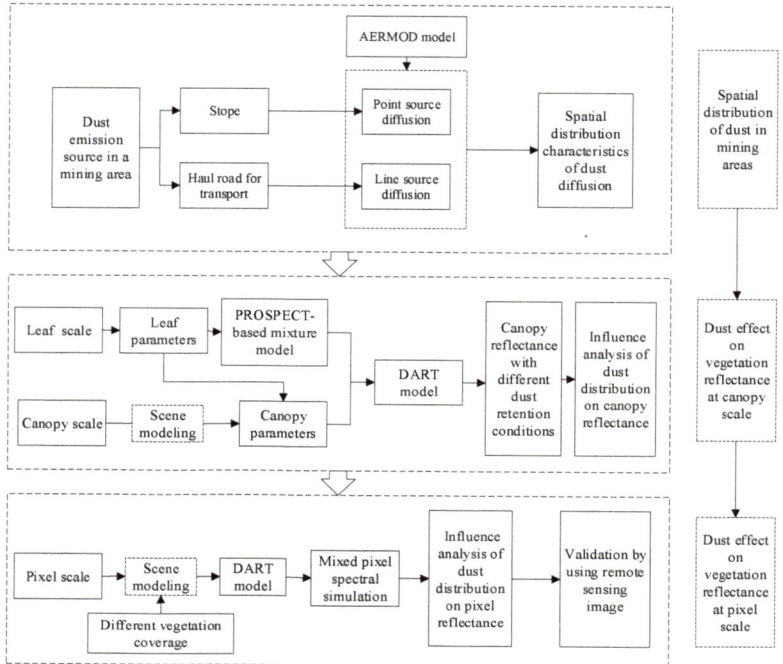

Figure 2. A general flowchart of this study.

2.3. AERMOD Modeling Dust Dispersion

The dust emission intensity in a mining area needs to be determined first because it is the input data of AERMOD. Then the dust dispersion can be described by using AERMOD.

2.3.1. Determining Dust Emission Intensity in the Mining Area

There are two dust sources in the mining area: stope and haul road.

(1) In the stope

A stope is a place for mining ore. The dust in the stope is mainly generated by the forklift when it works, and the dust emission intensity can be calculated with the following formula [30]:

$$Q = 0.0523 \times U^{1.30} \times H^{2.01} \times W^{-1.40} \times M \tag{1}$$

where Q is the dust emission intensity (kg/h), U is the wind speed (m/s), H is the unloading height of forklift (m), W is the water content of material (%), and M is the loading capacity per unit time (m^3/h). According to field monitoring data in the study area, U was 1.6 m/s, H was 4 m, W was 10%, and M was 50 m^3/h. The final dust emission intensity was 3.1 kg/h.

(2) Along the haul road

The haul road, linking a stope to a concentrator, is used for transporting iron ore and bulk and material. The formula of dust source intensity on the haul road is as follows [31]:

$$Q_{road} = 0.123 \times V/5 \times (M/6.8)^{0.85} \times P/0.5 \times 0.72L$$
$$Q = Q_{road} * 1.2 \tag{2}$$

where Q_{road} is the dust generated by vehicle (kg/vehicle), V is the driving speed of the vehicle (km/h), M is the vehicle load (t), p is the road surface material quantity (kg/m^2), and L is the road length (km). According to the actual conditions, the road was set with length 200 m and width 12 m. The height affected by dust was 1.5 m, the tail gas mixing height was 2 m, V = 50 km/h, M = 45 t, p = 0.3 kg /m^2, and the average transportation was 20 times per hour. The final emission intensity was 7.92 kg/km/vehicle or 44 g/km/s.

2.3.2. Dust Dispersion by Using AERMOD and Dustfall Amount Transformation

The two dust sources, stope and haul road, were classified into continuous point and line sources. The dust emission intensity of each source was set as the initial value of AERMOD (ver 18081) with a grain size of 75 µm, which was the mean grain size of the dust samples in the study area. Furthermore, the meteorological data in the mining area were also set as the initial values for running AERMOD. The average wind speed was 1.6 m/s, measured from 1 June to 12 June 2013, in the growing season. Based on the above data, the spatial dispersion of dust in the two sources was simulated by using AERMOD.

Daily average dust concentration is a direct result derived from AERMOD. For this study, it should be converted into dustfall amount according to the following relationship [32]:

$$C_{TSP} = K \cdot C_{DF} \qquad (3)$$

where C_{TSP} is the dust concentration (µg/m^3), C_{DF} is the dustfall amount (t/(km^2·30d)) and K is the correlation coefficient with the value of 11.630 (30d·(10^3 km))$^{-1}$. The transformed amount of dustfall is the cumulative value in 30 days, which would hardly deposit on the canopy totally. Considering the influence of wind and the canopy structure on the dust deposition. the dustfall amount was set as a cumulative value of 2 d.

2.4. Spectra Simulation of Dusty Vegetation by Using DART

DART is one of the most accurate and complete models that operate on different 3D scenes simulating radiative transfer from the visible to thermal infrared in the Earth landscapes and the atmosphere. It has been developed since 1992. It models optical signals at the entrance of imaging radiometers and laser scanners onboard satellites and airplanes, as well as the 3D radiative budget of urban and natural landscapes for any experimental configuration and instrumental specification [24]. The input parameters of the DART model mainly include two parts, one is the related parameters of objects in the simulation scene, and the other is the environment parameters of the simulated scene. The object parameters in the simulation scene are physical and biochemical parameters, structural parameters and scattering properties of the object. The scene environment parameters include solar zenith angle, azimuth angle, observation zenith angle and azimuth angle, scene size and resolution, etc.

In this study, maize was selected as an example of vegetation for its wide distribution in the mining area. Firstly, the reflectance at leaf scale was obtained for the DART input. For clean leaves, the reflectance was derived using the PROSPECT model with four input variables: N (leaf structure parameter), EWT (equivalent water thickness or water content), C_{ab} (chlorophyll a + b content) and C_m (dry matter content) [33]. The four input variables of maize leaf were adapted from the LOPEX93 dataset [33], where N = 1.34, EWT = 0.0137 cm, C_{ab} = 45.27 µg/cm^2, C_m = 0.0047 g/cm^2. For dusty leaves, the reflectance was derived using the PROSPECT-based mixture model [34,35] based on the dust dispersion along the haul road. To simulate the vertical distribution difference of deposited dust on maize canopy, the canopy model of maize was established with upper, middle and lower layers. Then the dusty leaf spectra under different dust amounts at different layers were imported into the DART model. In this case, the zenith angle and azimuth of the sun were set as 70° and 141° by referring to the Landsat 8 OLI imaging scene parameters (August 9, 2013) passing the study area for validation. Finally, the reflectance of maize canopy with different dust amounts was simulated.

Based on the canopy-scale result achieved, pixel-scale spectral data could be obtained for further analysis. At the pixel scale, it was supposed that the target pixel had only two endmembers, vegetation and soil. To explore the dust influence on the reflectance of mixed pixels with different vegetation covers, FVC was set as 89% (very high), 75% (high), 51% (medium) and 23% (low). The spectral differences of pixels with different FVCs could be compared with different dust amount. Taking pixel reflectance at 626 nm (red band) and 840 nm (NIR band) as examples, a quantitative analysis was made to reflect the dust effect on pixel spectra under different vegetation covers using the change rate. Its formula is as follows:

$$CR = \frac{R_d - R_0}{R_0} \quad (4)$$

where CR is the change rate of vegetation reflectance, R_d is the reflectance of dusty vegetation and R_0 is the reflectance of dust-free vegetation.

2.5. Validation for Simulation with Satellite Images

Two scenes of Landsat OLI images (path 122 and row 32), with 30 m spatial resolution, were used to validate the simulated results. One was acquired before a rainy day (24 July 2013) and set as the dusty scene, and the other was acquired after a rainy day (9 August 2013) and set as the dust-free scene. The two images were atmospherically corrected to surface reflectance by using the Fast Line-of-Slight Atmospheric Analysis of Spectral Hypercubes (FLAASH) module of ENVI software. Then, *FVC* was derived by using a dimidiate pixel model based on the dust-free reflectance image [36]. The calculation formula of FVC can be expressed as:

$$FVC = \frac{NDVI - NDVI_{soil}}{NDVI_{veg} - NDVI_{soil}} \quad (5)$$

where *NDVI* is the normalized difference vegetation index, $NDVI_{veg}$ is the *NDVI* value for "pure" vegetation pixel and $NDVI_{soil}$ is the *NDVI* value for "pure" bare soil pixel in the image. In this study, $NDVI_{veg}$ is 0.944, and $NDVI_{soil}$ is 0.173. Finally, the reflectance difference image (between the dusty scene and the dust-free scene) and *FVC* image were used to understand the simulated result with the DART model.

3. Results and Discussion

3.1. Spatial Characteristics of Dust Dispersion Based on the AERMOD Simulation

3.1.1. Dust Dispersion in the Stope (Point Source)

In the horizontal direction, the dust diffusion was consistent with the wind direction (Figure 3a). The dust concentration gradually increased first, reached a maximum of 94.50 mg/m^3, and then began to decline. The dust pollution range was about 20 m horizontally from the point source. In the vertical direction, the dust concentration increased firstly and then decreased within 1.5 m to the pollution source (Figure 3b).

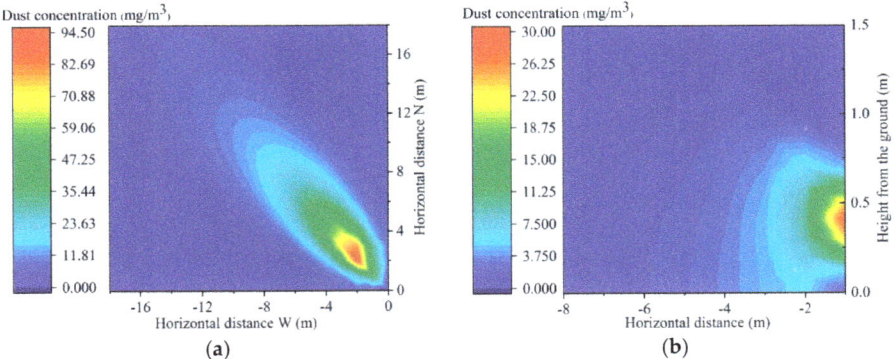

Figure 3. Spatial distribution of dust concentration in the stope shovel loading (**a**) is in horizontal direction, and (**b**) is in vertical direction.

3.1.2. Dust Dispersion along the Road Transport (Line Source)

Under the condition of 1.6 m/s wind, the dust was almost distributed symmetrically along the road, which meant that the wind had a slight influence on the dust dispersion. In the horizontal direction, the dust concentration was consistent on the line parallel to the road, and the dust concentration reached the maximum at 47.60 mg/m^3 on the road and then gradually decreased with the distance off the road (Figure 4a). In the vertical direction, the dust raised by the trucks formed continuous dust layers in a "saddle" shape with different concentrations (Figure 4b). According to the simulation results by AERMOD, the dust concentration decreased with the increasing height.

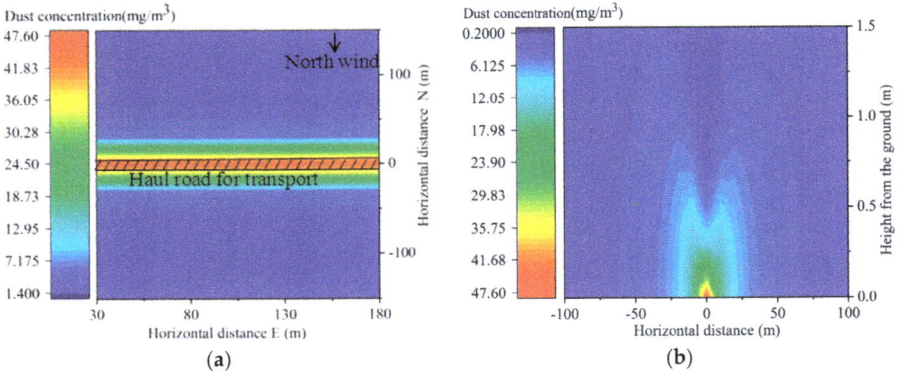

Figure 4. Spatial distribution of dust concentration in road transportation (**a**) is in horizontal direction, and (**b**) is in vertical direction).

Taken together, dust pollution along the haul road was more severe and more extensive than that in the stope. After conversion, the amount of dust deposited on the vegetation canopy at different heights (0.5 m, 1.0 m, and 1.5 m) and distances (from 10 m to 100 m) is shown in Figure 5. Generally, the amount of dust decreased with increasing height. For 1.0 m and 1.5 m height, amounts of dust increased firstly when the distance to the road was not greater than 30 m and then decrease with increasing distance. When the horizontal distance was greater than 60 m, the amounts of dust at different heights were close.

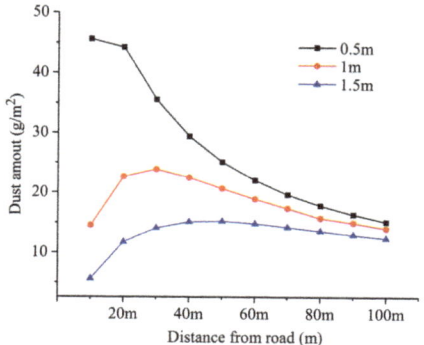

Figure 5. Changes of dustfall at different heights from the ground.

AERMOD is suitable for the near-field range (<50 km), so it could be applied in this study. According to AERMOD, the maximum dust concentration was 94.50 mg/m^3 in the stope and 47.60 mg/m^3 along the haul road, and the maximum dustfall amount was 45.0 g/m^2 at the 0.5 m height. The result was close to some reported results. For example, the maximum dust concentration was 10.78 mg/m^3 in Sistan, Iran during the summer dusty period [37], and the maximum dustfall amount was 14.98 g/m^2 in a limestone quarry in north Israel [38].

3.2. Canopy Spectra under Different Dust Retention Conditions Based on DART Simulation

According to the dust concentration along the road, the dust amount set on the vegetation canopy was 0–80 g/m^2, with a gradient level of 8 g/m^2. Dust falling on vegetation canopy has different effects on the reflectance of different bands (Figure 6). Generally, the reflectance decreased with the increasing amount of dust in the range of 0.7–1.4 μm and increased with increasing amount of dust outside the range. Under the same amount of dust, the spectral response of the canopy was different when the canopy was covered by dust on the upper, middle and lower layers. The spectral change rate increased gradually when the dust only deposited from lower to upper layers. In other words, the dust on the upper layer had a greater impact on the canopy spectra. Furthermore, the dust deposited on all the layers has had the greatest impact on the canopy.

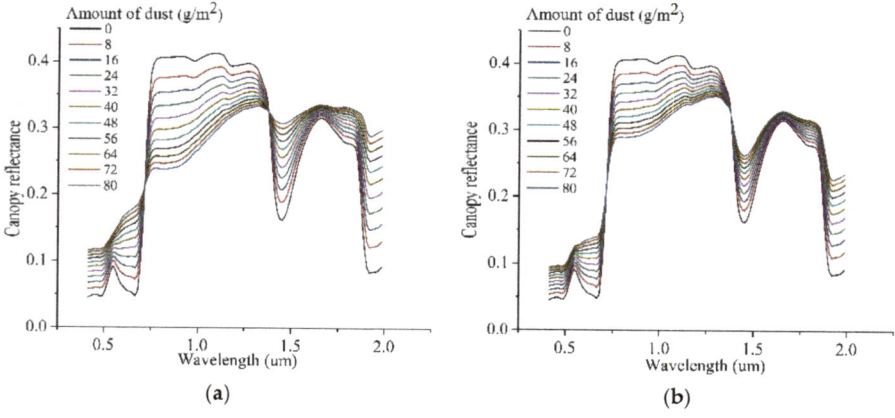

Figure 6. *Cont.*

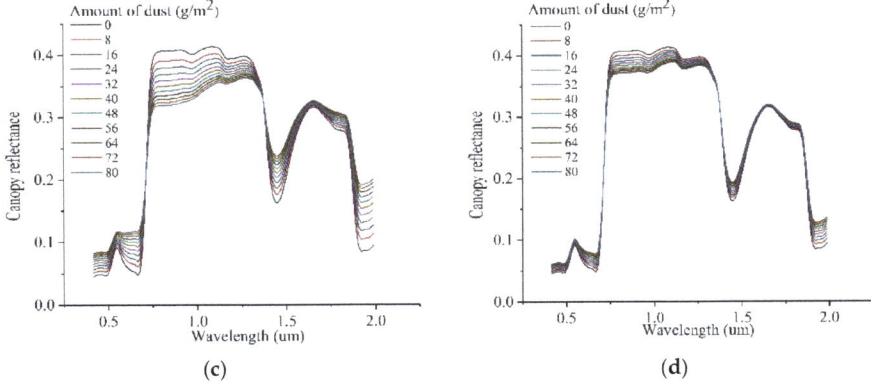

Figure 6. The simulated spectral results of maize canopy under different dust retention conditions (**a**) is dust retention on all the layers of canopy, (**b**) is dust retention only on the top layer of canopy, (**c**) is dust retention only on the middle layer of canopy, and (**d**) is dust retention only on the bottom layer of canopy).

3.3. Pixel Spectra Change under Different Dust Amount and FVCs

Figure 7 shows the spectra of dust-free mixed pixels when the FVC was set as 89%, 75%, 51%, 23%, 13% and 8% initially. The results show that when the vegetation cover decreased, the pixel reflectance gradually changed to the characteristics of soil reflectance. When vegetation cover was less than 23%, the pixel spectra were similar to the soil spectrum.

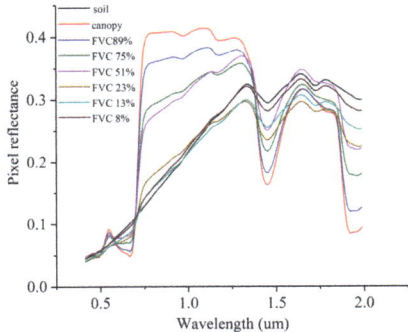

Figure 7. Mixed pixel spectra under different fractional vegetation covers (FVCs) (FVC of canopy is 100%).

Scenes with FVCs of 89%, 75%, 51% and 23% were selected to simulate the spectra of mixed pixels with different dust amount. The change trend of pixel spectra was the same as that at the canopy scale when the amount of dust increased. However, the lower the vegetation cover was, the less the impact of deposited dust on the pixel spectra (Figure 8). The result shows that the change rate decreased with the decrease of FVC at both 626 nm and 840 nm (Figure 9). When the vegetation cover was low, the soil mainly contributed to the pixel spectra. According to the measured soil spectra with different dust deposition levels, the dust effect on it could be ignored. Thus, the pixel spectrum influenced by the dust deposition was mainly determined by the vegetation FVC in the pixel.

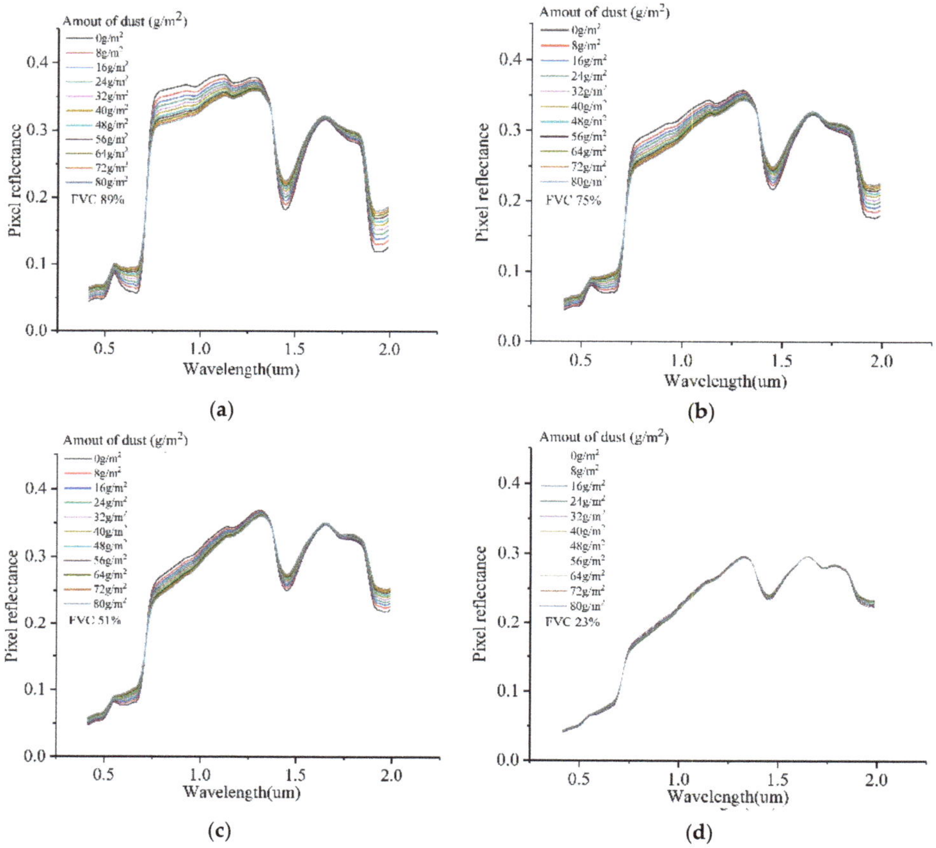

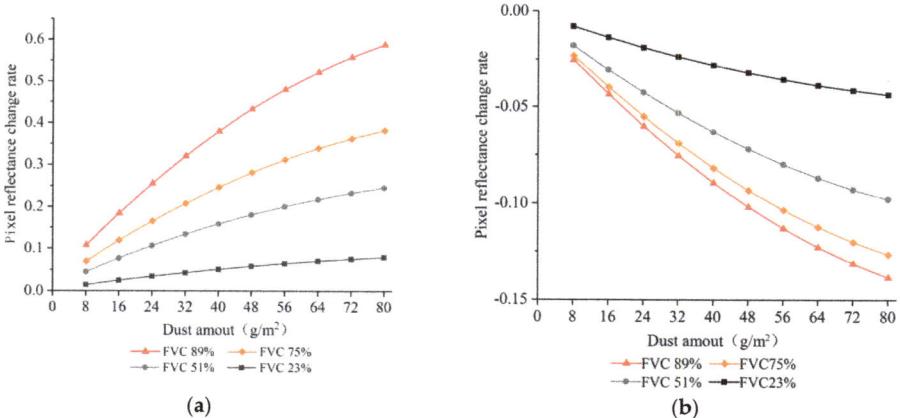

Figure 8. Effects of dust retention on vegetation spectra under different FVC conditions: (**a**) 89%, (**b**) 75%, (**c**) 51%, and (**d**) 23%.

Figure 9. Spectral change rates corresponding to different FVCs under dust retention conditions (**a**) is red band (626 nm), and (**b**) is NIR band (840 nm)).

3.4. Test by Using Satellite Images

With two scenes of remote sensing images, the pixel spectra on both sides of the haul road in the mining area were extracted and compared to test the above result created with simulation data. To comprehensively consider the pixel spectra influenced by dust absorption on vegetation under different FVCs, the difference in reflectance at the NIR band (845–885 nm) of total 302 pixels between dusty and dust-free remote sensing images was obtained to compare with FVC (Figure 10). The results showed that the difference in reflectance at the NIR band increased while FVC increased, which meant that the influence of dust retention on the spectrum of pixels became stronger. However, some errors may be caused because of the difference between the simulation and the validation. Vegetation type is one of the factors. Maize was selected as the experimental plant in the DART simulation, but the selected pixels on the Landsat image for validation may not only be occupied by maize because there were various types of vegetation in the study area. On the other hand, other parameters in the DART model, such as soil background, could also cause errors due to differences from the actual situation. These factors should be considered in future studies to obtain more accurate results.

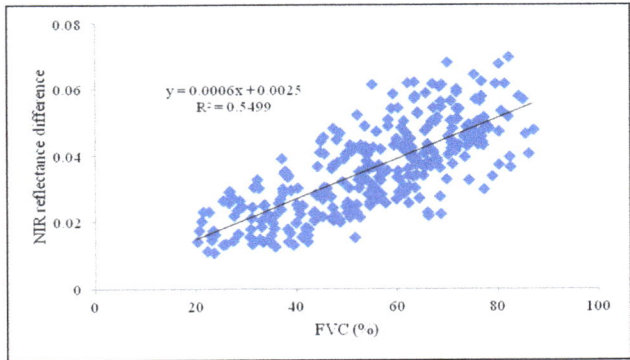

Figure 10. Reflectance difference in NIR band of Landsat 8 image corresponding to different FVCs (the pixels are spot in the map in Figure 1).

The tested result with the satellite images indicates that the influence of dust retention on the pixel spectra with different FVCs was consistent with that created with the simulated spectral data. From the perspective of application, this might be very significant for inversion using remote sensing images under dusty conditions. To remove dust influence accurately on inversion, vegetation cover factor must be considered for its effect on the reflectance. The accurate inversed vegetation parameters are important for sustainable management in mining areas [39].

4. Conclusions

Dust pollution is severe in some mining areas in China. In this study, the dust dispersion was studied by taking an iron mining area as an example. The effect of dust on vegetation spectra was investigated at both the canopy scale and the pixel scale using a spectral simulation method. The AERMOD simulation was conducted in a stope and along a haul road in a mining area. Along the haul road, dust pollution was more severe due to extensive and high-concentration distribution. According to DART simulation, the vertical distribution of dust deposition had an important influence on the canopy spectra. At the canopy scale, the higher the dust deposition on the vegetation canopy layer, the stronger the dust deposition on the vegetation influences. At the pixel scale, the spectra of pixel under dusty conditions varied with the vegetation cover. As the FVC decreased, the dust impact on the pixel spectra would be decreased. This simulated result was compared by the test result estimated using the satellite remote sensing images. The comparative result indicates that the influence of dust retention on the pixel spectra with different FVCs was consistent with that created with the simulated spectral data. These findings would be beneficial to decision-makers or researchers for the remote sensing application to mapping and assessing the dust effect in mining areas.

Author Contributions: Conceptualization, B.M. and X.L.; methodology, B.M.; software, X.L. and B.M.; validation, Z.J. and X.L.; formal analysis, B.M. and R.P.; investigation, B.M. and A.L.; resources, B.M.; data curation, X.L. and Z.J.; writing—original draft preparation, B.M.; writing—review and editing, B.M., R.P. and D.C.; funding acquisition, B.M. and D.C. All authors have read and agreed to the published version of the manuscript.

Funding: This research is jointly supported by the Fundamental Research Funds for the Central Universities (N2001020, N17241004, N160104006, and N170103009), and the National Natural Science Foundation of China (41871310, 41201359).

Acknowledgments: We would like to acknowledge the anonymous reviewers for their valuable suggestions.

Conflicts of Interest: The authors declare no conflict of interest.

References

1. Kaufman, Y.J.; Tanre, D.; Dubovik, O.; Karnieli, A.; Remer, L.A. Absorption of sunlight by dust as inferred from satellite and ground-based remote sensing. *Geophys. Res. Lett.* **2001**, *28*, 1479–1482. [CrossRef]
2. Fairlie, T.D.; Jacob, D.J.; Park, R.J. The impact of transpacific transport of mineral dust in the United States. *Atmos. Environ.* **2007**, *41*, 1251–1266. [CrossRef]
3. Tegen, I.; Werner, M.; Harrison, S.P.; Kohfeld, K.E. Relative importance of climate and land use in determining present and future global soil dust emission. *Geophys. Res. Lett.* **2004**, *31*, 4. [CrossRef]
4. Doronzo, D.M.; Khalaf, E.A.; Dellino, P.; De Tullio, M.D.; Dioguardi, F.; Gurioli, L.; Mele, D.; Pascazio, G.; Sulpizio, R. Local impact of dust storms around a suburban building in arid and semi-arid regions: Numerical simulation examples from Dubai and Riyadh, Arabian Peninsula. *Arab. J. Geosci.* **2014**, *8*, 7359–7369. [CrossRef]
5. Kok, J.F.; Parteli, E.J.R.; Michaels, T.I.; Karam, D.B. The physics of wind-blown sand and dust. *Rep. Prog. Phys.* **2012**, *75*, 106901. [CrossRef] [PubMed]
6. Kinuthia, J.; Nidzam, R. Towards zero industrial waste: Utilisation of brick dust waste in sustainable construction. *Waste Manag.* **2011**, *31*, 1867–1878. [CrossRef]
7. Gulia, S.; Goyal, P.; Goyal, S.K.; Kumar, R. Re-suspension of road dust: Contribution, assessment and control through dust suppressants—A review. *Int. J. Environ. Sci. Technol.* **2018**, *16*, 1717–1728. [CrossRef]
8. Steiner, D.; Lanzerstorfer, C. Investigation of dust resistivity for a fractioned biomass fly ash sample during poor combustion conditions with regard to electrostatic precipitation. *Fuel* **2018**, *227*, 59–66. [CrossRef]
9. Zhong, B.; Wang, L.; Liang, T.; Xing, B. Pollution level and inhalation exposure of ambient aerosol fluoride as affected by polymetallic rare earth mining and smelting in Baotou, north China. *Atmos. Environ.* **2017**, *167*, 40–48. [CrossRef]
10. Gan, Y.; Griffin, W.M. Analysis of life-cycle GHG emissions for iron ore mining and processing in China—Uncertainty and trends. *Resour. Policy* **2018**, *58*, 90–96. [CrossRef]
11. Chaulya, S.K.; Ahmad, M.; Singh, R.S.; Bandopadhyay, L.K.; Bondyopadhay, C.; Mondal, G.C. Validation of Two Air Quality Models for Indian Mining Conditions. *Environ. Monit. Assess.* **2003**, *82*, 23–43. [CrossRef] [PubMed]
12. Cimorelli, A.J.; Perry, S.G.; Venkatram, A.; Weil, J.C.; Paine, R.J.; Wilson, R.B.; Lee, R.F.; Peters, W.D.; Brode, R.W. AERMOD: A Dispersion Model for Industrial Source Applications. Part I: General Model Formulation and Boundary Layer Characterization. *J. Appl. Meteorol.* **2005**, *44*, 682–693. [CrossRef]
13. Lorber, M.; Eschenroeder, A.; Robinson, R. Testing the USA EPA's ISCST-Version 3 model on dioxins: A comparison of predicted and observed air and soil concentrations. *Atmos. Environ.* **2000**, *34*, 3995–4010. [CrossRef]
14. Levy, J.I.; Spengler, J.D.; Hlinka, D.; Sullivan, D.; Moon, D. Using CALPUFF to evaluate the impacts of power plant emissions in Illinois: Model sensitivity and implications. *Atmos. Environ.* **2002**, *36*, 1063–1075. [CrossRef]
15. Holmes, N.S.; Morawska, L. A review of dispersion modelling and its application to the dispersion of particles: An overview of different dispersion models available. *Atmos. Environ.* **2006**, *40*, 5902–5928. [CrossRef]

16. Huang, D.; Guo, H. Dispersion modeling of odour, gases, and respirable dust using AERMOD for poultry and dairy barns in the Canadian Prairies. *Sci. Total. Environ.* **2019**, *690*, 620–628. [CrossRef]
17. Farmer, A.M. The effects of dust on vegetation—A review. *Environ. Pollut.* **1993**, *79*, 63–75. [CrossRef]
18. Lin, W.; Li, Y.; Du, S.; Zheng, Y.; Gao, J.; Sun, T. Effect of dust deposition on spectrum-based estimation of leaf water content in urban plant. *Ecol. Indic.* **2019**, *104*, 41–47. [CrossRef]
19. Peng, J.; Wang, J.-Q.; Xiang, H.-Y.; Niu, J.-L.; Chi, C.-M.; Liu, W.-Y. Effect of Foliar Dustfall Content (FDC) on High Spectral Characteristics of Pear Leaves and Remote Sensing Quantitative Inversion of FDC. *Guang Pu Xue Yu Guang Pu Fen Xi = Guang Pu* **2015**, *35*, 1365–1369.
20. Zajec, L.; Gradinjan, D.; Klančnik, K.; Gaberščik, A. Limestone dust alters the optical properties and traits of Fagus sylvatica leaves. *Trees* **2016**, *30*, 2143–2152. [CrossRef]
21. Asner, G.; Martin, R.E. Spectral and chemical analysis of tropical forests: Scaling from leaf to canopy levels. *Remote Sens. Environ.* **2008**, *112*, 3958–3970. [CrossRef]
22. Mertens, K.C.; De Baets, B.; Verbeke, L.P.C.; Dewulf, R. A sub-pixel mapping algorithm based on sub-pixel/pixel spatial attraction models. *Int. J. Remote Sens.* **2006**, *27*, 3293–3310. [CrossRef]
23. Huesca, M.; García, M.; Roth, K.L.; Casas, A.; Ustin, S.L. Canopy structural attributes derived from AVIRIS imaging spectroscopy data in a mixed broadleaf/conifer forest. *Remote Sens. Environ.* **2016**, *182*, 208–226. [CrossRef]
24. Gastellu-Etchegorry, J.-P.; Yin, T.; Lauret, N.; Cajgfinger, T.; Gregoire, T.; Grau, E.; Feret, J.-B.; Lopes, M.; Guilleux, J.; Dedieu, G.; et al. Discrete Anisotropic Radiative Transfer (DART 5) for Modeling Airborne and Satellite Spectroradiometer and LIDAR Acquisitions of Natural and Urban Landscapes. *Remote Sens.* **2015**, *7*, 1667–1701. [CrossRef]
25. Duthoit, S.; Demarez, V.; Gastellu-Etchegorry, J.-P.; Martin, E.; Roujean, J.L. Assessing the effects of the clumping phenomenon on BRDF of a maize crop based on 3D numerical scenes using DART model. *Agric. For. Meteorol.* **2008**, *148*, 1341–1352. [CrossRef]
26. Ben Hmida, S.; Kallel, A.; Gastellu-Etchegorry, J.-P.; Roujean, J.-L. Crop Biophysical Properties Estimation Based on LiDAR Full-Waveform Inversion Using the DART RTM. *IEEE J. Sel. Top. Appl. Earth Obs. Remote Sens.* **2017**, *10*, 4853–4868. [CrossRef]
27. Morton, D.C.; Rubio, J.; Cook, B.D.; Gastellu-Etchegorry, J.-P.; Longo, M.; Choi, H.; Hunter, M.; Keller, M. Amazon forest structure generates diurnal and seasonal variability in light utilization. *Biogeosciences* **2016**, *13*, 2195–2206. [CrossRef]
28. Schneider, F.D.; Leiterer, R.; Morsdorf, F.; Gastellu-Etchegorry, J.-P.; Lauret, N.; Pfeifer, N.; Schaepman, M. Simulating imaging spectrometer data: 3D forest modeling based on LiDAR and in situ data. *Remote Sens. Environ.* **2014**, *152*, 235–250. [CrossRef]
29. Ma, B.; Pu, R.; Wu, L.; Zhang, S. Vegetation Index Differencing for Estimating Foliar Dust in an Ultra-Low-Grade Magnetite Mining Area Using Landsat Imagery. *IEEE Access* **2017**, *5*, 8825–8834. [CrossRef]
30. Wanjun, T. Study on Dust Distribution and Diffusion Mechanism in Open Pit Coal Mine. Ph.D. Thesis, China University of Mining and Technology, Xuzhou, China, 2018.
31. Wang, D.; Lin, G.; Xu, Z. *Practical Technology for Atmospheric Environmental Impact Assessment*; Standards Press of China: Beijing, China, 2010.
32. Tian, G.; Li, J.-M.; Li, G.; Huang, Y.-H.; Yan, B.-L. Correlation between dust fall and TSP from construction sites. *Environ. Sci.* **2007**, *28*, 1941–1943.
33. Fang, M.; Ju, W.; Zhan, W.; Cheng, T.; Qiu, F.; Wang, J. A new spectral similarity water index for the estimation of leaf water content from hyperspectral data of leaves. *Remote Sens. Environ.* **2017**, *196*, 13–27. [CrossRef]
34. Ma, B.; Li, X.; Liang, A.; Chen, Y.; Che, D. Experimental and Numerical Investigation of Dustfall Effect on Remote Sensing Retrieval Accuracy of Chlorophyll Content. *Sensors* **2019**, *19*, 5530. [CrossRef] [PubMed]
35. Chen, Y. Experimental Study of Dustfall Effect on Chlorophyll Retrieval Using Remote Sensing. Master's Thesis, Northeastern University, Shenyang, China, 2018.
36. Zhang, X.; Liao, C.; Li, J.; Sun, Q. Fractional vegetation cover estimation in arid and semi-arid environments using HJ-1 satellite hyperspectral data. *Int. J. Appl. Earth Obs. Geoinf.* **2013**, *21*, 506–512. [CrossRef]
37. Behrooz, R.D.; Esmaili-Sari, A.; Bahramifar, N.; Kaskaoutis, D. Analysis of the TSP, PM10 concentrations and water-soluble ionic species in airborne samples over Sistan, Iran during the summer dusty period. *Atmos. Pollut. Res.* **2017**, *8*, 403–417. [CrossRef]

38. Bluvshtein, N.; Mahrer, Y.; Sandler, A.; Rytwo, G. Evaluating the impact of a limestone quarry on suspended and accumulated dust. *Atmos. Environ.* **2011**, *45*, 1732–1739. [CrossRef]
39. Beadel, S.; Shaw, W.; Bawden, R.; Bycroft, C.; Wilcox, F.; McQueen, J.; Lloyd, K. Sustainable management of geothermal vegetation in the Waikato Region, New Zealand, including application of ecological indicators and new monitoring technology trials. *Geothermics* **2018**, *73*, 91–99. [CrossRef]

Publisher's Note: MDPI stays neutral with regard to jurisdictional claims in published maps and institutional affiliations.

© 2020 by the authors. Licensee MDPI, Basel, Switzerland. This article is an open access article distributed under the terms and conditions of the Creative Commons Attribution (CC BY) license (http://creativecommons.org/licenses/by/4.0/).

MDPI
St. Alban-Anlage 66
4052 Basel
Switzerland
Tel. +41 61 683 77 34
Fax +41 61 302 89 18
www.mdpi.com

Remote Sensing Editorial Office
E-mail: remotesensing@mdpi.com
www.mdpi.com/journal/remotesensing

www.ingramcontent.com/pod-product-compliance
Lightning Source LLC
LaVergne TN
LVHW070251100526
838202LV00015B/2209